BIOTECHNOLOGICAL APPROACHES FOR PEST MANAGEMENT AND ECOLOGICAL SUSTAINABILITY

BIOTECHNOLOGICAL APPROACHES FOR PEST MANAGEMENT AND ECOLOGICAL SUSTAINABILITY

 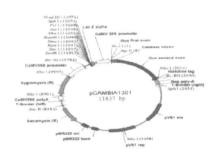

Hari C. Sharma

CRC Press
Taylor & Francis Group
Boca Raton London New York

CRC Press is an imprint of the
Taylor & Francis Group, an **informa** business

CRC Press
Taylor & Francis Group
6000 Broken Sound Parkway NW, Suite 300
Boca Raton, FL 33487-2742

First issued in paperback 2019

© 2009 by Taylor & Francis Group, LLC
CRC Press is an imprint of Taylor & Francis Group, an Informa business

No claim to original U.S. Government works

ISBN-13: 978-1-56022-163-0 (hbk)
ISBN-13: 978-0-367-38619-1 (pbk)

Library of Congress Cataloging-in-Publication Data

Sharma, H. C. (Hari Chand)
    Biotechnological approaches for pest management and ecological sustainability / Hari C. Sharma.
        p. cm.
    Includes bibliographical references and index.
    ISBN 978-1-56022-163-0 (alk. paper)
    1. Agricultural pests--Biological control. 2. Insect pests--Biological control. 3. Plants--Insect resistance--Genetic aspects. 4. Plant biotechnology. I. Title.

    SB950 .S43008
    632'.96--dc22   2008029544

Visit the Taylor & Francis Web site at
http://www.taylorandfrancis.com

and the CRC Press Web site at
http://www.crcpress.com

Contents

Foreword

There has been an unprecedented rise in food prices in recent times, taking some basic foods beyond the reach of the poor and imposing a crippling burden on the economies of the poorest countries. Rising temperatures and climate change have also emerged as serious challenges to crop production and food security. It is ironic, that, while on the one hand there is a need to feed a world population that is expected to exceed 8 billion by the year 2025, on the other, cropland availability has been showing a declining trend. The decrease in the availability of arable land is expected to be much greater in developing countries, where most of the increase in population is expected to occur, than in developed countries.

Unless crop productivity is maximized from the available arable land, meeting the increasing demand for food, feed, and fodder may no longer be possible. One of the areas where a substantial increase in food production can be realized is through the reduction of crop losses due to biotic stresses, which are now estimated at US$243.4 billion annually. Massive applications of pesticides to minimize losses due to insect pests, diseases, and weeds have resulted in high levels of pesticide residue in food and food products and has had an adverse effect on the beneficial organisms in the environment. A large number of insect species have now developed high levels of resistance to currently available insecticides, which has necessitated either the application of even higher doses or an increased frequency of insecticide application. The use of biotechnological tools to minimize losses due to insect pests has therefore become inevitable.

Though the Green Revolution led to significant advances in crop improvement and crop protection technologies, total food production and per capita availability of food have stagnated over the past two decades. There is an urgent need to examine how the tools of science can be used to increase crop productivity without any of the adverse impacts on the environment. A substantial increase in food production can be realized through the application of the modern tools of biotechnology for pest management. Significant progress has been made in the past two decades in using biotechnological tools to understand gene structure and function, and in introducing exotic genes into crop plants for resistance to insect pests. Toxin genes from the bacterium, *Bacillus thuringiensis* have been incorporated into several crops and insect-resistant genetically-engineered cotton, corn, and potato are now being cultivated over large areas of Asia, Africa, Australia, the Americas, and in some parts of Europe.

Large-scale deployment of insect-resistant transgenic crops has raised many concerns about their possible interaction with non-target organisms in the ecosystem, bio-safety of the food derived from genetically-engineered crops, and their likely impact on the environment. As a result, people in certain parts of the world have adopted a cautious approach to accepting products derived through the application of modern tools of biotechnology, although transgenes are not conceptually different from native genes or organisms used for increasing crop production through conventional technologies. Therefore, there is a need to take a critical look at the potential benefits of using modern tools of biotechnology for pest management, and the likely interaction of genetically modified plants with non-target organisms in the ecosystem. This will not only lead to informed decisions for development and deployment of transgenic crops with insect resistance for pest management, but will also help in planning appropriate strategies to deploy them for sustainable crop production.

This book, *Biotechnological Approaches for Pest Management and Ecological Sustainability*, is a comprehensive work that deals with a gamut of issues ranging from host plant resistance to insect pests, phenotyping transgenic plants and mapping populations for insect resistance, physico-chemical and molecular markers associated with insect resistance, potential of insect-resistant transgenic crops for pest management, and the use of biotechnological tools for diagnosis of insects and monitoring insect resistance to insecticides. It also covers the use of genetic engineering to produce robust natural enemies and more virulent strains of entomopathogenic microbes, bio-safety of food derived from genetically engineered plants, detection of transgene(s) in food and food products, and the potential application of the modern tools of biotechnology for pest management and sustainable crop production.

This valuable book comes at a time when alternative strategies are urgently needed to deal with biotic stresses to ensure a food secure future. It will serve as a useful source of information to students, scientists, NGOs, administrators, and research planners in the 21st century.

William D. Dar
Director General
International Crops Research Institute
for the Semi-Arid Tropics (ICRISAT)

Preface

Recombinant DNA technology has significantly enhanced our ability for crop improvement and crop protection to meet the increasing demand for food, feed, and fodder. Considerable progress has been made over the past two decades in manipulating genes from diverse sources and inserting them into crop plants to confer resistance to insect pests and diseases, tolerance to herbicides, drought, improved nutritional quality, increased effectiveness of bio-control agents, and a better understanding of the nature of gene action and metabolic pathways. Genes that confer resistance to insect pests have been inserted into several crops. Transgenic crops with insect resistance are now being grown in several countries worldwide. There has been a rapid increase in the area planted with transgenic crops from 1.7 million ha in 1996 to in excess of 100 million ha in 2007. Deployment of insect-resistant transgenic plants for pest control has resulted in a significant reduction in insecticide use, reduced exposure of farmers to insecticides, reduction of the harmful effects of insecticides on non-target organisms, and a reduction in the amount of insecticide residues in food and food products. Adoption of transgenic crops for pest management also offers the additional advantage of controlling insect pests that have become resistant to commonly used insecticides.

However, the products of biotechnology need to be commercially viable, environmentally benign, easy to use in diverse agro-ecosystems, and have a wide-spectrum of activity against the target insect pests, but harmless to non-target organisms. There is a need to pursue a pest management strategy that takes into account the insect biology, insect plant interactions, and their influence on the non-target organisms in the eco-system. There is a need to combine exotic genes with conventional host plant resistance, and with traits that confer resistance to other insect pests and diseases of importance in the target regions. It is important to devise and follow bio-safety regulations, and to make the products of biotechnology available to farmers who cannot afford the high cost of seeds and chemical pesticides. Use of molecular techniques for diagnosis of insect pests and their natural enemies, and for gaining an understanding of their interactions with their host plants will provide a sound foundation for the development of insect-resistant cultivars in future. Genetic engineering can also be used to produce robust natural enemies, and more stable and virulent strains of entomopathogenic bacteria, fungi, viruses, and nematodes for use in integrated pest management. Molecular markers can also be used for the identification of newer insecticide molecules with different modes of action and for monitoring insect resistance to insecticides. Molecular marker-assisted selection promises to accelerate the pace of development of insect-resistant cultivars for integrated pest management. While there has been a general acceptance of the medicinal products derived through the application of the tools of biotechnology, the response to food derived from genetically modified plants has been based on caution. Rapid and cost effective development and adoption of biotechnology-derived products will depend on developing a full understanding of the interaction of genes within their genomic environment and with the environment in which their conferred phenotype must interact in the ecosystem. It is in this context that this book, *Biotechnological Approaches for Pest Management and Ecological Sustainability*, will serve as a useful source of information for students, scientists, administrators, and research planners.

It would not have been possible to undertake this gigantic task without the inspiration and encouragement from Dr. W.D. Dar (Director General), Dr. D.A. Hoisington (Deputy Director General), and Dr. C.L.L. Gowda (Theme Leader, Crop Improvement at ICRISAT). I am grateful to M.K. Dhillon, G. Pampapathy, Surekha, K.K. Sharma, T.G. Hash, G.V. Ranga Rao, Rajan Sharma, O.P. Rupela, R. Wadaskar, S. Deshpande, P. Lava Kumar, Farid Waliyar, T. Napolean, Rajiv Varshney, S. Senthilvel, and D.A. Hoisington for reviewing different chapters of this book. I also thank S.R. Venkateswarlu for help in the preparation of the manuscript. I am highly thankful to Mr. V. Venkatesan, for his prompt and timely help with literature search during the course of preparation of this manuscript. Last, but not least, for their patience, understanding, and support during the course of the preparation of this manuscript, I am extremely grateful to my wife, Veena Sharma and our daughters Anu and Ankita. The help rendered by the editors and the staff of CRC Press, who have done an excellent job in bringing out this book, is much appreciated.

Hari C. Sharma
Principal Scientist—Entomology
International Crops Research Institute
for the Semi-Arid Tropics (ICRISAT)

1

Pest Management and the Environment

Introduction

Low productivity in agriculture is one of the major causes of poverty, food insecurity, and malnutrition in developing countries, where agriculture is the driving force for broad-based economic growth. By the year 2020, the world population will exceed 7.5 billion. Nearly 1.2 billion people live in a state of absolute poverty [Food and Agriculture Organization (FAO), 1999; Pinstrup-Andersen and Cohen, 2000]. The availability of land for food production is decreasing over time, and such a decrease is expected to be much greater in the developing than in the developed countries. Mexico, Eucador, Nigeria, and Ethiopia had a per-capita cropland availability of 0.25 ha in 1990 compared to 0.10 ha in Egypt, Kenya, Bangladesh, Vietnam, and China. By 2025, per-capita cropland availability will be below 0.10 ha in countries such as Peru, Tanzania, Pakistan, Indonesia, and Philippines (Myers, 1999). Such a decrease in availability of cropland will have major implications for food security. The fate of small farm families in the short term will depend on precision agriculture, which involves the use of right inputs at the right time. Therefore, accelerated public investments are needed to facilitate agricultural growth through high-yielding varieties resistant to biotic and abiotic stresses, environmentally friendly production technology, availability of reasonably priced inputs, dissemination of information, improved infrastructure and markets, primary education, and health care. These investments need to be supported by good governance and an environment friendly policy for sustainable management of natural resources.

As a result of using high-yielding varieties, irrigation, fertilizers, and pesticides, crop productivity has increased five times over the past five decades. Productivity increases in agriculture led by research and development formed the basis for rapid economic growth and poverty reduction (McCalla and Ayers, 1997). Advances in crop improvement have led to the "Green Revolution" becoming one of the scientifically most significant events in the history of mankind. Productivity increases in rice, wheat, and maize helped to surpass in a decade the production accomplishments of the past century (Swaminathan, 2000).

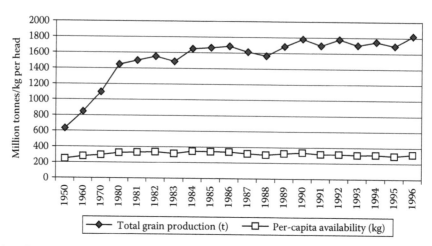

FIGURE 1.1   Grain production and per capita availability of grain between 1950 to 1990.

Grain production has shown a remarkable increase from 1950 to 1980, but only a marginal increase was recorded from 1980 onwards (Figure 1.1). Thereafter, the grain production has remained almost static. The rate of increase in food production decreased to 1% per annum in the 1990s as compared to a 3% increase in the 1970s. After the mid-1980s, there has been a slow and steady decline in per-capita availability of food grains (Dyson, 1999). By 2010, the number of people facing malnutrition will be 30% in Sub-Saharan Africa, 10% in West Asia and North Africa, 6% in East Asia, 12% in South Asia, and 7% in Latin America. As land and water are diminishing resources, there is no option than to increase crop productivity per unit area. There is a need to examine how science can be used to raise biological productivity without the associated ecological costs. Some of this increase in crop productivity can be achieved through the application of modern tools of biotechnology in integrated gene management, integrated pest management, and efficient postharvest management. Biotechnological approaches in agriculture and medicine can provide a powerful tool to alleviate poverty and improve the livelihoods of the rural poor (Sharma et al., 2002).

Pest-Associated Crop Losses and the Need for Pest Management

One of the practical means of increasing crop production is to minimize the pest-associated losses (Sharma and Veerbhadra Rao, 1995), currently estimated at 14% of the total agricultural production (Oerke, 2006). There are additional costs in the form of pesticides applied for pest control, valued at $10 billion annually. Insect pests, diseases, and weeds cause an estimated loss of US$243.4 billion in eight major field crops out of total attainable production of US$568.7 billion worldwide. Among these, insects cause an estimated loss of US$90.4 billion, diseases US$76.8 billion, and weeds US$64.0 billion. The actual losses have been estimated at 51% in rice, 37% in wheat, 38% in maize, 41% in potato, 38% in cotton, 32% in soybeans, 32% in barley, and 29% in coffee (Figure 1.2). Massive application of pesticides to minimize the losses due to insect pests, diseases, and weeds has resulted in adverse effects to the beneficial organisms, pesticide residues in the food and food products, and environmental pollution. As a result, the chemical control of insect pests is under increasing pressure. This has necessitated the use of target specific compounds with low persistence, and an increase in emphasis on integrated pest management (IPM). Although the benefits to agriculture from pesticide use to prevent insect-associated losses cannot be overlooked,

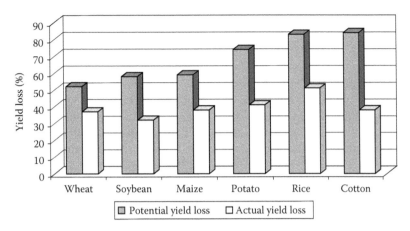

FIGURE 1.2 Extent of losses due to insect pests in major crops.

there is now a greater need to develop alternative technologies that will allow a rational use of pesticides for sustainable crop production. IPM has historically placed great hopes on host plant resistance. However, conventional host plant resistance to insects involves quantitative traits at several loci, and as a result, the progress has been slow and at times difficult to achieve.

What Is Available in the Basket and What Can We Do?

Crop production is severely constrained by increasing difficulties in controlling the damage by insect pests because of the development of insect resistance to insecticides. Therefore, there is a need to adopt pest management strategies to reduce over-dependence on synthetic pesticides. Natural enemies, biopesticides, natural plant products, and pest-resistant varieties offer a potentially safe method of managing insect pests. Unlike synthetic pesticides, some of these technologies (insect-resistant varieties, natural enemies, *Bacillus thuringiensis* (*Bt*) Berliner, nucleopolyhedrosis viruses [NPVs], entomopathogenic fungi, and nematodes) have the advantage of replicating themselves or their effect in the field, and thus having a cumulative effect on pest populations. Despite being environmentally friendly, the alternative technologies have some serious limitations, such as: (1) mass production, (2) slow rate of action, (3) cost effectiveness, (4) timely availability, and (5) limited activity spectrum. Some of the natural enemies such as *Trichogramma, Cotesia, Bracon, Chrysoperla,* and *Coccinella,* and the biopesticides such as *Bt* and NPVs, are being produced commercially. Strains of *Pseudomonas, Beauveria,* and *Metarhizium* are also effective in controlling insects. Natural plant products from neem, *Azadirachta indica* A. Juss., custard apple, *Annona squamosa* L., and *Pongamia pinnata* (L.) Pierre., have been recommended for pest control. Several insect-resistant varieties have been developed, but very few are cultivated by the farmers on a large scale because of lack of sustained seed supply (Sharma and Ortiz, 2002). However, many alternative technologies are not as effective as the synthetic insecticides and, therefore, have not been widely adopted by the farming community on a large scale. There is, therefore, an urgent need to improve:

- Mass production and the delivery system of natural enemies.
- Bio-efficacy and formulations of biopesticides and natural plant products.
- Production and distribution of insect-resistant cultivars.

Pest Management Components

Management of insect pests on high-value crops relies heavily on insecticides, often to the exclusion of other methods (Sharma and Veerbhadra Rao, 1995). With an increasing restraint on insecticide use due to development of resistance in insect populations and environmental contamination, integration of several management techniques has become necessary to reduce the reliance on insecticides and prolong the utility of important molecules (Reddy et al., 1997). In order to overcome the toxic and chronic effects of pesticides, as well as pest resurgence, intensive research efforts are needed to develop a balanced program for IPM. Various components of IPM are discussed below.

Economic Thresholds

The present methodology for assessing insect damage to undertake control measures is cumbersome, and the grassroot-level field workers and farmers are unable to understand and practice them. Simple methods to assess insect damage and density would be useful for timely application of appropriate control measures. Economic threshold levels (ETLs) are available for a limited number of insect species. ETLs developed without taking into consideration the potential of naturally occurring biological control agents and levels of resistance in the cultivars grown to the target pests are of limited value. The ETLs have to be developed for specific crop-pest-climatic situations. The ETLs developed in one region are not applicable in other areas where the crop-pest and socioeconomic conditions are different. Simple methods of assessing ETLs could help avoid unnecessary pesticide applications.

Biological Control

A large number of parasites, predators, bacteria, fungi, and viruses regulate populations of insect pests under natural conditions. However, only a few biocontrol agents have been exploited successfully for controlling insect pests. Identification of potential biocontrol agents would help to launch a successful battle against the crop pests. A more pragmatic approach would be the conservation of biocontrol agents. Major improvements in biological control of insect pests can be made through habitat management. Increasing genetic diversity has been proposed as a means of augmenting natural enemy populations. However, the response of natural enemies to genetic diversity varies across crops and cropping systems (Andow, 1991). Hedgerows, cover crops, and weedy borders provide nectar, pollen, and refuge to the natural enemies. Mixed planting and provision of flowering plants at the field borders can increase the diversity of the habitat, and provide more effective shelter and alternative food sources to predators and parasites. Inter- or mixed-cropping, which involve simultaneous growing of two or more crops on the same piece of land, is one of the oldest and most common cultural practices in tropical countries for risk aversion and pest management. Densities of natural enemies have been found to be greater in 52.7% of the species in polycultures, while 9.3% species had lower densities (Andow, 1991). Predators and parasites have been found to result in higher mortality of herbivore arthropods in polycultures in nine studies, lower rates of mortality in two studies, and no differences in four studies (Russell, 1989). For biological control to be successful, it is important to ensure that essential parasitoid resources and hosts coincide in time and space.

Further, technology available for mass rearing of some of the potential parasites, predators, and pathogens is not satisfactory. An appropriate mass rearing or multiplication technology would help realize the potential of biocontrol agents in pest management. A few studies have focused on insect population dynamics and key mortality factors under field conditions (Zalucki et al., 1986; Fitt, 1989). Early-stage mortality is invariably the most severe, although its causes and extent vary greatly, and comparable data sets are too few to identify the factors responsible for population regulation across regions. Varieties with moderate levels of resistance that allow the pest densities to remain below ETLs are best suited for use in IPM in combination with natural enemies. Restless behavior and prolonged developmental period of the immature stages increases the susceptibility of the target pest to the natural enemies. However, plant morphological characteristics and secondary plant substances sometimes reduce the effectiveness of natural enemies for pest management. The use of insect-pest-resistant varieties and biological control brings together unrelated mortality factors, which reduce the pest population's genetic response to selection pressure from plant resistance and natural enemies. Acting in concert, they provide a density-independent mortality at times of low pest density, and density-dependent mortality at times of high pest density (Bergman and Tingey, 1979).

Parasitoids

A diverse range of parasitoids lay their eggs on or in the body of an insect host, which is then used as a food by the developing larvae. The most important parasitoid groups are trichogrammatids, ichneumonids, braconids, chalcids, and tachinids. The trichogrammatids parasitize the eggs of several insect species, and have been used extensively in biological control. The ichneumonids and braconids prey mainly on the larvae of butterflies and moths. The chalcid wasps parasitize the eggs and larvae of insects. The most important and widely used parasitoids for biological control of insects are the egg parasitoids such as *Trichogramma*, *Chelonus*, and *Telenomus*. The larval parasitoids such as *Cotesia*, *Encarsia*, *Gonatocerus*, *Campoletis*, *Bracon*, *Enicospilus*, *Palexorista*, *Carcelia*, *Sturmiopsis*, etc., have also been used in several countries for biological control of insects.

Cropping systems can be altered successfully to augment and enhance the effectiveness of natural enemies (Andow and Risch, 1985; Altieri, Wilson, and Schmidt, 1985). Optimal microclimatic conditions, nectar sources, and alternate hosts may exist in some cropping systems, but not others. Physicochemical characteristics of the host plants also play an important role in host specificity of both the insect hosts and their parasitoids (Sharma, Pampapathy, and Sullivan, 2003). Host-plant-mediated differences in the activity and abundance of natural enemies have been recorded in the case of *Helicoverpa armigera* (Hubner) (Pawar, Bhatnagar, and Jadhav, 1986; Zalucki et al., 1986; Manjunath et al., 1989). The average rates of parasitism of the eggs of *H. armigera* (mainly by *Trichogramma* spp.) have been found to be 33% on sorghum, 15% on groundnut, and 0.3% on pigeonpea, and little or no parasitism was observed on chickpea (Pawar, Bhatnagar, and Jadhav, 1986). Manjunath et al. (1989) observed up to 98% parasitism of *H. armigera* eggs by *Trichogramma chilonis* Ishii on tomato, potato, and lucerne, but no egg parasitism was recorded on chickpea, probably because of the acid exudates secreted by the leaves. Therefore, due consideration should be given to the host plant and the species of the parasitoid involved while planning for biological control of insect pests.

Predators

In general, predators have received much less attention than parasites as natural control agents. They exercise greater control on pest populations in a diverse array of crops and cropping systems. The most common predators include *Chrysopa, Chrysoperla, Nabis, Geocoris, Orius, Polistes,* and the species belonging to Pentatomidae, Reduviidae, Coccinellidae, Carabidae, Formicidae, and Araneida (Zalucki et al., 1986; King and Coleman, 1989; Romeis and Shanower, 1996). Some predators have been used in augmentative release studies, notably *Chrysoperla carnea* (Stephen). Although effective in large numbers, the high cost of large-scale production precludes their economic use as a biological control (King and Coleman, 1989). Naturally occurring predators play a major role in keeping the insect pest populations below ETLs. Coccinellids and chrysophids have been used successfully for biological control of insect pests under greenhouse conditions, and in some situations, under field conditions.

Increasing genetic diversity also helps to increase the abundance and effectiveness of generalist predators (Sunderland and Samu, 2000; Schmidt et al., 2004). Some natural enemies may be more abundant in polycultures because of the greater availability of nectar, pollen, and diversity of prey (Bugg, Ehler, and Wilson, 1987) for a longer period of time (Topham and Beardsley, 1975). Populations of coccinellid beetles (*Coccinella transversalis* Fab. and *Adalia bipunctata* L.), lace wings (*Chrysopa* spp.), reduviid and pirate bugs [*Coranus triabeatus* (Horvath)], and spiders (*Lycosa* spp. and *Araneus* spp.) have been found to be significantly greater in maize-cowpea intercrop than on cotton alone. Greater numbers of *Geocoris* spp. and other predators have been recorded on knotweed than on other weed species, which was attributed to availability of floral nectar and of alternate prey (Bugg, Ehler, and Wilson, 1987). Greenbug, *Schizaphis graminum* (Rondani), on strips of sorghum interplanted in cotton has been found to support large numbers of the coccinellid predator, *Hippodamia convergens* Guerin-Meneville and other predators (Fye, 1972). Abundance of the predatory mite, *Metaseiulus occidentalis* (Nesbitt) was greater in plots adjacent to alfalfa intercropped in cotton (Corbett and Plant, 1993). Mulching of the soil surface with crop residue also increases the abundance of the generalist predators (Altieri, Wilson, and Schmidt, 1985; Schmidt et al., 2004).

Entomopathogenic Bacteria

Several entomopathogenic bacteria play a major role in controlling insect pests under natural conditions. Formulations based on *B. thuringiensis* have been marketed since the 1950s. There are 67 registered *Bt* products with more than 450 formulations. The major boost to the production and use of *Bt* products came with the discovery of the HD-1 strain of *Bt* subspecies *kurstaki*, which is effective against a large number of insect species (Dulmage, 1970). Several commercial products such as Thuricide®, Dipel®, Trident®, Condor®, and Biobit® are marketed worldwide. There are several subspecies of this bacterium, which are effective against lepidopteran, dipteran, and coleopteran insects. Formulations based on *Bt* account for nearly 90% of the total biopesticide sales worldwide (Neale, 1997), with annual sales of nearly US$90 million (Lambert and Peferoen, 1992). *Bacillus israeliensis* L. has been used extensively for the control of mosquitoes. Narrow host range, necessity to ingest the *Bt* toxins by the target insects, ability of insect larvae to avoid lethal doses of *Bt* by penetrating into the plant tissue, inactivation by sunlight, and effect of plant surface chemicals on its toxicity limit its widespread use in crop protection (Navon, 2000). There may be a limitation on the use of *Bt*-based products in crops or areas where transgenic plants with *Bt* toxin genes have been deployed as a strategy for resistance management.

Baculoviruses

Baculoviruses are regarded as safe and selective pesticides. They have been used against many insect species worldwide, mainly against lepidopteran insect pests. The NPVs exist as populations in nature, with a wide variation in virulence. Movement within and from soil is basic to the long-term survival and effectiveness of NPVs. The NPVs have amenalistic interactions with other biotic agents. Their use and effectiveness is highly dependent on the environment (Fuxa, 2004). The NPVs can be used for the control of some difficult to control insect pests such as *H. armigera* (Pokharkar, Chaudhary, and Verma, 1999). The most successful examples have been the use of NPV of soybean caterpillar, *Anticarcia gemmatalis* (Hubner) and of *Heliothis/Helicoverpa* (Moscardi, 1999). Narrow host range, slow rate of insect mortality, difficulties in mass production, stability under sunlight, and farmers' attitude have limited the use of NPVs as commercial pesticides. Addition or tank mixing of chemical pesticides and genetic engineering can be used to overcome some of the shortcomings of baculoviruses. Jaggery (uncleaned sugar syrup) (0.5%), sucrose (0.5%), egg white (3%), and chickpea flour (1%) are effective in increasing the activity of NPVs (Sonalkar et al., 1998). Adjuvants such as liquid soap (0.5%), indigo (0.2%), urea (1%), and cottonseed extract are useful in increasing the stability of NPVs to UV rays of light. A significant and negative correlation has been observed between insect mortality due to NPVs and foliar pH, phenols, tannins, and protein binding capacity (Ramarethinam et al., 1998). Much remains to be done to develop effective formulations of baculoviruses for effective control of insect pests.

Entomopathogenic Fungi

Entomopathogenic fungi have been recognized as important natural enemies of insect pests. Species pathogenic to insect pests are *Metarhizium anisopliae* (Metsch.), *M. flavoviride* (Metsch.), *Nomuraea rileyi* (Farlow) Samson, *Beauveria bassiana* (Balsamo), *Paecilomyces farinosus* (Holm ex Gray) Brown & Smith (Hajeck and St.-Leger, 1994; Saxena and Ahmad, 1997). Mass production of different entomopathogenic fungi may not be difficult. Glucose, yeast extract, basal salts, agar, and carrot medium can be used for the multiplication of *M. anisopliae*. Zapek Dox Broth (containing 2% chitin and 3% molasses) promotes growth and sporulation of most entomopathogenic fungi (Srinivasan, 1997). For commercial production, a solid-state fermentation system may be more effective. Adhesion of fungal spores to host cuticle and their germination is a prerequisite for efficacy of fungal pathogens. High relative humidity (RH > 90%) is required for germination of fungal spores, and is a big handicap in the widespread use of entomopathogenic fungi. However, special formulations in oil can overcome this problem by creating high RH microclimates around the spores, enabling entomopathogenic fungi to function in low RH environments (Bateman et al., 1993).

Entomopathogenic Nematodes

Entomopathogenic nematodes of the genera *Steinernema* and *Heterorhabditis* have emerged as excellent candidates for biological control of insect pests. Entomopathogenic nematodes are associated with the bacterium *Xenorhabdus* and are quite effective against a wide range of soil-inhabiting insects. The relationship between the nematodes and the bacterium is symbiotic because the nematodes cannot reproduce inside the insects without the bacterium, and the bacterium cannot enter the insect hemocoel without the nematode and cause the infection (Poinar, 1990). Broad host range, virulence, safety to nontarget organisms, and effectiveness have made them ideal biological control agents (Georgis, 1992). Liquid formulations and application strategies have allowed nematode-based products to be quite competitive for pest management in high-value crops. Entomopathogenic nematodes are generally more expensive to produce than the insecticides, and their effectiveness is limited to certain niches and insect species. There is a need to improve culturing techniques, formulations, quality, and the application technology.

Cultural Control

The need for ecologically sound, effective, and economic methods of pest control has prompted renewed interest in cultural methods of pest control. The merit of many of the traditional farm practices has been confirmed by learning why farmers do what they do. But some practices still remain to be thoroughly investigated and understood. A number of cultural practices, such as selection of healthy seeds, synchronized and timely sowing, optimum spacing, removal of crop residues, optimum fertilizer application, and regulation of irrigation, help in minimizing the pest incidence. A number of crop husbandry practices that help reduce pest damage can be quite effective under subsistence farming conditions and these involve no additional costs to the farmers, and do not disturb natural enemies of the insect pests and the environment.

Date of Sowing and Planting Density

Sowing time considerably influences the extent of insect damage. Normally, farmers plant with the onset of rains. Synchronous and timely or early sowing of cultivars with similar maturity over large areas reduces population build up of insect pests and the damage they cause. In Tamil Nadu, India, there is an old adage among the farmers, "inform your neighbor before you plant sorghum lest his crop be destroyed by shoot fly [*Atherigona soccata* (Rondani)] and head bugs [*Calocoris angustatus* (Lethiery)]." Early and uniform sowing of sorghum over large areas has resulted in reducing the damage by shoot fly and sorghum midge [*Stenodiplosis sorghicola* (Coqillett)] in Maharashtra, India. Early planting of pigeonpea results in reduced damage by *H. armigera* (Dahiya et al., 1999). The traditional practice of using a high seeding rate helps to maintain optimum plant stand and reduce insect damage in cereals (Gahukar and Jotwani, 1980). Shoot fly and midge damage in sorghum is higher when plant densities are low (Sharma, 1985). Timely thinning of the crop also helps to reduce pest damage.

Nutrient Management

The extent and nature of fertilizer application influence the crop susceptibility to insects. In some instances, high levels of nutrients increase the level of insect resistance, and in others they increase the susceptibility. An increase in nitrogenous and phosphatic fertilizers decreases shoot fly, *A. soccata*, and spotted stem borer, *Chilo partellus* (Swinhoe), infestation in sorghum (Chand, Sinha, and Kumar, 1979), possibly by increasing plant vigor (Narkhede, Umrani, and Surve, 1982). Plants treated with K and NK also suffer low shoot fly and borer damage in sorghum (Balasubramanian et al., 1986). Shoot fly and stem borer damage has been found to be greater in plots treated with cattle manure. This may be due to the attraction of shoot flies to the odors emanating from organic manure. Application of biofertilizer (*Azospirillum* sp.) increases the phenolic content of sorghum seedlings and results in a decrease in shoot fly damage (Mohan et al., 1987). *Azospirillum* also increases the effectiveness of carbofuran for shoot fly control (Mote, 1986). Application of potash decreases the incidence of top shoot borer, *Scirpophaga excerptalis* (Walker) in sugarcane. High levels of nitrogen lead to greater damage by the cotton jassid, *Amrasca biguttula biguttula* Ishida. A change in nutrient supply also affects the resistance to greenbug,

S. graminum, in sorghum (Schweissing and Wilde, 1979). Increase in nitrogen in potato leaves increased the development and survival of serpentine leaf miner, *Liriomyza trifolii* Burgess (Facknath and Lalljee, 2005). Potassium and phosphorus, on the other hand, decreased the host suitability of potato plants to *L. trifolii*, and were detrimental to the pest.

Intercropping and Crop Rotations

Crop rotation is another means of reducing insect infestation. It breaks the continuity of the food chain of oligophagous pests. Sorghum is generally rotated with cotton, groundnut, sunflower, or sugarcane to reduce the damage by *A. soccata*, *S. sorghicola*, and *C. angustatus*. A carefully selected cropping system (intercropping or mixed cropping) can be used to reduce pest incidence, and minimize the risks involved in monocultures. Sorghum shoo-fly, *A. soccata*, and midge, *S. sorghicola*, damage is reduced when sorghum is intercropped with leguminous crops. Intercropping sorghum with cowpea or lablab reduced the damage by spotted stem borer, *C. partellus* by 50%, and increased the grain yield by 10% to 12% over a single crop of sorghum (Mahadevan and Chelliah, 1986). Intercropping sorghum with pigeonpea reduces the damage by *H. armigera* in pigeonpea (Hegde and Lingappa, 1996). Intercropping chickpea with mustard or linseed (Das, 1998) reduces the damage by *H. armigera*. Sesame, sunflower, marigold, and carrot can be used as trap crops for *H. armigera* (Sharma, 2001, 2005). Carrot intercropped with lucerne has been shown to suffer less damage by the rust fly, *Psila rosae* F. (Ramert, 1993). Intercropping bean with collards decreases flea beetle, *Phyllotreta cruciferae* Goeze densities on the collards and minimizes the leaf damage (Altieri, van Schoonhoven, and Doll, 1977). Intercropping red clover with maize also reduces the damage by the European corn borer, *Ostrinia nubilalis* (Hubner) (Lambert et al., 1987).

Field Sanitation and Tillage

Collecting and burning of stubbles and chaffy panicles reduces the carryover of spotted stem borer, *C. partellus*, and midge, *S. sorghicola*, in sorghum. Stalks from the previous season should be fed to cattle or burnt before the onset of monsoon rains to reduce the carryover of *C. partellus* (Gahukar and Jotwani, 1980). Piling and burning of trash at dusk in the field attracts the adults of white grubs, *Holotrichia consanguinea* (Blanchard), and the red hairy caterpillar, *Amsacta moorei* Butler, and kills them. This helps to reduce the oviposition and damage by these insects. Ploughing the fields after crop harvest and before planting reduces the abundance and carryover of white grubs, grasshoppers, hairy caterpillars, and stem borers by exposing them to parasites, predators, and adverse weather conditions (Gahukar and Jotwani, 1980). Timely weeding reduces the extent of damage by some insects (Sharma et al., 2004). Many common weeds also act as hosts for oviposition, and provide a better ecological niche for the insects to hide, thus shielding them from natural enemies and insecticide sprays. However, many weed hosts also sustain the natural enemies of insect pests, and thus may help in increasing the efficiency of natural enemies in population suppression of insect pests. Flooding of the fields at the time of pupation reduces the survival of *H. armigera* (Murray and Zalucki, 1990).

Chemical Control

Insecticides are the most powerful tool in pest management. Insecticides are highly effective, rapid in action, adaptable to most situations, flexible enough to meet the changing agronomic requirements, and economical. Insecticides are the most reliable means of reducing crop damage when the pest populations exceed ETLs. When used properly based on ETLs, insecticides provide a dependable tool to protect the crop from insect pests. Despite their effectiveness, much insecticide use has been unsound, leading to problems such as pest resurgence, development of resistance, pesticide residues, nontarget effects, and direct hazards to human beings (Smith et al., 1974). Despite several advantages of insecticides for pest control, their use often results in direct toxicity to natural enemies (Sharma and Adlakha, 1981), and also through the poisoned prey (Sharma and Adlakha, 1986), and a consideration of these is essential for optimizing their use in pest management. There is substantial literature on the comparative efficacy of different insecticides against insect pests. Most insecticide applications are targeted at the larval stages. Control measures directed at adults, eggs, and neonate larvae are most effective in minimizing insect damage. Spray decisions based on egg counts could destroy both invading adults and eggs, and leave a residue to kill future eggs and the neonate larvae. Young larvae are difficult to find, and at times burrow into the plant parts where they become less accessible to contact insecticides. Ultra low volume (ULV) applicators have been found to be more effective than the other types of sprayers (Parnell et al., 1999).

Development of Resistance to Insecticides and Strategies for Resistance Management

Excessive and indiscriminate use of insecticides not only has resulted in development of insecticide-resistant insect populations, but also decimated useful parasites and predators in the ecosystem. There are several reports that substantiate the development of resistance to insecticides in insects of public health importance, stored grains, and field crops. A large number of insects have shown resistance to insecticides belonging to different groups, and 645 cases of resistance have been documented (Rajmohan, 1998). Most reports of resistance development pertain to organophosphates (250), followed by synthetic pyrethroids (156), carbamates (154), and others (including chlorinated hydrocarbons) (85). Many species (about 85) of insects have developed resistance to more than two groups of insecticides. The highest numbers of insects and mites showing resistance to pesticides have been recorded in vegetables (48), followed by those infesting fruit crops (25), cotton (21), cereals (15), and ornamentals (13). *Heliothis/Helicoverpa* (which are the most serious pests on cotton, legumes, vegetables, and cereals) have shown resistance to several groups of insecticides (Figure 1.3). This has resulted in widespread failure of chemical control, resulting in extreme levels of debt for farmers, at times even causing them to commit suicide. The cotton white fly, *Bemisia tabaci* (Genn.) has shown resistance to insecticides in cotton, brinjal, and okra; while the tobacco caterpillar, *Spodoptera litura* (F.) has been found to be resistant to insecticides on cotton, cauliflower, groundnut, and tobacco. Green peach and potato aphid, *Myzus persicae* Sulzer, cotton aphid, *Aphis gossypii* Glover, mustard aphid, *Lipaphis erysimi* Kalt., and diamond back moth, *Plutella xylostella* (L.), have also been found to exhibit resistance to insecticides in several crops. Development of resistance to insecticides has necessitated the application of higher dosages of the same pesticide or an increased number of pesticide applications. The farmers often resort to application of insecticide mixtures to minimize the insect damage. This has not only increased the cost of pest control, but also resulted in insecticidal hazards and pollution of the environment. It is in this context that the use of integrated pest management becomes all the more important.

Insecticide resistance management strategies have aimed either at preventing the development of resistance or to contain it (Forrester, 1990). All rely on a strict temporal restriction in the use of certain insecticides such as pyrethroids and their alteration with

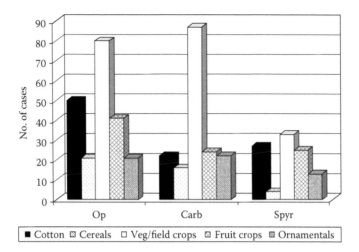

FIGURE 1.3 Number of insect species that have developed resistance to different insecticides. OP, organophosphates; Carb, carbamates; and Spyr, synthetic pyrethroids.

other insecticide groups to minimize selection for resistance (Sawicki and Denholm, 1987). Because of economic advantages and low toxicity to mammals and to some parasites and predators, much effort has been directed towards developing management strategies aimed at prolonging the use of synthetic pyrethroids (King and Coleman, 1989). In Australia, the strategy is dictated by the cotton growing season, which is divided into three stages: I, early growth; II, the peak squaring, flowering, and boll formation; and III, end of season. The use of pyrethroids is restricted to stage II, which corresponds to a 42-day period for a single *H. armigera* generation, with no more than three applications. Endosulfan should not be used during stage III to decrease the chances of reselection for resistance to cyclodiene insecticides (Forrester, 1990).

The strategy depends for its success in dilution of resistance through interbreeding with immigrant populations of susceptible insects from unsprayed crops and wild hosts. However, if resistance is present, and a large proportion of the local over-wintered population is resistant, then this strategy is unlikely to be effective in the long term. There is some uncertainty as to whether insecticide resistance confers reduced fitness in pyrethroid-resistant strains. Increased incidence of the nerve insensitivity resistance mechanism, which does not confer the disadvantage of reduced fitness, implies a major threat to current resistance management strategies. Counter to the benefits derived from diluting influxes of susceptible insects, the capacity of certain insect species for long range movement has serious implications in relation to the spread of insecticide-resistant populations, particularly into unsprayed areas or crops (Daly and Gregg, 1985).

The efficacy of insecticides against target insect pests also depends on the formulation, type of application equipment, and technology for delivering the insecticides. Some of the application equipment does not give the desired performance for specific crop-pest, climatic, and topographic conditions. There is a need to devise suitable application equipment to meet the farmers' needs in rain-fed agriculture. Further, the types of insecticide formulations needed in rain-fed areas are different from those for irrigated areas. Dry areas need different types of pesticide formulations, which require a minimum amount of water. Hence, research efforts should be concentrated on developing the right type of plant protection equipment vis-à-vis insecticide formulations. There is a growing support for the use of growth-regulating substances that hamper the development of various stages of insects. Efforts should continue to search and identify newer compounds that can be used successfully in pest control programs.

Pest Resurgence

Resurgence of cotton whitefly, *B. tabaci*, has been reported from several crops as a result of overuse of synthetic pyrethroids. Resurgence of aphid, *A. gossypii*, and leafhopper, *A. biguttula biguttula*, in cotton, brown planthopper, *Nilaparvata lugens* (Stal), in rice, and mites, *Tetranychus*, in vegetables and apple have also been reported (Heinrichs and Mochida, 1984; Hoffmann and Frodsham, 1993; Hardin et al., 1995). Some of the major factors causing pest resurgence are:

- Application of high doses of nitrogenous fertilizers.
- Use of sublethal doses of insecticides.
- Reduction in population of natural enemies due to broad-spectrum pesticides.
- Favorable conditions for insect multiplication.

This problem not only leads to increased use of pesticides, but also increases the cost of cultivation, greater exposure of the operators to toxic chemicals, and failure of the crop in the event of poor control of the target insect pests.

Pesticide Residues in Food and Food Products

Pesticide residues find their way to human beings through consumption of commodities contaminated with pesticides. Many scientific studies have proved biomagnification of pesticide residues in human tissues, and products of animal origin. Chlorinated insecticides are more persistent in nature compared to organophosphates, carbamates, and synthetic pyrethroids. In order to regulate pesticide residues to safe levels, the Food and Agriculture Organization (FAO) and World Health Organization (WHO) have prescribed pesticide residue tolerance limits for agricultural commodities for 50 pesticides. Over 100,000 cases of accidental exposure to pesticides are reported every year, of which a large number are fatal. Pesticides are composed of active ingredients and inert material. The inert substances at times may be more toxic than the active ingredient. Some of the inert ingredients have been suspected to be carcinogenic, while others have been linked to disorders of the central nervous system, liver, and kidney, as well as birth defects and acute toxicity. All pesticides are designed to kill insect pests, but can also kill human beings and other nontarget organisms if ingested in sufficient amounts. A natural mix of pesticides and fertilizers can significantly affect the immune and neuroendocrine systems.

Contamination of Soil and Water

Most of the pesticides applied for pest control ultimately find their way into soil and water. It has been estimated that nearly 50% of the pesticides applied to crop foliage reach the soil either as spray drift or as runoff. Pesticide residues in soil find their way into the aquatic system or may accumulate in plants. Most lawn care chemicals have been associated with the death of birds. Fish in rivers and streams have often been found to contain residues of more than one pesticide. Pesticides are also responsible for a decline in the number of amphibians and are also associated with the elimination of many species that are important pollinators of plants. Therefore, it is important to strategize the use of insecticides in pest management. To achieve this objective, it is important to use alternative methods of pest control as and when feasible to avoid routine treatments. In addition, it should be borne in mind that complete control of a pest population is not necessary to prevent economic loss.

Pesticides of Plant Origin

A large number of plant products derived from neem, custard apple, tobacco, pyrethrum, etc., have been used as safer pesticides for pest management. Neem leaves and kernel powder have traditionally been used by farmers against pests of household, agricultural, and medical importance. Neem derivatives comprise a complex array of novel compounds with profound behavioral and physiological effects, such as repellence, phagodeterrance, growth disruption, and inhibition of oviposition (Sharma et al., 1984; Sankaram et al., 1988). Some of these effects have been attributed to azadirachtin, salannin, nimbin, zedunin, and meliantriol (Sharma et al., 1984; Shanker and Parmar, 1999; Sharma, Sankaram, and Nwanze, 1999). The complexity of the chemical structures of neem compounds precludes their synthesis on a practical scale. Therefore, use of neem leaf and seed kernel extract, and neem oil has been recommended for pest management. Although neem is active against a wide range of insect pests, it is known to have little or no effect against major groups of beneficial insects such as spiders, ladybird beetles, parasitic wasps, and predatory mites. Identification and promotion of pesticides of plant origin is one of the alternatives to overcome the ill effects of pesticides. At present, neem products are being marketed globally, although their production and use is limited by the availability of quality raw material. Efforts are needed to identify more molecules of plant origin so that they can be used successfully in pest management in the future.

Host Plant Resistance

With the development of insect resistance to insecticides, adverse effects of insecticides on natural enemies, and public awareness of environment conservation, there has been a renewed interest in the development of crop cultivars with resistance to insect pests. It is important to adopt pest control strategies that are: (1) ecologically sound, (2) economically practical, and (3) socially acceptable. Host plant resistance (HPR) along with natural enemies and cultural practices can play a major role in pest management (Painter, 1951; Smith, 1989; Sharma and Ortiz, 2002). Inspite of the importance of HPR as an important component of IPM, breeding for plant resistance to insects has not been as rapidly accepted as breeding of disease-resistant cultivars. This was partly due to the relative ease with which insect control is achieved with the use of insecticides, and the slow progress in developing insect-resistant cultivars because of the difficulties involved in ensuring adequate insect infestation for resistance screening.

High levels of plant resistance are available against a few insect species only. However, very high levels of resistance are not a prerequisite for use of HPR in IPM. Varieties with low to moderate levels of resistance or those that can avoid the pest damage can be deployed for pest management in combination with other components of pest management (Panda and Khush, 1995; Sharma and Ortiz, 2002). Deployment of pest-resistant cultivars should be aimed at conservation of the natural enemies and minimizing the number of pesticide applications. Use of insect-resistant cultivars also improves the efficiency of other pest management practices, including the synthetic insecticides (Sharma, 1993; Panda and Khush, 1995; Sharma et al., 2003). Host-plant resistance can be used as: (1) a principal component of pest control, (2) an adjunct to cultural, biological, and chemical control, and (3) a check against the release of susceptible cultivars.

HPR as a method of insect control in the context of IPM has a greater potential than any other method of pest suppression. In general, the use of pest-resistant varieties is not subjected to the vagaries of nature, unlike chemical and biological control methods. Use of insect-resistant varieties has contributed immensely to sustainable crop production worldwide (Smith, 1989; Panda and Khush, 1995). Plant resistance as a method of pest control offers many advantages, and in some cases, it is the only practical and effective method of pest management. However, there may be problems if we rely exclusively on plant resistance for insect control, for example, high levels of resistance may be associated with low yield potential or undesirable quality traits, and resistance may not be expressed in every environment wherever a variety is grown. Therefore, insect-resistant varieties need to be carefully fitted into the pest management programs in different agro-ecosystems. Insect-resistant varieties have been deployed for the control of a number of insect pests worldwide (Painter, 1951; Maxwell and Jennings, 1980; Smith, 1989; Sharma and Ortiz, 2002). Several insect pests have been kept under check through the use of insect resistant cultivars, for example, grapevine phylloxera, *Phylloxera vitifoliae* (Fitch.) (resistant rootstocks from the United States); cotton jassid, *A. biguttula biguttula* (Krishna, Mahalaxmi, Khandwa 2, and MCU 5); wooly apple aphid, *Eriosoma lanigerum* (Hausmann) (Northern Spy rootstocks); Hessian fly, *Mayetiola destructor* (Say) (Pawnee, Poso 42, and Benhur); rice gall midge, *Orseola oryzae* Wood-Mason (IR 36, Kakatiya, Surekha, and Rajendradhan); spotted alfalfa aphid, *Therioaphis maculata* (Buckton) (Lahontan, Sonora, and Sirsa); sorghum shoot fly, *A. soccata* (Maldandi, Swati, and Phule Yashoda); sorghum midge, *S. sorghicola* (ICSV 745, ICSV 88032, and ICSV 804); and sorghum head bug, *Eurystylus oldi* Poppius (*guineense* sorghums in West Africa) (Painter, 1951; Adkisson and Dyck, 1980; Maxwell and Jennings, 1980; Smith, 1989; Sharma, 1993; Sharma and Ortiz, 2002).

Integrated Pest Management

Mating Disruption and Mass Trapping

Mating disruption has been tried for controlling several insect pests, such as pink bollworm, *Pectinophora gossypiella* (Saunders), gypsy moth, *Lymantria dispar* L., codling moth, *Cydia pomonella* (L.), and the Guatemalan potato moth, *Tecia solanivora* Povolny (Carde and Minks, 1995). To develop an effective mating disruption program, a number of conditions need to be fulfilled. The target insect should be relatively immobile so that the females that have mated outside the treated area do not enter and lay eggs in the treated fields. Insects such as cotton bollworm, *H. armigera*, which is a highly mobile pest, are very difficult to control with mating disruption unless thousands of hectares are treated simultaneously. The pest should ideally be restricted to a single crop, otherwise all the target crops within an area need to be treated. The pheromone should be synthesized at an economically acceptable cost, for example, the spotted bollworm, *Earias vittella* (Fab.), can be readily controlled with mating disruption, but the method is not economically viable due to the high cost of the pheromone. The pheromone must be stable and formulated such that it releases the pheromone in a controlled manner in the crop habitat.

Mass trapping by pheromone- or kairomone-baited traps can be attempted to reduce insect infestations. It is important to understand that not all insects can be controlled by mass trapping. It is also better to think in terms of population suppression rather than control. The most promising candidates are insects that use aggregation pheromones, such as Spruce bark beetle, *Ips typographus* (L.). For this pest, trap densities of 20 to 30 traps per hectare have been used. Sex pheromones can be used for mass trapping of some insects. It is necessary to catch 95% of the male moths before there is any significant impact on the ability of the population to reproduce. Mobile insects such as *H. armigera* cannot be successfully controlled by mass trapping or mating disruption, as the females that have mated outside the treated area lay eggs in the area where the males may have been successfully removed. Mercury lamps spaced 300 m apart over a large number of contiguous cotton fields reduced the egg laying by 41.5%, and the frequency of pesticide application by two to three times in China (Zhao et al., 1999). However, the application and economics of such an approach need to be looked into critically. Compound traps having two lamps with sex pheromone or poplar branches have been used to control *H. armigera* in China. In comparison to the control plots, the numbers of eggs on cotton plants in plots with traps were reduced by 34.5% within 160 m. Mass trapping has been shown to work successfully for lepidopteran moths, which are relatively immobile, such as rice stem borers (Pyralidae), potato tuber moth [*Phthorimaea operculella* (Zeller)], diamondback moth, *P. xylostella*, and brinjal fruit and shoot borer, *Leucinodes orbonalis* (Guen.) (Howse, Stevens, and Jones, 1997). For pests such as these, trap densities of 10 to 20 traps per hectare have been shown to be effective at reducing damage levels.

Population Prediction Models and Early Warning Systems

Monitoring the movement of insect pests can provide early warning of pest invasion in an area or crop. Although work on long-distance movement using remote sensing, backtracking, and other techniques have indicated that some insects are able to cover large distances, their occurrence in significant numbers at a particular location can seldom be predicted with certainty. Pheromone-baited traps alone or in combination with other lures have been used for monitoring insect populations (Nesbitt et al., 1979), but the relationship between egg, larval, and insect catch in traps is closest only when insect densities are low at the beginning of the season. Trapping is useful as a qualitative measure indicating the initiation of infestation or migration (wave front), and the need to begin scouting for immature stages in the crop.

Models are conceptual or mathematical devices that aim to describe or simulate natural processes. They can be used to predict the outcome of hypothetical eventualities and as management tools to predict or establish the optimal tactics required to achieve a particular result within the constraints of the model. The population models are useful for developing appropriate pest management strategies, such as optimal timing of insecticide application (Apel, Herrmann, and Richter, 1999). The use of phenological or time parameters in predictive models is important to improve their performance. In Australia, the size of the second generation of *H. armigera* is linked to first generation, winter rainfall (positive effect) and spring rainfall (negative effect), which account for 96% of the variation in second generation (Maelzer and Zalucki, 1999).

SIRATAC, a computer-based pest management system, has been developed to rationalize insecticide use on cotton (Hearn et al., 1981). This system incorporates a temperature-driven cotton development model, including the natural fruiting habit of the plant, and submodels to incorporate damage relationships, the impact of natural enemies, and predetermined or dynamic thresholds for pests. The system gives management options, and the outcomes of using "soft" or "hard" insecticides. Significant improvements in this model were obtained by adjusting the threshold at specific times during the crop growth period, which roughly correspond to the three phases of an insecticide resistance management strategy (Cox et al., 1991). The model HEAPS incorporates modules based on adult movement, oviposition, development, survival, and host phenology, and estimates populations in each of a grid of simulation units into which a cotton-producing region is divided, taking into account both bollworm species in cotton as well as in other crops, and noncrop hosts in the region (Dillon and Fitt, 1990). A relatively simple simulation model of *H. armigera* on pigeonpea has been developed by Holt, King, and Armes (1990) to optimize insecticide use for the control of susceptible and resistant larvae of *H. armigera* on pigeonpea. The driving variable is the flowering phenology of the crop, on which oviposition time and survival strongly depend. The optimal time and application frequency to control the progenies of a wholly immigrant population were most sensitive to the time and duration of immigration, flowering time, moth age at immigration, and the development time of young larvae. Like the other models described, this model is also highly specific to the local ecology of the pest and the cropping system in question.

The IPM Practice

In view of the need to make use of and exploit the existing spectra of natural enemies to reduce excessive dependence on chemical control, particularly where there is resistance to insecticides, various IPM programs have been developed in which different control tactics are combined to suppress pest numbers below a threshold (FAO, 1995; Gopal and Senguttuvan, 1997). These vary from judicious use of insecticides based on ETLs and regular scouting to ascertain pest population levels to sophisticated systems, almost exclusively for cotton, using computerized crop and population models to assess the need, optimum timing, and selection of insecticides for sprays.

Classical integrated management programs for apple pests in Canada (Pickett and Patterson, 1953) and for cotton pests in Peru (Dout and Smith, 1971) provided some of the early models for successful implementation of IPM in the field. The FAO subsequently provided the coordination to spread the IPM concept in developing countries. The success of an IPM program in rice in South East Asia (FAO, 1995) was based on linking outbreaks of the brown planthopper, *N. lugens* with the application of broad-spectrum insecticides, and the realization of the fact that the brown planthopper populations were kept under check by the natural enemies in the absence of insecticide application. Much of the impact of this program was brought out through field demonstrations, training programs, and farmers' field schools. As a result, some of the broad-spectrum insecticides were also banned in some countries. The success of some of these programs has led to the collaboration of a global IPM facility under the auspices of FAO, the United Nations Development Program (UNDP), and the World Bank, which will serve as a coordinating and promoting entity for IPM worldwide. The establishment of International Agricultural Research Centers (IARCs) has also contributed significantly to IPM, particularly through the development and promotion of pest-resistant cultivars worldwide.

Is Genetic Engineering of Plants and Biocontrol Agents an Answer?

The promise of biotechnology as an instrument of development lies in its capacity to improve the quantity and quality of plants and biocontrol agents quickly and effectively.

Genetic engineering reduces the time required to combine favorable traits over the conventional methods. Increased precision also translates into improved predictability of the products. The application of biotechnology can create plants that are resistant to drought, insect pests, weeds, and diseases. Plant characteristics can also be altered for early maturity, increased transportability, reduced postharvest losses, and improved nutritional quality. Significant progress has been made over the past three decades in handling and introduction of exotic genes into plants, and has provided opportunities to modify crops to increase yields, impart resistance to biotic and abiotic stress factors, and improve nutrition. Genes from bacteria such as *B. thuringiensis* have been used successfully for pest control through transgenic crops on a commercial scale. Trypsin inhibitors, lectins, ribosome inactivating proteins, secondary plant metabolites, vegetative insecticidal proteins, and small RNA viruses can be used alone or in combination with *Bt* genes (Hilder and Boulter, 1999; Sharma et al., 2002).

In addition to widening the pool of useful genes, genetic engineering also allows the introduction of several desirable genes in a single event, thus reducing the time required to introgress novel genes into the elite background. Toxin genes from *Bt* have been inserted into crop plants since the mid-1980s. Since then, there has been a rapid growth in the area under transgenic crops in the United States, Australia, China, and India, among others. The area planted to transgenic crops increased from 1.7 million ha in 1996 to 100 million ha in 2006 (James, 2007). In addition to the reduction in losses due to insect pests, the development and deployment of transgenic plants with insecticidal genes will also lead to:

- A major reduction in insecticide sprays;
- Reduced exposure of farm labor and nontarget organisms to pesticides;
- Increased activity of natural enemies;
- Reduced amounts of pesticide residues in the food and food products; and
- A safer environment in which to live.

The benefits to growers would be higher yields, lower costs, and ease of management, in addition to reduction in the number of pesticide applications (Griffiths, 1998; Sharma et al., 2002). In the diverse agricultural systems such as those prevailing in the tropics, it is important to understand the biology and behavior of all the insect species in the ecosystem so that informed decisions can be made as to which crops to transform, and the toxins to be deployed. It is also important to consider the resistance management strategies, economic value, and environmental impact of the exotic genes in each crop, and whether a crop serves as a source or sink for the insect pests and their natural enemies.

Conclusions

Crop cultivars derived through conventional plant breeding or biotechnological approaches will continue to play a pivotal role in IPM in different crops and cropping systems. Cultural practices that help reduce the intensity of insect pests should be followed wherever feasible. The role of natural enemies as control agents is not very clear, although efforts should be made to increase their abundance through reducing pesticide application and adopting cropping practices that encourage their activity. Most of the studies have indicated that insecticide applications are more effective than the natural plant products, *Bt*, NPV, or the release of natural enemies. However, biopesticides can be applied in rotation or in combination with the synthetic insecticides. Scouting for eggs and young larvae is most critical for timely application of control measures. Control measures on most crops must start with the onset of infestation and must coincide with egg laying and presence of small larvae. Transgenics with different insecticidal genes can be exploited for sustainable crop production in the future.

References

Adkisson, P.L. and Dyck, V.A. (1980). Resistant varieties in pest management systems. In Maxwell, F.G. and Jennings, P.R. (Eds.), *Breeding Plants Resistant to Insects*. New York, USA: John Wiley & Sons, 233–251.

Altieri, M.A., van Schoonhoven, A. and Doll, J.A. (1977). The ecological role of weeds in insect pest management systems: A review illustrated by bean (*Phaseolus vulgaris*) cropping systems. *Pest Articles and News Summaries (PANS)* 23: 195–205.

Altieri, M.A., Wilson, R.C. and Schmidt, L.L. (1985). The effect of living mulches and weed cover on the dynamics of foliage and soil arthropod communities in three crop systems. *Crop Protection* 4: 2010–2213.

Andow, W. (1991). Vegetational diversity and arthropod population response. *Annual Review of Entomology* 36: 561–586.

Andow, D.A. and Risch, S.J. (1985). Predation in diversified agroecosystems: Relations between a coccinellid predator and its food. *Journal of Applied Ecology* 22: 357–372.

Apel, H., Herrmann, A. and Richter, O. (1999). A decision support system for integrated pest management of *Helicoverpa armigera* in the tropics and subtropics by means of a rule-based fuzzy-model. *Zeitschrift fur Agrarinformatik* 7: 83–90.

Balasubramanian, G., Balasubramanian, M., Sankran, K. and Manickam, T.S. (1986). Effect of fertilization on the incidence of shoot fly and stem borer on rainfed sorghum. *Madras Agricultural Journal* 73: 471–473.

Bateman, R., Carey, M., Moore, D. and Prior, C. (1993). The enhanced infectivity of *Metarhizium flavoviride* in oil formulations to desert locusts at low humidities. *Annals of Applied Biology* 122: 145–152.

Bergman, J.M. and Tingey, W.M. (1979). Aspects of interaction between plant genotypes and biological control. *Bulletin of Entomological Society of America* 25: 275–279.

Bugg, R.L., Ehler, L.E. and Wilson, L.T. (1987). Effect of common knotweed (*Polygonum viculare*) on abundance and efficiency of insect predators of crop pests. *Hilgardia* 55(7): 1–53.

Carde, R.T. and Minks, A.K. (1995). Control of moth pests by mating disruption: Successes and constraints. *Annual Review of Entomology* 40: 559–585.

Chand, P., Sinha, M.P. and Kumar, A. (1979). Nitrogen fertilizer reduces shoot fly incidence in sorghum. *Science and Culture* 45: 61–62.

Corbett, A. and Plant, R.E. (1993). Role of movement in the response of natural enemies to agroecosystem diversification: A theoretical evaluation. *Environmental Entomology* 22: 519–531.

Cox, P.G., Marsden, S.G., Brook, K.D., Talpaz, H. and Hearn, A.B. (1991). Economic optimization of *Heliothis* thresholds on cotton using a pest management model. *Agricultural Systems* 35: 157–171.

Dahiya, S.S., Chauhan, Y.S., Johansen, C. and Shanower, T.G. (1999). Adjusting pigeonpea sowing time to manage pod borer infestation. *International Chickpea and Pigeonpea Newsletter* 6: 44–45.

Daly, J.C. and Gregg, P. (1985). Genetic variation in *Heliothis* in Australia: Species identification and gene flow in the two species *H. armigera* (Hubner) and *H. punctigera* (Wallengren) (Lepidoptera: Noctuidae). *Bulletin of Entomological Research* 75: 169–184.

Das, S.B. (1998). Impact of intercropping on *Helicoverpa armigera* (Hub.): Incidence and crop yield of chickpea in West Nimar Valley of Madhya Pradesh. *Insect Environment* 4: 84–85.

Dillon, M.L. and Fitt, G.P. (1990). HEAPS: A regional model of *Heliothis* population dynamics. In *Proceedings of the Fifth Australian Cotton Conference*, 8–9 August 1990, Broadbeach, Queensland, Australia. Brisbane, Queensland, Australia: Australian Cotton Growers' Research Association, 337–344.

Dout, R.L. and Smith, R.F. (1971). The pesticide syndrome: Diagnosis and prophylaxis. In Huffaker, C.B. (Ed.), *Biological Control*. New York: Plenum Press, 3–15.

Dulmage, H.D. (1970). Insecticidal activity of HD-1, a new isolate of *Bacillus thuringiensis* var. *alesti*. *Journal of Invertebrate Pathology* 15: 232–239.

Dyson, T. (1999). World food trends and prospects to 2025. *Proceedings National Academy of Sciences USA* 96: 5929–5936.

Facknath, S. and Lalljee, B. (2005). Effect of soil-applied complex fertilizer on an insect–host plant relationship: *Liriomyza trifolii* on *Solanum tuberosum. Entomologia Experimentalis et Applicata* 115: 67–77.

Fitt, G.P. (1989). The ecology of *Heliothis* species in relation to agroecosystems. *Annual Review of Entomology* 34: 17–52.

Food and Agriculture Organization (FAO). (1995). *Inter-Country Program for the Development and Application of Integrated Pest Control in Rice in South and South-East Asia*. Rome, Italy: FAO Plant Protection Service.

Food and Agriculture Organization (FAO). (1999). *The State of Food Insecurity in the World*. Rome, Italy: Food and Agriculture Organization.

Forrester, N.W. (1990). Designing, implementing and servicing an insecticide resistance management strategy. *Pesticide Science* 28: 167–179.

Fuxa, J.R. (2004). Ecology of insect nucleopolyhedrosis viruses. *Agriculture, Ecosystems & Environment* 103: 27–43.

Fye, R.E. (1972). The interchange of insect parasites and predators between crops. *Pest Articles and News Summaries (PANS)* 18: 143–146.

Gahukar, R.T. and Jotwani, M.G. (1980). Present status of field pests of sorghum and millets in India. *Tropical Pest Management* 26: 138–151.

Georgis, R. (1992). Present and future prospects for entomopathogenic nematode products. *Biocontrol Science and Technology* 2: 83–99.

Gopal, S. and Senguttuvan, T. (1997). Integrated management of tomato fruit borer with insecticides, neem products and virus. *Madras Agricultural Journal* 84: 82–84.

Griffiths, W. (1998). Will genetically modified crops replace agrochemicals in modern agriculture? *Pesticide Outlook* 9: 6–8.

Hajeck, A.E. and St.-Leger, R.J. (1994). Interactions between fungal pathogens and insect hosts. *Annual Review of Entomology* 39: 293–322.

Hardin, M.R., Benrey, B., Coll, M., Lamp, W.O., Roderick, G.K. and Barbosa, P. (1995). Arthropod pest resurgence: An overview of potential mechanisms. *Crop Protection* 14: 3–18.

Hearn, A.B., Ives, P.M., Room, P.M., Thomson, N.J. and Wilson, L.T. (1981). Computer-based cotton pest management in Australia. *Field Crops Research* 4: 321–332.

Hegde, R. and Lingappa, S. (1996). Effect of intercropping on incidence and damage by *Helicoverpa armigera* in pigeonpea. *Karnataka Journal of Agricultural Sciences* 9: 616–621.

Heinrichs, E.A. and Mochida, O. (1984). From secondary to major pest status: The case of insecticide-induced rice brown planthopper, *Nilaparvata lugens*, resurgence. *Protection Ecology* 7: 201–218.

Hilder, V.A. and Boulter, D. (1999). Genetic engineering of crop plants for insect resistance: A critical review. *Crop Protection* 18: 177–191.

Hoffmann, M.P. and Frodsham, A.C. (1993). *Natural Enemies of Vegetable Insect Pests*. Ithaca, New York, USA: Cooperative Extension, Cornell University.

Holt, J., King, A.B.S. and Armes, N.J. (1990). Use of simulation analysis to assess *Helicoverpa armigera* control on pigeonpea in southern India. *Crop Protection* 9: 197–206.

Howse, P., Stevens, J.M. and Jones, O. (1997). *Insect Pheromones and Their Use in Pest Management*. New York, USA: Marcel Dekker.

James, C. (2007). Global Status of Commercialized Biotech/GM Crops: 2006. ISAAA Briefs no. 35. Ithaca, New York, USA: International Service for Acquisition on Agri-Biotech Applications (ISAAA). http://www.isaaa.org/resources/publications/briefs/35. Accessed June 15, 2007.

King, E.G. and Coleman, R.J. (1989). Potential for biological control of *Heliothis* species. *Annual Review of Entomology* 34: 53–75.

Lambert, B. and Peferoen, M. (1992). Insecticidal promise of *Bacillus thuringiensis*. *Bioscience* 42: 112–121.

Lambert, J.D.H., Arnason, J.T., Serratos, A., Philogene, B.J.R. and Faris, M.A. (1987). Role of inter-cropped red clover in inhibiting European corn borer (Lepidoptera: Pyralidae) damage to corn in eastern Ontario. *Journal of Economic Entomology* 80: 1192–1196.

Maelzer, D.A. and Zalucki, M.P. (1999). Analysis of long-term light-trap data for *Helicoverpa* spp. (Lepidoptera: Noctuidae) in Australia: The effect of climate and crop host plants. *Bulletin of Entomological Research* 89: 455–463.

Mahadevan, N.R. and Chelliah, S. (1986). Influence of intercropping legumes with sorghum on the infestation of the stem borer, *Chilo partellus* (Swinhoe) in Tamil Nadu in India. *Tropical Pest Management* 32: 162–163.

Manjunath, T.M., Bhatnagar, V.S., Pawar, C.S. and Sithanatham, S. (1989). Economic importance of *Heliothis* spp. in India and an assessment of their natural enemies and host plants. In King, E.G. and Jackson, R.D. (Eds.). *Proceedings of the Workshop on Biological Control of* Heliothis: *Increasing the Effectiveness of Natural Enemies*, November 1985. New Delhi, India: Far Eastern Regional Research Office, U.S. Department of Agriculture, 196–228.

Maxwell, F.G. and Jennings P.R. (Eds.) (1980). *Breeding Plants Resistant to Insects*. New York, USA: John Wiley & Sons.

McCalla, A.F. and Ayers, W.S. (1997). *Rural Development: From Vision to Action*. Washington, D.C., USA: The World Bank.

Mohan, S., Jayaraj, S., Purshothaman, D. and Rangarajan, A.V. (1987). Can the use of *Azospirillum* biofertilizer control sorghum shoot fly? *Current Science* 56: 723–725.

Moscardi, F. (1999). Assessment of the application of baculoviruses for control of Lepidoptera. *Annual Review of Entomology* 44: 257–289.

Mote, U.N. (1986). Effect of carbofuran and *Azospirillum* on shoot fly incidence and yield of rabi sorghum cultivars. *Current Research Reporter* 2: 118–121.

Murray, D.A.H. and Zalucki, M.P. (1990). Survival of *Helicoverpa punctigera* (Wallengren) and *H. armigera* (Hubner) (Lepidoptera: Noctuidae) pupae submerged in water. *Journal of Australian Entomological Society* 29: 191–192.

Myers, N. (1999). The next green revolution: Its environmental underpinnings. *Current Science* 76: 507–513.

Narkhede, P.L., Umrani, N.K. and Surve, S.P. (1982). Shoot fly incidence in relation to "P" fertilization to winter sorghum. *Sorghum Newsletter* 25: 79–90.

Navon, A. (2000). *Bacillus thuringiensis* insecticides in crop protection: Reality and prospects. *Crop Protection* 19: 669–676.

Neale, M.C. (1997). Bio-pesticides: Harmonization of registration requirements within EU directive 91-414. An industry view. *Bulletin OEPP* 27: 89–93.

Nesbitt, B.F., Beevor, P.S., Hall, D.R. and Lester, R. (1979). Female sex pheromone components of the cotton bollworm, *Heliothis armigera*. *Journal of Insect Physiology* 25: 535–541.

Oerke, E.C. (2006). Crop losses to pests. *Journal of Agricultural Science* 144: 31–43.

Painter, R.H. (1951). *Insect Resistance in Crop Plants*. New York, USA: MacMillan.

Panda N. and Khush, G.S. (1995). *Host Plant Resistance to Insects*. Wallingford, Oxon, UK: CAB International.

Parnell, M.A., King, W.J., Jones, K.A., Ketunuti, U. and Wetchakit, D. (1999). A comparison of motor-ized knapsack mistblower, medium volume application, and spinning disk, very low volume application, of *Helicoverpa armigera* nuclear polyhedrosis virus on cotton in Thailand. *Crop Protection* 18: 259–265.

Pawar, C.S., Bhatnagar, V.S. and Jadhav, D.R. (1986). *Heliothis* species and their natural enemies, with their potential for biological control. *Proceedings of the Indian Academy of Science (Animal Science)* 95: 695–703.

Pickett, A.D. and Patterson, N.A. (1953). The influence of spray programs on the fauna of apple orchards in Novo Scotia. IV. A review. *Canadian Entomologist* 85: 472–478.

Pinstrup-Andersen, P. and Cohen, M. (2000). Modern biotechnology for food and agriculture: Risks and opportunities for the poor. In Persley, G.J. and Lantin M.M. (Eds.), *Agricultural Biotechnology and the Poor*. Washington, D.C., USA: Consultative Group on International Agricultural Research, 159–172.

Poinar Jr., G.O. (1990). Taxonomy and biology of *Steinernematidae* and *Heterorhabtidae*. In Gaugler, R. and Kaya, H.K. (Eds.), *Entomopathogenic Nematodes in Biological Control*. Boca Raton, Florida, USA: CRC Press, 23–61.

Pokharkar, D.S., Chaudhary, S.D. and Verma, S.K. (1999). Utilization of nuclear polyhedrosis virus in the integrated control of fruit-borer (*Helicoverpa armigera*) on tomato (*Lycopersicon esculentum*). *Indian Journal of Agricultural Sciences* 69: 185–188.

Rajmohan, N. (1998). Pesticide resistance: A global scenario. *Pesticides World* 3: 34–40.

Ramarethinam, S., Marimuthu, S., Rajagopal, B. and Murugesan, N.V. (1998). Relative changes in the virulence and infectivity of the nuclear polyhedrosis virus in relation to the dietary habit of *Helicoverpa armigera* (Hubner) (Lep., Noctuidae). *Journal of Applied Entomology* 122: 623–628.

Ramert, B. (1993). Mulching with grass and bark and intercropping with *Medicago littoralis* against carrot fly [*Psila rosae* (F.)]. *Biological Agriculture and Horticulture* 9: 125–135.

Reddy, D.V.R., Sharma, H.C., Gaur, T.B. and Divakar, B.J. (Eds.). (1997). *Plant Protection and Environment*. Rajendranagar, Hyderabad, Andhra Pradesh, India: Plant Protection Association of India.

Romeis, J. and Shanower, T.G. (1996). Arthropod natural enemies of *Helicoverpa armigera* in India. *Biocontrol Science and Technology* 6: 481–508.

Russell, E.P. (1989). Enemies hypothesis: A review of the effect of vegetational diversity on predatory insects and parasitoids. *Environmental Entomology* 18: 590–599.

Sankaram, A.V.B., Murthy, M.M., Bhaskaraiah, K., Subramanyam, M., Sultana, N., Sharma, H.C., Leuschner, K., Ramaprasad, G., Sitaramaiah, S., Rukmini, C. and Rao, P.U. (1988). Chemistry, biological activity and utilization aspects of some promising neem extractives. In Schmutterer, H. and Ascher, K.R.S. (Eds.), *Natural Pesticides From Neem Tree and Other Tropical Plants*. Eschborn, Germany: German Society for Technical Cooperation (GTZ), 127–148.

Sawicki, R.M. and Denholm, I. (1987). Management of resistance to pesticides in cotton. *Tropical Pest Management* 33: 262–272.

Saxena, H. and Ahmad, R. (1997). Field evaluation of *Beauveria bassiana* (Balsamo) Vuillemin against *Helicoverpa armigera* (Hubner) infecting chickpea. *Journal of Biological Control* 11: 93–96.

Schmidt, M.H., Thewes, H., Thies, C. and Tscharntke, T. (2004). Aphid suppression by natural enemies in mulched cereals. *Entomologia Experimentalis et Applicata* 113: 87–93.

Schweissing, F.C. and Wilde, G. (1979). Temperature and plant nutrient effects on resistance of seedling sorghum to the greenbug. *Journal of Economic Entomology* 72: 20–23.

Shanker, J.S. and Parmar, B.S. (1999). Recent developments in botanicals and biopesticides. *Indian Journal of Plant Protection* 27: 139–154.

Sharma, H.C. (1985). Strategies for pest control in sorghum in India. *Tropical Pest Management* 31: 167–185.

Sharma, H.C. (1993). Host plant resistance to insects in sorghum and its role in integrated pest management. *Crop Protection* 12: 11–34.

Sharma, H.C. (2001). *Cotton Bollworm/Legume Pod Borer,* Helicoverpa armigera *(Hubner) (Noctuidae: Lepidoptera): Biology and Management. Crop Protection Compendium*. Wellingford, UK: Commonwealth Agricultural Bureau International.

Sharma, H.C. (Ed.). (2005). Heliothis/Helicoverpa *Management: Emerging Trends and Strategies for Future Research*. New Delhi, India: Oxford and IBH Publishing.

Sharma, H.C. and Adlakha, R.L. (1981). Selective toxicity of some insecticides to the adults of ladybird beetle, *Coccinella septempunctata* L. and cabbage aphid, *Brevicoryne brassicae* L. *Indian Journal of Entomology* 3: 92–99.

Sharma, H.C. and Adlakha, R.L. (1986). Relative toxicity of different insecticides to *Coccinella septempunctata* L. after predating upon poisoned cabbage aphid, *Brevicoryne brassicae* L. *Indian Journal of Entomology* 48: 204–211.

Sharma, H.C. and Ortiz, R. (2002). Host plant resistance to insects: An eco-friendly approach for pest management and environment conservation. *Journal of Environmental Biology* 23: 111–135.

Sharma, H.C. and Veerbhadra Rao, M. (Eds.). (1995). *Pests and Pest Management in India: The Changing Scenario*. Rajendranagar, Andhra Pradesh, India: Plant Protection Association of India.

Sharma, H.C., Crouch, J.H., Sharma, K.K., Seetharama, N. and Hash, C. T. (2002). Applications of biotechnology for crop improvement: prospects and constraints. *Plant Science* 163: 381–395.

Sharma, H.C., Leuschner, K., Sankaram, A.V.B., Gunasekhar, D., Marthandamurthi, K., Bhaskariah, M., Subramanyam, M. and Sultana, N. (1984). Insect antifeedants and growth inhibitors from *Azadirachta indica* and *Plumbago zeylanica*. In Schmutterer, H. and Ascher, K.R.S. (Eds.), *Natural Pesticides from the Neem Tree and Other Tropical Plants*. Eschborn, Germany: German Society for Technical Cooperation (GTZ), 291–320.

Sharma, H.C., Pampapathy, G. and Sullivan, D.J. (2003). Influence of host plant resistance on activity and abundance of natural enemies. In Ignacimuthu, S. and Jayaraj, S. (Eds.), *Biological Control of Insect Pests*. New Delhi, India: Phoenix Publishing House, 282–296.

Sharma, H.C., Sankaram, A.V.B. and Nwanze, K.F. (1999). Utilization of natural pesticides derived from neem and custard apple in integrated pest management. In Singh, R.P. and Saxena, R.C. (Eds.), *Azadirachta indica* A. Juss. New Delhi, India: Oxford & IBH Publishing, 199–211.

Sharma, H.C., Sullivan, D.J., Sharma, M.M. and Shetty, S.V.R. (2004). Influence of weeding regimes and pearl millet genotypes on parasitism of the Oriental armyworm, *Mythimna separata*. *BioControl* 49: 689–699.

Sharma, H.C., Taneja, S.L., Kameswara Rao, N. and Prasada Rao, K.E. (2003). *Evaluation of Sorghum Germplasm for Resistance to Insect Pests*. Information Bulletin no. 63. Patancheru 502 324, Andhra Pradesh, India: International Crops Research Institute for the Semi-Arid Tropics (ICRISAT).

Smith, C.M. (1989). *Plant Resistance to Insects*. New York, USA: John Wiley & Sons.

Smith, R.F., Huffaker, C.B., Adkisson, P.L. and Newsom, D.L. (1974). Progress achieved in the implementation of integrated control projects in the USA and tropical countries. *EPPO Bulletin* 4: 221–239.

Sonalkar, V.U., Deshmukh, S.D., Satpute, U.S. and Ingle, S.T. (1998). Efficacy of nuclear polyhedrosis virus in combination with adjuvants against *Helicoverpa armigera* (HBN). *Journal of Soils and Crops* 81: 67–69.

Srinivasan, T.R. (1997). *Studies on Pathogenicity and Virulence of* Metarhizium anisopliae *(Metsch.),* Metarhizium flavoride *(Metsch.) and* Nomuraea rileyi *(Farlow) Samson and Management of* Helicoverpa armigera *(Hübner) and* Spodoptera litura *(F.) in* Lycopersicon esculentum *(L.)*. Ph.D. thesis, Tamil Nadu Agricultural University, Coimbatore, Tamil Nadu, India.

Sunderland, K. and Samu, K. (2000). Effects of agricultural diversification on the abundance, distribution, and pest control potential of spiders: A review. *Entomologia Experimentalis et Applicata* 95: 1–13.

Swaminathan, M.S. (2000). Genetic engineering and food security: Ecological and livelihood issues. In Persley, G.J. and Lantin, M.M. (Eds.), *Agricultural Biotechnology and the Rural Poor*. Washington, D.C., USA: Consultative Group on International Agricultural Research, 37–44.

Topham, M. and Beardsley, J.W. Jr. (1975). Influence of nectar source plants on the New Guinea sugarcane weevil parasite, *Lixophaga sphenophori* (Villeneuve). *Proceedings of the Hawaiian Entomology Society* 22: 145–154.

Zalucki, M.P., Dalglish, G., Firempong, S. and Twine, P. (1986). The biology and ecology of *Heliothis armigera* (Hubner) and *H. punctigera* Wallengren (Lepidoptera: Noctuidae) in Australia: What do we know? *Australian Journal of Ecology* 34: 779–814.

Zhao, W.X., He, J.F., Zhang, G.Y., Li, J.S. and Liu, Q. (1999). A study on techniques to trap and kill cotton bollworm adults with high-voltage mercury lamps. *Acta Agriculturae Universitatis Henanensis* 33: 151–155.

2

Applications of Biotechnology in Agriculture: The Prospects

Introduction

There has been a remarkable increase in grain production over the past five decades, but only a marginal increase was realized after the 1990s (Myers, 1999). Much of the increase in grain production resulted from an increase in the area under cultivation, irrigation, better agronomic practices, and improved cultivars. Yields of several crops have already reached a plateau in developed countries and, therefore, most of the productivity gains in the future will have to be achieved in developing countries through better management of natural resources and crop improvement. Productivity gains are essential for long-term economic growth, but in the short term, these are even more important for maintaining adequate food supplies for the growing population. It is in this context that biotechnology will play an important role in increasing food production in the near future. There is a need to take a critical, but practical look at the prospects of biotechnological applications for increasing crop production and improving nutritional quality. Genetic engineering offers plant breeders access to an infinitely wide array of novel genes and traits, which can be inserted into high-yielding and locally adapted cultivars (Sharma, 2004). This approach offers rapid introgression of novel genes and traits into elite agronomic backgrounds. Biotechnological tool can be used to:

- Develop new hybrid crops based on genetic male-sterility;
- Exploit apomixis to fix hybrid vigor in crops;
- Increase resistance to insect pests, diseases, herbicides, and abiotic stress factors;
- Improve effectiveness of natural enemies and entomopathogenic bacteria, viruses, and fungi;
- Enhance nutritional value of crops through enrichment with vitamin A and essential amino acids;

- Improve shelf life and postharvest quality of produce;
- Increase efficiency of phosphorus uptake and nitrogen fixation;
- Improve adaptation to soil salinity and aluminum toxicity;
- Understand the nature of gene action and metabolic pathways;
- Increase photosynthetic activity, sugar, starch production, and crop yield; and
- Produce antibodies, pharmaceuticals, and vaccines.

New crop cultivars with resistance to insect pests and diseases combined with biocontrol agents should lead to a reduced reliance on pesticides, and thereby reduce farmers' crop protection costs, while benefiting both the environment and public health. Similarly, genetic modification for herbicide resistance to achieve efficient and cost-effective weed control can increase farm incomes, while reducing the labor demand for weeding and herbicide application. Labor released from agriculture can then be used for other profitable endeavours. By increasing crop productivity, agricultural biotechnology can substitute for the need to cultivate new land and thereby conserve biodiversity in areas that are marginal for crop production. The potential of these technologies has been tested extensively in model crop species of temperate and subtropical agriculture. However, there is an urgent need for an increased focus on crops relevant to the small farm holders and poor consumers in the developing countries of the humid and semiarid tropics. The promise of biotechnology can be realized by utilizing the information and products generated through research on genomics and genetic engineering to increase the productivity of crops through enhanced resistance to biotic and abiotic stress factors and improved nutritional quality (Sharma et al., 2002).

The Genomics Revolution

The last decade has seen the whole genome sequencing of model organisms such as human (TIHGMC, 2001), yeast (Piskur and Langkjaer, 2004), *Caenorhabditis elegans* White (Chalfie, 1998), *Arabidopsis thaliana* (L.) Heynh. (TAGI, 2000; Shoemaker et al., 2001), and rice, *Oryza sativa* L. (Palevitz, 2000). The whole genome sequencing is now being carried out for several other crop species, such as *Zea mays* L., *Sorghum bicolor* (L.) Moench., *Medicago sativa* L., and *Musa* spp. Systematic whole genome sequencing provides critical information on gene and genome organization and function, and has revolutionized our understanding of crop production, and ability to manipulate traits contributing to high crop productivity (Pereira, 2000). Similarly, advances in microarray technology will allow the simultaneous expression and analysis of vast numbers of genes that will elucidate gene function, and the complex multifaceted interactions between genes that result in different phenotypes under varying environmental conditions (Shoemaker et al., 2001). These studies need to be augmented by more specific investigations based on gene suppression, cosuppression, or antisensing of a defined sequence (Primrose and Twyman, 2003). This knowledge from model systems will increase our understanding of plant biology and thereby increase our ability to exploit genomic information in agriculture. Advances in these areas will fuel the mapping of QTL (quantitative trait loci) underlying agronomic traits in plants and other organisms. The use of QTL markers in crop improvement promises to revolutionize plant breeding in particular, facilitating the rapid and efficient utilization of novel traits from closely related wild species.

Marker-Assisted Selection

It takes five to six generations to transfer a trait within a species into high-yielding, locally adapted cultivars through conventional breeding, and one has to plant a large number of progenies to be able to select the plants with appropriate combination of traits (Figure 2.1). The improved lines developed then have to go through a set of multilocation tests, before a variety can be identified for cultivation by farmers. This process takes a minimum of seven to ten years. Recombinant DNA technologies, besides generating information on gene sequences and function, allows the identification of specific chromosomal regions carrying, genes contributing to traits of economic interest (Karp et al., 1997). The identification of DNA markers for traits of interest usually depends on making crosses between two genotypes with substantial and heritable differences in the trait(s) of interest. Depending on the crop and traits involved, mapping populations are then derived from the progeny of this cross by selfing once, many times, backcrossing to one of the parental genotypes (BC), or subjecting plants to tissue culture to generate doubled haploids (DH). The major advantage of most mapping populations is that each line is homozygous and therefore can be multiplied indefinitely through self-pollination. This then allows the population to be evaluated under many environments and seasons, facilitating a much more accurate estimate of phenotypic variation, on which to base the mapping exercise. Mapping populations also allow scientists from many diverse disciplines to study different aspects of the same trait in the same population. This approach can only be used when parental genotypes can be identified with opposing phenotypes for the trait of interest. Interspecific crosses can also be used to good effect, but linkage maps derived from such crosses may have limited relevance in crop improvement programs (Fulton et al., 1997).

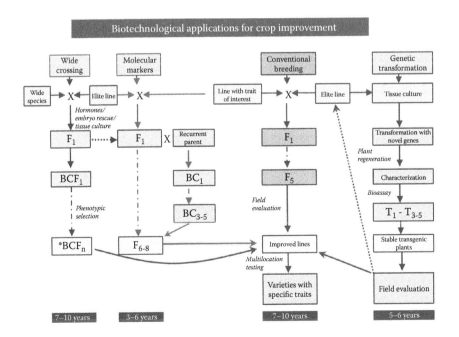

FIGURE 2.1 Application of tools of biotechnology in crop improvement.

For marker-assisted selection (MAS), the elite lines can be crossed with another line having trait(s) of interest. The F_1 hybrid is crossed with the recurrent parent (invariably the elite parent) (BC_1), and the gene transfer is monitored through MAS until BC_{3-5} (until the QTL or the gene of interest is transferred into the elite line). In wild relatives that are not easily crossable with the cultivated types, the F_1 hybrids may have to be produced through embryo rescue and tissue culture, and the progenies advanced as in the conventional backcross breeding approach (phenotypic selection) or through MAS, using cultivated species as the recurrent parent. Progenies from F_2 to F_{6-8} generations can also be advanced as per conventional pedigree breeding, and plants with appropriate combination of traits can be used as improved varieties or as donor parents in conventional breeding. The F_{6-8} progenies can also be used as random inbred lines (RILs) for mapping the trait(s) of interest. For RILs, 250 to 300 plants are selected at random in F_2s, and advanced by selfing the plants at random in each line in each generation. The plants obtained in BC_5 can be used as isogenic lines to study the inheritance or role of traits of interest. The MAS takes three to six years and thus speeds up the pace of transferring the traits of interest into the improved varieties, and it does not require large-scale planting of the progenies up to crop harvest, as only the plants showing the presence of the trait or QTL need to be maintained up to maturity. Wide hybridization may take seven to ten years or longer (BCF_n), depending on the success in transferring the trait(s) of interest into the elite line without other unwanted traits that influence the quality of the produce and productivity potential of the crop.

Once genomic regions contributing to the trait of interest have been assigned and the alleles at each locus designated, they can be transferred into locally adapted, high-yielding cultivars by making requisite crosses. The offspring with a desired combination of alleles can then be selected for further evaluation using MAS. Wild relatives of commercial crops contain alleles of importance for improving crop performance and resistance to biotic and abiotic stress factors, and these can be effectively incorporated into crop breeding programs through MAS (Xiao et al., 1996). DNA marker technology has been used in commercial plant breeding programs since the early 1990s, and has proved useful for the rapid and efficient transfer of the traits into agronomically desirable varieties and hybrids (Miflin, 2000). The development of genetic maps in a number of species has led to the realization of positional similarity of maps across species. This "synteny" will allow advances made in one species to be applied in others relatively easily (Gale and Devos, 1998). This information can also be used by biochemists and physiologists to understand the genetics of metabolic processes, analyze traits controlled by several QTLs, and identify favorable alleles at each locus. The alleles can be combined by simple crossing, and the most favorable combinations assembled in the same background.

The use of DNA markers for indirect selection offers greatest potential gains for quantitative traits with low heritability as these are the most difficult characters to work with in the field through phenotypic selection. However, these types of traits are also among the most difficult to develop effective MAS systems. The expression of these traits can be greatly affected by "genotype-by-environment interaction" and "epistasis," which can complicate the development of MAS systems to the same extent that they confound traditional field-based selection. The quality of a marker-assisted selection program can only be as good as the quality of the phenotypic data on which the development of that marker was based. It is therefore essential to use large mapping populations, which are precisely and accurately characterized in many locations across several years. The selective power of markers must then be verified in a range of populations representing the diversity of current breeding populations. Only then will it be possible to identify markers that can be applied effectively and robustly to assist the selection of complex characters.

Gene Sequence and Function

Genes can be discovered using a variety of approaches (Shoemaker et al., 2001; Primrose and Twyman, 2003), but a routine large-scale approach can commonly be followed by generating and sequencing a library of expressed genes. This library typically consists of thousands of strands of complementary DNA (cDNA) that represent abundantly expressed genes under particular environmental conditions at the sampled growth stage. When sequenced, these cDNAs are termed expressed sequence tags (EST). A large number of ESTs are now available in the public databases for several crops, such as *Z. mays, A. thaliana, O. sativa, S. bicolor,* and *Glycine max* (L.) Merr. A comparison of the EST databases from different plants can reveal the diversity in coding sequences between closely and distantly related species while mapping of ESTs may elucidate the synteny between those species. When a high level of sequence similarity is detected between an EST and a gene of known function in another species, it is possible to infer probable gene function in the species of interest. However, the emphatic elucidation of gene function still requires experimental verification. Only a small proportion of genes are abundantly transcribed in any particular environment or growth stage and, therefore, a complete picture can only be obtained by generating a range of cDNA libraries from plants grown under different environmental conditions and sampled at different growth stages or by sequencing an entire cDNA genome library. For understanding gene functions of a whole organism, functional genomics technology is now focused on high throughput methods using insertion mutant isolation, gene chips or microarrays, and proteomics. These and other high throughput techniques offer powerful new uses for the genes discovered through sequencing (Hunt and Livesey, 2000).

Metabolic Pathways

Knowledge of the changes in a specific plant function induced by different treatments has led to the development of methods to isolate genes involved in the metabolic pathways or their associated physiology (Zhang and Forde, 2000). The availability of tagged mutant populations and use of elegant screening systems based on knowledge of metabolism provide a relatively easy approach to isolating the genes for key steps. Many secondary plant metabolites such as alkaloids, terpenoids, and flavonoids have been implicated in several functions in plant physiology, including host plant resistance to biotic stress factors. Many compounds of the flavonoid biosynthetic pathway (flavanones, flavones, flavanols, and isoflavonoids) accumulate in response to biotic and abiotic stresses (Ebel, 1986; Sharma and Norris, 1991; Heller and Forkman, 1993). Chalcone synthase catalyzes the condensing of 3-malonyl-CoA and hydroxy cinnamoly-CoA ester to form the chalcone intermediate, and chalcone is converted into flavanone by chalcone isomerase. Flavanones are converted into flavones by flavone synthase. Dihydroflavanols are derived from flavanones, which are precursors for the production of flavonols and anthocyanins. Flavonoids and anthocyanins are associated with resistance to insect pests and diseases in several crops. Genetic engineering can be used to change the metabolic pathways to increase the amounts of secondary metabolites, which play an important role in host plant resistance to insect pests and diseases, for example, medicarpin and sativan in alfalfa, cajanol and stilbene in pigeonpea, deoxyanthocyanidin flavonoids (luteolinidin, apigenindin, etc.) in sorghum, and stilbene in chickpea (Heller and Forkman, 1993). The expression of phytoalexins in transgenic plants may be difficult due to complexities involved in their biosynthesis. However, stilbenes have been expressed in transgenic tobacco plants exhibiting various degrees of inhibition of fungal growth. Molecular mechanisms underlying the activation of defense genes implicated in phytoalexin biosynthesis are quite common in a large number of plant species. Biotechnology offers a great promise to increase the production of secondary metabolites in plants that are used in medicine, aromatic industry, and in host resistance to insect pests and diseases, or inhibit the production of toxic metabolites in crop produce meant for food, feed, and fodder.

Trait Analysis

Gene pools in plants may have begun diverging over 150 million years ago. The resultant diversity in genes has led to variation in expression of traits, and generation of completely new plant functions and phenotypes. Important traits in field crops can now be addressed from a general perspective through comparative gene function analysis using model plants, for example, the gene for leafy mutant phenotype in *A. thaliana* is a single gene determining initiation of flowering (Pereira, 2000). This type of approach has opened up an exciting new field of "allele mining" of germplasm collections. The known sequence of a gene can now be used to identify related genes in crop plants. Alternatively, DNA marker linkage maps can be used to analyze the genetic basis of traits and identify allelic variants. In this way, complex traits can be dissected into their component genes through intensive fine mapping. Map-based cloning of genes has been successful in a number of cases, and is becoming easier with the development of different genomic libraries of crops in a range of yeast- and bacteria-based vectors.

Genetic Transformation

Genetic transformation offers direct access to a vast pool of useful genes not previously available for crop improvement (Dunwell, 2000; Sharma et al., 2000, 2002). Current genetic engineering techniques allow the simultaneous use of several desirable genes in a single event, thus allowing coordinated approaches for introduction of novel genes or traits into the elite background. The priorities for applied transgenic research are similar to those of conventional plant breeding, aiming to selectively alter, add, or remove a specific character in order to address biotic and abiotic constraints to productivity. Genetic engineering also offers the possibility of introducing a desirable character from closely related plants without associated deleterious genes or from related species that do not readily cross with the crop of interest, or from completely unrelated species, even in other taxonomic phyla. In many species, the development of rapid, efficient, and routine transformation systems is still in progress, and this represents a bottleneck in the development of stable, high-yielding transgenic plants in many crop species. Development and deployment of transgenic plants is an important prerequisite for sustainable and economic use of biotechnology for crop improvement. As a result of advances in genetic transformation and gene expression during the last two decades, there has been rapid progress in using genetic engineering for crop improvement in terms of herbicide tolerance, insect resistance, and male-sterility systems (Hilder and Boulter, 1999; Sharma, Sharma, and Crouch, 2004). The potential of this technology has now been widely recognized and extensively adopted in the private sector for crop improvement.

Resistance to Insect Pests, Diseases, and Herbicides

The first transgenic plants with *Bacillus thuringiensis* (*Bt*) (Berliner) genes were produced in 1987 (Barton, Whiteley, and Yang, 1987; Vaeck et al., 1987). While most of the insect-resistant transgenic plants have been developed by using *Bt* δ-endotoxin genes, many studies are underway to use non-*Bt* genes, which interfere with the nutritional requirements of the insects. Such genes include protease inhibitors, chitinases, secondary plant metabolites, and lectins (Hilder and Boulter, 1999; Sharma, Sharma, and Crouch, 2004). Genes conferring resistance to insects have been inserted into crop plants such as maize, cotton, potato, tobacco, rice, broccoli, lettuce, walnuts, apple, alfalfa, and soybean. A number of transgenic crops have now been released for on-farm production or field testing (James, 2007). The first transgenic crop with resistance to insects was grown in 1994, and large-scale cultivation was undertaken in 1996 in the United States. Since then, there has been a rapid increase in the area sown with transgenic crops in the United States, Canada, Australia, Argentina, India, and China. Transgenic crops are now grown in over 25 countries in the world. Successful control of cotton bollworms has been achieved through transgenic cotton (Wilson et al., 1992; Benedict et al., 1996; Zhao et al., 1998). Cry type toxins from *Bt* are effective against cotton bollworms, *Heliothis virescens* F. and *Helicoverpa armigera* (Hubner), corn earworm, *Helicoverpa zea* (Boddie), the European corn borer, *Ostrinia nubilalis* (Hubner), and rice stem borer, *Scirpophaga incertulas* (Walker) (Armstrong et al., 1995; Benedict et al., 1996; Nayak et al., 1997; Mqbool et al., 1998; Alam et al., 1999). Successful expression of *Bt* genes against the lepidopterous insects has also been achieved in tomato (Delannay et al., 1989), potato (Jansens et al., 1995), brinjal (Kumar et al., 1998), groundnut (Singsit et al., 1997), and chickpea (Kar et al., 1997).

There will be tremendous benefits to the environment through the deployment of transgenic plants for pest management (Sharma and Ortiz, 2000). Deployment of insect-resistant crops has led to 1 million kg reduction of pesticides applied for pest control in the United States in 1999 as compared to 1998 (NRC, 2000). Papaya with transgenic resistance to ringspot virus (Gonsalves, 1998) has been grown in Hawaii since 1996. Rice yellow mottle virus (RYMV), which is difficult to control with conventional approaches, can now be controlled through transgenic rice, which will provide insurance from total crop failure. Globally, herbicide-resistant soybean and insect-resistant maize and cotton account for 85% of the total area under transgenic crops. The area planted to genetically improved crops has increased dramatically from less than 1 million ha in 1995 to 100 million ha in 2006 (James, 2007). Transgenic plants with insecticidal genes are set to feature prominently in pest management in both the developed and developing worlds in the future. Such an effort will play a major role in minimizing insect-associated losses, increase crop production, and improve the quality of life for the rural poor. Development and deployment of transgenic plants with insecticidal genes for pest control will lead to:

- Reduction in insecticide sprays.
- Increased activity of natural enemies.
- IPM of secondary pests.

Tolerance to Abiotic Stresses

Development of crops with an inbuilt capacity to withstand abiotic stresses would help stabilize the crop production and significantly contribute to food security in developing countries. Plants expressing trehalose phosphate synthase, and trehalose phosphate phosphatase have larger leaves, altered stem growth, and improved response to stress (Goddijn et al., 1997; Pilon Smits et al., 1998). Over-expression of various glutamate dehydrogenases (GDH) also improves plant growth and stress tolerance. Plants have been specifically transformed with genes encoding α- and β-subunits of the chloroplast-located GDH from the alga, *Chlorella sorokiniana* (Shihira and Krauss) (Schmidt and Miller, 1997). Similar improvements in plant performance have been reported for rice plants transformed with the barley late embryogenesis (LEA) gene (Wu and Ho, 1997). Plants with an ability to produce more citric acid in roots provide tolerance to aluminum in acid soils (De la Fuente et al., 1997). Introduction of functional calcineurin activity provides tolerance to salinity (Pardo, Hasegawa, and Bressan, 1999; McCourt et al., 1999) involving the introduction of a gene encoding a plant farnesyltransferse (Pei et al., 1998) and inhibitors of this enzyme when expressed in plants enhance drought tolerance, improve resistance to senescence, and modify the growth habit. A salt tolerance gene isolated from mangrove, *Avicennia marina* (Forssk.) Vierh., has been cloned, and can be transferred into other crop plants (Swaminathan, 2000). The *gutD* gene from *Escherichia coli* Escherich can also be used to provide salt tolerance (Liu et al., 1999). These genes hold a great potential for increasing crop production in marginal lands.

Increased Starch and Sugar Production

Sucrose phosphate synthase (SPS) is a key enzyme in the regulation of sucrose metabolism. Transgenic plants expressing the maize SPS under the control of a promoter from the small subunit of tobacco (*Rubisco*) has shown increased foliar sucrose or starch ratios in leaves, and decreased amounts of foliar carbohydrates when plants were grown with CO_2 enrichment (Signora et al., 1998). Modification of the activity of metabolites of the TCA (tricarboxylic acid) cycle by reducing the amount of the NAD-malic enzyme can also be used for increasing starch concentrations (Leaver et al., 1998). Introduction of the *E. coli* inorganic pyrophosphatase to alter the amount of sugar (Sonnewald and Willmitzer, 1996), and modification of hexokinases (Sheen, 1998), which affect the sugar-sensing capacities of a plant, and sucrose-binding proteins (Grimes and Chao, 1998), and a class of cupin protein (Dunwell, 1998) have been implicated in sugar unloading in developing legume seeds. This has opened up exciting possibilities for changing the chemical composition of the food grains to meet specific requirements.

Altering Senescence and Drought Resistance

It has long been argued that a reduction in senescence (Smart et al., 1996; De Nijs, Broer, and Van Doorn, 1997) would improve the performance of plants and thereby increase crop yield. Introduction of farnesyl transferase and the isopentenyl transferase genes delays senescence. Senescence associated promoters *SAG1* and *SAG2* will be useful for producing transgenic plants with improved performance (Amasino and Gan, 1997). Delayed leaf senescence is also associated with resistance to drought.

Increased Photosynthetic Efficiency, Crop Growth, and Yield

An exciting experimental approach to radically change plant metabolism is currently being investigated with respect to introducing the C_4 type of photosynthesis into C_3 plants such as *A. thaliana* (Ishimaru et al., 1997) and potato (Ishimaru et al., 1998). The C_3 photosynthesis suffers from O_2 inhibition due to the oxygenase reaction of ribulose 1,5-biophosphate carboxylase/oxygenase (Rubisco), and the subsequent loss of CO_2 from photorespiration. In contrast, C_4 plants such as maize have evolved a biochemical mechanism to overcome this inhibition. A key feature of this mechanism is the activity of phosphoenolpyruvate carboxylase (PEPC) (Grula and Hudspeth, 1999), which fixes atmospheric CO_2 in the cytosol of mesophyll cells. Using an *Agrobacterium*-mediated transformation system, the intact maize gene for PEPC has recently been transferred into the C_3 plants (Matsuoka et al., 1998a, 1998b; Ku et al., 1999). Physiologically, these plants exhibit reduced O_2 inhibition of photosynthesis and have photosynthetic rates comparable to those of control untransformed plants. Investigations into the manipulation of the key photosynthetic enzymes, Rubisco, pyruvate phosphate kinase (PPDK), and NADP malate dehydrogenase (NADP-MDH) in the C_4 dicotyledonous species, *Flaveria bidentis* (L.) Kuntze have also been reported (Furbank et al., 1997). An alternative strategy to reduce photorespiration by manipulating catalase amounts in tobacco has also been described (Brisson, Zelitch, and Havir, 1998). Appropriate manipulation of the enzymes involved in photosynthetic activity can be used to increase the productivity potential of C_3 plants.

Genes determining plant height in *A. thaliana* are orthologous (similar to) to dwarf genes in cereals, which have been used in conventional plant breeding in the Green Revolution (Pereira, 2000). These genes (*NORIN 10*) were introduced into Western wheat varieties in the 1950s and have now been isolated, and identical phenotypes reconstructed in other crops through genetic transformation (Peng et al., 1999). The dwarfing genes can now be routinely deployed in various crop species to increase crop productivity. Improved yield can also be achieved by manipulation of fructose-1,6-bisphosphate aldolase (FDA), an enzyme that reversibly catalyses the conversion of triosephosphate to fructose-1,6-bisphosphate. Leaves of transgenic plants expressing FDA from *E. coli* in the chloroplast show significantly enhanced starch accumulation, lower sucrose concentration, and higher root mass (Barry, Cheikh, and Kishore, 1998).

A more generic method for changing plant performance may be to modify plastid number (Osteryoung, 1998), and the expression of a hybrid protein comprising a yeast gene encoding 5-amino levulinic acid synthase and an N-terminal transit sequence for the small subunit of carboxydismutase (Grimm, 1998). Manipulation of chlorophyll a/b binding genes has also been used to modify chlorophyll amounts. Degreening of oilseed rape caused by sublethal freezing during seed maturation (Johnson-Flanagan et al., 1998) can be accomplished by antisense reduction of the type I chlorophyll a/b binding protein of light harvesting complex II (Kirchanski, 1998). Other nonphotosynthetic approaches to increase yield of both shoot and root include overexpression of a cyclin gene, preferably the *cycla* gene from *A. thaliana* (Doerner and Lamb, 1998).

Improved Nutrition

Several traits can be targeted to improve the nutritional status of crop produce. These include carbohydrates, proteins, oils, vitamins, iron, and essential amino acids. Transgenic rice with a capacity to produce beta-carotene can be used to overcome the deficiency of vitamin A (Ye et al., 2000). Similarly, transgenic rice with elevated levels of iron has been produced using genes involved in the production of an iron binding protein that facilitates iron availability in the human diet (Goto et al., 1999). Altering protein levels, composition of fatty acids, vitamins, and amino acids is being increasingly targeted for value addition. It is now possible to alter the composition of fatty acids so that polyunsaturated (e.g., linoleic acid) content is decreased while that of monounsaturated (e.g., oleic acid) content is increased to allow processing without the traditional use of hydrogenation and thus, avoid the undesirable trans-fatty acids. Amounts of essential amino acids such as lysine, methionine, threonine, and tryptophan can be increased to improve the nutritional quality of cereal grains. Transgenic modifications have also been used to alter the ratio of amylose to amylopectin in starch (McLaren, 1998). Decreasing the amounts of oligosaccharides (such as raffinose and stachyose) improves digestibility, and decreases the degree of flatulence during digestion. Transgenic technology can also be used to remove antinutritional factors (Kaufman et al., 1998).

Production of Pharmaceuticals and Vaccines

Microbial production of bioactive chemotherapeutics is a classical fermentation technology, and these molecules have a wide application in veterinary medicine, agriculture, and human health. Many of these technologies can now be improved through biotechnological applications, and open up newer vistas for production of many molecules in other organisms and plants. Plants and other organisms can also be used as factories for producing enzymes for industrial applications. Several vaccines can be produced in plants (Anderson, 1996). Vaccines against infectious diseases of the gastrointestinal tract have been produced in potatoes and bananas (Moffat, 1995; Thanavala et al., 1995; Artsaenko et al., 1998; Tacker et al., 1998). The antigen proteins produced by the transgenic plants retain the immunogenic properties upon purification, which can be used for production of antibodies when injected into mice. Mice eating the transgenic plants have shown an immune response. Such an immune response has been demonstrated for cholera toxin B (Arakawa, Chang, and Langridge, 1998). Anticancer antibodies expressed in rice and wheat could be useful in diagnosis and treatment of this disease (Stoger et al., 1999). There is also a great potential to increase the yield of medicines derived from plants (e.g., salicylic acid) through the use of transgenic technology.

Production of Antibodies

The ability of antibodies to interact with their cognate antigens with high specificity and affinity has been exploited in a wide range of applications in immunotherapy and immunodiagnostics. In recent years, plants have been shown to produce numerous antibody molecules, ranging from small antigen-binding fragments to large multimeric antibody complexes. The use of plants as a vehicle for antibody production opens up an inexpensive method for large-scale production of antibodies for immunotherapy and diagnostic applications (Owen et al., 1992; Fiedler and Conrad, 1995). A range of different antibody fragments, including those with affinity for phytochrome (Owen et al., 1992), abscisic acid (Artsaenko et al., 1995), fungal cutinase (Schouten et al., 1996), oxazalone (Fiedler and Conrad, 1995), and tobacco mosaic virus (TMV) coat protein (Zimmerman et al., 1998) have been stably expressed. One of the benefits of producing antibodies in plants is the potential to exploit their natural storage capabilities. Seeds are suitable for expression and storage of single chain antibodies (*scFvs*) (Fiedler and Conrad, 1995). Plants are also a suitable host for producing functional *sIgA* molecules (Ma et al., 1995). Rosso et al. (1996) described the transient expression of a functional *scFv* directed against salivary secretions of root-knot nematode, *Meloidogyne incognita* (Kofoid and White) Chitwood, in tobacco protoplasts, while Baum et al. (1996) expressed a functional whole antibody specific to stylet secretions of this nematode. This approach can be exploited in the future to express antibodies directed against specific functions of herbivorous arthropods.

Genes that are based on antibody technology can also be exploited for genetic transformation of crop plants. Single chain antibodies can be used to block the function of essential insect proteins, which serve as control agents against nematodes, pathogens, and viruses (Van Engelen et al., 1994; Rosso et al., 1996). This approach of controlling insects would offer the advantage of allowing some degree of selection for specificity effects, so that insect pests but not the beneficial organisms are targeted. The development of a delivery system from transgenic plants to the insect hemolymph will remove a key constraint in the transgenic approach to crop protection.

Genetic Improvement of Entomopathogenic Microorganisms

Genetic engineering can also be used to improve the efficacy of entomopathogenic microorganisms. Efforts to improve *B. thuringiensis* have largely been focused on increasing its host range and stability (van Frankenhuyzen, 1993). The toxin gene from *B. thuringiensis* has been inserted in *Pseudomonas fluorescens* Migula. The recombinant bacteria are fermented and then killed for application on crops for pest management. Work on baculoviruses is focused on incorporation of genes that produce the proteins that kill the insects at a faster rate (Bonning and Hammock, 1992), and on removal of the polyhedrin gene, which produces the protective viral-coat protein, and its persistence in the field (Corey, 1991). Spiders and scorpions produce powerful neurotoxins that have been expressed in transgenic organisms (Barton and Miller, 1991). Genes encoding neurotoxins from predatory mites (Tomalski and Miller, 1991) and scorpion (Stewart et al., 1991) have also been deployed in recombinant baculoviruses to increase their biological activity. Incorporation of benomyl resistance into the entomopathogenic fungus, *Metarhizium anisopliae* (Mets.) Sorokin could make this fungus more useful for use in integrated pest management (Goettel et al., 1990). The insect-specific neurotoxin AaIT from the venom of the scorpion, *Androctonus australis* Hector expressed in tobacco has shown insecticidal activity against *H. armigera* larvae (up to 100% mortality after 6 days) (Yao et al., 1996). Transgenic plants of tobacco have been obtained containing an insecticidal spider peptide gene, and some of these plants have shown resistance to *H. armigera* (Jiang, Zhu, and Chen, 1996). The role of neurotoxins from insects and spiders need to be studied in greater detail before they are deployed in other organisms and plants because of their possible toxicity to mammals.

Genetic Improvement of Natural Enemies

There is tremendous scope for developing natural enemies with genes for resistance to pesticides and ability to withstand adverse weather conditions (Hoy, 1992). One of the major problems in using natural enemies in pest control is the difficulty involved in mass rearing, and ability to withstand adverse conditions, and biotechnology has the promise to solve many of these problems. It can also help to understand the genetics and physiology of reproduction, and control of sex ratio in natural enemies. This information can be used to improve rearing of the natural enemies for biological control. Biotechnological interventions can also be used to broaden the host range of natural enemies or enable their production on artificial diet or nonhost insect species that are easy to multiply under laboratory conditions.

Application of Biotechnology in Biosystematics and Diagnostics

Diagnosis and characterization of insect pests and their natural enemies through biosystematics can be exploited in pest management in the future. Characterization of natural enemies is particularly challenging in the tropics, where there is enormous diversity (Waage, 1991). Biochemical and molecular markers have proved to be quite useful in distinguishing between pest and nonpest forms. Biotechnological tools based on molecular markers offer a great promise for improving biosystematics, identification of insect pests, and natural enemies across a range of taxonomic groups (Hawksworth, 1994), and can also be used for rapid diagnosis even by nontaxonomists in far-flung regions.

Exploitation of Male-Sterility and Apomixis

Genetic (GMS) or cytoplasmic (CMS) male-sterility leads to the suppression of production of viable pollen (Liedl and Anderson, 1993). Male-sterility has been observed in a wide variety of higher plants and is characterized by low levels of male-sterility or complete absence of pollen production. The male-sterility phenotype affects the pollen-producing organs because of a high energy requirement. The best known examples of this trait are the CMS observed in *Z. mays, S. bicolor, Pennisetum glaucum* (L.) R. Br., and *Helianthus annus* L. Both the GMS and CMS systems have been exploited for developing rice, maize, sorghum, and pearl millet hybrids (Williams and Levings, 1992). A general characteristic of CMS is the dysfunction of mitochondria in tapetal cells. Mitochondrial genomes encoding chimeric proteins are presumably present in all tissues of the plant. Mitochondrial dysfunction produced by a chimeric protein interferes with the organelle function, and affects pollen production. Biotechnological approaches can be used to transfer CMS from within a species or from one species into another.

Apomixis resulting from the development of asexual embryos produces a large number of nucellar offsprings, which are genetically similar to the female parent (Sharma and Thorpe, 1995; Savidan, 2000). Obligate apomixis offers an opportunity to clone plants through seed propagation and gene manipulation, and could be used as a potent tool in plant breeding. It provides uniformity in seed propagation of rootstocks and true breeding of F_1 hybrids. Genetic manipulation of apomixes has the potential to result in production of stable and superior hybrids. Some apomictic cultivars have already been released for citrus, Kentucky grass, and buffalo grass. Development of cross compatible apomictic plants will allow for hybridization to break the barriers in gene transfer. This will also help to fix heterosis and obtain nonsegregating populations from hybrids with a unique combination of characters from the parents. Genetic engineering of apomixes can be used for fixing the genetic variability to produce crops with high productivity and better food quality. This system has been well studied in citrus, sorghum, maize, turf grass, and other crop plants. Introduction of apomictic genes into crops will have revolutionary implications for plant breeding and agriculture, where social and economic benefits will exceed those of the Green Revolution. However, the dangers associated with genetic uniformity could be exacerbated by inappropriate use of this technology, and therefore its application would have to be considered on a case-by-case basis.

Conclusions

Availability of information and technology in developing countries, where the need to increase food production is most urgent, will be a key factor in the use of biotechnology for sustained food security. Predicted growth in world population and the likely effects of climate change will pose a serious challenge to crop production and food security, particularly in developing countries. The augmentation of conventional breeding with the use of marker-assisted selection and transgenic plants promises to facilitate substantial increases in food production. However, knowledge of the physiology and biochemistry of plants will be extremely important for interpreting the information from molecular markers and deriving new and more effective paradigms in plant breeding. The application of DNA marker technologies in exploiting the vast and largely under-utilized pool of favorable alleles existing in the wild relatives of crops will provide a huge new resource of genetic variation to fuel the next phase of crop improvement. In particular, tremendous benefits will be derived through the transfer of genes important for crop protection and crop quality. However, rapid and cost effective development, and adoption of biotechnology-derived products will depend on developing a full understanding of the interaction of genes within their genomic environment, and with the environment in which their conferred phenotype must interact.

References

Alam, M.F., Datta, K., Abrigo, E., Oliva, N., Tu, J., Virmani, S.S. and Datta, S.K. (1999). Transgenic insect-resistant maintainer line (IR68899B) for improvement of hybrid rice. *Plant Cell Reports* 18: 572–575.

Amasino, R.M. and Gan, S. (1997). Transgenic plants with altered senescence characteristics. *European Patent* 804066 (Europe, Mexico, USA).

Anderson, J. (1996). *Evolution Is Not What it Used to Be*. New York, USA: W.H. Freeman and Company.

Arakawa, T., Chang, K.X. and Langridge, W.H.R. (1998). Efficacy of a food plant based oral cholera toxin B subunit vaccine. *Nature Biotechnology* 16: 292–297.

Armstrong, C.L., Parker, G.B., Pershing, J.C., Brown, S.M., Sanders, P.R., Duncan, D.R., Stone, T., Dean, D.A., DeBoer, D.L., Hart, J., Howe, A.R., Morrish, F.M., Pajeau, M.E., Petersen, W.L., Reich, B.J., Rodriguez, R., Santino, C.G., Sato, S.J., Schuler, W., Sims, S.R., Stehling, S., Tarochione, L.J. and Fromm, M.E. (1995). Field evaluation of European corn borer control in progeny of 173 transgenic corn events expressing an insecticidal protein from *Bacillus thuringiensis*. *Crop Science* 35: 550–557.

Artsaenko, O., Keltig, B., Feidler, U., Conrad, U. and Dureng, K. (1998). Potato tubers as a biotechnology for recombinant antibiotics. *Molecular Breeding* 4: 313–319.

Artsaenko, O., Peisker, M., zur Neiden, U., Feidler, U., Weiler, E.W., Muntz, K. and Conrad, U. (1995). Expression of a single chain antibody against abscisic acid creates a wilty phenotype in transgenic tobacco. *Plant Journal* 8: 745–750.

Barry, G.F., Cheikh, N. and Kishore, G. (1998). Expression of fructose 1,6 bisphosphate aldolase in transgenic plants. *Patent Application* WO 98/58069 (PCT, several countries).

Barton, K.A. and Miller, M.J. (1991). Insecticidal toxins in plants. *European Patent* EP0431829.

Barton, K., Whiteley, H. and Yang, N.S. (1987). *Bacillus thuringiensis* δ-endotoxin in transgenic *Nicotiana tabacum* provides resistance to lepidopteran insects. *Plant Physiology* 85: 1103–1109.

Baum, T.J., Hiatt, A., Parrott, W.A., Pratt, L.H. and Hussey, R.S. (1996). Expression in tobacco of a functional monoclonal antibody specific to stylet secretions of the root-knot nematode. *Molecular Plant Microbe Interactions* 9: 382–387.

Benedict, J.H., Sachs, E.S., Altman, D.W., Deaton, D.R., Kohel, R.J., Ring, D.R. and Berberich, B.A. (1996). Field performance of cotton expressing CryIA insecticidal crystal protein for resistance to *Heliothis virescens* and *Helicoverpa zea* (Lepidoptera: Noctuidae). *Journal of Economic Entomology* 89: 230–238.

Bonning, B.C. and Hammock, B.D. (1992). Development and potential of genetically engineered viral insecticides. *Biotechnology and Genetic Engineering Reviews* 10: 455–489.

Brisson, L.F., Zelitch, I. and Havir, E.A. (1998). Manipulation of catalase levels produces altered photosynthesis in transgenic tobacco plants. *Plant Physiology* 116: 259–269.

Chalfie, M. (1998). Genome sequencing. The worm revealed. *Nature* 396: 620–621.

Corey, J.S. (1991). Releases of genetically modified viruses. *Medical Virology* 1: 79–88.

De la Fuente, J.M., Ramírez Rodríguez, V., Cabera Ponce, J.L. and Herrera Estrella, L. (1997). Aluminum tolerance in transgenic plants by alteration of citrate synthesis. *Science* 276: 1566–1568.

De Nijs, J.J.M., Broer, J. and Van Doorn, J.E. (1997). Plants with delayed or inhibited ripening or senescence. *Patent Application* EP 784423 (PCT, several countries).

Delannay, X., LaVallee, B.J., Proksch, R.K., Fuchs, R.L., Sims, S.K., Greenplate, J.T., Marrone, P.G., Dodson, R.B., Augustine, J.J., Layton, J.G. and Fischhoff, D.A. (1989). Field performance of transgenic tomato plants expressing *Bacillus thuringiensis* var *kurstaki* insect control protein. *Biotechnology* 7: 1265–1269.

Doerner, P.W. and Lamb, C.J. (1998). Method of increasing growth and yield in plants. *Patent Application* WO 98/03631 (USA).

Dunwell, J.M. (1998). Novel food products from genetically modified crop plants: Methods and future prospects. *International Journal of Food Science and Technology* 33: 205–213.

Dunwell, J.M. (2000). Transgenic approaches to crop improvement. *Journal of Experimental Botany* 51: 487–496.

Ebel, J. (1986). Phytoalexin synthesis: The biochemical analysis of the induction process. *Annual Review of Phytopathology* 24: 235–264.

Fiedler, U. and Conrad, U. (1995). High-level production and long term storage of engineered antibodies in transgenic tobacco seeds. *Biotechnology* 13: 1090–1093.

Fulton, T., Beck Bunn, T., Emmatty, D., Eshed, Y., Lopez, J., Petiard, V., Uhlig, J., Zamir, D. and Tanksley, S.D. (1997). QTL analysis of an advanced backcross of *Lycopersicon peruvianum* to the cultivated tomato and comparisons with QTLs found in other wild species. *Theoretical and Applied Genetics* 95: 881–894.

Furbank, R.T., Chitty, J.A., Jenkins, C.L.D., Taylor, W.C., Trenanion, S.J., Von Coammererer, S. and Asthon, A.R. (1997). Genetic manipulation of key photosynthetic genes in the C_4 plant *Flaveria bidentis. Australian Journal of Plant Physiology* 24: 477–485.

Gale, M.D. and Devos, K.M. (1998). Comparative genetics in the grasses. *Proceedings National Academy of Sciences USA* 95: 1971–1974.

Goettel, M.S., St. Leger, R.J., Bhairi, S., Jung, M.K., Oakley, B.R., Roberts, D.W. and Staples, R.C. (1990). Pathogenicity and growth of *Metarhizium anisopliae* stably transformed to benomyl resistance. *Current Genetics* 17: 129–132.

Goddijn, O.J.M., Pen, J., Smeekens, J.C.M. and Smits, M.T. (1997). Regulating metabolism by modifying the level of trehalose-6-phosphate by inhibiting endogenous trehalase levels. *Patent Application* WO 97/42326 (PCT, several countries).

Gonsalves, D. (1998). Control of papaya ringspot virus in papaya: A case study. *Annual Review of Phytopathology* 36: 415–437.

Goto, F., Yoshihara, T., Shigemoto, N., Toki, S. and Takaiwa, F.T. (1999). Iron fortification of rice seed by the soybean ferritin gene. *Nature Biotechnology* 173: 282–286.

Grimes, H.D. and Chao, W.S. (1998). Sucrose-binding proteins. *Patent Application* WO 98/53086 (PCT, several countries).

Grimm, B. (1998). How to affect the chlorophyll biosynthesis in plants. *Patent Application* WO 98/24920. (PCT, several countries).

Grula, J.W. and Hudspeth, R.L. (1999). Promoters derived from the maize phosphoenolpyruvate carboxylase gene involved in C_4 photosynthesis. *U.S. Patent* 5856177 (USA).

Hawksworth, D.L. (Ed.). (1994). *The Identification and Characterization of Pest Organisms*. Wallingford, Oxon, UK: CAB International.

Heller, W. and Forkman, G. (1993). Biosynthesis of flavonoids. In J.B. Harborne (Ed.), *The Flavonoids, Advances in Research Since 1986*. London, UK: Chapman and Hall, 499–535.

Hilder, V.A. and Boulter, D. (1999). Genetic engineering of crop plants for insect resistance: A critical review. *Crop Protection* 18: 177–191.

Hoy, M.A. (1992). Biological control of arthropods: Genetic engineering and environmental risks. *Biological Control* 2: 166–170.

Hunt, S.P. and Livesey, F.J. (2000). *Functional Genomics: A Practical Approach*. Oxford, UK: Oxford University Press.

Ishimaru, K., Ichikawa, H., Matsuoka, M. and Ohsugi, R. (1997). Analysis of a C_4 pyruvate, orthophosphate dikinase expressed in C_3 transgenic *Arabidopsis* plants. *Plant Science* 129: 57–64.

Ishimaru, K., Okhawa, Y., Ishige, T., Tobias, D.J. and Ohsugi, R. (1998). Elevated pyruvate orthophosphate dikinase (PPDK) activity alters carbon metabolism in C_3 transgenic potatoes with a C_4 maize PPDK gene. *Physiologica Plantarum* 103: 340–346.

James, C. (2007). *Global Status of Commercialized Biotech/GM Crops: 2006*. ISAAA Briefs no. 35. Ithaca, New York, USA: International Service for Acquisition of Agri-Biotech Applications (ISAAA). http://www.isaaa.org/resources/publications/briefs/35.

Jansens, S., Cornelissen, M., Clercq de R., Reynaerts, A. and Peferoen, M. (1995). *Phthorimaea operculella* (Lepidoptera: Gelechiidae) resistance in potato by expression of *Bacillus thuringiensis* Cry IA(b) insecticidal crystal protein. *Journal of Economic Entomology* 88: 1469–1476.

Jiang, H., Zhu, Y.X. and Chen, Z.L. (1996). Insect resistance of transformed tobacco plants with a gene of a spider insecticidal peptide. *Acta Botanica Sinica* 38: 95–99.

Johnson-Flanagan, A.M., Singh, J., Robert, L.S. and Morisette, J.C.P. (1998). Anti-sense RNA for CAB transcript to reduce chlorophyll content in plants. *U.S. Patent* 5773692 (USA).

Kar, S., Basu, D., Das, S., Ramkrishnan, N.A., Mukherjee, P., Nayak, P. and Sen, S.K. (1997). Expression of cryIA(c) gene of *Bacillus thuringiensis* in transgenic chickpea plants inhibits development of pod borer (*Heliothis armigera*) larvae. *Transgenic Research* 6: 177–185.

Karp, A., Edwards, K.J., Bruford, M., Funk, S., Vosman, B., Morgante, M., Seberg, O., Kremer, A., Boursot, P., Arctander, P., Tautz, D. and Hewitt, G.M. (1997). Molecular technologies for biodiversity evaluation: Opportunities and challenges. *Nature Biotechnology* 15: 625–628.

Kaufman, P.B., Cseke, L.J., Warber, S., Duke, J.A. and Brielman, H.L. (1998). *Nutritional Products from Plants*. Boca Raton, Florida, USA: CRC Press.

Kirchanski, S.J. (1998). Phytochrome regulated transcription factor for control of higher plant development. *Patent Application* WO 98/48007 (PCT, several countries).

Ku, M.S.B., Agarie, S., Nomura, M., Fukayama, H., Tsuchida, H., Ono, K., Hirose, S., Toki, S., Miyao, M. and Matsuoka, M. (1999). High-level expression of maize phosphoenolpyruvate carboxylase in transgenic rice plants. *Nature Biotechnology* 17: 76–80.

Kumar, P.A., Mandaokar, A., Sreenivasu, K., Chakrabarti, S.K., Bisaria, S., Sharma, S.R., Kaur, S. and Sharma, R.P. (1998). Insect-resistant transgenic brinjal plants. *Molecular Breeding* 4: 33–37.

Leaver, C.J., Hill, S.A., Jenner, H.L. and Winning, B.M. (1998). Transgenic plants having increased starch content. *Patent Application* WO 98/23757 (PCT, several countries).

Liedl, B.E. and Anderson, N.O. (1993). Reproductive barriers: Identification, uses and circumvention. *Plant Breeding Reviews* 11: 11–154.

Liu, Y., Wang, G., Liu, J., Peng, X., Xie, Y., Dai, J., Guo, S. and Zhang, F. (1999). Transfer of *E. coli* gutD gene into maize and regeneration of salt-tolerant transgenic plants. *Life Sciences, China* 42: 90–95.

Ma, J.K.C., Hiatt, A., Hein, M., Vine, D.N., Wang, F., Stabila, P., van Dollerweerd, C., Mostov, K. and Lehner, T. (1995). Generation and assembly of secretory antibodies in plants. *Science* 268: 716–719.

Matsuoka, M., Nomura, M., Agarie, S., Tokutomi, M. and Ku, M.S.B. (1998a). Evolution of C_4 photosynthetic genes and overexpression of maize C_4 genes in rice. *Journal of Plant Research* 111: 333–337.

Matsuoka, M., Tokutomi, M., Toki, S. and Ku, M.S.B. (1998b). C_3 plants expressing photosynthetic enzymes of C_4 plants. *Patent Application* EP 874056 (USA).

McCourt, P., Ghassemian, M., Cutler, S. and Bonetta, D. (1999). Stress tolerance and delayed senescence in plants. *Patent Application* WO 99/06580 (PCT, several countries).

McLaren, J.S. (1998). The success of transgenic crops in the USA. *Pesticide Outlook* 9: 36–41.

Miflin, B. (2000). Crop improvement in the 21st century. *Journal of Experimental Botany* 51: 1–8.

Moffat, A.S. (1995). Exploring transgenic plants as a new vaccine source. *Science* 268: 656–658.

Mqbool, S.B., Husnain, T., Raizuddin, S. and Christou, P. (1998). Effective control of yellow rice stem borer and rice leaf folder in transgenic rice *indica* varieties Basmati 370 and M 7 using novel δ-endotoxin cry2A *Bacillus thuringiensis* gene. *Molecular Breeding* 4: 501–507.

Myers, N. (1999). The next Green Revolution: Its environmental underpinnings. *Current Science* 76: 507–513.

Nayak, P., Basu, D., Das, S., Basu, A., Ghosh, D., Ramakrishnan, N.A., Ghosh, M. and Sen, S.K. (1997). Transgenic elite *indica* rice plants expressing CryIAc delta-endotoxin of *Bacillus thuringiensis* are resistant against yellow stem borer (*Scirpophaga incertulas*). *Proceedings National Academy of Sciences USA* 94: 2111–2116.

NRC (National Research Council). (2000). *Genetically Modified Pest-Protected Plants: Science and Regulation.* Washington, D.C., USA: National Research Council.

Osteryoung, K.W. (1998). Plant plastid division genes. *Patent Application* WO 98/00436 (PCT, several countries).

Owen, M., Gandecha, A., Cockburn, W. and Whitelam, G.C. (1992). Synthesis of a functional anti-phytochrome single-chain Fv protein in transgenic tobacco. *Biotechnology* 10: 790–794.

Palevitz, B.A. (2000). Rice genome gets a boost. Private sequencing effort yields rough draft for public. http://www.the-scientist.com/yr2000/may/palevitz-pl-000501.html.

Pardo, J.M., Hasegawa, P.M. and Bressan, R.A. (1999). Transgenic plants tolerant of salinity stress. *Patent Application* WO 99/05902 (PCT, several countries).

Pei, Z.M., Ghassemian, M., Kwak, C.M., McCourt, P. and Schroeder, J.I. (1998). Role of farnesyl transferase in ABA regulation of guard cell anion channels and plant water loss. *Science* 282: 287–290.

Peng, J., Richards, D.E., Hartley, N.M., Murphy, G.P., Devos, K.M., Flintham, J.E., Beales, J., Fish, L.J., Worland, A.J., Pelica, F., Sudhakar, D., Christou, P., Snape, J.W., Gale, M.D. and Harberd, N.P. (1999). "Green Revolution" genes encode mutant gibberellin response modulators. *Nature* 400: 256–261.

Pereira, A. (2000). Plant genomics is revolutionising agricultural research. *Biotechnology Development Monitor* 40: 2–7.

Pilon Smits, E.A.H., Terry, N., Sears, T., Kim, H., Zayed, A., Hwang, S.B., Van Dun, K., Voogd, E., Verwoerd, T.C., Krutwagen, R.W.H.H. and Goddijn, O.J.M. (1998). Trehalose-producing transgenic tobacco plants show improved growth performance under drought stress. *Journal of Plant Physiology* 152: 525–532.

Piskur, J. and Langkjaer, R.B. (2004). Yeast genome sequencing: The power of comparative genomics. *Molecular Microbiology* 53: 381–389.

Primrose, S.B. and Twyman, R. (2003). *Principles of Genome Analysis and Genomics.* Oxford, UK: Blackwell Science.

Rosso, M.N., Schouten, A., Roosien, J., Borst-Vrenssen, T., Hussey, R.S., Gommers, F.J., Bakker, J., Schots, A. and Abad, P. (1996). Expression and functional characterization of a single chain FV antibody directed against secretions involved in plant nematode infection process. *Biochemistry and Biophysics Research Communications* 220: 255–263.

Savidan, Y. (2000). Apomixes: Genetics and breeding. *Plant Breeding Review* 18: 13–86.

Schmidt, R.R. and Miller, P. (1997). Novel polypeptides and polynucleotides relating to the α- and β-subunits of glutamate dehydrogenases and methods of use. *Patent Application* WO 97/12983 (PCT, several countries).

Schouten, A., Roosien, J., van Engelen, F.A., de Jong, G.A.M., Borst-Vrenssen, A.W.M., Zilverentant, J.F., Bosch, D., Stiekema, W.J., Gommers, F.J., Schots, A. and Bakker, J. (1996). The C-terminal KDEL sequence increases the expression level of a single-chain antibody designed to be targeted to both the cytosol and the secretory pathway in transgenic tobacco. *Plant Molecular Biology* 30: 781–793.

Sharma, H.C. (2004). Biotechnological approaches for crop improvement with special reference to host plant resistance to insects. In Chhillar, B.S., Singh, R., Bhanot, J.P. and Ram, P. (Eds.), *Recent Advances in Host Plant Resistance to Insects*. Hisar, Haryana, India: Center for Advanced Studies, Department of Entomology, CCS Haryana Agricultural University, 230–244.

Sharma, H.C. and Norris, D.M. (1991). Chemical basis of resistance in soybean to cabbage looper, *Trichoplusia ni. Journal of Science of Food and Agriculture* 55: 353–364.

Sharma, H.C. and Ortiz, R. (2000). Transgenics, pest management, and the environment. *Current Science* 79: 421–437.

Sharma, H.C., Sharma, K.K. and Crouch, J.H. (2004). Genetic engineering of crops for insect control: Effectiveness and strategies for gene deployment. *CRC Critical Reviews in Plant Sciences* 23: 47–72.

Sharma, H.C., Crouch, J.H., Sharma, K.K., Seetharama, N. and Hash, C.T. (2002). Applications of biotechnology for crop improvement: Prospects and constraints. *Plant Science* 163: 381–395.

Sharma, H.C., Sharma, K.K., Seetharama, N. and Ortiz, R. (2000). Prospects for using transgenic resistance to insects in crop improvement. *Electronic Journal of Biotechnology* 3, no 2. http://www.ejborg/content/vol 3/issue 2/full/20.

Sharma, K.K. and Thorpe, T.A. (1995). Asexual reproduction in asexual vascular plants in nature. In Thorpe, T.A. (Ed.), *Embryogenesis in Plants*. Dordrecht, The Netherlands: Kluwer Academic Publishers, 17–72.

Sheen, J. (1998). Stress-protected transgenic plants. *Patent Application* WO 98/26045 (PCT, several countries).

Shoemaker, D.D., Schadt, E.E., Armour, Y.D., He, P., Garrett Engel, P.D. and McDonayl, P.M. (2001). Experimental annotation of the human genome using microarray technology. *Nature* 409: 922–927.

Signora, L., Galtier, N., Skot, L., Lukas, H. and Foyer, C.H. (1998). Overexpression of phosphate synthase in *Arabidopsis thaliana* results in increased foliar sucrose/starch ratios and favors decreased foliar carbohydrate accumulation in plants after prolonged growth with CO_2 enrichment. *Journal of Experimental Botany* 49: 669–680.

Singsit, C., Adang, M.J., Lynch, R.E., Anderson, W.F., Wang, A., Cardineau, G. and Ozias Akins, P. (1997). Expression of a *Bacillus thuringiensis* cryIA(c) gene in transgenic peanut plants and its efficacy against lesser cornstalk borer. *Transgenic Research* 6: 169–176.

Smart, C., Thomas, H., Hosken, S., Schuch, W.W., Drake, C.R., Grierson, D., Farrell, A., John, I. and Greaves, J.A. (1996). Regulation of senescence. *Patent Application* EP 719341 (PCT, several countries).

Sonnewald, U. and Willmitzer, L. (1996). Plasmids for the production of transgenic plants that are modified in habit and yield. *U.S. Patent* 5492820 (USA).

Stewart, L.M.D., Hirst, M., Ferber, M.L., Merryweather, A.T., Cayley, P.J. and Possee, R.D. (1991). Construction of an improved baculovirus insecticide containing an insect-specific toxin gene. *Nature* 352: 85–88.

Stoger, E., Williams, S., Christou, P., Down, R.E. and Gatehouse, J.A. (1999). Expression of the insecticidal lectin from snowdrop (*Galanthus nivalis*, agglutin; GNA) in transgenic wheat plants: Effects on predation by the grain aphid, *Sitobion avenae. Molecular Breeding* 5: 65–73.

Swaminathan, M.S. (2000). Genetic engineering and food security: Ecological and livelihood issues. In Persley, G.J. and Lantin, M.M. (Eds.), *Agricultural Biotechnology and the Rural Poor*. Washington, D.C., USA: Consultative Group on International Agricultural Research, 37–42.

Tacker, C.O., Mason, H.S., Losonsky, G., Clements, J.D., Lavine, M.M. and Arntzen, L.J. (1998). Immunogenecity in bananas of recombinant bacterial antigen delivered in a transgenic potato. *Nature Medicine* 4: 607–609.

TAGI (The *Arabidopsis* Genome Initiative). (2000). Analysis of the genome sequence of the flowering plant *Arabidopsis thaliana*. *Nature* 408: 796–815.

Thanavala, Y., Yang, Y.F., Lyons, P., Mason, H.S. and Arntzen, C. (1995). Immunogenicity of transgenic plant-derived hepatitis B surface antigen. *Proceedings National Academy of Sciences USA* 92: 3358–3361.

TIHGMC (The International Human Genome Mapping Consortium). (2001). A physical map of the human genome. *Nature* 409: 934–941.

Tomalski, M.D. and Miller, L.K. (1991). Insect paralysis by baculovirus-mediated expression of a mite neurotoxin gene. *Nature* 352: 82–85.

Vaeck, M., Reynaerts, A., Hofte, H., Jansens, S., DeBeuckleer, M., Dean, C., Zabeau, M., Van Montagu, M. and Leemans, J. (1987). Transgenic plants protected from insect attack. *Nature* 327: 33–37.

Van Engelen, F.A., Schouten, A., Molthoff, J.W., Roosien, J., Dirske, W.G., Schots, A., Bakker, J., Gommers, F.J., Jongsma, M.A., Bosch, D. and Stiekma, W.J. (1994). Coordinate expression of antibody subunit genes yields high levels of functional antibodies in roots of transgenic tobacco. *Plant Molecular Biology* 26: 1707–1710.

Van Frankenhuyzen, K. (1993). The challenge of *Bacillus thuringiensis*. In Entwistle, P.F., Cory, J.S., Bailey, M.J. and Higgs, S. (Eds.), *Bacillus thuringiensis: An Environmental Biopesticide—Theory and Practice*. Chichester, UK: John Wiley & Sons, 1–35.

Waage, J.K. (1991). Biodiversity as a resource for biological control. In Hawksworth, D.L. (Ed.), *The Biodiversity of Microorganisms and Invertebrates: Its Role in Sustainable Agriculture*. Wallingford, Oxon, UK: CAB International, 149–161.

Williams, M.E. and Levings, C.S. III. (1992). Molecular biology of cytoplasmic male-sterility. *Plant Breeding Reviews* 10: 23–53.

Wilson, W.D., Flint, H.M., Deaton, R.W., Fischhoff, D.A., Perlak, F.J., Armstrong, T.A., Fuchs, R.L., Berberich, S.A., Parks, N.J. and Stapp, B.R. (1992). Resistance of cotton lines containing a *Bacillus thuringiensis* toxin to pink bollworm (Lepidoptera: Gelechiidae) and other insects. *Journal of Economic Entomology* 85: 1516–1521.

Wu, R.J. and Ho. T.H.D. (1997). Production of water stress or salt stress tolerant transgenic cereals plants. *Patent Application* WO 97/13843 (PCT, several countries).

Xiao, J., Grandillo, S., Ahn, S.N.K., McCouch, S.R., Tanksley, S.D., Li, J. and Yuan, L. (1996). Genes from wild rice improve yield. *Nature* 384: 223–224.

Yao, B., Wu, C.J., Zhao, R.M. and Fan, Y.L. (1996). Application of insect-specific neurotoxin AaIT gene in baculovirus and plant genetic engineering. *Rice Biotechnology Quarterly* 26: 24.

Ye, X., Al Babili, S., Kloti, A., Zhang, J., Lucca, P., Beyer, P. and Potrykus, I. (2000). Engineering the provitamin A (beta-carotene) biosynthetic pathway into (carotenoid-free) rice endosperm. *Science* 287: 303–305.

Zhang, H. and Forde, B.G. (2000). Regulation of *Arabidopsis* root development by nitrate availability. *Journal of Experimental Botany* 51: 51–59.

Zhao, J., Zhao, Z., Lu, K.J., Fan, M.G. and Guo, S.D. (1998). Interactions between *Helicoverpa armigera* and transgenic Bt cotton in North China. *Scientia Agricultura Sinica* 31: 1–6.

Zimmermann, S., Schillberg, S., Liao, Y.C. and Fischer, R. (1998). Intracellular expression of a TMV-specific single chain Fv fragment leads to improved virus resistance in *Nicotiana tabacum*. *Molecular Breeding* 4: 369–379.

3

Evaluation of Transgenic Plants and Mapping Populations for Resistance to Insect Pests

Introduction

Because of the development of insect resistance to insecticides, adverse effects of insecticides on natural enemies, and environmental pollution, there has been a renewed interest in the development of insect-resistant cultivars for pest management (N. Panda and Khush, 1995; Sharma and Ortiz, 2002; Smith, 2005). Many crop species possess genetic variation, which can be exploited to produce varieties that are less susceptible to insect pests. However, the development of varieties resistant to insects has not been as rapid as for disease resistance. Slow progress in developing insect-resistant cultivars has been due mainly to the difficulties involved in ensuring adequate insect infestation for resistance screening in addition to low levels of resistance to certain insect species in the cultivated germplasm. Under natural conditions, the insect infestation is either too low or too high and, as a result, it becomes difficult to make a meaningful selection. Therefore, it is important to develop techniques to screen for resistance to insect pests under optimum levels of infestation and under similar environmental conditions. Insects reared on artificial diets can be used to test the material under uniform infestation. However, insect-rearing programs are expensive, technology development requires several years, and may not produce the behavioral or metabolic equivalent of an insect population in nature.

Although screening under artificial infestation is a necessity for most insect pests, it may not be an absolute requirement for insects that occur in epidemic proportions over years or are endemic to particular areas, for example, cotton bollworms, *Pectinophora gossypiella* (Saunders), *Heliothis virescens* (F.), *Helicoverpa armigera* (Hubner), and *Earias vittella* (Fab.), and jassids, *Amrasca biguttula biguttula* (Ishida) in cotton; aphids, *Lipaphis erysimi* (Kalt.) and *Brevicoryne brassicae* L. in cruciferous crops; pod borer, *H. armigera* in pigeonpea and chickpea; corn earworm, *Helicoverpa zea* (Boddie) in maize; and shoot fly, *Atherigona soccata* (Rondani) and sorghum midge, *Stenodiplosis sorghicola* (Coquillett) in sorghum. There is a need to refine techniques to screen for resistance to insect pests in several crops, and develop uniform and standardized procedures for assessing insect populations and damage. The following techniques can be used to evaluate transgenic plants and mapping populations for resistance to insect pests in different crops.

Techniques to Screen for Resistance to Insects Under Natural Infestation

Rarely is a researcher able to grow a set of genotypes and evaluate insect damage accurately under natural infestation. Either there are insufficient insect numbers to cause adequate damage or insects occur at an inappropriate phenological stage of crop growth (Sharma, Singh, and Ortiz, 2001; Sharma, 2005). Field infestations are normally used to evaluate a large number of genotypes at an early stage of the resistance-breeding program. However, field evaluations are influenced by nontarget insects, which may interfere with the damage caused by the target insect, or the insect populations are kept under check by the natural enemies. As a result, it is difficult to achieve dependable screening of the test material under field conditions. Managed or augmented insect density ensures uniform distribution of insects under field conditions, but the insects may be influenced by biotic and abiotic population regulation factors. Major advantages of screening under natural infestation are the convenience and low cost. However, seasonality, unpredictability, interference by nontarget insects, and uneven distribution makes screening under natural conditions unreliable, time consuming, and less effective.

Knowledge concerning the periods of maximum insect abundance and hot spots is the first step to initiate work on screening and breeding for resistance to insect pests. Delayed plantings of the crop and use of infester rows of a susceptible cultivar of the same or of a different species have been used to increase insect infestation under natural conditions (Sharma et al., 1992; Smith, Khan, and Pathak, 1994; Sharma, 2005). Sowing time should be adjusted such that the most susceptible stage of the crop is exposed to maximum insect infestation. Generally, no insecticide should be applied in the resistance screening nursery, but plant protection measures may be adopted if necessary to control other insects that interfere with screening for resistance to the target insect species. Several procedures have been employed to obtain adequate insect pressure for resistance screening under field conditions (Table 3.1). The objective of all these approaches is to have an optimum insect density-to-damage ratio that allows the researcher to observe maximum differences between the resistant and susceptible genotypes. Screening for resistance to insects under natural conditions is a long-term process because of the variations in insect populations across seasons and locations. In addition, there are large differences in the flowering times of different genotypes. Genotypes flowering at the beginning and the end of the season are exposed to low insect infestation, whereas those flowering during the mid-season suffer heavy damage. As a result, it is difficult to identify reliable and stable sources of resistance under natural infestation.

Use of Hot-Spot Locations

Hot spots are the locations where the insects are known to occur regularly in optimum numbers across seasons. Hot-spot locations can be used for large-scale screening of the germplasm, segregating breeding material, and multilocational testing of the transgenic plants. Several insect species are known to occur in high numbers every year at several

TABLE 3.1

Techniques to Screen for Resistance to Insects under Field Conditions

Crop	Insect Species	Remarks	References
Sorghum	Shoot fly, *Atherigona soccata*	Infester rows of a susceptible cultivar planted 20 to 25 days earlier than the test material, and spreading moistened fishmeal in the infester rows.	Doggett, Starks, and Eberhart (1970); Taneja and Leuschner (1985)
	Sorghum midge, *Stenodiplosis sorghicola*	Planting infester rows 20 days earlier than the test material, grouping the test material according to maturity, and spreading midge-infested panicles in infester rows.	Sharma, Vidyasagar, and Leuschner (1988a)
	Head bug, *Calocoris angustatus*	Plant infester rows 20 days earlier than the test material, split plantings, and group material according to maturity.	Sharma and Lopez (1992)
Pigeonpea	Pod borer, *Helicoverpa armigera*	Adjusting planting date, grouping material according to maturity, and tagging the inflorescences at flowering.	Sharma et al. (2005a)
Soybean	Velvetbean caterpillar, *Anticarsia gemmatalis* (Hubner)	Plot size or shape does not affect insect density or defoliation. Estimates of larval density and defoliation are better in plots of two or more rows than in one-row plots.	Funderburk et al. (1989)
Cowpea	Legume pod borer, *Maruca vitrata* (Geyer)	Visual examination of flower injury or larval presence and/or a pod evaluation index can be used as a selection criteria.	Jackai (1982); Oghiakhe, Jackai, and Makanjuola (1992)
Egg plant	Spotted beetle, *Henosepilachna vigintioctopunctata* (Fab.)	The damage index calculated by multiplying pest incidence and intensity/100 can be used as a selection criterion.	Rajendran and Gopalan (1997)
Brachiaria spp.	Spittlebugs, *Zulia colombiana* (Lall.) and *Aeneolamia reducta* (Lall.)	Screening is more reliable if established in old fields of a spittlebug susceptible grass such as *Brachiaria decumbens*.	Lapointe, Arango, and Sotelo (1989)
Wheat	Russian wheat aphid, *Diuraphis noxia* (Kurdj.)	Embryo and colony counts are suitable alternatives to nymphal counts for measuring antibiosis.	Scott, Worrall, and Frank (1990)
Maize	Fall armyworm, *Spodoptera frugiperda* (J.E. Smith)	Feeding lesions in the whorl and unfolded leaves at 7 and 14 days after infestation were effective to distinguish between resistant and susceptible genotypes.	Davis, Ng, and Williams (1992)
Cotton	Tarnished plant bug, *Lygus lineolaris*	Evaluation of anthers in white flowers of medium-sized squares for discoloration is a good indicator of genotypic resistance.	Maredia et al. (1994)
Soybean	Defoliators	Visual and quantitative assessments were highly correlated.	Bowers et al. (1999)
	Stem fly, *Melanagromyza sojae* (Zehnt.)	Maximin-minimax approach involving the yield component and response to entire insect-pest complex was useful to identify resistant genotypes with good yield potential.	Sharma (1996)
Potato	Tuber moth, *Phthorimaea operculella* (Zeller)	Oviposition preference over a period of 24 h can be used to evaluate potato lines for resistance to potato tuber moth.	Gyawali (1989)
Fragaria spp.	Aphid, *Chaetosiphon fragaefolii* (Cock)	Honeydew production, survival, and production of nymphs can be used as indicators of aphid resistance.	Shanks and Grath (1992)

locations. As far as possible, efforts should be made to select locations where appropriate attention can be given to managing the trials and evaluation of the test material for resistance at the appropriate stage.

Adjusting Planting Date

The test material should be planted such that the most susceptible stage of the crop is exposed to maximum or optimum levels of insect infestation. Fortnightly or monthly plantings of a susceptible cultivar during the cropping season can also be used to determine the optimum time for planting the test material. Most of the crops planted 20 to 25 days later than the normal planting times are exposed to heavy insect damage. The periods of maximum insect abundance can also be determined through monitoring of insect populations through light or pheromone traps. Sorghum crop planted during the second week of July in southern India is exposed to heavy damage by shoot fly, *A. soccata*, sorghum midge, *S. sorghicola*, and head bug, *Calocoris angustatus* (Lethiery) (Sharma et al., 1992). Most of the crops (cotton, maize, pigeonpea, tomato, sunflower, and sorghum) planted during July in South Asia are exposed to heavy *H. armigera* infestations during the rainy season, while October is the optimum time to plant the test material (chickpea, tomato, and safflower) during the post-rainy season for resistance screening (Sharma et al., 2005a). Similar planting times should be determined for different insect pests and their crop hosts in each region to maximize the chances for obtaining adequate levels of insect infestation for resistance screening.

Manipulation of Cultural Practices

Cultural practices can be manipulated to increase the activity of certain insects in the field (Hollingsworth and Berry, 1982). Closer spacing leads to greater incidence of rice gall midge, *Orseolia oryzae* Wood-Mason (Prakasa Rao, 1975). Lower planting densities lead to greater infestation of sorghum shoot fly, *A. soccata*, and sorghum midge, *S. sorghicola* (Sharma et al., 1992). A drought-stressed sorghum crop suffers greater severity of spotted stem borer, *Chilo partellus* (Swinhoe) damage, while sprinkler irrigation during the flowering stage increases the severity of sorghum midge and head bug, *C. angustatus* (Sharma et al., 1992). Fertilizer application increases the damage by planthopper, *A. biguttula biguttula* and bollworms, *E. vittella* and *H. armigera* in cotton.

Planting Infester Rows

Planting susceptible cultivars as infester rows along the field borders or at regular intervals in the field can be used to increase insect infestation in the test material. The infester rows may be planted 20 to 25 days earlier, or an early flowering crop or cultivar can be used as infester rows so that the flowering in the infester rows occurs earlier than the test material in the case of insects feeding on the reproductive parts (Figure 3.1). There should be sufficient time for the insect to multiply on the infester rows, and then move to the test material (Sharma, Vidyasagar, and Leuschner, 1988a; Smith, Khan, and Pathak, 1994). Insects collected from nearby fields can also be released in the infester rows. Insects can also be attracted to the infester rows by kairomones (e.g., fishmeal for sorghum shoot fly) (Figure 3.2). Plant material with diapausing insects (e.g., chaffy sorghum panicles with midge larvae or stalks of cereal crops with diapausing larvae of stem borers) can also be spread in the infester rows to increase the severity of insect damage. The infester rows

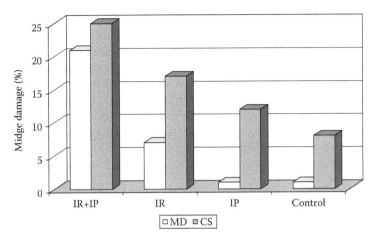

FIGURE 3.1 Effect of planting infester rows (IR) inoculated with sorghum panicles having diapausing larvae (IP) on sorghum midge, *Stenodiplosis sorghicola* damage under field conditions. MD, Spikelets with midge larvae (%); CS, midge-damaged chaffy spikelets (%).

should be removed after infestation of the test material has taken place so that it does not compete with the test material.

Trap crops that are attractive to the target insect species can also be used to increase insect infestation in the test material (Stern, 1969). The *H. armigera* infestations can be increased by planting *Solanum vairum* Dunal in tomato (Talekar, Hau, and Chang, 1999), and sesame and pigeonpea in cotton (Sharma et al., 2005a). Mustard planted in cabbage and cauliflower can be used to increase diamondback moth, *Plutella xylostella* (L.), infestation. *Brassica juncea* (L.) Czern. can be planted in cotton to increase the infestation by *Lygus lineolaris* (Palisat de Beauvois) (Laster and Meredith, 1974). When the main crops reach the susceptible stage, the trap crops are either allowed to senesce or are chopped off so that the insects migrate to the test material.

Grouping the Material According to Maturity and Height

To screen for resistance to insect species that feed on reproductive structures, it is important to group the test material according to maturity and height as there are large

FIGURE 3.2 Infester row fishmeal technique to screen sorghum for resistance to shoot fly, *Atherigona soccata*.

differences in the flowering times of different genotypes within a crop. Because of fluctuations in insect populations over the crop-growing season, it becomes difficult to obtain uniform insect pressure on genotypes flowering at different times. Genotypes flowering at the beginning and end of the cropping season escape insect damage, while those flowering in the mid-season are exposed to heavy insect damage. As a result, it becomes difficult to make meaningful comparisons between the resistant and susceptible genotypes. To overcome this problem, it is important to group the test material according to maturity and height. It is equally important to include resistant and susceptible checks of appropriate maturity in different trials for proper comparison (Sharma et al., 1999, 2005a; B. Singh and Yadav, 1999a, 1999b).

Sequential Plantings

Even with the best of insect forecasting models, it is not possible to pinpoint the periods of maximum abundance of an insect under field conditions. At the same time, it is not possible to take up timely planting of the test material because of the variation in the onset of rainfall in different seasons and locations. As a result, it becomes difficult to coincide the susceptible stage of the crop with maximum insect abundance because of changes in weather conditions, and the effect of environment on both the insect and the crop. Therefore, the test material should be planted two to three times at an interval of 15 to 20 days so that one of the plantings is exposed to adequate insect damage. Such an approach also helps to reduce the chances of escape from insect damage as the late-flowering genotypes in the first planting and the early-flowering genotypes in the second planting are most likely to flower during the mid-season, and are exposed to maximum insect abundance (Sharma, Singh, and Nwanze, 1997).

Selective Control of Nontarget Insects

Resistance screening trials are quite often rendered useless because of interference by the nontarget insect pests. Selective insecticides can be used either to control other insects infesting the crop at the same or at different times or to kill the natural enemies of the target insect species to allow population build up of the target insect (Sharma, Vidyasagar, and Leuschner, 1988a). It is important to protect the cotton crop from jassid damage during the seedling stage to evaluate the test material for resistance to bollworms. Sorghum at the seedling stage (5 to 10 days old) can be sprayed with insecticides to control the shoot fly to avoid interference with screening for resistance to spotted stem borer. Similarly, sorghum and pigeonpea can be sprayed at the milk stage to control the sucking pests to avoid interference in screening for resistance to sorghum midge and pod borer, respectively. Insecticide application at times also leads to resurgence of some insect species, for example, brown planthopper, *Nilaparvata lugens* (Stal), in rice (Chelliah and Heinrichs, 1980; Heinrichs et al., 1982). Pesticide application kills the natural enemies and the insects then multiply at a faster rate in the absence of natural enemies. Selective insecticides can also be used as powerful tool in conserving and enhancing insect populations (Shepard, Carner, and Turnipseed, 1977).

Augmentation of Insect Populations

Insect infestation in the screening nursery can be increased by collecting indigenous insect populations from the surrounding areas and releasing them in the test plots. Insect abundance in the screening nursery can also be augmented by placing nondestructive light, pheromone, or kairomone traps in the field. Light traps are effective in attracting brown leaf hopper, *N. lugens*, green leaf hopper, *Nephotettix virescens* (Distant), and gall midge, *O. oryzae* in rice (Prakasa Rao, 1975); stem borers of cereals (Sharma et al., 1992); and *H. armigera* in cotton and food legumes (Sharma et al., 2005a). Kairomones present in the leaves of susceptible pigeonpea varieties, which are attractive to egg-laying females of *H. armigera*, can be used to increase insect abundance in the screening nursery. Moistened fishmeal is highly attractive to the sorghum shoot fly, *A. soccata*, and is used in the interlard fishmeal technique to screen for resistance to this pest (Doggett, Starks, and Eberhart, 1970; Sharma et al., 1992).

Labelling the Plants or Inflorescences Flowering at the Same Time

The test material flowering at the same time can be labelled with a ribbon or marked with paint. Genotypes flowering at different times can be marked with different colored labels. This enables the comparison of the test material flowering at the same time with the resistant and susceptible controls of similar maturity. For comparisons to be meaningful, the portion of the inflorescence (30 to 45 cm) can also be marked with twine or colored labels. The data on insect damage should be recorded in the marked plants or portion of inflorescence and comparisons made among the genotypes flowering during the same period. For this purpose, three to five inflorescences may be tagged in each plot, depending on the stage of screening of the material, and the resources available.

Techniques to Screen for Resistance to Insects under Artificial Infestation

Mass Rearing

Insects can be reared in large numbers on natural or artificial diets in the laboratory or collected from the field to screen the test material in the field or in the greenhouse under uniform insect infestation (Smith, 1989; Sharma et al., 1992; Smith, Khan, and Pathak, 1994). Many insect species can be reared on natural diets. Greenbug, *Schizaphis graminum* (Rondani) (Starks and Burton, 1977), brown planthopper, *N. lugens* (Heinrichs, Medrano, and Rupasus, 1985), and many species of aphids and leafhoppers can be reared in the greenhouse on their natural host plants. Protocols for mass production of insects on artificial diets for infestation of the test material under greenhouse or field conditions have been described for several insect species (P. Singh and Moore, 1985; Armes, Bond, and Cooters, 1992; Sharma et al., 1992; Smith, Khan, and Pathak, 1994; N. Panda and Khush, 1995) (Figure 3.3).

Insect rearing should be planned such that the appropriate stage of the test insect is available in adequate numbers for infesting the test material during the susceptible stage of the crop. One of the major constraints to large-scale production of insects on artificial diets is the high cost of infrastructure, particularly when special equipment and facilities with appropriate control over temperature, relative humidity, and photoperiod are required. Continuous production of insects on artificial diet also decreases their genetic diversity (Berenbaum, 1986), which might lead to outcomes dissimilar to that of the natural populations (Guthrie et al., 1974). This problem can be overcome by infusing insects from the field periodically or fresh culture can be initiated at the beginning of each season. This will also

FIGURE 3.3 Screening for resistance to brown planthopper, *Nilaparvata lugens* in rice under greenhouse conditions.

avoid the build up of pathogen infections such as bacteria, fungi, protozoa, and viruses in the insect colony.

Infestation Techniques

Standardization of techniques to infest the material at the susceptible stage with uniform insect density is essential for successful evaluation of the test material under greenhouse or field conditions. Several techniques have been used for infestation and evaluation of the test material in the field (Smith, Khan, and Pathak, 1994; N. Panda and Khush, 1995) (Table 3.2). While devising techniques to infest the test material, it is important to take into consideration the:

- Stage of the insect and application procedure;
- Number of insects required and time of infestation;
- Number of infestations;
- Cannibalism among the individuals of the same species; and
- Susceptible stage of the crop and the site of infestation.

Efforts should be made to obtain uniform infestation at the most susceptible stage of the crop in a manner closer to the natural infections. The amount of food available to the insects and the insect density also influence the expression of resistance to insects, and therefore, efforts should be made to use optimum and uniform infestation to result in maximum differences between the resistant and susceptible genotypes (Sharma and Lopez, 1993). Adults, eggs, or the first-instar larvae can be spread uniformly in the test material by hand or by mixing them with an inert carrier (Smith, Khan, and Pathak, 1994). Adults can also be released on the test material inside cages (Sharma et al., 1992). Eggs are released into the plant whorl or egg masses stapled on to the underside of the leaves (Davis, 1985). Eggs can also be mixed in a liquid carrier such as agar-water and dispensed

TABLE 3.2

Techniques to Screen for Resistance to Insect Pests under Artificial Infestation in the Field

Crop	Insect Species	Remarks	References
Maize	Asian corn borer, *Ostrinia furnacalis* (Guen.)	Artificial infestation of 15- and 20-day-old plants with 100 and 200 eggs per plant resulted in leaf damage rating of 7.3 and 7.6, respectively.	Legacion and Gabriel (1988)
	European corn borer, *Ostrinia nubilalis* (Hubner)	Infesting the plants with 10 to 15 larvae per plant using *Bazooka* applicator.	Mihm (1982)
	Fall armyworm, *Spodoptera frugiperda*	Infesting the plants with 10 to 15 larvae per plant using *Bazooka* applicator.	Mihm (1983)
	Corn earworm, *Helicoverpa zea*	Releasing 10 larvae per plant with *Bazooka*.	Mihm (1982)
	Sugarcane stem borer, *Diatraea* spp.	Plants infested with 20 to 30 neonate larvae.	Mihm (1985)
	Pink borer, *Sesamia calamistis* (Hmps.)	Plants infested with 10 and 15 larvae resulted in 52 and 72.9% stand loss, respectively.	Nwosu (1992)
	African corn borer, *Busseola fusca*	Diapausing larvae collected from stubbles can be used to start insect colony for artificial infestation.	Van Rensburg and Van Rensburg (1993)
Sorghum	Spotted stem borer, *Chilo partellus*	Infestation of 18- to 20-day-old seedlings with 5 to 7 larvae per plant using *Bazooka* applicator.	Sharma et al. (1992)
	Midge, *Stenodiplosis sorghicola*	Infest the panicles at half-anthesis with 40 midge females for two consecutive days in a headcage.	Sharma, Vidyasagar, and Leuschner (1988b)
	Head bug, *Calocoris angustatus*	Infest the sorghum panicles with 10 pairs of bugs at the pre-anthesis stage in a headcage.	Sharma and Lopez (1992)
	African head bug, *Eurystylus oldi* (Poppius)	Infest sorghum panicles with 20 pairs of adults or 50 nymphs at the post-anthesis stage in a headcage.	Sharma, Doumbia, and Diorisso (1992)
Pearl millet	Head miner, *Heliocheilus albipunctella* (de Joannis)	Panicles infested with 30 to 40 larvae suffered 51 to 60% damage. Forty eggs per panicle resulted in 51 to 80% damage, and provided consistent infestation.	Youm, Yacouba, and Kumar (2001)
Sugarcane	Sugarcane borer, *Eldana saccharina* (Walker)	Infesting water-stressed potted plants with eggs is effective to screen for resistance.	Leslie and Nuss (1992)

through a pressurized syringe, for example, *Diabrotica virgifera virgifera* (L.) eggs in maize (Sutter and Branson, 1980). Eggs suspended in 0.2% agar-agar solution can also be spread on plants in controlled amounts through hypodermic syringes or pressure applicators.

Hall et al. (1980) tested three methods of artificial infestation of cotton plants with tobacco budworm, *H. virescens*. Eggs suspended in corn meal and distributed with a dispensing device, eggs suspended in xanthane gum solution and distributed with a hand sprayer, and first-instars suspended in corn meal and distributed with a dispenser. The latter two approaches resulted in desirable infestation, but infestation with the first-instars was preferred since it overcomes the problems associated with egg viability. Manual infestation with neonate larvae is quite effective, but is cumbersome and time consuming. A *Bazooka* applicator has been developed at the International Maize and Wheat Improvement Center

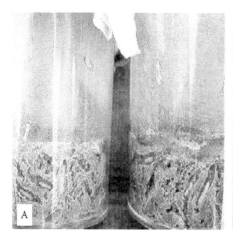

FIGURE 3.4   Screening for resistance to spotted stem borer, *Chilo partellus* in sorghum. (A) Rearing stem borer larvae on artificial diet, and (B) infestation of sorghum seedlings using *Bazooka* applicator.

(CIMMYT) for infesting the maize plants with the neonate larvae (Mihm, 1982). This technique has been used effectively with several insect species (Mihm, 1982, 1983, 1985; Smith, Khan, and Pathak, 1994). The *Bazooka* applicator has been modified at the International Crop Reseach Institute for the Semi-Arid Tropics (ICRISAT) to infest a large number of sorghum genotypes with neonate larvae of spotted stem borer, *C. partellus* (Sharma et al., 1992) (Figure 3.4). The larvae can be mixed with a carrier (corn cob grits, poppy seed, or any other lightweight material, such as foxtail and finger millet seed) in such a manner that each stroke of the *Bazooka* applicator releases known numbers of larvae on each plant (generally 5 to 10 larvae per plant). The larval density can be manipulated by changing the ratio between the numbers of larvae and the amount of carrier used.

   The *Bazooka* applicator is quite efficient for infesting a large number of plants under field conditions. The applicator should be standardized to deliver a uniform number of larvae to distinguish between the resistant and susceptible plants. While infesting the plants, the *Bazooka* should be rotated constantly to ensure that the larvae are distributed uniformly in the carrier. Care should be taken that the neonate larvae are not killed due to abrasive action of the carrier. Infestations should be carried out early in the morning or late in the evenings. If there is heavy rainfall immediately after infestation, it is important to repeat the infestation or check the plants for feeding symptoms, and re-infest the plants that do not show any signs of insect feeding. Field infestation should be carried out at the most susceptible stage of the crop. In cereal crops, it may be necessary to infest the crop when there is no excess water in the leaf whorls. However, in cotton, grain legumes, and vegetable crops, which do not have leaf whorls or distinct sites where the larvae can be released, it may be useful to infest the crop after the rain or when there is enough morning dew. The crop can also be sprayed with water before infestation. The larvae should be dispensed into the leaf whorls of cereals, into the ears of sorghum, silks in corn, and the inflorescence in food legumes and cotton. The following protocol may be followed for evaluation of the test material under artificial infestation.

- Raise the crop in the field, and do not apply any insecticides, except to control non-target insects with selective insecticides. Space the insect infestation and insecticide application so that there is no effect of the insecticide on the target insect.

- Infest the crop with eggs or neonate larvae at the susceptible stage of the crop as appropriate.
- After one week, check the plants for insect infestation. Record data on plant damage and insect numbers as appropriate.
- At maturity, record data on number of fruiting bodies, insect damage (e.g., number of pods or bolls infested), and yield of the infested and uninfested plants or plots.
- Select the genotypes based on insect numbers/damage, and grain yield, in comparison to the resistant and susceptible checks of similar maturity.

Artificial infestation in the field provides useful information on plant resistance based on antibiosis and tolerance components of resistance. However, it does not account for antixenosis for oviposition, which is an important component of resistance to insects in several crops. This technique has been used extensively in cereal crops, but has not been tried on a large scale in grain legumes, fiber, and vegetable crops. One of the reasons for this may be the absence of a pocket where the larvae can be released through the *Bazooka* applicator. Use of sticky materials is limited to eggs, which upon drying may pose problems for egg hatching. Therefore, there is a need to refine this technique for large-scale application in legumes, vegetables, and fruit crops.

Planning the Rearing Schedule and Egg and Pupal Storage

Studies on plant resistance should ideally be carried out by infesting the plants or plant parts of the same age and growth stage as subjected to insect attack under natural conditions. Therefore, efforts should be made to have the laboratory-reared insects available for infestation at the most susceptible stage of the crop. For this purpose, planning for multiplication of insects and planting of the test material should be undertaken two to three months in advance. In case there is a mismatch between the availability of insects and the crop growth, one can hasten or slow-down insect development in the laboratory. If the insect rearing is delayed, the temperature in the rearing room can be increased by 2 to 3°C to speed up insect development. If the insect culture is ready earlier than the test material, one can delay the insect development by: (1) reducing the temperature in the rearing room, (2) storing the eggs or pupae at low temperatures (e.g., spotted stem borer, *C. partellus* and pod borer, *H. armigera* eggs can be stored at 10 to 12°C at the black head stage (two days after eggs-laying) for nearly 10 days in the refrigerator (Dhillon and Sharma, 2007). The eggs can be taken out four to six hours before use in laboratory or field infestations. The pupae of *Heliothis/Helicoverpa* can also be stored for two to three months in deep-freeze, and taken out as and when the insects are required for bioassay. This information has been used effectively for large-scale screening of the test material for resistance to insects.

Caging the Plants with Insects

Caging the test plants with insects in the greenhouse or in the field is another dependable method of screening for insect resistance. In this method, considerable control is exercised over maintaining a uniform insect pressure on the test entries, and plants are infested at the same phenological stage. This protects the insects from natural enemies and also prevents the insects from migrating away from the test plants. Cages can be designed to cover the whole plants, such as those of chickpea, safflower, sunflower, and tomato, or to cover only the plant parts that are damaged by the insects, such as panicles of sorghum

FIGURE 3.5   Headcage technique to screen sorghum for resistance to sorghum midge, *Stenodiplosis sorghicola*. (A) sorghum panicle at anthesis infested with sorghum midge females, (B) reaction to midge of a susceptible sorghum cultivar, CSH 1, and (C) reaction to midge of a resistant sorghum cultivar, ICSV 745.

(Figure 3.5) or inflorescences of cotton, pigeonpea, etc. (Sharma et al., 1992, 2005a; Sharma, Pampapathy, and Kumar, 2005). Cage size and shape are determined by the type and number of test plants needed for evaluation. For valid comparisons, resistant and susceptible checks of appropriate maturity should also be included and infested at the same time as the test genotypes.

Greenhouse Screening

Screening for resistance to some insects can be carried out at the seedling stage in the greenhouse under no-choice, dual-choice, or multi-choice conditions, and has been utilized successfully in case of sorghum (Starks and Burton, 1977; Nwanze and Reddy, 1991; Sharma et al., 1992), rice (Medrano and Heinrichs, 1985; Heinrichs, Medrano, and Rapusas, 1985), wheat (Webster and Smith, 1983), soybean (Fehr and Caviness, 1977; All, Boerma, and Todd, 1989), alfalfa (Sorenson and Horber, 1974), pearl millet (Sharma and Youm, 1999; Sharma and Sullivan, 2000), groundnut (Sharma, Pampapathy, and Kumar, 2002), and chickpea (Sharma et al., 2005a, 2005b; Sharma, Pampapathy, and Kumar, 2005) (Table 3.3). Screening can be carried out under no-choice (Figures 3.6 and 3.7) or multi-choice conditions. The test material can also be infested with insects under multi-choice, dual-choice, or no-choice conditions using cages of appropriate size. The test material can also be infested with insects at the susceptible stage without a cage. This approach is more appropriate to evaluate seedlings or small-sized plants, such as those of chickpea, safflower, cruciferous vegetables, etc. However, it may be a bit cumbersome to use the cage technique on plants with a large canopy, such as those of cotton and pigeonpea. The following procedure may be adopted to screen for resistance to insects under greenhouse conditions.

- Raise the plants in the greenhouse as per recommended package of practices. Infest the test plants at the susceptible stage of the crop, for example, for chickpea, the plants can be infested with neonate larvae of *H. armigera* at 30 days after seedling emergence or at the 50% flowering and pod setting stages; while for sorghum, the infestation should be carried out at 10 days after seedling emergence for *A. soccata*, and 20 days after seedling emergence with *C. partellus*, *Mythimna Separeta* (Walk.),

TABLE 3.3

Techniques to Screen for Resistance to Insect Pests under Greenhouse Conditions

Crop	Insect Species	Remarks	References
Rice	Yellow stem borer, *Scirpophaga incertulas* (Walker)	Oil drum pots were found to be simple and quick to screen deepwater rice for resistance to stem borer.	Catling, Islam, and Pattrasudhi (1988);
		Larvae of the stem borer were not able to grow on the callus of *Oryza ridleyi*, but normal growth was observed on susceptible variety, Rexero.	Caballero et al. (1988)
	Leafhoppers, *Nilaparvata lugens* *Sogatella furcifera* (Horvath)	Five 2nd or 3rd instar nymphs per seedling at 3.5 to 4 leaf stage can be used to identify resistant genotypes using seed box technique. Measurement of plant damage and honeydew were similar.	Wu, Li, and Zhang (1989); Jeyarani and Velusamy (2002); Liu et al. (2002)
	Green leafhopper, *Nephotettix virescens*	Screening first in open free-choice test, then verifying the resistance in no-choice test.	Karim and Saxena (1990)
	African rice gall midge, *Orseolia oryzivora*	Using midge culture to start infestation in a tunnel screenhouse in field was similar to insect cages.	C.T. Williams et al. (2001)
Wheat	Russian wheat aphid, *Diuraphis noxia*	Initial and final plant height, initial and final *D. noxia* infestation, damage rating, leaf area, and dry plant mass can be used to evaluate for resistance.	Tolmay, van der Westhuizen, and van Deventer (1999)
Maize	Rice weevil, *Sitophilus oryzae* (L.)	Eight pairs of 7-day-old adults per 20 grains in plastic tubes can be used for resistance screening.	de Sarup (1989)
	Corn earworm, *Helicoverpa zea*	Callus feeding can be used to assess antibiosis mechanism of resistance.	W.P. Williams, Buckley, and Davis (1987)
Sorghum	Sorghum shoot fly, *Atherigona soccata*	Screening under multi-, double-, or no-choice conditions. Shoot fly females (12 to 16) enclosed with 7- to 10-day-old seedlings for 24 h.	Sharma et al. (1992)
	Spotted stem borer, *Chilo partellus*	The tray method required 250 egg masses in 15 g of carrier to infest 1000 plants.	Nwanze and Reddy (1991)
	Sorghum midge, *Stenodiplosis sorghicola*	Five midge females released per 50 sorghum spikelets for 6 h to measure antixenosis for oviposition.	Franzmann (1996)
	Maize weevil, *Sitophilus zeamais* (Mots.)	Infest 50 grains with three 16- to 20-day-old females for 14 days.	Larrain, Araya, and Paschke (1995)
Pearl millet	Oriental armyworm, *Mythimna separata*	Fifteen-day-old seedlings infested with 10 neonate larvae per 5 plants.	Sharma and Sullivan (2000)
Bermuda grass	Fall armyworm, *Spodoptera frugiperda*	There were differences in callus feeding, but the resistance ratings were not the same as in leaf screening tests.	Croughan and Quisenberry (1989)
Soybean	*Helicoverpa zea*, *Heliothis virescens*, *Pseudoplusia includens* (Walker)	Neonate larvae placed on 12- to 16-day-old potted plants with free-choice among the plants for 14 days within a replicate.	All, Boerma, and Todd (1989)
Groundnut	Tobacco caterpillar, *Spodoptera litura*	Thirty-day-old seedlings infested with 20 neonate larvae per plant	Sharma, Pampapathy, and Kumar (2002)

continued

TABLE 3.3   (continued)

Crop	Insect Species	Remarks	References
Cowpea	Spotted pod borer, *Maruca vitrata*	Screening test based on nonpreference for oviposition and larval feeding, final-instar larval weight, pupal period, and adult size.	Echendu and Akingbohungbe (1990)
	Pod sucking bug, *Clavigralla tomentosicollis*	Dry cowpea seeds were better than fresh seeds or fresh pods to identify protracted nymphal development and high cohort mortality.	Jackai (1990)
Pigeonpea	Spotted pod borer, *Maruca vitrata*	Plants infested with 10 neonate larvae at 50% flowering.	Sharma and Franzmann (2000); Sharma, Saxena, and Bhagwat (1999)
Chickpea	Pod borer, *Helicoverpa armigera*	Plants infested with 20 neonate larvae at 30 days after planting or at the flowering stage.	Sharma, Pampapathy, and Kumar (2005)
Okra	Cotton leafhopper, *Amrasca biguttula biguttula*	One-week-old plants can be used for rapid screening. Inverse relationship between nymphs and trichome density.	Mahal, Lal, and Singh (1993)
Lentil	Weevil, *Sitona crinitus* Herbst.	Infesting 10 lentil seedlings with 6 pairs caused high damage (visual damage score of 7.0 on a 1 to 9 scale).	El Damir, El Bouhssini, and Al Salti (1999)
Brassica	Mustard aphid, *Lipaphis erysimi*	*Brassica* germplasm can be screened for resistance to aphid at cotyledonary stage using 10 and 15 apterae of *L. erysimi*, and aphid clones from a common host.	Kher and Rataul (1992)
Mustard	Flea beetle, *Phyllotreta cruciferae* (Geoze)	Test arenas without borders and evaluating damage when 50% of the cotyledon area in the standard check was damaged.	Palaniswamy and Lamb (1992)
Sitka spruce	White pine weevil, *Pissodes strobi* (Peck.)	Caging seedlings under choice conditions is a useful tool for mass screening of clonal material.	Klimaszewski et al. (2000)

and *Spodoptera litura* (F.). The plants can be covered with cages, a muslin cloth, or a nylon bag.

- The experiment may be terminated when the differences between the resistant and susceptible checks are maximum, and more than 80% plants or leaf area/pods are damaged in the susceptible check. Data may be recorded on plant damage, egg laying, larval survival, and larval weights as appropriate. Insect damage can also be evaluated visually on a 1 to 9 scale (where 1 = <10% damage, and 9 = >80% damage).

Use of Excised Plant Parts

Artificial infestation in the field and in the greenhouse requires a large number of insects, and there is no control over the environmental conditions. Differences in flowering times of the test material add another variable to the screening process. Cage screening in the field and in the greenhouse at times may be cumbersome as a number of large-sized cages may be required to complete the screening process. To overcome some of these problems and to screen large numbers of lines rapidly, leaf discs and excised plant parts can be used successfully to evaluate the test material for resistance to insects (Sams, Laver, and Redcliffe, 1975; Fitt, Mares, and Llewellyn, 1994; Fitt et al., 1998; Olsen and Daly, 2000; Sharma et al., 2005b) (Table 3.4). These techniques can also be used to study the consumption

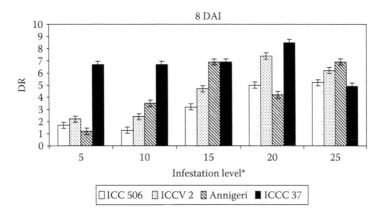

FIGURE 3.6 Reaction of three chickpea genotypes for resistance to *Helicoverpa armigera* under no-choice cage screening in the greenhouse. DR, damage rating, 1 = <10% leaf area damaged, and 9 = >80% leaf area damaged; DAI, days after infestation. *Number of larvae per plant.

and utilization of food by insects to gain a better understanding of the mechanisms of resistance. Marfa et al. (2002) used whole plants or small pieces of leaves or stems of transgenic plants to determine the degree of insect resistance in transgenic rice plants (transformed with either the *cry1B* or maize proteinase inhibitor [*mpi*] genes) to rice stem borer, *Chilo suppressalis* (Walker). The bioassay consisting of rice seeds germinated *in vitro* using Murashige and Skoog medium in test tubes, and then infestation of each 3- to 4-cm long seedling with one neonate larva obtained from surface-sterilized eggs of rice stem borer, *C. suppressalis* has been found to be quite effective. On nontransformed plants, the larvae developed normally, while those fed on plants transformed with *cry1B* died after a few days of infestation. On transgenic plants with the *mpi* gene, the neonate larvae did not die, but their growth was slower as compared to that on the nontransgenic control plants.

Detached Leaf Assay

The detached leaf assay has been used successfully in several crops in which a leaf with a petiole, a small branch, or an inflorescence can be placed into water, 3% agar-agar, 1% sugar solution, Arnon, or Hoagland solution in a plastic or a glass jar of appropriate capacity to keep the leaves in a turgid condition for five to seven days (Figure 3.8).

FIGURE 3.7 No-choice cage technique to screen chickpea for resistance to *Helicoverpa armigera*. ICCC 37, Susceptible check and ICC 506, Resistant check.

TABLE 3.4

Use of Excised Plant Parts to Evaluate Genotypic Resistance to Insect Pests

Crop	Insect Species	Remarks	References
Cotton	Cotton bollworm, *Helicoverpa armigera*	A 48 h technique using VI instar larvae proved as reliable as the standard technique, but was faster and more energy efficient.	McColl and Noble (1992)
		Detached leaf petiole embedded in 3% agar in 250 mL cup, and infested with 10 to 20 neonate larvae can be used to evaluate for resistance to insects.	Olsen and Daly (2000); Sharma et al. (2005b)
	Whitefly, *Bemisia tabaci* (Genn.)	First fully expanded leaf with petiole placed in water and infested with 10 white fly females can be used to screen for resistance.	Ranjith and Mohansundram (1992)
Rice	Striped stem borer, *Chilo suppressalis*	Orientation and settling responses of the third and fourth instars on the stem pieces differed significantly, and were correlated with feeding ratio.	Saeb, Nouri Ghonbalani, and Rajabi (2001)
Pigeonpea	Pod borer, *Helicoverpa armigera*	Detached trifoliate petiole embedded in 3% agar in a 250 mL cup, and infested with 20 neonate larvae can be used to screen for resistance.	Sharma et al. (2005b)
Chickpea	Pod borer, *Helicoverpa armigera*	Detached 5 to 7 cm terminal branch embedded in 3% agar in a 250 mL cup, and infested with 10 neonate larvae can be used to screen for resistance.	Sharma et al. (2005b)
Groundnut	Pod borer, *Helicoverpa armigera*	Detached leaf petiole embedded in 3% agar in a 250 mL cup, and infested with 20 neonate larvae.	Sharma et al. (2005b)
Peas	Pea weevil, *Bruchus pisorum* L.	Flat and swollen pods (>20 mm long) can be used to evaluate peas for resistance to pea weevil. Dual-choice and no-choice tests can be used to identify genotypes resistant to *B. pisorum*.	Hardie and Clement (2001)
Soybean	Twospotted spider mite, *Tetranychus urticae* Koch	Excised leaf and intact whole plant bioassays did not differ for the variables measured. Excised leaf bioassay was easier to set up and monitor, and took 75% less space in the growth chamber.	Elden (1999)
Brachiaria spp.	Spittle bug, *Aeneolamia varia* (Dist.)	A small plant growth unit supporting a single stem cutting (vegetative propagule) infested with 6 adults or 10 nymphs can be used to screen for resistance to *A. varia*.	Cardona, Miles, and Sotelo (1999)
Lettuce	Banded cucumber beetle, *Diabrotica balteata* Le conte	Whole plants were more useful to evaluate lettuce cultivars for resistance to *D. balteata* than excised plant parts.	Huang et al. (2003)
Brassicas	Mustard aphid, *Lipaphis erysimi*	Placing leaves in Hoagland's solution was better for screening of *Brassica* germplasm for aphid resistance.	H. Singh, Rohilla, and Singh (2001)
Strawberry	Two-spotted mite, *Tetranychus urticae*	Detached leaf in water supported by a floating polystyrene raft, with the lower side of the leaf facing downwards maintained the leaves in turgid condition for 4 weeks.	Sonneveld, Wainwright, and Labuschagne (1997)

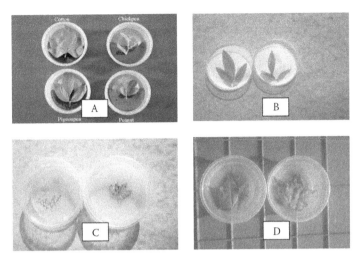

FIGURE 3.8 Detached leaf assay to screen for resistance to insect pests. (A) Uninfested leaves of cotton, chickpea, pigeonpea, and groundnut. (B), (C), and (D) are reactions of pigeonpea, chickpea, and cotton to neonate larvae of *Helicoverpa armigera* using detached leaf assay.

The detached leaves are infested with neonate or third-instar larvae, nymphs, or adults. The test is easy and quick, and can be carried out with different parts of the same plant at different growth stages without sacrificing the plant. The reactions of test genotypes invariably are similar to those under natural infestation (Figure 3.9). However, excision of plant parts at times leads to a change in the physico-chemical properties of the plants, and the magnitude of resistance can either decrease (Thomas, Sorensen, and Painter,

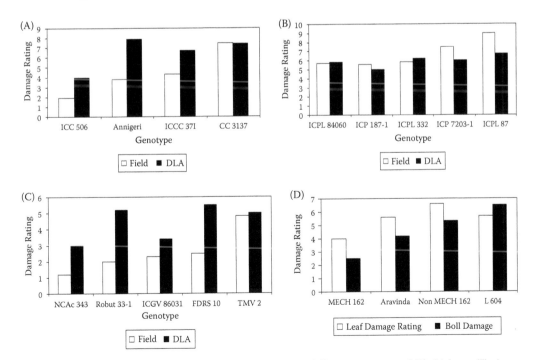

FIGURE 3.9 Expression of resistance to *Helicoverpa armigera* in different genotypes of (A) chickpea, (B) pigeonpea, (C) groundnut, and (D) cotton in detached leaf assay (DLA) and under natural infestation in the field.

1966) or increase (van Emden and Bashford, 1976), depending on the nature of insect plant interactions in different crops. These techniques can also be used to study consumption and utilization of food by insects. The following procedure may be followed to screen the test material using detached leaf assay.

- Detach a leaf with a petiole (first fully expanded leaf of cotton, pigeonpea, groundnut, sunflower, or tomato) or a 7- to 10-cm growing terminal branch of chickpea, pigeonpea, safflower, or an inflorescence (15 cm long) of pigeonpea, chickpea, cotton, or tomato, and place it immediately in agar-agar (3%), or insert the petiole or the inflorescence in sugar or Arnon solution in a plastic or conical flask of appropriate capacity (250 to 1000 mL) or insert the leaf petiole or the branch into 3% agar-agar (10 to 15 mL) solution poured on to one side of the plastic cup.
- Release an adequate number of neonate or third-instar larvae that result in maximum differences between the resistant and susceptible genotypes.
- Record data on leaf/pod feeding, larval survival, and larval weights at four to six days after infestation, when the differences between the resistant and susceptible checks are maximum, and more than 80% of the leaf area/pods are damaged in the susceptible check. Evaluate leaf/pod damage on a 1 to 9 scale (where 1 = <10% leaf area/pods damaged, and 9 = >80% leaf area/pods damaged).

Leaf Disc or Pod/Boll Assay

Leaf discs of appropriate size or pods and bolls at the susceptible stage can also be used to assess the genotypic resistance to insects. The leaf discs and pods/bolls can be offered to the larva in no-choice, dual-choice, or multi-choice assays. The bioassays can be carried out as follows.

- Place a filter paper at the bottom of the Petri dish (7.5 cm diameter), and another filter paper inside the lid, and moisten the filter paper inside the lid with 2 to 3 mL of water. This keeps the excised plant parts in a turgid condition.
- Take a leaf disc of appropriate size (2.5 to 5 sq. cm) from the mid-portion of the first fully expanded leaf, a pod, or a boll (7 to 10 days old), and place in a Petri-dish arena in no-choice, dual-choice, or multi-choice conditions.
- Release neonate or third-instar larvae in the Petri dish arena, as appropriate.
- Record data on extent of leaf/pod feeding, larval survival, and larval weights at 48 or 72 hours after infestation, when the differences between the resistant and susceptible genotypes are maximum.

Oviposition Nonpreference

Antixenosis for oviposition is a major component of resistance to insects and, therefore, it can be used as one of the criteria to evaluate relative resistance of test genotypes to insect pests. For this purpose, the plants can be grown in pots in the greenhouse or 20-cm long branches or inflorescences can be brought from the field and placed in a conical flask containing water, sugar (1%), Hoagland, or Arnon solution. The oviposition preference can be studied under no-choice, dual-choice, or multi-choice conditions.

- Raise plants in pots in the greenhouse or take inflorescences from the field and place them in a conical flask (250 mL capacity) containing water or 1% sucrose or

Arnon solution. Place a cotton swab around the stem at the neck of the conical flask to keep the inflorescences in upright condition.

- Place the plants or inflorescences inside a cage (30 × 30 × 30 or 50 × 50 × 50 cm) under no-choice, dual-choice, or multi-choice conditions. Known resistant or susceptible checks should also be placed inside the cage in dual-choice and multi-choice tests. Release 5 to 20 pairs of insects inside the cage for three days, and repeat the experiment 5 to 10 times.

- Record data on eggs laid after three days. The number of eggs can also be recorded daily, and the plants changed every day. Compute the oviposition nonpreference in relation to the number of eggs laid on the resistant or the susceptible check.

Diet Impregnation Assay to Assess Antibiosis

Antibiosis to insects in general is mainly because of poor nutritional quality or secondary plant substances, for example, gossypol and tannins in cotton (Lukefahr, Martin, and Meyer, 1965; Sharma and Agarwal, 1983b), tridecanone and tomatine in tomato (Campbell and Duffey, 1979; W.G. Williams et al., 1980), maysin in maize (Waiss et al., 1979), oxalic and malic acids in chickpea (Yoshida, Cowgill, and Wightman, 1997), and flavonoids in pigeonpea (Stevenson et al., 2005). Antibiosis effects of secondary metabolites can be measured in terms of survival and development of insects on host plant tissues incorporated into inert materials or artificial diet (Table 3.5). To assess the antibiosis component of

TABLE 3.5

Artificial Diet Impregnation Assay to Assess Antibiosis Component of Resistance to Insect Pests

Crop	Insect Species	Remarks	References
Cotton	*Spodoptera littoralis* (Boisd.)	Gossypol incorporated in artificial diet.	Meisner et al. (1977)
Pigeonpea	Pod borer, *Helicoverpa armigera*	Ten grams of leaf powder mixed in 250 mL of artificial diet differentiates resistant and susceptible genotypes.	Kumari, Sharma, and Jagdishwar Reddy (2008)
Chickpea	Pod borer, *Helicoverpa armigera*	Ten grams of leaf powder mixed in 250 mL of artificial diet can be used to assess antibiosis component of resistance.	Sharma et al. (2005a)
Maize	European corn borer, *Ostrinia nubilalis*	Bioassays for antibiotic effects should utilize a diet with limited contaminants, incorporate tissue on a weight basis, and focus on silk tissue.	Warnock et al. (1997)
	Southwestern corn borer, *Diatraea grandiosella* (Dyar)	Add 15 g lyophilized husk tissue collected within 3 days of silk emergence to artificial diet. Weigh the larvae after 21 days.	W.P. Williams and Buckley (1996)
Sorghum	Fall armyworm, *Spodoptera frugiperda*	Greater antibiosis was observed when the dried milk stage florets were mixed with diet without pinto bean.	Diawara, Wiseman, and Isenhour (1991)
	Spotted stem borer, *Chilo partellus*	Lyophilized leaf powder (10 to 20 g per 250 mL of diet) can be used to assess antibiosis component of resistance.	Kishore Kumar, Sharma, and Dharma Reddy (2005)
Soybean	Corn earworm, *Helicoverpa zea*	Lyophilized pod powder in artificial diet can be used to assess antibiosis component of resistance.	Dougherty (1976)

resistance through diet impregnation assay, the test material can be grown in the greenhouse or field conditions (Kumari, Sharma, and Jagdishwar Reddy, 2008). No insecticide should be applied on the test material, but care should be taken that there is no heavy incidence of any disease or nontarget insect species.

- The plant material (leaves at the beginning of flowering in chickpea, pigeonpea, and cotton; flowers at petal opening; or pods/bolls at 10 to 15 days after flowering) should be harvested and immediately placed in an icebox.
- The test material should be brought to the laboratory and placed in a freeze-drier until the material is completely dry. If freeze-drying equipment is not available, the test material can be placed in paper bags and placed in an oven at 55°C and dried to a constant weight (in about four to five days).
- The dried material should be powdered in a pestle and mortar or in a Willey mill, placed in sealed packets or paper bags, and stored in desiccators until used in diet incorporation assay.
- Take 10 to 20 g of the powdered material (10 g leaf or pod powder in case of chickpea and pigeonpea) and add it to diet ingredients to prepare 250 mL of diet. Pour 7 to 10 mL of diet in 15- to 20-mL glass vials or plastic cups, allow the diet to solidify, and release neonate larvae singly or in groups in each vial. Keep the vials at 27 ± 2°C or at room temperature.
- Record data on larval survival and larval weight at 10 days after initiating the experiment, number of larvae pupated, adult emergence, fecundity, larval and pupal development periods, and pupal weights.

A relative assessment of the adverse effects of different genotypes on the survival and development of insects will give an idea of the antibiosis component of resistance because of the presence or absence of secondary metabolites and/or poor nutritional quality of food.

Bioassay of Transgenic Plants for Resistance to Insects

Evaluation of transgenic plants should be carried out in contained facilities (P_2 level greenhouse or in the laboratory) with the approval of the institute's biosafety committee or as per guidelines of the national regulatory bodies, with proper control over pollen movement, and disposal of plant residues, water, and soil used for growing the plants. The transgenic events along with the nontransgenic plants of the same cultivar should be grown under similar conditions. If possible, susceptible and resistant checks (identified through conventional host plant resistance evaluations) of the same crop should also be included as controls. The material can be tested using leaf disc assay, detached leaf assay, no-choice cage screening in the greenhouse, feeding preference assays, and consumption and utilization of food. The bioassay can be performed with neonate larvae or nymphs, third-instar larvae, and adults as per the feeding behavior of the insect, and the reliability of the testing procedure. There should be 5 to 10 replications for each transgenic plant or event depending on the availability of the plant material and the experimental requirements.

FIGURE 3.10 A contained field trial for evaluating transgenic plants for insect resistance under artificial infestation.

The experiments can be performed using completely randomized design or any other appropriate system. The plant residue and the insects used in bioassay should be properly disposed of or burnt. The seeds of the transgenic material should be stored in a cool and dry place under proper safeguards.

Transgenic events selected for insect resistance with optimum expression of the transgene are tested under contained field trials, after obtaining necessary permission from the institutional and national biosafety regulatory bodies. The experimental area should be fenced all around, and the plants covered with a nylon mesh to prevent the movement of pollinators and the pollen from transgenic plants to nontransgenic plants (Figure 3.10). The transgenic events to be tested along with the nontransgenic plants of the same cultivar should be planted in a randomized block design with three to five replications. A plot size of 10 to 50 square meters should be maintained, depending on the resources and the material to be tested. The nontransgenic plants of the same cultivar should be planted around the test plots to monitor the movement of pollen, if any, through molecular techniques (RT-PCR, ELISA, Southern blot, etc.). A nonhost crop should be planted all around the boundary of the test plots to prevent movement of pollen from the test plants to the non-transgenic plants of the same crop or closely related wild relatives of the transgenic plants being tested. The test material is infested with a uniform number of neonates or third-instar larvae at a density sufficient to differentiate resistant and susceptible plants. The plant material can also be brought to the laboratory and subjected to the bioassays described above. The test material can also be infested with adults. In the case of lepidopteran insects that are affected by light, infestation with adults does not lead to uniform infestation of the test material. Once the transgenic events have been tested in contained field trials for two to three seasons for their efficacy to control the target pests, yield potential, nutritional quality, and environmental safety may be tested as per national biosafety guidelines. The promising events thus identified are tested in the target areas in an open field in replicated trials. The test material can be subjected to natural infestation or infested artificially with insects reared in the laboratory. Once the effectiveness of the transgene, and the agronomic, nutritional, and biosafety desirability of an event is proved across seasons and locations, the institutions/scientists concerned can apply for the deregulation of the event for use by farmers as a component of integrated pest management.

Phenotyping Mapping Populations for Resistance to Insects

Phenotyping for identification of quantitative trait loci (QTL) or genes associated with insect resistance can be carried out under field, greenhouse, or laboratory conditions. Nearly 250 to 500 F_{6-10} recombinant inbred lines (RILs) derived from a single cross, double haploids, backcross progenies, or F_2s should be tested under field conditions in a lattice square or balanced alpha design with three to five replications. A plot size of two rows, 2 to 4 m long, should be maintained. The plot size and number of replications to be used will depend on the resources available and field variability. Evaluation for resistance to insects can be carried out under natural or artificial infestation. Data should be recorded on insect survival and development, plant damage, and yield, with utmost precision. The precision of the phenotypic data determines the quality of the genetic linkage map and the robustness of the QTL or markers associated with resistance to insects. The mapping populations should be tested across seasons and locations to assess the genotype-environment interactions. The mapping populations can also be evaluated using cage screening in the field or in the greenhouse, excised plant parts, diet impregnation assay, or monitoring of physicochemical traits associated with resistance to insects. The information generated on insect survival, development, and plant damage is used to develop the genetic linkage maps and identify the QTLs or genes associated with resistance to insects.

Measurement of Host Plant Resistance to Insects

Precise assessment of insect damage or insect numbers at the initial stage is most important to screen for resistance to insect pests. Data should be recorded at the appropriate stage on extent of leaf, pod, boll, or plant damage, depending on the nature of the damage (Sharma, Singh, and Ortiz, 2001). Some polyphagous insect pests, such as pod borers and bollworms (*Heliothis*, *Helicoverpa*, and *Maruca*), feed on several plant parts, such as leaves, flower buds, flowers, pods, bolls, and the seeds, whereas the cereal stem borers (*C. partellus*, *Busseola fusca* Fuller, and *Sesamia inferens* Wlk.) feed on leaves, leaf sheaths, stems, and the peduncle, leading to leaf scarification, stem tunnelling, peduncle breakage, or partial seed set. Damage to vegetative plant parts results in indirect loss, and generally gets compensated as a result of regrowth in the affected plants or plant parts. However, damage to the reproductive parts, particularly to flowers and developing seeds, results in direct loss and there is little chance to recover from insect damage. Hence, the level of infestation during the flowering and fruiting phases is quite important for assessing insect damage to quantify genotypic resistance to insects feeding on reproductive structures. Numbers of fruiting bodies and the extent of damage is the final outcome of complex interactions involving the insects and their host plants. Percentage damage to plants and plant parts, such as leaves, bolls, fruits, and pods, are commonly used for determining the genotypic susceptibility to pod borers and bollworms. However, this criterion often leads to unreliable results due to variations in insect population, damage to the foliage (which is not reflected in the damage to the fruiting bodies), damage to flowers, shedding of the reproductive plant parts as a result of insect infestation, and the genotypic ability to produce a second flush in case the first flush is lost due to insect damage. At times, the second flush may escape insect damage, and thus give an erroneous estimate of genotypic susceptibility to insect pests based on damage to bolls or pods. To overcome these problems, the test material can be evaluated on a visual damage rating scale of 1 to 9, taking into consideration the extent of leaf damage or recovery from insect feeding during the vegetative stage (on seedlings and leaves), numbers of fruiting bodies and their distribution on the plant, and the proportion of fruiting bodies or grain damaged by insects.

Data can also be recorded on insect numbers, survival and development, fecundity, consumption and utilization of food, host selection and feeding behavior, numbers of the diapausing insects, etc. The following methods can be used to assess insect damage to evaluate the germplasm, segregating breeding material, mapping populations, and transgenic plants for resistance to insects.

Visual Damage Rating

1. Plants with little damage during the vegetative stage and damaged plants that show good recovery from insect damage, and/or fruiting bodies uniformly distributed throughout the plant canopy, and <10% of the fruiting bodies with insect damage.

2. Plants with 10 to 20% damage during the vegetative stage and showing good recovery from insect damage, and/or large number of fruiting bodies retained on the plant with uniform distribution, and 10 to 20% of the fruiting bodies with insect damage.

3. Plants with 20 to 30% damage during the vegetative stage, and showing good recovery from insect damage, and/or a large number of fruiting bodies retained on the plant and 20 to 30% of the fruiting bodies with insect damage.

4. Plants with 30 to 40% damage during the vegetative stage with moderate recovery from insect damage, and/or moderate number of fruiting bodies retained on the plant, and 30 to 40% of the fruiting bodies with insect damage.

5. Plants with 40 to 50% damage during the vegetative stage with moderate recovery, and/or nearly 50% of the fruiting bodies retained on the plant, and 40 to 50% of the fruiting bodies with insect damage.

6. Plants with 50 to 60% damage during the vegetative stage, with low to moderate recovery, and/or nearly 40% of the fruiting bodies retained on the plant, and 50 to 60% of the fruiting bodies with insect damage.

7. Plants with 60 to 70% damage during the vegetative stage, with poor recovery from insect damage, and/or nearly 30% of the fruiting bodies retained on the plant, and 60 to 70% of the fruiting bodies with insect damage.

8. Plants with 70 to 80% damage during the vegetative stage with very poor recovery from insect damage, and/or nearly 20% of the fruiting bodies retained on the plant, and 70 to 80% of the fruiting bodies with insect damage.

9. Plants with >80% damage during the vegetative stage with little recovery from insect damage, and/or <20% of the fruiting bodies retained on the plant, and >80% of the fruiting bodies with insect damage.

Indirect Feeding Injury

Measurements of insect damage to plants are more useful than insect numbers, insect growth, or development. Plant damage and the resulting reduction in yield or quality are important while establishing the goals of a crop improvement program. Measurements of yield reduction indicate direct insect feeding injury to plants. Plant damage can also be determined by measuring the number of plants with insect damage, incidence of tissue necrosis, fruit abscission, and stem damage. The quality of produce can also be used to measure the effect of insect damage. Leaf defoliation or damage to fruiting bodies is usually determined by rating scales that make use of visual estimates of plant damage based on percentages or numerical ratings as described above. Several such rating scales have been developed to assess insect damage in crop plants (Sharma et al., 1992; Smith, Khan, and Pathak, 1994). Direct measurements of leaf area are also used to measure insect damage. Indirect feeding injury measurements, such as plant growth, photosynthetic rates, transpiration rates, ethylene production, and respiratory rates can also be used to assess genotypic resistance to insects. In some cases, the test genotypes can be first evaluated visually for insect damage on a scale of 1 to 9, and then percentage damage to the fruiting bodies or seeds can be recorded by counting the total number of fruiting bodies with and without insect damage in five plants selected at random in the center of each plot. Additional information on susceptibility to other insects and diseases should also be recorded at the appropriate stage of crop growth, as and when feasible.

Simulated Feeding Injury

Insect feeding injury can be simulated by mechanical defoliation. This is particularly useful in the case of leaf-feeding insects such as grasshoppers, beetles, and leaf-feeding caterpillars. However, plants respond differently to artificial defoliation than to actual insect feeding. Therefore, the relationship between artificial defoliation and natural insect feeding should be determined before results on artificial defoliation are accepted as a measure of insect damage-yield-loss relationship. Insect injury can also be measured by injection of toxic insect secretions into plant tissues, for example, the application of crude extract of greenbug in sorghum.

Association of Physico-Chemical Characteristics and Molecular Markers with Insect Resistance

Plant resistance can also be assessed by measuring the concentrations of allelochemicals or the density, size, and distribution of morphological structures, and by using molecular markers associated with resistance to insects (Smith, Khan, and Pathak, 1994; Sharma and Nwanze, 1997). This permits a rapid identification of plants potentially resistant to the target insect species. Use of biochemical (Table 3.6) and morphological markers (Table 3.7) to measure host plant resistance to insects also overcomes the variation associated with insect infestation, and the effect of environmental factors on the expression of resistance to insects. Molecular markers associated with insect resistance can also be used to monitor the presence of QTLs associated with resistance to insects in the test material using association mapping (Dhillon et al., 2006). Molecular markers can be used to track genes from both conventional sources and transgenic plants for gene pyramiding.

Sampling Insect Populations

Plant resistance to insects can be measured in terms of insect numbers on the test genotypes. Insect numbers can be estimated by sampling at the plant site where damage has taken place, and at the appropriate phenological stage and time. The population of small-sized immobile insects can be measured visually, but this method is subjected to variations in

TABLE 3.6

Biochemical Markers Associated with Resistance to Insect Pests

Crop	Insect Species	Remarks	References
Cotton	Tobacco budworm, *Heliothis virescens*	Gossypol and cyanidin-3-glucoside amounts are highly correlated with antibiosis to bollworms.	Hedin et al. (1983)
	Bollworm, *Helicoverpa armigera*	Gossypol amounts are associated with bollworm damage.	Tang and Wang (1996)
	Spotted bollworm, *Earias vittella*	Gossypol and tannin contents are linked to spotted bollworm resistance.	Sharma and Agarwal (1983b)
	Cotton jassid, *Amrasca buguttula biguttula*	Condensed tannins and proline contents are associated with resistance/susceptibility.	Sharma and Agarwal (1983a)
Rice	Striped stem borer, *Chilo suppressalis*	GUS histochemical assay is convenient to screen transgenic plants for resistance to this pest.	Wu et al. (2000); Shen et al. (2003)
		Pentadecanal acts as oviposition deterrent.	Dhaliwal and Pathak (1988)
	Brown planthopper, *Nilaparvata lugens*	Low amounts of asparagine and glutamic acid are associated with resistance/susceptibility.	Sogawa and Pathak (1970)
Maize	European corn borer, *Ostrinia nubilalis*	DIMBOA content is associated with resistance to European corn borer and corn earworm.	Klun and Robinson (1969)
	Corn earworm, *Helicoverpa zea*	Maysin, DIMBOA, and MBOA contents are associated with earworm resistance.	Reed, Bhendley, and Showers (1972); Waiss et al. (1979); Campos et al. (1988)
Wheat	Wheat blossom midge, *Sitodiplosis mosellana* (Gehin)	The concentration of ferulic acid in kernels is correlated with midge infestation.	Abdel et al. (2001)
Sorghum	Sorghum midge, *Stenodiplosis sorghicola*	Tannin content of sorghum grain is associated with resistance to midge damage.	Sharma, Vidyasagar, and Leuschner (1990)
	Shoot fly, *Atherigona soccata*	Low lysine content in resistant varieties. Dhurrin acts as antifeedant.	S.P. Singh and Jotwani (1980); B.U. Singh and Rana (1986)
Pigeonpea	Pod borer, *Helicoverpa armigera*	Stilbene, quercetin, and quercetrin, and quercetrin–3-methyl ether are associated with susceptibility to pod borer.	Green et al. (2002, 2003)
		Protease inhibitors contribute to insect resistance.	Giri and Kachole (1998)
Chickpea	Pod borer, *Helicoverpa armigera*	Oxalic acid and malic acid associated resistance to pod borer damage. Mackiain, judaicin, and stilbene-2-carboxylic acid act as antifeedants.	Yoshida, Cowgill, and Wightman (1995); Simmonds and Stevenson (2001)
Soybean	Cabbage looper, *Trichoplusia ni* (Hub.)	Coumesterol, diadzein, and glyceollins contribute to insect resistance.	Sharma and Norris (1991)
Tomato	Tomato fruit worm, *Helicoverpa zea*	2-Tridecanone imparts resistance to fruitworm.	Kennedy and Yamamoto (1979); W.G. Williams et al. (1980)
	Colorado potato beetle, *Leptinotarsa decemlineata* (Say)	α-tomatine associated with resistance.	Barbour and Kennedy (1991)

continued

TABLE 3.6   (continued)

Crop	Insect Species	Remarks	References
Groundnut	Tobacco caterpillar, *Spodoptera litura*	High cholorogenic and caffeoylquinic acid contents associated with resistance.	Kimmins, Padgham, and Stevenson (1995)
Sunflower	Sunflower midge, *Contarinia schulzi* (Sch.)	2,4-dichlorophenoxyacetic acid (2,4-D) injected in flower buds resulted in damage similar to sunflower midge.	Brewer, Anderson, and Urs (1994)
Sweet potato	Sweet potato weevil, *Cylas formicarius elegantulus* (Fab.)	Boehmeryl acetate and its alcohol, boehmerol act as oviposition stimulants.	Son, Severson, and Kays (1991)
Lettuce	Lettuce root aphid, *Pemphigus bursarius* (L.)	Visualizing phenolic acids by using fluorescence microscopy.	Cole (1987)
Carrot	Carrot root fly, *Psila rosae* (F.)	Visualizing phenolic acids by using fluorescence microscopy.	Cole (1987)
		Cholorogenic acid imparts resistance.	Ellis (1999)
Bitter gourd	Tobacco caterpillar, *Spodoptera litura*	Momordicine III acts as antifeedant.	Yasui, Kato, and Yazawa (1998)
	Melon fruit fly, *Bactrocera cucurbitae* (Coquillett)	Moisture and potassium contents of the fruit are positively correlated, while the reducing sugars are negatively associated with fruit fly infestation.	Dhillon et al. (2005b)

TABLE 3.7

Morphological Markers Associated with Resistance to Insect Pests

Crop	Insect Species	Remarks	References
Rice	Striped stem borer, *Chilo suppressalis*	Nonglandular trichomes impart resistance. Silica content associated with resistance.	Patanakamjorn and Pathak (1967); Pathak et al. (1971)
	Brown planthopper, *Nilaparvata lugens*	Nonglandular trichomes impart resistance.	Petinez (1986)
Maize	Earworm, *Helicoverpa zea*	Trichomes associated with oviposition preference.	Widstrom, McMillian, and Wiseman (1979)
	Spotted stem borer, *Chilo partellus*	Nonglandular trichomes impart resistance.	Kumar (1992)
Wheat	Hessian fly, *Mayetiola destructor* (Say)	Nonglandular trichomes impart resistance.	Roberts et al. (1979)
	Barley aphid, *Rhopalosiphum padi* (L.)	Nonglandular trichomes impart resistance.	Roberts and Foster (1983)
Sorghum	Shoot fly, *Atherigona soccata*	Trichomes, leaf glossiness, and leaf sheath pigmentation are associated with oviposition nonpreference. Silica deposition imparts resistance.	Bothe and Pokharkar (1985); Sharma and Nwanze (1997); Dhillon et al. (2005a)
	Stem borer, *Chilo partellus*	Ligular hairs, leaf glossiness, and panicle initiation are associated with resistance.	Sharma and Nwanze (1997)
	Sorghum midge, *Stenodiplosis sorghicola*	Short and tight glumes, and faster rate of grain development impart resistance.	Sharma, Vidyasagar, and Leuschner (1990)

continued

TABLE 3.7   (continued)

Crop	Insect Species	Remarks	References
	African head bug, *Eurystylus oldi*	Longer covering of the grain by the glumes and grain hardness impart resistance.	Sharma et al. (1994)
Cotton	Bollworms, *Helicoverpa, Heliothis, Earias*	Leaf pubescence associated with oviposition preference.	Sharma and Agarwal (1983b); Navasero and Ramaswamy (1991); Butter and Singh (1996)
	Cotton jassid, *Amrasca biguttula biguttula*	Hairy varieties are resistant to jassid damage.	Sharma and Agarwal (1983a)
	White fly, *Bemisia tabaci*	Hairy varieties are susceptible to white fly.	Navon et al. (1991)
Pigeonpea	Pod borer, *Helicoverpa armigera*	Nonglandular trichomes associated with resistance.	Romeis, Shanower, and Peter (1999)
Chickpea	Pod borer, *Helicoverpa armigera*	Glandular trichomes contribute to pod borer resistance.	Yoshida, Cowgill, and Wightman (1997)
Cowpea	Pod borer, *Maruca vitrata*	Trichomes inhibit feeding by the neonate larvae.	Jackai and Oghiakhe (1989)
	Pod sucking bug, *Clavigralla tomentosicollis*	Nonglandular trichomes impart resistance.	Chiang and Singh (1988)
Soybean	Cabbage looper, *Trichoplusia ni*	Trichomes inhibit feeding by the neonate larvae.	Khan, Ward, and Norris (1986)
	Stem flies, *Melanagromyza, Ophiomyia*	Nonglandular trichomes impart resistance.	Chiang and Norris (1983)
Alfalfa	Alfalfa aphid, *Therioaphis maculata* (Buckton)	Nonglandular trichomes impart resistance.	Manglitz and Kehr (1984)
Tomato	Fruitworm, *Helicoverpa armigera*	Glandular trichomes associated with resistance.	Farrar and Kennedy (1987)
		Leaf pubescence associated with oviposition preference.	AVRDC (1987)
Brinjal	Fruit borer, *Leucinodes orbonalis* (Guen.)	Nonglandular trichomes impart resistance.	R.N. Panda and Das (1974)
Okra	Leafhopper, *Amrasca biguttula biguttula*	Nonglandular trichomes impart resistance.	R. Singh and Agarwal (1988)
Bittergourd	Melon fruit fly, *Bactrocera cucurbitae*	Flesh thickness and fruit diameter are associated with susceptibility to fruit fly.	Dhillon et al. (2005b)

colony size and pattern of insect distribution. Shaking the plants, use of sampling nets, use of traps, or actual counts are used to obtain an estimate of insect numbers per plot or plant. Insect populations can also be sampled for host preference or antixenosis by counting the insects that settle on a host plant for oviposition, for example, numbers of head bugs or midge females on sorghum panicles at the panicle emergence and flowering stages, respectively (Sharma et al., 1992). Counting the numbers of eggs laid on different genotypes can also be used as a measure of host plant preference by the adults. Population build up can be measured at an appropriate stage as a measure of antixenosis and antibiosis or host suitability for the insects. Numbers of insects entering in diapause at the end of the cropping season, such as diapausing larvae of cereal stem borers and pupae of *Heliothis/Helicoverpa*, can also be used a criterion to measure host plant resistance to insects.

Host plant resistance to insects can be measured in terms of eggs laid or relative preference for egg laying by the females as in *H. virescens* on cotton (Lukefahr, Haughtaling, and Graham, 1971; Robinson, Wolfonbarger, and Doday, 1980); *H. zea* in tomato (Gosenza and Green, 1979) and maize (Widstrom, McMillian, and Wiseman, 1979); *H. armigera* in soybean (N. Panda and Daugherty, 1975), chickpea (Cowgill and Lateef, 1996), pigeonpea (Sharma et al., 2005a), and cotton (Venugopal Rao, Tirumala Rao, and Reddy, 1991); and *A. soccata* in sorghum (Sharma et al., 1992). It is important to record egg laying during the most susceptible stage of the crop and just before egg hatching, for example, at the flowering stage in cotton, chickpea, and pigeonpea for *H. armigera*, and at the seedling stage for sorghum shoot fly, *A. soccata*.

Measurements of Yield and Quality

Yield reduction measures direct insect feeding injury to plants. Measurements of quality of produce can also be used to measure the effect of insect damage in sorghum, tomato, seed cotton, etc. For example, head bug, *C. angustatus*, damage in sorghum during grain development affects both grain quality and seedling emergence (Sharma, Soman, and Subramanian, 1995). The test entries can be planted under protected and unprotected conditions, and percentage reduction in grain yield under unprotected conditions can be used as an index of relative susceptibility or resistance to insects. This method also takes into account the tolerance or recovery component of resistance, and the stability of resistance across seasons and locations. Different levels of insect infestation can also be created using different spray regimes or through artificial infestation. Genotypes with high yield and low insect damage and regression coefficients are selected in comparison to the susceptible genotypes.

Measurements of Insect Survival and Development

Effects of antibiosis on insect survival and development are expressed in terms of larval mortality, decreased larval and pupal weights, prolonged larval and pupal development period, failure to pupate, and reduced fecundity and egg viability. Insect survival and development can be measured to assess plant resistance to insects if antibiosis is the principal component of resistance. Effects of plant resistance on survival and development are also measured in terms of amount of food consumed per unit of body weight per day and food utilization (Beck and Reese, 1976). Measurement of feces production (Kasting and McGinnis, 1962) or honeydew production (Paguia, Pathak, and Heinrichs, 1980; Pathak, Saxena, and Heinrichs, 1982) can also be used to assess food utilization and antibiosis mechanism of resistance to insects. Antibiosis effects are also expressed in terms of weight and size of insects, sex ratio, and proportion of insects entering into diapause. Plant extracts or specific chemicals that confer resistance to insects can also be bioassayed to determine the levels of resistance to insects in different crops. These can range from simple antifeedant or phago-stimulant tests to sophisticated behavioral and metabolic effects on insects. These assays normally include presenting plant fractions or chemicals to the test insects on host or nonhost tissue or substrates by surface coating or impregnation, inert substrates such as filter paper, glass fiber discs, or incorporation into artificial diets.

Consumption and Utilization of Food

Several indices of consumption and utilization of food can be used to determine the level of plant resistance to insects (Waldbauer, 1968). Effects of plant resistance on insect feeding and development are measured in terms of amount of food consumed per unit of body weight per day (consumption index, CI) or leaf area consumed, larval growth rates (GR), approximate digestibility (AD), efficiency of conversion of ingested food (ECI) into body matter, and the efficiency of conversion of digested food (ECD) into body matter. The CI is calculated as: $CI = F/TA$. Where F = weight of food ingested, T = duration (in days) of feeding period, and A = mean weight of the insect during the feeding period. Mean larval weight is determined by averaging the initial and final weights of the larvae (Soo Hoo and Fraenkel, 1966). The approximate digestibility (AD) of food is calculated as:

$$AD = \frac{\text{Weight of food ingested} - \text{weight of feces}}{\text{Weight of food ingested}} \times 100$$

Growth rate (GR) is calculated as percentage increase in larval weight per unit of time. GR = Weight gain/Initial weight of larva $\times$ 100
Larval efficiency in converting the ingested food into body matter (ECI) is calculated as:

$$ECI = \frac{\text{Weight gained by the larva}}{\text{Weight of food ingested}} \times 100$$

The efficiency with which the larvae convert the digested food into body matter (ECD) is calculated as:

$$ECD = \frac{\text{Weight gained by the larvae}}{\text{Weight of food ingested} - \text{Weight of feces}} \times 100$$

Measurements of Insect Behavior

Several techniques have been developed for studying insect behavior inside plant tissue or to quantify the effects of allomones and kairomones on insect behavior (Table 3.8).

TABLE 3.8

Physical Methods for Detecting and Monitoring Insect Behavior and Measuring Plant Resistance to Insects

Crop	Insect Species	Remarks	References
Rice	Rice bloodworm, *Chironomus tepperi* Skuse	Computer-based image analysis can be used to assess rice cultivars for resistance.	Stevens et al. (2000)
Maize	Leaf feeding insects	Pulsed x-ray can be used to examine silica deposition in the leaf blades.	Cheng and Kim (1989)
Sorghum	Sorghum midge, *Stenodiplosis sorghicola*	X-ray monitoring of the diapausing midge larvae can be used as an indicator for midge infestation.	Harris (1971)
		Sorghum midge females show differential responses to odors from wild relatives of sorghum.	Sharma and Franzmann (2001)
Sugarcane	African sugarcane borer, *Eldana saccharina*	Use NIR in screening the material for resistance. Stalk surface wax (alcohols and carbonyls) contributed towards resistance.	Rutherford and Van Staden (1996)
Cowpea	Bruchid, *Callosobruchus maculatus* (Fab.)	Use biomonitor to measure activity of internally feeding insects. The activity of larvae can be recorded for 24 h starting 14 days after oviposition.	Devereau et al. (2003)
Lettuce	Lettuce root aphid, *Pemphigus bursarius*	Feeding behavior monitored on resistant and susceptible varieties.	Cole, Riggall, and Morgan (1993)

The responses of insects to volatile stimuli have been studied for several insects. Several designs of olfactometers have been used to observe insect behavior (Traynier, 1967; Sharma and Franzmann, 2001). Olfactory responses are also studied physiologically by electro-antenograms (Visser, Van Straten, and Maarse, 1979; Guerin and Visser, 1980; Guerin, Stadler, and Buser, 1983; Saxena and Okech, 1985; Khan, Ciepiela, and Norris, 1987), elec-troretinograms, and by electronic feeding monitors (McLean and Weigt, 1968).

Selection Indices

In addition to low insect damage and/or low insect numbers on the resistant genotypes, several crops possess the ability to recover from insect damage and produce the same or more yield under insect infestation, for example, in the case of *H. armigera* damage in chickpea, pigeonpea, and cotton, certain genotypes have the ability to produce a second flush when the first flush is heavily damaged. In the case of damage by shoot fly, *A. soccata* and spotted stem borer, *C. partellus* in sorghum, the plants have the ability to produce axial tillers when the main plant is damaged, depending on the availability of nutrients and water in the soil. In some cases, the resistant genotypes have the ability to withstand higher insect numbers or damage. Several indices can be used to measure plant resistance or tolerance to insects based on insect numbers or damage and yield under infested and uninfested conditions. Some of the indices that can be used to select genotypes based on one or more damage parameters, and yield under infested and uninfested conditions are discussed below.

Tolerance Index

Tolerance to insect damage can be measured in terms of plant damage per unit number of insects.

$$\text{Tolerance rating} = \frac{\text{No. of insects per plant}}{\text{Damage rating}}.$$

This system identifies the cultivars that suffer relatively low insect damage despite having a high insect population. N. Panda and Heinrichs (1983) calculated tolerance and antibiosis indices based on insect dry weight.

$$\text{Tolerance} = \frac{\text{Insect dry weight on the test cultivar}}{\text{Insect dry weight on the susceptible cultivar}}.$$

Loss in Grain Yield

Tolerance to insect damage can also be measured in terms of loss in grain yield. Loss in grain yield of a test genotype under insect infestation relative to that under uninfested conditions is measured as:

$$\text{Loss in grain yield (\%)} = \frac{[\text{GYI}_{UN} - \text{GYI}_{IN}]}{\text{GYI}_{UN}} \times 100$$

where GYI_{UN} and GYI_{IN} are the grain yields of the *I*th genotype under uninfested (GYI_{UN}) and infested (GYI_{IN}) conditions, respectively.

Selection based on loss in grain yield favors genotypes with low grain yield potential. High loss in grain yield indicates more sensitivity to insect infestation.

Relative Efficiency Index

The Relative Efficiency Index (REI) can be calculated to classify genotypes based on grain yield (Graham, 1984; Rosales-Serna et al., 2000) under infested and uninfested conditions.

$$REI = \left(\frac{GY_{I\,(IN)}}{GY_{(IN)}}\right) \times \left(\frac{GY_{I\,(UN)}}{GY_{(UN)}}\right)$$

where $GY_{I\,(IN)}$ and $GY_{(IN)}$ are the grain yields of the *I*th genotype and average grain yield of the test genotypes, respectively, under insect infestation, while $GY_{I\,(UN)}$ and $GY_{(UN)}$ are the grain yields of the *I*th genotype and grain yield of all the test genotypes, respectively, under uninfested conditions. REI favors the genotypes with high yield potential and tolerance to insect damage.

Fischer and Maurer's Stress Susceptibility Index

The Fischer and Maurer's Stress Susceptibility Index (FMSSI) has been used for selecting genotypes for drought tolerance (Fischer and Maurer, 1978), and can be used to categorize the test genotypes for insect resistance as follows:

$$FMSSI = \frac{1 - (GY_{I\,(IN)}/GY_{I\,(UN)})}{1 - (GY_{IN}/GY_{UN})}$$

where $GY_{I\,(IN)}$ and $GY_{I\,(UN)}$ are the average grain yields of the *I*th genotype under infested and uninfested conditions, respectively, whereas GY_{IN} and GY_{UN} are the mean grain yields of all the test genotypes under infested and uninfested conditions, respectively. Selections based on FMSSI favors genotypes with low grain yield potential and high yield response under insect infestation. FMSSI values <1.0 or >1.0 indicate high or low tolerance to insect damage, respectively.

Fernandez Stress Tolerance Index

The Fernandez Stress Tolerance Index (FSTI) is based on grain yield reduction adjusted to the severity of insect damage in a particular environment (Fernandez, 1993).

$$FSTI = \left(\frac{GY_{I\,(IN)}}{GY_{(IN)}}\right) \times \left(\frac{GY_{I\,(UN)}}{GY_{(UN)}}\right) \times \left(\frac{GY_{(IN)}}{GY_{(UN)}}\right)$$

where $GY_{I(IN)}$ and $GY_{(IN)}$ are the grain yields of the *I*th genotype, and all the test genotypes, respectively, under insect infestation, while $GY_{I(UN)}$ and $GY_{(UN)}$ are average grain yield responses of the *I*th genotype, and all the test genotypes, respectively, under uninfested conditions. High FSTI values indicate high tolerance to insect damage and high grain yield potential.

Conclusions

Variations in insect populations in space and time, influence of environmental factors on behavior and biology of insects, plant growth, and biochemical composition are major constraints in screening and breeding for resistance to insect pests. Difficulties in mass production of a large number of insect species for artificial infestation at the susceptible stage of the crop hampers the development of crop cultivars with resistance to insect pests. Techniques to screen for resistance to insects under field conditions, artificial infestation, greenhouse and laboratory conditions (no-choice cage screening, detached leaf assay, diet impregnation assay, oviposition nonpreference, survival and development, consumption and utilization of food, and monitoring physico-chemical traits and insect behavior) have been developed for many insects. However, there is a need for standardizing protocols for artificial rearing of insects and precise infestation and damage evaluation procedures to generate quality data for evaluation of germplasm, segregating breeding materials, mapping populations, and transgenic plants for resistance to insect pests.

References

Abdel Aal, E.S.M., Hucl, P., Sosulski, F.W., Graf, R., Gillott, C. and Pietrzak, L. (2001). Screening spring wheat for midge resistance in relation to ferulic acid content. *Journal of Agricultural and Food Chemistry* 49: 3559–3566.

All, J.N., Boerma, H.R. and Todd, J.W. (1989). Screening soybean genotypes in the greenhouse for resistance to insects. *Crop Science* 29: 1156–1159.

Armes, N.J., Bond, G.S. and Cooters, R.J. (1992). *The Laboratory Culture and Development of* Helicoverpa armigera. Natural Resources Institute Bulletin No.57. Chatham, UK: Natural Resources Institute.

AVRDC (Asian Vegetable Research and Development Center). (1987). *Screening Hairy Tomatoes for Resistance to Tomato Fruit Worm.* Progress Report. Shanhua, Taiwan: Asian Vegetable Research and Development Center.

Barbour, J.D. and Kennedy, G.G. (1991). Role of steroidal glycoalkaloid α-tomatine in host plant resistance of tomato to Colorado potato beetle. *Journal of Chemical Ecology* 60: 289–300.

Beck, S.D. and Reese, J.C. (1976). Insect-plant interactions: Nutrition and metabolism. In Wallace, J.W. and Mansell, R.L. (Eds.), *Biochemical Interaction between Plants and Insect. Recent Advances in Phytochemistry*, vol. 10. New York, USA: Plenum Press, 41–92.

Berenbaum, M. (1986). Post ingestive effects of phytochemicals on insects. In Miller, T.A. and Miller, J. (Eds.), *Insect-Plant Interactions*. New York, USA: Springer-Verlag, 121–153.

Bothe, N.N. and Pokharkar, R.N. (1985). Role of silica content in sorghum for reaction to shoot fly. *Journal of Research, Maharashtra Agricultural Universities* 10(3): 338–339.

Bowers, G.R., Kenty, M.M., Way, M.O., Funderburk, J.E. and Strayer, J.R. (1999). Comparison of three methods for estimating defoliation in soybean breeding programs. *Agronomy Journal* 91: 242–247.

Brewer, G.J., Anderson, M.D. and Urs, N.V.R.R. (1994). Screening sunflower for tolerance to sunflower midge using the synthetic auxin 2,4-dichlorophenoxyacetic acid. *Journal of Economic Entomology* 87: 245–251.

Butter, N.S. and Singh, S. (1996). Ovipositional response of *Helicoverpa armigera* to different cotton genotypes. *Phytoparasitica* 24(2): 97–102.

Caballero, P., Singh, D.H., Khan, Z.R., Saxena, R.C., Juliano, B.O. and Zapata, F.J. (1988). Use of tissue culture to evaluate rice resistance to lepidopterous pests. *International Rice Research Newsletter* 13(5): 14–15.

Campbell, B.C. and Duffey, S.S. (1979). Tomatine and parasitic wasps: Potential incompatibility of plant antibiotic with biological control. *Science* 205: 700–705.

Campos, F., Atkinson, J., Arnason, J.T., Philogene, B.J.R., Morand, P., Werstiuk, N.H. and Timmins, G. (1988). Toxicity and toxicokinetics of 6-methoxybenzoxazolinone (MBOA) in the European corn borer, *Ostrinia nubilalis* (Hubner). *Journal of Chemical Ecology* 14: 989–1002.

Cardona, C., Miles, J.W. and Sotelo, G. (1999). An improved methodology for massive screening of *Brachiaria* spp. genotypes for resistance to *Aeneolamia varia* (Homoptera: Cercopidae). *Journal of Economic Entomology* 92: 490–496.

Catling, H.D., Islam, Z. and Pattrasudhi, R. (1988). New methods of screening deepwater rice for yellow stem borer resistance. In *Proceedings of the International Deepwater Rice Workshop*. Manila, Philippines: International Rice Research Institute, 539–549.

Chelliah, S. and Heinrichs, E.A. (1980). Factors affecting insecticide-induced resurgence of the brown planthopper, *Nilaparvata lugens* on rice. *Environmental Entomology* 9: 773–777.

Cheng, C. and Kim, H.G. (1989). The use of X-ray contact microradiography in the study of silica deposition in the leaf blade. *Maize Genetics Cooperation News Letter* 63: 45–47.

Chiang, H.S. and Norris, D.M. (1983). Morphological and physiological parameters of soybean resistance to agromyzid bean flies. *Environmental Entomology* 12: 260–265.

Chiang, H.S. and Singh, S.R. (1988). Pod hairs as a factor in *Vigna vexillata* resistance to the pod sucking bug, *Clavigralla tomentosicollis*. *Entomologia Experimentalis et Applicata* 47: 195–199.

Cole, R.A. (1987). Intensity of radical fluorescence as related to the resistance of seedlings of lettuce to the lettuce root aphid and carrot to the carrot fly. *Annals of Applied Biology* 111: 629–639.

Cole, R.A., Riggall, W. and Morgan, A. (1993). Electronically monitored feeding behaviour of the lettuce root aphid (*Pemphigus bursarius*) on resistant and susceptible lettuce varieties. *Entomologia Experimentalis et Applicata* 68: 179–185.

Cowgill, S.E. and Lateef, S.S. (1996). Identification of antibiotic and antixenotic resistance to *Helicoverpa armigera* (Lepidoptera: Noctuidae) in chickpea. *Journal of Economic Entomology* 89: 224–229.

Croughan, S.S. and Quisenberry, S.S. (1989). Evaluation of cell culture as a screening technique for determining fall armyworm (Lepidoptera: Noctuidae) resistance in Bermuda grass. *Journal of Economic Entomology* 82: 232–235.

Davis, F.M. (1985). Entomological techniques and methodologies used in research programmes on plant resistance to insects. *Insect Science and Its Application* 6: 391–400.

Davis, F.M., Ng, S.S. and Williams, W.P. (1992). *Visual Rating Scales for Evaluating Whorl-Stage Corn for Resistance to Fall Armyworm*. Technical Bulletin 186. Starkville, Mississippi, USA. Mississippi Agriculture and Forestry Experiment Station.

de Sarup, P. (1989). Standardization of technique for evaluating maize varieties for resistance to adults of *Sitophilus oryzae* Linn. in storage. *Journal of Entomological Research* 20: 1–5.

Devereau, A.D., Gudrups, I., Appleby, J.H. and Credland, P.F. (2003). Automatic, rapid screening of seed resistance in cowpea, *Vigna unguiculata* (L.) Walpers, to the seed beetle *Callosobruchus maculatus* (F.) (Coleoptera: Bruchidae) using acoustic monitoring. *Journal of Stored Products Research* 39: 117–129.

Dhaliwal, G.S. and Pathak, M.D. (1988). Pentadecanal: A semiochemical from rice. In *Proceedings XVIII International Congress of Entomology*. Vancouver, Canada: Entomological Society of Canada, 178.

Dhillon, M.K. and Sharma, H.C. (2007). Effect of storage temparature and duration on viability of eggs of *Helicoverpa armigera* (Lepidoptera: Noctuidae). *Bulletin of Entomological Research* 97: 55–59.

Dhillon, M.K., Sharma, H.C., Folkerstma, R.T. and Chandra, S. (2006). Genetic divergence and molecular characterization of sorghum hybrids and their parents for reaction to shoot fly, *Atherigona soccata*. *Euphytica* 149: 199–210.

Dhillon, M.K., Sharma, H.C., Singh, R. and Naresh, J.S. (2005a). Mechanisms of resistance to shoot fly, *Atherigona soccata* in sorghum. *Euphytica* 144: 301–312.

Dhillon, M.K., Singh, R., Naresh, J.S. and Sharma, N.K. (2005b). Influence of physico-chemical traits of bittergourd, *Mimordica charantia* L. on larval density and resistance to melon fruit fly, *Bactrocera cucurbiate* (Coquilellett). *Journal of Applied Entomology* 129: 393–399.

Diawara, M.M., Wiseman, B.R. and Isenhour, D.J. (1991). Bioassay for screening plant accessions for resistance to fall armyworm (Lepidoptera: Noctuidae) using artificial diets. *Journal of Entomological Science* 26: 367–374.

Doggett, H., Starks, K.J. and Eberhart, S.A. (1970). Breeding for resistance to the sorghum shoot fly. *Crop Science* 10: 528–531.

Dougherty, D.E. (1976). *Pinitol and Other Soluble Carbohydrates in Soybean as Factors in Facultative Parasite Nutrition.* Ph.D. thesis, University of Georgia, Athens, GA.

Echendu, T.N.C. and Akingbohungbe, A.E. (1990). Intensive free-choice and no-choice cohort tests for evaluating resistance to *Maruca testulalis* (Lepidoptera: Pyralidae) in cowpea. *Bulletin of Entomological Research* 80: 289–293.

El Damir, M., El Bouhssini, M. and Al Salti, N. (1999). A simple screening technique of lentil germplasm for resistance to *Sitona crinitus* H. (Coleoptera: Curculionidae) under artificial infestation. *Arab Journal of Plant Protection* 17: 33–35.

Elden, T.C. (1999). Laboratory screening techniques for evaluation of soybean germplasm for resistance to twospotted spider mite (Acari: Tetranychidae). *Journal of Entomological Science* 34: 132–143.

Ellis, P.R. (1999). The identification and exploitation of resistance in carrots and wild *Umbelliferae* to the carrot fly *Psila rosae* (F.). *Integrated Pest Management Review* 4: 1607–1616.

Farrar, R. Jr. and Kennedy, G.G. (1987). 2-Undecanone, a constituent of the glandular trichomes of *Lycopersicon hirsutum* f. *glabratum.* Effects on *Heliothis zea* and *Manduca sexta* growth and survival. *Entomologia Experimentalis et Applicata* 43: 17–23.

Fehr, W.R. and Caviness, C.E. (1977). *Stages of Soybean Development.* Iowa State University Special Report 80. Ames, Fowa, USA: Iowa State University.

Fernandez, C.G.J. (1993). Effective selection criteria for assessing plant stress tolerance. In George Kuo, C. (Ed.), *Adoption of Food Crops to Temperature and Water Stress: Proceedings of an International Symposium,* 13–18 August, 1992. AVRDC Publication No. 93–410. Shanhua, Taiwan: Asian Vegetable Research and Development Center (AVRDC), 93–410.

Fischer, R.A. and Maurer, R. (1978). Drought resistance in spring wheat cultivars. I. Grain yield responses. *Australian Journal of Agricultural Research* 29: 897–912.

Fitt, G.P., Mares, C.L. and Llewellyn, D.J. (1994). Field evaluation and potential ecological impact of transgenic cottons (*Gossypium hirsutum*) in Australia. *BioControl Science and Technology* 4: 535–548.

Fitt, G.P., Daly, J.C., Mares, C.L. and Olsen, K. (1998). Changing efficacy of transgenic Bt cotton: Patterns and consequences. In Zalucki, M.P., Drew, R.A.I. and White, G.G. (Eds.), *Pest Management: Future Challenges, Proceedings, 6th Australasian Applied Entomology Research Conference,* vol. 1. Brisbane, Queensland, Australia: University of Queensland Press, 189–196.

Franzmann, B.A. (1996). Evaluation of a laboratory bioassay for determining resistance levels to sorghum midge, *Contarinia sorghicola* (Coquillett) (Diptera: Cocidomyiidae). *Australian Journal of Entomology* 35: 119–123.

Funderburk, J.E., Soffes, A.R., Barnett, R.D., Herzog, D.C. and Hinson, K. (1989). Plot size and shape in relation to soybean resistance for velvetbean caterpillar (Lepidoptera: Noctuidae). *Journal of Economic Entomology* 83: 2107–2110.

Giri, A.P. and Kachole, M.S. (1998). Amylase inhibitors of pigeonpea (*Cajanus cajan*) seeds. *Phytochemistry* 47: 197–202.

Gosenza, G.W. and Green, H.B. (1979). Behaviour of tomato fruitworm, *Heliothis zea* (Boddie) on susceptible and resistant lines of tomatoes. *Horticulture Science* 14: 171–173.

Graham, R.D. (1984). Breeding for nutritional characteristics in cereals. *Advances in Plant Nutrition* 1: 57–102.

Green, P.W.C., Stevenson, P.C., Simmonds, M.S.J. and Sharma, H.C. (2002). Can larvae of the pod borer, *Helicoverpa armigera* (Lepidoptera: Noctuidae), select between wild and cultivated pigeonpea [*Cajanus* sp. (Fabaceae)]. *Bulletin of Entomological Research* 92: 45–51.

Green, P.W.C., Stevenson, P.C., Simmonds, M.S.J. and Sharma, H.C. (2003). Phenolic compounds on the pod surface of pigeonpea, *Cajanus cajan,* mediate feeding behavior of larvae of *Helicoverpa armigera. Journal of Chemical Ecology* 29: 811–821.

Guerin, P.M., Stadler, E. and Buser, H.R. (1983). Identification of host plant attractants for the carrot fly, *Psila rosae*. *Journal of Chemical Ecology* 9: 843–861.

Guerin, P.M. and Visser, J.H. (1980). Electroantennogram responses of the carrot fly, *Psila rosae*, to volatile plant components. *Physiological Entomology* 5: 111–119.

Guthrie, W.D., Rathore, Y.S., Cox, D.F. and Reed, G.L. (1974). European corn borer: Virulence on crop plants of larvae reared for different generations on a meridic diet. *Journal of Economic Entomology* 67: 605–606.

Gyawali, B.K. (1989). Simple technique for screening potato lines against tuber moth, *Phthorimaea* (*Gnorimoschema*) *operculella* Zeller. *Quarterly Newsletter, Asia and Pacific Plant Protection Commission* 32: 34–35.

Hall, P.K., Parrott, W.L., Jenkins, J.N. and McCarty, J.C. Jr. (1980). Use of tobacco budworm eggs and larvae for establishing field infestations on cotton. *Journal of Economic Entomology* 73(3): 393–395.

Hardie, D.C. and Clement, S.L. (2001). Development of bioassays to evaluate wild pea germplasm for resistance to pea weevil (Coleoptera: Bruchidae). *Crop Protection* 20: 517–522.

Harris, K.M. (1971). X-ray detection of *Contarinia sorghicola* (Coq.) larvae and pupae in sorghum spikelets. *Bulletin of Entomological Research* 60: 379–382.

Hedin, P.A., Jenkins, J.N., Collum, D.H., White, W.H., Parrot, W.L. and MacGown, M.W. (1983). Cyanidin-3-β-glucoside, a newly recognized basis for resistance in cotton to the tobacco budworm *Heliothis virescens* (Fab.) (Lepidoptera: Noctuidae). *Experientia* 39: 799–801.

Heinrichs, E.A., Medrano, F.G. and Rapusas, H.R. (1985). *Genetic Evaluation for Insect Resistance in Rice*. Los Banos, Philippines, International Rice Research Institute.

Heinrichs, E.A., Aquino, G.B., Chelliah, S., Valencia, S.L. and Reissig, W.H. (1982). Resurgence of *Nilaparvata lugens* (Stal) populations as influenced by method and timing of insecticide applications in lowland rice. *Environmental Entomology* 11:78–84.

Hollingsworth, C.S. and Berry, R.E. (1982). Two spotted spider mite (Acari: Tetranychidae) in peppermint: population dynamics and influence of cultural practices. *Environmental Entomology* 11: 1280–1284.

Huang, J., Nuessly, G.S., McAuslane, H.J. and Nagata, R.T. (2003). Effect of screening methods on expression of Romaine lettuce resistance to adult banded cucumber beetle, *Diabrotica balteata* (Coleoptera: Chrysomellidae). *Florida Entomologist* 86: 194–198.

Jackai, L.E.N. (1982). A field screening technique for resistance of cowpea (*Vigna unguiculata* W.) to the pod borer, *Maruca testulalis* (Geyer) (Lepidoptera: Pyralidae). *Bulletin of Entomological Research* 72: 145–156.

Jackai, L.E.N. (1990). Screening of cowpeas for resistance to *Clavigralla tomentosicollis* Stal (Hemiptera: Coreidae). *Journal of Economic Entomology* 83: 300–305.

Jackai, L.E.N. and Oghiakhe, S. (1989). Pod wall trichomes and resistance of two wild cowpea, *Vigna vexillata*, accessions to *Maruca testulalis* (Geyer) (Lepidoptera: Noctuidae) and *Clavigralla tomentosicollis* Stal (Hemiptera: Coreidae). *Bulletin of Entomological Research* 79: 595–605.

Jeyarani, S. and Velusamy, R. (2002). Genetic evaluation for resistance to rice mealy bug in planthopper resistant rice varieties. *Journal of Ecobiology* 14: 51–56.

Karim, A.N.M. and Saxena, R.C. (1990). Resistance of green leafhopper (GLH) *Nephotettix virescens* in free-choice and no-choice tests. *International Rice Research Newsletter* 11: 15–16.

Kasting, R. and McGinnis, A.J. (1962). Quantitative relationship between consumption and excretion of dry matter by the larvae of pale western cutworms, *Agrotis orthogonia* Morr. (Lepidoptera: Noctuidae). *Canadian Entomologist* 94: 441–443.

Kennedy, G.G. and Yamamoto, R.T. (1979). A toxic factor causing resistance in a wild tomato to the tobacco hornworm and some other insects. *Entomologia Experimentalis et Applicata* 26: 121–126.

Khan, Z.R., Ciepiela, A. and Norris, D.M. (1987). Behavioral and physiological responses of cabbage looper, *Trichoplusia ni* (Hubner), to steam distillates from resistant versus susceptible soybean plants. *Journal of Chemical Ecology* 13: 1903–1915.

Khan, Z.R., Ward, J.T. and Norris, D.M. (1986). Role of trichomes in resistance to cabbage looper, *Trichopulsia ni*. *Entomologia Experimentalis et Applicata* 42: 109–117.

Kher, S. and Rataul, H.S. (1992). Screening technique in *Brassica* crops for aphid-resistance. *Indian Journal of Entomology* 54: 181–189.

Kimmins, F.M., Padgham, D.E. and Stevenson, P.C. (1995). Growth inhibition of the cotton bollworm (*Helicoverpa armigera*) larvae by caffeoylquinic acids from the wild groundnut, *Arachis paraguariensis*. *Insect Science and Its Application* 16: 363–368.

Kishore Kumar, V., Sharma, H.C. and Dharma Reddy, K. (2005). Antibiosis mechanism of resistance to spotted stem borer, *Chilo partellus* in sorghum, *Sorghum bicolor*. *Crop Protection* 25: 66–72.

Klimaszewski, J., Bernier Cardou, M., Cyr, D., Alfaro, R. and Lewis, K. (2000). Screening of sitka spruce (*Picea sitchensis*) seedlings for resistance to the white pine weevil (*Pissodes strobi*) in a caging experiment. *Belgian Journal of Entomology* 2: 273–286.

Klun, J.A. and Robinson, J.F. (1969). Concentration of two 1,4 benzoxazinones in dent corn at various stages of development of the plant and its relation to resistance to the host plant to the European corn borer. *Journal of Economic Entomology* 62: 214–220.

Kumar, H. (1992). Inhibition of ovipositional responses of *Chilo partellus* (Lepidoptera: Pyralidae) by the trichomes on the lower leaf surface of a maize cultivar. *Journal of Economic Entomology* 85: 1736–1739.

Kumari, A.D., Sharma, H.C. and Jagdishwar Reddy, D. (2008). Incorporation of lyophilized leaves and pods into artificial diet to assess antibiosis component of resistance to *Helicoverpa armigera* in pigeonpea. *Journal of Food Legumes* (in press).

Lapointe, L., Arango, G. and Sotelo, G. (1989). A methodology for evaluation of host plant resistance in *Brachiaria* to spittlebug species (Homoptera: Cercopidae). *Proceedings of the XVI International Grassland Congress*, 4–11 October, 1989. Nice, France. Versailles, France: Association Francaise pour la Production Fourragere, Centre National de Recherche Agronomique, 731–732.

Larrain, P.I., Araya, J.E. and Paschke, J.D. (1995). Methods of infestation of sorghum lines for the evaluation of resistance to the maize weevil, *Sitophilus zeamais* Motschulsky (Coleoptera: Curculionidae). *Crop Protection* 14: 561–564.

Laster, M.L. and Meredith, W.R. Jr. (1974). Evaluating the response of cotton cultivars to tarnished plant bug injury. *Journal of Economic Entomology* 67: 686–688.

Legacion, D.M. and Gabriel, B.P. (1988). Techniques for artificial infestation of corn plants in identifying sources of leaf-feeding resistance to the Asiatic corn borer, *Ostrinia furnacalis* (Guenee). *Philippine Agriculturist* 71: 371–374.

Leslie, G. and Nuss, K. (1992). Screening sugarcane for resistance to *Eldana*. *South African Sugar Journal* 76: 168–170.

Liu, G.J., Fu, Z.H., Shen, J.H. and Zhang, Y.H. (2002). Comparative study on evaluation methods for resistance to rice planthoppers (Homoptera: Delphacidae) in rice. *Chinese Journal of Rice Science* 16: 52–56.

Lukefahr, M.J., Haughtaling, J.E. and Graham, H.M. (1971). Suppression of *Heliothis* populations with glabrous cotton strains. *Journal of Economic Entomology* 64: 486–488.

Lukefahr, M.J., Martin, D.F. and Meyer, J.R. (1965). Plant resistance to five Lepidoptera attacking cotton. *Journal of Economic Entomology* 58: 516–518.

Mahal, M.S., Lal, H. and Singh, R. (1993). Standardization of a technique for screening okra germplasm for resistance against cotton jassid, *Amrasca biguttula* (Ishida). II. Ovipositional preference of adults. *Journal of Insect Science* 6: 223–225.

Manglitz, G.R. and Kehr, W.R. (1984). Resistance to spotted alfalfa aphid (Homoptera: Aphididae) in alfalfa seedlings of two plant introductions. *Journal of Economic Entomology* 77: 357–359.

Maredia, K.M., Tugwell, N.P., Waddle, B.A. and Bourland, F.M. (1994). Technique for screening cotton germplasm for resistance to tarnished plant bug, *Lygus lineolaris* (Palisat de Beauvois). *Southwestern Entomologist* 19: 63–70.

Marfa, V., Mele, E., Vassal, J.M. and Messeguer, J. (2002). *In vitro* insect-feeding bioassay to determine the resistance of transgenic rice plants transformed with insect resistance genes against striped stem borer (*Chilo suppressalis*). *In Vitro Cellular and Developmental Biology-Plant* 38: 310–315.

McColl, A.L. and Noble, R.M. (1992). Evaluation of a rapid mass-screening technique for measuring antibiosis to *Helicoverpa* spp. in cotton cultivars. *Australian Journal of Experimental Agriculture* 32: 1127–1134.

McLean, D.L. and Weigt, W.A. Jr. (1968). An electronic system to record aphid salivation and ingestion. *Annals of Entomological Society of America* 61: 180–185.

Medrano, F.G. and Heinrichs, E.A. (1985). A simple technique of rearing yellow stem borer, *Scirpophaga incertulas* (Walker). *International Rice Research Newsletter* 10: 14–15.

Meisner, J., Navon, A., Zur, M. and Ascher, K.R.S. (1977). The response of *Spodoptera littoralis* larvae to gossypol incorporated in an artificial diet. *Environmental Entomology* 6: 243–244.

Mihm, J.A. (1982). Techniques for mass rearing and infestation in screening for host plant resistance to corn earworm, *Heliothis zea*. In *Proceedings, International Workshop on* Heliothis *Management*, 15–20 November, 1981. Patancheru, Andhra Pradesh, India: International Crops Research Institute for the Semi-Arid Tropics, 255–266.

Mihm, J.A. (1983). *Techniques for Efficient Mass Screening and Infestation of Fall Armyworm*, Spodoptera frugiperda *J.E. Smith for Plant Resistance Studies*. El Batan, Mexico: International Maize and Wheat Improvement Center.

Mihm, J.A. (1985). Methods of artificial rearing with *Diatraea* species and evaluation of stem borer resistance in sorghum. In *Proceedings International Sorghum Entomology Workshop*, 15–21 July, 1984, Texas A & M University, College Station, Texas, USA. Patancheru, Andhra Pradesh, India: International Crops Research Institute for the Semi-Arid Tropics, 169–174.

Navasero, R.C. and Ramaswamy, S.B. (1991). Morphology of leaf surface trichomes and its influence on egg laying by *Heliothis virescens*. *Crop Science* 31: 342–353.

Nwanze, K.F. and Reddy, Y.V.R. (1991). A rapid method for screening sorghum for resistance to *Chilo partellus* (Swinhoe) (Lepidoptera: Pyralidae). *Journal of Agricultural Entomology* 8: 41–49.

Navon, A., Melamed Madjar, V., Zur, M. and Ben Moshe, E. (1991). Effects of cotton cultivars on feeding of *Heliothis armigera* and *Spodoptera littoralis* larvae and on oviposition of *Bemisia tabaci*. *Agriculture, Ecosystem and Environment* 35: 73–80.

Nwosu, K.I. (1992). Optimum larval population of *Sesamia calamistis* Hmps. (Lepidoptera: Noctuidae) for artificial infestation of maize plants. *Insect Science and Its Application* 13: 369–371.

Oghiakhe, S., Jackai, L.E.N. and Makanjuola, W.A. (1992). A rapid visual field screening technique for resistance of cowpea (*Vigna unguiculata*) to the legume pod borer *Maruca testulalis* (Lepidoptera: Pyralidae). *Bulletin of Entomological Research* 82: 507–512.

Olsen, K.M. and Daly, J.C. (2000). Plant-toxin interactions in transgenic Bt cotton and their effect on mortality of *Helicoverpa armigera* (Lepidoptera: Noctuidae). *Journal of Economic Entomology* 93: 1293–1299.

Paguia, P., Pathak, M.D. and Heinrichs, E.A. (1980). Honey dew excretion measurement techniques for determining differential feeding activity of biotypes of *Nilaparvata lugens* on rice varieties. *Journal of Economic Entomology* 73: 35–40.

Palaniswamy, P. and Lamb, R.J. (1992). Screening for antixenosis resistance to flea beetles, *Phyllotreta cruciferae* (Goeze) (Coleoptera: Chrysomelidae), in rapeseed and related crucifers. *Canadian Entomologist* 124: 895–906.

Panda, N. and Daugherty, D.M. (1975). Note on the antibiosis factor of resistance to corn earworm in pubescent genotypes of soybean. *Indian Journal of Agricultural Sciences* 45: 68–72.

Panda, N. and Heinrichs, E.A. (1983). Levels of tolerance and antibiosis in rice varieties having moderate resistance to the brown planthopper, *Nilaparvata lugens*. *Environmental Entomology* 12: 1204–1214.

Panda, N. and Khush, G.S. (1995). *Host Plant Resistance to Insects*. Wallingford, Oxon, UK: Commonwealth Agricultural Bureau, International.

Panda, R.N. and Das, R.C. (1974). Ovipositional preference of shoot and fruit borer (*Leucinodes orbonalis* Guen.) to some varieties of brinjal. *South Indian Horticulture* 22: 46–50.

Patanakamjorn, S. and Pathak, M.D. (1967). Varietal resistance of rice to the Asiatic rice borer, *Chilo suppressalis* (Lepidoptera: Crambidae) and its association with various plant characters. *Annals of Entomological Society of America* 60: 287–292.

Pathak, P.K., Saxena, R.C. and Heinrichs, E.A. (1982). Parafilm jacket for measuring honey dew excretion by *Nilaparvata lugens* in rice. *Journal of Economic Entomology* 75: 194–195.

Pathak, M.D., Andres, F., Galacgae, N. and Raros, R. (1971). *Resistance of Rice Varieties to Striped Rice Borers*. Technical Bulletin 11. Manila, Philippines: International Rice Research Institute.

Petinez, M.T. (1986). Field screening of different systemic insecticides and varietal resistance against rice green leafhoppers (*Nephotettix cincticeps, Nephotettix apicalis*) and brown planthoppers (*Nilaparvata lugens*). *Scientific Journal* 5(2): 155.

Prakasa Rao, P.S. (1975). Some methods of increasing field infestation of rice gall midge. *Rice Entomology Newsletter* 2: 16–17.

Rajendran, B. and Gopalan, M. (1997). An improved technique for screening and grading of egg plant, *Solanum melongena* L., accessions for resistance to spotted beetle, *Henosepilachna vigintioctopunctata* (Fab.). *Journal of Entomological Research* 21: 245–251.

Ranjith, A.M. and Mohanasundaram, M. (1992). A new method for rapid leaf screening of cotton germplasm for the whitefly, *Bemisia tabaci* Gennadius. *Madras Agricultural Journal* 79: 218–219.

Reed, G.L., Bhendley, T.A. and Showers, W.B. (1972). Influence of resistant corn leaf tissue on the biology of European corn borer. *Annals of Entomological Society of America* 65: 658–662.

Roberts, J.J. and Foster, T.E. (1983). Effect of leaf pubescence in wheat on bird cherry oat aphid (Homoptera: Aphididae). *Journal of Economic Entomology* 76: 1320–1322.

Roberts, J.J., Gallun, R.L., Patterson, F.L. and Foster, J.E. (1979). Effects of wheat leaf pubescence on the Hessian fly. *Journal of Economic Entomology* 72: 211–214.

Robinson, S.H., Wolfonbarger, D.A. and Doday, R.H. (1980). Antixenosis of smooth leaf cotton to the ovipositional response of tobacco budworm. *Crop Science* 20: 646–649.

Romeis, J., Shanower, T.G. and Peter, A.J. (1999). Trichomes on pigeonpea (*Cajanus cajan* (L.) Millsp.) and two wild *Cajanus* spp. *Crop Science* 39: 564–569.

Rosales-Serna, R., Ramirez-Vallejo, P., Acosta-Gallegos, J.A., Castillo-Gonzalez, F. and Kelly, J.D. (2000). Grain yield and drought tolerance of common bean under field conditions. *Agrociencia* 34: 153–165.

Rutherford, R.S. and Van Staden, J. (1996). Towards a rapid near-infrared technique for prediction of resistance to sugarcane borer *Eldana saccharina* Walker (Lepidoptera: Pyralidae) using stalk surface wax. *Journal of Chemical Ecology* 22: 681–694.

Saeb, H., Nouri Ghonbalani, G. and Rajabi, G. (2001). Laboratory assays for evaluating varieties of rice for resistance to striped stem borer, *Chilo suppressalis* (Walker). *Iranian Journal of Agricultural Sciences* 32: 757–763.

Sams, D.W., Laver, F.I. and Redcliffe, E.B. (1975). An excised leaflet test for evaluating resistance to green peach aphid in water bearing *Solanum* germplasm. *Journal of Economic Entomology* 68: 607–609.

Saxena, R.C. and Okech, S.H. (1985). Role of plant volatiles in resistance of selected rice varieties to brown planthopper. *Nilaparvata lugens* (Stal) (Homoptera: Delphacidae). *Journal of Chemical Ecology* 1: 1601–1616.

Scott, R.A., Worrall, W.D. and Frank, W.A. (1990). Comparison of three techniques for measuring antibiosis to Russian wheat aphid. *Southwestern Entomologist* 15: 439–446.

Shanks, C.H. and Garth, J.K.L. (1992). Honeydew production, survival and reproduction by *Chaetosiphon fragaefolii* (Cockerell) (Homoptera: Aphididae) on susceptible and resistant clones of *Fragaria* spp. *Scientia Horticulturae* 50: 71–77.

Sharma, A.N. (1996). Comparison of two screening procedures and classification of soybean genotypes into insect-resistant groups. *International Journal of Pest Management* 42: 307–310.

Sharma, H.C. (Ed.). (2005). Heliothis/Helicoverpa *Management: Emerging Trends and Strategies for Future Research*. New Delhi, India: Oxford & IBH Publishers.

Sharma, H.C. and Agarwal, R.A. (1983a). Role of some chemical components and leaf hairs in varietal resistance in cotton to jassid, *Amrasca biguttula biguttula* Ishida. *Journal of Entomological Research* 7: 145–149.

Sharma, H.C. and Agarwal, R.A. (1983b). Factors affecting genotypic susceptibility to spotted bollworm (*Earias vittella* Fab.) in cotton. *Insect Science and Its Application* 4: 363–372.

Sharma, H.C., Doumbia, Y.O. and Diorisso, N.Y. (1992). A headcage technique to screen sorghum for resistance to mirid head bug, *Eurystylus immaculatus* Odh. in West Africa. *Insect Science and Its Application* 13: 417–427.

Sharma, H.C. and Franzmann, B.A. (2000). Biology of the legume pod borer, *Maruca vitrata* and its damage to pigeonpea and adzuki bean. *Insect Science and Its Application* 20: 99–108.

Sharma, H.C. and Franzmann, B.A. (2001). Host plant preference and oviposition responses of the sorghum midge, *Stenodiplosis sorghicola* (Coquillett) (Dipt., Cecidomyiidae) towards wild relatives of sorghum. *Journal of Applied Entomology* 125: 109–114.

Sharma, H.C. and Lopez, V.F. (1992). Screening for plant resistance to sorghum head bug, *Calocoris angustatus* Leth. *Insect Science and Its Application* 13: 315–325.

Sharma, H.C. and Lopez, V.F. (1993). Influence of panicle size, infestation levels, and environment on genotypic resistance in sorghum to head bug, *Calocoris angustatus* Lethiery. *Entomologia Experimentalis et Applicata* 71: 101–110.

Sharma, H.C. and Norris, D.M. (1991). Chemical basis of resistance in soybean to cabbage looper, *Trichoplusia ni. Journal of Science of Food and Agriculture* 55: 353–364.

Sharma, H.C. and Nwanze, K.F. (1997). *Mechanisms of Resistance to Insects in Sorghum and Their Usefulness in Crop Improvement.* Information Bulletin no. 45. Patancheru, Andhra Pradesh, India: International Crops Research Institute for the Semi-Arid Tropics.

Sharma, H.C. and Ortiz, R. (2002). Host plant resistance to insects: An eco-friendly approach for pest management and environment conservation. *Journal of Environmental Biology* 23: 111–135.

Sharma, H.C., Pampapathy, G. and Kumar, R. (2002). Technique to screen peanuts for resistance to the tobacco armyworm, *Spodoptera litura* (Lepidoptera: Noctuidae) under no-choice cage conditions. *Peanut Science* 29: 35–40.

Sharma, H.C., Pampapathy, G. and Kumar, R. (2005). Standardization of cage techniques to screen chickpeas for resistance to *Helicoverpa armigera* (Lepidoptera: Noctuidae) under greenhouse and field conditions. *Journal of Economic Entomology* 98: 210–216.

Sharma, H.C., Saxena, K.B. and Bhagwat, V.R. (1999). *Legume Pod Borer,* Maruca vitrata*: Bionomics and Management.* Information Bulletin no. 55. Patancheru, Andhra Pradesh, India: International Crops Research Institute for the Semi-Arid Tropics.

Sharma, H.C., Singh, F. and Nwanze, K.F. (Eds.). (1997). *Plant Resistance to Insects in Sorghum.* Patancheru, Andhra Pradesh, India: International Crops Research Institute for the Semi-Arid Tropics.

Sharma, H.C., Singh, B.U. and Ortiz, R. (2001). Host plant resistance to insects: Measurement, mechanisms and insect-plant-environment interactions. In Ananthakrishnan, T.N. (Ed.), *Insects and Plant Defense Dynamics.* New Delhi, India: Oxford and IBH Publishing, 133–159.

Sharma, H.C., Soman, P. and Subramanian, V. (1995). Effect of host plant resistance and chemical control of head bug, *Calocoris angustatus* Leth., on grain quality and seedling establishment in sorghum. *Annals of Applied Biology* 126: 131–142.

Sharma, H.C. and Sullivan, D.J. (2000). Screening for plant resistance to the Oriental armyworm, *Mythimna separata* (Lepidoptera: Noctuidae) in pearl millet, *Pennisetum glaucum. Journal of Agricultural and Urban Entomology* 17: 125–134.

Sharma, H.C., Vidyasagar, P. and Leuschner, K. (1988a). Field screening for resistance to sorghum midge (Diptera: Cecidomyiidae). *Journal of Economic Entomology* 81: 327–334.

Sharma, H.C., Vidyasagar, P. and Leuschner, K. (1988b). No-choice cage technique to screen for resistance to sorghum midge (Diptera: Cecidomyiidae). *Journal of Economic Entomology* 81: 415–422.

Sharma, H.C., Vidyasagar, P. and Leuschner, K. (1990). Components of resistance to the sorghum midge, *Contarinia sorghicola. Annals of Applied Biology* 116: 327–333.

Sharma, H.C. and Youm, O. (1999). Host plant resistance in integrated pest management. In Khairwal, I.S., Rai, K.N., Andrews, D.J. and Harinarayana, H. (Eds.), *Pearl Millet Improvement.* New Delhi, India: Oxford and IBH Publishing, 381–415.

Sharma, H.C., Ahmad, R., Ujagir, R., Yadav, R.P., Singh, R. and Ridsdill-Smith, T.J. (2005a). Host plant resistance to cotton bollworm/legume pod borer, *Helicoverpa armigera.* In Sharma, H.C. (Ed.), Heliothis/Helicoverpa *Management: Emerging Trends and Strategies for Future Research.* New Delhi, India: Oxford and IBH Publishing, 167–208.

Sharma, H.C., Doumbia, Y.O., Haidra, M., Scheuring, J.F., Ramaiah, K.V. and Beninati, N.F. (1994). Sources and mechanisms of resistance to sorghum head bug, *Eurystylus immaculatus* Odh. in West Africa. *Insect Science and Its Application* 15: 39–48.

Sharma, H.C., Pampapathy, G., Dhillon, M.K. and Ridsdill-Smith, T.J. (2005b). Detached leaf assay to screen for host plant resistance to *Helicoverpa armigera*. *Journal of Economic Entomology* 98: 568–576.

Sharma, H.C., Singh, B.U., Hariprasad, K.V. and Bramel-Cox, P.J. (1999). Host-plant resistance to insects in integrated pest management for a safer environment. *Proceedings, Academy of Environmental Biology* 8: 113–136.

Sharma, H.C., Taneja, S.L., Leuschner, K. and Nwanze, K.F. (1992). *Techniques to Screen Sorghums for Resistance to Insects*. Information Bulletin no. 32. Patancheru, Andhra Pradesh, India: International Crops Research Institute for the Semi-Arid Tropics.

Shen, S.Q., Wu, D.X., Cui, H.R., Xia, Y.W. and Shu, Q.Y. (2003). Study on some problems in breeding insect-resistant *Bt* transgenic Japonica rice. *Journal of Zhejiang University, Agriculture and Life Sciences* 29: 499–503.

Shepard, M., Carner, G.R. and Turnipseed, S.G. (1977). Colonization and resurgence of insect pests of soybean in response to insecticides and field isolation. *Environmental Entomology* 6: 501–506.

Simmonds, M.S.J. and Stevenson, P.C. (2001). Effects of isoflavonoids from *Cicer* on larvae of *Helicoverpa armigera*. *Journal of Chemical Ecology* 27: 965–977.

Singh, B. and Yadav, R.P. (1999a). Location of sources of resistance amongst chickpea (*Cicer arietinum* L.) genotypes against gram pod borer (*Heliothis armigera* Hub.) under normal sown conditions using new parameters. *Journal of Entomological Research* 23: 19–26.

Singh, B. and Yadav, R.P. (1999b). Field screening of chickpea (*Cicer arietinum* L.) genotypes against gram pod borer (*Heliothis armigera* Hub.) under late sown conditions. *Journal of Entomological Research* 23(2): 133–140.

Singh, B.U. and Rana, B.S. (1986). Resistance in sorghum to the shoot fly, *Atherigona soccata* Rondani. *Insect Science and Its Application* 7: 577–587.

Singh, H., Rohilla, H.R. and Singh, V. (2001). A technique of rearing crucifer aphid (*Lipaphis erysimi*) under laboratory conditions. *Indian Journal of Agricultural Sciences* 71: 346–347.

Singh, P. and Moore, R.F. (1985). *Handbook of Insect Rearing*, vol. I and II. New York, USA: Elsevier.

Singh, R. and Agarwal, R.A. (1988). Influence of leaf-veins on ovipositional behaviour of jassid, *Amrasca biguttula biguttula* (Ishida). *Journal of Cotton Research and Development* 2(1): 41–48.

Singh, S.P. and Jotwani, M.G. (1980). Mechanisms of resistance to shoot fly. III. Biochemical basis of resistance. *Indian Journal of Entomology* 42: 551–556.

Smith, C.M. (1989). *Plant Resistance to Insects*. New York, USA: John Wiley & Sons.

Smith, C.M. (2005). *Plant Resistance to Arthropods: Molecular and Conventional Approaches*. Berlin, Germany: Springer Verlag.

Smith, C.M., Khan, Z.R. and Pathak, M.D. (1994). *Techniques for Evaluating Insect Resistance in Crop Plants*. Boca Raton, Florida, USA: CRC Press.

Sogawa, K. and Patthak, M.D. (1970). Mechanisms of brown planthopper resistance in Mudgo variety of rice (Hemiptera: Delphacidae). *Applied Entomology and Zoology* 5: 145–158.

Son, K.C., Severson, R.F. and Kays, S.J. (1991). A rapid method for screening sweet potato genotypes for oviposition stimulants to the sweet potato weevil. *HortScience* 26: 409–410.

Sonneveld, T., Wainwright, H. and Labuschagne, L. (1997). Two methods for determining the resistance of strawberry cultivars to two-spotted mite (Acari: Tetranychidae). *Proceedings of the Third International Strawberry Symposium*, 29 April–4 May, 1996, Veldhoven, Netherlands. Volume I. *Acta Horticulturae* 439: 199–204.

Soo Hoo, C.F. and Fraenkel, G. (1966). The consumption, digestion, and utilization of food plants by a phytophagous insect, *Prodenia eridania* (Cramer). *Journal of Insect Physiology* 12: 71–73.

Sorenson, E.L. and Horbor, E. (1974). Selecting alfalfa seedlings to resist the potato leafhopper. *Crop Science* 14: 85–86.

Starks, K.J. and Burton, R.L. (1977). Greenbugs: Determining biotypes, culturing, and screening for plant resistance. *USDA-ARS Technical Bulletin* 1556.

Stern, V.M. (1969). Interplanting alfalfa in cotton to control lygus bugs and other insect pests. Tall Timbers Conference. *Ecology and Animal Control Habitat Management* 1: 55–69.

Stevens, M.M., Fox, K.M., Warren, G.N., Cullis, B.R., Coombes, N.E. and Lewin, L.G. (2000). An image analysis technique for assessing resistance in rice cultivars to root-feeding chironomid midge larvae (Diptera: Chironomidae). *Field Crops Research* 66: 25–36.

Stevenson, P.C., Green, P.W.C., Simmonds, M.S.J. and Sharma, H.C. (2005). Physical and chemical mechanisms of plant resistance to *Helicoverpa armigera*: Recent research on chickpea and pigeonpea. In Sharma, H.C. (Ed.), Heliothis/Helicoverpa *Management: Emerging Trends and Strategies for Future Research*. New Delhi, India: Oxford and IBH Publishers, 209–222.

Sutter, G.R. and Branson, T.F. (1980). A procedure for artificially infesting field plots with corn rootworm eggs. *Journal of Economic Entomology* 73: 135–137.

Talekar, N.S., Hau, T.B.H. and Chang, W.C. (1999). *Solanum viarum*, a trap crop for *Helicoverpa armigera*. *Insect Environment* 5: 142.

Taneja, S.L. and Leuschner, K. (1985). Resistance screening and mechanisms of resistance in sorghum to shoot fly. In *Proceedings, International Sorghum Entomology Workshop*, 15–21 July, 1984, Texas A & M University, College Station, Texas, USA. Patancheru, Andhra Pradesh, India: International Crops Research Institute for the Semi-Arid Tropics, 115–131.

Tang, D.L. and Wang, W.G. (1996). Influence of contents of secondary metabolism substances in cotton varieties on the growth and development of cotton bollworm. *Plant Protection* 22: 6–9.

Thomas, J.G., Sorensen, E.L. and Painter, R.H. (1966). Attached vs. excised trifoliates for evaluation of resistance in alfalfa to the spotted alfalfa aphid. *Journal of Economic Entomology* 59: 444–448.

Tolmay, V.L., van der Westhuizen, M.C. and van Deventer, C.S. (1999). A six week screening method for mechanisms of host plant resistance to *Diuraphis noxia* in wheat accessions. *Euphytica* 107: 79–89.

Traynier, R.M.M. (1967). Effect of host plant odour on behaviour of the adult cabbage root fly, *Erioischia brassicae*. *Entomologia Experimentalis et Applicata* 10: 321–328.

Van Emden, H.F. and Bashford, M.A. (1976). The effect of leaf excision on performance of *Myzus persicae* and *Brevicoryne brassicae* in relation to nutrient treatment of plants. *Physiological Entomology* 1: 67–71.

Van Rensburg, J.B.J. and Van Rensburg, G.D.J. (1993). Laboratory production of *Busseola fusca* (Fuller) (Lepidoptera: Noctuidae) and techniques for the detection of resistance in maize plants. *African Entomology* 1: 25–28.

Venugopal Rao, N., Tirumala Rao, K. and Reddy, A.S. (1991). Ovipositional and larval development sites of gram caterpillar (*Helicoverpa armigera*) in pigeonpea. *Indian Journal of Agricultural Sciences* 61: 608–609.

Visser, J.H., Van Straten, S. and Maarse, T. (1979). Isolation and identification of volatiles in the foliage of potato, *Solanum tuberosum*, a host plant of the Colorado potato beetle, *Leptinotarsa decemlineata*. *Journal of Chemical Ecology* 5: 11–23.

Waiss, A.C. Jr., Chan, B.G., Elliger, C.A., Wiseman, B.R., McMillian, W.W., Widstrom, N.W., Zuber, M.S. and Keaster, A.J. (1979). Maysin, a flavone glycoside from corn silks with antibiotic activity toward corn earworm. *Journal of Economic Entomology* 72: 256–258.

Waldbauer, G.P. (1968). The consumption and utilization of food by insects. *Advances in Insect Physiology* 5: 229–288.

Warnock, D.F., Hutchison, W.D., Kurtti, T.J. and Davis, D.W. (1997). Laboratory bioassays for evaluating sweet corn antibiosis on European corn borer (Lepidoptera: Pyralidae) larval development. *Journal of Entomological Science* 32: 342–357.

Webster, J.A. and Smith, D.H. Jr. (1983). Developing small grains resistant to the cereals leaf beetle. *USDA Technical Bulletin* 1673.

Widstrom, N.W., McMillian, W.W. and Wiseman, B.R. (1979). Ovipositional preference of the corn earworm and the development of trichomes on the exotic corn selections. *Environmental Entomology* 8: 833–839.

Williams, C.T., Ukwungwu, M.N., Singh, B.N. and Okhidievbie, O. (2001). Assessment of host plant resistance in *Oryza sativa* to the African rice gall midge, *Orseolia oryzivora* Harris and Gagne (Dipt., Cecidomyiidae), with a description of a new method for screening under artificial infestation. *Journal of Applied Entomology* 125: 341–349.

Williams, W.G., Kennedy, G.G., Yamamoto, R.T., Thacker, J.D. and Bordner, J. (1980). 2-Tridecanone: A naturally occurring insecticide from the wild tomato, *Lycopersicon hirsutum* f. *glabratum*. *Science* 207: 888–889.

Williams, W.P. and Buckley, P.M. (1996). Southwestern corn borer growth on laboratory diets containing lyophilized corn husks. *Crop Science* 36: 462–464.

Williams, W.P., Buckley, P.M. and Davis, F.M. (1987). Feeding response of corn earworm (Lepidoptera: Noctuidae) to callus and extracts of corn in the laboratory. *Environmental Entomology* 16: 532–534.

Wu, G., Cui, H.R., Shu, Q.Y., Ye, G.Y. and Xia, Y.W. (2000). GUS histochemical assay: A rapid way to screen striped stem borer (*Chilo suppressalis*) resistant transgenic rice with a cry1Ab gene from Bt (*Bacillus thuringiensis*). *Journal of Zhejiang University, Agriculture and Life Sciences* 26: 141–143.

Wu, J.T., Li, G.X. and Zhang, L.Y. (1989). Studies on methods of evaluating moderate resistance to the brown planthopper, *Nilaparvata lugens* (Homoptera: Delphacidae) in rices, *Oryza* spp. *Journal of South China Agricultural University* 10: 72–78.

Yasui, H., Kato, A. and Yazawa, M. (1998). Antifeedants to armyworms. *Spodoptera litura* and *Pseudaletia separata*, from bitter gourd leaves *Momordica charantia*. *Journal of Chemical Ecology* 24: 803–813.

Yoshida, M., Cowgill, S.E. and Wightman, J.A. (1995). Mechanisms of resistance to *Helicoverpa armigera* (Lepidoptera: Noctuidae) in chickpea: Role of oxalic acid in leaf exudates as an antibiotic factor. *Journal of Economic Entomology* 88: 1783–1786.

Yoshida, M., Cowgill, S.E. and Wightman, J.A. (1997). Roles of oxalic and malic acids in chickpea trichome exudate in host-plant resistance to *Helicoverpa armigera*. *Journal of Chemical Ecology* 23: 1195–1210.

Youm, O., Yacouba, M. and Kumar, K.A. (2001). An improved infestation technique using eggs of the millet head miner (*Heliocheilus albipunctella*) (Lepidoptera: Noctuidae) in millet resistance screening. *International Journal of Pest Management* 47: 289–292.

4

Host Plant Resistance to Insects: Potential and Limitations

Introduction

There is a large gap between potential yields and actual yields harvested by farmers. Average yields for most crops, particularly in the developing countries, are only one-third or less of the potential yield, and insect pests are one of the major constraints in crop production. Insect pests have high reproductive rates, a fast generation turnover, wide genetic diversity, and an ability to withstand, metabolize, and avoid toxic chemicals. As a result, it is difficult to control several insect species through currently available insecticides. To harvest the potential yields of high-yielding cultivars, the farmers resort to heavy use of insecticides. However, even if 90% of the insects are killed as a result of insecticide application, the remaining population multiplies at a much faster rate in the absence of natural enemies (which are killed by the insecticides) (Knipling, 1979), and the farmers have to apply insecticides more frequently and at higher doses, which finally results in failure of control operations and environmental pollution. Failure to control insects has forced farmers to give up cultivation of some crops in different countries. Indiscriminate use of insecticides has resulted in adverse effects on nontarget organisms, insecticide residues in food and food products, pest resurgence, development of resistance, and environmental pollution. The current sensitivities about environmental pollution, human health hazards, and pest resurgence as a consequence of improper use of synthetic insecticides has led to greater emphasis on alternative methods of insect control.

Host plant resistance (HPR), natural plant products, biopesticides, natural enemies, and agronomic practices offer a potentially viable option to control insect pests, as they are relatively safe to the nontarget beneficial organisms and human beings. Heavy insecticide use has also led to an exponential increase in the number of insect species resistant to insecticides (Georghiou, 1986). The most effective strategy for pest management is to use selective insecticides at a low dosage in combination with plant resistance to slow the rate of evolution of insecticide-resistant insect populations. Improving plant resistance to insect pests through conventional breeding, wide hybridization, marker-assisted selection, and genetic transformation will significantly contribute to sustainable crop protection and environmental conservation. Varieties with adequate levels of resistance to insect pests will encourage farmers to reduce insecticide application, and thus minimize the environmental hazards. Such varieties will provide safer farm environments and produce with lower insecticide residues. Development of crop cultivars with resistance to insects would provide an effective complementary approach in integrated pest management (IPM). Resistant or less susceptible cultivars would also provide an equitable, environmentally sound, and sustainable pest management tool. Therefore, we need to make a concerted effort to transfer insect resistance genes into genotypes with desirable agronomic characteristics, and with adaptation to different agroecosystems.

In spite of the importance of HPR as a component of IPM, breeding for resistance to insects has not been as successful as breeding for disease resistance (Sharma, 2002). This is largely because of the relative ease with which insect control is achieved through insecticide use, and the slow progress in developing insect-resistant cultivars as a result of the difficulties involved in ensuring adequate insect infestation for resistance screening. To breed for resistance to insect pests, it is important to have optimum levels of natural infestation or ability to rear the insects on artificial diets in the laboratory. Insect-rearing programs are expensive and, in some cases, it may not produce the behavioral or metabolic equivalent of an insect population in nature. However, with the development of insect resistance to insecticides, adverse effects of insecticides on natural enemies, and public awareness of environmental conservation, there has been a renewed interest in the development of crop cultivars with resistance to insect pests in the national programs. The establishment of International Agricultural Research Centers (IARCs), and the collection and evaluation of germplasm for insect resistance in several crops has given a renewed impetus to the identification and use of HPR in pest management.

With the domestication of plants, farmers harvested the seeds from the plants that were able to withstand adverse environmental factors, including insect pests and diseases. The plants that were resistant to insect pests survived until crop harvest where the herbivore pressure resulted in plant mortality, or their proportion decreased over time where the herbivore pressure did not result in plant mortality. This process led to natural selection of plants with resistance to biotic and abiotic stresses prevalent in an ecosystem. Because of this unintentional but continuous selection of plants with resistance to insect pests, several varieties with resistance to insects were selected by the farmers (Sharma, 2002). The best examples of this process are shoot fly-resistant sorghum landraces cultivated during the post-rainy season in India, and head bug resistance in *guineense* sorghums in West Africa (Sharma, 1993, 1996; Sharma et al., 1997, 2003b). Resistance of plants to insects enables a plant to avoid or inhibit host selection, oviposition, and feeding, reduce insect survival, retard development, and tolerate or recover from injury from insect populations that would otherwise cause greater damage to other genotypes of the same species under similar environmental conditions (Smith, 1989). Resistance of plants to insects is the consequence of heritable plant characteristics that result in a plant being relatively less damaged than the plant without these characters. The ability of plants to resist insect damage is based on morphological and biochemical characteristics of the plants, which affect the behavior and biology of insects, and thus influence the extent of damage caused by the insect pests. Resistance traits are preadaptive and genetically inherited characters of the plant that increase their chances of survival and reproduction. Plant resistance to insects is relative, and the level of resistance is expressed in relation to resistant and susceptible genotypes of the same species under similar environmental conditions. Conventional host plant resistance to insects will play a major role in application of tools of biotechnology for integrated pest management and sustainable crop production. Sources of resistance with diverse mechanisms (genes) will be the key for marker assisted transfer of insect resistance genes into high yielding cultivars. Conventional host plant resistance can also be deployed along with the novel genes to make transgenic plants an effective weapon for pest management. Host plant resistance to insects may be classified into the following categories.

Pseudo-resistance: Pseudo-resistance or false resistance through avoidance of insect infestation.

Constitutive resistance: Constitutive resistance is independent of environmental factors and is due to physico-chemical characteristics of the host plant that affect the host selection and feeding behavior, survival, development, and fecundity of insect pests, and thus, the extent of insect damage.

Inducible resistance: Inducible resistance is due to the influence of temperature, photoperiod, plant-water potential, chemicals, and pathogen or insect damage on the production and accumulation of secondary plant substances (phytoalexins) or due to their effect on nutritional quality of the host plant.

Associate resistance: Associate resistance is due to the presence of resistant or nonhost plants in the vicinity. Associate resistance also occurs in multilines or synthetics as a result of diversion or delaying actions of mixture of plants resulting in slow development of an insect biotype that is capable of damaging the resistant cultivars.

Identification and Utilization of Resistance

Extensive screening of germplasm has been carried out to identify sources of resistance to insect pests in several crops (Painter, 1951; Panda and Khush, 1995; Clement and Quisenberry, 1999; Sharma et al., 2003b; Smith, 2005). Sources of resistance have been identified against the following insect pests.

- Corn earworm, *Helicoverpa zea* (Boddie), corn borer, *Ostrinia nubilalis* (Hubner), sugarcane borer, *Diatraea grandiosella* (Dyar), fall armyworm, *Spodoptera frugiperda* (J.E. Smith), and spotted stem borer, *Chilo partellus* (Swinhoe), in maize (Mihm, 1982, 1985; Swarup, 1987).

- Brown planthopper, *Nilaparvata lugens* (Stal) (Figure 4.1), gall midge, *Orseolea oryzae* Wood-Mason, and stem borers, *Scirpophaga incertulas* (Walker) and *Chilo suppressalis* (Walker), in rice (Saxena, 1986; Kalode, Bentur, and Srinivasan, 1989; Smith, Khan, and Pathak, 1994).

- Hessian fly, *Mayetiola destructor* (Say), and greenbug, *Schizaphis graminum* (Rondani), in wheat (Starks and Merkle, 1977; Roberts et al., 1979; Smith, 2005).

- Sorghum shoot fly, *Atherigona soccata* (Rondani) (Figure 4.2), spotted stem borer, *C. partellus*, sorghum midge, *Stenodiplosis sorghicola* (Coquillett), and head bug, *Calocoris angustatus* (Lethiery) in sorghum (Sharma et al., 1992, 2003b; Sharma, Vidyasagar, and Leuschner, 1988a, 1988b; Sharma, 1996; Padma Kumari, Sharma, and Reddy, 2000).

FIGURE 4.1   Brown planthopper, *Nilaparavata lugens* damage in different rice cultivars.

- Oriental armyworm, *Mythimna separata* (Walker), in pearl millet (Sharma and Sullivan, 2000).

- Tobacco budworm, *Heliothis virescens* (F.), cotton bollworm, *Helicoverpa armigera* (Hubner), and leafhopper, *Amrasca biguttula biguttula* Ishida, in cotton (Hall et al., 1980; Sharma and Agarwal, 1983a; Nanthagopal and Uthamasamy, 1989; Sundramurthy and Chitra, 1992; Rao and Prasad, 1996; Wu, Cai, and Zhang, 1997; Sharma, 2005; Sharma et al., 2005b).

- Mexican bean beetle, *Epilachna varivestis* Mulsant, in soybean (Kogan, 1982).

- Groundnut leaf miner, *Aproaerema modicella* DuV., aphids, *Aphis craccivora* Koch., jassids, *Empoasca kerri* Pruthi, *H. zea*, and tobacco caterpillar, *Spodoptera litura* (F.), in groundnut (Beland and Hatchett, 1976; Wightman et al., 1990; Sharma, Pampapathy, and Kumar, 2002).

- Legume pod borer, *H. armigera*, in chickpea (Figure 4.3) and pigeonpea (Srivastava and Srivastava, 1990; Cowgill and Lateef, 1996; Shanower, Yoshida, and Peter, 1997; Sharma et al., 2005b, 2005d; Sharma, Pampapathy, and Kumar, 2005).

FIGURE 4.2   Shoot fly, *Atherigiona soccata* damage in resistant (IS 18551) and susceptibe (CSH 1) cultivars of sorghum.

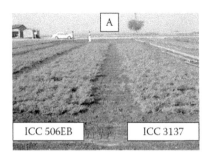

FIGURE 4.3 *Helicoverpa armigera* damage at the vegetative (A) and reproductive (B) stages in resistant (ICC 506EB) and susceptible (ICC 3137) genotypes of chickpea.

- Spotted pod borer, *Maruca vitrata* (Geyer), in pigeonpea and cowpea (Sharma, Saxena, and Bhagwat, 1999; Sharma and Franzmann, 2000).
- Pea weevil, *Bruchus pisorum* L., in pea (Clement et al., 1994).

Moderate levels of resistance have been discovered (1) against the Colorado potato beetle, *Leptinotarsa decemlineata* (Say), potato tuber moth, *Phthorimaea opercullela* Zeller, and green peach potato aphid, *Myzus persicae* (Sulzer) in potato (Sams, Lauer, and Redcliffe, 1975; Sinden et al., 1986; Ortiz et al., 1990); (2) against the spotted alfalfa aphid, *Therioaphis maculata* (Buckton) in alfalfa (Thomas, Sorensen, and Painter, 1966); (3) carrot fly, *Psila rosae* (Fab.) in carrot (Stadler and Buser, 1984); (4) fruit borers, *H. zea* and *H. armigera* in tomato (Cosenza and Green, 1979; Kashyap and Verma, 1987); (5) cotton aphid, *Aphis gossypii* Glover in muskmelon (Kennedy, Kishaba, and Bohn, 1975); and (6) cabbage maggot, *Hylemyia* sp. in turnip (Varis, 1958). Insect-resistant cultivars with desirable agronomic backgrounds have been developed in several crops, and cultivars with multiple resistance to insect pests and diseases will be in greater demand in the future for sustainable crop production. This requires a concerted effort from scientists involved in crop improvement programs worldwide. At the moment, there is a great amount of variability in:

- Emphasis placed on plant resistance in crop improvement programs.
- Availability of cost-effective and reliable resistance screening techniques.
- Progress made in identification and utilization of sources of resistance to insect pests.
- Multilocational testing to understand genotype-environment interactions.
- Emphasis given to insect resistance in identifying and releasing new crop cultivars.
- Efforts to spread and popularize insect-resistant cultivars.

Wild Relatives of Crops as Sources of Resistance to Insects

Wild species of crops are important sources of genes for resistance to biotic and abiotic constraints (Stalker, 1980; Plucknett et al., 1987; Xiao et al., 1996). In cotton, resistance to *H. armigera* has been reported in several wild species, such as *Gossypium thurberi* Todaro,

Cajanus sericeus

Cajanus scarabaeoides

FIGURE 4.4   Wild relatives of pigeonpea (*Cajanus sericeus* and *Cajanus scarabaeoides*) that have shown high levels of resistance to pod borer, *Helicoverpa armigera.*

G. somalense (Guerke), *G. armourianum* Kearny, *G. gossypiodes* (Ulbrich) Standley, *G. capitis viridis* Mauer, *G. raimondii* Ulbrich, *G. trilobum* (DC) Skovsted, *G. latifolium* (Murray) Roberty, and *G. barbosanum* L.L. Phillips & D. Clement, and in the wild races of *G. hirsutum* (P. Singh and Narayanan, 1994). *Gossypium raimondii* is resistant to pink bollworm, *Pectinophora gossypiella* (Saunders) and *H. armigera,* while *G. armourianum* and *G. raimondii* are resistant to white fly, *Bemisia tabaci* (Genn.) and the leafhopper, *A. biguttulla biguttulla.* In potato, the wild relative, *Solanum brachistotrichum* (Bitter) Rydberg is resistant to green peach potato aphid, *M. persicae.* In cultivated tomato, insect resistance is rare, but it is more prevalent in wild accessions of *Lycopersicon esculentum* var. *cerasiforme* (Mill. & Alef.) (Eigenbrode, Trumble, and Jones, 1993). Accessions belonging to *L. hirsutum* f. *glabratum* C. H. Müll, and *L. hirsutum* Dunal are resistant to *H. armigera* (Kashyap et al., 1990).

Wild relatives of pigeonpea, such as *Cajanus scarabaeoides* (L.) Thouars, *C. sericeus* (Benth. ex Bak.) van der Maesen, *C. acutifolius* (F.v. Muell.) van der Maesen, *C. albicans* (W. & A.) van der Maesen, *Rhynchosia aurea* (Willd.) DC, and *R. bracteata* Benth. ex Bak. are highly resistant to *H. armigera* (Sharma et al., 2001; Green et al., 2006) (Figure 4.4). Some of the wild relatives of pigeonpea have also shown resistance to pod fly, *Melanagromyza obtusa* Malloch and pod wasp, *Tanaostigmodes cajaninae* La Salle (Sharma, Pampapathy, and Reddy, 2003). Of these, *C. scarabaeoides, C. sericeus, C. acutifolius,* and *C. albicans* can be easily crossed with cultivated pigeonpea. Wild chickpea species, such as *Cicer bijugum* Rech., *C. reticulatum* (Lad.), *C. judaicum* Boissier., *C. pinnatifidum* Jaub. & Sp., *C. microphyllum* Benth., and *C. cuneatum* A. Rich., have shown high levels of resistance to *H. armigera* (Sharma et al., 2005e, 2005f) (Figure 4.5). Accessions belonging to *C. bijugum, C. pinnatifidum,* and *C. echinospermum* (Davis) have also shown resistance to the bruchid, *Callosobruchus chinensis* L. (K.B. Singh, Ocampo, and Robertson, 1998). Accessions belonging to wild relatives of groundnut, such as *Arachis cardenasii* Krapov. & WC Gregory, *A. duranensis* Krapov. & WC Gregory, *A. kempff-mercadoi* Krapov, WC Gregory & CE Simpson, *A. monticola* Krapov. & Rigoni, *A. stenosperma* Krapov. & WC Gregory, *A. paraguariensis* Chodat & Hassl., *A. pusilla* Benth., and *A. triseminata* Krapov. & WC Gregory, have shown multiple resistances to leaf miner, *A. modicella, H. armigera,* and *E. kerri* (Sharma et al., 2003a). *Arachis cardenasii* (ICG 8216), *A. ipaensis* Krapov. & WC Gregory (ICG 8206), *A. paraguariensis* (ICG 8130), and *A. appressipila* Krapov. & WC Gregory (ICG 8946) exhibit antibiosis to *S. litura* under no-choice conditions. Wild relative of pea, *Pisum fulvum* (Sibth. & Sm.) is resistant to the bruchid, *Bruchus pisorum* L. (Clement, Hardie, and Elberson, 2002), while the wild relative of cowpea, *Vigna vexillata* (L.) Benth is resistant to spotted pod borer, *M. vitrata* and the pod sucking bug, *Clavigralla tomentosicollis* Stal (Jackai and Oghiakhe, 1989). Lines showing high levels of resistance to insects can be used in wide hybridization to increase the levels and diversify the basis of resistance to the target insects.

FIGURE 4.5 Wild relatives of chickpea (*Cicer bijugum*, *Cicer judaicum*, and *Cicer microphyllum*) that have shown high levels of resistance to pod borer, *Helicoverpa armigera*.

In sorghum, accessions belonging to *Sorghum laxiflorum* Bailey, *S. australiense* Garber & Snyder, *S. brevocallosum* Garber, *S. dimidiatum* Stapf., *S. matarkense* Garber, *S. nitidum* (Vahl) Pers., *S. purpureosericeum* (Hochst. Ex A. Rich.), *S. timorense* (Kunth) Buse, *S. versicolor* Andersson, *S. angustum* S.T. Blake, *S. ecarinatum* Lazarides, *S. exstans* Lazarides, *S. interjectum* Lazarides, and *S. intrans* F. Mull. ex Benth. are highly resistant to sorghum shoot fly, *A. soccata*. *Sorghum laxiflorum*, *S. australiense*, *S. brevocallosum*, *S. dimidiatum*, *S. matarkense*, *S. nitidum*, *S. purpureosericeum*. *S. timorense*, *S. versicolor*, *S. angustum*, *S. ecarinatum*, *S. exstans*, *S. interjectum*, *S. stipoideum* (Ewart & Jean White) C. Gardener and C.E. Hubb., and *S. intrans* are resistant to spotted stem borer, *C. partellus* (Venkateswaran, 2003). Sorghum midge, *S. sorghicola*, females did not lay any eggs in the spikelets of *S. angustum*, *S. amplum* Lazarides, and *S. bulbosum* Lazarides compared to 30 eggs in spikelets of *S. halepense* (L.) Pers. under no-choice conditions (Sharma and Franzmann, 2001). Odors from the panicles of *S. halepense* are more attractive to the females of sorghum midge than the odors from panicles of *S. stipoideum*, *S. brachypodum* Lazarides, *S. angustum*, *S. macropsermum* Garber, *S. nitidum*, *S. laxiflorum*, and *S. amplum*. In the case of rice, *Oryza eichingeri* A. Peter has shown high levels of resistance to brown planthopper, *N. lugens* and the green leafhopper, *Nephotettix* sp.; *O. brachyantha* A. Chev. et Roehr., and *O. minuta* J.S. Presl. ex C.B. Presl. to yellow stem borer, *C. suppressalis*; and *O. alta* Swallen. and *O. ridleyi* Hook. f. to striped rice stem borer, *S. incertulas* (Khush, 1977; Khush and Brar, 1991; Brar and Khush, 1997). In the case of wheat, the wild relatives *Aegilops tauschii* (Coss.) Schmal., *Triticum ventricosum* Ces. (Syn. *Aegilops ventricosa* Tausch.), and *T. turgidum* L. are resistant to Hessian fly, *M. destructor*; and *T. monococcum* L. and *T. turgidum* to Russian wheat aphid, *Diuraphis noxia* (Kurdj.) (Clement, 2002).

Inducible Resistance

Induced resistance is the qualitative and/or quantitative enhancement of a plant's defense mechanisms against insect pests in response to extrinsic physical or chemical stimuli. Induced resistance results in changes in a plant that produce a negative effect on herbivores (Karban and Baldwin, 1997). Earlier research in host-pathogen systems documented that protection could be effective for a week or more (McIntyre and Dodds, 1979; Chaudary,

Schwarzbach, and Fischbeck, 1983). Recent reports on the duration of induced resistance in host-pathogen systems have been documented in barley (Pelcz and Wolffgang, 1986), cotton (Liu, Chen, and Wang, 1990), and cucumber (Dalisay and Kuc, 1995). Insect herbivory has been shown to induce resistance in soybean plants (Lin, Kogan, and Fischer, 1990; Fischer, Kogan, and Paxton, 1990; Lin and Kogan, 1990) against the soybean looper, *Pseudoplusia includens* (Walker) and the Mexican bean beetle, *E. varivestis*. Cross-resistance to *P. includens* due to previous *Cerotoma trifurcata* (Forster) injury in the form of induced resistance is beneficial to the plant. Bean leaf beetle, *C. trifurcata* herbivory also affects larval growth rates and reduced the suitability of foliage to the corn earworm, *H. zea* in soybean (Felton, Summers, and Mueller, 1994).

Chemically induced expression systems or "gene switches" enable temporal, spatial, and quantitative control of genes introduced into plants or those that are already present in the plants to impart resistance to insects. This approach has provided opportunities for management of development of resistance in insect pest populations in transgenic crops. In addition to insect or pathogen attack, resistance can also be induced by suboptimal concentrations of potassium iodide, copper, and herbicides. Effectiveness of the chemical injury inducer Actigard™ in providing resistance to various insect pests and pathogens in the tomato has been demonstrated by Inbar et al. (1998). Induced resistance against *E. varivestis* lasted three days after damage in soybean (Underwood, 1998). Proteinase inhibitors and oxidative enzymes such as polyphenol oxidase, peroxidase, and lipoxygenase persist for at least 21 days after induction in damaged tomato leaflets (Stout, Workman, and Duffey, 1996).

A wide range of inducible genes has been identified in plants based on endogenous chemical signals, such as phytohormones, responses to insect and pathogen attack, or wounding. The best-studied system utilizes the *PR1-a* promoter from tobacco, which is induced during systemic resistance response following pathogen infection (Uknes et al., 1993). The *PR1-a* mRNA levels can also be induced by exogenous application of salicylic acid or 1,2,3-benzothaidiazole-7-carbothioic acid S-methyl ester (BTH) (Ward et al., 1991; Gatz, 1997). The latter has been commercialized as a plant immunization chemical. However, *PR1-a* use may be limited due to its responsiveness to exogenous signals such as pathogens, UV-B, and pollutants (Gatz, 1997). Safeners used for detoxification of certain herbicides also induce a range of metabolic enzymes, such as glutathione-S transferase and cytochrome P 450 (Jepson, Martinez, and Sweetruan, 1998). Safener-induced cDNA clones (*In 2-2* and *GST-27*) have been isolated from maize and *Arabidopsis*. These promoters can be induced by M-(aminocarbonyl)-2-chlorobenzenesulfonamide and the herbicide sulfonylurea. Tetracycline and lactose represser operator systems from *Escherichia coli* Escherich have also been used to control gene expression (Jepson et al., 1994). Tetracycline (*TetR*) repressor protein has been used to regulate the expression from *CaMV 35S* promotor (Gatz and Quail, 1998). Another system in yeast uses a copper-dependent transcriptional activation system, which consists of *ace1* gene encoding a metalloresponsive factor expressed constitutively. Activation of the reporter gene (*ace1*) is achieved in the presence of copper (Mett, Lochhead, and Reynolds, 1993). Copper has traditionally been used as a fungicide, but application of copper for activation of the reporter gene may not be ideal in all crops. Another transcriptional activation system (*Alc*) is based on ethanol (Caddick et al., 1998). The *AlcR* transcription factor interacts with ethanol or related inducers. It has a low activity, rapid and reversible induction, and sufficiently high levels of induction to generate phenotypic effects. Glucocorticoid receptor (GR) (Wang et al., 1997) and ecdysone receptor (Wing, Slawecki, and Carlson, 1988; Jepson, Martinez, and Sweetruan, 1998) also have the potential for use as switches for inducible resistance to insects.

Octadecanoid and the salicylic acid pathways are involved in the induced attraction of the parasitoid wasp, *Cotesia rubecula* (Marshall) by *Arabidopsis thaliana* (L.) Heyn. infested with the herbivore, *Pieris rapae* L. (van Poecke and Dicke, 2002). Besides exogenous application of jasmonic acid or salicylic acid, use is also made of transgenic *A. thaliana* that does not show induced jasmonic acid levels after wounding (S-12) and transgenic *A. thaliana* that does not accumulate salicylic acid (NahG). Treatment of *A. thaliana* with jasmonic acid resulted in an increased attraction of parasitoid wasps compared with untreated plants, whereas treatment with salicylic acid did not. Transgenic plants impaired in the octadecanoid or the salicylic acid were less attractive than wild-type plants (van Poecke and Dicke, 2002).

Factors Affecting Expression of Resistance to Insects

The most desirable form of insect resistance is the one that is stable across locations and seasons. However, several climatic and edaphic factors influence the level and nature of resistance to insect pests (Kogan, 1982). Inherited characters, especially those involving physiological characteristics, are influenced by the environmental factors. Some of the factors that influence plant resistance are discussed below.

Soil Moisture

Moisture stress alters the plant's reaction to insect damage, leading either to an increase or decrease in susceptibility to insect pests. Populations of *Aphis fabae* (Scop.) have lower rates of reproduction on water-stressed plants (McMutry, 1962). High levels of water stress also reduce damage by sorghum shoot fly, *A. soccata* (Soman et al., 1994). However, water-stressed plants of sorghum suffer greater damage by the spotted stem borer, *C. partellus* and sugarcane aphid, *Melanaphis sacchari* (Zehntner) (Sharma et al., 2005c). Atmospheric humidity also interferes with insect-plant interactions (Sharma et al., 1999a). High humidity increases the detection of odors and, thus, may influence host finding by the insects. In cotton, frequent irrigation increases vegetative growth and subsequent damage by *H. armigera* (B. Singh et al., 2005).

Plant Nutrition

Nutrients play an important role in plant resistance to insects. In some instances, high levels of nutrients increase the level of plant resistance to insects, and in others they may increase the susceptibility. Application of nitrogenous fertilizers decreases the damage by shoot fly, *A. soccata*, and spotted stem borer, *C. partellus*, in sorghum (Reddy and Narasimha Rao, 1975; Chand, Sinha, and Kumar, 1979). A decrease in shoot fly damage has also been observed after application of phosphatic fertilizers (Channabasavanna, Venkat Rao, and Rajagopal, 1969; Sharma, Singh, and Nwanze, 1997). Changes in nutrient supply also affect the sorghum resistance to greenbug, *S. graminum* (Schweissing and Wilde, 1979). Application of potash decreases the incidence of the top borer, *Scirpophaga excerptalis* (Walker), in sugarcane. High levels of nitrogen lead to greater damage by the cotton jassid, *A. biguttula biguttula* (Purohit and Deshpande, 1992).

Temperature

Temperature is one of the most important factors affecting the behavioral and physiological interactions of insects and their host plants (Benedict and Hatfield, 1988). Temperature-induced stress changes the levels of biochemicals, enzymes, morphological defenses, or nutritional quality of the host plant. Temperature affects not only plant growth, but also the biology, behavior, and population dynamics of the insects (Tingey and Singh, 1980). In general, low temperatures have a negative effect on plant resistance to insects (Kogan, 1982). However, lower temperatures resulting in reduced rate of growth of the developing grain increase sorghum susceptibility to midge, *S. sorghicola* (Sharma et al., 1999a). Differences in genoptypic suscepti-bility to greenbug in sorghum increase with an increase in temperature (Schweissing and Wilde, 1978). In alfalfa, the level of resistance to pea aphid, *Acyrthosiphon pisum* Harris and alfalfa aphid, *T. maculata* is enhanced at higher temperatures (Kogan, 1982). In sorghum, expres-sion of resistance to sorghum midge, *S. sorghicola* is influenced by temperature and the relative humidity (Sharma, Venkateswarulu, and Sharma, 2003). There is considerable variation in the influence of temperature on expression of resistance to insects, and such interactions need to be kept in mind while identifying sources of resistance to insect pests for use in crop improve-ment programs.

Photoperiod

Photoperiod influences plant growth and physicochemical characteristics of crop plants and, thus, influences the interaction between insects and crop plants. Failure or inability to grow certain crop plants during the off season at times is largely associated with increased susceptibility to insects and diseases. Intensity and quality of light influences the biosynthesis of phenylpropanoids (Hahlbrock and Grisebach, 1979) and anthocyanins (Carew and Krueger, 1976). Prolonged exposure to high-intensity light induces susceptibil-ity in PI 227687 soybean plants to the cabbage looper, *Trichoplusia ni* (Hubner) (otherwise resistant) (Khan et al., 1986). Susceptibility in sorghum to midge, *S. sorghicola*, increases under long day length in Kenya near the Equator (Sharma et al., 1999a).

Insect Biotypes

Biotypes are populations of insects capable of damaging and surviving on cultivars known to be resistant to other populations of the same species (Kogan, 1982). The term biotype is used for a group of insects primarily distinguishable on the basis of their interaction with genetically stable varieties or clones of host plants. Most biotypes have been recorded in aphids because of parthenogenesis. Even a single mutant aphid capable of feeding on a resistant genotype can build up into a new biotype (Pathak, 1970). Instances of emergence of new biotypes have been documented in the case of gall midge, *O. oryzae* in rice (Bentur, Srinivasan, and Kalode, 1987), Hessian fly, *M. destructor* in wheat, and brown planthopper, *N. lugens* in rice (Smith, 2005). To overcome the problem of biotypes, genotypes with differ-ent mechanisms of resistance should be utilized in a breeding program to have stable and high levels of resistance against a number of prevalent insect biotypes in a region.

Influence of HPR on Pest Population Dynamics and Economic Injury Levels

Economic injury levels (EILs) can be used to determine the levels of host plant resistance that can be practically attainable and economically rewarding in crop improvement

(Sharma and Teetes, 1995). Studies on the effect of insect-resistant cultivars on EILs will also be useful in assessing the contribution of insect-resistant germplasm in regulating pest populations, avoiding excessive insecticide use, and determining the levels of insect resistance needed in the newly developed cultivars, as well as the effectiveness of insect-resistant cultivars in IPM for sustainable crop production. Because the term resistance conveys different expectations to different people, and as the farmers desire no additional risk, the use of EILs to define the level of resistance of a newly released crop cultivar is critical in pest management. Insect-resistant cultivars not only decrease the density of pest populations, but also delay the time required by the insect pests to attain the EILs, depending on the level and the mechanism of resistance. If the EIL is based on insect damage (e.g., percentage of deadhearts for shoot fly and stem borer in sorghum, number of leaves damaged by aphids, or percentage of leaf area consumed by armyworms), and the insect population increases over the crop growing season, then a susceptible cultivar will suffer economic loss in July, a moderately resistant cultivar in August, and a resistant cultivar can withstand the insect density until the end of the season under Indian conditions (Sharma, 1993). In the case of insects in which the damage is limited to a particular stage and a short span of time (e.g., deadheart formation due to sorghum shoot fly), a cultivar can be planted up to a period when insect density is expected to be below EIL. If EIL is based on adults, which is a nondamaging stage of the insect (e.g., sorghum midge adults or number of *Helicoverpa* or *Spodoptera* moths caught in pheromone or light traps), the EIL will increase with an increase in the level of insect resistance.

The relationship between insect density and grain damage for sorghum midge changes with the level of resistance to this insect (Hallman, Teetes, and Johnson, 1984; Sharma, Vidyasagar, and Nwanze, 1993). In the case of sorghum midge, the EIL is 1 adult per panicle on a susceptible cultivar CSH 1, 25 adults per panicle on a moderately resistant cultivar ICSV 745, and >50 adults per panicle on a highly resistant cultivar ICSV 197 (Sharma, Vidyasagar, and Nwanze, 1993) (Table 4.1). If the EIL is based on adults, which also cause damage (e.g., head bugs in sorghum), and the resistance is based on nonpreference and antibiosis (which will decrease the initial infestation and the rate of population increase), then the time taken by the insect to attain the EIL will also be extended by the resistant cultivars. The EILs for sorghum head bug *C. angustatus* have been determined to be 0.2 to 0.9 bugs per panicle on CSH 11, a highly susceptible sorghum; 0.5 to 1.3 bugs on IS 21443, a moderately resistant sorghum; and 10 to 15 bugs on IS 17610, a resistant sorghum (Sharma and Lopez, 1993). On chickpea, larval days for *H. armigera* needed to justify insecticide

TABLE 4.1

Economic Injury Levels (EILs)* for Sorghum Midge, *Stenodiplosis sorghicola* on Resistant and Susceptible Genotypes of Sorghum (ICRISAT Center, 1989–90 Post-Rainy Season)

| Cultivar | EILs at Different Infestation Levels (No. of Midges per Panicle) | | | | |
	1	5	10	20	40
ICSV 197-R	3.1	333.3	2.9	12.5	100.0
ICSV 745-R	20.0	12.2	2.2	3.3	6.7
ICSV 1-MR	0.8	2.2	2.3	0.6	0.2
ICSV 112-S	0.6	2.9	0.8	0.6	0.2
CSH 1-S	0.3	0.5	0.5	0.3	0.2

* Number of midges per panicle. R, resistant; MR, moderately resistant; S, susceptible.

application have been estimated to be 3.7 to 3.8 on Annigeri and ICCC 37 (susceptible cultivars) as compared to 2.7 larval days on ICC 506, a resistant cultivar (Wightman et al., 1995). HPR has a great influence on EILs, and this information is very important in cropping systems involving insect resistant cultivars as a component of pest management. One of the first and most important adjustments to crop management recommendations that must be made relates to economic thresholds or action thresholds in relation to host plant resistance (Sharma and Teetes, 1995). In some cases, there are several cultivars of a crop with different levels of resistance. Experimental and empirical data should be generated to determine the level of resistance of a cultivar, which is critical in deciding the nature and timing of the intervention (insecticide application or release of natural enemies) needed to suppress an increasing pest population.

Host Plant Resistance in Integrated Pest Management

Host plant resistance as a method of insect control in the context of IPM has a greater potential than any other method of pest suppression. In general, the use of insect-resistant varieties is not subjected to the vagaries of nature, unlike chemical and biological control methods. HPR along with natural enemies and cultural practices is a central component of any pest management strategy (Painter, 1951, 1958; Maxwell and Jennings, 1980; Smith, 1989; Sharma, 2002; Sharma and Ortiz, 2002). Plant resistance as a method of pest control offers many advantages, and in some cases, it is the only practical and effective method of pest management. However, there may be some problems if we rely exclusively on plant resistance for insect control, for example, high levels of resistance may be associated with low yield potential or undesirable quality traits, and resistance may not be expressed in every environment where a variety is grown. Therefore, insect-resistant varieties need to be carefully fitted into the pest management programs in different agroecosystems. The nature of deployment, alone or in combination with other methods of insect control, depends on the level and mechanisms of resistance, and the cropping system (Kennedy et al., 1987).

High levels of plant resistance are available against a few insect species only. However, very high levels of resistance are not a prerequisite for use of HPR as a component in integrated pest management. Varieties with low to moderate levels of resistance or those that can avoid insect damage can be deployed for pest management in combination with other components of pest management. Deployment of insect-resistant cultivars should be aimed at conservation of the natural enemies and minimizing the number of insecticide applications. Use of insect-resistant cultivars also improves the efficiency of other pest management practices, including the synthetic insecticides (Adkisson and Dyck, 1980; Heinrichs, 1988; Sharma, 1993; Panda and Khush, 1995). HPR can be used as:

- A principal component of pest control.
- An adjunct to cultural, biological, and chemical control.
- A check against the release of susceptible cultivars.

HPR as a Principal Method of Insect Control

HPR has often been used for the management of several insect species. However, only a few insect species can be controlled by the use of resistant varieties alone. Insect-resistant varieties have been deployed for the control of a number of insect pests worldwide (Painter, 1951; Maxwell and Jennings, 1980; Smith, 1989, 2005; Sharma and Ortiz, 2002). Several insect pests have been kept under check through the use of insect-resistant cultivars. The major examples include:

- Grapevine rootstocks from the United States, resistant to phylloxera, *Phylloxera vitifoliae* (Fitch.) (Painter, 1951; Adkisson and Dyck, 1980).
- Cotton cultivars Krishna, Mahalaxmi, Khandwa 2, and MCU 5, resistant to leaf-hopper, *A. biguttula biguttula* (Sundramurthy and Chitra, 1992).
- Northern Spy rootstocks of apple, resistant to wooly apple aphid, *Eriosoma lanigerum* (Hausm.) (Martin, 1973).
- Wheat varieties Pawnee, Poso 42, and Benhur, resistant to Hessian fly, *M. destructor* (Maxwell, Jenkins, and Parrot, 1972).
- Rice varieties IR 36, Kakatiya, Surekha, and Rajendradhan, resistant to gall midge, *O. oryzae* (Kalode, 1987).
- Alfalfa varieties Lahontan, Sonora, and Sirsa 9, resistant to spotted alfalfa aphid, *T. maculata* (Howe and Smith, 1957; Hunt et al., 1966).
- Sorghum varieties Maldandi, Swati, ICSV 705, ICSV 700, CISV 708, IS 18551, and SFCR 151, resistant to sorghum shoot fly, *A. soccata* (Figure 4.6); DJ 6514, TAM 2577, AF 28, ICSV 745 (Figure 4.7) and ICSV 88032, resistant to sorghum midge, *S. sorghicola* (Figure 4.8) (Agrawal, Sharma, and Leuschner, 1987; Sharma et al., 1994a, 1994b, 2005g; Sharma, 2001; Agrawal et al., 2005); and CSM 388 and Malisor 84-7, resistant to head bug, *Eurystylus oldi* (Pop.) (Sharma, 1993; Sharma, Lopez, and Vidyasagar, 1994; Sharma et al., 2005g).
- Pigeonpea varieties ICPL 332 and ICPL 88039, resistant to *H. armigera*.
- Chickpea varieties ICC 506 and ICCV 10, resistant to *H. armigera* (Sharma et al., 2005b).

The benefits of HPR depend on the pattern of insect invasion, for example, many insects, such as aphids, whiteflies, and mites, invade the crop in low numbers, and their abundance increases over several generations before reaching the economic threshold levels.

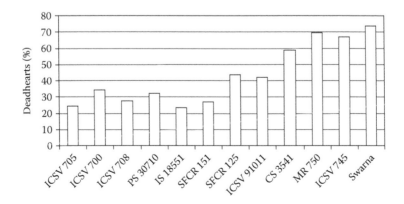

FIGURE 4.6 Expression of resistance to sorghum shoot fly, *Atherigona soccata* in sorghum.

FIGURE 4.7 Sorghum cultivar ICSV 745 with high levels of resistance to midge, *Stenodiplosis sorghicola*, which can be deployed as a principal component of controlling this pest.

For such insects, even low levels of antixenotic and antibiotic resistance would be useful in delaying the time required to reach the damaging levels. However, for insects that invade a crop in large numbers due to immigration, such as *Heliothis/Helicoverpa* and the army-worms *M. separata*, *Spodoptera exempta* (Walker), and *S. frugiperda*, grass hoppers, and locusts, the effects of HPR in suppressing insect populations and damage may not be apparent in the first few generations. The influence of insect-resistant varieties on insect populations can be demonstrated by making use of the simple insect models of Knipling (1979) as adopted by Adkisson and Dyck (1980), and Sharma (1993). Kennedy et al. (1987) demonstrated the usefulness of even low levels of resistance through simulation models. They suggested that maize with antixenotic resistance might not reduce the damage by *H. zea*, while maize with antibiotic type of resistance could cause nearly 50% mortality in the first- and second-instar larvae. However, it will take the insect nearly 23 generations to overcome the antixenotic resistance by 50%, while it will require only seven generations for overcoming the antibiotic type of resistance. A combination of antixenotic and antibiotic resistance is more durable (Gould, Kennedy, and Johnson, 1991). When the two types of resistances are combined, the insect would take 32 generations to overcome the antibiotic

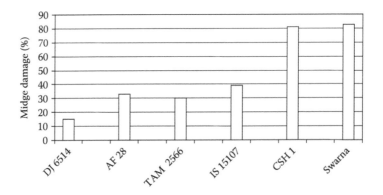

FIGURE 4.8 Sorghum midge, *Stenodiplosis sorghicola*, damage in different genotypes of sorghum.

TABLE 4.2

Population Increase of Sorghum Midge, *Stenodiplosis sorghicola*, on Three Sorghum Genotypes*

Generation		No. of Midge Flies ha^{-1} [a]		
		CSH 1	IS 12664C	DJ 6514
First year	P_1	100	100	100
	F_1	600	300	100
	F_2	3600	900	100
	F_3	21600	1700	100
	F_4	129600	8100	100
Diapause population[b] (1%)	[c]	1554	120	4
Second year	P_2	1554	120	4
	F_1	9324	360	4
	F_2	55944	1080	4
	F_3	335664	3240	4
	F_4	2113984	9720	4
Diapause population[b] (1%)		25149	142	1

* CSH 1, susceptible; IS 12664C, moderately resistant; and DJ 6514, resistant. A hypothetical example.

[a] Midge population multiplies by 6 times on CSH 1 and 3 times on IS 12664C as compared to that on DJ 6514. The midge population at the beginning of the season is assumed to be 100 flies ha^{-1} (P_1).

[b] In each generation (F_1–F_4), 1% of the total population enters diapause.

resistance, and over 100 generations to overcome the antixenotic resistance. If the insects are exposed to toxin-producing plants, a biotype capable of overcoming the resistance can emerge quickly. Addition of 10% and 30% susceptible plants in the field can delay the development of a new biotype by 150 and 500 generations, respectively.

The impact of growing a highly resistant, a moderately resistant, and a susceptible variety on insect populations over a period of time has been explained for sorghum midge, *S. sorghicola* in Table 4.2. At the end of four generations during the first year, there would be 129,600, 8100, and 100 midges ha^{-1} on CSH 1, susceptible; IS 12664C, moderately resistant; and DJ 6514, a highly resistant cultivar, respectively. By the second year, there would be 2,113,984, 9,720, and 4 midges ha^{-1} in areas planted to CSH 1, IS 12664C, and DJ 6514, respectively. More importantly, there would be 25,149, 142, and one diapausing midge larvae ha^{-1} in areas planted to CSH 1, IS 12664C, and DJ 6514, respectively. Even moderate levels of resistance, such as those available for the polyphagous pest *H. armigera* in chickpea, can have considerable effect on the population increase of this pest (Table 4.3). The effect of growing a resistant cultivar would be similar for other insects depending on the level of resistance and the mechanisms involved. If the mortality of *H. armigera* larvae is 15% on ICCC 37, 35% on ICCV 2, and 40% on ICC 506, and assuming that there are 10 female moths per hectare in the beginning of the season, each female moth lays an average of 500 eggs, and there are three generations in a cropping season, then there will be 191,914,063 moths in an area planted with the susceptible cultivar ICCC 37 as compared to 85,820,313 moths in the area planted with the moderately resistant cultivar ICCV 2, and 67,500,000 moths in the area planted with the resistant cultivar ICC 506. Based on rates of insect multiplication, there would be 2.84 and 1.27 times as many insects in areas planted to ICCC 37 and ICCV 2, respectively, as compared with areas cropped with ICC 506. Thus, even moderate levels of plant resistance exercise considerable influence on insect populations, which is cumulative over time. These models can also be used to explain the situations where

TABLE 4.3

Population Dynamics of *Helicoverpa armigera* on a Susceptible (ICCC 37), a Moderately Resistant (ICCV 2), and a Resistant (ICC 506) Chickpea Cultivar*

Generation		No. of *Helicoverpa armigera* Moths ha^{-1}		
		ICCC 37	**ICCV 2**	**ICC 506**
Parent generation	P1	10	10	10
First generation	F1	4250	3250	3000
Second generation	F2	903125	528125	450000
Third generation	F3	191914063	85820313	67500000
Population ratio in relation to the resistant check (ICC 506)		2.84	1.27	1.00

* A hypothetical example based on the model proposed by Knipling (1979).

Note: It has been assumed that each female moth lays an average of 500 eggs, and the sex ratio is 1:1. There are three generations in a cropping season. The *Helicoverpa armigera* population at the beginning of the season is assumed to be 10 female moths ha^{-1} (P$_1$). In each generation (F$_1$–F$_3$), the larval mortality is 15% in ICCC 37, 35% in ICCV 2, and 40% in ICC 506.

minor insect pests become very serious with the introduction of newly developed high-yielding insect-susceptible cultivars.

HPR and Biological Control

Plant resistance and biological control are the key components of integrated pest management (Starks, Muniappan, and Eikenbary, 1972). HPR to insects, in general, is compatible with the natural enemies for pest management. Varieties with moderate levels of resistance that allow the insect densities to remain below economic threshold levels are best suited for use in pest management in combination with natural enemies. The natural enemies not only help to control the target pests, but also reduce the population densities of other insects within their host range (Maxwell, 1972). Insect-resistant varieties also increase the effectiveness of the natural enemies because of a favorable ratio between the densities of the target pest and its natural enemies. Such a combination is more effective in crops with a tolerance mechanism of resistance (Kogan, 1982).

The use of HPR and biological control brings together unrelated mortality factors and thus reduces the insect population's genetic response to selection pressure from either plant resistance or from the natural enemies. Acting in concert, they provide a density-independent mortality at times of low insect density, and density-dependent mortality at times of pest abundance (Bergman and Tingey, 1979). In addition to the direct and indirect effects of plant resistance on insect pests, the selection pressure imposed by natural enemies can also result in magnification of the effects of plant resistance on insect density (van Emden, 1991). In general, the rate of insect adaptation to a resistant cultivar is lower when the suppression is achieved by the combined action of plant resistance and natural enemies than by high levels of plant resistance alone (Gould, Kennedy, and Johnson, 1991).

Restless behavior of the insects on the resistant varieties also increases their vulnerability to the natural enemies (Pathak, 1970). A prolonged developmental period of the immature stages also increases the susceptibility period of the target insect species to the natural enemies or result in synchronization of the insect developmental stages with the peak activity and abundance of the natural enemies. Moderate levels of plant resistance in chickpea in combination with natural enemies (e.g., *Campoletis chlorideae* Uchida) can exercise considerable effect on population dynamics of the pod borer, *H. armigera*. If we assume that there are 10 female moths per hectare in the beginning of the season, each *H. armigera* female lays an average of 500 eggs, and there are three generations in a cropping season, the larval parasite, *C. chlorideae* results in 20% *H. armigera* mortality in each generation (Romeis and Shanower, 1996), and the *H. armigera* larval mortality is 15% in ICCC 37, 35% in ICCV 2, and 40% in ICC 506; then there would be 98,260,000 moths in an area planted with ICCC 37, 43,940,000 moths in the area planted to ICCV 2, and 34,560,000 moths in the area planted to the resistant cultivar, ICC 506. As compared to the resistant check, ICC 506, there would be 2.84 and 1.27 times more moths in the areas planted with ICCC 37 and ICCV 2, respectively. There would be 5.55 times more insects in an area planted with a susceptible cultivar alone as compared to the areas having a resistant cultivar in combination with the parasitoid, *C. chlorideae*. The ratio between *H. armigera* moth populations (across cultivars) in areas with different cultivars plus control with the natural enemies as compared to those with the cultivars alone would be 0.51. Thus, natural enemies in combination with plant resistance can have a dramatic effect on the population dynamics of insect pests.

Tritrophic Interactions

Biological control processes involve the tritrophic interactions between the plants, the target pests, and the natural enemies. These interactions are not limited to the impact of plant chemicals on the target insect and the effect of insect fitness on natural enemies, but also include the direct effects of plants on the natural enemies. The physical or chemical characteristics of the host plant that influence the activity and abundance of natural enemies involve:

Physical interactions

- Physical protection to herbivores from natural enemies.
- Effect of morphological characteristics of the plants on the activity of natural enemies.
- Effect of plant phenology on the searching ability of natural enemies.

Chemical interactions

- Effects on fitness of the insect host.
- Effects on the fitness of the natural enemies.
- Effects on the acceptability of the insect host to the natural enemies.
- Effects on the ratio between insects and the natural enemies.
- Plant chemicals acting as kairomones or allomones to the natural enemies.

It is very likely that plants have evolved mechanisms to attract the natural enemies to reduce the extent of insect damage, for example, the female parasitoid wasp, *Campoletis sonorensis* (Cam.) responds to the volatiles of cotton plant over a short distance, while searching for its prey, *H. virescens*. It is easier for the wasp to find the host habitat first and then the prey within the vicinity of the host plant. Tobacco, cotton, and maize plants produce distinct volatile blends in response to damage by *H. virescens* and *H. zea* (de Moraes et al., 1998). The parasitic wasp, *Cardiochiles nigriceps* (Vier.) exploits these differences to distinguish the plant infestation by *H. virescens* from that by *H. zea*, and selectively prefers the plants damaged by *H. virescens*. Therefore, the nature of insect-host plant interactions is critical in determining the extent of parasitization by the natural enemies. This type of strategy is compatible with biological control (Williams, Elzen, and Vinson, 1988). In contrast, plant secondary compounds such as tomatine in the insect host's diet may affect the parasitization. In other cases, changes in host suitability due to the insect host's diet can influence the developmental rate, size, survival, parasitization success, sex ratio, fecundity, and life span of the parasitoids (Vinson and Barbosa, 1987).

Compatibility of Plant Resistance and Biological Control

A synergism between plant resistance and biological control is a valuable phenomenon in the development of practical insect pest management. The trichogrammatid egg parasitoids of *H. armigera* are highly active on sorghum, but nearly absent on chickpea and pigeonpea, whereas the larval parasitoid, *C. chlorideae* is active in sorghum, pearl millet, and chickpea, but inactive in pigeonpea (Figure 4.9) (Pawar, Bhatnagar, and Jadhav, 1986). Pigeonpea genotypes with resistance to the pod borer, *H. armigera* have no adverse affects on larval parasitization and mortality under field conditions (Sharma, Pampapathy, and Wani, 2006).

Pearl millet genotypes also have no adverse effects on the parasitization of the larvae of Oriental armyworm, *M. separata* (Sharma et al., 2004). Insects feeding on a resistant plant generally have a slower growth and extended developmental period, and poorly developed insects are more vulnerable to natural enemies for a longer period, and this increases the probability of mortality due to biotic and abiotic factors. Insects developing slowly on resistant varieties are more effectively regulated by the predators than those feeding on the susceptible varieties (Price et al., 1980). In certain cases, the plant secondary compounds imparting resistance to insects are compatible with the performance of natural enemies, for example, *Cotesia congregata* (Say) [monophagous parasitoid of *Manduca sexta* (L.)] shows no detrimental effects on exposure to nicotine in tobacco (Barbosa et al., 1986). Gossypol in the pigment glands of cotton ingested by *H. virescens* has no adverse effect on *C. sonorensis*. Larger adults of *C. sonorensis* are produced on insect larvae fed on low concentrations of

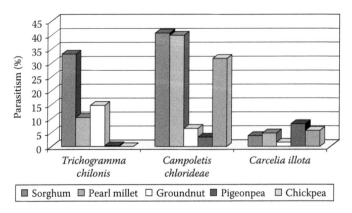

FIGURE 4.9 Parasitism of *Helicoverpa armigera* by *Trichogramma chilonis* (eggs), *Campoletis chlorideae* (larvae), and *Carcelia illota* (larvae).

gossypol than those fed on gossypol-free diets (Williams, Elzen, and Vinson, 1988). Plant resistance and biological control are compatible in the case of greenbug, *S. graminum* (Starks, Muniappan, and Eikenbary, 1972). The movement of the greenbugs on the resistant sorghum cultivars increases the parasitization (Starks and Burton, 1977). Since insecticides are generally not used on resistant cultivars, the natural enemies of the greenbug are also protected. *Aphidius matricariae* Hal. is effective in controlling *M. persicae* on aphid-resistant chrysanthemum (Wyatt, 1970). Levels of parasitization of sorghum midge, *S. sorghicola* by *Neotrichoporoides* (*Tetrastichus*) *diplosidis* Crawford were higher on the moderately resistant genotypes IS 10132 and PM 9760 compared to those on the highly resistant genotypes, such as IS 3461, IS 7005, IS 9807, IS 19512, and AF 28 (with colored grain and high tannin content) (Sharma, 1994; Sharma, Singh, and Nwanze, 1997). Higher levels of parasitization [by *Cotesia flavipes* (Cam.)] have been recorded on stem borer, *C. partellus* resistant genotypes than on susceptible ones, irrespective of the time and method of infestation (Duale and Nwanze, 1997). Reduced rate of multiplication of aphids on the moderately resistant varieties can magnify the plant resistance in the presence of natural enemies (van Emden and Wearing, 1965). Voraciousness of a given predator fails to control the aphid population daily multiplying by a factor of 1.2 on a susceptible cultivar, but adequately controls the aphid population if the growth rate decreases to 1.15 on a resistant cultivar (Bombosch, 1963). Low levels of plant resistance have a positive interaction with the predators of cabbage aphid, *Brevicoryne brassicae* L. on Brussels sprouts (Dodd, 1973). A moderate level of resistance in rice varieties IR 46 and Utri Rajapan effectively increases the predation rate by the spiders *Lycosa* spp. on the brown planthopper, *N. lugens* (Kartohardjono and Heinrichs, 1984).

A number of predators and parasitoids, including *C. sonorensis* and *C. nigriceps*, attack early instars of *H. virescens*, but generally do not attack bigger larvae. Because of a low level of antibiosis in moderately resistant plants, the larvae of *H. virescens* remain in early instars for longer periods and are more likely to be parasitized (Danks, Rabb, and Southern, 1979). The antixenotic factor may decrease or increase the fitness due to density and/or frequency-dependent predation. In avoiding plants or plant parts with antixenotic resistance, young larvae become aggregated at a limited number of feeding sites, and this situation is more favorable for natural enemies, while searching for their prey (Gould, Kennedy, and Johnson, 1991). Compatibility between moderate levels of plant resistance and natural enemies has also been reported in the case of sorghum midge, *C. sorghicola* (Sharma, 1994).

Incompatibility of Plant Resistance and Biological Control

Plant morphological characteristics and secondary plant substances sometimes may have an adverse effect on the natural enemies. Leaf hairs and trichomes offer an effective plant defence mechanism against arthropods, but it can interfere with the mobility of natural enemies (Stipanovic, 1983). Genotypes of tobacco with glandular trichomes severely limit the parasitization of the eggs of tobacco hornworm, *M. sexata* by *Telenomus* sp. and *Trichogramma minutum* Riley (Rabb and Bradley, 1968). In the case of pigeonpea and chickpea, trichomes and trichome exudates hamper the parasitization of eggs of *H. armigera* by the egg parasitoid, *Trichogramma chilonis* Ishii (Romeis, Shanower, and Zebitz, 1997). Increased trichome density on cotton leaves reduces the ability of the parasitoid, *T. pretiosum* Riley and *Chrysopa rufilabris* (Burm.) to find and parasitize corn earworm, *H. zea* eggs (Treacy, Zummo, and Benedict, 1985). Secondary plant substances can prolong post-embryonic development (Isenhour and Wiseman, 1987, 1989) or reduce the number of adult parasitoids arriving on the host (van Emden, 1978). The development rate is impaired in parasitoids fed on corn earworm larvae reared on artificial diets containing nicotine, tomatine, and gossypol (Vinson and Barbosa, 1987). Some of the midge-resistant sorghum lines with colored grain and high tannin content have lower levels of parasitization by *N. diplosidis* than the midge-resistant or susceptible sorghum genotypes having white grain and low or no tannins (Sharma, 1994). Tomatine also affects the egg predators, *Coleomegilla maculata* (DeGeer) and *Geocoris punctipes* (Say) adversely when *H. zea* is fed on the foliage of wild tomato line, PI 134417 (Barbour, Farrar Jr., and Kennedy, 1993). In tomato, tomatine is absorbed by the endoparasitoid, *Hyposoter exiguae* (Vier.) from its host, *H. zea*, which prolongs the parasitoid larval period, reduces pupal eclosion and adult size, and shortens its longevity and fecundity (Campbell and Duffey, 1979; Herzog and Funderburk, 1985).

Plant Resistance–Insect Pathogen Interaction

Insect pathogens can be more effective in a pest management program if antibiosis factors of host resistance are compatible with insect pathogens. Plant resistance may improve or reduce the effectiveness of insect pathogens depending on the nature of plant resistance (Schultz, 1983; Barbosa, 1988). The effectiveness of *Bacillus thuringiensis* (Berliner) is greater on insect pests adapted to high tannin content (with a gut pH of 8.0 to 9.5). Secondary plant substances alter the pathogenicity of *B. thuringiensis* on *M. sexta* and *T. ni* (Krischik, Barbosa, and Reichelderfer, 1988). *Manduca sexta* can tolerate higher levels of toxins such as nicotine and gain protection from entomogenous bacteria. However, when *T. ni* is reared on diets with higher concentrations of *B. thuringiensis*, the presence of dietary nicotine increases the insect mortality. The pathogenicity of the fungus, *Nomuraea rileyi* (Farlow) is reduced if corn earworm larvae ingest tomatine from tomato plants (Gallardo et al., 1990). A synergistic interaction has been observed between maize cultivars resistant to the fall armyworm, *S. furgiperda* and the nuclear polyhedrosis virus (NPV) (Hamm and Wiseman, 1986). Rutin and chlorogenic acid inhibit the infectivity of NPV against *H. zea* (Felton and Duffey, 1990).

Manipulation of Plant Characteristics to Increase the Effectiveness of Natural Enemies

Plant characteristics can also be manipulated to increase the effectiveness of the natural enemies. For example, the hairiness of cucumbers interferes with the biological control of the greenhouse whitefly, *Trialeurodes vaporariorum* (West.) by *Encarsia formosa* (Gahan). Movement of *E. formosa* is 30% higher on cucumber hybrids with half the number of hairs (van Lenteren, 1991). Similarly, development of pigeonpea lines without glandular trichomes may lead to greater parasitization of *H. armigera* eggs by *Trichogramma* spp. (Sharma, H.C., unpublished). Brussels sprouts with glossy leaves are more attractive to aphid predators than the cultivars with waxy foliage (Way and Murdie, 1965), while predation by *Hippodamia convergens* (G.M.) on adults of *Plutella xylostella* (L.) larvae is significantly greater on lines with glossy leaves as compared to lines with normal wax in *Brassica oleracea* L. (Eigenbrode and Kabalo, 1999). Therefore, efforts should be made to identify insect-resistant genotypes that are compatible or hospitable to the natural enemies.

HPR and Cultural Control

Cultural practices cause specific physiological changes that reduce the suitability of host plants for phytophagous insects (Hare, 1983). Most of these practices are compatible with other pest control tactics, including HPR, and have long been associated with subsistence farming. Insect-resistant cultivars, including those that can escape pest damage, are highly useful in pest management in combination with cultural practices. This will have the same effect on the population dynamics of the pest species in question as the combined action of insecticides and insect-resistant cultivars. Cultural control by itself may not reduce the pest populations below economic threshold levels, but aids in reducing the losses through interaction with plant resistance (Glass, 1975). Plant resistance in concert with cultural control can drastically reduce the need for insecticide application. For example, late planting of sorghum cultivars M 35-1 and Phule Yashoda resistant to shoot fly, *A. soccata*, can reduce the deadheart formation substantially during the post rainy season. Cultural control is a powerful tool to suppress insect pests in different agroecosystems. This technique involves two basic approaches:

- Making the environment less favorable to the pest; and
- Making the environment more favorable to the pest's natural enemies.

Asynchrony Between Plant Growth and Insect Populations

Insect-resistant varieties in combination with early-planting, early-maturity, defoliation, destruction of stalks, and deep-ploughing can be used effectively to control boll weevil, *Anthonomus grandis* (Boh.) and bollworms, *H. virescens* and *P. gossypiella*, in cotton (Adkisson and Gaines, 1960). This not only reduces the pest damage but also decreases the over-wintering populations of these pests, and thus results in reduced crop loss in the following seasons. The nectarless cotton varieties reduce pink bollworm infestation by 50% (Wilson and Shaver, 1973), and this in combination with cultural practices can reduce the pink bollworm infestation by 16-fold (Adkisson and Dyck, 1980). Through careful planning, the cropping pattern can be adjusted such that the most susceptible stage of the crop avoids the peak periods of insect activity. A combination of plant resistance and short-duration cotton varieties has been quite effective for controlling the bollworms (Walker and Niles, 1971). Short-season and rapid-fruiting cotton varieties mature two to three weeks earlier than the long-duration varieties. Early harvest coupled with area-wide stalk destruction reduces the over-wintering population of diapausing insects by 90% (Adkisson and Dyck, 1980). Such a system not only suppresses the pest population but also restores the biological control and significantly reduces the need for insecticide application.

Genetic Diversity

Genetic diversity, through its influence on herbivores and on the natural enemies, can play a key role in pest management. Polyculture (growing more than one crop in the same area) is one way of increasing crop diversity. Polycultures are ecologically complex because of interspecific and intraspecific competition with the insects and the natural enemies (van Emden, 1965). Elimination of alternate habitats leads to decreased predator and parasite populations and an increase in insect abundance (Southwood, 1975). Population densities of insect pests are frequently lower in polycultures (Risch, Ow, and Altieri, 1983), because of associational resistance or resource concentration (Roots, 1973), and the action of natural enemies (Russell, 1989). Specialist insects are generally less abundant in diverse habitats because of low concentration of their food in the habitat and increased activity of natural enemies. Plant diversity may also provide important resources for the natural enemies, such as alternative prey, nectar, pollen, and breeding sites. In diverse plant communities, a specialist insect is less likely to find its host because of the presence of confusing or masking effects of chemical stimuli from the nonhost plants, physical barriers to movement, and changes in the micro-environment of the target insect. Consequently, insect survival may be lower in polycultures than in the monocultures (Baliddawa, 1985). The population density of melon worm, *Diaphania hyalinata* (L.) is lower in polyculture (maize-cowpea-squash) than in monoculture (squash alone) (Letourneau, 1986). The total crop yield in polycultures is greater when estimated as a land equivalent ratio. However, the role of polycultures or plant diversity should be carefully assessed for its effects on insect damage, and the activity and abundance of natural enemies. Also, the effects of plant diversity on pest damage can change over time and locations, depending on the herbivore diversity and interactions among the harmful and beneficial insects.

Multilines/Synthetics

The feeding preference of polyphagous insects can be altered by including genetically different genotypes with similar maturity and height. Biological control of the cereal leaf beetle, *Oulema melanopus* (L.) has been achieved with mixed cropping of beetle-resistant and beetle-susceptible wheat varieties on an area-wide basis (Casagrande and Haynes, 1976). Simulated growth of aphid predators on the susceptible plants in variety mixtures also slows down the rate of development of virulent aphid biotypes (Wilhoit, 1991). The combined effect of varietal mixtures and natural enemies is quite effective in suppressing populations of insect pests.

Trap Crops

Trap crops attract insect pests or other organisms so that pest incidence on the target crop is minimized. Reduction in pest damage is achieved either by preventing the insect pests from infesting the target crop or by concentrating them in a certain part of the field where they can be easily destroyed (Hokkanen, 1991). The principle is similar to associational resistance, in which the insect pests show a distinct preference for certain plant species, cultivars, or a crop stage. Crop stands can be manipulated in time and space so that attractive host plants are offered to the insect pests at a critical stage of insect development. The insects concentrate at the desired site on the trap crop, and as a result, the main crop seldom needs to be treated with insecticides and thus the natural control of insect pests remains operational in most of the field. Trap cropping is particularly important in subsistence farming in the developing countries, and its application in cotton and soybean has been very successful (Newsom et al., 1980). In cotton/sesame intercrop, row strips of sesame (constituting 5% of the total area) can be used as a trap crop to attract *Heliothis/Helicoverpa* species from the main crop of cotton. Sesame, which is highly attractive to *Heliothis* species (from the seedling stage to senescence), attracts large numbers of insects away from the cotton (Sharma, 2005). It also attracts the parasitoid, *C. sonorensis*, which parasitizes large numbers of *H. virescens* larvae (Pair, Laster, and Martin, 1982).

Nutrient Application and Plant Resistance

Crop susceptibility to different insect pests changes with the amount and type of fertilizers applied. Therefore, care should be taken to apply appropriate combinations of nutrients to minimize pest damage and realize maximum crop yield in combination with insect-resistant cultivars. Expression of resistance to insects changes with the availability of nutrients. Sorghum shoot fly incidence decreases with an increase in application of nitrogenous fertilizers (Reddy and Narashima Rao, 1975; Sharma, 1985).

HPR and Chemical Control

Insecticides are most effective to obtain immediate control of pest outbreaks. The suppressive action of insecticides is independent of the pest density. The same level of mortality can be expected from a given dose whether the pest population is high or low, and this characteristic is important in applying pest management, that is, the use of insecticides to reduce populations to levels that can subsequently be controlled by other pest control techniques. However, their broad-spectrum mode of action destroys the beneficial insects as well, and thus leads to adverse effects in the environment. Therefore, it is important to integrate other methods of insect control with different modes of action to minimize insecticide use for sustainable crop protection. Plant resistance enhances the effectiveness of the insecticides through:

- Better insecticide coverage of the plant parts to be protected through modified plant canopy, for example, loose panicles in sorghum and frego-bract in cotton;
- Imbalanced nutrition or toxic substances having an adverse effect on insect growth and development, which may increase the insect susceptibility to insecticides; and
- Easy access to parasites and predators through changes in plant canopy.

The most common form of integrated control involves the use of insect-resistant cultivars and insecticides. The pest numbers are reduced in each generation, and this process slows the rate of population growth of the target insects (Painter, 1951). Even a moderately resistant cultivar in combination with insecticides can bring about a substantial reduction in pest numbers, and minimize the losses in grain yield. If insecticide application results in 80% larval mortality of *H. armigera* larvae, and the larval survival is 85% in ICCC 37, 65% on ICCV 2, and 60% on ICC 506; then there would be 2.84 times more moths in the area planted with ICCC 37 and 1.27 times more moths in the area planted with ICCV 2 as compared to the area planted with ICC 506. There would be 14.22 times more insects in an area planted with a susceptible cultivar alone as compared to the areas having a resistant cultivar plus one insecticide application in the first generation. The ratio between *H. armigera* moth populations (across cultivars) in areas planted with different cultivars plus one insecticide spray in the first generation compared to those with cultivars alone would be 0.20.

Rice varieties with resistance to stem borer, *C. suppressalis* and brown planthopper, *N. lugens* are widely grown in South India and South East Asia, and require fewer insecticidal applications than the varieties susceptible to these pests (Pathak, Beachell, and Andres, 1973; Panda and Khush, 1995). Several other varieties of rice with resistance to different insects have been used in integrated pest management in rice, cotton, sorghum, groundnut, tobacco, soybean, etc. (Kalode and Sharma, 1995; Panda and Khush, 1995; Sharma and Ortiz, 2002). Plant resistance may also enhance the effectiveness of the insecticides through better penetration of the insecticides to the target insect through modified plant morphology, for example, loose panicles in sorghum (Sharma, Lopez, and Vidyasagar, 1994) and open canopy in cotton increases the effectiveness of insecticides. Such traits also allow an easy access to parasites and predators. Moderate resistance based on imbalanced nutrition or toxic substances may increase the susceptibility of insects to insecticides. Insect-resistant corn hybrids require less insecticide application than the susceptible ones (McMillian et al., 1972).

Interaction between Antixenotic Mechanism of Resistance and Chemical Control

Chemicals and morphological traits conferring antixenosis to insects may be effective either at the egg-laying or feeding stages of the insect. In frego-bract cottons, the square has a rolled, twisted, and open bract (unlike in normal cotton where the bract is flat and encloses the square). Insecticide application is not required for boll weevil control on frego-bract cotton varieties, where up to 94% of the cotton boll weevil, *A. grandis*, population was suppressed. This also reduces the over-wintering population of the weevil (Jenkins and Parrot, 1971). A high level of oviposition suppression can be very useful in eradication programs. As boll weevils feed and oviposit on the cotton buds, the exposed buds in the frego-bract cotton can ideally be covered with insecticides. When sprayed with methyl-parathion, frego-bract buds have seven times more deposits of insecticide residue than those with normal bracts. A combination of the okra-leaf and frego-bract improves the efficiency of chemical control in cotton (Jones et al., 1986). In sorghum, better control of head bugs and head caterpillars can be obtained in genotypes with open and semicompact panicles (Sharma, Lopez, and Vidyasagar, 1994).

A two-locus model has been developed for resistance avoidance to illustrate that insect resistance can be managed in the field by the use of insecticides having insect-repellent properties (Gould, 1984). Insecticides reduce pest damage and decrease the population of the target pest, whereas repellency is likely to be effective in limiting pest damage to treated crops as it may keep the pests away from the target crop and reduce the pest incidence. Repellency is equivalent to using low doses of insecticides along with the repellent properties of the host plant. Insecticide formulations having noninsecticidal compounds with repellent properties (naturally derived or synthetically developed products) could lower the rate of insect adaptation to an insecticide. Insects lack the sensory apparatus to respond to most of the insecticides, but they have evolved the ability to respond to the volatile secondary plant compounds produced by the plants. Therefore, the selective pressure due to insecticides without antixenosis is likely to be much stronger than in populations faced with a resistant host plant with antixenosis mechanism of resistance (Pluthero and Singh, 1984).

Interaction between Antibiosis Mechanism of Resistance and Chemical Control

Plant resistance reduces the rate of insect population increase, which is cumulative over-time. Insecticides kill the target pest while the antibiosis component of resistance aims at decreasing the rate of population increase through reduced vigor, longer development period, and reduced fecundity. This prevents the insect population from reaching economic threshold levels. The toxicity of an insecticide is influenced by the insect body weight, and a lower concentration is needed to control the insects feeding on a resistant variety than those feeding on a susceptible variety. Nymphs of wheat grain aphid, *Sitobion avenae* (F.) reared on a resistant wheat variety (Altar) possessing the antibiosis compound DIMBOA (2,4-dihydroxy-7-methoxy 1,4-benzoxazin-3-one) were significantly more susceptible to deltamethrin than nymphs reared on the susceptible wheat variety (Dollar bird). The LD_{50} value (adjusted for weight) was reduced by 91% for nymphs reared on the aphid-resistant cultivar (Nicol et al., 1993). In concentrations as low as one-hundredth of those affecting the survival and reproduction of the aphid, DIMBOA strongly inhibited detoxifying enzymes (peroxidase, polyphenol oxidase, *N*-demethylase, glutathione *S*-transferase, and UDP-glucose transferase) (Leszczynski et al., 1983).

Moderate Levels of Plant Resistance and Chemical Control

Moderate levels of resistance to the target pest can be used effectively in combination with insecticides to keep the pest density below economic threshold levels. A rational combination of moderate levels of plant resistance and insecticides reduces the chances of insect resurgence and helps to conserve the natural enemies, preserve environmental quality, slow down the rate of selection for insecticide-resistant insect strains, and increase the profitability of crop production (Adkisson and Dyck, 1980). Population of *M. persicae* on partially resistant varieties has been observed to be about 85% of that on the susceptible varieties of Brussels sprouts. However, LD_{50} of malathion for the aphid on the partially resistant cultivar was about 55% compared to that of the aphids on the susceptible variety. Insecticide requirement was much less on the partially resistant variety than that on the susceptible variety (Muid, 1977). The increased susceptibility to malathion on the partially resistant variety appeared to be due to the interaction of insecticide with insects of low vitality. Even with small levels of plant resistance, insecticide concentration can be reduced to one-third of that required on a susceptible variety (van Emden, 1990). A half dose of chlorfenvinphos gave equal or better control of the turnip root fly, *Delia floralis* (Fall.) on resistant cultivars of Swede (Cruciferae) S 7790 than the full dose on the susceptible cultivar, Ruta. The reduction in indices of root fly damage and the increase in percentage of marketable produce were more pronounced on the resistant than on the susceptible cultivar (Taksdal, 1993).

Insecticide application did not result in a substantial increase in grain yield of chickpea variety, ICC 506 with moderate levels of resistance to *H. armigera*, while four to five sprays were needed to realize the maximum yield potential of the susceptible varieties such as Annigeri and ICCC 37 (Wightman et al., 1995). In the case of pigeonpea, insecticide (endosulfan) and the *Helicoverpa*-resistant cultivar, ICPL 332 (see Figure 4.10) has a substantial effect on pod damage and grain yield (Figure 4.11). Pod damage in the *Helicoverpa*-susceptible cultivar, ICPL 87 was 35.8% in endosulfan-treated plots compared to 70.8% in untreated control plots. Grain yield was 970.22 kg ha^{-1} in plots treated with endosulfan (0.07%) compared to 277.4 kg ha^{-1} in untreated plots (Sharma and Pampapathy, 2004). In the *Helicoverpa*-resistant cultivar, ICPL 332 the pod borer damage ranged from 4.3% in endosulfan-treated plots compared to 14.3% in the untreated control plots. Grain yield was 1642.8 kg ha^{-1} in the untreated control plots compared to 1833.6 kg ha^{-1} in plots treated with endosulfan. The *Helicoverpa*-resistant cultivar ICPL 332 resulted in a significant decrease in *H. armigera* damage, and thus can be used as a component in integrated pest management in pigeonpea.

FIGURE 4.10 Pigeonpea cultivar ICPL 332, with moderate levels of resistance to *Helicoverpa armigera*, which can be deployed for IPM in combination with other components of pest management.

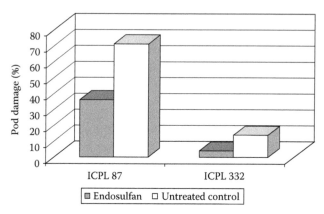

FIGURE 4.11   Effect of insecticide application (one spray of 0.07% endosulfan) on pod damage by *Helicoverpa armigera*, and grain yield in a susceptible (ICPL 87) and resistant (ICPL 332) cultivars of pigeonpea.

Reduced insecticide use on vegetables and fruits for controlling insects not only benefits the agroecosystem and the natural enemies, but also results in lower insecticide residues in the food. The susceptibility of white backed planthopper, *Sogatella furcifera* (Horvath) and brown planthopper *N. lugens* to insecticides is affected by the level of plant resistance in the rice cultivars. Planthoppers reared on moderately resistant rice varieties had lower LD_{50} values than those reared on the susceptible varieties (Heinrichs et al., 1984). A combination of moderate levels of plant resistance and insecticide application can be used for effective control of insect pests. Application of carbofuran granules results in a significant reduction in deadheart formation due to shoot fly in sorghum cultivars with moderate levels of resistance (M 35-1 and ICSV 705), but there is no effect of carbofuran application on shoot fly damage on the susceptible cultivar, CSH 1 (Figure 4.12). Maximum grain yield in sorghum has been realized with four sprays of demeton-*S*-methyl against sorghum head bug, *C. angustatus* on IS 9692 and CSH 11, whereas only one to two sprays are sufficient to realize the maximum yield potential of the head bug-resistant genotypes,

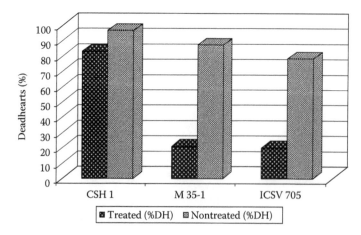

FIGURE 4.12   Effect of carbofuran treatment on deadheart formation due to sorghum shoot fly, *Atherigona soccata* in susceptible (CSH 1) and moderately resistant (M 35-1 and ICSV 705) cultivars of sorghum.

TABLE 4.4

Effect of Different Levels of Protection on Insect-Resistant and Insect-Susceptible Cultivars of Sorghum on Insect Damage and Grain Yield

Genotype	Shoot Fly DH (%)		Head Bugs[a] (Milk Stage)		Midge Score[b]		Grain Yield (kg ha⁻¹)	
	P	UP	P	UP	P	UP	P	UP
ICSV 197-MR	29.7	98.2	83	197	1.3	1.3	5,000	1,333
ICSV 700-SBR	7.6	72.5	4	163	5.3	9.0	1,000	0.000
ICSV 745-MR	30.1	98.4	4	101	1.0	1.7	5,000	1,000
PS 30710-SFR	8.0	73.7	6	55	3.7	8.7	4,333	667
Swarna-S	24.5	97.9	0	80	4.7	9.0	2,000	0.000
SE ±		3.00		61		0.73		433

MR, midge resistant; SBR, stem borer resistant; SFR, shoot fly-resistant; S, susceptible check; P, protected; UP, unprotected; and DH, dead hearts.

[a] Number of head bug nymphs and adults per panicle.

[b] Midge score (1 = <10% spikelets with midge damage, and 9 = >80% spikelets with midge damage).

IS 17610 and IS 21443 (Sharma and Lopez, 1993). Head bug-resistant genotypes not only reduce the rate of population increase of the insect, but are also able to tolerate greater head bug densities.

High Levels of Plant Resistance and Chemical Control

High levels of resistance can be quite effective in controlling certain insect pests. However, there may be need to protect such varieties against secondary pests. HPR in combination with insecticide application (use of carbofuran granules) has a dramatic effect in reducing shoot fly (*A. soccata*) damage in sorghum (Sharma et al., 1999b) (Table 4.4). However, HPR alone is quite effective for the control of sorghum midge, *S. sorghicola*. Insecticide application alone is ineffective in reducing the shoot fly damage on the susceptible cultivars. The corn earworm-resistant untreated maize hybrid, 471-U6 × 81-1 had 48% more undamaged ears under artificial infestation (heavy population pressure) than the susceptible hybrid treated with seven applications of tetrachlorvinphos (Gardona[R]). However, one insecticide application on the resistant hybrid gives earworm control equal to that achieved with seven insecticide applications on the susceptible hybrid (McMillian et al., 1972; Wiseman, Harrell, and McMillian, 1973). One of the most successful examples of drastic reduction in insecticide use with resistant varieties is the large-scale planting of rice gall midge-resistant varieties in India (Panda and Khush, 1995), and sorghum midge-resistant hybrids in Australia.

Advantages of HPR

Utilization of plant resistance as a control strategy has enormous practical relevance and additional emotional appeal (Davies, 1981; Sharma, 1996). It is in this context that HPR assumes a central role in our efforts to increase the production and productivity of crops. Plant resistance to insects is the backbone of any pest management system because of:

Specificity: It is specific to the target pest or a group of pests, and generally has no adverse effects on the nontarget natural enemies.

Cumulative effects: The effects of plant resistance on insect population density are cumulative over successive generations of the target pest because of reduced survival, delayed development, and reduced fecundity.

Persistence: Most of the insect-resistant varieties express moderate to high levels of resistance to the target insect pests throughout the crop-growing season, except under certain environmental conditions or occasional occurrence of the new biotypes or of high pest densities. In contrast, the insecticides have to be applied repeatedly to achieve satisfactory control of the pest populations.

Compatibility: It is compatible with other methods of pest control, and also improves the efficiency of other methods of pest management.

Environmental safety: There are no harmful effects of HPR on nontarget organisms, humans, and the environment.

Ease of adoption: It does not involve any costs to the farmers. Also, the farmers do not have to have knowledge of any application techniques.

Very high levels of resistance may neither be attainable nor required. A variety capable of reducing the pest population by 50% in each generation can be quite useful in reducing the damage below economic threshold levels within a few generations (Painter, 1951). The cumulative and persistent effects of plant resistance are a strong contrast to the explosive effects of insecticides, where the insect population multiplies at a much faster rate after the insecticide application because of the absence of natural enemies. One of the notable examples of use of HPR is the near elimination of the Hessian fly, *M. destrutor* by the use of resistance varieties in the United States, and that of gall midge, *O. oryzae* and brown planthopper, *N. lugens* in rice in Asia (Panda and Khush, 1995).

Limitations of HPR

Plant resistance is not a panacea for solving all pest problems. Certain limitations and problems will always be associated any insect control program, and HPR is no exception. Development of plant varieties resistant to insect pests takes a long time. It took 15 years to develop the sorghum midge-resistant variety ICSV 745 (Sharma, 1993). In contrast, only three to five years were needed to develop the spotted alfalfa aphid-resistant Cody, Moapa, and Lahontan varieties (Panda and Khush, 1995). Developing insect-resistant crop varieties requires a great deal of expertise and resources. It is usually necessary to organize a well-planned multidisciplinary team of entomologists and plant breeders. Commitment of relatively long-term funding is a critical factor in the ultimate success of plant resistance programs. Some mechanisms of plant resistance may involve the diversion of some resources by the plant to extra structures or production of defense chemicals at the expense of other physiological processes, including those contributing to yield (Mooney, Gulmon, and

Johnson, 1983). Although concentration of defense chemicals responsible for resistance is low in plant tissues, the total amount per hectare may be quite high. The production cost of 34 kg of gossypol in terms of glucose equivalent in cotton will be 70.7 kg of glucose ha^{-1} (Mitra and Bhatia, 1982). The genetic, biochemical, and physiological mechanisms involve the initial cost of resistance as well as the extent to which they can be modified by screening and selection. Some mechanisms are less costly, whereas others involving high costs cannot be modified. More information is needed on mechanisms of resistance, the genetic regulation of resistance traits, and biochemical pathways and their physiological effects (Simms, 1992). One might expect a negative correlation between the potential yield of a cultivar and its level of resistance to the target pest. This is illustrated by the failure to evolve insect-resistant varieties in soybean, pigeonpea, chickpea, etc. Although HPR promises to contribute a great deal to pest management in several crops, progress has been slow mainly because of low yield potential of the resistant varieties (Newsom et al., 1980, Sharma, 1993). Yield of promising soybean lines at the advanced stages of testing usually falls short of the performance of commercial varieties. The growers are not willing to plant an insect-resistant variety unless it can yield about as much without insecticide treatment as susceptible varieties with insecticide treatment. However, the fundamental objective of breeding for insect resistance in crop plants is to reduce the amount of insecticides needed to achieve satisfactory control of the target pests, and an acceptable level of sustainable resistance, compatible with yield and quality of the produce. Despite many dramatic successes in HPR, there still are cases where plant resistance to one insect leads to increased susceptibility to another insect. Cotton varieties with high levels of gossypol are resistant to *H. virescens* and blister beetles, but gossypol at concentrations found in cotton plants is attractive to the boll weevil, *A. grandis* (Bell et al., 1987).

Secondary plant substances have a negative effect on the fitness of herbivores. However, many herbivores possess remarkable potential for utilizing or metabolizing the toxic plant chemicals and, therefore, their role as plant defense chemicals is not sacrosanct. Insects also evolve into biotypes to overcome antibiosis resistance. However, partially resistant varieties would probably last longer in the field than those with high levels of resistance (Lamberti, Waller, and van der Graaff, 1983). The current theories on plant defense strategies do not take into account the complexity of tritrophic interactions. Plant resistance based on antibiosis may not always be compatible with biological control (Boethel and Eikenbary, 1986). Elucidation of these interactions can further the understanding, and provide greater potential for manipulation of these systems to specific crop species and varieties. The possibility of using compounds from plants to reduce herbivore damage and increase the effectiveness of biological control agents is quite attractive. Ideally, plant resistance should strive to reduce substances attractive to herbivores, while increasing the substances attractive to the natural enemies.

The chemical basis of plant resistance to insects at times can modify the toxicity of insecticides to insects, for example, 2-tridecanone in wild tomato reduces the toxicity of carbaryl to *H. virescens* (Brattesten, 1988). Some plant defense chemicals also affect the food quality. Gossypol and related compounds that confer resistance to insects in cotton are toxic to nonruminant vertebrates (Lambou et al., 1966). Rutin, cholorogenic acid, tomatine, and phenols have toxic effects on humans. Some of these compounds may also be carcinogenic and mutagenic. Insect-resistant tomato plants when ploughed into the fields after crop harvest may lead to build up of allelopathic compounds in the soil and affect the growth of the subsequent crop. The accumulation of such chemicals in the soil may alter the microflora in the rhizosphere. Therefore, all such interactions should be kept in mind while developing and deploying insect-resistant cultivars for pest management. The major limitations of plant resistance are:

Time for development: It takes a long time to identify and develop insect-resistant cultivars. This method is not suitable to solve sudden or localized pest problems. It takes 5 to 15 years to identify sources of resistance and transfer the resistance traits into cultivars with high yield potential and desirable quality traits.

Genetic limitations: Absence of adequate levels of resistance in the available germplasm may deter the use of plant resistance for managing certain pests. Such limitations can now be overcome through the use of interspecific hybridization, mutations, and genetic transformation.

Occurrence of insect biotypes: Occurrence of new biotypes of the target pest may limit the use of certain insect-resistant varieties in time and space. In such situations, one has to go for polygenic resistance or continuously search for new genes, and transfer them into high yielding varieties.

Conflicting resistance traits: Certain plant characteristics may confer resistance to one pest, but render such plants more susceptible to other pests, for example, hairiness in cotton confers resistance to jassids, *A. biguttula biguttula* (Sharma and Agarwal, 1983a), but such varieties are preferred for oviposition by the spotted bollworm, *Earias vittella* (Fab.) (Sharma and Agarwal, 1983b). Also, pubescence in soybean confers resistance to leafhoppers, but pod borer, *Grapholitha glycinivorella* Matsumura, prefers pubescent varieties for oviposition (Nishijima, 1960). Varieties with high levels of resistance to sorghum midge are susceptible to head bugs, shoot fly, and stem borer (Sharma, 1993).

Plant resistance at times may be associated with low yield or factors resulting in poor or unacceptable produce, for example, sorghum genotypes with high tannin content are resistant to sorghum midge (*S. sorghicola*) and birds, but such a grain has poor nutritional quality. Similarly, gossypol and other terpenoids in cotton confer resistance to bollworms (Sharma and Agarwal, 1982a, 1982b), but high gossypol content spoils the quality of cottonseed oil. In such situations, one has to break the linkage between the factors conferring resistance to the target insects and the low yield potential or arrive at a threshold levels for the resistant traits (secondary metabolites) that result in reduced pest susceptibility, and at the same time do not have an adverse effect on the quality of the produce.

Conclusions

Considerable progress has been made in identification and utilization of plant resistance to insects. Multilocational testing of the identified sources and breeding material need to be strengthened to identify stable and diverse sources of resistance or establish the presence of new insect biotypes. Resistance to insects should be given as much emphasis as yield to identify new varieties and hybrids. Insect-resistant varieties exercise a constant and cumulative effect on insect populations over time and space, have no adverse effects on the environment, reduce the need to use insecticides, have no extra cost to the farmers, and do not require inputs and application skills by the farmers. Therefore, plant resistance to insects should form the backbone of pest management programs for integrated pest management. Molecular markers can be used to accelerate the pace of introgression of insect resistance genes into high yielding cultivars. Conventional plant resistance to insects can also be deployed in combination with novel genes to make plant resistance an effective weapon for pest management and sustainable crop production.

References

Adkisson, P.L. and Dyck, V.A. (1980). Resistant varieties in pest management systems. In Maxwell, F.G. and Jennings, P.R. (Eds.), *Breeding Plants Resistant to Insects*. New York, USA: John Wiley & Sons, 233–251.

Adkisson, P.L. and Gaines, J.G. (1960). *Pink Bollworm Control as Related to the Total Cotton Insect Control Program of Central Texas*. Miscellaneous Publication, 444. College Station, Texas, USA: Texas Agriculture Experiment Station.

Agrawal, B.L., Sharma, H.C., Abraham, C.V. and Stenhouse, J.W. (2005). Registration of ICSV 88032 midge-resistant sorghum variety. *International Sorghum and Millets Newsletter* 46: 43–46.

Agrawal, B.L., Sharma, H.C. and Leuschner, K. (1987). Registration of ICSV 197 midge resistant sorghum cultivar. *Crop Science* 27: 1312–1313.

Baliddawa, C.W. (1985). Plant species diversity and crop pest control. *Insect Science and Its Application* 6: 479–487.

Barbosa, P. (1988). Natural enemies and herbivore-plant interactions: Influence of plant allelochemicals and host specificity. In Barbosa, P. and Letourneau, D.K. (Eds.), *Novel Aspects of Insect-Plant-Interactions*. New York, USA: John Wiley & Sons, 201–229.

Barbosa, P., Saunders, J.A., Kemper, J., Trumble, R., Olechno, J. and Martinat, P. (1986). Plant allelochemicals and insect parasitoids: Effects of nicotine on *Cotesia congregata* and *Hyposoter annulipes*. *Journal of Chemical Ecology* 12: 1319–1328.

Barbour, J.D., Farrar, R.R. Jr. and Kennedy, G.G. (1993). Interaction of *Manduca sexta* resistance in tomato with insect predators of *Helicoverpa zea*. *Entomologia Experimentalis et Applicata* 68: 143–155.

Beland, G.C. and Hatchett, J.H. (1976). Expression of antibiosis to the bollworm in two soybean genotypes. *Journal of Economic Entomology* 6: 557–560.

Bell, A.A., Stipanovic, R.D., Elzen, G.W. and Williams, H.J. Jr. (1987). Structural and genetic variation of natural pesticides in pigment glands of cotton (*Gossypium*). In Waller, G.R. (Ed.), *Allelochemicals: Role in Agriculture and Forestry*. ACS Symposium Series 330. Washington, D.C., USA: American Chemical Society.

Benedict, J.H. and Hatfield, J.L. (1988). Influence of temperature induced stress on host plant suitability to insects. In Heinrichs, E.A. (Ed.), *Plant Stress Insect Interactions*. New York, USA: John Wiley & Sons, 139–165.

Bentur, J.S., Srinivasan, T.E. and Kalode, M.B. (1987). Occurrence of a virulent rice gall midge (GM) *Orseolia oryzae* Wood–Mason biotype (?) in Andhra Pradesh, India. *International Rice Research Newsletter* 12: 33–34.

Bergman, J.M. and Tingey, W.M. (1979). Aspects of interaction between plant genotypes and biological control. *Bulletin of Entomological Society of America* 25: 275–279.

Boethel, D.J. and Eikenbary, R.D. (Eds.). (1986). *Interactions of Plant Resistance and Parasitoids and Predators of Insects*. Chichester, UK: Ellis Harwood.

Bombosch, S. (1963). Untersuchungen zur Vermchrung von *Aphis fabae* Scop. in Samenrubenbestanden unter besonderer Berucksichtigung der Schwebfliegen (Diptera, Syrphidae). *Zietschrift fur Angewande Entomologie* 52: 105–141.

Brar, D.S. and Khush, G.S. (1997). Alien introgression in rice. *Plant Molecular Biology* 35: 35–47.

Brattesten, L.B. (1988). Potential role of plant allelochemicals in the development of insecticide resistance. In Barbosa, P. and Letourneau, D.K. (Eds.), *Novel Aspects of Insect-Plant-Interactions.* New York, USA: John Wiley & Sons, 313–355.

Caddick, M.X., Greenland, A.J., Jepson, I., Krause, K.P., Qu, N., Riddell, K.V., Salter, M.G., Schuch, W. and Tomsett, A.B. (1998). An ethanol-inducible gene switch for plants used to manipulate carbon metabolism. *Nature Biotechnology* 16: 177–180.

Campbell, B.C. and Duffey, S.S. (1979). Tomatine and parasitic wasps: Potential incompatibility of plant antibiotics with biological control. *Science* 205: 700–705.

Carew, D.P. and Krueger, J. (1976). Anthocyanidins of *Catharanthus callus* cultures. *Phytochemistry* 15: 442.

Casagrande, R.A. and Haynes, D.L. (1976). The impact of pubescent wheat on the population dynamics of the cereal leaf beetle. *Environmental Entomology* 5: 133–159.

Chand, P., Sinha, M.P. and Kumar, A. (1979). Nitrogen fertilizer reduces shoot fly incidence in sorghum. *Science and Culture* 45: 6162.

Channabasavanna, G.P., Venkat Rao, B.V. and Rajagopal, G.K. (1969). Preliminary studies on the effect of incremental levels of phosphatic fertilizer on the incidence of jowar shoot fly. *Mysore Journal of Agriculture Sciences* 3: 253–255.

Chaudary, R.C., Schwarzbach, E. and Fischbeck, G. (1983). Quantitative studies of resistance induced by avirulent cultures of *Erysiphe graminis* f. sp. *hordei* in barley. *Phytopathology Zeitschrift fur Angewandte* 108: 80–87.

Clement, S.L. (2002). Insect resistance in wild relatives of food legumes and wheat. In *Plant Breeding for 11th Millennium, Proceedings of the 12th Australian Plant Breeding Conference*, 15–20 September, 2002. Perth, Australia: Australian Plant Breeding Association, 287–293.

Clement, S.L., Hardie, D.C. and Elberson, L.R. (2002). Variation among accessions of *Pisum fulvum* for resistance to pea weevil. *Crop Science* 42: 2167–2173.

Clement, S.L. and Quisenberry, S.S. (Eds.). (1999). *Global Plant Genetic Resources for Insect-Resistant Crops.* Boca Raton, Florida, USA: CRC Press.

Clement, S.L., Sharaf El-Din, N., Weigand, S. and Lateef, S.S. (1994). Research achievements in plant resistance to insect pests of cool season food legumes. *Euphytica* 73:41–50.

Cosenza, G.W. and Green, H.P. (1979). Behaviour of tomato fruit worm, *Heliothis zea* on susceptible and resistant lines of processing tomatoes. *Horticulture Science* 14: 171.

Cowgill, S.E. and Lateef, S.S. (1996). Identification of antibiotic and antixenotic resistance to *Helicoverpa armigera* (Lepidoptera: Noctuidae) in chickpea. *Journal of Economic Entomology* 89: 224–229.

Dalisay, R.F. and Kuc, J.A. (1995). Persistence of induced resistance and enhanced peroxidase and chitinase activities in cucumber plants. *Physiological Plant Pathology* 47: 315–327.

Danks, H.V., Rabb, R.L. and Southern, P.S. (1979). Biology of insect parasites of *Heliothis* larvae in North Carolina. *Journal of Georgia Entomological Society* 14: 36–64.

Davies, J.C. (1981). Pest losses and control of damage on sorghum in developing countries. The realities and myths. In House, L.R., Mughogo, L.K. and Peacock, J.M. (Eds.), *Sorghum in the Eighties.* Patancheru, Andhra Pradesh, India: International Crops Research Institute for the Semi-Arid Tropics (ICRISAT), 215–224.

de Moraes, C.M., Lewis, W.J., Pare, P.W., Alban, H.T. and Tumilson, J.H. (1998). Herbivore infested plants selectively attract parasitoids. *Nature* 393: 570–573.

Dodd, G.D. (1973). *Integrated Control of the Cabbage Aphid,* Brevicoryne brassicae *(L).* Ph.D. thesis, University of Reading, Reading, UK.

Duale, A.H. and Nwanze, K.F. (1997). Effects of plant resistance to insects on the effectiveness of natural enemies. In Sharma, H.C., Singh, F. and Nwanze, K.F. (Eds.), *Plant Resistance to Insects in Sorghum.* Patancheru, Andhra Pradesh, India: International Crops Research Institute for the Semi-Arid Tropics (ICRISAT), 161–167.

Eigenbrode, S.D. and Kabalo, N.N. (1999). Effects of *Brassica oleracea* wax blooms on predation and attachment by *Hippodamia convergens*. *Entomologia Experimentalis et Applicata* 91: 125–130.

Eigenbrode, S.D., Trumble, J.T. and Jones, R.A. (1993). Resistance to beet armyworm [*Spodoptera exigua* (Hubner)], hemipterans, and *Liriomyza* spp. in *Lycopersicon. Journal of American Society of Horticultural Science* 118: 525–530.

Felton, G.W. and Duffey, S.S. (1990). Inactivation of baculovirus by quinines formed in insect-damaged plant tissues. *Journal of Chemical Ecology* 16: 1221–1236.

Felton, G.W., Summers, C.B. and Mueller, A.J. (1994). Oxidative responses in soybean foliage to herbivory by bean leaf beetle and three-cornered alfalfa hopper. *Journal of Chemical Ecology* 20: 639–650.

Fischer, D.C., Kogan, M. and Paxton, J. (1990). Effect of glyceollin, a soybean phytoalexin, on feeding by three phytophagous beetles (Coleoptera: Coccinellidae and Chrysomelidae): dose vs. response. *Environmental Entomology* 19: 1278–1282.

Gallardo, F., Boethel, D.J., Fuxa, J.R. and Richter, A. (1990). Susceptibility of *Heliothis zea* (Boddie) larvae to *Nomuraea rileyi* (Farlow) Samson. Effects of λ-tomatine at the third trophic level. *Journal of Chemical Ecology* 16: 1751–1759.

Gatz, C. (1997). Chemical control of gene expression. *Annual Review of Plant Physiology and Plant Molecular Biology* 48: 89–108.

Gatz, C. and Quail, P.H. (1998). *Tn10*-encoded *Tet* repressor can regulate an operator-containing plant promoter. *Proceedings National Academy of Sciences USA* 85:1394–1397.

Georghiou, G.P. (1986). The magnitude of resistance problems. In *Pesticide Resistance, Strategies and Tactics for Management.* Washington, D.C., USA: National Academy Press, 14–43.

Glass, E.H. (1975). *Integrated Pest Management Rationale, Potential, Needs and Implementation.* College Park, Maryland, USA: Entomological Society of America.

Gould, F. (1984). Role of behaviour in the evolution of insect adaptation to insecticides and resistant host plants. *Bulletin of Entomological Society of America* 30: 33–41.

Gould, F., Kennedy, G.G. and Johnson, M.T. (1991). Effects of natural enemies on the rate of herbivore adaptation to resistant host plants. *Entomologia Experimentalis et Applicata* 58: 1–14.

Green, P.W.C., Sharma, H.C., Stevenson, P.C. and Simmonds, M.S.J. (2006). Susceptibility of pigeonpea and some of its wild relatives to predation by *Helicoverpa armigera*: Implications for breeding resistant cultivars. *Australian Journal of Agricultural Research* 57: 831–836.

Hahlbrock, K. and Grisebach, H. (1979). Enzymatic controls in the biosynthesis of lignins and flavonoids. *Annual Review of Plant Physiology* 30: 105–130.

Hall, P.K., Parrot, W.L., Jenkins, J.N. and McCarty, J.C. Jr. (1980). Use of tobacco budworm eggs and larvae for establishing field infestations in cotton. *Journal of Economic Entomology* 73: 393–395.

Hallman, G.J., Teetes, G.L. and Johnson, J.W. (1984). Relationship of sorghum midge (Diptera: Cecidomyiidae) density to damage to resistant and susceptible sorghum hybrids. *Journal of Economic Entomology* 77: 83–87.

Hamm, J.J. and Wiseman, B.R. (1986). Plant resistance and nuclear polyhedrosis virus for suppression of the fall armyworm (Lepidoptera: Noctuidae). *Florida Entomologist* 69: 541–549.

Hare, J.D. (1983). Manipulation of host suitability for herbivore pest management. In Denno, R.F. and McClure, M.S. (Eds.), *Variable Plants and Herbivores in Natural and Managed Systems.* New York, USA: Academic Press, 655–679.

Heinrichs, E.A. (1988). Role of insect-resistant varieties in rice IPM systems. In Teng, P.S. and Heong, K.L. (Eds.), *Pesticide Management and Integrated Pest Management in Southeast Asia.* College Park, Maryland, USA: Consortium for International Crop Protection.

Heinrichs, E.A., Fabellar, L.T., Basilio, R.P., Tu, C.W. and Medrano, F. (1984). Susceptibility of rice plant hoppers, *Nilaparvata lugens* and *Sogatella furcifera* (Homoptera: Delphacidae) to insecticides as influenced by level of resistance in the host plant. *Environmental Entomology* 13: 455–458.

Herzog, D.C. and Funderburk, J.E. (1985). Plant resistance and cultural practice interactions with biological control. In Hoy, M.A. and Herzog, D.C. (Eds.), *Biological Control in Agricultural IPM Systems.* London, UK: Academic Press, 67–88.

Hokkanen, H.M.T. (1991). Trap cropping in pest management. *Annual Review of Entomology* 36: 119–138.

Howe, W.L. and Smith, O.F. (1957). Resistance to the spotted alfalfa aphid in Lehontan alfalfa. *Journal of Economic Entomology* 50: 320–324.

Hunt, O.J., Peadon, R.N., Crahan, H.L. and Lieberman, F.V. (1966). Registration of Washoe alfalfa. *Crop Science* 6: 160.

Inbar, M., Doodstar, H., Sonoda, R.M., Leibee, G.L. and Mayer, R.T. (1998). Elicitors of plant defensive systems reduce insect densities and disease incidence. *Journal of Chemical Ecology* 24: 135–149.

Insenhour, D.J. and Wiseman, B.R. (1987). Foliage consumption and development of fall armyworm (Lepidoptera: Noctuidae) as affected by the interactions of a parasitoid, *Campoletis sonorensis* (Hymenoptera: Ichneumonidae), and resistant corn genotypes. *Environmental Entomology* 16: 1181–1184.

Insenhour, D.J. and Wiseman, B.R. (1989). Parasitism of the fall armyworm (Lepidoptera: Noctuidae) as effected by host feeding on silks of *Zea mays* v. Zaplote Chico. *Environmental Entomology* 18: 394–397.

Jackai, L.E.N. and Oghiakhe, S. (1989). Podwall trichomes and resistance of two wild cowpea, *Vigna vexillata*, accessions to *Maruca testulalis* (Geyer) (Lepidoptera: Pyralidae) and *Clavigralla tomentosicollis* Stal (Hemiptera: Coreidae). *Bulletin of Entomological Research* 79: 595–605.

Jenkins, J.N. and Parrot, W.L. (1971). Effectiveness of frego bract as a boll weevil resistance character in cotton. *Crop Science* 11: 739–743.

Jepson, I., Martinez, A. and Sweetruan, P. (1998). Chemical inducible gene expression systems for plants: A review. *Pesticide Science* 54: 360–367.

Jepson, I., Lay, V.J., Holt, D.C., Bright, S.W.J. and Greenland, A.J. (1994). Cloning and characterization of maize herbicide safener-induced cDNAs encoding subunits of glutathione S-transferase isoforms I, II, IV. *Plant Molecular Biology* 26: 1855–1866.

Jones, J.E., James, D., Sistler, F.E. and Stringer, S.J. (1986). Spray penetration of cotton canopies as affected by leaf and bract isolines. *Louisiana Agriculture* 29: 15–17.

Kalode, M.B. (1987). Insect pests of rice and their management. In Veerabhadra Rao, M. and Sithanantham, S. (Eds.), *Plant Protection in Field Crops*. Rajendranagar, Hyderabad, Andhra Pradesh, India: Plant Protection Association of India, 61–74.

Kalode, M.B., Bentur, J.S. and Srinivasan, T.E. (1989). Screening and breeding rice for stem borer resistance. In Nwanze, K.F., (Ed.), *International Workshop on Sorghum Stem Borers*. Patancheru, Andhra Pradesh, India: International Crops Research Institute for the Semi-Arid Tropics (ICRISAT), 153–157.

Kalode, M.B. and Sharma, H.C. (1995). Host plant resistance to insects: Progress, problems, and future needs. In Sharma, H.C. and Veerbhadra Rao, M. (Eds.), *Pests and Pest Management in India: The Changing Scenario*. Rajendranagar, Hyderabad, Andhra Pradesh, India: Plant Protection Association of India, 229–243.

Karban, R. and Baldwin, I.T. (1997). *Induced Responses to Herbivory*. Chicago, Illinois, USA: University of Chicago Press.

Kartohardjono, A. and Heinrichs, E.A. (1984). Populations of the brown planthopper, *Nilaparvata lugens* (Stal) (Homoptera: Delphacidae), and its predators on rice varieties with differing levels of resistance. *Environmental Entomology* 13: 359–365.

Kashyap, R.K. and Verma, A.N. (1987). Development and survival of fruit borer, *Heliothis armigera* (Hübner), on resistant and susceptible tomato genotypes. *Journal of Plant Diseases and Protection* 94: 14–21.

Kashyap, R.K., Banerjee, M.K., Kalloo, G. and Verma, A.N. (1990). Survival and development of fruit borer, *Heliothis armigera* (Hubner) (Lepidoptera: Noctuidae) on *Lycopersicon* spp. *Insect Science and Its Application* 11: 877–881.

Kennedy, G.G., Kishaba, A.N. and Bohn, G.W. (1975). Response of several pest species to *Cucumis melo* L. lines resistant to *Aphis gossypii* Glover. *Environmental Entomology* 4: 653–657.

Kennedy, G.G., Gould, F., de Ponting, O.M.B. and Stinner, R.E. (1987). Ecological, agricultural, genetic and commercial considerations in the development of insect resistant germplasms. *Environmental Entomology* 16: 327–338.

Khan, Z.R., Norris, D.M., Chiang, H.S. and Oosterwyk, A.S. (1986). Light induced susceptibility in soybean to cabbage looper, *Trichoplusia ni* (Lepidoptera: Noctuidae). *Environmental Entomology* 15: 803–808.

Khush, G.S. (1977). Disease and insect resistance in rice. *Advances in Agronomy* 29: 265–341.

Khush, G.S. and Brar, D.S. (1991). Genetics of resistance to insects in crop plants. *Advances in Agronomy* 45: 223–274.

Knipling, E.F. (1979). The basic principles of insect population suppression. *Bulletin of Entomological Society of America* 12: 7–15.

Kogan, M. (1982). Plant resistance in pest management. In Metcalf, R.L. and Luckmann, W.H. (Eds.), *Introduction to Insect Pest Management*, 2nd Edition. New York, USA: John Wiley & Sons, 93–134.

Krischik, V.A., Barbosa, P. and Reichelderfer, C.F. (1988). Three trophic level interactions: Allelochemicals, *Manduca Sexta*, and *Bacillus thuringiensis* var. *kurstaki*. *Environmental Entomology* 17: 476–482.

Lamberti, F., Waller, J.M. and van der Graaff, N.A. (1983). *Durable Resistance in Crops*. New York, USA: Plenum Press.

Lambou, M.G., Shaw, R.L., Decossas, K.M. and Vix, H.L.E. (1966). Cottonseeds role in a hungry world. *Economic Botany* 20: 256–267.

Leszczynski, B., Urbanska, A., Matok, H. and Dixon, A.F.G. (1983). Detoxifying enzymes of the grain aphid. *Bulletin OILB/SORB* 16: 165–172.

Letourneau, D.K. (1986). Associational resistance in squash monocultures and polycultures in tropical Mexico. *Environmental Entomology* 15: 285–292.

Lin, H.C. and Kogan, M. (1990). Influence of induced resistance in soybean on the development and nutrition of the soybean looper and the Mexican bean beetle. *Entomologia Experimentalis et Applicata* 55: 131–138.

Lin, H.C., Kogan, M. and Fischer, D. (1990). Induced resistance in soybean to the Mexican bean beetle (Coleoptera: Coccinellidae): Comparisons of inducing factors. *Environmental Entomology* 19: 1852–1857.

Liu, X.Z., Chen, C.Y. and Wang, C.J. (1990). Selection of inducer of resistance to cotton Fusarium wilt and the optimal interval duration. *Acta Phytopathologica Sinica* 20: 123–126.

Martin, H. (1973). *The Scientific Principles of Crop Protection*. London, UK: Edward Arnold.

Maxwell, F.G. (1972). Host plant resistance to insects: Nutritional and pest management relationships. In Rodriguez, J.G. (Ed.), *Insect and Mite Nutrition*. Amsterdam, The Netherlands: North-Holland Publishing Company.

Maxwell, F.G. and Jennings, P.R. (Eds.). (1980). *Breeding Plants Resistant to Insects*. New York, USA: John Wiley & Sons.

Maxwell, F.G., Jenkins, J.N. and Parrot, W.L. (1972). Resistance of plants to insects. *Advances in Agronomy* 24: 187–265.

McIntyre, J. and Dodds, J.A. (1979). Induction of localized and systemic protection against *Phytophthora parasitica* var. *nicotianae* by tobacco mosaic virus of tobacco, hypersensitive to the virus. *Physiological Plant Pathology* 15: 321–330.

McMillian, W.W., Wiseman, B.R., Widstrom, N.W. and Harrell, E.A. (1972). Resistant sweet corn hybrid plus insecticide to reduce losses from corn earworms. *Journal of Economic Entomology* 65: 229–231.

McMutry, J.A. (1962). Resistance of alfalfa to spotted alfalfa aphid in relation to environmental factors. *Hilgardia* 32: 501–539.

Mett, V.L., Lochhead, L.P. and Reynolds, P.H.S. (1993). Copper-controllable gene expression system for whole plants. *Proceedings National Academy of Sciences USA* 90: 4567–4571.

Mihm, J.A. (1982). Techniques for mass rearing and infestation in screening for host plant resistance to corn earworm, *Heliothis zea*. In *Proceedings, International Workshop on Heliothis Management*, 15–20 November, 1981. Patancheru, Andhra Pradesh, India: International Crops Research Institute for the Semi-Arid Tropics, 255–266.

Mihm, J.A. (1985). Methods of artificial rearing with *Diatraea* species and evaluation of stem borer resistance in sorghum. In *Proceedings, International Sorghum Entomology Workshop*, 15–21 July, 1984. College Station, Texas, USA: Texas A & M University; Patancheru, Andhra Pradesh, India: International Crops Research Institute for the Semi-Arid Tropics, 169–174.

Mitra, R. and Bhatia, C.R. (1982). Bioenergetic considerations in breeding for insect and pathogen resistance in plants. *Euphytica* 31: 429–437.

Mooney, H.A., Gulmon, S.L. and Johnson, N.D. (1983). Physiological constraints on plant chemical defenses. In P.A. Hedin (Ed.), *Plant Resistance to Insects*. ACS Symposium Series 208. Washington, D.C., USA: American Chemical Society, 21–36.

Muid, B. (1977). *Host Plant Modification of Insecticide Resistance in* Myzus persicae. Ph.D. thesis, University of Reading, Reading, UK.

Nanthagopal, R. and Uthamasamy, S. (1989). Life tables for American bollworm, *Heliothis armigera* (Hübner) on four species of cotton under field conditions. *Insect Science and Its Application* 10: 521–530.

Newsom, L.D., Kogan, M., Miner, F.D., Rabb, R.L., Turnipseed, S.G. and Whitcomb, W.H. (1980). General accomplishments toward better pest control in soybean. In Huffaker, C.B. (Ed.), *New Technology of Pest Control*. New York: John Wiley & Sons, 51–98.

Nicol, D., Wratten, S.D., Eaton, N. and Copaja, S.V. (1993). Effects of DIMBOA levels in wheat on the susceptibility of the grain aphid *Sitobion avenae* to deltamethrin. *Annals of Applied Biology* 122: 427–433.

Nishijima, Y. (1960). Host plant preference of the soybean pod borer, *Grapholitha glycinivorella* Matsumura (Lep., Eucomidae). I. Oviposition site. *Entomologia Experimentalis et Applicata* 3: 38–47.

Ortiz, R., Raman, K.V., Iwanaga, M. and Palacios, M. (1990). Breeding for resistance to potato tuber moth, *Phthorimaea opercullela* (Zeller) in diploid potatoes. *Euphytica* 50: 119–125.

Padma Kumari, A.P., Sharma, H.C. and Reddy, D.D.R. (2000). Components of resistance to sorghum head bug, *Calocoris angustatus*. *Crop Protection* 19: 385–392.

Painter, R.H. (1951). *Insect Resistance in Crop Plants*. New York, USA: MacMillan.

Painter, R.H. (1958). Resistance of plants to insects. *Annual Review of Entomology* 3: 267–290.

Pair, S.D., Laster, M.L. and Martin, D.F. (1982). Parasitoids of *Heliothis* spp. (Lepidoptera: Noctuidae) larvae in Mississippi associated with sesame interplanting on cotton. 1971–1974: Implications of host-habitat interaction. *Environmental Entomology* 11: 509–512.

Panda, N. and Khush, G.S. (1995). *Host Plant Resistance to Insects*. Wallingford, Oxon, UK: CAB International.

Pathak, M.D. (1970). Genetics of plants in pest management. In Rabb, R.L. and Guthrie, F.E. (Eds.), *Concepts of Pest Management*. Raleigh, North Carolina, USA: North Carolina State University, 138–157.

Pathak, M.D., Beachell, H.M. and Andres, F. (1973). IR 20, a pest and disease-resistant high yielding rice. *International Rice Communication Newsletter* 92 (3): 1–8.

Pawar, C.S., Bhatnagar, V.S. and Jadhav, D.R. (1986). *Heliothis* species and their natural enemies, with their potential for biological control. *Proceedings of the Indian Academy of Science* (*Animal Science*) 95: 695–703.

Pelcz, J. and Wolffgang, H. (1986). Duration of induced resistance and susceptibility changes in the host-parasite combination barley/powdery mildew (*Hordeum vulgare* L./*Erysiphe graminis* f. sp. *hordei* March). *Archives of Phytopathology and Plant Protection, East Germany* 22: 459–464.

Plucknett, D.L., Smith, N.J.H., Williams, J.T. and Murthi, A.N. (1987). *Genebanks and the World's Food*. Princeton, New Jersey, USA: Princeton University Press.

Pluthero, F.G. and Singh, R.S. (1984). Insect behavioral responses to toxins: Practical and evolutionary considerations. *Canadian Entomologist* 116: 57–68.

Price, P.W., Button, C.E., Gross, P., McPheron, B.A., Thompson, J.N. and Weis, A.E. (1980). Interactions among three trophic levels: Influence of plants on interactions between insect herbivores and natural enemies. *Annual Review of Ecology and Systematics* 11: 41–65.

Purohit, M. and Deshpande, A.D. (1992). Effect of nitrogenous fertilizer application on cotton leafhopper, *Amrasca biguttula biguttula* (Ishida). In *National Seminar on Changing Scenario in Pests and Pest Management in India*, 31 January–1 February, 1992. Rajendranagar, Hyderabad, Andhra Pradesh, India: Plant Protection Association of India.

Rabb, R.L. and Bradley, J.R. (1968). The effect of host plants on parasitism of eggs of tobacco hornworm. *Journal of Economic Entomology* 61: 1249–1252.

Rao, C.N. and Prasad, V.D. (1996). Comparative population growth rates of *Helicoverpa armigera* (Hub.) on certain cultivars of cotton, *Gossypium hirsutum* L. *Annals of Plant Protection Sciences* 4: 138–141.

Reddy, K.S. and Narasimha Rao, D.V. (1975). Effect of nitrogen application on shoot fly incidence and grain maturity in sorghum. *Sorghum Newsletter* 18: 23–24.

Risch, S.J., Ow, D. and Altieri, M.A. (1983). Agroecosystem diversity and pest control: Data, tentative conclusions and new research directions. *Environmental Entomology* 12: 625–629.

Roberts, J.J., Gallun, R.L., Patterson, F.L. and Foster, J.E. (1979). Effects of wheat leaf pubescens on Hessian fly. *Journal of Economic Entomology* 72: 211–214.

Romeis, J. and Shanower, T.G. (1996). Arthropod natural enemies of *Helicoverpa armigera* (Hubner) (Lepidoptera: Noctuidae) in India. *BioControl Science and Technology* 6: 481–508.

Romeis, J., Shanower, T.G. and Zebitz, C.P.W. (1997). Plant volatile infochemicals mediate plant preference of *Trichogramma chilonis. Journal of Chemical Ecology* 23: 2455–2465.

Roots, R.B. (1973). Organization of a plant-arthropod association in simple and diverse habitats. The fauna of collards (*Brassica oleracea*). *Ecological Monograph* 43: 95–124.

Russell, E.P. (1989). Enemies hypothesis: A review of the effect of vegetational diversity on predatory insects and parasitoids. *Environmental Entomology* 18: 590–599.

Sams, D.W., Lauer, F.I. and Redcliffe, E.B. (1975). An excised leaflet test for evaluating resistance to green peach aphid in water bearing *Solanum* germplasm. *Journal of Economic Entomology* 68: 607–609.

Saxena, R.C. (1986). Biochemical basis of insect resistance in rice varieties. In Green, M.B. and Hedin, P.A. (Eds.), *Natural Resistance of Plants to Pests*. Washington, D.C.: American Chemical Society, 142–159.

Schultz, J.C. (1983). Impact of variable plant defensive chemistry on susceptibility of insects to natural enemies. In Hedin, P.A. (Ed.), *Plant Resistance to Insects*. ACS Symposium Series 208. Washington, D.C., USA: American Chemical Society, 37–54.

Schweissing, F.C. and Wilde, G. (1978). Temperature influence on greenbug resistance of crops in the seedling stage. *Environmental Entomology* 7: 831–834.

Schweissing, F.C. and Wilde, G. (1979). Temperature and plant nutrient effects on resistance of seedling sorghum to the greenbug. *Journal of Economic Entomology* 72: 20–23.

Shanower, T.G., Yoshida, M. and Peter, A.J. (1997). Survival, growth, fecundity and behavior of *Helicoverpa armigera* (Lepidoptera: Noctuidae) on pigeonpea and two wild *Cajanus* species. *Journal of Economic Entomology* 90: 837–841.

Sharma, H.C. (1985). Strategies for pest control in sorghum in India. *Tropical Pest Management* 31: 167–185.

Sharma, H.C. (1993). Host plant resistance to insects in sorghum and its role in integrated pest management. *Crop Protection* 12: 11–34.

Sharma, H.C. (1994). Effect of insecticide application and host plant resistance on parasitization of sorghum midge. *Contarinia sorghicola* Coq. *BioControl Science and Technology* 4: 53–60.

Sharma, H.C. (1996). Role of host resistance to insects in integrated pest management. In Upadhyaya, R.K., Mukerji, K.G. and Razak, R.L. (Eds.), *IPM System in Agriculture, vol 1. Principles and Perspectives*. New Delhi, India: Aditya Books Private Ltd., 317–333.

Sharma, H.C. (2001). Host plant resistance to sorghum midge, *Stenodiplosis sorghicola* (Coquillett): A sustainable approach for integrated pest management and environment conservation. *Journal of Eco-Physiology and Occupational Health* 1: 1–34.

Sharma, H.C. (2002). Host plant resistance to insects: Principles and practices. In Sarath Babu, B., Varaprasad, K.S., Anitha, K., Prasada Rao, R.D.V.J. and Chandurkar, P.S. (Eds.), *Resources Management in Plant Protection*, vol. 1. Rajendarnagar, Hyderabad, Andhra Pradesh, India: Plant Protection Association of India, 37–63.

Sharma, H.C. (Ed.). (2005). Heliothis/Helicoverpa *Management: Emerging Trends and Strategies for Future Research*. New Delhi, India: Oxford and IBH Publishers.

Sharma, H.C. and Agarwal, R.A. (1982a). Consumption and utilization of bolls of different cotton genotypes by larvae of *Earias vittella* F. and effect of gossypol and tannins on food utilization. *Zietschrift fur Angewandte Zoologie* 68: 13–38.

Sharma, H.C. and Agarwal, R.A. (1982b). Effect of some antibiotic compounds in *Gossypium* on the post-embryonic development of spotted bollworm (*Earias vittella* F.). *Entomologia Experimentalis et Applicata* 31: 225–228.

Sharma, H.C. and Agarwal, R.A. (1983a). Role of some chemical components and leaf hairs in varietal resistance in cotton to jassid, *Amrasca biguttula biguttula* Ishida. *Journal of Entomological Research* 7: 145–149.

Sharma, H.C. and Agarwal, R.A. (1983b). Ovipositional behavior of spotted bollworm, *Earias vittella* Fab. on some cotton genotypes. *Insect Science and Its Application* 4: 373–376.

Sharma, H.C. and Franzmann, B.A. (2000). Biology of the legume pod borer, *Maruca vitrata* and its damage to pigeonpea and adzuki bean. *Insect Science and Its Application* 20: 99–108.

Sharma, H.C. and Franzmann, B.A. (2001). Host plant preference and oviposition responses of the sorghum midge, *Stenodiplosis sorghicola* (Coquillett) (Dipt., Cecidomyiidae) towards wild relatives of sorghum. *Journal of Applied Entomology* 125: 109–114.

Sharma, H.C. and Lopez, V.F. (1993). Comparison of economic injury levels for sorghum head bug, *Calocoris angustatus* on resistant and susceptible genotypes at different stages of panicle development. *Crop Protection* 12: 259–266.

Sharma, H.C., Lopez, V.F. and Vidyasagar, P. (1994). Effect of panicle compactness and host plant resistance in sequential plantings on population increase of panicle feeding insects in *Sorghum bicolor* L. (Moench). *International Journal of Pest Management* 40: 216–221.

Sharma, H.C. and Ortiz, R. (2002). Host plant resistance to insects: An eco-friendly approach for pest management and environment conservation. *Journal of Environmental Biology* 23: 111–135.

Sharma, H.C. and Pampapathy, G. (2004). Effect of natural plant products, brassinolide, and host plant resistance in combination with insecticides on *Helicoverpa armigera* (Hubner) damage in pigeonpea. *Indian Journal of Plant Protection* 32: 40–44.

Sharma, H.C., Pampapathy, G. and Kumar, R. (2002). Technique to screen peanuts for resistance to the tobacco armyworm, *Spodoptera litura* (Lepidoptera: Noctuidae) under no-choice cage conditions. *Peanut Science* 29: 35–40.

Sharma, H.C., Pampapathy, G. and Kumar, R. (2005). Standardization of cage techniques to screen chickpeas for resistance to *Helicoverpa armigera* (Lepidoptera: Noctuidae) under greenhouse and field conditions. *Journal of Economic Entomology* 98: 210–216.

Sharma, H.C., Pampapathy, G. and Reddy, L.J. (2003). Wild relatives of pigeonpea as a source of resistance to the pod fly (*Melanagromyza obtusa* Malloch) and pod wasp (*Tanaostigmodes cajaninae* La Salle). *Genetic Resources and Crop Evolution* 50: 817–824.

Sharma, H.C., Pampapathy, G. and Wani, S.P. (2006). Influence of host plant resistance on biocontrol of *Helicoverpa armigera* in pigeonpea. *Indian Journal of Plant Protection* 34: 129–131.

Sharma, H.C., Saxena, K.B. and Bhagwat, V.R. (1999b). *Legume Pod Borer,* Maruca vitrata: *Bionomics and Management.* Information Bulletin no. 55. Patancheru, Andhra Pradesh, India: International Crops Research Institute for the Semi-Arid Tropics (ICRISAT).

Sharma, H.C., Singh, F. and Nwanze, K.F. (Eds.). (1997). *Plant Resistance to Insects in Sorghum.* Patancheru, Andhra Pradesh, India: International Crops Research Institute for the Semi-Arid Tropics (ICRISAT).

Sharma, H.C. and Sullivan, D.J. (2000). Screening for plant resistance to the Oriental armyworm, *Mythimna separata* (Lepidoptera: Noctuidae) in pearl millet, *Pennisetum glaucum. Journal of Agricultural and Urban Entomology* 17: 125–134.

Sharma, H.C. and Teetes, G.L. (1995). Yield loss assessments and economic injury levels for panicle feeding insects of sorghum. In *Proceedings, International Working Group Meeting on Panicle Feeding Pests of Sorghum and Pearl Millet,* 4–8 October, 1993. ICRISAT Sahelian Center, Niamey, Niger. Patancheru, Andhra Pradesh, India: International Crops Research Institute for the Semi-Arid Tropics, 125–134.

Sharma, H.C., Venkateswarulu, G. and Sharma, A. (2003). Environmental factors influence the expression of resistance to sorghum midge, *Stenodiplosis sorghicola. Euphytica* 130: 365–375.

Sharma, H.C., Vidyasagar, P. and Leuschner, K. (1988a). Field screening for resistance to sorghum midge (Diptera: Cecidomyiidae). *Journal of Economic Entomology* 81: 327–334.

Sharma, H.C., Vidyasagar, P. and Leuschner, K. (1988b). No-choice cage technique to screen for resistance to sorghum midge (Diptera: Cecidomyiidae). *Journal of Economic Entomology* 81: 415–422.

Sharma, H.C., Vidyasagar, P. and Nwanze, K.F. (1993). Effect of host-plant resistance on economic injury levels for the sorghum midge, *Contarinia sorghicola*. *International Journal of Pest Management* 39: 435–444.

Sharma, H.C., Agrawal, B.L., Abraham, C.V., Vidyasagar, P., Nwanze, K.F. and Stenhouse, J.W. (1994a). Registration of nine lines with resistance to sorghum midge: ICSV 692, ICSV 729, ICSV 730, ICSV 731, ICSV 736, ICSV 739, ICSV 744, ICSV 745, and ICSV 748. *Crop Science* 34: 1425–1426.

Sharma, H.C., Agrawal, B.L., Abraham, C.V., Stenhouse, J.W. and Aung, T. (2005a). Registration of sorghum varieties ICSV 735, ICSV 758, and ICSV 804 resistant to sorghum midge, *Stenodiplosis sorghicola*. *International Sorghum and Millets Newsletter* 46: 46–49.

Sharma, H.C. Ahmad, R., Ujagir, R., Yadav, R.P., Singh, R. and Ridsdill-Smith, T.J. (2005b). Host plant resistance to *Helicoverpa*: The prospects. In Sharma, H.C. (Ed.), *Helicoverpa/Heliothis Management: Emerging Trends and Strategies for Future Research*. New Delhi, India: Oxford and IBH Publishers, 171–213.

Sharma, H.C., Dhillon, M.K., Kibuka, J. and Mukuru, S.Z. (2005c). Plant defense responses to sorghum spotted stem borer, *Chilo partellus* under irrigated and drought conditions. *International Sorghum and Millets Newsletter* 46: 49–52.

Sharma, H.C., Mukuru, S.Z., Manyasa, E. and Were, J. (1999a). Breakdown of resistance to sorghum midge, *Stenodiplosis sorghicola*. *Euphytica* 109: 131–140.

Sharma, H.C., Pampapathy, G., Dhillon, M.K. and Ridsdill-Smith, T.J. (2005d). Detached leaf assay to screen for host plant resistance to *Helicoverpa armigera*. *Journal of Economic Entomology* 98: 568–576.

Sharma, H.C., Pampapathy, G., Dwivedi, S.L. and Reddy, L.J. (2003a). Mechanisms and diversity of resistance to insect pests in wild relatives of groundnut. *Journal of Economic Entomology* 96: 1886–1897.

Sharma, H.C., Pampapathy, G., Lanka, S.K. and Ridsdill-Smith, T.J. (2005e). Potential for exploitation of wild relative of chickpea, *Cicer reticulatum* for imparting resistance to *Helicoverpa armigera*. *Journal of Economic Entomology* 98: 2246–2253.

Sharma, H.C., Pampapathy, G., Lanka, S.K. and Ridsdill-Smith, T.J. (2005f). Antibiosis mechanism of resistance to legume pod borer, *Helicoverpa armigera* in wild relatives of chickpea. *Euphytica* 142: 107–117.

Sharma, H.C., Reddy, B.V.S., Dhillon, M.K., Venkateswaran, K., Singh, B.U., Pampapathy, G., Folkerstma, R., Hash, C.T. and Sharma, K.K. (2005g). Host plant resistance to insects in sorghum: Present status and need for future research. *International Sorghum and Millets Newsletter* 46: 36–43.

Sharma, H.C., Reddy, B.V.S., Stenhouse, J.W. and Nwanze, K.F. (1994b). Host plant resistance to sorghum midge, *Contarinia sorghicola*. *International Sorghum and Millets Newsletter* 35: 30–41.

Sharma, H.C., Singh, B.U., Hariprasad, K.V. and Bramel-Cox, P.J. (1999b). Host plant resistance to insects in integrated pest management for a safer environment. *Proceedings, Academy of Environmental Biology* 8: 113–136.

Sharma, H.C., Stevenson, P.C., Simmonds, M.S.J and Green, P.W.C. (2001). *Identification of* Helicoverpa armigera *(Hübner) Feeding Stimulants and the Location of their Production on the Pod-Surface of Pigeonpea* [Cajanus cajan *(L.) Millsp*.]. Final Technical Report. DfiD Competitive Research Facility Project [R 7029 (C)]. Patancheru, Andhra Pradesh, India: International Crops Research Institute for the Semi-Arid Tropics (ICRISAT).

Sharma, H.C., Sullivan, D.J., Sharma, M.M. and Shetty, S.V.R. (2004). Influence of weeding regimes and pearl millet genotypes on parasitism of the Oriental armyworm, *Mythimna separata*. *BioControl* 49: 689–699.

Sharma, H.C., Taneja, S.L., Leuschner, K. and Nwanze, K.F. (1992). *Techniques to Screen Sorghums for Resistance to Insects*. Information Bulletin no. 32. Patancheru, Andhra Pradesh, India: International Crops Research Institute for the Semi-Arid Tropics (ICRISAT).

Sharma, H.C., Taneja, S.L., Kameswara Rao, N. and Prasada Rao, K.E. (2003b). *Evaluation of Sorghum Germplasm for Resistance to Insect Pests*. Information Bulletin no. 63. Patancheru, Andhra Pradesh, India: International Crops Research Institute for the Semi-Arid Tropics (CRISAT).

Simms, E.L. (1992). Costs of plant resistance to herbivory. In Fritz, R.S. and Simms, E.L. (Eds.), *Plant Resistance to Herbivores and Pathogens: Ecology, Evolution and Genetics*. Chicago: University of Chicago Press, 392–425.

Sinden, S.C., Sanford, L.C., Cantelo, W.W. and Deahl, K.L. (1986). Leptine glycoalkaloids and resistance to the Colorado potato beetle (Coleoptera: Chrysomelidae) in *Solanum chacoense*. *Environmental Entomology* 15: 1057–1062.

Singh, B., Singh, B., Jalota, S., Zalucki, M.P. and Dilwari, V.K. (2005). Role of water management on incidence of *Helicoverpa armigera* (Hubner) in upland cotton. *Indian Journal of Ecology* 32: 24–28.

Singh, K.B., Ocampo, B. and Robertson, L.D. (1998). Diversity for abiotic and biotic stress resistance in the wild annual *Cicer* species. *Genetic Resources and Crop Evolution* 45: 9–17.

Singh, P. and Narayanan, S.S. (1994). Breeding for resistance to biotic stress in cotton. *Journal of Cotton Research and Development* 8: 48–66.

Smith, C.M. (1989). *Plant Resistance to Insects*. New York: John Wiley & Sons.

Smith, C.M. (2005). *Plant Resistance to Arthropods: Molecular and Conventional Approaches*. Dordrecht, The Netherlands: Springer Verlag.

Smith, C.M., Khan, Z.R. and Pathak, M.D. (1994). *Techniques for Evaluating Insect Resistance in Crop Plants*. Boca Raton, Florida, USA: CRC Press.

Soman, P., Nwanze K.F., Laryea, K.B., Butler, D.R. and Reddy, Y.V.R. (1994). Leaf surface wetness in sorghum and resistance to shoot fly, *Atherigona soccata*: role of soil and plant water potentials. *Annals of Applied Biology* 124: 97–108.

Southwood, T.R.E. (1975). The dynamics of insect populations. In Pimentel, D. (Ed.), *Insects, Science, and Society*. New York, USA: Academic Press, 151–199.

Srivastava, C.P. and Srivastava, R.P. (1990). Antibiosis in chickpea (*Cicer arietinum*) to gram pod borer, *Heliothis armigera* (Hubner) Noctuidae: Lepidoptera in India. *Entomon* 15: 89–94.

Stadler, E. and Buser, H.R. (1984). Defense chemicals in the leaf surface wax synergistically stimulate oviposition by phytophagous insects. *Experientia* 40: 1157–1159.

Stalker H.T. (1980). Utilization of wild species for crop improvement. *Advances in Agronomy* 23: 111–147.

Starks, K.J. and Burton, R.L. (1977). Greenbugs: A comparison of mobility on resistant and susceptible varieties of four small grains. *Environmental Entomology* 6: 331–332.

Starks, K.J. and Merkle, O.G. (1977). Low level resistance in wheat to greenbug. *Journal of Economic Entomology* 70: 305–306.

Starks, K.J., Muniappan, R. and Eikenbary, R.D. (1972). Interaction between plant resistance and parasitism against the greenbug on barley and sorghum. *Annals of Entomological Society of America* 65: 650–655.

Stipanovic, R.D. (1983). Function and chemistry of plant trichomes and glands in insect resistance: Protective chemicals in plant epidermal glands and appendages. In Hedin, P.A. (Ed.), *Plant Resistance to Insects*. ACS Symposium Series 208. Washington, D.C., USA: American Chemical Society, 69–100.

Stout, M.J., Workman, J. and Duffey, S.S. (1996). Differential induction of tomato foliar proteins by arthropod herbivores. *Journal of Chemical Ecology* 20: 2575–2594.

Swarup, P. (1987). Insect pest management in maize. In Veerbhadra Rao, M. and Sithanantham, S. (Eds.), *Plant Protection in Field Crops*. Rajendranagar, Hyderabad, Andhra Pradesh, India: Plant Protection Association of India, 105–112.

Sundramurthy, V.T. and Chitra, K.L. (1992). Integrated pest management in cotton. *Indian Journal of Plant Protection* 20: 1–17.

Taksdal, G. (1993). Resistance in swedes to the turnip root fly and its relation to integrated pest management. *Bulletin of OILB/SROP* 16: 13–20.

Thomas, J.G., Sorensen, E.L. and Painter, R.H. (1966). Attached vs. excised trifoliates for evaluation of resistance in alfalfa to the spotted alfalfa aphid. *Journal of Economic Entomology* 59: 444–448.

Tingey, W.M. and Singh, S.R. (1980). Environmental factors affecting the magnitude and expression of resistance. In Maxwell, F.G. and Jennings, P.R. (Eds.), *Breeding Plants Resistant to Insects*. New York, USA: John Wiley & Sons, 87–113.

Treacy, M.F., Zummo, G.R. and Benedict, J.H. (1985). Interactions of host-plant resistance in cotton with predators and parasites. *Agriculture, Ecosystems, and the Environment* 13: 151–157.

Uknes, S., Dincher, S., Friedrich, L., Negrotto, D., Williams, S., Thompson-Taylor, H., Potter, S., Ward, E. and Ryals, J. (1993). Regulation of pathogenesis-related protein-1a gene expression in tobacco. *Plant Cell* 5: 159–169.

Underwood, N.C. (1998). The timing of induced resistance and induced susceptibility in the soybean-Mexican bean beetle system. *Oecologia* 114: 376–381.

van Emden, H.F. (1965). The role of uncultivated land in the biology of crop pests and beneficial insects. *Science of Horticulture* 17: 121–136.

van Emden, H.F. (1978). Insects and secondary plant substances, an alternative viewpoint with special reference to aphids. In Harborne, J.B. (Ed.), *Biochemical Aspects on Plant and Animal Coevolution*. London, UK: Academic Press, 310–323.

van Emden, H.F. (1990). The interaction of host plant resistance with other control measures. *Proceedings, Brighton Crop Protection Conference* 3: 939–949.

van Emden, H.F. (1991). The role of host plant resistance in insect pest mismanagement. *Bulletin of Entomological Research* 81: 123–126.

van Emden, H.F. and Wearing, C.H. (1965). The role of the aphid host plant in delaying economic damage levels in crops. *Annual Review of Entomology* 14: 197–270.

van Lenteren, J.C. (1991). Biological control in a tritrophic system approach. In Peters, D.C. and Webster, J.A. (Eds.), *Aphid-Plant Interactions: Populations to Molecules*. Stillwater, Oklahoma, USA: Oklahoma State University Press, 3–28.

van Poecke, R.M.P. and Dicke, M. (2002). Induced parasitoid attraction by *Arabidopsis thaliana*: Involvement of the octadecanoid and the salicylic acid pathway. *Journal of Experimental Botany* 53: 1793–1799.

Varis, A.L. (1958). On the susceptibility of the different varieties of big-leafed turnip to damage caused by cabbage maggots (*Hylemyia* sp.). *Journal of Science, Agriculture Society of Finland* 30: 271–275.

Venkateswaran, K. (2003). *Diversity Analysis and Identification of Sources of Resistance to Downy Mildew, Shoot Fly and Stem Borer in Wild Sorghums*. Ph.D. thesis, Department of Genetics, Osmania University, Hyderabad, Andhra Pradesh, India.

Vinson, S.B. and Barbosa, P. (1987). Interrelationships of nutritional ecology of parasitoids. In Slansky Jr., F. and Rodriguez, J.G. (Eds.), *Nutritional Ecology of Insects, Mites, Spiders and Related Invertebrates*. New York, USA: John Wiley & Sons, 673–695.

Walker Jr., J.K. and Niles, G.A. (1971). Population dynamics of the boll weevil and modified cotton types: Implication for pest management. *Texas Agriculture Experimentation Station Bulletin* 1109.

Wang, Y., DeMayo, F.J., Tsai, S.Y. and O'Mally, B.W. (1997). Ligand-induced and liver expression in trans mice. *Nature Biotechnology* 15:239–243.

Ward, E.R., Uknes, S.J., Williams, S.C., Dincher, S.S., Wiederhold, D.L., Alexander, D.C., Ahl-Goy, P., Metraux, J.P. and Ryals, J.A. (1991). Coordinate gene activity in response to agents that induce systemic acquired resistance. *Plant Cell* 3: 1085–1094.

Way, M.J. and Murdie, G. (1965). An example of varietal variations in resistance of brussels sprouts. *Annals of Applied Biology* 56: 326–328.

Wightman, J.A., Anders, M.M., Rameshwar Rao, V. and Mohan Reddy, L. (1995). Management of *Helicoverpa armigera* (Lepidoptera: Noctuidae) in chickpea in southern India: Thresholds and economics of host plant resistance and insecticide application. *Crop Protection* 14: 37–46.

Wightman, J.A., Dick, K.M., Ranga Rao, G.V., Shanower, T.G. and Gold, C.G. (1990). Pests of groundnut in the semi-arid tropics. In Singh, S.R. (Ed.), *Insect Pests of Legumes*. New York, USA: Longman and Sons, 243–322.

Wilhoit, L.R. (1991). Modeling the population dynamics of different aphid genotypes in plant variety mixtures. *Ecological Modeling* 55: 257–283.

Williams, H.J., Elzen, G.W. and Vinson, S.B. (1988). Parasitoid-host-plant interactions, emphasizing cotton (*Gossypium*). In Barbosa, P. and Letourneau, D.K. (Eds.), *Novel Aspects of Insect-Plant-Interactions*. New York, USA: John Wiley & Sons, 171–200.

Wilson, F.D. and Shaver, T.N. (1973). Glands, gossypol content, and tobacco budworm development in seedlings and floral parts of cotton. *Crop Science* 13: 107–110.

Wing, K.D., Slawecki, P. and Carlson, G.R. (1988). RH 5849, a non-steroidal ecdysone agonist: Effects on larval Lepidoptera. *Science* 241: 470–472.

Wiseman, B.R., Harrell, E.A. and McMillian, W.W. (1973). Continuation of tests of resistant sweet corn hybrid plus insecticides to reduce losses from corn earworm. *Environmental Entomology* 2: 919–920.

Wu, L.F., Cai, Q.N. and Zhang, Q.W. (1997). The resistance of cotton lines with different morphological characteristics and their F_1 hybrids to cotton bollworm. *Acta Entomologica Sinica* 40: 103–109.

Wyatt, I.J. (1970). The distribution of *Myzus persicae* (Sulz.) on year round chrysanthemums. II. Winter season: The effect of parasitism by *Aphidius matricariae* Hal. *Annals of Applied Biology* 65: 41–42.

Xiao, J., Grandillo, S., Ahn, S.N.K., McCouch, S.R., Tanksley, S.D., Li, J. and Yuan, L. (1996). Genes from wild rice improve yield. *Nature* 384: 223–224.

5

Mechanisms and Inheritance of Resistance to Insect Pests

Introduction

Evaluation of germplasm collections have resulted in identification of several sources of resistance to insect pests in different crops (Panda and Khush, 1995; H.C. Sharma and Ortiz, 2002; Smith, 2005). However, screening of thousands of germplasm accessions probably has resulted in missing many germplasm accessions with moderate levels of resistance, but with different genes for insect resistance (Clement and Quisenberry, 1999). The identified sources of resistance have not been used widely because of low heritability or linkage drag. Varieties with resistance to insect pests have been identified and released for cultivation in different crops (Panda and Khush, 1995). However, the levels of resistance in most of the varieties released for cultivation are low to moderate. Therefore, there is a need to increase the levels and diversify the basis of resistance through exploitation of resistance sources in cultivated germplasm and wild relatives of crops with different mechanisms of resistance. The progress in developing crop cultivars with resistance to insects has been quite slow because of lack of information on the mechanisms that contribute to insect resistance, the numbers of genes involved, and the nature of gene action (Smith, 2005). Lack of such information reduces the efficiency of breeding for insect resistance and confounds the development of effective marker-assisted selection systems. There is a need to understand the mechanisms and inheritance of resistance to insects to identify molecular markers associated with different mechanisms of resistance. Such an information will also be useful to plan appropriate strategies for marker-assisted introgression of insect resistant genes into high yielding cultivars, understand the nature of gene action, number of genes involved, and pyramiding of resistance genes to develop cultivars with stable and durable resistance to insect pests.

Mechanisms of Resistance to Insects

Knowledge of insect-plant relationships and the mechanisms that contribute to insect resistance is critical for developing germplasm with high yield and durable resistance (H.C. Sharma, 1994). In view of limited success in the past in developing crop cultivars with resistance to insect pests by using known sources of resistance, there is a need to identify genotypes with diverse mechanisms (genes), and pyramid the resistance genes to increase the levels and diversify the bases of resistance. Information on different mechanisms of resistance to insects, such as antixenosis or nonpreference, antibiosis, and tolerance is discussed below.

Antixenosis

Antixenosis or nonpreference is expressed in terms of unsuitability of the host plant for oviposition or feeding. The word antixenosis is derived from the Greek, *xeno*, meaning "guest." It describes the inability of a plant to serve as a host to a herbivore insect, and forces the insect to change its host plant for feeding and oviposition. Antixenotic (nonpreference) resistance reduces the rate of both initial and successive insect population buildup. Antixenotic resistance may also shift the insect population to other fields of the same crop or to other host plants of the insect. It is due to physico-chemical characteristics of the host plant that affect insect behavior adversely, resulting in selection of an alternative host plant. Absence of physicochemical stimuli that are involved in selection of the host plant or presence of repellents, deterrents, and antifeedants contribute to the antixenosis mechanism of resistance. Sensory cues that mediate host selection for oviposition include visual, tactile, and chemical stimuli (Thompson and Pellmyr, 1991). Antixenosis is an important component of resistance to the Hessian fly, *Mayetiola destructor* (Say) in wheat (Roberts et al., 1979), the onion fly, *Hylemyia antiqua* (Meigen) in onion, the carrot fly, *Psila rosae* (Fab.) in carrot (Stadler and Buser, 1984), and the Asiatic stem borer, *Chilo suppressalis* (Walker) in rice (Saxena, 1986). Antixenosis for oviposition is a major component of resistance to the shoot fly, *Atherigona soccata* (Rondani) and midge, *Stenodiplosis sorghicola* (Coquillett) in sorghum (Figure 5.1) (H.C. Sharma and Vidyasagar, 1994; H.C. Sharma and Nwanze, 1997; H.C. Sharma, Franzmann, and Henzell, 2002). Cotton genotypes with red and okra leaf

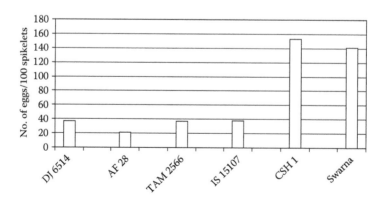

FIGURE 5.1 Oviposition by the females of sorghum midge, *Stenodiplosis sorghicola*, on different sorghum genotypes under no-choice conditions in the headcage.

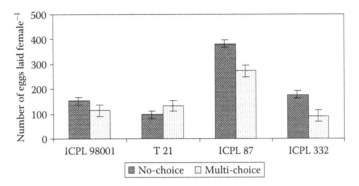

FIGURE 5.2 Oviposition preference by the *Helicoverpa armigera* females on pigeonpea under no-choice and multichoice conditions.

and frego-bract are nonpreferred by the cotton bollworm, *Helicoverpa armigera* (Hubner) for egg laying than the genotypes with normal leaves (Wu, Cai, and Zhang, 1997). Oviposition nonpreference is also a component of resistance to *H. armigera* in chickpea (Cowgill and Lateef, 1996) and pigeonpea (Figure 5.2) (H.C. Sharma et al., 2001; Kumari, Sharma, and Reddy, 2005), and to *Helicoverpa zea* (Boddie) in tomato (Cosenza and Green, 1979). Both antixenosis and antibiotic types of resistance have been observed against the pea weevil, *Bruchus pisorum* L. (Clement et al., 1994). Nonpreference for feeding and nutritional antibiosis are major components of resistance in soybean to *Epilachna varivestis* Mulsant (Kogan, 1982).

Antibiosis

Antibiosis is expressed in terms of larval mortality, decreased larval and pupal weights, prolonged larval and pupal development, and reduced fecundity. Antibiosis effects are also expressed in terms of sex ratio and proportion of insects entering diapause. The antibiotic type of resistance prolongs the generation time, and can reduce the insect population within a few generations through a cumulative effect on insect survival and development (Coaker, 1959; Knipling, 1979; Jayaraj, 1982; H.C. Sharma, 1993). Both chemical and morphological factors mediate antibiosis (Panda and Khush, 1995; H.C. Sharma and Nwanze, 1997; Smith, 2005). Lethal effects may be acute, often affecting young larvae, while chronic effects lead to mortality of older larvae, pupae, and adults (H.C. Sharma, Vidyasagar, and Subramanian, 1993; H.C. Sharma and Nwanze, 1997; H.C. Sharma et al., 2001; Smith, 2005; Stevenson et al., 2005). Individuals surviving the direct effects of antibiosis may have reduced body weight, a prolonged period of development, and reduced fecundity. Antibiosis may also include subchronic effects, such as slow growth rates, poor utilization of food, and reduced fecundity (Reese and Beck, 1976).

Antibiosis is an important component of resistance to *H. armigera* in cotton. Cotton genotypes Suvin (*Gossypium barbandense* L.) and MCU 9 (*G. hirsutum* L.) are more suitable than TKHe 44 (*G. arboreum* L.) for the growth and development of *H. armigera* (Nanthagopal and Uthamasamy, 1989). Net reproductive rate, intrinsic rate of increase, and fecundity have been found to be relatively lower on cotton genotypes LK 861 and LPS 141 as compared to MCU 5 (Rao and Prasad, 1996). There is a large variation in larval survival, larval and pupal weights, egg viability, and longevity of *H. armigera* adults when reared on different chickpea genotypes (Srivastava and Srivastava, 1990). Life table analysis suggested that there is considerable variation in net reproductive rate (142.1 to 268.6), mean generation

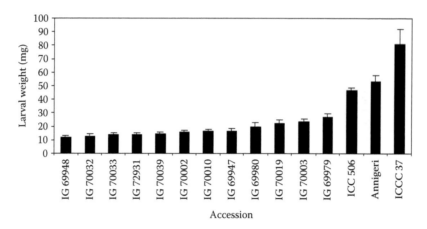

FIGURE 5.3 Expression of antibiosis to *Helicoverpa armigera* larvae in wild relatives of chickpea (larval weights at 10 days after releasing the larvae on the leaves of different wild relatives of chickpea). ICC 506, Annigeri, and ICCC 37 are the resistant, commercial, and susceptible checks, respectively, of the cultivated chickpea.

time (39.1 to 45.2 days), intrinsic rate of increase (0.12 to 0.14), finite rate of increase (1.13 to 1.15), and multiplication rate per week (2.57 to 3.02) on different genotypes (R.P. Sharma and Yadav, 2000).

Antibiosis is one of the important components of resistance to *H. armigera* in wild relatives of chickpea (Figure 5.3) (H.C. Sharma et al., 2005b). The fecundity of the females emerging from insects reared on resistant genotypes of chickpea is also affected adversely (Figure 5.4). Larval mortality and prolongation of the larval period are the main components of resistance to *H. armigera* in wild relatives of pigeonpea (Shanower, Yoshida, and Peter, 1997). Larval and pupal weights and developmental period are all adversely affected when *H. armigera* larvae were fed on the flowers of wild relatives of pigeonpea, such as *Cajanus cajanifolius* (Hains) Maesen and *C. sericeus* (F. Muell. ex-Benth.) F. Muell.; and only few larvae survived to maturity (Dodia et al., 1996). A significant reduction in fecundity has also been observed when the larvae were reared on resistant varieties.

In groundnut, larvae of *H. zea* fed on PI 229358 showed a significant reduction in larval growth (Beland and Hatchett, 1976). Some of the groundnut accessions suffer heavy leaf

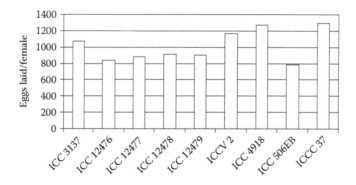

FIGURE 5.4 Fecundity of *Helicoverpa armigera* females emerging from insects reared on different chickpea genotypes.

feeding by *Spodoptera litura* (F.), but result in slow growth of the larvae because of poor nutritional quality of the food or the presence of secondary plant substances (Stevenson et al., 1993). Larval mortality is the principal component of resistance to *H. armigera* in wild tomato, *Lycopersicon hirsutum* f. sp. *glabratum* Mull. (Tingey and Sinden, 1982; Kashyap and Verma, 1987). Antibiosis is also an important component of resistance to European corn borer, *Ostrinia nubilalis* (Hubner) and corn earworm, *H. zea* in maize (Klun, Tipton, and Brindley, 1967), and Colorado potato beetle, *Leptinotarsa decemlineata* (Say) (Sinden et al., 1986) and potato tuber moth, *Phthorimaea operculella* (Zeller) in potato. The larvae of cabbage looper, *Trichoplusia ni* (Hubner) are more efficient in food utilization on the soybean susceptible genotype, Davis, than on the resistant genotype, PI 227687 (H.C. Sharma and Norris, 1993).

Antibiosis expressed in terms of larval mortality, slow growth, and delayed development is one of the components of resistance to spotted stem borer, *Chilo partellus* (Swinhoe) (V.K. Kumar, Sharma, and Reddy, 2005), and the sorghum midge, *S. sorghicola* (H.C. Sharma, Vidyasagar, and Subramanian, 1993; H.C. Sharma, Franzmann, and Henzell, 2002) in sorghum. Antibiosis is also expressed in terms of reduced fecundity in the sorghum head bug, *Calocoris angustatus* (Lethiery) (H.C. Sharma, Lopez, and Nwanze, 1993).

Tolerance

Tolerance or recovery resistance enables a plant to withstand damage from an insect population that is injurious to another cultivar without a tolerance mechanism of resistance. Expression of tolerance is determined by the inherent capability of a genotype to overcome an insect infestation or to recover from insect damage and/or add new plant growth after insect damage. Plants with an ability to tolerate insect damage at times may produce more yield than the plants of a nontolerant susceptible cultivar at the same level of insect infestation. Tolerance often occurs in combination with antixenosis and antibiosis components of resistance. Production of tillers in sorghum following damage to the main plant by sorghum shoot fly, *A. soccata*; and stem borer, *C. partellu* (H.C. Sharma and Nwanze, 1997), serves as a component of recovery resistance (Figure 5.5). Fertility status and moisture availability in the soil influence tiller production in plants damaged by shoot fly and stem borers in sorghum (H.C. Sharma, 1993). Increase in grain mass in panicles partially damaged by sorghum midge, *S. sorghicola* (H.C. Sharma, 1997; H.C. Sharma, Abraham, and Stenhouse, 2002), and less grain damage per unit number of head bugs, *C. angustatus* (H.C. Sharma and Lopez, 1990, 1993; Padma Kumari, Sharma, and Reddy, 2000) serve as a component of recovery resistance to insect feeding and damage in sorghum. Temperature (Schweissing and Wilde, 1978) and nutrients (Schweissing and Wilde, 1979) affect the tolerance of sorghum seedlings to damage by greenbug, *Schizaphis graminum* (Rondani). Tolerance to insect damage has also been observed in alfalfa to weevil, *Hypera postica* (Gyllen.) (Dogger and Hanson, 1963), in maize to corn earworm, *H. zea* (Wiseman, McMillian, and Widstrom, 1972), in rice to brown planthopper, *Nilaparvata lugens* (Stal) (Panda and Heinrichs, 1983; Heinrichs et al., 1984), in wheat to greenbug, *S. graminum* (Starks and Merkle, 1977), in muskmelon to the aphid, *Aphis gossypii* Glover (Kennedy, Kishaba, and Bohn, 1975), and in turnip to cabbage maggot, *Hylemyia* sp. (Varis, 1958).

Tolerance is an important component of resistance to *H. armigera* in cotton (Balasubramanian, Gopalan, and Subramanian, 1977; Murthy et al., 1998). It is expressed in terms of rejuvenation potential, healthy leaf growth, flowering compensation potential, and plant vigor. The cotton genotype JK 276-4, possessing quick rejuvenation and higher fruiting efficiency, suffers relatively less yield loss by *H. armigera* (Murthy et al., 1998).

FIGURE 5.5 Production of auxiliary tillers in sorghum as a result of damage to the main shoot by the sorghum shoot fly, *Ahterigona soccata*, which serves as a mechanism of recovery resistance.

The ability to recover from *H. armigera* damage is also an important component of resistance in chickpea and pigeonpea (Figure 5.6) (Lateef, 1985; Srivastava and Srivastava, 1989; Lateef and Sachan, 1990; H.C. Sharma et al., 2005a). Effects of tolerance are cumulative as a result of inter- and intraplant growth compensation, mechanical strength of tissues and organs, and growth regulation and partitioning (Tingey, 1981). Plants with a tolerance mechanism of resistance have a great value in pest management as such plants also prevent evolution of new insect biotypes, and help in maintaining the populations of the natural enemies.

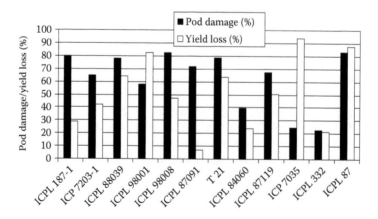

FIGURE 5.6 Recovery resistance to *Helicoverpa armigera* in pigeonpea. The genotypes suffering high pod damage and low yield loss (ICPL 187-1, ICP 7203-1, ICPL 87091, and ICPL 84060 have good recovery resistance to damage by pod borer).

Escape

Avoidance of insect damage through early or late flowering during periods of low insect abundance is another mechanism of resistance, called pseudo-resistance (Kogan, 1982; Smith, 2005). It is not a true mechanism of resistance, but serves as an effective means of avoiding pest damage. Many of the landrace cultivars have evolved such that they flower during the periods of low insect activity. This mechanism can be exploited to minimize insect damage, depending on the length of the cropping season, and the cultivars grown.

Breeding for Resistance to Insect Pests

Several methods have been used to transfer insect resistance genes into high-yielding cultivars (Panda and Khush, 1995; Smith 2005). Mass selection, pure line selection, recurrent selection, and backcross methods have been used to incorporate insect resistance genes into the crop plants. In self-pollinated crops, backcross, bulk, and pedigree breeding have been used to incorporate resistance genes into agronomically desirable cultivars, while mass selection and recurrent selection methods have been used in cross pollinated crops.

Mass Selection

Mass selection involves selecting individual insect-resistant plants in a population in each generation (cycle of selection), combining the seed of the resistant plants, and growing progenies in the following generation. It allows selection of several insect-resistant plants in each cycle of selection. Maximum improvement for resistance is achieved during initial selection, followed by two to five additional cycles of selection. Mass selection has been used effectively to breed for resistance to the potato leafhopper, *Empoasca fabae* (Harris) (Sanford and Ladd, 1983) and to identify genotypes with resistance to sugarbeet root maggot, *Tetanops myopaeformis* Roder (Theurer et al., 1982). Pure line selection is another form of mass selection that involves the selection of individual resistant plants that are advanced separately. In each cycle of selection, resistant selections are selfed and advanced. In cross-pollinated crops, individual selections are interplanted in a later selection cycle to form a composite of all the insect-resistant plants/lines.

Recurrent Selection

Recurrent selection has been used to concentrate insect resistance genes from diverse sources into the same background through several cycles of intermating and selection. In each cycle, resistant plants are selected among the progeny produced through a previous mating of resistant individuals. Recurrent selection allows introduction of resistance to several insects and diseases from different sources in later selection cycles. Recurrent selection has been used to increase the levels of resistance in maize to European corn borer *O. nubilalis* (Russell, Lawrence, and Guthrie, 1979; Klenke, Russell, and Guthrie, 1986). The corn borer damage decreased significantly after four cycles of selection. Variability for resistance to *O. nubilalis* also decreased in each cycle of selection. Recurrent selection has also been used to increase the level of resistance in maize populations to corn earworm, *H. zea*, and fall armyworm, *Spodoptera frugiperda* (J.E. Smith) (Widstrom, 1989; Widstrom et al.,

1992; Butron et al., 2002), and to the maize weevil, *Sitophilus zeamais* Motso. (Widstrom, 1989). Recurrent selection has also been used to breed for resistance to sorghum shoot fly, *A. soccata*, and sorghum midge, *S. sorghicola* (Agrawal and Abraham, 1985; H.C. Sharma et al., 2005c). In potato, recurrent selection has been used to improve resistance to *E. fabae*. Seven cycles of selection resulted in a major reduction in *E. fabae* damage (Sanford and Ladd, 1983). Both recurrent selection and pedigree breeding have been used to breed for resistance to *E. fabae* in alfalfa (Elden and Elgin, 1987). The level of resistance in red clover, *Trifolium pratense* L. to *Acyrthosiphon pisum* (Harris) and yellow clover aphid, *Therioaphis trifolii* (Monell) has also been increased by recurrent selection (Gorz, Manglitz, and Haskins, 1979). Five cycles of selection for *T. trifolii* and three cycles of selection for *A. pisum* have been used to develop the synthetic cultivar N-2. Recurrent selection is useful for gene pyramiding and for developing cultivars with multiple resistance to several insect pests and diseases.

Pedigree Breeding

Pedigree breeding involves selection of individual plants in segregating populations on the basis of insect resistance with desirable agronomic plant types. The best F_2 plants are selected, and planted as F_3 families. In the F_3 generation, selection for resistance is made within the family. Selected F_4 resistant families are planted, and evaluated for resistance. In the F_5 and F_6 generations, the material is subjected to more rigorous screening in replicated trials, and yield tests are also conducted to eliminate the resistant families with poor agronomic desirability. In the later generations, selections are made for families with insect resistance, high yield, and other agronomic characters. The advantage of pedigree breeding is that it eliminates unacceptable plant material early in the breeding program, allowing detailed evaluation of selected lines over a period of time. The major disadvantages are that its use is limited to self-pollinated crops, and that only a limited number of lines can be processed due to extra time required for planting, harvesting, and data collection. Pedigree breeding has been used for increasing the levels of resistance in rice to green leafhopper, *Nephotettix virescens* (Distant), the brown plant hopper *N. lugens*, and the rice gall midge, *Orseolia oryzae* Wood-Mason (Khush, 1980). Pedigree breeding has been used to incorporate resistance to shoot fly, *A. Soccata*, midge, *S. sorghicola*, and spotted stem borer, *C. partellus* into elite lines in sorghum (H.C. Sharma, Singh and Nwanze, 1997), and to *H. armigera* in chickpea and pigeonpea (Dua et al., 2005). The bulk breeding method is also used to incorporate arthropod resistance into self-pollinated crops. Bulk breeding is similar to the pedigree breeding method, but selection normally does not occur until the F_5 generation.

Backcross Breeding

It involves the use of recurring backcrosses to one of the parents (recurrent parent) of a hybrid, accompanied by selection for resistance to the target insect species. The nonrecurrent parent is a source with a higher level of resistance than that used in the previous backcross. Backcross breeding can be used as a rapid way to incorporate insect resistance into agronomically desirable cultivars that are susceptible to insects. After each cross, selections are made for agronomically desirable insect-resistant plants. Several high-yielding cultivars of rice and soybean with insect resistance have been developed using backcross breeding (Khush and Brar, 1991; Smith, 2005). Backcross breeding has not been found to be useful in breeding maize for resistance to insects. Reciprocal translocation studies have shown that at least 12 genes are involved in European corn borer, *O. nubilalis* resistance in maize (G.E. Scott, Dicke, and Pesho, 1966; Onukogu et al., 1978). For this reason, maize hybrid development has been conducted using other breeding methods.

Development of F₁ Hybrids Using Cytoplasmic Male Sterility

Male-sterility results from the inability of plants to produce functional pollen due to reproductive deficiency in hermaphrodite flowers. Cytoplasmic male-sterility occurs due to the mutation of mitochondria or some other cytoplasmic factors outside the nucleus, which results in the transformation of fertile cytoplasm into sterile cytoplasm. The male-sterile line is maintained by pollinating it with pollen from the maintainer line, which differs from the A line only for male-sterility. The F_1 hybrids for cultivation are produced by crossing the male-sterile line with a pollinator that carries the fertility restoration genes. A pollinator line that results in maximum heterosis for grain yield and other traits of interest is used for hybrid production. Cytoplasmic male-sterility has been exploited in several crops to develop hybrids for increasing crop productivity, particularly in cereal crops such as rice, maize, sorghum, and pearl millet. However, large-scale cultivation of hybrids based on a single source of male sterility may pose a serious challenge to sustainable crop production because of decreased genetic diversity and increased susceptibility to insect pests. Therefore, there is a need to develop male-sterile and restorer lines with resistance to insects to develop insect-resistant hybrids for cultivation by the farmers. To produce insect-resistant hybrids involving a cytoplasmic male-sterility system, it is important to transfer insect resistance genes into both male-sterile and restorer lines to produce hybrids with resistance to insects. Sorghum hybrids with resistance to shoot fly and sorghum midge have been developed based on cytoplasmic male sterility (H.C. Sharma et al., 1996, 2005c; Dhillon et al., 2006a; H.C. Sharma, Dhillon, and Reddy, 2006). Hybrids with resistance to sorghum midge have been widely deployed in Australia for controlling sorghum midge, *S. sorghicola* (Henzell et al., 1997).

Genetic Basis of Resistance

Information on mechanisms and inheritance of resistance to insects can be utilized in selecting parents with diverse mechanisms or with different genes for resistance, selection of appropriate breeding methodology (pedigree, backcross, or population improvement) depending on the number of genes involved and nature of gene action, and developing isolines and mapping populations for resistance to insects. Resistance to insects may be oligogenic, polygenic, or cytoplasmic.

Oligogenic Resistance

Oligogenic resistance is controlled by one (monogenic or vertical resistance) or a few major genes, and the gene effects are easy to detect. Such resistance can be transferred into the elite lines through pedigree or backcross breeding. Resistance to several insects in rice, sweet potato, corn, and sorghum is monogenic in nature (Painter, 1958; M.D. Pathak, 1970; Kogan, 1982; H.C. Sharma, 1993). The chances of evolution of insect biotypes capable of overcoming the monogenic resistance are quite high, and such resistance breaks down over a period of time. Some of the notable examples of vertical resistance are resistance in wheat to Hessian fly, *M. destructor* (Smith, 2005), and rice resistance to brown planthopper, *N. lugens* (Khush and Brar, 1991). The vertical type of resistance for tolerance or moderate levels of resistance provides stable resistance to insect pests (Panda and Khush, 1995). Since availability of new genes for resistance may be a limiting factor, it is important to manage known resistant genes to maximize their effect and durability through: (1) recycling and sequential release of resistance genes, (2) gene pyramiding, (3) regional biotype-specific deployment of resistance genes, and (4) development of synthetics involving multilines with different genes for resistance.

Polygenic Resistance

Polygenic resistance is controlled by several genes. It is also called horizontal resistance. Inheritance of such genes is usually complex, and it takes a long time to transfer polygenic resistance into improved cultivars. Pedigree breeding involving multiple crosses or population improvement approaches can be used to breed for horizontal resistance. Because of the additive effect of several genes, the insects take a longer time to overcome such resistance and thus the chances of evolution of new insect biotypes are low or minimal. Resistance to several insect species in crops such as cotton, legumes, maize, rice, and wheat is under polygenic control (Panda and Khush, 1995; Smith, 2005). Progress in breeding for insect resistance involving a horizontal or additive type of gene action has been slow, and sufficient levels of resistance have not been achieved in cultivars with desirable agronomic backgrounds. The horizontal type of resistance does not exert a selection pressure on pest populations, and is quite durable. However, strong environmental influence and genetic variability in insect populations complicate the identification and transfer of resistance genes into high-yielding cultivars.

Selection for the additive type of gene action should be deferred until later generations, when a desired level of homozygosity is achieved. In the early generations, selection may be exercised against lines highly susceptible to the target insect pests. Indirect selection using physico-chemical traits associated with insect resistance can also be practiced in the early generations, for example, glossy leaf trait for resistance to shoot fly, *A. soccata* and short and tight glumes for resistance to midge, *S. sorghicola* in sorghum (H.C. Sharma and Nwanze, 1997), trichomes for *H. zea* resistance in tomato (Farrar and Kennedy, 1987), leaf hairs for resistance to leafhopper, *Amrasca biguttula biguttula* Ishida in cotton (H.C. Sharma and Agarwal, 1983), and nonglandular trichomes in wild relatives of pigeonpea, *Cajanus scarabaeoides* (L.) F. Muell. for resistance to *H. armigera* (Romeis et al., 1999; Rupakala et al., 2005). Molecular markers linked with insect resistance genes or the factors associated with insect resistance can be used to improve the efficiency of screening and breeding for resistance to insects, including the additive type of gene action (H.C. Sharma, Abraham, and Stenhouse, 2002).

Cytoplasmic Effects

Cytoplasmic resistance is due to the factors in the cytoplasm of the host plant. Cytoplasmic inheritance is maternal, and can be detected by making reciprocal crosses. The first cytoplasmic male-sterile line in rice was developed by substituting nuclear genes of the *indica* variety, Taichung Native 1 (Athwal and Virmani, 1972), and since then, a large number of cytoplasmic male-sterile (CMS) lines have been developed by exploiting both intra- and interspecific CMS systems (Virmani and Wan, 1988). The male sterile cytoplasm affects rice plant reactions to pathogens, as the WA male sterile cytoplasm in rice is less susceptible to *Pyricularia oryzae* Cav. and *Xanthomonas oryzae* pv. *oryzae* (Swings et al.) than the fertile cytoplasm (R.C. Yang et al., 1989). In maize, the use of CMS-T cytoplasm for hybrid production has been stopped because of its susceptibility to southern corn leaf blight, *Bipolaris maydis* (Nisikado and Miyake), which severely damaged maize crop in the United States in the 1970s (Tatum, 1971). In pearl millet, most of the hybrids are based on A_1 cytoplasm (Burton and Athwal, 1967), and the hybrids based on 111A CMS line are less susceptible to the Oriental armyworm, *Mythimna separata* (Walker) than those based on 5141A and 5054A (H.C. Sharma, Bhatnagar, and Davies, 1982). Most of the sorghum hybrids are based on *Milo* (A_1) cytoplasm (Stephens and Holland, 1954), and a large number of these hybrids are highly susceptible to insect pests. Sorghum lines KS 34 to 39 are as susceptible as CKA (Combine Kafir-based CMS lines) to greenbug, *S. graminum* (Ross and Kofoid, 1979). Cytoplasmic male-sterile lines are more susceptible to insects than the corresponding maintainer lines, suggesting that resistance to insects is influenced by the interaction of factors in the cytoplasm of the maintainer lines and the nuclear genes (H.C. Sharma, 2001; H.C. Sharma et al., 1994, 2004). Resistance to sorghum shoot fly, *A. soccata*, sorghum midge, *S. sorghicola*, sugarcane aphid, *Melanaphis sacchari* (Zehntner), and shoot bug, *Peregrinus maidis* Ashmead, in F_1 hybrids of sorghum is influenced by CMS lines, while resistance to spotted stem borer, *C. partellus* is influenced by the restorer lines (H.C. Sharma et al., 2004; Dhillon et al., 2006a, 2006b). Hybrids based on insect-resistant male-sterile and restorer lines are resistant, while those based on insect susceptible CMS, and resistant, or susceptible restorer lines are susceptible to shoot fly, sugarcane aphid, shoot bug, head bug, and midge (H.C. Sharma et al., 1996, 2000a; Dhillon et al., 2006a, 2006b). A_4M and A_4VzM cytoplasms in sorghum are less susceptible to shoot fly than the A_1, A_2, and A_3 cytoplasms (Dhillon, Sharma, and Reddy, 2005; Dhillon et al., 2005; H.C. Sharma et al., 2005c). To overcome the problem of a narrow genetic base based on a single source of male sterility, efforts should be made to transfer the insect resistance genes into diverse genetic backgrounds or discover new sources of male-sterility that are less susceptible to insect pests.

Genetics and Inheritance of Resistance to Insect Pests

Information on genetics and inheritance of resistance to insect pests is important in crop improvement, which indicates the degree of ease or difficulty involved in incorporating the resistance genes into the improved cultivars. Resistance to insects may be dominant or recessive. The general combining ability (GCA) of a cultivar to transmit resistance to the progeny is determined from the average resistance levels of the F_1 and F_2 plants in all crosses involving a particular genotype, while the specific combining ability (SCA) is a measure of the amount of resistance transferred by a genotype in a single cross to a particular parent. Heritability or variation observed in the progeny of a cross is another measure of inheritance of resistance, which may be affected by several factors. When several different alleles from genes in resistant plants contribute to variation for resistance to an insect species, these are called additive effects. Epistatic effects of alleles may also contribute to variation. The genetics and inheritance of resistance to insects has been discussed by Khush and Brar (1991), Panda and Khush (1995), and Smith (2005). Information on inheritance of resistance to insects in different crops is discussed below.

Rice

Many genes have been identified in rice that contribute for resistance to brown planthopper, *N. lugens*, green leafhoppers, *Nephotettix cincticeps* Uhler and *N. virescens*, gall midge,

O. oryzae, whitebacked planthopper, *Sogatella furcifera* (Horvath), and yellow stem borer, *Scirpophaga incertulas* (Walker) (Khush and Brar, 1991). The genes *Glh2, Glh3, Glh5, Glh6,* and *Glh7* are inherited as dominant traits for resistance to green leafhopper, while *glh4* and *glh8* are recessive (Athwal and Pathak, 1972; Rezaul, Kamin, and Pathak, 1982; M.D. Pathak and Khan, 1994). Four genes condition resistance to *N. cincticeps*. The gene *Grh1* (green rice leaf-hopper) B, from the cultivar Norin PL2 (Kobayashi et al., 1980) is located on rice chromosome 5 (Tamura et al., 1999), while *Grh2* from DV S5 is located on chromosome II (Wang et al., 2004). *Grh3* is located on chromosome 6, and on chromosome 3 (Saka et al., 1997; Fukuta et al., 1998; Yazawa et al., 1998). Seven genes control the expression of resistance to *N. virescens* (Chelliah, 1986).

Thirteen *Bph* (brown planthopper) genes control resistance to *N. lugens* (Khush and Brar, 1991). The wild rice, *Oryza australiensis* Domin. and *O. officinalis* Wall. ex Watt. are the sources of *bph10* and *bphl3* genes, respectively (Ishii et al., 1994; Renganayaki et al., 2002). The genes *Bph1, 3, 6, 9, 10,* and *13* are inherited as dominant traits, while *bph2, 4, 5, 7, 8, 11,* and *12* are inherited as recessive traits. *Bph1* and *bph2* segregate independently of *bph3* and *bph4*. The genes *bph3* and *bph4*, and *bph1* and *bph2* are linked closely (Ikeda and Kaneda, 1981). Resistance to gall midge, *O. oryzae* is controlled by several dominant genes (Sastry and Prakasa Rao, 1973). Two of these genes were designated as *Gm* (gall midge) 1 and *Gm2* (Chaudhary et al., 1986). Both of these genes segregate independently of one another. Six additional dominant genes (*Gm4, 5, 6, 7, 8,* and *9*) and a recessive gene (*gm3*) have also been documented (Srivastava et al., 1993; D. Yang et al., 1997; A. Kumar, Shrivastava, and Sahu, 1998; A. Kumar, Shrivastava, and Shukla, 2000; A. Kumar et al., 2000; Katiyar et al., 2001). Both dominant and recessive genes control the inheritance of resistance to the white backed planthopper, *S. furcifera*. The genes *Wbph* (white-backed planthopper) *1* in N 22 and *Wbph2* in ARC 10239 are inherited as dominant traits (Sidhu, Khush, and Medrano, 1979; Angeles, Khush, and Heinrichs, 1981). *Wbph3* in AOR 52 and *wbph4* in Podiwi A8 are inherited as dominant and recessive traits, respectively (Hernandez and Khush, 1981). Resistance in rice to stem borer, *S. incertulas* is polygenic, and exists in many genotypes (Khush and Brar, 1991).

Wheat and Barley

Identification of germplasm for resistance to the Hessian fly, *M. destructor* began in the early 1900s, and 29 genes from rye, wheat, durum wheat, *Triticum turgidum* var. *durum* L., *Aegilops tauschii* Coss., or *Aegilops triuncialis* L. that control resistance to this insect have been identified (Smith, 2005). All the genes (except *h4*) are inherited as dominant or partially dominant traits. Genes *HI* to *H5, H7, H8,* and *Hl2* have been derived from wheat; *H6, H9* to *H11, H14* to *H20, H28,* and *H29* from durum wheat; *H21* and *H25* from rye; and *H22* to *H24* and *H26* from *A. tauschii* (Berzonsky et al., 2003). *H30* is a single dominant gene from *A. triuncialis* (Martin-Sanchez et al., 2003). The genes *H3, H6,* and *H9* occur on wheat chromosome 5A, with *H3* linked to *H6* and *H9*. *H5* is inherited independently of *H9* (Gallun and Patterson, 1977; Stebbins, Patterson, and Gallun, 1983). *Hl0* and *Hl2* are also located on wheat chromosome 5A (Ohm et al., 1995). Several of these genes have been deployed in response to evolution of Hessian fly, *M. destructor* biotypes.

Eleven genes control the expression of resistance to greenbug, *S. graminum* in *A. speltoides* (Tausch), *A. tauschii*, rye, tall wheatgrass, *Agropyron elongatum* (Host.) Beauv., and wheat. Resistance to biotypes A, B, and C derived from CI 17609 has been introgressed from rye as a dominant gene (*Gb2*) on wheat chromosome IA (Hollenhorst and Lappa, 1983). The dominant genes *Gb3* and *Gb4* on chromosome 7D originated from *A. tauschii*, and have been introduced into wheat germplasm CI 17895 and CI 17959, respectively (Harvey, Martin, and Livers, 1980; Hollenhorst and Lappa, 1983; Weng and Lazar, 2002). The gene *Gb5* located on wheat chromosome 7A in CI 17882 was introgressed from *A. speltoides* (Hollenhorst and Lappa, 1983; Tyler, Webster, and Merkle, 1987), while *Gb6* was identified in GRS 1201 (Porter, Webster, and Friebe, 1994). *Gbx* and *Gbz* from *A. tauschii* are allelic or tightly linked to *Gb3* (Zhu et al., 2004), and are inherited as single dominant genes (Zhu et al., 2004). *Gby* is located on the wheat chromosome 7A (Boyko, Starkey, and Smith, 2004). *Rsgla* confers inducible *S. graminum* resistance in barley, and is triggered by feeding of avirulent *S. graminum* biotype (Carver et al., 1988; Hays et al., 1999).

Twelve genes confer resistance to the Russian wheat aphid, *Diuraphis noxia* (Kurdj.) in barley, rye, and wheat. The dominant genes *Dnl* and *Dn2* were identified in *Triticum aestivum* L. accessions PI 137739 (*Dn1*) and PI 262660 (*Dn2*) from Iran and Azerbaijan, respectively (du Toit, 1987, 1988, 1989). The recessive gene *dn3* is present in an amphiploid wheat parent derived from crosses between *A. tauschii* and durum wheat. The *Dn5* was identified in PI 294994 (du Toit 1987; Saidi and Quick, 1996; Zhang, Quick, and Liu, 1998), while *Dn4* and *Dn6* originated from PI 372129 and PI 243781, respectively (Nkongolo et al., 1989, 1991a, 1991b; Saidi and Quick, 1996). The gene *Dn7* from rye has been transferred to the wheat cultivar Gamtoos (Marais and du Toit, 1993; Marais, Horn, and du Toit, 1994; Marais, Wessels, and Horn, 1998). The genes *Dn8* and *Dn9* are coexpressed with *Dn5* in PI 294994 (Liu et al., 2001), while *Dnx* from PI 220127 is inherited as a dominant trait (Harvey and Martin, 1990; Liu et al., 2001). Resistance to *D. noxia* biotype I in barley line STARS-930IB (derived from PI 573080) is controlled by dominant alleles at two loci (Momhinweg, Porter, and Webster, 1995). An incompletely dominant allele at the *Rdnl* locus and a dominant allele at the *Rdn2* locus confer a high level of resistance to *D. noxia* in barley.

Maize

Several genes condition resistance to the European corn borer, *O. nubilalis* (G.E. Scott and Guthrie, 1967; Chiang and Hudson, 1973). Different genes condition resistance to first and second broods, but some genes condition resistance to both broods (Jennings, Russell, and Guthrie, 1974). Resistance to ear damage by *H. zea* in sweetcorn involves multiple genes, and is controlled by epistatic as well as additive-dominance effects (Warnock, Davis, and Gingera, 1998). Resistance to *H. zea* and *S. zeamais* is largely controlled by additive effects rather than dominance or epistatic effects (Widstrom, 1972; Widstrom, Wiseman, and McMillian, 1972; Widstrom and McMillian, 1973). Both GCA and SCA effects explain significant amounts of variation in different maize populations for resistance to fall armyworm, *S. frugiperda* and the sugarcane borer, *Diatraea grandiosella* (Dyar) (Williams, Buckley, and Davis, 1995; Williams, Davis, and Buckley, 1998). More than one pair of genes control maize silk resistance to *H. zea*, and some of these genes interact in a nonallelic manner (Wiseman and Bondari, 1992, 1995). Inheritance of maysin content in maize, which imparts resistance to *H. zea*, is governed by the presence of a major modifier gene (Widstrom and Snook, 2001). The locus for maysin production is governed by a single modifier gene in GT 114 × GT 119 (Widstrom and Snook, 1994). However, some accessions of maize, such as PI 245138, which have low levels of maysin, are also resistant to *H. zea* (Wilson, Wiseman, and Snook, 1995). Stalk resistance to the stem borer, *Sesamia nonagrioides* (leFebvre) is inherited quantitatively (Cartea et al., 1999, 2001), and additive, dominant, and epistatic effects control the gene action. Both additive and dominant effects explain the variation in expression of resistance to the corn leaf aphid, *Rhopalosiphum maidis* (Fitch.) (Bing and Guthrie, 1991), and to the spotted stem borer, *C. partellus* (R.S. Pathak, 1991).

Sorghum

Resistance to the sorghum shoofly, *A. soccata* is controlled by additive effects, and is expressed as a partially dominant trait at low to moderate levels of infestation (Rana, Jotwani, and Rao, 1981; Dhillon et al., 2006a). The additive component increases at high infestation, but the dominance component is unaffected (Borikar and Chopde, 1980). However, P.T. Gibson and Maiti (1983) reported that resistance is expressed as a recessive trait conditioned by a single gene. Additive effects control the expression of resistance to the spotted stem borer, *C. partellus* (R.S. Pathak and Olela, 1983). Additive gene effects govern resistance to leaf feeding and stem tunneling, while resistance to deadheart formation is controlled by a nonadditive type of gene action (R.S. Pathak, 1990). Additive type of gene action controls leaf feeding and deadheart resistance to spotted stem borer, *C. partellus*, while resistance to tiller production and exit holes is governed by additive and dominance effects, while resistance to stem tunneling is governed by dominance type of gene actions (Sharma et al., 2007). Sorghum resistance to greenbug, *S. graminum* biotype C, was first detected in *Sorghum virgatum* (Hack.) Stapf. Resistance in the sorghum genotype SA 7536 is inherited as an incompletely dominant trait (Weibel et al., 1972). Resistance to biotypes C, E, F, and I is inherited as an incompletely dominant trait controlled by a few major genes (Weibel et al., 1972; Puterka and Peters, 1995; Tuinstra, Wilde, and Krieghauser, 2001). Resistance to the sorghum midge, *S. sorghicola* from AF 28 is inherited as a recessive trait, and is controlled by two or more loci (Boozaya-Angoon et al., 1984, Rossetto and Igue, 1983), and by additive gene effects (Widstrom, Wiseman, and McMillian, 1984; H.C. Sharma et al., 1996, 2000), which varies across locations (H.C. Sharma, Mukuru, and Stenhouse, 2004). However, resistance to sorghum head bug, *C. angustatus*, is inherited as a partially dominant trait controlled by both additive and nonadditive gene action (H.C. Sharma et al., 2000b), while resistance to the African head bug, *Eurystylus oldi* (Poppius) is largely controlled by the additive type of gene effects (Ratnadass et al., 2002; Aladele and Ezeaku, 2003). Expression of resistance in the F_1 hybrids is influenced by cytoplasmic male sterility, and resistance is needed in both parents to produce sorghum hybrids with resistance to shoot fly, *A. soccata*, sugarcane aphid, *M. sacchari*, midge, *S. sorghicola*, and head bug, *C. angustatus* (H.C. Sharma et al., 1996, 2005c, 2004; Dhillon et al., 2006a, 2006b).

Cotton

Most of the characters associated with resistance to bollworm, *H. armigera*, in cotton are governed by oligogenes, and can be transferred into locally adapted cultivars. Growth and development of *H. armigera* was considerably reduced on some of the male sterile lines of cotton (Natarajan et al., 1985). Inheritance of gossypol-containing glands in cotton, which is associated with resistance to bollworms is due to the G 13 allele (Calhoun, 1997). Diallel analysis has indicated that the additive type of gene effects account for approximately 90% of the total genetic variance in cotton for resistance to the tobacco budworm, *Heliothis virescens* (Fab.), and number of gossypol glands (Wilson and Lee, 1971; Wilson and Smith, 1977). Wilson and George (1979, 1983) evaluated the combining ability for resistance to seed damage by pink bollworm, *Pectinophora gossypiella* (Saunders), and observed that gene action contributing to resistance in progeny was additive, and only a few genes conditioned resistance. Trichome density in *Gossypium* species, which is associated with resistance to the leaf hopper, *A. biguttula biguttula*, has been associated with five genes referred to as t_1–t_5 (Lacape and Nguyen, 2005).

Oilseeds

Resistance in soybean to defoliating insects is polygenic. Heritability estimates for resistance to the cabbage looper, *T. ni* (Luedders and Dickerson, 1977) and *E. varivestis* (Sisson et al., 1976) suggest quantitative inheritance. The F_2 plants from a cross between the parents resistant and susceptible to the soybean looper, *P. includens* also exhibit partial dominance or a quantitative inheritance (Kilen, Hatchett, and Hartwig, 1977). Johnson and Beard (1977) observed that a phytomelanin (achene) layer conferring resistance to sunflower moth, *Homoeosoma electellum* (Hulst.), in different *Helianthus* species is inherited as a dominant trait.

Alfalfa

A single dominant gene in alfalfa, *Medicago sativa* L., and sweet clover, *Melilotus infesta* Guss., controls resistance to the pea aphid, *A. pisum* (D.V. Glover and Stanford, 1966), and the sweet clover aphid, *Therioaphis riehmi* (Bomer) (Manglitz and Gorz, 1968). Resistance to spotted alfalfa aphid, *Therioaphis maculata* (Buckton), is controlled by several genes (E. Glover and Melton, 1966), indicating that resistance is quantitative. Combining ability effects in alfalfa for resistance to *E. fabae* are significant and additive in nature (Soper, Mcintosh, and Elden, 1984; Elden, Elgin, and Soper, 1986).

Potato

Resistance to the green peach potato aphid, *Myzus persicae* (Sulzer) in cultivated and wild *Solanum* species is expressed as a partially dominant trait (Sams, Lauer, and Radcliffe, 1976). One dominant gene controls glandular trichome-mediated resistance to *M. persicae* in *Solanum tarijense* Hawkes and *S. berthaultii* Hawkes, but in *S. phureja* (Juz et Buk.) × *S. berthaultii* crosses, two genes control the expression of resistance (R.W. Gibson, 1979). Resistance to *M. persicae* is mediated by a complex interaction between trichome density and droplet size of exudates, and is a quantitatively inherited trait (Mehlenbacher, Plaisted, and Tingey, 1983, 1984). Resistance to potato tuber moth, *P. operculella*, has been derived from *Solanum sparsipilum* (Bitter) Juz. & Buk., and is controlled by a few major genes (Ortiz et al., 1990).

Grain Legumes

Resistance in Lima bean, *Phaseolus lunatus* L., to the leafhopper, *Empoasca kraemeri* Ross & Moore, is due to the quantitative effect of several genes, and is inherited as a recessive trait (Lyman and Cardona, 1982). Both additive and dominant gene effects are responsible for *E. kraemeri* resistance in cultivars of the common bean, *Phaseolus vulgaris* L. (Kornegay and Temple, 1986). Inheritance of hooked trichomes is complex, and is controlled by additive, dominant, and epistatic gene effects (Park et al., 1994). There is also evidence for transgressive segregation (levels of resistance greater than that of the resistant parent) in some progenies from crosses between resistant and susceptible bean cultivars. Resistance in wild strains of *P. vulgaris* to *Zabrotes subfasciatus* (Boheman) is controlled by the toxic seed protein, arcelin. The presence of arcelin is inherited as a dominant trait (Romero Andreas, Yandell, and Bliss, 1986; Kornegay, Cardona, and Posso, 1993). Resistance to bean weevil, *Acanthoscelides obtectus* (Say), is derived from a wild *Phaseolus* accession, and is inherited as a complementary effect of two recessive genes (Kornegay and Cardona, 1991). Resistance in mungbean, *Vigna radiata* (L.) R. Wilczek to the Azuki bean weevil, *Callosobruchus chinensis* L., and the cowpea weevil, *C. maculatus* (F.), is derived from the wild mungbean, *Vigna radiata* var. *sublobata* (L.) Wilczer (Roxb.) Verdc., and is inherited as a simple dominant trait (Tomaka et al., 1992). In *desi* chickpea, highly significant variances have been observed for pod borer, *H. armigera* damage, suggesting the involvement of the additive type of gene action (Singh et al., 1991), but there was a preponderance of the nonadditive type of gene action in the *kabuli* types. From line × tester studies, it was concluded that resistance to *H. armigera* is controlled by multiple genes. In most studies, gene action was found to be predominantly additive, although the nonadditive type of gene action was also observed (Singh et al., 1991; Gowda et al., 2005). In pigeonpea, combining ability studies have indicated the preponderance of the nonadditive type of gene action for resistance to *H. armigera* and *Maruca vitrata* (Geyer) (Lal, Singh, and Vishwajeet, 1989). Verulkar, Singh, and Bhattacharya (1997) indicated the involvement of a single dominant gene in the antixenosis mechanism of resistance in *C. scarabaeoides* to *H. armigera*.

Nonglandular trichomes, which are associated with resistance to *H. armigera* in *C. scarabaeoides*, are inherited as a dominant trait (Rupakala et al., 2005). Resistance to the cowpea aphid, *A. craccivora* is inherited as a monogenic dominant trait (Bala et al., 1987). Resistance to *C. maculatus* is controlled by a combination of major and minor genes expressed as a recessive trait (Redden, Dobie, and Gatehouse, 1983; Redden, Singh, and Luckefahr, 1984), and is controlled by both additive and dominance effects (Fatunla and Badaru, 1983). Cowpea resistance to the cowpea curculionid beetle, *Chalcodermus aeneus* Boheman is controlled by one pair of genes, and is controlled by the additive type of gene action (Ferry and Cuthbert, 1975).

Vegetables

The level of resistance in carrot to western plant bugs, *Lygus hesperus* Knight and *Lygus elisus* van Duzee, has been increased through self-pollination for three generations (D.R. Scott, 1977). de Ponti (1979) increased resistance in cucumber to the two spotted mite, *Tetranychus urticae* Koch by crossing several moderately resistant cultivars, indicating polygenic inheritance of resistance. A single gene controls the inheritance of cucurbitacin (a feeding deterrent) in cucurbits. However, two or three gene pairs control resistance to the spotted cucumber beetle, *Diabrotica undecimpunctata howardi* Barber. Factors other than cucurbitacin also condition the expression of resistance to insects (G.C. Sharma and Hall, 1971). Resistance in lettuce to the leaf aphid, *Nasonovia ribisnigri* (Mosley), has been transferred from *Lactuca virosa* L. to *Lactuca sativa* L. by interspecific crossing (Eenink, Groenwold, and Dieleman, 1982). Resistance is monogenic and inherited as a dominant trait (Eenink, Dieleman, and Groenwold, 1982).

Vine size in tomato is associated with resistance to *H. zea* and there are large pleiotropic effects (Ferry and Cuthbert, 1973). Antixenosis in tomato Entry 38 is under polygenic control. In *L. hirsutum* f. *glabratum*, resistance is controlled by more than one factor (Kalloo et al., 1989). Acyl sugars present in *Lycopersicon pennellii* (Correll) D'Arcy are responsible for high levels of resistance to the spider mite, *Tetranychus evansi* Baker & Pritchard. High acyl sugar content in the cross *L. esculentum* × *L. pennellii* is inherited as a recessive trait, and is controlled by a single locus (Resende et al., 2001).

In okra, resistance to *A. biguttula biguttula* is controlled by dominant genes (B.R. Sharma and Gill, 1982). Both additive and dominance gene effects were significant, but additive and dominance × dominance type of interactions appear to be more important than other effects.

Fruits

Resistance in fruits to several species of aphids is controlled by a single dominant gene, for example, rosy apple aphid, *Dysaphis plantaginea* Pass., and rosy leaf-curling aphid, *Dysaphis devecta* Walk., in apple (Alston and Briggs, 1977). In raspberry, *Rubus phoenicolasius* Maxim., two dominant genes control resistance to the raspberry aphid, *Amphorophora idaei* Bomer (Daubeny, 1966; Jones, McGavin, and Birch, 2000). Monet and Massonie (1994) identified a single gene in peach, *Prunus persica* (L.), (Batsch.) that controls resistance to peach potato aphid, *M. persicae*.

Conclusions

Sources of resistance to insect pests have been identified in different crops. However, the identified sources of resistance have not been used widely because of low heritability or linkage drag. The progress in developing crop cultivars with resistance to insects has been quite slow because of lack of information on the mechanisms that contribute to insect resistance, the number of genes involved, and the nature of gene action. Levels of resistance to insect pests in most of the varieties released for cultivation are low to moderate and, therefore, there is a need to diversify the bases of resistance through gene pyramiding from cultivated germplasm and closely related wild relatives of crops. Resistance to insects is largely due to antixenosis, antibiosis, and tolerance, and is controlled by many genes. Resistance may be dominant or recessive. Pedigree and backcross approaches can be used to transfer insect resistance genes into the cultivars with desirable agronomic backgrounds, while recurrent selection is useful for gene pyramiding for the same or different insect pests. Cytoplasmic male-sterility can be exploited for developing insect-resistant hybrids, and resistance is needed in both the parents to produce insect-resistant hybrids. Information on mechanisms and inheritance of resistance will be useful for marker-assisted introgression of resistance genes into high yielding cultivars.

References

Agrawal, B.L. and Abraham, C.V. (1985). Breeding sorghum for resistance to shoot fly and midge. In *Proceedings, International Sorghum Entomology Workshop,* July 15–21, 1984, Texas A&M University, College Station, Texas, USA. Patancheru, Andhra Pradesh, India: International Crops Research Institute for the Semi-Arid Tropics (ICRISAT), 371–384.

Aladele, S.E. and Ezeaku, I.E. (2003). Inheritance of resistance to head bug (*Eurystylus oldi*) in grain sorghum (*Sorghum bicolor*). *African Journal of Biotechnology* 2: 202–205.

Alston, F.H. and Briggs, J.B. (1977). Resistance genes in apple and biotypes of *Dysaphis devecta*. *Annals of Applied Biology* 87: 75–81.

Angeles, E.R., Khush, G.S. and Heinrichs, E.A. (1981). New genes for resistance to whitebacked planthopper in rice. *Crop Science* 21: 47–50.

Athwal, D.S. and Pathak, M.D. (1972). Genetics of resistance to rice insects. In *Rice Breeding*. Manila, Philippines: International Rice Research Institute, 375–386.

Athwal, D.S. and Virmani, S.S. (1972). Cytoplasmic male sterility and hybrid breeding in rice. In *Rice Breeding*. Manila, Philippines: International Rice Research Institute, 615–620.

Bala, H.D., Singh, B.B., Singh, S.R. and Ladeinde, T.A.O. (1987). Inheritance of resistance to aphid in cowpea. *Crop Science* 27: 892–894.

Balasubramanian, G., Gopalan, M. and Subramanian, T.R. (1977). Resistance to leafhopper in upland cotton. *Indian Journal of Agricultural Sciences* 47: 82–86.

Beland, G.C. and Hatchett, J.H. (1976). Expression of antibiosis to the bollworm in two soybean genotypes. *Journal of Economic Entomology* 6: 557–560.

Berzonsky, W.A., Ding, H., Haley, S.D., Harris, M.O., Lamb, R.J., McKenzie, R.I.H., Ohm, H.W., Patterson, F.L., Peairs, F.B, Poner, D.R., Ratcliffe, R.H. and Shanower, T.G. (2003). Breeding wheat for resistance to insects. *Plant Breeding Reviews* 22: 221–296.

Bing, J.W. and Guthrie, W.D. (1991). Generation mean analysis for resistance in maize to the corn leaf aphid (Homoptera: Aphididae). *Journal of Economic Entomology* 84: 1080–1082.

Boozaya-Angoon, D., Starks, K.J., Weibel, D.E. and Teetes, G.L. (1984). Inheritance of resistance in sorghum, *Sorghum bicolor*, to the sorghum midge, *Contarinia sorghicola* (Diptera: Cecidomyiidae). *Environmental Entomology* 13: 1531–1534.

Borikar, S.T. and Chopde, P.R. (1980). Inheritance of shoot-fly resistance under three levels of infestation in sorghum. *Maydica* 25: 175–183.

Boyko, E.V., Starkey, S.R. and Smith, C.M. (2004). Molecular genetic mapping of *Gby*, a new green bug resistance gene in bread wheat. *Theoretical and Applied Genetics* 109: 1230–1236.

Burton, G.W. and Athwal, D.S. (1967). Two additional sources of cytoplasmic male-sterility in pearl millet and their relationship to Tift 23A. *Crop Science* 7: 209–211.

Butron, A., Widstrom, N.W., Snook, M.E. and Wiseman, B.R. (2002). Recurrent selection for corn earworm (Lepidoptera: Noctuidae) resistance in three closely related corn southern synthetics. *Journal of Economic Entomology* 95: 458–462.

Calhoun, D.S. (1997). Inheritance of high glanding, an insect resistance trait in cotton. *Crop Science* 37: 1181–1186.

Cartea, M.E., Malvar, R.A., Butron, A., Vales, M.I. and Ordas, A. (1999). Inheritance of antibiosis to *Sesamia nonagrioides* (Lepidoptera: Noctuidae) in maize. *Journal of Economic Entomology* 92: 994–998.

Cartea, M.E., Malvar, R.A., Vales, M.I., Butron, A. and Ordas, A. (2001). Inheritance of resistance to ear damage caused by *Sesamia nonagrioides* (Lepidoptera: Noctuidae) in maize. *Journal of Economic Entomology* 94: 277–283.

Carver, B.F., Morgan, G.H., Edwards, L.H. and Webster, J.A. (1988). Registration of four pairs of greenbug-resistant vs. susceptible near-isolines of winter barley germplasms. *Crop Science* 28: 1034–1035.

Chaudhary, B.P., Srivastava, P.S., Shrivastava, M.N. and Khush, G.S. (1986). Inheritance of resistance to gall midge in some cultivars of rice. In *Rice Genetics*. Manila, Philippines: International Rice Research Institute, 523–528.

Chelliah, S. (1986). Genetics of resistance in rice to planthoppers and leafhoppers. In *Rice Genetics. Proceedings, International Rice Genetics Symposium*, 27–31 May, 1985. Manila, Philippines: Island Publications Co., 513–522.

Chiang, M.S. and Hudson, M. (1973). Inheritance of resistance to the European corn borer in grain corn. *Canadian Journal of Plant Science* 53: 779–782.

Clement, S.L. and Quisenberry, S.S. (Eds.). (1999). *Global Plant Genetic Resources for Insect-Resistant Crops*. Bota Racon, Florida, USA: CRC Press.

Clement, S.L., Sharaf El-Din, N., Weigand, S. and Lateef, S.S. (1994). Research achievements in plant resistance to insect pests of cool season food legumes. *Euphytica* 73:41–50.

Coaker, T.H. (1959). Investigations on *Heliothis armigera* in Uganda. *Bulletin of Entomological Research* 50: 487–506.

Cosenza, G.W. and Green, H.P. (1979). Behaviour of tomato fruit worm, *Heliothis zea* on susceptible and resistant lines of processing tomatoes. *Horticulture Science* 14: 171.

Cowgill, S.E. and Lateef, S.S. (1996). Identification of antibiotic and antixenotic resistance to *Helicoverpa armigera* (Lepidoptera: Noctuidae) in chickpea. *Journal of Economic Entomology* 89: 224–229.

Daubeny, H.A. (1966). Inheritance of immunity in the red raspberry to the North American strain of the aphid, *Amphorophora rubi* (Kalt.). *Proceedings of American Society for Horticulture Science* 88: 346–351.

de Ponti, O.M.B. (1979). Resistance in *Cucumis sativus* L. to *Tetranychus urticae* Koch. 5. Raising the resistance level by the exploitation of transgression. *Euphytica* 28: 569–577.

Dhillon, M.K., Sharma, H.C. and Reddy, B.V.S. (2005). Agronomic characteristics of different cytoplasmic male-sterility systems and their reaction to sorghum shoot fly, *Atherigona soccata*. *International Sorghum and Millets Newsletter* 46: 52–55.

Dhillon, M.K., Sharma, H.C., Reddy, B.V.S., Singh, R. and Naresh, J.S. (2006a). Inheritance of resistance to sorghum shoot fly, *Atherigona soccata*. *Crop Science* 46: 1377–1383.

Dhillon, M.K., Sharma, H.C., Pampapathy, G. and Reddy, B.V.S. (2006b). Cytoplasmic male-sterility affects expression of resistance to shoot bug, *Peregrinus maidis*, sugarcane aphid, *Melanaphis sacchari*, and spotted stem borer, *Chilo partellus* in sorghum. *International Sorghum and Millets Newsletter* 47: 66–68.

Dhillon, M.K., Sharma, H.C., Reddy, B.V.S., Singh, R., Naresh, J.S. and Kai, Z. (2005). Relative susceptibility of different male-sterile cytoplasms in sorghum to shoot fly, *Atherigona soccata*. *Euphytica* 144: 275–283.

Dodia, D.A., Patel, A.J., Patel, I.S., Dhulia, F.K. and Tikka, S.B.S. (1996). Antibiotic effects of pigeonpea wild relatives on *Helicoverpa armigera*. *International Chickpea and Pigeonpea Newsletter* 3: 100–101.

Dogger, J.R. and Hanson, C.H. (1963). Reaction of alfalfa varieties and strains to alfalfa weevil. *Journal of Economic Entomology* 56: 192–197.

du Toit, F. (1987). Resistance in wheat (*Triticum aestivum*) to *Diuraphis noxia* (Hemiptera: Aphididae). *Cereal Research Communications* 15: 175–179.

du Toit, F. (1988). Another source of Russian wheat aphid (*Diuraphis noxia*) resistance in *Triticum aestivum*. *Cereal Research Communications* 16: 105–106.

du Toit, F. (1989). Inheritance of resistance in two *Triticum aestivum* lines to Russian wheat aphid (Homoptera: Aphididae). *Journal of Economic Entomology* 82: 1251–1253.

Dua, R.P., Gowda, C.L.L., Kumar, S., Saxena, K.B., Govil, J.N., Singh, B.B., Singh, A.K., Singh, R.P., Singh, V.P. and Kranthi, S. (2005). Breeding for resistance to *Heliothis/Helicoverpa*: Effectiveness and limitations. In Sharma, H.C. (Ed.), *Heliothis/Helicoverpa Management: Emerging Trends and Strategies for Future Research*. New Delhi, India: Oxford and IBH Publishers, 223–242.

Eenink, A.H., Groenwold, R. and Dieleman, F.L. (1982). Resistance of lettuce (*Lactuca*) to the leaf aphid, *Nasonovia ribisnigri*. I. Transfer of resistance from *L. virosa* to *L. sativa* by interspecific crosses and selection of resistant breeding lines. *Euphytica* 31: 291–300.

Eenink, A.H., Dieleman, F.L. and Groenwold, R. (1982). Resistance of lettuce (*Lactuca*) to the leaf aphid, *Nasonovia ribisnigri*. II. Inheritance of the resistance. *Euphytica* 31: 301–304.

Elden, T.C. and Elgin, J.H. Jr. (1987). Recurrent seedling and individual plant selection for potato leafhopper (Homoptera: Cicadellidae) resistance in alfalfa. *Journal of Economic Entomology* 80: 690–695.

Elden, T.C., Elgin, J.H. Jr. and Soper, J.F. (1986). Inheritance of pubescence in selected clones from two alfalfa populations and relationship to potato leafhopper resistance. *Crop Science* 26: 1143–1146.

Farrar, R.R. and Kennedy, G.G. (1987). Growth, food consumption and mortality of *Heliothis zea* larvae on foliage of the wild tomato, *Lycopersicon hirsutum* f. *glabratum*: and the cultivated tomato, *L. esculentum*. *Entomologia Experimentalis et Applicata* 44: 213–219.

Fatunla, T. and Badaru, K. (1983). Inheritance of resistance to cowpea weevil (*Callosobruchus maculatus* Fabr.). *Journal of Agriculture Science, Cambridge* 101: 423–426.

Ferry, R.L. and Cuthbert, F.P. Jr. (1973). Factors affecting evaluation of fruitworm resistance in the tomato. *Journal of American Society of Horticulture Science* 98: 457–459.

Ferry, R.L. and Cuthbert, F.P. (1975). A tomato fruitworm antibiosis in *Lycopersicon*. *Horticulture Science* 10: 146.

Fukuta, Y., Tamura, K., Hirac, M. and Oya, S. (1998). Genetic analysis of resistance to green rice leaf-hopper (*Nephotettix cincticeps* Uhler) in rice parental line Norin-PL6, using RFLP markers. *Breeding Science* 48: 243–249.

Gallun, R.L. and Patterson, F.L. (1977). Monosomic analysis of wheat for resistance to Hessian fly. *Journal of Heredity* 68: 223–226.

Gibson, P.T. and Maiti, R.K. (1983). Trichomes in segregating generations of sorghum matings. I. Inheritance of presence and density. *Crop Science* 23: 73–75.

Gibson, R.W. (1979). The geographical distribution, inheritance and pest-resisting properties of sticky-tipped foliar hairs on potato species. *Potato Research* 22: 223–237.

Glover, D.V. and Stanford, E.H. (1966). Tetrasomic inheritance of resistance in alfalfa to the pea aphid. *Crop Science* 6: 161–165.

Glover, E. and Melton, B. (1966). Inheritance patterns of spotted alfalfa aphid resistance in *Zea* plants. *New Mexico Agriculture Experimentation Station Research Report* 127: 1–40.

Gorz, H.J., Manglitz, G.R. and Haskins, A. (1979). Selection for yellow clover aphid and pea aphid resistance in clover. *Crop Science* 19: 257–260.

Gowda, C.L.L., Ramesh, S. Chandra, S., and Upadhyaya, H.D. (2005). Genetic basis of pod borer (*Helicoverpa armigera*) resistance and grain yield in *desi* and *Kabuli* chickpea (*Cicer arietinum* L.) under unprotected conditions. *Euphytica* 145: 199–214.

Harvey, T.L. and Martin, T.J. (1990). Resistance to Russian wheat aphid, *Diuraphis noxia*, in wheat (*Triticum aestivum*). *Cereal Research Communications* 18: 127–129.

Harvey, T.L., Martin, T.J. and Livers, R.W. (1980). Resistance to biotype C greenbug in synthetic hexaploid wheats derived from *Triticum tauschii*. *Journal of Economic Entomology* 73: 387–389.

Hays, D.B., Porter, D.R., Webster, J.A. and Carver, B.F. (1999). Feeding behavior of biotypes E and H greenbug (Homoptera: Aphididae) on previously infested near-isolines of barley. *Journal of Economic Entomology* 92: 1223–1229.

Heinrichs, E.A., Fabellar, L.T., Basilio, R.P., Tu, C.W. and Medrano, F. (1984). Susceptibility of rice plan-thoppers, *Nilaparvata lugens* and *Sogatella furcifera* (Homoptera: Delphacidae) to insecticides as influenced by level of resistance in the host plant. *Environmental Entomology* 13: 455–458.

Henzell, R.G., Peterson, G.C., Teetes, G.L., Franzmann, B.A., Sharma, H.C., Youm, O., Ratnadass, A., Toure, A., Raab, J. and Ajayi, O. (1997). Breeding for resistance to panicle pests of sorghum and pearl millet. In *Proceedings, the International Conference on Genetic Improvement of Sorghum and Pearl Millet*, 23–27 September, 1996. Lubbock, Texas, USA: Texas A&M University, 255–280.

Hernandez, J.E. and Khush, G.S. (1981). Genetics of resistance to whitebacked planthopper in some rice (*Oryza sativa* L.) varieties. *Oryza* 18: 44–50.

Hollenhorst, M.M. and Lappa, L.R. (1983). Chromosomal location of genes for resistance to greenbug in "Largo" and "Amigo" wheats. *Crop Science* 23: 91–93.

Ikeda, R. and Kaneda, C. (1981). Genetic analysis of resistance to brown planthopper, *Nilaparvata lugens* Stal in rice. *Japanese Journal of Plant Breeding* 31: 279–285.

Ishii, T., Brar, D.S., Multani, D.S. and Khush, G.S. (1994). Molecular tagging of genes for brown plan-thopper resistance and earliness from *Oryza australiensis* into cultivated rice, *O. sativa. Genome* 37: 217–221.

Jayaraj, S. (1982). Biological and ecological studies of *Heliothis*. In Reed, W. and Kumble, V. (Eds.), *Proceedings of the International Workshop on* Heliothis *Management*, 15–20 November, 1981. Patancheru, Andhra Pradesh, India: International Crops Research Institute for the Semi-Arid Tropics (ICRISAT), 17–28.

Jennings, C.W., Russell, W.A. and Guthrie, W.D. (1974). Genetics of resistance in maize to first- and second-brood European corn borer. *Crop Science* 14: 394–398.

Johnson, A.L. and Beard, B.H. (1977). Sunflower moth damage and inheritance of the phytomelanin layer in sunflower achenes. *Crop Science* 17: 369–372.

Jones, A.T., McGavin, W.J. and Birch, A.N.E. (2000). Effectiveness of resistance genes to the large raspberry aphid, *Amphorophora idaei* Bomer, in different raspberry (*Rubus idaeus* L.) genotypes and under different environmental conditions. *Annals of Applied Biology* 136: 107–113.

Kalloo, G., Banerjee, M.K., Kashyap, R.K. and Yadav, A.K. (1989). Genetics of resistance to fruitworm in *Lycopersicon*. *Plant Breeding* 102: 173–175.

Kashyap, R.K. and Verma, A.N. (1987). Development and survival of fruit borer, *Heliothis armigera* (Hübner), on resistant and susceptible tomato genotypes. *Journal of Plant Diseases and Protection* 94(1): 14–21.

Katiyar, S.K., Tan, Y., Huang, B., Chandel, G., Xu, Y., Zhang, Y., Xie, Z. and Bennett, J. (2001). Molecular mapping of gene *Gm-6(l)* which confers resistance against four biotypes of Asian rice gall midge in China. *Theoretical and Applied Genetics* 103: 953–961.

Kennedy, G.G., Kishaba, A.N. and Bohn, G.W. (1975). Response of several pest species to *Cucumis melo* L. lines resistant to *Aphis gossypii* Glover. *Environmental Entomology* 4: 653–657.

Khush, G.S. (1980). Breeding rice for multiple disease and insect resistance. In *Rice Improvement in China and Other Asian Countries*. Los Banos, Philippines: International Rice Research Institute, 219–238.

Khush, G.S. and Brar, D.S. (1991). Genetics of resistance to insects in crop plants. *Advances in Agronomy* 45: 223–274.

Kilen, T.C., Hatchett, J.H. and Hartwig, E.E. (1977). Evaluation of early generation soybeans for resistance to soybean looper. *Crop Science* 17: 397–398.

Klenke, J.R., Russell, W.A. and Guthrie, W.D. (1986). Distributions for European corn borer (Lepidoptera: Pyralidae) ratings of S_1 lines from "BS9" corn. *Journal of Economic Entomology* 79: 1076–1081.

Klun, J.A., Tipton, C.C. and Brindley, T.A. (1967). 2-4 dihydoxy-7-methodoxy-1,4-benoxazin-3-zone (DIMBOA) as active agent in the resistance of maize to the European corn borer. *Journal of Economic Entomology* 60: 1529–1533.

Knipling, E.F. (1979). The basic principles of insect population suppression. *Bulletin of Entomological Society of America* 12: 7–15.

Kobayashi, A., Kaneda, C., Ikeda, R. and Ikehashi, H. (1980). Inheritance of resistance to green rice planthopper, *Nephotettix cincticeps* in rice. *Japanese Journal of Breeding* 30(1): 56–57.

Kogan, M. (1982). Plant resistance in pest management. In Melcalf, R.L. and Luckmann, W.H. (Eds.), *Introduction to Insect Pest Management*, 2nd Edition. New York, USA: John Wiley & Sons, 93–134.

Kornegay, J.L. and Cardona, C. (1991). Inheritance of resistance to *Acanthoscelides obtectus* in a wild common bean accession crossed to commercial bean cultivars. *Euphytica* 52: 103–111.

Kornegay, J., Cardona, C. and Posso, C.E. (1993). Inheritance of resistance to Mexican bean weevil in common bean, determined by bioassay and biochemical tests. *Crop Science* 33: 589–594.

Kornegay, J.L. and Temple, S.R. (1986). Inheritance and combining ability of leafroller defense mechanisms in common bean. *Crop Science* 26: 1153–1158.

Kumar, A., Shrivastava, M.N. and Sahu, R.K. (1998). Genetic analysis for gall midge resistance: A reconsideration. *Rice Genetics Newsletter* 17: 83–84.

Kumar, A., Shrivastava, M.N. and Shukla, B.C. (2000). Genetic analysis of gall midge (*Orseolia oryzae* Wood-Mason) biotype I resistance in the rice cultivar RP 2333-156-8. *Oryza* 37: 79–80.

Kumar, A., Bhandarkar, S., Pophlay, D.J. and Shrivastava, M.N. (2000). A new gene for gall midge resistance in rice accession Jhilpiti. *Rice Genetics Newsletter* 15: 142–153.

Kumar, V.K., Sharma, H.C. and Reddy, K.D. (2005). Antibiosis component of resistance to spotted stem borer, *Chilo partellus* in sorghum, *Sorghum bicolor*. *Crop Protection* 25: 66–72.

Kumari, A.D., Sharma, H.C. and Reddy, D.J. (2005). Oviposition non-preference as component of resistance to pod borer, *Helicoverpa armigera* in pigeonpea. *Journal of Applied Entomology* 130: 10–14.

Lacape, J.M. and Nguyen, T.B. (2005). Mapping quantitative trait loci associated with leaf and stem pubescence in cotton. *Journal of Heredity* 96: 441–446.

Lal, S.S., Singh, I.P. and Vishwa Jeet. (1989). Heterosis and combining ability for resistance to pod fly and lepidopteran borers in pigeonpea (*Cajanus cajan*). *Indian Journal of Agricultural Sciences* 69(11): 786–788.

Lateef, S.S. (1985). Gram pod borer [*Heliothis armigera* (Hub.)] resistance in chickpea. *Agriculture, Ecosystems, and Environment* 14: 95–102.

Lateef, S.S. and Sachan, J.C. (1990). Host-plant resistance to *Helicoverpa armigera* (Hub.) in different agroecological contexts. In *Chickpea in the Nineties, Proceedings of the Second International Workshop on Chickpea*, 4–8 December, 1989. Patancheru, Andhra Pradesh, India: International Crops Research Institute for the Semi-Arid Tropics/International Center for Agricultural Research in the Dry Areas, 181–190.

Liu, X., Smith, C.M., Gill, B.S. and Tolmay, V. (2001). Microsatellite markers linked to six Russian wheat aphid resistance genes in wheat. *Theoretical and Applied Genetics* 102: 504–510.

Luedders, V.D. and Dickerson, W.A. (1977). Resistance of selected soybean genotypes and segregating populations to cabbage looper feeding. *Crop Science* 17: 395–396.

Lyman, J.M. and Cardona, C. (1982). Resistance in lima beans to a leafhopper, *Empoasca kraemeri*. *Journal of Economic Entomology* 75: 281–286.

Manglitz, G.R. and Gorz, H.J. (1968). Inheritance of resistance in sweetclover to the sweetclover aphid. *Journal of Economic Entomology* 61: 90–94.

Marais, G.F. and du Toit, F.A. (1993). A monosomic analysis of Russian wheat aphid resistance in the common wheat PI294994. *Plant Breeding* 111: 246–248.

Marais, G.F., Horn, M. and du Toit, F. (1994). Intergeneric transfer (rye to wheat) of a gene(s) for Russian wheat aphid resistance. *Plant Breeding* 113: 265–271.

Marais, G.F., Wessels, W.G. and Horn, M. (1998). Association of a stem rust resistance gene (*Sr45*) and two Russian wheat aphid resistance genes (*Dn5* and *Dn7*) with mapped structural loci in common wheat. *South African Journal of Plant and Soil* 15: 67–71.

Martin-Sanchez, J.A., Gomez-Colmenarejo, M., Del Moral, J., Sin, E., Montes, M.J., Gonzalez-Belinchon, C., Lopez Brana, I. and Delibes, A. (2003). A new Hessian fly resistance gene (H30) transferred from wild grass *Aegilops triuncialis* to hexaploid wheat. *Theoretical and Applied Genetics* 106: 1248–1255.

Mehlenbacher, S.A., Plaisted, R.L. and Tingey, W.M. (1983). Inheritance of glandular trichomes in crosses with *Solanum berthaultii*. *American Potato Journal* 60: 699–708.

Mehlenbacher, S.A., Plaisted, R.L. and Tingey, W.M. (1984). Heritability of trichome density and droplet size in interspecific potato hybrids and relationship to aphid resistance. *Crop Science* 24: 320–322.

Momhinweg, D.W., Porter, D.R. and Webster, J.A. (1995). Inheritance of Russian wheat aphid resistance in spring barley. *Crop Science* 35: 1368–1371.

Monet, R. and Massonie, G. (1994). Genetic determination of resistance to green aphid (*Myzus persicae*) in peach. *Agronomie* 2: 177–182.

Murthy, J.S.V.S., Rajasekhar, P., Venkataiah, M. and Ranganathacharyulu, N. (1998). Evaluation of some cotton genotypes for resistance to bollworm (*Helicoverpa armigera* Hub.). *Annals of Agricultural Research* 19: 30–33.

Nanthagopal, R. and Uthamasamy, S. (1989). Life tables for American bollworm *Heliothis armigera* (Hübner) on four species of cotton under field conditions. *Insect Science and Its Application* 10(4): 521–530.

Natarajan, K., Krishnaswami, R., Regupathy, A. and Jayaraj, S. (1985). Development and behaviour of gram pod borer, *Heliothis armigera* (Hubner) on cytoplasmic male sterile and fertile cotton lines. In Raghupathy, A. and Jayaraj, S. (Eds.), *Behavioral and Physiological Approaches in Pest Management*. Coimbatore, Tamil Nadu, India: Tamil Nadu Agricultural University, 200–202.

Nkongolo, K.K., Quick, J.S., Limin, A.E. and Fowler, D.B. (1991a). Sources and inheritance of resistance to Russian wheat aphid in *Triticum* species amphiploids and *Triticum tauschii*. *Canadian Journal of Plant Science* 71: 703–708.

Nkongolo, K.K., Quick, J.S., Meyers, W.L. and Peairs, F.B. (1989). Russian wheat aphid resistance of wheat, rye, and triticale in greenhouse tests. *Cereal Research Communications* 17: 227–232.

Nkongolo, K.K., Quick, J.S., Peairs, F.B. and Meyers, W.L. (1991b). Inheritance of resistance of PI 373129 wheat to the Russian wheat aphid. *Crop Science* 31: 905–906.

Ohm, H.W., Sharma, H.C., Patterson, F.L., Ratcliffe, R.H. and Obani, M. (1995). Linkage relationships among genes on wheat chromosome 5A that condition resistance to Hessian fly. *Crop Science* 35: 1603–1607.

Onukogu, F.A., Guthrie, W.D., Russell, W.A., Reed, G.L. and Robbins, J.C. (1978). Location of genes that condition resistance in maize to sheath collar feeding by second-generation European corn borer. *Journal of Economic Entomology* 71: 1–4.

Ortiz, R. Iwanaga, M., Raman, K.V. and Palacios, M. (1990). Breeding for resistance to potato tuber moth, *Phthorimaea operculella* (Zeller), in diploid potatoes. *Euphytica* 50: 119–125.

Padma Kumari, A.P., Sharma, H.C. and Reddy, D.D.R. (2000). Components of resistance to sorghum head bug, *Calocoris angustatus. Crop Protection* 19: 385–392.

Painter, R.H. (1958). Resistance of plants to insects. *Annual Review of Entomology* 3: 267–290.

Panda, N. and Heinrichs, E.A. (1983). Levels of tolerance and antibiosis in rice varieties having moderate resistance to the brown planthopper, *Nilaparvata lugens. Environmental Entomology* 12: 1204–1214.

Panda, N. and Khush, G.S. (1995). *Host Plant Resistance to Insects.* Wallingford, Oxon, UK: CAB International.

Park, S.J., Timmins, P.R., Quiring, D.T. and Jul, P.Y. (1994). Inheritance of leaf area and hooked trichome density of the first trifoliolate leaf in common bean (*Phaseolus vulgaris* L). *Canadian Journal of Plant Science* 74: 235–240.

Pathak, M.D. (1970). Genetics of plants in pest management. In Rabb, R.L. and Guthrie, F.E. (Eds.), *Concepts of Pest Management.* Raleigh, North Carolina, USA: North Carolina State University, 138–157.

Pathak, M.D. and Khan, Z.R. (1994). *Insect Pests of Rice.* Manila, Philippines: International Rice Research Institute.

Pathak, R.S. (1990). Genetics of sorghum, maize, rice and sugarcane resistance to the cereal stem borer, *Chilo* spp. *Insect Science and Its Application* 11: 689–699.

Pathak, R.S. (1991). Genetic expression of the spotted stem borer, *Chilo partellus* (Swinhoe) resistance in three maize crosses. *Insect Science and Its Application* 12: 147–151.

Pathak, R.S. and Olela, J.C. (1983). Genetics of host plant resistance in food crops with special reference to sorghum stem borers. *Insect Science and Its Application* 4: 127–134.

Porter, D.R., Webster, J.A. and Friebe, B. (1994). Inheritance of greenbug biotype G resistance in wheat. *Crop Science* 34: 625–628.

Puterka, G.J. and Peters, D.C. (1995). Genetics of greenbug (Homoptera: Aphididae) virulence to resistance in sorghum. *Journal of Economic Entomology* 88: 421–429.

Rana, B.S., Jotwani, M.G. and Rao, N.G.P. (1981). Inheritance of host plant resistance to the sorghum shoot fly. *Insect Science and Its Application* 2: 105–110.

Rao, C.N. and Prasad, V.D. (1996). Comparative population growth rates of *Helicoverpa armigera* (Hub.) on certain cultivars of cotton, *Gossypium hirsutum* L. *Annals of Plant Protection Sciences* 4: 138–141.

Ratnadass, A., Chantereau, J., Coulibaly, M.F. and Cilas, C. (2002). Inheritance of resistance to the panicle feeding bug, *Eurystylus oldi*, and sorghum midge, *Stenodiplosis sorghicola* in sorghum. *Euphytica* 123: 131–138.

Redden, R.J., Dobie, P. and Gatehouse, A. (1983). The inheritance of seed resistance to *Callosobruchus maculatus* F. in cowpea [*Vigna unguiculata* (L.) Walp.]. I. Analyses of parental, F_1, F_2 and backcross seed generation. *Australian Journal of Agricultural Research* 34: 681–695.

Redden, R.J., Singh, S.R. and Luckefahr, M.J. (1984). Breeding for cowpea resistance to bruchids at IITA. *Protection Ecology* 7: 291–303.

Reese, J.C. and Beck, S.D. (1976). Effects of certain allelochemics on the black cutworm, *Agrotis ipsilon*: effects of catechol, dopamine, and chlorogenic acid on larval growth, development and utilization of food. *Annals of Entomological Society of America* 69: 68–72.

Renganayaki, K., Fritz, A.K., Sadasivam, S., Pammi, S., Harrington, S.E., McCouch, S.R., Kumar, S.M. and Reddy, A.S. (2002). Mapping and progress toward map-based cloning of brown planthopper biotype-4 resistance gene introgressed from *Oryza officinalis* into cultivated rice, *O. sativa. Crop Science* 42: 2112–2117.

Resende, J.T.V., Maluf, W.R., Cardoso, M.G., Nelson, D.L. and Faria, M.V. (2001). Inheritance of acyl-sugar contents in tomatoes derived from an interspecific cross with the wild tomato, *Lycopersicon pennelii* and their effect on spider mite repellence. *Genetics and Molecular Research* 1: 106–116.

Rezaul Kamin, A.N.M. and Pathak, M.D. (1982). New genes for resistance to green leafhopper, *Nephotettix virescens* (Distant) in rice, *Oryza sativa* L. *Crop Protection* 1: 483–490.

Roberts, J.J., Gallun, R.L., Patterson, F.L. and Foster, J.E. (1979). Effects of wheat leaf pubescens on Hessian fly. *Journal of Economic Entomology* 72: 211–214.

Romeis, J., Shanower, T.G. and Peter, A.J. (1999). Trichomes of pigeonpea and two wild *Cajanus* species. *Crop Science* 39: 1–5.

Romero Andreas, J., Yandell, B.S. and Bliss, F.A. (1986). Bean arcelin. I. Inheritance of a novel seed protein of *Phaseolus vulgaris* L. and its effect on seed composition. *Theoretical and Applied Genetics* 72: 123–128.

Ross, W.M. and Kofoid, K.D. (1979). Effect of non-milo cytoplasms on the agronomic performance of sorghum. *Crop Science* 19: 267–270.

Rossetto, C.J. and Igue, T. (1983). Heranea de resistencia variedade de sorgo AF 28 a *Contarinia sorghicola* Coquillett. *Bragantia* 42: 211–219.

Rupakala, A., Rao, D., Reddy, L.J., Upadhyaya, H.D. and Sharma, H.C. (2005). Inheritance of trichomes and resistance to pod borer (*Helicoverpa armigera*) and their association in interspecific crosses between cultivated pigeonpea (*Cajanus cajan*) and its wild relative *C. scarabaeoides*. *Euphytica* 145: 247–257.

Russell, W.A., Lawrence, G.D. and Guthrie, W.D. (1979). Effects of recurrent selection for European corn borer resistance on other agronomic characters in synthetic cultivars of maize. *Maydica* 24: 33–47.

Saidi, A. and Quick, J.S. (1996). Inheritance and allelic relationships among Russian wheat aphid resistance genes in winter wheat. *Crop Science* 36: 256–258.

Saka, N., Toyama, T., Tuji, T., Nakamae, H. and Izawa, T. (1997). Fine mapping of green ricehopper resistant gene *Grh-3 (t)* and screening of *Grh-3 (t)* among green ricehopper resistant and green leafhopper resistant cultivars in rice. *Breeding Science* 47 (suppl.l): 1–55.

Sams, D.W., Lauer, F.I. and Radcliffe, E.B. (1976). Breeding behavior of resistance to green peach aphid in tuber-bearing *Solanum* germplasm. *American Potato Journal* 53: 23–29.

Sanford, L.L. and Ladd, T.L. Jr. (1983). Selection for resistance to potato leafhopper in potatoes. III. Comparison of two selection procedures. *American Potato Journal* 60: 653–659.

Sastry, M.V.S. and Prakasa Rao, P.S. (1973). Inheritance of resistance to rice gall midge, *Pachydiplosis oryzae* Wood-Mason. *Current Science* 42: 652–653.

Saxena, R.C. (1986). Biochemical basis of insect resistance in rice varieties. In Green, M.B. and Hedin, P.A. (Eds.), *Natural Resistance of Plants to Pests*. Washington, D.C., USA: American Chemical Society, 142–159.

Schweissing, F.C. and Wilde, G. (1978). Temperature influence on greenbug resistance of crops in the seedling stage. *Environmental Entomology* 7: 831–834.

Schweissing, F.C. and Wilde, G. (1979). Temperature and plant nutrient effects on resistance of seedling sorghum to the greenbug. *Journal of Economic Entomology* 72: 20–23.

Scott, D.R. (1977). Selection for *Lygus* bug resistance in carrot. *HortScience* 12: 452.

Scott, G.E., Dicke, F.F. and Pesho, G.R. (1966). Location of genes conditioning resistance in corn to leaf feeding of the European corn borer. *Crop Science* 4: 444–446.

Scott, G.E. and Guthrie, W.D. (1967). Reactions of permutations of maize double crosses to leaf feeding of European corn borers. *Crop Science* 7: 233–235.

Shanower, T.G., Yoshida, M. and Peter, A.J. (1997). Survival, growth, fecundity and behavior of *Helicoverpa armigera* (Lepidoptera: Noctuidae) on pigeonpea and two wild *Cajanus* species. *Journal of Economic Entomology* 90: 837–841.

Sharma, B.R. and Gill, B.S. (1982). Genetics of resistance to cotton jassid, *Amrasca biguttula biguttula* (Ishida) in okra. *Euphytica* 33: 215–220.

Sharma, G.C. and Hall, C.V. (1971). Cucurbitacin B and total sugar inheritance in *Cucurbita pepo* L. related to spoiled cucumber beetle feeding. *Journal of American Society of Horticulture Science* 96: 750–754.

Sharma, H.C. (1993). Host plant resistance to insects in sorghum and its role in integrated pest management. *Crop Protection* 12: 11–34.

Sharma, H.C. (1994). Insect plant relationships. In Gujar, G.T. (Ed.), *Recent Advances in Insect Physiology and Toxicology*. New Delhi, India: Agricole Publishing Company, 1–13.

Sharma, H.C. (1997). Influence of panicle size on midge damage and compensation in grain mass in sorghum. *International Sorghum and Millets Newsletter* 38: 85–87.

Sharma, H.C. (2001). Cytoplasmic male-sterility and source of pollen influence the expression of resistance to sorghum midge, *Stenodiplosis sorghicola*. *Euphytica* 122: 391–395.

Sharma, H.C., Abraham, C.V. and Stenhouse, J.W. (2002). Compensation in grain weight and volume in sorghum is associated with expression of resistance to sorghum midge, *Stenodiplosis sorghicola*. *Euphytica* 125: 245–254.

Sharma, H.C. and Agarwal, R.A. (1983). Role of some chemical components and leaf hairs in varietal resistance in cotton to jassid, *Amrasca biguttula biguttula* Ishida. *Journal of Entomological Research* 7: 145–149.

Sharma, H.C., Bhatnagar, V.S. and Davies, J.C. (1982). Studies on *Mythimna separata* at ICRISAT. *Sorghum Entomology Progress Report 6*. Patancheru, Andhra Pradesh, India: International Crops Research Institute for the Semi-Arid Tropics (ICRISAT).

Sharma, H.C., Dhillon, M.K., Pampapathy, G. and Reddy, B.V.S. (2007). Inheritance of resistance to spotted stem borer, *Chilo partellus* in sorghum, *Sorghum bicolor*. *Euphytica* 156: 117–128.

Sharma, H.C., Dhillon, M.K. and Reddy, B.V.S. (2006). Expression of resistance to *Atherigona soccata* in F_1 hybrids involving shoot fly resistant and susceptible cytoplsmic male-sterile and restorer lines of sorghum. *Plant Breeding* 125: 473–477.

Sharma, H.C., Franzmann, B.A. and Henzell, R.G. (2002). Mechanisms and diversity of resistance to sorghum midge, *Stenodiplosis sorghicola* in *Sorghum bicolor*. *Euphytica* 124: 1–12.

Sharma, H.C. and Lopez, V.F. (1990). Mechanisms of resistance in sorghum to head bug, *Calocoris angustatus*. *Entomologia Experimentalis et Applicata* 57: 285–294.

Sharma, H.C. and Lopez, V.F. (1993). Survival of *Calocoris angustatus* (Hemiptera: Miridae) nymphs on diverse sorghum genotypes. *Journal of Economic Entomology* 86: 607–613.

Sharma, H.C., Lopez, V.F. and Nwanze, K.F. (1993). Genotypic effects of sorghum accessions on fecundity of sorghum head bug, *Calocoris angustatus* Lethiery. *Euphytica* 65: 167–175.

Sharma, H.C., Mukuru, S.Z. and Stenhouse, J.W. (2004). Variation in inheritance of resistance to sorghum midge, *Stenodiplosis sorghicola* across locations in India and Kenya. *Euphytica* 138: 219–225.

Sharma, H.C. and Norris, D.M. (1993). Innate differences in consumption and utilization of soybean leaves by the cabbage looper, *Trichoplusia ni* Hübner (Lepidoptera: Noctuidae). *Zietschrift fur Angewandte Entomologie* 116: 527–531.

Sharma, H.C. and Nwanze, K.F. (1997). *Mechanisms of Resistance to Insects in Sorghum and Their Usefulness in Crop Improvement*. Information Bulletin no. 45. Patancheru, Andhra Pradesh, India: International Crops Research Institute for the Semi-Arid tropics (ICRISAT).

Sharma, H.C. and Ortiz, R. (2002). Host plant resistance to insects: An eco-friendly approach for pest management and environment conservation. *Journal of Environmental Biology* 23: 111–135.

Sharma, H.C., Singh, F. and Nwanze, K.F. (Eds.) (1997). *Plant Resistance to Insects in Sorghum*. Patancheru, Andhra Pradesh, India: International Crops Research Institute for the Semi-Arid Tropics (ICRISAT).

Sharma, H.C. and Vidyasagar, P. (1994). Antixenosis to sorghum midge, *Contarinia sorghicola* in *Sorghum bicolor*. *Annals of Applied Biology* 124: 495–507.

Sharma, H.C., Vidyasagar, P. and Subramanian, V. (1993). Antibiosis component of resistance in sorghum to sorghum midge, *Contarinia sorghicola*. *Annals of Applied Biology* 123: 469–483.

Sharma, H.C., Abraham, C.V., Vidyasagar, P. and Stenhouse, J.W. (1996). Gene action for resistance in sorghum to midge, *Contarinia sorghicola*. *Crop Science* 36: 259–265.

Sharma, H.C., Ahmad, R., Ujagir, R., Yadav, R.P., Singh, R. and Ridsdill-Smith, T.J. (2005a). Host plant resistance to *Helicoverpa*: The prospects. In Sharma, H.C. (Ed.), Helicoverpa/Heliothis *Management: Emerging Trends and Strategies for Future Research*. New Delhi, India: Oxford & IBH Publishers, 171–213.

Sharma, H.C., Dhillon, M.K., Naresh, J.S., Singh, R., Pampapathy, G. and Reddy, B.V.S. (2004). Influence of cytoplasmic male-sterility on the expression of resistance to insects in sorghum. In Fisher, T., Turner, N., Angus, J., McIntyre, L., Robertson, M., Borrell, A. and Lloyd, D. (Eds.), *New Directions for a Diverse Planet: Proceedings of the 4th International Crop Science Congress*, 25 September–1 October, 2004, Brisbane, Queensland, Australia. http://www.cropscience.org.au.

Sharma, H.C., Mukuru, S.Z., Gugi, H. and King, S.B. (2000a). Inheritance of resistance to sorghum midge and leaf diseases in Kenya. *International Sorghum and Millets Newsletter* 41: 37–42.

Sharma, H.C., Pampapathy, G., Lanka, S.K. and Ridsdill-Smith, T.J. (2005b). Antibiosis mechanism of resistance to legume pod borer, *Helicoverpa armigera* in wild relatives of chickpea. *Euphytica* 142: 107–117.

Sharma, H.C., Reddy, B.V.S., Dhillon, M.K., Venkateswaran, K., Singh, B.U., Pampapathy, G., Folkerstma, R., Hash, C.T. and Sharma, K.K. (2005c). Host plant resistance to insects in sorghum: Present status and need for future research. *International Sorghum and Millets Newsletter* 46: 36–43.

Sharma, H.C., Satyanarayana, M.V., Singh, S.D. and Stenhouse, J.W. (2000b). Inheritance of resistance to head bugs and its interaction with grain molds in *Sorghum bicolor*. *Euphytica* 112: 167–173.

Sharma, H.C., Stevenson, P.C., Simmonds, M.S.J and Green, P.W.C. (2001). *Identification of* Helicoverpa armigera *(Hübner) Feeding Stimulants and the Location of Their Production on the Pod-Surface of Pigeonpea* [Cajanus cajan *(L.) Millsp.*]. Final Technical Report. DfiD Competitive Research Facility Project [R 7029 (C)]. Patancheru, Andhra Pradesh, India: International Crops Research Institute for the Semi-Arid Tropics (ICRISAT).

Sharma, H.C., Vidyasagar, P., Abraham, C.V. and Nwanze, K.F. (1994). Influence of cytoplasmic male-sterility on host plant interaction with sorghum midge, *Contarinia sorghicola*. *Euphytica* 74: 35–39.

Sharma, R.P. and Yadav, R.P. (2000). Construction of life tables to establish antibiosis resistance to the gram pod borer, *Heliothis armigera* (Hüb.) among chickpea genotypes. *Journal of Entomological Research* 24(4): 365–368.

Sidhu, G.S., Khush, G.S. and Medrano, F.G. (1979). A dominant gene in rice for resistance to white-backed planthopper and its relationship to other plant characteristics. *Euphytica* 28: 227–232.

Sinden, S.C., Sanford, L.C., Cantelo, W.W. and Deahl, K.L. (1986). Leptine glycoalkaloids and resistance to the Colorado potato beetle (Coleoptera: Chrysomellidae) in *Solanum chacoense*. *Environmental Entomology* 15: 1057–1062.

Singh, O., Gowda, C.L.L., Sethi, S.C. and Lateef, S.S. (1991). Inheritance and breeding for resistance to *Helicoverpa armigera*, pod borer in chickpea. In *Golden Jubilee Symposium on Genetic Research and Evaluation. Current Trends and Next Fifty Years*, 12–15 February, 1991. New Delhi, India: Indian Society of Genetics and Plant Breeding, Indian Agricultural Research Institute, 121.

Sisson, V.A., Miller, P.A., Campbell, W.V. and VanDuyn, J.W. (1976). Evidence of inheritance of resistance to the Mexican bean beetle in soybeans. *Crop Science* 16: 835–837.

Smith, C.M. (2005). *Plant Resistance to Arthropods: Molecular and Conventional Approaches*. Dordrecht, The Netherlands: Springer Verlag.

Soper, J.F., McIntosh, M.S. and Elden, T.C. (1984). Diallel analysis of potato leafhopper resistance among selected alfalfa clones. *Crop Science* 24: 667–670.

Srivastava, C.P. and Srivastava, R.P. (1989). Screening for resistance to gram pod borer, *Heliothis armigera* (Hubner) in chickpea (*Cicer arietinum* L.) genotypes and observations on its mechanism of resistance in India. *Insect Science and Its Application* 10: 255–258.

Srivastava, C.P. and Srivastava, R.P. (1990). Antibiosis in chickpea (*Cicer arietinum*) to gram pod borer, *Heliothis armigera* (Hubner) Noctuidae: Lepidoptera) in India. *Entomon* 15: 89–94.

Srivastava, M.N., Kumar, A., Shrivastava, S.K. and Sahu, R.K. (1993). A new gene for resistance to rice gall midge in rice variety Abhaya. *Rice Genetics Newsletter* 10: 79–80.

Stadler, E. and Buser, H.R. (1984). Defense chemicals in the leaf surface wax synergistically stimulate oviposition by phytophagous insects. *Experientia* 40: 1157–1159.

Starks, K.J. and Merkle, O.G. (1977). Low level resistance in wheat to greenbug. *Journal of Economic Entomology* 70: 305–306.

Stebbins, N.B., Patterson, F.I. and Gallun, R.I. (1983). Inheritance of resistance of PI94587 wheat to biotypes B and D of Hessian fly. *Crop Science* 23: 251–253.

Stephens, J.C. and Holland, R.F. (1954). Cytoplasmic male sterility for hybrid sorghum seed production. *Agronomy Journal* 46: 20–23.

Stevenson, P.C., Anderson, J.C., Blaney, W.M. and Simmonds, M.S.J. (1993). Developmental inhibition of *Spodoptera litura* (Fab.) larvae by a novel caffeoyl quinic acid from the wild groundnut *Arachis paraguariensis* (Chodat & Hassl.). *Journal of Chemical Ecology* 19: 2917–2933.

Stevenson, P.C., Green, P.W.C., Simmonds, M.S.J. and Sharma, H.C. (2005). Physical and chemical mechanisms of plant resistance to *Helicoverpa*: Recent research on chickpea and pigeonpea. In Sharma, H.C. (Ed.), Helicoverpa/Heliothis *Management: Emerging Trends and Strategies for Future Research.* New Delhi, India: Oxford & IBH Publishers, 215–228.

Tamura, K., Fukuta, Y., Hirae, M., Oya, S., Ashikawa, I. and Vagi, T. (1999). Mapping of the Grh/locus for green rice leafhopper resistance in rice using RFLP markers. *Breeding Science* 49: 11–14.

Tatum, L.A. (1971). The southern corn leaf blight epidemic. *Science* 171: 1113–1116.

Theurer, J.C., Blickenstaff, C.C., Mahrt, G.G. and Doney, D.L. (1982). Breeding for resistance to the sugarbeet root maggot. *Crop Science* 22: 641–645.

Thompson, J.N. and Pellmyr, O. (1991). Evolution of oviposition behavior and host preference in Lepidoptera. *Annual Review of Entomology* 36: 65–89.

Tingey, W.M. (1981). The environmental control of insects using plant resistance. In Pimentel, D. (Ed.), *CRC Handbook of Pest Management in Agriculture.* Boca Raton, Florida, USA: CRC Press, 175–197.

Tingey, W.M. and Sinden, S.L. (1982). Glandular pubescence, glycoalkaloid composition and resistance to the green peach potato aphid, potato leaf hopper, and potato flea hopper in *Solanum berthaultii. American Potato Journal* 59: 95–106.

Tomaka, N., Lairungreang, C., Nakeeraks, P., Egawa, Y. and Thavarasook, C. (1992). Development of bruchid-resistant mungbean line using wild mungbean germplasm in Thailand. *Plant Breeding* 109: 60–66.

Tuinstra, M.R., Wilde, G.E. and Krieghauser, T. (2001). Genetic analysis of biotype I resistance in sorghum. *Euphytica* 121: 87–91.

Tyler, J.M., Webster, J.A. and Merkle, O.G. (1987). Designations for genes in wheat germplasm conferring greenbug resistance. *Crop Science* 27: 526–527.

Varis, A.L. (1958). On the susceptibility of the different varieties of big-leafed turnip to damage caused by cabbage maggots (*Hylemyia* sp.). *Journal of Science, Agriculture Society of Finland* 30: 271–275.

Verulkar, S.B., Singh, D.P. and Bhattacharya, A.K. (1997). Inheritance of resistance to podfly and pod borer in the interspecific cross of pigeonpea. *Theoretical and Applied Genetics* 95: 506–508.

Virmani, S.S. and Wan, B.H. (1988). Development of CMS lines in hybrid rice breeding. In *Rice Breeding.* Manila, Philippines: International Rice Research Institute, 103–114.

Wang, C., Yasui, H., Yoshimura, A., Zhai, H. and Wan, J. (2004). Inheritance and QTL mapping of antibiosis to green leafhopper in rice. *Crop Science* 44: 389–393.

Warnock, D.F., Davis, D.W. and Gingera, G.R. (1998). Inheritance of ear resistance to European corn-borer in "Apache" sweet corn. *Crop Science* 38: 1451–1457.

Weibel, D.E., Starks, K.J., Wood, E.A. Jr. and Morrison, R.D. (1972). Sorghum cultivars and progenies rated for resistance to greenbugs. *Crop Science* 12: 334–336.

Weng, Y. and Lazar, M.D. (2002). Amplified fragment length polymorphism and simple sequence repeat-based molecular tagging and mapping of greenbug resistance gene *GbJ* in wheat. *Plant Breeding* 121: 218–223.

Widstrom, N.W. (1972). Reciprocal differences and combining ability for corn earworm injury among maize single crosses. *Crop Science* 12: 245–247.

Widstrom, N.W. (1989). Breeding methodology to increase resistance in maize to corn earworm, fall armyworm, and maize weevil. In *Toward Insect Resistant Corn for the Third World: Proceedings of the International Symposium on Methodologies for Developing Host Plant Resistance to Corn Insects.* Mexico D.F., Mexico: International Wheat and Maize Research Center (CIMMYT), 209–219.

Widstrom, N.W. and McMillian, W.W. (1973). Genetic factors conditioning resistance to earworm in maize. *Crop Science* 13: 459–461.

Widstrom, N.W. and Snook, M.E. (1994). Inheritance of maysin content in silks of maize inbreds resistant to corn earworm. *Plant Breeding* 112: 120–126.

Widstrom, N.W. and Snook, M.E. (2001). Recurrent selection for maysin, a compound in maize silks, antibiotic to earworm. *Plant Breeding* 120: 357–359.

Widstrom, N.W., Wiseman, B.R. and McMillian, W.W. (1972). Resistance among some maize inbreds and single crosses to fall armyworm injury. *Crop Science* 12: 290–292.

Widstrom, N.W., Wiseman, B.R. and McMillian, W.W. (1984). Patterns of resistance in sorghum to the sorghum midge. *Crop Science* 24: 791–793.

Widstrom, N.W., Williams, W.P., Wiseman, B.R. and Davis, F.M. (1992). Recurrent selection for resistance to leaf feeding by fall armyworm in maize. *Crop Science* 32: 1171–1174.

Williams, W.P., Buckley, P.F. and Davis, F.M. (1995). Combining ability in maize for fall armyworm and southwestern corn borer resistance based on a laboratory bioassay for larval growth. *Theoretical and Applied Genetics* 90: 275–278.

Williams, W.P., Davis, F.M. and Buckley, P.F. (1998). Resistance to southwestern corn borer in corn after anthesis. *Crop Science* 38: 1514–1517.

Wilson, F.D. and George, B.W. (1979). Combining ability in cotton for resistance to pink bollworm. *Crop Science* 19: 834–836.

Wilson, F.D. and George, B.W. (1983). A genetic and breeding study of pink bollworm resistance and agronomic properties in cotton. *Crop Science* 23: 1–4.

Wilson, F.D. and Lee, J.A. (1971). Genetic relationships between tobacco budworm feeding response and gland number in cotton seedlings. *Crop Science* 11: 419–421.

Wilson, F.D. and Smith, J.N. (1977). Variable expressivity and gene action of gland-determining alleles in *Gossypium hirsutum* L. *Crop Science* 17: 539–543.

Wilson, R.L., Wiseman, B.R. and Snook, M.E. (1995). Evaluation of pure red pericarp and eight selected maize accessions for resistance to corn earworm (Lepidoptera, Noctuidae) silk feeding. *Journal of Economic Entomology* 88: 755–758.

Wiseman, B.R. and Bondari, K. (1992). Genetics of antibiotic resistance in corn silks to the corn earworm (Lepidoptera: Noctuidae). *Journal of Economic Entomology* 85: 293–298.

Wiseman, B.R. and Bondari, K. (1995). Inheritance of resistance in maize silks to the corn earworm. *Entomologia Experimentalis et Applicata* 77: 315–321.

Wiseman, B.R., McMillian, W.W. and Widstrom, N.W. (1972). Tolerance as a mechanism of resistance in corn to the corn earworm. *Journal of Economic Entomology* 65: 835–837.

Wu, L.F., Cai, Q.N. and Zhang, Q.W. (1997). The resistance of cotton lines with different morphological characteristics and their F_1 hybrids to cotton bollworm. *Acta Entomologica Sinica* 40: 103–109.

Yang, D., Parco, A., Nandi, S., Subudhi, P., Zhu, Y., Wang, G. and Huang, H.N. (1997). Construction of a bacterial artificial chromosome (BAC) library and identification of overlapping BAC clones with chromosome 4-specific RFLP markers in rice. *Theoretical and Applied Genetics* 95: 1147–1154.

Yang, R.C., Lu, H.R., Zhang, X.B., Xia, Y.H., Li, W.M., Liang, K.J., Wang, N.Y. and Chen, Q.H. (1989). Study of the susceptibility of CMS WA cytoplasm in rice to blast and bacterial blight. *Acta Agronomica Sinica* 15: 310–318.

Yazawa, S., Yasui, H., Yoshimura, A. and Iwata, N. (1998). RFLP mapping of genes for resistance to green rice leafhopper (*Nephotettix cincticeps* Uhler) in rice cultivar DV85 using near isogenic lines. *Science Bulletin - Faculty of Agriculture, Kyushu University* 52: 169–175.

Zhang, Y.J., Quick, S. and Liu, S. (1998). Genetic variation in PI294994 wheat for resistance to Russian wheat aphid. *Crop Science* 38: 527–530.

Zhu, L.C., Smith, C.M., Fritz, A., Boyko, E.V. and Flinn, M.B. (2004). Genetic analysis and molecular mapping of a wheat gene conferring tolerance to the greenbug (*Schizaphis graminum* Rondani). *Theoretical and Applied Genetics* 109: 289–293.

6

Physico-Chemical and Molecular Markers
for Resistance to Insect Pests

Introduction

Because of environmental and human health problems associated with excessive use of pesticides, there has been an increased emphasis on alternative methods of controlling insect pests, including host plant resistance to insects, which can play a pivotal role in integrated pest management. Sources of resistance to insects have been identified long ago, but these have not been used effectively in crop improvement, because the levels of resistance are either too low or it is not possible to screen the test material under uniform and optimum insect infestation levels to identify lines combining desirable agronomic traits and resistance to insect pests (Sharma and Ortiz, 2002; Sharma et al., 2005). Thus, there is a need for improving the accuracy and precision of phenotyping for insect resistance. Once the phenotyping systems are developed, molecular markers can be used to accelerate the pace of development of insect-resistant cultivars, dissecting the genetic basis of resistance, identifying the location of underlying genes, and understanding the nature of gene action (Sharma et al., 2002).

Screening of germplasm collections has resulted in identification of accessions with moderate to high levels of resistance against several insect species in different crops (Smith, 1989, 2005; Panda and Khush, 1995; Sharma and Ortiz, 2002; Sharma et al., 2005). However, large-scale screening of thousands of germplasm accessions for insect resistance probably has resulted in missing many germplasm accessions with diverse genes for insect resistance (Clement and Quisenberry, 1999). Many of the identified sources of resistance have not been used widely because the levels of resistance are either low or it has not been possible to transfer the genes for insect resistance into high-yielding cultivars because of linkage drag. To overcome these problems, marker-assisted selection can be used as a powerful tool in terms of pyramiding resistance genes, and identifying segregants not carrying undesirable traits (provided there are no pleiotropic effects of the insect resistance genes).

Varieties with resistance to insect pests have been identified and released for cultivation in different crops (Panda and Khush, 1995; Sharma and Ortiz, 2002). However, the levels of resistance in some of the varieties released for cultivation are low to moderate. Therefore, there is a need to increase the levels and diversify the basis of resistance to insect pests through exploitation of resistance sources in the cultivated germplasm, wild relatives of crops, and genetic engineering of novel genes from unrelated species to make host plant resistance an effective weapon for pest management.

The last decade has seen rapid progress in molecular biology, with whole genome sequencing of model organisms such as humans, *Saccharomyces*, *Arabidopsis*, and *Oryza* (Chalfie, 1998; Sherman, 1998; Palevitz, 2000; Shoemaker et al., 2001; Piskur and Langkjaer, 2004). Systematic whole genome sequencing will provide critical information on gene and genome organization and function, which will revolutionize our understanding of crop production and ability to manipulate traits contributing to crop productivity (Pereira, 2000; Crouch et al., 2005). The advances in genome sequencing in major crops will have substantial spillover effects on lesser-studied crops. Recombinant DNA technologies will allow identification of specific chromosomal regions carrying the genes associated with resistance to the target insect pest (Karp et al., 1997). There are many types of DNA markers, and each of these have a different set of advantages for any particular application in linkage mapping and marker-assisted selection (MAS) for resistance to insect pests. Once genomic regions contributing to the trait of interest have been identified and the alleles at each locus designated by molecular markers, they can be transferred into locally adapted, high-yielding cultivars by crossing, and following the marker(s) through subsequent generations of inbreeding or backcrossing. Wild relatives of crops also contain alleles of importance for resistance to insect pests. Since these alleles are often recessive, they can only effectively be utilized in crop breeding programs through MAS (Xiao et al., 1996; Miflin, 2000). Molecular markers can be used to:

- Estimate genetic variances (Bai, Michaels, and Pauls, 1998).
- Predict hybrid performance (Bohn et al., 1997).
- Estimate the number of genes in which the parents differ (Kisha, Sneller, and Diers, 1997).
- Identify quantitative trait loci (QTL) associated with resistance to insect pests (Peirera, 2000).

Genetic maps based on recombination frequencies are an important tool in crop improvement (Karp et al., 1997; Mohan et al., 1997). At times, there may be discrepancies in physical and genetic maps and, therefore, it is important to correlate genetic and physical maps for mapping and isolating the genes of interest. Physical or genetic mapping can be accomplished by using:

- Terminal deficiencies (B.Y. Lin et al., 1997).
- Translocation lines (Kunzel, Korzum, and Meister, 2000).
- Pulsed field gel electrophoresis (Bonnema et al., 1996).
- YAC and BAC contiguous DNA sequences (Kurata et al., 1997; Dunford et al., 2002).
- Genomic introgression (Humphreys et al., 1998).
- Fluorescence *in situ* hybridization (FISH) (J. Jiang and Gill, 1994; Kim et al., 2004).

Physical maps can be used for gene isolation through microdissection and micro-cloning. Gene structure and function can also be elucidated by comparing the mapped positions of genes with their physical locations, and physical distances between genes can be used to gain an understanding of the mechanism of chromosome recombination and rearrangements. High-density genetic linkage maps have been developed for barley, *Hordeum vulgare* L.; maize, *Zea mays* L; potato, *Solanum tuberosum* L.; rye, *Secale cereale* L.; sorghum, *Sorghum bicolor* (L.) Moench; soybean, *Glycine max* (L.) Merr.; tomato, *Lycopersicon esculentum* Mill; and wheat, *Triticum aestivum* L. (Paterson, Tanksley, and Sorrells, 1991; Hernandez et al., 2001; Korzun et al., 2001; Boyko et al., 2002; Sharopova et al., 2002; Song et al., 2004; Somers, Isaac, and Edwards, 2004). Molecular markers in many of these crops have also been linked to genes expressing resistance to several insect pests. Developments in DNA marker technology can be used to accelerate the process of transferring insect resistance into improved cultivars. Once markers linked closely to the resistance genes are identified, MAS can be practiced in early generations at early stages of plant development, and speed up the selection process. The MAS can also be used for pyramiding resistance genes from diverse sources. Location of the marker away from the gene of interest may lead to cross over between the gene of interest and the marker, and the marker identified for a gene in one cross may not be useful in another cross unless the marker is linked to the resistance gene (Mohan et al., 1997). The MAS takes 3 to 6 years, and thus speeds up the pace of transferring the traits of interest into improved varieties, and it does not require large-scale planting of the segregating material up to crop harvest, as only the plants showing the presence of markers associated with QTLs linked to resistance alleles need to be maintained up to maturity.

Mapping Populations

It takes five to six generations to transfer insect resistance traits into the high-yielding cultivars through conventional breeding, while gene transfer from wild relatives may take a considerably longer time due to the complexity of achieving interspecific hybrids on a sufficiently large scale to identify stable progeny with an acceptable combination of traits. In either case, MAS can dramatically speed up the process by reducing the number of generations and the size of the populations required to identify individuals with appropriate combination of genes, while having the minimal amount of linkage drag from the wild relatives. The improved lines with insect resistance thus developed will need to be tested across seasons and locations, before a variety can be identified for cultivation by farmers. This process takes 7 to 10 years. In MAS programs, the elite breeding line or cultivars can be crossed with the source of resistance, and the F_1 hybrid recrossed with the recurrent parent (invariably the elite parent) (BC_1), and the gene transfer can be monitored through MAS up to BC_{3-5} (until a line with the QTL or the gene of interest in the genomic background of the elite line with a minimum of donor parent genome is identified).

Near isogenic lines (NILs), F_2 and backcross populations, doubled haploids, and recombinant inbred lines (RILs) can be used for gene mapping in many crops (Mohan et al., 1994, 1997). Mapping populations from interspecific crosses are often used for genetic linkage studies due to high levels of detectable polymorphism, but linkage maps derived from such crosses may have limited relevance in crop breeding programs due to different recombination patterns (Fulton et al., 1997). However, markers developed from such maps may be valuable tools for introgression of genes of interest from the wild relatives into the cultigen. The F_{6-8} progenies of crosses involving a resistance source from the wild relatives and the cultivated types can also be used as RILs for mapping insect resistance (provided the population has been advanced through the generations in a correct manner).

Linkage between a resistance gene and a linked marker may vary greatly. The two may be completely linked, where no crossing over occurs between the resistance gene and the marker during meiosis, and the gene and marker are always transferred together from one generation to another. The resistance gene and a molecular marker at times are incompletely linked, and crossing over may occur between the gene and the marker during meiosis, thus breaking the linkage between the marker allele and the resistance allele of the parents. Estimates of the recombination between the resistance gene and the linked marker are measured as the recombination frequency (RF), which is measured among backcross progenies of segregating F_2s, $F_{2:3}$ families, or RILs by matching the phenotype and genotype of each progeny (Lander et al., 1987). Mapping simultaneously estimates all recombination frequencies for markers segregating as dominant, recessive, and codominant traits in the mapping population. The linkage between QTL and marker loci is based on the distribution patterns for the resistance characters linked to insect resistance genes and the molecular marker at each locus (Lincoln, Daly, and Lander, 1993).

To estimate recombination frequency between a resistance gene and a molecular marker, researchers often analyze between 100 and 500 $F_{2:3}$ progenies or RILs derived from crosses of parents with known resistance or susceptibility. DNA is collected from the resistant and susceptible parents, as well as from each F_2 plant or several plants in F_6 RILs. Different DNA markers from a variety of chromosome locations are screened to identify those detecting polymorphisms between the parents. If the parent polymorphisms are apparent in the bulked segregant DNA samples, the marker is referred to as a putatively linked marker, and DNA of all F_2 plants or RILs families is also evaluated. At this point, QTL mapping software packages such as Mapmaker, Joinmap, QTL cartographer, etc., are used to correlate the phenotype and genotype of the plants in a mapping population to develop the genetic linkage map. Linkage mapping software packages contain mapping functions such as the Haldane's (1919) and Kosambi's (1944) functions to correct any under-estimation.

Markers associated with insect resistance can be physico-chemical factors of the plants (plant hairs or trichomes, flower and seed color, leaf size and shape, thickness of the cell wall, protein, amino acid, sugar, and fatty acid profiles, secondary metabolites such as alkaloids, terpenoids, flavonoids, isozymes, and protease inhibitors) or DNA based [RAPDs (random-amplified polymorphic DNA), RFLPs (restriction fragment length polymorphisms), AFLPs (amplified fragment length polymorphic DNA), SCARs (sequence characterized amplified regions), STS (sequence tagged sites), SSRs (simple sequence repeats), ALPs (amplicon length polymorphisms), and DArTs (diversity array technology)] (Crouch et al., 2005).

Physico-chemical Markers Associated with Resistance to Insects

Morphological Markers

Plants have acquired several physicochemical characteristics that contribute to insect resistance in different crops (Panda and Khush, 1995; Smith, 2005) (Table 6.1). The role of various physico-chemical components in host plant resistance to insects is discussed below.

TABLE 6.1

Morphological Markers Associated with Resistance to *Heliothis/Helicoverpa*

Crop	Morphological Trait	Remarks	References
Cotton	Nectarless	Nonpreference	Lukefahr et al. (1960)
	Leaf pubescence	Oviposition preference	Butter et al. (1996)
	Low bract teeth	Resistance	Rao et al. (1996)
	Red leaf	Resistance	Ansingkar et al. (1984)
	Okra leaf	Resistance	Wu, Cai, and Zhang (1997)
	Frego-bract	Resistance	Wu, Cai, and Zhang (1997)
	Erect growth habit	Resistance	Navasero and Ramaswamy (1991)
	Glandless	Susceptibility	Narayanan and Singh (1994)
	Tough bolls	Resistance	Mohan et al. (1995)
Pigeonpea	Nonglandular trichomes	Resistance	Romeis, Shanower, and Peter (1999)
	Colored seeds	Resistance	Nanda, Sasmal, and Mohanty (1996)
	Long and wide pods	Susceptibility	Nanda, Sasmal, and Mohanty (1996)
	Clustered podding	Susceptibility	Raut et al. (1993)
	Indeterminate growth habit	Resistance	Lateef and Pimbert (1990)
Chickpea	Glandular trichomes	Resistance	Yoshida, Cowgill, and Wightman (1997)
	Long and wide pods	Susceptibility	Gururaj et al. (1993)
Tomato	Glandular trichomes	Resistance	Farrar and Kennedy (1988), Sivaprakasam (1996)
	Pubescence	Oviposition preference	AVRDC (1987)
	Calyx texture and flowering	Resistance	Cosenza and Green (1979)
	Fruit skin	Resistance	Cosenza and Green (1979)
	Bigger vine size	Resistance	Ferry and Cuthbert (1973)
Soybean	Trichomes	Resistance	Panda (1979)

Visual Stimuli

Insects use vision to recognize the shape of, and chemical stimuli to perceive, the host plant. Host finding from a distance is the first step in host plant selection and colonization by insects. It involves the visual stimuli emitted by the host plant, for example, aphids are attracted to yellow-white surfaces (Moericke, 1969; Prokopy and Owens, 1983), cotton white fly, *Bemisia tabaci* (Genn.) is attracted to yellow-green and yellow-red (Hussain and Trehan, 1940), and sorghum midge, *Stenodiplosis sorghicola* (Coquillett) to white-yellow colors (Sharma, Leuschner, and Vidyasagar, 1990; Sharma and Franzmann, 2002). Grasshoppers orient themselves towards vertical lines or objects (Williams, 1954). The glossy trait of cabbage, *Brassica oleracea* var *capitata* L. leaves confers antixenosis to the cabbage aphid, *Brevicoryne brassicae* L. (Singh and Ellis, 1993). Sorghum genotypes with glossy leaf trait (light, yellow, shiny, and erect leaves) (Figure 6.1) are nonpreferred for oviposition by sorghum shoot fly, *Atherigona soccata* (Rondani) as compared to genotypes with dark green, drooping, and nonshiny leaves (Sharma and Nwanze, 1997). Plants having yellowish green leaves are nonpreferred for oviposition by pod borer, *Helicoverpa armigera* (Hubner) as

FIGURE 6.1   Glossy leaf trait in sorghum associated with resistance to sorghum shoot fly, *Atherigona soccata*.

compared to the cultivars with dark green leaves (Sundramurthy and Chitra, 1992; Jing et al., 1997). Red strains of *Gossypium arboreum* L. in cotton are resistant to spotted bollworm, *Earias vittella* (Fab.) (Sharma, Agarwal, and Singh, 1982). In groundnut, genotypes with dark green and smaller leaflets are less damaged by *H. armigera* than those with longer shoots, and larger and light green leaflets (Arora, Kaur, and Singh, 1996).

Phenological Traits

Rice varieties with tight leaf sheaths are less susceptible to the Asiatic stem borer, *Chilo suppressalis* (Walker) (Patanakamjorn and Pathak, 1967), while spring wheat varieties with tight leaf sheaths are resistant to leaf miner, *Hydrellia griseola* (Fall) (Zhu, 1981). In sorghum, varieties with short and light glumes are resistant to sorghum midge, *S. sorghicola* (Sharma, Vidyasagar, and Leuschner, 1990), while those with longer glume covering of the grain are less susceptible to the African head bug, *Eurystylus oldi* (Poppius) (Sharma et al., 1994). Sorghum genotypes with compact panicles are more susceptible to *H. armigera* than those with loose panicles (Sharma, Lopez, and Vidyasagar, 1994). Silica deposition on the fourth and fifth leaves of sorghum is associated with resistance to shoot fly, *A. soccata* (Ponnaiya, 1951; Bothe and Pokharkar, 1985), while high silica content has been reported to interfere with feeding by the Asiatic stem borer, *C. suppressalis* (Pathak et al., 1971) in rice, and shoot borer, *Chilotraea infuscatellus* Snellen in sugarcane (Rao, 1967).

In cotton, earliness, smaller leaves, and nonclustered bolls provide a mechanism to avoid damage by bollworms. Smooth and okra leaves, frego-bract, nectarlessness, open canopy, and naked seed characteristics are associated with resistance to *H. armigera* (Khalifa, 1979). A combination of frego-bract, okra leaf, red leaf, and nectarlessness contribute to maximum reduction of damage by *H. armigera* (Bhat and Jayaswal, 1988; Wu, Cai, and Zhang, 1997). The nectarless trait in cotton is associated with reduced numbers of tobacco budworm, *Heliothis virescens* (F.), and pink bollworm, *Pectinophora gossypiella* (Saunders) (Lukefahr, Houghtaling, and Graham, 1975).

Husk tightness in maize imparts resistance to corn earworm, *Helicoverpa zea* (Boddie) (Wiseman and Isenhour, 1994; Widstrom et al., 2003b), and has also been found to result in reduced levels of aflatoxin B_1 (Widstrom et al., 2003a). Pearl millet varieties with a dense covering of anthers and without awns are more susceptible to *H. armigera* than the varieties without anther cover and awns (Sharma and Youm, 1999). Pigeonpea genotypes with determinate growth habit, clustered pods, and dense plant canopy are more susceptible to pod borers, *H. armigera* and *Maruca vitrata* (Geyer) than genotypes with nonclustered pods (Sharma, Bhagwat, and Saxena, 1997), while the genotypes with smaller pods, pod wall tightly fitting to the seeds, and a deep constriction between the seeds are less susceptible to *H. armigera* (Nanda, Sasmal, and Mohanty, 1996). Several morphological traits such as pod shape, pod wall thickness, and crop duration influence *H. armigera* damage in chickpea (Ujagir and Khare, 1988). Main stem thickness, leaflet shape and length, leaf hairiness, and peg length are associated with resistance or susceptibility to *H. armigera*, and tobacco leaf caterpillar, *Spodoptera litura* (F.), in wild relatives of groundnut (Sharma et al., 2003). The *H. zea* damage in tomato is associated with vine size and fruit number (Ferry and Cuthbert, 1973). Calyx texture, flowering, and fruit skin thickness are also associated with resistance to *H. zea* (Cosenza and Green, 1979). Pubescence on the leaf tip is associated with reduced defoliation by *H. zea*, *Spodoptera exigua* (Hubner), and soybean looper, *Pseudoplusia includens* (Walker) in soybean (Hulburt, Boerma, and All, 2004).

Leaf Hairs

Leaf hairs (that do not produce glandular secretions) play an important role in host plant resistance to insects. Hairy cotton varieties are preferred by the cotton leaf roller, *Sylepta derogata* Fab. (Mahal, Dhawan, and Singh, 1980), cotton white fly, *B. tabaci* (Ilyas, Puri, and Rote, 1991), and tobacco budworm, *H. virescens* (Lukefahr, Haughtaling, and Graham, 1971; Robinson, Wolfonbarger, and Doday, 1980). Glabrous cottons have 60% fewer eggs of *H. virescens* than the hairy ones. Some of the eggs laid on glabrous surface are also dislodged. Females of *H. armigera* also lay more eggs on hairy varieties of cotton compared to the glabrous ones (Natarajan, 1990). However, leaf hairs in cotton confer resistance to leafhopper, *Amrasca biguttula biguttula* Ishida (Sharma and Agarwal, 1983a) (Figure 6.2), and western tarnished plant bug, *Lygus hesperus* Knight (Benedict, Leigh, and Hyer, 1983). Similarly in soybean, *H. armigera* females laid three times more eggs on the pubescent lines as compared to the nonpubescent ones (Panda and Daugherty, 1975), but pubescent varieties are highly resistant to leafhopper, *Empoasca fabae* Harris (Kogan, 1982). Wild relatives of pigeonpea such as *Cajanus scarabaeoides* (L.) F. Muell. and *C. acutifolius* (F. Muell. Ex Benth.) Maesen with nonglandular trichomes are not preferred by *H. armigera* females for egg laying (Sharma et al., 2001). Rice varieties with hairy upper leaf lamina are less susceptible to the Asiatic rice stem borer, *C. suppressalis* (Patanakamjorn and Pathak, 1967), while wheat varieties with pubescent leaves are resistant to the cereal leaf beetle, *Oulema melanopus* (L.) (Wallace, McNeal, and Berg, 1974). Leaf hairs also confer resistance to the jassid, *A. biguttula biguttula* in brinjal (Gaikwad, Darekar, and Chavan, 1991), and to the mustard aphid, *Lipaphis erysimi* Kalten. in *Brassica juncea* L. (Lal, Singh, and Singh, 1999).

FIGURE 6.2 Leaf hairiness in cotton associated with resistance to leafhopper, *Amrasca biguttula biguttula*.

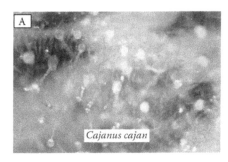

Cajanus cajan

Cajanus scarabaeoides

FIGURE 6.3   Glandular (A) and nonglandular (B) trichomes in pigeonpea, *Cajanus cajan*, and its wild relative, *Cajanus scarabaeoides*, are associated with susceptibility and resistance, respectively, to *Helicoverpa armigera*.

Trichomes

Trichomes are hair-like glandular outgrowths on the epidermis of plants. These are found on leaves, shoots, and roots, and occur in several shapes and sizes (Jeffree, 1986). They affect locomotion, attachment, shelter, feeding, and survival of insects. The glandular trichomes in dicotyledons produce sticky exudates that may contain chemicals acting as phagostimulants (Figure 6.3), antifeedants, and/or repellents (Figure 6.4), and disrupt development and survival of insects (Peter, Shanower, and Romeis, 1995; Romeis, Shanower, and Peter, 1999; Rupakala et al., 2005). Trichomes act as a defense mechanism of plants to insects, although in some instances, they also provide a suitable substrate for egg laying and attachment, or interfere with the activity of natural enemies (Kauffman and Kennedy, 1989; Sharma, Pampapathy, and Sullivan, 2003). Some insect species have also developed adaptations to neutralize the effect of trichomes or their exudates (Gregory et al., 1986; Kennedy et al., 1987).

Hooked trichomes on bean, *Phaseolus vulgaris* L., impair the movement of the aphid, *Aphis craccivora* Koch (Johnson, 1953) and potato leafhopper, *E. fabae* (Pillemer and Tingey, 1978). Glandular exudates in *Solanum neocardenasii* Hawkes and Hjerting provide resistance

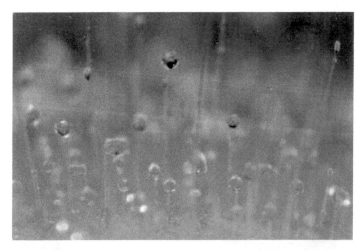

FIGURE 6.4   Acid exudates on the leaf surface of chickpea which act as antifeedants to pod borer, *Helicoverpa armigera*.

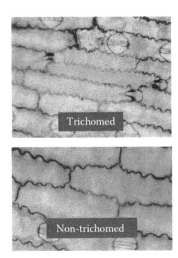

FIGURE 6.5 Trichomes on the under surface of sorghum leaf associated with resistance to sorghum shoot fly, *Atherigona soccata.*

against *Myzus persicae* Sulzer (Lapointe and Tingey, 1984). Cotton white fly, *B. tabaci*, adults are trapped by the trichome exudates in tomato (Kisha, 1984). Presence of trichomes is also associated with oviposition nonpreference in wheat to Hessian fly, *Mayetiola destructor* (Say) (Roberts et al., 1979) and to the cereal leaf beetle, *O. melanoplus* (Lampert et al., 1983), to the spotted stem borer, *Chilo partellus* (Swinhoe) in maize (Kumar, 1992), and shoot fly *A. soccata* in sorghum (Sharma and Nwanze, 1997) (Figure 6.5).

Glandular trichomes in pigeonpea are linked to susceptibility to *H. armigera* (Peter, Shanower, and Romeis, 1995; Romeis, Shanower, and Peter, 1999; Sharma et al., 2001). Trichomes and their exudates in chickpea influence the movement and feeding of neonate larvae of *H. armigera* (Stevenson et al., 2005), and influence the feeding by larvae of the spotted pod borer, *M. vitrata* in cowpea, *Vigna unguiculata* (L.) Walp. (Jackai and Oghiakhe, 1989), and cabbage looper, *Trichoplusia ni* (Hubner), in soybean (Khan, Ward, and Norris, 1986). Trichomes on the pods of *Vigna vexillata* (L.), a wild relative of cowpea, are partly responsible for resistance to the pod sucking bug, *Clavigralla tomentosicollis* Stal (Chiang and Singh, 1988). Removal of glandular trichomes in wild potato, *Solanum berthaulti* Hawkes, results in increased mortality of the Colorado potato beetle, *Leptinotarsa decemlineata* (Say) (Neal, Steffens, and Tingey, 1989), and decreases the feeding by green peach potato aphid, *M. persicae* (Lapointe and Tingey, 1984). Glandular trichomes in tomato and its wild relative, *L. hirsutum* f *glabratum* L. confer resistance to *H. zea* (Farrar and Kennedy, 1987) and the aphid, *Macrosiphum euphorbiae* (Thomas) (Musetti and Neal, 1997). Trichomes obstruct the feeding by black wine weevil, *Otiorhynchus sulcatus* (Fab.) in strawberry, *Fragaria chiloensis* (L.) Duch. (Doss et al., 1987), and *Epilachna verivestis* (Mulsant) in soybean (Gannon and Beach, 1996).

Plant Growth Responses

Hypersensitive growth responses of plants, such as rapidly growing tissues of cotton bolls, may kill larvae of the bollworm, *H. armigera*. Rapidly growing tissue of cotton bolls also kill the larvae of boll weevil, *Anthonomus grandis* (Boheman) (Hinds, 1906) and spotted bollworm, *E. vittella* (Sharma and Agarwal, 1983b). Wild accession of *Pisum sativum* L. ssp.

humile Poir. respond to pea weevil, *Bruchus pisorum* L., eggs by forming callus (Clement, Hardie, and Elberson, 2002). Mustard plants with a hypersensitive response to insect feeding produce a necrotized zone around the base of eggs of the cabbage worm, *Artogeia rapae* L., leading to desiccation of the eggs (Shapiro and DeVay, 1987).

Biochemical Markers

Secondary plant substances and poor nutritional quality of the host plant are important components of resistance to insects. Antibiosis to insects in general is due to secondary plant substances, for example, gossypol and tannins in cotton (Lukefahr, Houghtaling, and Graham, 1975; Sharma, 1982; Sharma and Agarwal, 1983a, 1983b), tridecanone and tomatine in tomato (Farrar and Kennedy, 1988, 1990), maysin and DIMBOA in maize (Klun, Tipton, and Brindley, 1967; Waiss et al., 1979), oxalic and malic acids in chickpea (Yoshida, Cowgill, and Wightman, 1995), and isoflavonoids in pigeonpea (Sharma et al., 2001; Green et al., 2002, 2003) (Table 6.2). The role of secondary plant substances as biochemical markers in plant resistance to insects is discussed below.

Attractants

Chemical cues involved in long-range identification of the host plant are called attractants, although the final acceptance or rejection of a host plant depends on contact with the host plant for feeding and oviposition. Dipropyl disulfide is involved in host finding by the onion fly, *Delia antigua* (Meigen) (Stadler, 1986). Carrot fly, *Psila rosae* (F.), is guided to the host plant by a combination of compounds with varying degrees of volatility (Stadler and Buser, 1984). Glucosinates and isothiocyanates play an important role in host finding by the insects feeding on cruciferous plants (Renwick and Radke, 1988). Plant odors emanating from the sorghum panicles at anthesis are involved in host finding by the sorghum midge, *S. sorghicola* (Sharma, Leuschner, and Vidyasagar, 1990). In olfactometer tests, some cotton genotypes are less attractive to the larvae of spotted bollworm, *E. vittella* (Sharma and Agarwal, 1981).

Repellents

Plant defense chemicals that prevent or reduce the contact between the insect herbivores and their host plants are called repellents. Steam distillates of rice and soybean varieties that are resistant to insects are repellent to brown planthopper, *Nilaparvata lugens* (Stal) (Saxena and Okech, 1985) and cabbage looper, *T. ni* (Khan, Ciepla, and Norris, 1987). Thymol and carvacresol are repellent to rice leaf folder, *Cnaphalocrocis medinalis* Gn.; α-pinine and 3-carene are repellent to pine beetle, *Blastophagus piniperda* (L.) (Oksanen, Pertlunen, and Kangas, 1970), while tomatine and capsicin are repellent to Colorado potato beetle, *L. decemlineata* (Schreiber, 1958). Essential oils are repellent to corn earworm, *H. zea* (Starks et al., 1965), and tannic acid to alfalfa weevil, *Hypera postica* (Gyllen.) (Bennett, 1965). Genetic engineering can be used to increase the production of repellents as a strategy for developing crop plants with better resistance to insects through changes in the metabolic pathways.

Phagostimulants

The chemicals that stimulate feeding by insects are called phagostimulants. Several chemicals on the pod surface of pigeonpea that are absent from the pods of wild relatives influence host plant selection for feeding by the larvae of *H. armigera* (Sharma et al., 2001).

TABLE 6.2

Biochemical Markers Associated with Resistance to *Heliothis/Helicoverpa*

Crop	Biochemical Marker	Remarks	References
Cotton	High gossypol	Resistance	Lukefahr et al. (1975), Hedin et al. (1983), Mohan et al. (1994)
	High heliocides (H_1–H_4)	Resistance	Chan et al. (1978), Ramnath (1990), Khan et al. (1999)
	High tannins	Resistance	Narayanan and Singh (1994)
	High phenols	Resistance	Rajeswari et al. (1999)
	High bud sucrose	Susceptibility	Narayanan and Singh (1994)
	Sugars and proteins	Susceptibility	Tang and Wang (1996)
Pigeonpea	Water extract	Antifeedant	Shanower, Yoshida, and Peter (1997), Green et al. (2002)
	Acetone extract	Phagostimulant	Shanower, Yoshida, and Peter (1997), Green et al. (2002)
	Low amino acids, sugars, proteins, and high phenol content	Resistance	Sahoo and Patnaik (2003)
	High tannins and polyphenols	Resistance	Dodia et al. (1998)
	High isoquercetrin, quercetin, and quercetin-3-methyl ether	Phagostimulant	Green et al. (2002, 2003)
	High stilbene content	Resistance	Green et al. (2002, 2003)
	Sesquiterpenes in lipid layers	Attractant	Langenheim (1994)
	High trypsin and amylase inhibitors	Antibiosis	Giri and Kachole (1998), Mullimani and Sudhendra (2002)
Chickpea	High oxalic acid	Antibiosis	Yoshida, Cowgill, and Wightman (1995)
	High malic acid	Antifeedant	Rembold (1981)
	Hexanal, β-pinine-1-acetate, limonene, α-cedrene methyl-cyclopentene, 2,3,4-trimethyl hexane	Attractants	Rembold and Winter (1982), Rembold et al. (1989, 1990)
	Low amino acid and soluble sugar content	Resistance	Chhabra et al. (1990)
	Lectins	Antibiosis	Shukla et al. (2005)
	High polyphenols, cellulose, hemicellulose, and lignins	Resistance	Chhabra et al. (1990)
	High activity of phenol ammonia lyase	Resistance	Bhatnagar et al. (2000)
	High judaicin-7-O-glucoside, 2-methoxy judaicin, maakiain, and cholorogenic acid	Antibiosis	Simmonds and Stevenson (2001)
Tomato	More glandular exudates	Resistance	Juvik and Stevens (1982)
	High L-tomatine, 2-tridecanone, Fe, and zinc	Resistance	Ferry and Cuthbert (1975), Dimock and Kennedy (1983), Farrar and Kennedy (1988, 1990)
	High phenolics	Resistance	Isman and Duffey (1982a, 1982b), Kashyap (1983)
	High ascorbic acid and low reducing sugars	Resistance	Banerjee and Kalloo (1989)
	Sequiterpenes: (+)-(E)-endo-santelon-12-oic acid, (+)-(E)-endo-β-bergamoten-12-oic acid	Oviposition stimulants	Coates et al. (1988)
	Chlorogenic acid, caryophyllene, and humulene	Antibiosis	Frelichowski and Juvik (2001) Farrar and Kennedy (1988, 1990)
	Acylglucoses	Resistance	Felton, Broaduray, and Duffey (1989) Juvik et al. (1994)
Maize	High maysin and isomaysin	Resistance	Wiseman et al. (1992)
Soybean	High pinitol	Resistance	Reese et al. (1982)
Groundnut	High caffeoylquinic acid	Antibiosis	Kimmins et al. (1995)

Acetone extracts of pods of pigeonpea and its wild relative, *Cajanus platycarpus* (Benth.) Maesen, have a significant feeding stimulant effect on *H. armigera* larvae, whereas the extracts from *C. scarabaeoides* pods have no such effect (Shanower, Yoshida, and Peter, 1997). Quercetin, quercetrin, and guercetin-3-methyl ether (present in pod surface exudates) play an important role in food selection behavior of *H. armigera* larvae in pigeonpea (Green et al., 2002, 2003). Sterols and soybean leaf extractables in combination with sucrose are phagostimulant to the larvae of the cabbage looper, *T. ni* (Sharma and Norris, 1994a).

Antifeedants

Chemicals that inhibit feeding are called phagodeterrents. High acidity in the leaf exudates of chickpea is associated with resistance to *H. armigera* (Srivastava and Srivastava, 1989; Bhagwat et al., 1995). Chickpea exudates have malate and oxalate as the main components, and there are characteristic differences among genotypes depending on diurnal cycles and growth stage. Varieties with the highest amounts of malic acid are resistant to *H. armigera* (Rembold, 1981; Rembold et al., 1990). Malic acid acts as an antifeedant to the *H. armigera* larvae (Bhagwat et al., 1995). Oxalic acid inhibits the growth of *H. armigera* larvae when incorporated into artificial diet, while malic acid shows no growth inhibition (Yoshida, Cowgill, and Wightman, 1995, 1997). The chickpea flavonoids judaicin 7-O-glucoside, 2 methoxy judaicin, judaicin, and maakiain present in wild relatives of chickpea (*Cicer bijugum* Rech. and *C. judaicum* Boiss.) have shown antifeedant activity towards the larvae of *H. armigera* (Simmonds and Stevenson, 2001). Stilbene, a phytoalexin, occurs at high concentrations in pigeonpea cultivars with resistance to *H. armigera* (Green et al., 2003).

Growth Inhibitors

Alkaloids, ketones, terpenoids, flavonoids, and organic acids produced by the plants are toxic to insects. Gossypol, heliocide 1, heliocide 2, gossypolone, hemigossypolone, lactone, condensed tannins, and volatile terpenes in cotton confer resistance to bollworms (Hedin et al., 1983; Sharma, 1982; Sharma and Agarwal, 1982a, 1982b; Tang and Wang, 1996). High density of gossypol glands on squares rather than leaves, calyxes, and bracts contributes to *H. armigera* resistance in cotton (Rajarajeswari and Subba Rao, 1997). Tannins and terpenoid aldehydes gradually increase from the cotyledon stage to the late-bloom stage of cotton (Zummo, Segers, and Benedict, 1984). However, there is a sudden drop in tannins at the one-third growth stage, which provides a window of increased vulnerability to *H. zea*. Phenolics have also been reported to impart resistance to bollworms. Catechin, chrysanthemin, isoquercitrin, quercetin, condensed tannins, cyanidin, and delphinidin have been tested in laboratory bioassays against *H. armigera*, and found to exert a weight inhibition of 50% at variable concentrations. The polar solvent extractables of the soybean genotype PI 227687, resistant to the cabbage looper, *T. ni*, contains diadzien, coumesterol, sojagol, and glyceollins. These compounds reduce feeding, survival, and development of *T. ni* (Sharma and Norris, 1991, 1994b).

In tomato, L-tomatine, 2-tridecanone, and phenolics are responsible for antibiosis to *H. zea* (Ferry and Cuthbert, 1975; Dimock and Kennedy, 1983; Farrar and Kennedy, 1988, 1990). The allelochemical 2-tridecanone is highly toxic to *H. zea* (Kennedy and Yamamoto, 1979). High phenol content has also been reported to be associated with resistance to *H. armigera* (Banerjee and Kalloo, 1989). Acylglucoses exuded by trichomes of the wild tomato species, *Lycopersicon pennellii* (Corr.) D'Arcy act as feeding or oviposition deterrents or both. Exposure to these compounds in artificial diet or as sprays on cultivated tomato leaves resulted in retarded growth rates, extended development period, and reduced survival of *H. zea* and *S. exigua* (Juvik et al., 1994). In the case of *H. virescens* resistance in tobacco, larval weight gain is negatively correlated with nicotine content at various leaf positions. Larvae survived better and grew faster on the bud leaves, which have lower nicotine levels than the leaves further down the stalk where alkaloid levels are high (Jackson, Johnson, and Stephenson, 2002). Luteolin-C-glycosides, maysin, and isomaysin, which occur in maize silks, confer resistance to *H. zea* (Waiss et al., 1979). In soybean, pinitol confers resistance to *H. zea* (Dougherty, 1976), while long chain fatty acids in sunflower act as growth inhibitors for *H. virescens* and *H. zea* (Elliger et al., 1976). In potato, concentrations of the glycoalkaloids, solanine and chaconine are associated with resistance to *E. fabae* (Sanford et al., 1992).

Nonprotein Amino Acids

Nonprotein or unusual amino acids are found in a number of plant species, and afford protection against herbivores. The protective effect is elicited through their structural analogy to the commonly occurring essential amino acids. The harmful effects on insects occur as these get incorporated into proteins that are toxic to insects. Among these, L-canavanine, azetidine-2-caboxylic acid, 2,4-diamino butyric acid, minosine, and 3-hydroxyproline have significant growth inhibition effects on insects (Parmar and Walia, 2001). L-canavanine is a structural homologue of L-arginine, and occurs in over 1,500 leguminous plant species. Some of the nonprotein amino acids also act as enzyme inhibitors. Canaline, a hydrolytic product of canavanine, inhibits pyridoxal phosphate-dependent enzymes by forming a covalent bond. Minosine is a strong inhibitor of rat liver cytothionine synthetase and cytothionase, and inhibits the growth of rust red flour beetle, *Tribolium castaneum* (Herbst) (Ishaaya et al., 1991).

Nutritional Factors

For a plant to serve as a host for an insect, it should be able to provide holistic nutrition to support growth, development, and reproduction (Beck, 1972). It should also be suitable for digestion, assimilation, and conversion into energy and structural compounds required by the insects. Some of the nutrients such as amino acids, phospholipids, fatty acids, steroids, and ascorbic acid also serve as phagostimulants for different insect species (Dadd, 1970). Variation in the amounts of these compounds results in various degrees of host plant acceptance and damage by the herbivorous insects. Imbalanced ascorbic acid content in maize (Penny, Scott, and Guthrie, 1967), amino acid content in pea (Auclair, Maltais, and Cartier, 1957), low quantities of glutamic acid and asparagine in rice (Sogawa and Pathak, 1970), low lysine content in sorghum (Singh and Jotwani, 1980), and high proline content in cotton (Sharma and Agarwal, 1983a) are associated with resistance to insects. Total soluble sugars in the pod wall have a significant and negative correlation with pod damage by *H. armigera* in pigeonpea. Pea varieties deficient in certain amino acids are resistant to the pea aphid, *Acyrthosiphon pisum* (Harris) (Auclair, 1963).

In cotton, higher amounts of reducing sugars are associated with resistance to the leafhopper, *A. biguttula biguttula* (Singh and Agarwal, 1988), and amino acid content to stem weevil, *Pempherulus affinis* (Kalt.) (Parameswaran, 1983). Rice varieties resistant to thrips, *Stenchaetothrips biformis* (Bagnall), have lower amounts of sugars and free amino acids (Thayumananan et al., 1990), while low amounts of amino acids are associated with resistance to brown planthopper, *N. lugens* (Sogawa and Pathak, 1970; Bharathi, 1989).

However, rice varieties resistant to gall midge, *Orseolia oryzae* Wood-Mason have higher amounts of free amino acids and low amounts of sugars in the shoot tips, while stem borer, *Scirpophaga incertulas* (Walker) resistant varieties have lower amounts of amino acids and sugars than the susceptible ones (Vidyachandra, Roy, and Bhaskar, 1981). Chhabra et al. (1990) suggested that high amounts of nonreducing sugars and low amounts of starch in chickpea variety GL 645 might be responsible for its low susceptibility to *H. armigera*. Mungbean varieties with high sugar and amino acid content in leaves are resistant to whitefly, *B. tabaci* and the jassid, *Empoasca kerri* (Prathi) (Chhabra et al., 1988). High amounts of vitamin C in guava fruits are associated with resistance to the fruitfly, *Bactrocera correcta* Bezzi (Arora et al., 2000). Leafhopper, *A. biguttula biguttula*, injury in okra, *Abelmoschus esculentus* (L.) Moench, is negatively associated with calcium, but positively correlated with phosphorous content (Uthamasamy, 1985).

Enzymes

Amylase and protease inhibitors in pigeonpea have been shown to have an adverse effect on growth and development of *H. armigera* (Giri and Kachole, 1998). There is considerable variation in *H. armigera* gut protease inhibitory activity in developing seeds of chickpea (Patankar et al., 1999), and proteinase inhibitors from the nonhost plants (groundnut, winged bean, and potato) are more efficient in inhibiting the gut proteinases of *H. armigera* larvae than those from its favored host plants such as chickpea, pigeonpea, and cotton (Harsulkar et al., 1999). Protease inhibitors and chlorogenic acid are associated with aphid resistance in tomato (Felton, Broaduray, and Duffey, 1989). Common bean genotypes with resistance to bean weevil, *Zabrotes subfasciatus* (Boh.) did not show a significant inhibition of amylase. These lines contained arcelin, a major compound in seed protein, which is not digestible by *Z. subfasciatus*, and hence its amino acids were not available to this insect (Minney et al., 1990).

Molecular Markers

Simultaneous occurrence of a trait in the same population of two or more discontinuous variants or genotypes is called genetic polymorphism. A wide variety of techniques have been developed in the past few years for detecting DNA sequence polymorphism (Paterson, Tanksley, and Sorrells, 1991; Staub, Serquen, and Gupta, 1996). DNA fingerprinting or profiling is based on combined use of several single locus detection systems for understanding various aspects of plant genomes. These include characterization of genetic diversity, genome fingerprinting, genome mapping, gene localization, genome evolution, population genetics, taxonomy, plant breeding, and diagnostics. A good marker for use in genomic studies should (1) be polymorphic in nature, (2) be codominant in inheritance, (3) occur frequently in the genome, (4) be selective neutral in behavior, (5) possess easy access and fast assay, (6) be highly reproducible, and (7) provide ease of exchange of data between laboratories.

Several types of molecular markers have been used to evaluate DNA polymorphism and are generally classified as hybridization-based markers and polymerase chain reaction (PCR)-based markers. In the former, DNA profiles are visualized by hybridizing the restriction enzyme-digested DNA to a labeled probe, which is a DNA fragment of known origin or sequence. Ever since thermostable DNA polymerase was introduced, the use of PCR in genomic research has increased tremendously (Saiki et al., 1988; Mullis, 1990). The primer sequences are chosen to allow base-specific binding to the template in reverse orientation. PCR is extremely sensitive and operates at a very high speed. Several types of molecular markers have been used for developing genetic linkage maps of different crops, and to identify quantitative trait loci (QTLs) associated with resistance to insects (Terry et al., 2000; Yencho, Cohen, and Byrne, 2000; Smith, 2005). Different types of markers used to detect QTLs associated with insect resistance are discussed below.

Restriction Fragment Length Polymorphisms

The restriction fragment length polymorphisms (RFLPs) markers detect the differences between genotypic DNA of different genotypes when restriction enzymes from bacteria cut genomic DNA at specific nucleotide binding sites and yield variable sizes of DNA fragments (Karp et al., 1997; Primrose, 1998; Srivastava and Narula, 2004). Restriction enzymes are also called restriction endonucleases because they cut the DNA molecule at a particular nucleotide sequence, and different restriction enzymes cut DNA into pieces of different sizes. The digested DNA is separated by electrophoresis according to size, normally on agarose or polyacrylamide gel. The separated DNA fragments are then transferred to a nylon membrane and exposed to a particular DNA probe in a procedure referred to as Southern blotting. Complementary binding between the probe DNA and membrane DNA provides information about the possible location of the gene for insect resistance. The differences in DNA banding patterns on an autoradiogram of a Southern blot indicate the presence of one or more restriction sites in a sequence. The sequence containing a restriction site is one allele, while the corresponding sequence missing the restriction site is the other allele. When restriction sites are compared between two genotypes, one genotype may have the site, while the other does not. If differences exist, they are referred to as polymorphisms between the two genotypes. RFLP probes allow mapping of loci linked to resistance genes, and hundreds of RFLP loci have been mapped in several crop genomes. RFLP markers usually behave in a codominant manner and thus can detect heterozygotes and more allelic variation in natural plant populations. RFLPs, being codominant markers, can detect the coupling phase of DNA molecules, as DNA fragments from all homologous chromosomes are detected. They can easily determine if a linked trait is present in a homozygous or heterozygous state in an individual. Such information is highly desirable for recessive traits. Their inability to detect single base changes restricts their use in detecting point mutations occurring within the regions at which they are detecting polymorphism. Use of RFLP markers needs more time to complete the analysis (7 to 10 days), and involves the use of radioactive isotopes. RFLP markers have been used effectively to map insect resistance genes in several crops (Yencho, Cohen, and Byrne, 2000; Smith 2005). The RFLP probes from one species can often be used to detect homologous regions in the genomes of related species and, thus, are widely used to identifiy syntenic genomic regions (Devos and Gale, 2000).

Sequence-Tagged Sites

RFLP probes specifically linked to a desired trait can be converted into PCR-based sequence-tagged sites (STS) markers based on nucleotide sequence of the probe giving polymorphic band pattern to obtain specific amplicon (Karp et al., 1997; Srivastava and Narula, 2004). Using this technique, tedious hybridization procedures involved in RFLP analysis can be overcome. However, in many cases, the power to detect polymorphism is reduced compared to the original RFLP probes. Nonetheless, this approach is extremely useful for studying the relationship between various species. When these markers are linked to some specific traits, they can be easily integrated into plant breeding programs for marker-assisted selection of the trait of interest (Yencho, Cohen, and Byrne, 2000).

Expressed Sequence Tags

The expressed sequence tags (ESTs) are obtained by partial sequencing of random cDNA clones (Powell et al., 1996; Karp et al., 1997). Once generated, they are useful in cloning specific genes of interest and synteny mapping of functional genes in related organisms. The ESTs are useful for full genome sequencing and mapping programs to identify active genes and, thus, help in identification of diagnostic markers. An EST that appears to be unique can be used to isolate new genes. Development of EST markers is dependent on extensive sequence data on regions of the genome, which are expressed. However, once developed, they provide high quality consistent results because they are limited to expressed regions of the genome, and the markers themselves are directly associated with functional genes.

Single Strand Conformation Polymorphisms

The single strand conformation polymorphisms (SSCPs) are a powerful and rapid technique for gene analysis, particularly for detection of point mutations and DNA polymorphism (Srivastava and Narula, 2004). The SSCPs can identify heterozygosity of DNA fragments of the same molecular weight and can detect changes of a few nucleotide bases as the mobility of the single-stranded DNA changes with change in its GC content due to conformational change. To overcome problems of reannealing and complex banding patterns, an improved technique called asymmetric-PCR SSCP has been developed wherein the denaturation step has been eliminated and a large-sized sample can be loaded for gel electrophoresis, making it a potential tool for high throughput analysis (Sunnucks et al., 2000). It is not well developed for use in plants, although its application in discriminating progenies can be exploited, once suitable primers are designed for agronomically important traits.

Microsatellites

Microsatellites are multilocus probes creating complex banding patterns, are non-species specific, occur ubiquitously, and belong to the repetitive DNA family (Karp et al., 1997; Cook et al., 2004). Fingerprints generated by the probes are known as oligonucleotide fingerprints. The methodology has been derived from RFLP, and specific fragments are visualized by hybridization with a labeled micro- or minisatellite probe. Microsatellite primers generate high levels of polymorphism and detect patterns of codominant inheritance in populations of segregating progeny. Using microsatellites, some instances of semicodominance have been identified for Russian wheat aphid, *Diuraphis noxia* (Kurdj.), resistance (X.M. Liu et al., 2001). Microsatellite markers have been found to be useful in many crops for identification of insect resistance genes (Yencho, Cohen, and Byrne, 2000; X.M. Liu et al., 2001; Miller, Altinkut, and Lapitan, 2001; X.M. Liu, Smith, and Gill, 2002; Malik et al., 2003; Cook et al., 2004; Zhu et al., 2004).

Simple Sequence Repeats

The simple sequence repeats (SSRs) are tandem arrays of 2 to 5 base repeat units (particularly dinucleotide repeats) that are widely distributed in eukaryotic DNA (Powell et al.,

1996; Burns et al., 2001; Sharopova et al., 2002; Zhang et al., 2002). The SSRs provide high-quality and consistent results, but the markers are expensive to develop as they require extensive sequence data from the species of interest. The sequence data can be generated from libraries of cloned DNA (with or without enrichment for a specific SSR motif). Existing EST resources for a specific crop can also be used. EST-derived SSRs are less polymorphic than those derived from genomic libraries, but both are useful as they are PCR-compatible and typically inherited in a codominant manner. Polymorphic SSRs have been used to map resistance to shoot fly, *A. soccata*, in sorghum (Folkertsma et al., 2003; Hash et al., 2003); greenbug, *Schizaphis graminum* (Rondani), resistance in wheat (Weng and Lazar, 2002); and for covering the soybean genome (Narvel et al., 2000).

Sequence-tagged microsatellite markers (STMs) primers are complementary to the flanking regions of SSR loci, and yield highly polymorphic amplification products. Polymorphism appears because of variation in the number of tandem repeats (VNTR loci) of a given repeat motif. Tri- and tetranucleotide microsatellites are more popular for STMS analysis (based on PCR product separation on PAGE gels) because they present a clear banding pattern after PCR and gel electrophoresis. However, dinucleotides are abundant in genomes and have been used as markers. The di- and tetranucleotide repeats are present mostly in the noncoding regions of the genome, while 57% of trinucleotide repeats reside in or around the genes. A good relationship has been observed between the number of alleles detected and the total number of SSRs within the targeted microsatellite DNA of a reference genotype. Thus, the larger the repeat number in the VNTR locus, the greater the number of alleles that are expected to be detected in a large population of diverse accessions.

Randomly Amplified Polymorphic DNA

The randomly amplified polymorphic DNAs (RAPDs) are based on PCR, and use arbitrary primers for initiating amplification of random pieces of the sampled plant DNA, and requires no knowledge of the genome to be screened. PCR-based markers involve *in vitro* amplification of particular DNA sequences or loci with the help of specifically or arbitrarily chosen oligonucleotide sequences (primers) and a thermostable DNA polymerase enzyme. The amplified fragments are separated electrophoretically and banding patterns detected by staining or autoradiography. PCR is a versatile technique invented during the mid-1980s (Saiki et al., 1985). However, the results are inconsistent across populations and laboratories (Powell et al., 1996; Staub, Serquen, and Gupta, 1996). The PCR primers used for RAPD analysis are short random DNA sequences approximately 10 nucleotides long that amplify homologous genomic DNA sequences during the PCR process. Differences in the DNA sequences of resistant and susceptible plants result in different primer binding sites, which result in differences in PCR products that allow the visualization of polymorphisms between the DNA in the resistant and susceptible plants. Insect resistance genes in apple, rice, and wheat have been mapped using RAPD primers (Dweikat et al., 1994; Nair et al., 1995, 1996; Roche et al., 1997; Botha and Venter, 2000; Venter and Botha, 2000; Selvi et al., 2002; Jena et al., 2003). Reproducibility of RAPD markers has been quite difficult in many studies. In addition, DNA banding patterns based on RAPD primer amplification do not reveal heterozygotes as they are inherited in a dominant manner.

Inter Simple Sequence Repeat

In this technique, primers based on microsatellites are utilized to amplify inter-simple sequence repeat (SSR) DNA sequences (Staub, Serquen, and Gupta, 1996, Srivastava and Narula, 2004). Here, various microsatellites anchored at the 3' end are used for amplifying genomic DNA, which increases their specificity. These are dominant markers, though occasionally a few of them exhibit codominance. An unlimited number of primers can be synthesized for various combinations of di-, tri-, tetra-, and pentanucleotides, with an anchor made up of a few bases, and can be exploited for a broad range of applications in plant species.

Sequence Characterized Amplified Regions

RAPD-generated polymorphic DNA bands can be sequenced and the information used to design location specific sequence characterized amplified regions (SCARs) (Hernandez, Martin, and Dorado, 1999). These are similar to STS markers in construction and application. The presence or absence of the band indicates variation in sequence. They have a better reproducibility than RAPDs. SCARs are usually dominant markers; however, some of them can be converted into codominant markers by digesting the PCR products with restriction enzymes and polymorphism can be deduced by either denaturing gel electrophoresis or SSCP. SCARs are more informative for genetic mapping than dominant RAPDs, and can be used for map-based cloning. SCARs also allow comparative mapping or homology studies among related species, thus making them extremely adaptable. SCARs have been used to identify and map genes for resistance to rice gall midge, *O. oryzae* (Sardesai et al., 2002) and brown planthopper, *N. lugens* (Renganayaki et al., 2002).

Amplified Fragment Length Polymorphisms

The amplified fragment length polymorphisms (AFLP) markers are based on selective PCR amplification of restriction enzyme-digested DNA fragments (Vos et al., 1995). The DNA generated in each amplification contains molecular markers of random origin, but the process results in a large number of amplified DNA bands from one amplification. Sample DNA is digested with different restriction enzymes and restriction enzyme adaptors are then annealed to the restriction products. Restriction digests are then preselected by PCR amplification with general restriction enzymes attached to unique oligonucleotide primers and preselected PCR products, and then selectively amplified using specific 3" oligonucleotide primers. Amplified fragments are denatured and then separated using polyacrylamide gel electrophoresis, and the gel exposed to radiographic film to view the AFLP polymorphic banding patterns. The AFLP markers detect many DNA polymorphisms and have been used successfully to identify and map resistance to brown planthopper, *N. lugens* (Sharma et al., 2001) and gall midge, *O. oryzae* (Sardesai et al., 2002) in rice; and pod borer, *H. armigera* resistance in chickpea (Lawlor et al., 1998).

Single Nucleotide Polymorphisms

The vast majority of allelic differences between DNA sequences of individuals are point mutations due to single nucleotide polymorphisms (SNPs) (Brookes, 1999). As such, there are a vast number of potential SNP markers in all species. Considerable amounts of sequence data are required to develop SNP markers. However, their advantage lies in the potential to screen for allelic differences using methods that do not involve electrophoresis, as in microarrays.

Diversity Array Technology

Diversity array technology (DArT) is a sequence-independent, high-throughput method, able to detect polymorphism at hundreds of marker loci in a single experiment (Akbari et al.,

2006; Yang et al., 2006). DArT markers are typed in parallel, using high throughput plat-forms, with a low cost. DArT fingerprints will be useful for accelerating plant breeding, and for the characterization and management of genetic diversity (Wittenberg et al., 2005).

Molecular Markers Linked to Insect Resistance in Different Crops

The linkage between quantitative trait loci (QTL) and molecular marker loci is determined the same way as phenotypic resistance is linked with the segregation of genes for resis-tance to insects. The QTL analyses allow the researcher to identify the loci from a group of polymorphic segregating molecular makers that contribute most significantly to explain the phenotypic variation for biochemical or biophysical characters mediating insect resis-tance. A key component of QTL analysis is the calculation of a LOD (logarithm of the odds to base 10) score, a statistical estimate of the likelihood of recombination between two loci due to chance alone. The LOD scores indicate whether the two loci are likely to be near one another on a chromosome (at least in terms of recombination based genetic map, if not the physical map of that character) and are therefore likely to be inherited together. LOD scores of three or more indicate that the two loci are close to one another on a chromosome, that is, there is <1 in 1,000 probability of allelic association in the segregating population that occurs due to chance alone. The progress in identifying genomic regions associated with resistance to insects in different crops is discussed below.

Cotton

A molecular linkage map of *Gossypium hirsutum* L. has been developed with 58 doubled and haploid plants based on 624 marker loci (510 SSRs and 114 RAPDs). The 489 loci assem-bled into 43 linkage groups and covered 3,314.5 cM (Zhang et al., 2005). Another map has been developed based on RFLP markers in four mapping populations of *G. hirsutum* (Ulloa et al., 2002), which comprised of 284 loci mapped to 47 linkage groups with an average distance of 5.3 cM, covering 1,502.6 cM (approximately 31% of the total recombinational length of the cotton genome). The linkage groups contained 2 to 54 loci each, and ranged in distance from 1.0 to 142.6 cM. A total of 53 polymorphic fragments and 32 polymorphic loci, representing five linkage groups, have been identified in two F_2 populations of *G. hirsutum* [HS 46 × MARCABUCAG8US-1-88 (MAR) and HS 46 × Pee Dee 5363 (PD 5363)] by using RFLPs (Shappley et al., 1996).

The prominent effects of "Okra-leaf" locus (which is associated with resistance to bollworms) on chromosome 15 are modified by QTLs on several other chromosomes (C. Jiang et al., 2000). Among the 62 possible QTLs (at LOD > 2.0) for 14 morphological traits, 38 (61.3%) mapped to D-subgenome chromosomes, suggesting that D-subgenome of tetraploid cotton has been subjected to relatively greater evolution than the A-subgenome. Expression of the dense glanding (*dg*) (which contributes to bollworm, *H. armigera* and *H. virescens* resistance) mutant is confined to expanding leaves, internodes, and bracteoles (Vroh Bi et al., 1999). The ability to trace DNA segments of known chromosomal locations from the donor *G. sturtianum* Willis through segregating generations can be a starting point to map low-gossypol seed and high-gossypol traits. However, there is no evi-dence of association of dense glanding with 13 other mutant markers in *G. barbadense* L. (Percy, 1999).

Rice

The complete genetic linkage map of rice was published by the end of the twentieth century (Kurata et al., 1994, 1997; Palevitz, 2000), can be used for identification of genomic regions associated with resistance to insects. Considerable progress has been made in identification of molecular markers/QTLs associated with resistance to insect pests in rice. Several RAPD markers linked to yellow stem borer, *S. incertulas* resistance have been identified; the chromosome location of these genes is unknown (Selvi et al., 2002).

Much progress has been made in tagging and mapping the genes conferring resistance to rice gall midge, *O. oryzae*. Resistance to *O. oryzae* is linked with *RG329* (1.3 cM) and *RG476* (3.4 cM) markers on chromosome 4 (Mohan et al., 1994). RAPD fragment $F8_{1700}$ is associated with susceptibility and $F10_{600}$ with resistance to *O. oryzae* (Nair et al., 1995), while RAPD fragment $E20_{570}$ is associated with resistance and $E20_{583}$ with susceptibility to gall midge (Nair et al., 1996). Behura et al. (2000) reported an AFLP marker, which is amplified in biotypes 1, 2, and 5 of rice gall midge, but not in biotype 4. The former are avirulent on hosts bearing the *Gm2* resistance gene (e.g., Phalguna), while biotype 4 is avirulent to hosts with *Gm2* gene. Avirulence to *Gm2* gene is sex-linked. An AFLP marker *SA598* linked to *Gm7*, conferring resistance to biotypes 1, 2, and 4, has also been identified (Sardesai et al., 2002). *Gm7* is a dominant gene and is nonallelic to *Gm2*. A dominant resistance gene, *Gm1*, has been tagged and mapped on chromosome 9 (Biradar et al., 2004). The SSR markers *RM316*, *RM444*, and *RM219* located on chromosome 9 are linked to *Gm1* at genetic distances of 8.0, 4.9, and 5.9 cM, respectively. *Gm8* has been tagged and mapped on the rice chromosome 8 (Jain et al., 2004). Two fragments, *AR257* and *AS168*, were linked to the resistant and susceptible phenotypes, respectively. Another resistant phenotype-specific marker, *AP19(587)* was also identified using RAPDs. There is a tight linkage between the markers and the *Gm8* locus. Many of these markers can be used in MAS to develop rice cultivars with resistance to gall midge.

For brown planthopper, *N. lugens*, genomic regions on chromosomes 3 and 4 contain genes for resistance (Z. Huang et al., 2001). Two QTLs are associated with resistance to *N. lugens* (Alam and Cohen, 1998), of which one QTL is linked to antixenosis and second to tolerance. In another population, Xu et al. (2002) found that a main effect QTL for *N. lugens* resistance maps to the vicinity of a major rice gene controlling leaf and stem pubescence, suggesting that this QTL may explain antixenosis to *N. lugens*. The *N. lugens* resistance genes *bph11* and *Bph13* map to chromosome 3 (Kawaguchi et al., 2001; Renganayaki et al., 2002). The *N. lugens* resistance genes *Bph1*, *bph2*, *bph9*, and *Bph10* have been mapped to a 25 cM block on rice chromosome 12 (Ishii et al., 1994; Hirabayashi and Ogawa, 1995; Murata et al., 2000; P.N. Sharma et al., 2003), while *bph4* and *bph12* map to chromosomes 6 and 4, respectively (Hirabayashi and Ogawa, 1999; Kawaguchi et al., 2001). A major resistance gene, *Bph18(t)*, has been identified in the introgression lines IR 65482-7-216-1-2, which had been derived from the wild rice, *Oryza australiensis* Domin. (Jena et al., 2006). It is nonallelic to *Bph10* and is located on the long arm terminal region of chromosome 12, flanked by SSR marker *RM463* and STS marker *S15552*.

QTLs linked to green rice leafhopper, *Nephotettix cincticeps* (Uhler) resistance genes in rice have been identified in progeny from doubled haploid populations involving *indica* and *japonica* rice crosses. *Oryza sativa* L. × *O. officinalis* Wall ex Watt. crosses have also been used to identify QTLs for resistance to green leafhopper (N. Huang et al., 1997; Y. Fukuta et al., 1998; Z. Huang et al., 2001). Chromosomal location of the gene linked to early panicle initiation and resistance to green leafhopper resistance [*Grh3(t)*] have been used to identify rice lines possessing both early maturity and insect resistance (Tsuji, Saka, and Izawa, 2003). The *indica* genotype was maintained in the upstream region of the *Grh3(t)* gene located on chromosome 6 of Aichi 66. Four genes condition resistance in rice to *N. cincticeps*. Gene *Grh1* (green rice leafhopper) from the cultivar Norin-PL2 (Kobayashi et al., 1980), is located on rice chromosome 5 (Tamura et al., 1999). *Grh3* is located on chromosome 6, and *Grh4* on chromosome 3 (Saka et al., 1997; Fukuta et al., 1998; Yazawa et al., 1998). QTLs identified by Fujita, Yoshimura, and Yashi (2003) coincide with each of these four resistance genes. The *Grh5* gene located on the distal region of the long arm of chromosome 8 is tightly linked to markers *RM3754* and *RM3761*, and can be used in MAS to breed for resistance to this insect (Fujita et al., 2006). The QTLs for *Nephotettix virescens* (Dist.) resistance on chromosomes 3 and 11 are very near to *Grh2* and *Grh4*. The near-isogenic lines (NIL) containing both *Grh2* and *Grh4* express resistance to *N. virescens* (Wang et al., 2003, 2004). RFLP markers have been used to identify resistance genes in pairs of near-isogenic lines (NIL), each containing a single gene for resistance to the whitebacked planthopper, *Sogatella furcifera* (Horvath) or to *Xanthomonas oryzae* pv. *oryzae* (Ishiyama), and 78 F_3 families from the cross, TN1/IR36, segregating for five single resistance genes (McCouch, Khush, and Tanksley, 1991). Location of the resistance genes in NIL pairs was confirmed by using RFLP markers from putative positive regions as probes on segregating F_2 populations from crosses involving resistant isoline × susceptible recurrent parent. Two RFLP markers were linked to *Wbph1* for resistance to the *S. furcifera*, although the chromosomal location of the gene was unclear, since both linked markers were detected using multiple-copy clones as probes.

Wheat

Monosomic analysis has been used to locate a single, dominant, Hessian fly, *M. destructor*, resistance gene (*H13*) in the D genome of common wheat germplasm KS81H1640HF derived from *Aegilops squarrosa* L. (Gill, Hatchett, and Raupp, 1987). Telocentric analysis was used to map the *H13* gene on 6Dq (long arm) 35.0 ± 8.0 recombination units from the centromere. Dweikat et al. (2004) used a set of near-isogenic wheat lines, each carrying a resistance gene at 1 of 11 loci (*H3, H5, H6, H9, H10, H11, H12, H13, H14, H16,* or *H17*). These were developed by backcrossing to Hessian fly susceptible wheat cultivar "Newton." Eleven RAPD markers linked to the 11 resistance genes were identified. Several of these markers can be used to determine the presence/absence of specific Hessian fly resistance genes in wheat lines that have 1 or more genes for resistance. Linkage relationships among genes on wheat chromosome 5A that condition resistance to *M. destructor* have been identified (Ohm et al., 1995). Testcross analyses indicated that six of these genes appear to occupy a single linkage block on wheat chromosome 5A in the order *H9* to *H15, H10, H17, H16,* and *H12*. Gene *H14* did not appear to be within the linkage block *H9* to *H12*. Linkage analysis identified one sequence tagged site (STS) marker, STS-*Pm3*, and eight microsatellite markers (*Xbarc263, Xcfa2153, Xpsp2999, Xgwm136, Xgdm33, Xcnl76, Xcnl117,* and *Xwmc24*) near the *H9* locus on the distal region of the short arm of chromosome 1A, contrary to the previously reported location of *H9* on chromosome 5A (Kong et al., 2005). Locus *Xbarc263* was 1.2 cM distal to *H9*, which itself was 1.7 cM proximal to loci *Xcfa2153, Xpsp2999,* and *Xgwm136*. Marker alleles at loci *Xgwm136, Xcfa2153,* and SOPO05$_{909}$ were shown to be specific to *H9* and not diagnostic to several other Hessian fly resistance genes, and therefore should be useful for pyramiding *H9* with other Hessian fly resistance genes in a single genotype.

RFLP markers linked to two Hessian fly resistance genes in wheat from *Triticum tauschii* (Coss.) Schmal. have also been identified (Ma et al., 1993). A new Hessian fly resistance gene (*H30*) transferred from the wild grass, *Aegilops triuncialis* L. to hexaploid wheat, has also been identified (Martin-Sanchez et al., 2003). Five SSR markers (*Xgwm136, Xcfa2153, Xpsp2999, Xgwm33,* and *Xbarc263*) are linked to the Hessian fly resistance gene, *Hdic* on the short arm of the wheat chromosome 1A, which is in the same region as *H9, H10,* and *H11* (X.M. Liu et al., 2005). The flanking markers *Xgwm333* and *Xcfa2153* have been mapped at 0.6 cM and 1.4 cM, respectively. *Hdic* confers a high level of antibiosis to Hessian fly biotypes GP and L, and the strains *vH9* and *vH10*. Allelic tests indicated that resistance genes in SW 8 and SW 34 may be allelic to *H26* and *H13*, or correspond to paralogs at both loci, respectively (X.M. Liu, Smith, and Gill, 2005; Wang et al., 2006). The genes *H26* and *H13* have been assigned to chromosomes 3D and 6D, respectively. *H13* maps to the distal cluster of genes on chromosome 6DS (X.M. Liu, Gill, and Chen, 2005). Hessian fly resistance from TA 2473 (wild diploid goat grass, *T. tauschii*) is governed by one dominant gene located on chromosome 4D (Cox and Hatchett, 1994). This gene, designated as *H26*, is the only Hessian fly resistance gene known to be located on 4D. Gene *H26* conditions a high level of antibiosis to Biotype L and has been transferred into a germplasm line, KS92WGRC26, which has most of the desirable agronomic traits of its recurrent parent, Karl. A new Hessian fly resistance gene from *Aegilops ventricosa* Tausch. present in 4D(4M^v) substitution line H-93-33 derived from the cross (*Triticum turgidum* Desf. H-1-1 × *A. ventricosa* no. 11) × *Triticum aestivum*, H-10-15), confers resistance to the Spanish population of *M. destructor*, and is inherited as a single dominant trait (Delibes et al., 1997). The resistance gene *H27* is allelic to that of *A. ventricosa* no. 10, but nonallelic to *H3* and *H6*. A new source of resistance to the highly virulent and widespread biotype L of Hessian fly has been identified in an accession of tetraploid durum wheat, *T. turgidum* (Williams et al., 2003). Genetic analysis and deletion mapping revealed that this line was homozygous for a single locus for resistance, *H31*, residing at the terminus of chromosome 5BS. *H31* is the first Hessian fly resistance gene to be placed on 5BS. AFLP analysis identified a marker, linked to the resistance allele, which retained its polymorphism after being converted to a highly specific sequence-tagged site marker, *Xupw4148*. This dominant 128-bp marker maps 3.3 cM from the *H31* resistance locus and amplifies in resistant plants only.

Ma et al. (1998) mapped resistance genes *Dn2* and *Dn4* conferring resistance to the Russian wheat aphid, *D. noxia*. The genes *Dnl, Dn2, Dn5, Dn6,* and *Dnx* are located on the short arm of wheat chromosome 7D (X.M. Liu et al., 2001; X.M. Liu, Smith, and Gill, 2002). They are allelic or a cluster of completely linked resistance genes, as there is no segregation for susceptibility to Russian wheat aphid among progeny from crosses involving plants containing each of the six genes in all possible combinations (X.M. Liu, Smith, and Gill, 2005).

QTLs for resistance to greenbug, *S. graminum* in wheat have also been identified (Castro et al., 2004). Weng and Lazar (2002) used AFLP and SSR markers to tag greenbug resistance gene, *Gb3* in wheat. The *Gb* gene conferring resistance *to S. graminum*, is inherited as a single dominant trait (Zhu et al., 2005). Microsatellite markers *Xwmc157* and *Xgdm150* flank *Gbx1* at 2.7 and 3.3 cM, respectively (Weng, Devkota, and Rudd, 2005). *Xwmc671* is linked to *Gba, Gbb, Gbc,* and *Gbd* at 34.3, 5.4, 13.7, and 7.9 cM, respectively; while *Xgdm150* is distal to *Gbc* at 17.7 cM. *Gbd* appears to be a new resistance gene, and is different from *Gbx1* or *Gbz*. The genes *Gbx1, Gba, Gbc,* and *Gbd* are either allelic or linked to *Gb3*. *Gby* is another greenbug resistance gene in the wheat line 'Sando's selection 4040', and is inherited as a single semidominant gene located on wheat chromosome 7A (Boyko, Starkey, and Smith, 2004). *Gby* was mapped to the area in the middle of the "island" of putative defense response genes that are represented by RFLP markers (*Xpsr119, XZnfp, Xbcd98,* and *Pr1b*), previously mapped to the distal part of the short arm of wheat chromosome group 7. The selection accuracy when the RFLP markers *Xhcd98*, *Xpsrll9*, or *XZnfp* and *Pr1b* flanking *Gby* are used together to tag *Gby* is 99.78%, suggesting that they can be used successfully in MAS (Boyko, Starkey, and Smith, 2004).

In barley, resistance to cereal aphids, *Rhopalosiphum padi* (L.) and *R. maidis* (Fitch.) (Moharramipour et al., 1997), and the Russian wheat aphid, *D. noxia* (Nieto-Lopez and Blake, 1994) has also been mapped.

Maize

A linkage map with 932 ESTs has been developed in maize (Davis et al., 1999). A small number of QTLs have been found to account for 50% of the genetic variation for resistance to sugarcane borer, *Diatraea saccharalis* Fab. and southwestern corn borer, *Diatraea grandiosella* (Dyar) (Khairallah et al., 1997). Several chromosomal regions were involved in the resistance to stem borers, and some QTLs were common to both the species. Genetic diversity decreased in a maize population during selection for resistance to pink stem borer, *Sesamia nonagrioides* LeFevbre (as expected since random genetic drift as well as selection could reduce genetic variability), but not significantly (Butron et al., 2005). Changes in the frequency of five alleles were significantly greater than expected. Linear trend of the departure from the random genetic drift model was significant for some allelic versions of two SSR markers, *umc1329* and *phi076*, indicating that directional selection was acting on these loci. The significant effect of directional selection on those markers suggested the presence of QTLs for tunnel length and/or for yield under artificial infestation with *S. nonagrioides* on the long arm of chromosome 4.

Genes controlling the synthesis of DlMBOA, the maize leaf organic acid with antibiotic effects on European corn borer, *Ostrinia nubilalis* (Hubner) leaf feeding, occur on maize chromosome 4. Resistance to stem boring by *O. nubilalis* in temperate maize and to *Diatraea* species in tropical maize appears to be controlled by QTLs on chromosomes 2, 3, 5, 7, and 9 (Cardinal et al., 2001). Of the 12 QTL for neutral detergent fiber (NDF) and acid detergent fiber (ADF) in leaf-sheaths, five for each trait were at or near QTL for European corn borer tunneling (Cardinal and Lee, 2005). Four of the eight leaf-sheath acid detergent lignin (ADL) QTL were detected in the same genomic regions as ECB QTL. Resistance to European corn borer may be associated with a subset of the QTLs observed for cell wall components and ADF and starch concentration in the stalk (Krakowsky, Lee, and Holland, 2007). QTLs affecting resistance to *O. nubilalis* have been identified in genomic regions of maize chromosomes 2, 3, 5, and 9 (Shon et al., 1993; Cardinal et al., 2001; Jampatong et al., 2002; Krakowsky et al., 2002). Comparison of QTLs linked to stem boring and leaf feeding resistance have shown that there are few common elements in the genetic control of the two types of resistances (Jampatong et al., 2002). However, QTL alleles on maize chromosomes 2, 5, 7, and 9 play a major role in resistance to stem boring by *O. nubilalis*, southwestern corn borer, *D. grandiosella* and the sugarcane borer, *D. saccharalis* (McMullen and Simcox, 1995; Bohn et al., 1996, 1997; Groh et al., 1998a, 1998b; Khairallah et al., 1998; Cardinal et al., 2001). The major QTLs for production of silk maysin and apimaysin, which control resistance to *H. zea*, occur on chromosomes 5 and 9. The relationship between the QTLs for different types of resistance on chromosomes 5 and 9 is not fully understood (Cardinal et al., 2001). Selection in maize for rind penetrometer resistance (RPR), linked to stalk resistance to European corn borer, has been found to be useful in enhancing germplasm for stalk strength, and therefore, improving stalk lodging resistance. Multilocus models, including single-effect QTLs and epistatic interactions, accounted for 33.4, 44.7,

48.4, and 58.7% of the total phenotypic variation in four populations, respectively [MoSCSSS-High (selection for high RPR), MoSCSSS-Low (selection for low RPR), MoSQB-Low (selection for low stalk crushing strength), and inbred lines Mo47 and B73] (Flint-Garcia, McMullen, and Darrah, 2003). One chromosomal region contained a QTL in all four populations, while two QTLs were common among three of the four populations, and five QTLs in two populations. Candidate genes that overlap QTL confidence intervals include those involved in lignin synthesis, the phenylpropanoid pathway, and the timing of vegetative phase change are linked to resistance to European corn borer, *O. nubilalis*. QTLs have been identified for RPR and second generation *O. nubilalis* resistance (Flint-Garcia et al., 2003). Phenotypic recurrent selection increases the frequency of resistance alleles over cycles of selection. Phenotypic selection for both high and low RPR was more effective than MAS in two populations, while MAS was more effective than phenotypic selection in another population. The MAS was more effective in selecting for increased susceptibility, but not in increasing resistance to damage by *O. nubilalis*.

QTL linked to maysin production (a glycosyl flavone that controls antibiosis to *H. zea* larvae) have been identified (Guo et al., 2001). Butron et al. (2001) identified two major QTLs for synthesis of maysin and related compounds, the already known *p1*, on the short arm of chromosome 1, and a novel one in the interval *csu1066-umc176* on genomic region 2C-2L. A QTL for husk tightness was located near *p1*. The functional allele for *p1* and the favorable allele for husk tightness were in repulsion linkage. In a marker-assisted selection program for increasing resistance to corn earworm, *H. zea* markers for silk antibiotic synthesis should be accompanied by markers for husk tightness. The maize chromosome 1 locus *p1*, which activates transcription of parts of the flavonoid pathway, explains 58% of the variance for maysin content (Byrne et al., 1996). A second QTL on chromosome 9, which is dominant for low maysin levels and interacts with *p1*, is *rem* (recessive enhancer of maysin) *1*. When a functional *p1* allele is present, *rem1* nearly doubles the maysin concentration. The *p1* locus is highly significant in explaining the variation for both *H. zea* larval weight reduction and increased silk maysin concentration (Byrne et al., 1997). Additional loci on chromosomes 1 and 9 also explain significant variation for *H. zea* larval weight and maysin concentration. Both maysin and apimaysin are closely related glycosyl flavones. A QTL for maysin on maize chromosome 9 (*rem1*) explained 55% of the variance for maysin synthesis, while the QTL for apimaysin from the *pr1* region of chromosome 5, explained 64% of the variance for apimaysin synthesis (Lee et al., 1998). Neither QTL affects the other, indicating that synthesis of maysin and apimaysin occurs independently. However, *rem1* accounted for only 14.1% of the *H. zea* antibiosis and *pr1* accounted for 14.7% of the antibiosis, suggesting that other antibiotic compounds may contribute to *H. zea* antibiosis in maize. Chlorogenic acid, a maize phenylpropanoid metabolite with an adjacent hydroxyl ring structure similar to maysin, has also been implicated in *H. zea* resistance (Duffey and Stout, 1996). Bushman et al. (2002) detected a QTL in maize silks corresponding to the *p1* locus that increases both chlorogenic acid and total flavone content. Chlorogenic acid accumulation is probably due to the *p1* induction of chlorogenic acid synthesis, and induction of flavonoid genes to increase phenylpropanoid pathway substrate availability (Bushman et al., 2002).

Sorghum

QTL analyses have been used to document resistance in sorghum to the greenbug, *S. graminum* (Agrama et al., 2002; Katsar et al., 2002). Several QTLs in linkage groups of

S. graminum-resistant sorghum germplasm are associated with RFLP loci in regions syntenic to locations of the wheat *Dn* genes controlling resistance to Russian wheat aphid, *D. noxia* (Katsar et al., 2002). Nagaraj et al. (2005) identified three QTLs associated with resistance to *S. graminum* biotype I, and five QTLs associated with resistance to biotype K. The amount of phenotypic variation explained by these QTLs ranged from 9% to 19.6%. Tao et al. (2003) identified two different mechanisms for resistance to sorghum midge, *S. sorghicola*, through QTL mapping. Genetic regions located on two separate linkage groups were associated with the antixenosis mechanism of resistance, and explained 12% and 15% of the total variation in egg laying, respectively. One region was associated with antibiosis resistance to midge, and explained 34.5% of the variation in egg and pupal counts. Two RIL mapping populations (*BT* × 623 × IS 15881-3 and 296B × IS 18551) have been phenotyped and genotyped for resistance to sorghum shoot fly, *A. soccata*, and one population (ICSV 745 × PB 15001-3) for resistance to spotted stem borer, *C. partellus* and sorghum midge, *S. sorghicola* SSR-based genetic linkage maps for these populations have been constructed to identify QTLs associated with resistance to these insects. Polymorphic SSRs associated with resistance to shoot fly and/or phenotypic traits associated with resistance to this insect have been identified (Folkertsma et al., 2003; Hash et al., 2003). Markers *Xtxp258* (bp 190/230) and *Xtxp289* (bp 270/294) are linked to trichome density; *Xgap1* (bp 180/254) and *Xtxp141* (bp 154/169) to deadheart incidence, leaf glossiness, and trichome density; *Xisp328* (bp 144/166) and *Xisp264* (bp 153/207) to leaf glossiness; and *Xisp258* (bp 170/193) and *Xtxp65* (bp 125/134) to deadheart incidence and leaf glossiness. These QTLs are now being transferred into locally adapted hybrid parental lines via SSR-based MAS.

Potato

Bonierbale et al. (1994) mapped RFLP loci on progeny of crosses between cultivated and wild potato, *Solanum berthaultii* Hawkes, for genes controlling glandular trichome-based resistance to Colorado potato beetle, *L. decemlineata*. Several QTLs explained various components of resistance based on potato leaf trichome type, density, and exudates. Yencho et al. (1996) mapped *L. decemlineata* resistance genes in reciprocal backcrosses between *S. tuberosum* and *S. berthaultii* using RFLP markers to identify QTLs linked to reduced oviposition or feeding. Three QTLs on three chromosomes in BCB (backcross to *S. berthaultii*), and two QTLs on two chromosomes in BCT (backcross to *S. tuberosum*) influenced resistance to *L. decemlineata* involving *S. berthaulti* as a source of resistance. The QTLs were additive, but one instance of epistasis was also observed. Each QTL accounted for 4% to 12% of the phenotypic variation. In the more resistant BCB population, a three-QTL model explained nearly 20% of the variation in oviposition. When alleles at the three QTLs were from *S. berthaultii*, the egg laying was reduced by 60% compared to that on the heterozygotes. The QTLs for resistance to *L. decemlineata* coincided with loci associated with the glandular trichomes, confirming the importance of the glandular trichomes in mediating resistance to this insect. However, a relatively strong and consistent QTL for insect resistance in both BCB and BCT (backcross to *S. tuberosum*) on chromosome 1 was also observed, but was not associated with any trichome traits, suggesting that the trichomes may not account for all of the resistance observed in these progenies. Chen et al. (2004) constructed two BAC libraries for wild Mexican diploid potato, *Solanum pinnatisectum* Dunal. Fifteen BAC clones harbored PPO (polyphenol oxidase) loci for *L. decemlineata* resistance. Development of these BAC libraries will be useful for BAC contig construction and map-based cloning of insect resistance genes.

Tomato

Resistance to tomato fruitworm, *H. zea* in tomato is mediated by 2-tridecanone (2-TD), and direct selection for RFLP loci increased the frequency of 2-TD-mediated resistance (Nienhuis et al., 1987). RAPD primers were identified as giving parent-specific bands when screened with a set of introgression lines containing introgressed regions of *L. pennellii*, that encompass five QTLs linked to production of acylsugars—the compounds associated with insect resistance (McNally and Mutschler, 1997). Primers giving *L. pennellii*-specific bands have been mapped to identify bands affiliated with the QTLs and flanking regions using subsets of 7 to 16 F_2 individuals, which contained small overlapping segments (zones) of the *L. pennellii* genome spanning those regions. Seventeen RAPD primers, *agt*-related primers and an *agt* clone, have been used in mapping the complete F_2 population of 144 individuals, which resulted in the identification of RAPD markers for three of the five QTLs, and construction of an integrated genomic map for tomato (*L. esculentum* × *L. pennellii* LA716) of 111 RAPD and eight acylglucose transferase-related markers added to a map of 150 RFLP markers. QTLs for tomato resistance to the leaf miner, *Liriomyza trifolii* (Burgess) have also been identified (Moreira, Mollema, and van Heusden, 1999).

Chickpea

Many studies (Gaur and Slinkard, 1990a, 1990b; Ahmad, Gaur, and Slinkard, 1992; Kazan et al., 1993; Simon and Muehlbauer, 1997; Winter et al., 1999, 2000; Huttel et al., 1999; Santra et al., 2000; Tekeoglu et al., 2000; Sharma and Crouch, 2004) have used interspecific mapping populations for developing genetic linkage maps of chickpea. The preliminary linkage map was based on interspecific crosses of *Cicer arietinum* L. × *C. reticulatum* (Ladz.) and *C. echinospermum* (Davis), and intraspecific crosses of *C. reticulatum* (Gaur and Slinkard, 1990a, 1990b). Winter et al. (1999) developed the first genomic map of chickpea based on 90 RILs derived from a cross of *C. reticulatum* (PI 489777) and the cultivated chickpea, ICC 4958, using 120 STMS markers. This map was then augmented using 118 STMS, 96 DAFs, 70 AFLP, 37 ISSR, 17 RAPD, 2 SCAR, 3 cDNA, and eight isozyme markers screened across 130 RIL from the same cross (Winter et al., 2000). Santra et al. (2000) used an RIL population from an interspecific cross of *C. arietinum* × *C. reticulatum* to generate a map of nine linkage groups with 116 markers (isozymes, RAPDs, and ISSRs) covering a map distance of 981.6 cM with an average distance of 8.4 cM between markers. The RIL population derived from a cross between a wilt-resistant kabuli variety (ICCV 2) and a wilt-susceptible desi variety (JG 62) has been used to develop the first molecular map of chickpea based on an intraspecific cross (Cho et al., 2002). This map consists of 58 STMS, 20 RAPD, and 4 SSR markers assigned to 14 linkage groups covering 458 cM with an average distance of 5.3 cM between markers.

Mapping complex traits such as resistance to pod borer, *H. armigera* in chickpea is only just beginning (Lawlor et al., 1998). A mapping population of 126 F_{13} RILs of ICCV 2 × JG 62, has been evaluated for resistance to *H. armigera*. The overall resistance score (1 = <10 leaf area and/or pods damaged, and 9 = >80% leaf area and/or pods damaged) varied from 1.7 to 6.0 in the RIL population compared to 1.7 in the resistant check, ICC 506EB, and 5.0 in the susceptible check, ICCV 96029. There were 4 to 31 larvae per 10 plants in the mapping population compared to 10 larvae on ICC 506EB and 18 on ICCV 96029. These results indicated that there is considerable variation in this mapping population for susceptibility to *H. armigera*. Another RIL mapping population from the cross Vijay (susceptible) × ICC 506EB (resistant) has been evaluated for resistance to *H. armigera*. Efforts are also underway to evaluate interspecific mapping populations based on the crosses ICC 3137 (*C. arietinum*) × IG 72933 (*C. reticulatum*) and ICC 3137 × IG 72953 (*C. reticulatum*) for resistance to pod borer to identify QTLs linked to various components of resistance to *H. armigera* (Sharma, Gaur, and Hoisington, 2005).

Pigeonpea

A few studies have been conducted to investigate polymorphism in pigeonpea and its wild relatives (Sharma and Crouch, 2004; Sharma, Gaur, and Hoisingtom, 2005). Screening of ten allozymes across one Zambian and 20 Indian genotypes of cultivated pigeonpea detected limited polymorphism (Boehringer, Lebot, and Aradhya, 1991), while Nadimpalli et al. (1993) used nuclear RFLPs to determine phylogenetic relationships among 12 species in four genera (*Cajanus, Dunbaria, Eriosema,* and *Rhynchosia*). Fifteen random genomic probes and six restriction enzymes revealed limited variation within each species, while considerable polymorphism was observed between the species. *Cajanus cajan* (L.) Millsp. was found to be closer to *C. scarabaeoides* than to *C. cajanifolius* (Haines). Ratnaparkhe et al. (1995) studied RAPD polymorphism in cultivated pigeonpea and its 13 wild relatives. The level of polymorphism among the wild species was very high, while little polymorphism was detected within the cultivated species. Low levels of genetic diversity were also observed in the cultivated pigeonpea using DArTs (Yang et al., 2006). Only 64 markers were polymorphic among the cultivated pigeonpeas. DArT markers also revealed genetic relationships among the accessions of different species consistent with the available information and systematic classification.

Variations in length and restriction sites of ribosomal DNA have also been studied among eight *Cajanus* species (Parani et al., 2000). The six genotypes of *C. cajan* did not show polymorphism in any of the enzyme-probe combinations, whereas RFLPs were readily detected among the eight species in all enzyme-probe combinations. The cultigen was found to be closely related to *C. scarabaeoides*. The studies indicated that isozyme, RAPD, and RFLP markers may not be adequate to develop a genomic map of pigeonpea based on intraspecific mapping populations. However, recently developed microsatellite markers have detected polymorphism in diverse pigeonpea germplasm using manual slab gel systems (Burns et al., 2001). Six of these markers have detected extensive diversity within and between cultivated pigeonpea accessions using capillary electrophoresis. Thus, it appears that SSR markers will readily detect polymorphism in breeding populations, although the number currently available is a limitation to their application. For this reason, a major SSR marker development program has been initiated in pigeonpea. Panguluri et al. (2006) used AFLP markers to detect polymorphism in cultivated pigeonpea and two of its wild relatives *Cajanus volubilis* Lour. and *Rhynchosia bracteata* Benth. ex Bak. The two wild species shared only 7.15% of the bands with the pigeonpea, whereas 86.71% common bands were observed among the pigeonpea cultivars. Similarly, 62.08% bands were polymorphic between *C. volubilis* and pigeonpea in comparison to 63.33% of polymorphic bands between *R. bracteata* and pigeonpea, and 13.28% polymorphic bands among pigeonpea cultivars. High levels of resistance to pod borer, *H. armigera*, and pod fly, *Melanagromyza obtusa* (Malloch), have been identified in wild relatives of pigeonpea such as *C. scarabaeoides, C. sericeus* van der Maesen, and *C. acutifolius* (Sharma et al., 2001, 2003b), which can be easily crossed with the cultivated pigeonpea. A mapping population based on *C. cajan* × *C. scarabaeoides* is under development, and will be evaluated for resistance to *H. armigera* to identify QTLs linked for resistance to these insects (Sharma and Crouch, 2004).

Cowpea

The development of RFLP map of cowpea has allowed the investigation of association between genes of interest (Fatokun et al., 1992; Young et al., 1992a, 1992b; Myers, Fatokun, and Young, 1996; Fatokun, Young, and Myers, 1997; Menendez, Hall, and Gepts, 1997). A cross between an aphid, *A. craccivora*-resistant cultivated cowpea, IT 84S-2246-4, and aphid-susceptible wild cowpea, NI 963, has been evaluated for aphid resistance and RFLP marker segregation (Myers, Fatokun, and Young, 1996). One RFLP marker, *bg4D9b*, has been found to be tightly linked to aphid resistance gene (*Rac1*), and several flanking markers in the same linkage group (linkage group 1) were also identified. The close association of *Rac1* and *bg4D9b* presents an opportunity for cloning this insect resistance gene. Githiri, Kimani, and Pathak (1996) studied the linkage of aphid resistance gene, *Rac* with various polymorphic loci controlling morphological traits and aspartate amino-transferase isozyme (*AAT*) to identify simply inherited and easily identifiable markers for aphid resistance, and to distinguish between *Rac1* and *Rac2*. The F_2 and F_2-derived F_3 populations from crosses IT 87S-1459 × TVu 946 × ICV 5 and IT 84S-2246 × TVu 946 segregating for *Rac1*, and the cross ICV 12 × TVu 946 segregating for *Rac2* have been evaluated for various polymorphic morphological traits. Locus *pd*, controlling peduncle color, was found to be linked to both *Rac1* and *Rac2*. The recombination frequencies estimated by the maximum likelihood method were 26% ± 8.3% and 35% ± 7.5% for *Rac1-pd* and *Rac2-pd* co-segregation, respectively, indicating that *Rac1* and *Rac2* were not different from one another. No linkage was detected between aphid resistance genes and the genes controlling morphological traits or *AAT* isozyme.

Common Bean

A genetic linkage map of common bean for agronomic traits has been developed by Tar'an, Michaels, and Pauls (2002). Schneider, Brothers, and Kelley (1997) used seven markers for MAS, and improved yield under stress conditions by 11%. Common bean near-isogenic lines differing for the recessive bean common mosaic virus (BCMV) resistance allele *bc-3* have been screened to identify RAPD markers linked to BCMV (Haley, Afanador, and Kelly, 1994). Categorization of the *bc-3* genotypes in the F_2 population revealed that selection against the repulsion-phase RAPD, as opposed to selection for the coupling-phase RAPD, provided a greater proportion of homozygous resistant selections (81.8% versus 26.3%), and a lower proportion of both segregating (18.2% versus 72.5%) and homozygous susceptible (0.0% versus 1.2%) selections. Selection of individual plants based on phenotype of both RAPD markers was identical to selection based solely on the repulsion-phase RAPD alone. Bulk segregant analysis and quantitative trait analysis identified eight markers associated with resistance to potato leafhopper, *E. fabae*, and four markers that were associated with resistance to *E. kraemeri* Ross and Moore (Murray et al., 2004). Three markers were associated with resistance to both species. Composite interval mapping identified QTL for resistance to the leafhopper on core-map linkage groups B1, B3, and B7. QTL for seed weight were close to the locus controlling testa color and the phaseolin gene. Mesoamerican bean lines BAT 881 and G 21212 showed transgressive segregation for resistance to thrips, *Thrips palmi* Karny, in the field. Correlations between damage and reproductive adaptation scores were significant within and between seasons (Frei et al., 2005). A major QTL (*Tpr6.1*) for thrips resistance located on LG *b06* explained up to 26.8% of variance for resistance in a single season. Joint interval mapping across seasons revealed various QTLs on LGs *b02*, *b03*, *b06*, and *b08*, some of which were located in regions containing genes encoding for disease resistance.

Greengram

A gene from TC 1966 conferring resistance to bruchid, *Callosobruchus* sp., has been mapped using RFLP markers (Young et al., 1992b). Fifty-eight F_2 progenies from a cross between TC 1966 and a susceptible mungbean cultivar were analyzed with 153 RFLP markers. Resistance was mapped to a single locus on linkage group VIII, approximately 3.6 cM from the nearest RFLP marker. Based on RFLP analysis, an individual plant was identified in the F_2 population that retained the bruchid resistance gene within a tightly linked double crossover. The RAPDs have also been used to identify markers linked to the bruchid resistance in mungbean (Villareal, Hautea, and Carpena, 1998). The technique was utilized in conjunction with NIL and RIL mapping populations. The resistant NILs were B4P3-3-23, B4P 5-3-10, B4Gr3-1, and DHK 2-18; carrying the bruchid resistance gene in four genetic backgrounds, Pagasa 3, Pagasa 5, Taiwan Green, and VC 1973A, respectively. The source of resistance to this bruchid was TC 1966, an accession of *Vigna radiata* var. *sublobata* (L.) Wilczer (Roxb.) Verdc. The bruchid resistance gene mapped 14.6 cM from the nearest RAPD marker *Q04*, and 13.7 cM from the nearest RFLP marker *pM151b*. The gene was 25 cM from *pM151a*. When *pM151a* and *pM151b* were considered as alleles of the same locus, the bruchid resistance gene was located 11.9 cM from the nearest RAPD marker *Q04* sub 900 and 5.6 cM from *pM151*. Yang et al. (1998) used MAS in backcross breeding for introgression of bruchid resistance in greengram, while Kaga and Ishimoto (1998) studied genetic localization of a bruchid resistance gene and its relationship to alkaloids, the vignatic acids in greengram.

Soybean

There has been limited success in developing soybean cultivars with resistance to insects because of the quantitative nature of resistance and linkage drag from resistance donor parents. J.J. Lin et al. (1996) developed a linkage map of soybean using RFLP, RAPD, and AFLP markers, while Narvel et al. (2000) used SSR makers to cover the soybean genome. Song et al. (2004) published an integrated genetic linkage map of soybean. Rector et al. (1998) used 139 RFLPs to construct a genetic linkage map of soybean to identify the QTLs associated with resistance to corn earworm, *H. zea* in a population of 103 F_2-derived lines from Cobb (susceptible) × PI 229358 (resistant). The genetic linkage map consisted of 128 markers, which formed 30 linkage groups covering approximately 1325 cM. One major and two minor QTLs were identified for resistance to *H. zea*. The major QTL was linked to the RFLP marker *A584* on linkage group (LG) *M*. The minor QTLs were linked to the RFLP markers *R249* (LG 'H') and *Bng047* (LG *D1*). The heritability (h^2) for resistance was estimated to be 64%. Another RFLP map based on Cobb × PI 171451 and Cobb × PI 227687 has been developed by Rector et al. (1999). Among the three resistant genotypes (PI 171451, PI 227687, and PI 229358), a QTL on LG *H* was shared among all three resistant genotypes, and a major QTL on LG *M* was shared between PI 171451 and PI 229358. A minor QTL on LG *C2* was unique to PI 227687, and a minor QTL on LG *D1* was unique to PI 229358. In addition, a QTL was detected on LG *F* in the susceptible genotype, Cobb. This QTL is in a region of the soybean genome that has previously been associated with a cluster of soybean pathogen-resistance loci. Using RFLP markers, Narvel et al. (2001) identified QTLs associated with insect resistance from PI 229358 and PI 171451. Marker analysis defined intervals of 5 cM or less for a QTL on linkage group *D1b* (*SIR-D1b*), and for *SIR-G*, *SIR-H*, and *SIR-M*. At least 13 of the 15 *SIR* genotypes studied had introgressed *SIR-M*. Only a few genotypes possessed *SIR-G* or *SIR-H*, and no genotype possessed *SIR-D1b*. Resistance to defoliating insects in soybean is expressed as a combination of antibiosis and antixenosis mechanisms of resistance. Both of these resistance modes are inherited quantitatively (Rector et al., 2000). RFLP maps based on F_2 populations segregating for antibiosis against *H. zea* indicated that heritability estimates for antibiosis were 54%, 42%, and 62% in Cobb × PI 171451, Cobb × PI 227687, and Cobb × PI 229358, respectively. An antibiosis QTL on linkage group *LG M* was detected in both Cobb × PI171451 and Cobb × PI 229358. An antixenosis QTL was also significant at this location in these two crosses. This is the only insect-resistance QTL that has been detected for both antibiosis and antixenosis. Antibiosis QTL were also detected on LG *F* and *B2* in Cobb × PI 227687, and LGs *G*, and *J* in Cobb × PI 229358. Antibiosis was conditioned by the PI (resistant parent) allele at the QTL on LGs *G, M,* and *B2*, whereas the susceptible parent, Cobb, provided antibiosis alleles at the QTL on LGs *F* and *J*.

A genetic map based on more than 500 markers on 240 RILs derived from nonresistant parents [Minsoy from China and Noir 1 from Hungary (MN population)] has shown transgressive segregation for resistance to *H. zea*, and soybean looper, *P. includens* (Terry et al., 1999). The two QTLs affected larval development, while another QTL affected only a single trait each, that is, larval weight, pupal weight, developmental rate, nutritional efficiency, or survival. Increased range of defensive effects among the segregant RILs was due to recombination among several parental genes that together quantitatively control plant defensive traits. QTLs have also been found on five LGs in the MN population and four in the MA population (Minsoy × Archer) (Terry et al., 2000). The QTL on LG *U2* is associated with major effects on larval development in both the MN and the MA populations. All other QTLs had lesser effects. The *U2* QTL associated with resistance to insects is of major importance in that: (1) it has been identified in different genetic backgrounds, (2) it is associated with several larval growth parameters, and (3) it explains a large proportion of the phenotypic variation.

Groundnut

The first linkage map of groundnut with a total map distance of nearly 1063 cM has been constructed using an F_2 population derived from two related diploid species [*Arachis stenosperma* (Karpov. & W.C. Gregory) and *A. cardenasii* (Karpov. & W.C. Gregory)] (Halward, Stalker, and Kochert, 1993). The first RFLP-based genetic linkage map of cultivated groundnut (derived from a BC_1 population) contained 350 RFLP loci distributed across 22 linkage groups, with a total map distance of approximately 2,700 cM (Burow et al., 1999). RAPD (*RKN 229, RKN 410,* and *RKN 440*) and RFLP (*R2430E, R2545E,* and *S1137E*) markers linked with root-knot nematode resistance have been identified (Burow et al., 1996; Choi et al., 1999). Resistant and susceptible alleles for RFLP loci *R2430E* and *R2545E* are quite distinct, and may be useful for identifying individuals homozygous for resistance in segregating populations (Choi et al., 1999). RAPD, SCAR, and RFLP markers have also been used to follow introgression of wild species chromosome segments with nematode resistance in *A. hypogaea* L. from *A. cardenasii* (Garcia et al., 1996). There is a need to convert these RFLP markers into PCR-based markers to make it easier to exploit the marker-trait relationships (Dwivedi et al., 2003). Resistance to the aphid vector, *A. craccivora,* has been identified in the breeding line ICG 12991 and is controlled by a single recessive gene (Herselman et al., 2004), which was mapped on linkage group 1 at 3.9 cM from a marker originating from the susceptible parent, explaining 76.1% of the phenotypic variation for aphid resistance.

Gene Synteny

Genes can be discovered using a variety of approaches (Primrose, 1998; Shoemaker et al., 2001). The development of genetic maps in a number of crop species having positional similarity will lead to better understanding of crop evolution and functioning of genes. Gene "synteny" will allow advances made in one species to have spillover impacts in other species (Gale and Devos, 1998). A comparison of EST databases from different plants can reveal the diversity in coding sequences between closely and distantly related plants, while mapping of ESTs may elucidate the synteny between those species. It can also allow development of SNP and/or SSR markers detected with primer pairs that amplify poly-morphic introns located between conserved exons (Bertin, Zhu, and Gale, 2005; Feltus et al., 2006; Huyen et al., 2006). For understanding gene functions of a whole organism, functional genomics is now using insertion mutant isolation, gene chips, or microarrays, and proteomics. This information can also be used to understand the genetics of metabolic processes, analyze traits controlled by several QTLs, and identify favorable alleles at each locus. The favorable alleles can be combined by simple crossing, and the most favorable combinations assembled in the same background using MAS.

There has been a considerable interest in exploiting gene synteny by using SSR markers identified in intensively studied crops such as pea, soybean, and *Medicago* in lesser-studied crops such as chickpea, pigeonpea, and lentil. A comparison of the linkage maps of *Cicer, Pisum, Lens,* and *Vicia* has revealed that these legumes share many common linkage groups (Gaur and Slinkard 1990a, 1990b; Weeden, Muehlbauer, and Ladizinsky, 1992; Kazan et al., 1993; Simon and Muehlbauer, 1997; Weeden et al., 2000). The extent of conservation of linkage arrangement may be as much as 40% of the genome (Weeden et al., 2000). The high level of conservation of linkage groups among *Cicer, Pisum, Lens,* and *Vicia* suggests that these genera are very closely related. There is a nearly 60% chance that microsatellites isolated in pea will amplify in chickpea (Edwards et al., 1996), although there is less than a 20% chance in the reverse direction (Pandian, Ford, and Taylor, 2000). Aubert et al. (2002) reported 41 new links between *P. sativum* and *Medicago truncatula* Geart. maps. Huyen et al. (2006) compared ESTs from the phylogenetically distant species, *M. truncatula, Lupinus albus* Lin., and *G. max,* to produce 500 intron-targeted amplified polymorphic markers (ITAPs). These markers were used to generate comparative genetic maps of lentil, *Lens culinaris* Medik., and white lupin, *L. albus.* The results showed that 90% of the ITAP markers amplified genomic DNA in *M. truncatula,* 80% in *L. albus,* and 70% in *L. culinaris.* Although a direct and simple syntenic relationship was observed between *M. truncatula* and *L. culinaris* genomes, where there is evidence of moderate chromosomal rearrangement, a more complicated pattern among homologous blocks was apparent between the *L. albus* and *M. truncatula* genomes. Based on taxonomic distance, it is expected that a similar trend may be observed between soybean and pigeonpea. Gualtieri (2002) studied the level of microsynteny between the *SYM2* region of pea and the orthologous region in *M. truncatula.* The *SYM2*-containing region in pea and the *SYM2*-orthologous region in *M. truncatula* share conserved gene content, and should provide the basis for cloning *SYM2.* Combining empirical lab-based approaches with bioinformatic strategies will be helpful in developing efficient systems for screening the vast public domain sequence databases of soybean and *Medicago* to liberate sequences of most value for molecular breeding in chickpea and pigeonpea. Information on conserved gene sequences among these genera will also facilitate prediction of gene location in crops based on its location in other genera.

Molecular Markers and Metabolic Pathways

There is considerable scope for changing the products of secondary metabolites that are associated with resistance to insect pests through biotechnological approaches. Harnessing synteny may have maximum benefit where entire metabolic pathways are dissected and studied in detail in model systems, thereby identifying the key genes for manipulating that trait, which can then be traced in the species of interest. Many secondary plant metabolites such as flavonoids, alkaloids, and terpenoids have been implicated in host plant resistance to insect pests. Many compounds of the flavonoid biosynthetic pathway (flavanones, flavones, flavanols, and isoflavonoids) accumulate in response to insect damage (Ebel, 1986; Sharma and Norris, 1991; Heller and Forkman, 1993). Molecular breeding and genetic engineering can be used to change the metabolic pathways to increase the amounts of various flavonoids conferring resistance to insect pests, for example, medicarpin and sativan in alfalfa, cajanol and stilbene in pigeonpea, and stilbene in chickpea (Heller and Forkman, 1993). Stilbenes have been expressed in transgenic tobacco plants, exhibiting various degrees of inhibition of fungal growth (Heller and Forkman, 1993). Maysin, a glycosyl flavone in maize silk, is associated with resistance to corn earworm, *H. zea* (Waiss et al., 1979). Most of the phenotypic variation in maysin concentration in maize silk is accounted for by the *p1* locus, the transcription activator of the portion of the flavonoid pathway leading to maysin synthesis. Reduced function *p1* allele results in decreased transcription of genes encoding enzymes of the *p1*-controlled portion of the pathway, and thus reduced maysin synthesis. The marker *umc105a* corresponds to the brown pericarp (*bp1*) locus. The *p1* and chromosome *9S* regions are the major QTLs controlling silk antibiosis to the corm earworm (Byrne et al., 1997). Composite interval mapping has shown a major QTL in the *asg20-whp1* interval of chromosome 2, and another near the *wx1* locus on chromosome 9 (Byrne et al., 1998). A gene that encodes chalcone synthase (*whp1*) on chromosome 2 and a silk specific gene (*sm1*) on chromosome 6 affect silk maysin concentration and resistance to corn earworm in maize (Byrne et al., 1998). The extra chromosome 5A of *Allium cepa* L. plays an important role in flavonoid biosynthesis (Masuzaki, Shigyo, and Yamauchi, 2006). The flavonoid 3'-hydroxylase (F3'H) gene controlling quercetin synthesis from kaempferol is located on chromosome 7A, and an anonymous gene involved in glucosidation of quercetin is located on chromosome 3A or 4A.

The Transgenic Approach and Gene Pyramiding through MAS

Genetic engineering offers the advantage of rapid introgression of novel genes and traits into elite agronomic backgrounds (Mohan et al., 1997). Transgenic resistance to insects has been demonstrated in plants expressing insecticidal genes such as δ-endotoxins from *Bacillus thuringiensis* (*Bt*) Berliner, protease inhibitors, enzymes, secondary plant metabolites, and plant lectins (Sharma, Sharma, and Crouch, 2004). While transgenic plants with introduced *Bt* genes have been deployed in several crops on a global scale, the alternative genes have received considerably less attention. The potential of some of the alternative genes can only be realized by deploying them in combination with conventional host plant resistance and *Bt* genes (Sharma et al., 2002). Many of the candidate genes used in genetic transformation of crops are highly specific or are only mildly effective against the target insect pest(s). In addition, crops frequently suffer from a number of primary herbivores. This suggests that single and multiple transgenes will need to be combined in the same variety with other sources, mechanisms, and targets of insect pest resistance in order to generate highly effective and sustainable seed-based technologies. From an evolutionary point of view, the development of plants with several genes for insect resistance will be expected to decrease the probability of insect pests to overcome newly deployed seed-based resistance technologies and thereby increase the effectiveness and prolong the life of such new varieties (Hadi, McMullen, and Finer, 1996; Karim, Riazuddin, and Dean, 1999), for example, the activity of *Bt* genes in transgenic plants is enhanced by the serine protease inhibitors (MacIntosh et al., 1990; Zhao et al., 1997) and tannic acid (Gibson et al., 1995). However, this may have some metabolic cost to the plant in some cases, and different resistance gene products may also have deleterious or nullifying interactions.

Combining transgene- and QTL-mediated resistance can be used as a viable strategy for insect control. A QTL conditioning maize earworm resistance in soybean PI 229358 and the *cry1Ac* transgene from the recurrent parent Jack-*Bt* has been pyramided into BC_2F_3 plants by MAS (Walker et al., 2002). Segregating individuals were genotyped through SSR markers linked to an antibiosis/antixenosis QTL on LG *M*, and tested for the presence of *cry1Ac*. MAS was used during and after the two backcrosses to develop a series of BC_2F_3 plants with or without the *cry1Ac* transgene and the QTL conditioning insect resistance. The BC_2F_3 plants homozygous for parental alleles at markers on LG *M*, and which either had or lacked Cry1Ac, were assigned to one of four possible genotype classes. These plants were used in no-choice detached leaf feeding bioassays with corn earworm, *H. zea* and soybean looper, *P. includens*. Fewer larvae of either species survived on leaves expressing the Cry1Ac protein. Though not as great as the effect of Cry1Ac, the PI 229358-derived LG *M* QTL also had a detrimental effect on larval weights of both species, and on defoliation by maize earworm, but did not reduce defoliation by the soybean looper. Weights of soybean looper larvae fed on foliage from transgenic plants with the PI-derived QTL were lower than those of larvae fed transgenic tissue with the corresponding Jack chromosomal segment (Walker et al., 2002). Therefore, combining transgene- and QTL-mediated resistance to lepidopteran insects may be a viable strategy for insect control.

Marker-Assisted versus Phenotypic Selection

Expression of physicochemical markers is influenced by the environment, and this makes them less reliable than molecular markers. A number of methods have been used for mapping QTLs associated with the traits of interest (Knot and Hailey, 1992; Karp et al., 1997). Among the molecular markers, RFLPs are the most reliable (Caetano-Anolles, Bassam, and Gresshoff, 1991; J.J. Lin et al., 1996), and can be converted into SCARs by sequencing the two ends of the DNA clones and designing oligonucleotide primers based on end sequences (Williams et al., 1991). Molecular markers and QTLs associated with insect resistance can be used to transfer the resistance genes from sources of resistance into high-yielding cultivars. It is important that the marker cosegregates with the gene, and is closely linked (1 cM or less) with the trait of interest. There are several advantages of using molecular markers to breed for insect resistance. Some types of molecular markers behave in a codominant manner to detect heterozygotes in segregating populations. Morphological markers typically behave in a dominant or recessive manner and do not detect heterozygotes (Staub, Serquen, and Gupta, 1996). Molecular markers are:

- Unaffected by the environment;
- Phenotype neutral; and
- Detectable at all stages of plant growth.

Marker-assisted selection can be used to accelerate the pace and accuracy of transferring resistance genes into improved cultivars. The MAS takes three to six years, thus speeding up the pace of transferring the traits of interest into the improved varieties, and it does not require large-scale planting of the segregating progenies up to crop harvest, as only the plants with marker allele indicating the presence of the trait or QTL need to be maintained up to maturity. Narvel et al. (2001) used microsatellite markers to identify soybean QTLs for resistance to foliar feeding lepidopteran insects, to determine the degree to which different QTLs have been transferred into soybean cultivars over a 30-year period. Very few resistant genotypes possessed multiple QTLs from different soybean linkage groups, and MAS was suggested as a means of introgressing minor genes linked to soybean resistance to insects into elite soybean cultivars. MAS in barley for resistance to cereal cyst nematode, *Heterodera avenae* Woll. could be accomplished approximately 30 times faster and for 75% lower cost compared to phenotypic selection (Kretschmer et al., 1997). Similar cost and labor savings have also been documented for microsatellite markers linked to cyst nematode, *Heterodera glycines* Ichinohe resistance in soybean (Mudge et al., 1997).

In contrast to the markers linked to resistance genes inherited as simple dominant traits, the improvement of polygenic traits (QTLs) through MAS is difficult due to involvement of a number of genes, and their interactions (epistatic effects). Resolving these problems often involves multiple field tests across several environments. However, such experiments often display significant QTL-environment interactions. Several studies on QTLs linked to stem borer resistance in maize underscore the problems involved in using QTLs in MAS. The relative efficiency of phenotypic and MAS has been found to be similar (Groh et al., 1998a, 1998b; Willcox et al., 2002). However, phenotypic selection was more favorable due to lower costs. MAS and phenotypic selection for leaf feeding resistance to *D. grandiosella* and *D. saccharalis* improved the efficiency of selection by 4%, indicating that MAS is less efficient than phenotypic evaluation (Bohn et al., 2001). Maximum progress has been made in breeding for insect resistance in common bean by using a combination of phenotypic performance and QTL-based index, followed by QTL based index, and conventional selection (Tar'an et al., 2003). Although the cost of MAS is approximately 90% less than the cost of conventional selection, accurate identification of QTL position and the cost to generate initial data for use in MAS make conventional selection more cost effective. However, marker-assisted selection of large DNA segments can still be highly effective. Stromberg, Dudley, and Rufener, (1994) did not get a better response to MAS than to conventional selection for resistance to southwestern corn borer, *D. grandiosella*. Three putative QTLs accounted for 28% of the phenotypic variance, and no significant differences were observed for leaf damage ratings or larval weights between lines selected by the two methods.

The use of DNA-based markers for indirect selection offers the greatest potential gains for quantitative traits with low heritability, as these are the most difficult characters to work with through conventional phenotypic selection. However, it is also difficult to develop effective markers for such traits. The expression of such traits is influenced by genotype-environment interaction and epistasis, which in addition to difficulties involved in accurately and precisely phenotyping such traits, which confound the development of effective MAS systems. The quality of an MAS program can only be as good as the quality of the phenotypic data on which the development of that marker was based. Therefore, it is essential to use large mapping populations characterized across seasons and locations, using well-defined phenotyping protocols. Nevertheless, when confidence limits are calculated for the QTL positions, they might cover several intervals or entire chromosome arms, if the heritability of the trait is low (Hyne et al., 1995). Fine mapping of such large QTL intervals, by phenotypic screening of several hundred individuals exhibiting molecular marker evidence of genetic recombination in the interval of interest, is then required to obtain tightly linked flanking markers that can be exploited in MAS.

Conclusions

A good beginning has been made in developing genetic linkage maps of many crops. However, the accuracy and precision of phenotyping for resistance to insect pests remains a critical constraint in many crops. There is a need to focus on developing innovative solutions to this problem. Improved phenotyping systems will have substantial impact on both conventional and MAS to breed for resistance to insect pests, in addition to the more strategic research that feeds into these endeavors. MAS has had a dramatic impact, particularly in the private sector, in breeding for disease resistance and quality traits where simply inherited components could be readily identified. The same potential may be achieved in the case of more complex traits such as resistance to insect pests and abiotic stresses. However, the practical and logistical demands for developing and implementing molecular breeding systems for these traits are quite complex. There is a need to use MAS to develop cultivars with improved resistance to insect pests and to strengthen *Bt* transgenic crops through introgression of other components of resistance through MAS. There are very few reports concerning the application of MAS for resistance to insect pests. However, those available fail to demonstrate an increase in efficiency of MAS over conventional breeding approaches, although combining MAS with conventional approaches has given better results. Thus, not only is there a need for precise mapping of the QTLs associated with resistance to insects, but also the development of a new paradigm in breeding based on re-engineering breeding programs to make best use of molecular marker data. Only a combination of conventional and molecular approaches can accelerate the progress in developing cultivars with insect resistance to increase crop productivity and improve livelihoods of the rural poor.

References

Agrama, H.A., Widle, G.E., Reese, J.C., Campbell, L.R. and Tuinstra, M.R. (2002). Genetic mapping of QTLs associated with greenbug resistance and tolerance in *Sorghum bicolor*. *Theoretical and Applied Genetics* 104: 1373–1378.

Ahmad, F., Gaur, P.M. and Slinkard, A.E. (1992). Isozyme polymorphism and phylogenetic interpretations in the genus *Cicer* L. *Theoretical and Applied Genetics* 83: 620–627.

Akbari, M., Wenzl, P., Caig, V., Carling, J., Xia, L., Yang, S., Uszynski, G., Mohler, V., Lehmensiek, A., Kuchel, H., Hayden, M.J., Howes, N., Sharp, P., Vaughan, P., Rathmell, B., Huttner, E. and Kilian, A. (2006). Diversity arrays technology (DArT) for high-throughput profiling of the hexaploid wheat genome. *Theoretical and Applied Genetics* 113: 1409–1420.

Alam, S.N. and Cohen, M.B. (1998). Detection and analysis of QTLs for resistance to the brown planthopper, *Nilaparvata lugens*, in a doubled-haploid rice population. *Theoretical and Applied Genetics* 97: 1370–1379.

Arora, R., Kaur, S. and Singh, M. (1996). Groundnut genotype reaction to *Helicoverpa armigera* in India. *International* Arachis *Newsletter* 16: 35–37.

Arora, P.K., Kaur, N., Thind, S.K., Aulakh, P.S. and Kaur, N. (2000). Metabolites of some commercial cultivars of guava in relation to incidence of fruit fly. *Pest Management in Horticulture Ecosystem* 6(1): 61–62.

Aubert, G., Morin, J., Jacquin, F., Loridon, K., Quillet, M.C., Petit, A., Rameau, C., Lejeune-Hénaut, I., Huguet, T. and Burstin, J. (2002). Functional mapping in pea as an aid to the candidate gene selection and for investigating synteny with the model legume *Medicago truncatula*. *Theoretical and Applied Genetics* 112: 1024–1041.

Auclair, J.L. (1963). Aphid feeding and nutrition. *Annual Review of Entomology* 8: 439–490.

Auclair, J.L., Maltais, J.B. and Cartier, J.J. (1957). Factors in resistance of peas to the pea aphid, *Acyrthosiphon pisum* (Harris) (Homoptera: Aphididae). II. Amino acids. *Canadian Entomologist* 10: 457–464.

Bai, Y., Michaels, T.E. and Pauls, K.P. (1998). Determination of genetic relationships among *Phaseolus vulgaris* populations in a conical cross from RAPD marker analyses. *Molecular Breeding* 4: 395–406.

Banerjee, M.K. and Kalloo, G. (1989). Role of glands in resistance to tomato leaf curl virus, *Fusarium* wilt, and fruit borer in *Lycopersicon*. *Current Science* 52: 575–576.

Beck, S.D. (1972). Nutrition, adaptation and environment. In J.G. Rodriguez (Ed.), *Insect and Mite Nutrition: Significance and Implications in Ecology and Pest Management*. Amsterdam, The Netherlands: North-Holland, 1–6.

Behura, S.K., Sahu, S.C., Nair, S. and Mohan, M. (2000). An AFLP marker that differentiates biotypes of the Indian gall midge (*Orseolia oryzae*, Wood-Mason) is linked to sex and avirulence. *Molecular and General Genetics* 263: 328–334.

Benedict, J.H., Leigh, T.F. and Hyer, A.H. (1983). *Lygus hesperus* (Heteroptera: Miridae) oviposition behaviour, growth and survival in relation to cotton trichome density. *Environmental Entomology* 12: 331–335.

Bennett, S.E. (1965). Tannic acid as a repellent and toxicant to alfalfa weevil larvae. *Journal of Economic Entomology* 58: 372.

Bertin, I., Zhu, J. H. and Gale, M.D. (2005). SSCP-SNP in pearl millet—a new marker system for comparative genetics. *Theoretical and Applied Genetics* 110: 1467–1472.

Bhagwat, V.R., Aherker, S.K., Satpute, V.S. and Thakre, H.S. (1995). Screening of chickpea (*Cicer arietinum* L.) genotypes for resistance to *Helicoverpa armigera* (Hb.) and its relationship with malic acid in leaf exudates. *Journal of Entomological Research* 19: 249–253.

Bharathi, M. (1989). *Mechanisms and Genetics of Resistance in Rice to the Brown Planthopper*, Nilaparvata lugens (*Stal*) (*Homoptera: Delphacidae*). Ph.D. thesis, Tamil Nadu Agricultural University, Coimbatore, Tamil Nadu, India.

Bhat, M.G. and Jayaswal, A.P. (1988). A study on factors of bollworm resistance in cotton (*Gossypium hirsutum* L.) using isogenic lines. *Journal of the Indian Society for Cotton Improvement* 13: 149–153.

Biradar, S.K., Sundaram, R.M., Thirumurugan, T., Bentur, J.S., Amudhan, S., Shenoy, V.V., Mishra, B., Bennet, J. and Sharma, P.N. (2004). Identification of flanking SSR markers for a major rice gall midge resistance gene *Gm1* and their validation. *Theoretical and Applied Genetics* 109: 1468–1473.

Boehringer, A., Lebot, V. and Aradhya, M. (1991). Isozyme variation in twenty-one perennial pigeonpea genotypes. *International Pigeonpea Newsletter* 14: 6–7.

Bohn, M., Groh, S., Khairallah, M.M., Hoisington, D.A., Utz, H.F. and Melchinger, A.E. (2001). Re-evaluation of the prospects of marker-assisted selection for improving insect resistance against *Diatraea* spp. in tropical maize by cross validation and independent validation. *Theoretical and Applied Genetics* 103: 1059–1067.

Bohn, M., Khairallah, M.M., Gonzalez-deLeon, D., Hoisington, D.A., Utz, H.F.J., Deutsch, A., Jewell, D.C., Mihm, J.A. and Melchinger, A.E. (1996). QTL mapping in tropical maize. I. Genomic regions affecting leaf feeding resistance to sugarcane borer and other traits. *Crop Science* 36: 1352–1361.

Bohn, M., Khairallah, M.M., Jiang, C., Gonzalez-de-Leon, D., Hoisington, D.A., Utz, H.F., Deutsch, J.A., Jewell, D.C., Mihm, J.A. and Melchinger, A.E. (1997). QTL mapping in tropical maize. II. Comparison of genomic regions for resistance to *Diatraea* spp. *Crop Science* 37: 1892–1902.

Bonierbale, M.W., Plaisted, R.L., Pineda, O. and Tanksley, S.D. (1994). QTL analysis of trichome-mediated insect resistance in potato. *Theoretical and Applied Genetics* 87: 973–987.

Bonnema, G., Hontelez, J., Verkerk, R., Zhang, Y.Q., van Daelen, R., van Kammen, A. and Zabel, P. (1996). An improved method of partially digesting plant megabase DNA suitable for YAC cloning: Application to the construction of a 5.5 genome equivalent YAC library of tomato. *Plant Journal* 9: 125–133.

Botha, A.M. and Venter, E. (2000). Molecular marker technology linked to pest and pathogen resistance in wheat breeding. *South African Journal of Science* 96: 233–240.

Bothe, N.N. and Pokharkar, R.N. (1985). Role of silica content in sorghum for reaction to shoot fly. *Journal of Research, Maharashtra Agricultural Universities* 10(3): 338–339.

Boyko, E.V., Starkey, S.R. and Smith, C.M. (2004). Molecular genetic mapping of *Gby*, a new greenbug resistance gene in bread wheat. *Theoretical and Applied Genetics* 109: 1230–1236.

Boyko, E.V., Kalendar, R., Korzun, V., Korol, A., Schulman, A. and Gill, B.S. (2002). A high density genetic map of *Aegilops tauschii* includes genes, retro-transposons and microsatellites which provide unique insight into cereal chromosome structure and function. *Plant Molecular Biology* 48: 767–790.

Brookes, A.J. (1999). The essence of SNPs. *Gene* 234: 177–186.

Burns, M.J., Edwards, K.J., Newbury, H.J., Ford-Lloyd, B.V. and Baggott, C.D. (2001). Development of simple sequence repeat (SSR) markers for the assessment of gene flow and genetic diversity in pigeonpea (*Cajanus cajan*). *Molecular Ecology Notes* 1: 283–285.

Burow, M.D., Simpson, C.E., Paterson, A.H. and Starr, J.L. (1996). Identification of peanut (*Arachis hypogaea* L.) RAPD markers diagnostic of root-knot nematode [*Meloidogyne arenaria* (Neal) Chitwood] resistance. *Molecular Breeding* 2: 369–379.

Burow, M.D., Simpson, C.E., Starr, J.L. and Paterson, A.H. (1999). Generation of a molecular marker map of the cultivated peanut, *Arachis hypogaea* L. In *Proceedings, Plant and Animal Genome VII*, 17–21 January, 1999, San Diego, California, USA.

Bushman, B.S., Snook, M.E., Gerke, J.P., Szalma, S.J., Berbow, M.A., Houchins, K.E. and McMullen, M.D. (2002). Two loci exert major effects on chlorogenic acid synthesis in maize silks. *Crop Science* 42: 1669–1678.

Butron, A., Li, R.G., Guo, B.Z., Widstrom, N.W., Snook, M.E., Cleveland, T.E. and Lynch, R.E. (2001). Molecular markers to increase corn earworm resistance in a maize population. *Maydica* 46: 117–124.

Butron, A., Tarrio, R., Revilla, P., Ordas, A. and Malvar, R.A. (2005). Molecular changes in the maize composite EPS12 during selection for resistance to pink stem borer. *Theoretical and Applied Genetics* 110: 1044–1051.

Byrne, P.F., McMullen, M.D., Snook, M.E., Musket. T.A., Theuri, J.M., Widstrom, N.W., Wiseman, B.R. and Coe, E.H. (1996). Quantitative trait loci and metabolic pathways: Genetic control of the concentration of maysin, a corn earworm resistance factor, in maize silks. *Proceedings National Academy of Sciences USA* 93: 8820–8825.

Byrne, P.F., McMullen, M.D., Snook, M.E., Musket, T.A., Theuri, J.M., Widstrom, N.W., Wiseman, B.R. and Coe, E.H. (1997). Identification of maize chromosome regions associated with antibiosis to corn earworm larvae (Lepidoptera: Noctuidae). *Journal of Economic Entomology* 90: 1039–1045.

Byrne, P.F., McMullen, M.D., Wiseman, B.R., Snook, M.E., Musket, T.A., Theuri, J.M., Widstrom, N.W. and Coe, E.H. (1998). Maize silk maysin concentration and corn earworm antibiosis: QTLs and genetic mechanisms. *Crop Science* 38: 461–471.

Caetano-Anolles, G., Bassam, B.J. and Gresshoff, P.M. (1991). DNA amplification fingerprinting using very short arbitrary oligonucleotide primers. *BioTechnology* 9: 553–557.

Cardinal, A.J. and Lee, M. (2005). Genetic relationships between resistance to stalk-tunneling by the European corn borer and cell-wall components in maize population B73 × B52. *Theoretical and Applied Genetics* 111: 1–7.

Cardinal, A.J., Lee, M., Sharopova, N., Woodman-Clikeman, W.L. and Long, M.I. (2001). Genetic mapping and analysis of quantitative trait loci for resistance to stalk tunneling by the European corn borer in maize. *Crop Science* 41: 835–845.

Castro, A.M., Vasicek, A., Ellerbrook, C., Gimenez, D.O., Tocho, E., Tacaliti, M.S. Clua, A. and Snape, J.W. (2004). Mapping quantitative trait loci in wheat for resistance against greenbug and Russian wheat aphid. *Plant Breeding* 123: 361–365.

Chalfie, M. (1998). Genome sequencing. The worm revealed. *Nature* 396: 620–621.

Chen, Q., Sun, S., Ye, Q., McCuine, S., Huff, E. and Zhang, H.B. (2004). Construction of two BAC libraries from the wild Mexican diploid potato, *Solanum pinnatisectum*, and the identification of clones near the late blight and Colorado potato beetle resistance loci. *Theoretical and Applied Genetics* 108: 1002–1009.

Chhabra, K.S., Kooner, B.S., Sharma, A.K. and Saxena, A.K. (1988). Sources of resistance in mungbean (*Vigna radiata*) to insect pests and mungbean yellow mosaic virus. *Proceedings of II International Symposium on Mungbean*, 16–20 November, 1987, Bangkok, Thailand, 308–314.

Chhabra, K.S., Sharma, A.K., Saxena, A.K. and Kooner, B.S. (1990). Sources of resistance in chickpea: Role of biochemical components of the incidence of gram pod borer, *Helicoverpa armigera* (Hubner). *Indian Journal of Entomology* 52: 423–430.

Chiang, H.S. and Singh, S.R. (1988). Pod hairs as a factor in *Vigna vexillata* resistance to the pod-sucking bug, *Clavigralla tomentosicollis*. *Entomologia Experimentalis et Applicata* 47: 195–199.

Cho, S., Kumar, J., Shultz, J., Anupama, K., Tefera, F. and Muehlbauer, F.J. (2002). Mapping genes for double podding and other morphological traits in chickpea. *Euphytica* 128: 285–292.

Choi, K., Burow, M.D., Church, G., Burow, G., Paterson, A.H., Simpson, C.E. and Starr, J.L. (1999). Genetics and mechanism of resistance to *Meloidogyne arenaria* in peanut germplasm. *Journal of Nematology* 31: 283–290.

Clement, S.L., Hardie, D.C. and Elberson, L.R. (2002). Variation among wild *Pisum* accessions for pod and seed resistance to pea weevil (Coleoptera: Bruchidae). *Crop Science* 42: 2167–2173.

Clement, S.L. and Quisenberry, S.S. (Eds.). (1999). *Global Plant Genetic Resources for Insect-Resistant Crops*. Boca Raton, Florida, USA: CRC Press.

Cook, J.P., Wichman, D.M., Martin, J.M., Bruckner, P.L. and Talbert, L.E. (2004). Identification of microsatellite markers associated with a stem solidness locus in wheat. *Crop Science* 44: 1397–1402.

Cosenza, G.W. and Green, H.P. (1979). Behaviour of tomato fruit worm, *Heliothis zea* on susceptible and resistant lines of processing tomatoes. *Horticulture Science* 14: 171.

Cox, T.S. and Hatchett, J.H. (1994). Hessian fly-resistance gene *H26* transferred from *Triticum tauschii* to common wheat. *Crop Science* 34: 958–960.

Crouch, J.H., Gaur, P.M., Buhariwalla, H.K., Barman, P. and Sharma, H.C. (2005). Towards molecular breeding of *Helicoverpa* resistance in grain legumes. In Sharma, H.C. (Ed.), Heliothis/Helicoverpa *Management: Emerging Trends and Strategies for Future Research*. New Delhi, India: Oxford and IBH Publishing, 307–328.

Dadd, R.H. (1970). Arthropod nutrition. In Florkin, M. and Scheer, B.T. (Eds.), *Chemical Ecology*. New York, USA: Academic Press, 35–95.

Davis, G.L., McMullen, M.D., Baysdorfer, C., Musket, T., Grant, D., Staebell, M., Xu, G., Polacco, M., Koster, L., Melia-Hancock, S., Houchins, K., Chao, S. and Cae, E.H. Jr. (1999). A maize map standard with sequenced core markers, grass genome reference points, and 932-ESTs in a 1736-locus map. *Genetics* 152: 1137–1172.

Delibes, A., Del Moral, J., Martin-Sanchez, J.A., Mejias, A., Gallego, M., Casado, D., Sin, E. and Lopez-Brana, I. (1997). Hessian fly-resistance gene transferred from chromosome 4M^v of *Aegilops ventricosa* to *Triticum aestivum*. *Theoretical and Applied Genetics* 94: 858–864.

Devos, K.M. and Gale, M.D. (2000). Genome relationships: The grass model in current research. *Plant Cell* 12: 637–646.

Dimock, M.B. and Kennedy, G.G. (1983). The role of glandular trichomes in the resistance of *Lycopersicon hirsutum* f. *glabratum* to *Heliothis zea*. *Entomologia Experimentalis et Applicata* 33: 757–760.

Doss, R.P., Shanks, C.H. Jr., Chamberlain, J.D. and Garth, J.K.L. (1987). Role of leaf hairs in resistance of a clone of beach strawberry, *Fragaria chiloensis*, to feeding by adult black vine weevil, *Otiorhynchus sulcatus* (Coleoptera: Curculionidae). *Environmental Entomology* 16: 764–768.

Dougherty, D.E. (1976). *Pinitol and Other Soluble Carbohydrates in Soybean as Factors in Facultative Parasite Nutrition*. Ph.D. thesis, Tifton, University of Georgia, USA.

Duffey, S.S. and Stout, M.J. (1996). Antinutritive and toxic components of plant defense against insects. *Archives of Insect Biochemistry* 32: 3–37.

Dunford, R.P., Yano, M., Kurata, N., Sasaki, T., Huestis, G., Rocheford, T. and Laurie, D.A. (2002). Comparative mapping of the barley *Ppd-H1* photoperiod response gene region, which lies close to a junction between two rice linkage segments. *Genetics* 161: 825–834.

Dweikat, I., Ohm, H., MacKenzie, S., Patterson, F., Cambron, S. and Ratcliffe, R. (1994). Association of a DNA marker with Hessian fly resistance gene *H9* in wheat. *Theoretical and Applied Genetics* 89: 964–968.

Dweikat, I., Ohm, H., Patterson, F. and Cambron, S. (2004). Identification of RAPD markers for 11 Hessian fly resistance genes in wheat. *Theoretical and Applied Genetics* 94: 419–423.

Dwivedi, S.L., Crouch, J.H., Nigam, S.N., Ferguson, M.E. and Paterson, A.H. (2003). Molecular breeding of groundnut for enhanced productivity and food security in the semi-arid tropics: Opportunities and challenges. *Advances in Agronomy* 80: 153–221.

Ebel, J. (1986). Phytoalexin synthesis: The biochemical analysis of the induction process. *Annual Review of Phytopathology* 24: 235–264.

Edwards, K.J., Barker, J.H.A., Daly, A., Jones, C. and Karp, A. (1996). Microsatellite libraries enriched for several microsatellite sequences in plants. *Biotechniques* 20: 758–760.

Elliger, C.A., Zinkel, D.F., Chen, B.G. and Waiss, A.C. Jr. (1976). Diterpene acids as larval growth inhibitors. *Experientia* 32: 1364–1366.

Farrar, R.R. and Kennedy, G.G. (1987). Growth, food consumption and mortality of *Heliothis zea* larvae on foliage of the wild tomato, *Lycopersicon hirsutum* f. *glabratum*: and the cultivated tomato, *L. esculentum*. *Entomologia Experimentalis et Applicata* 44: 213–219.

Farrar, R.R. and Kennedy, G.G. (1988). 2-undecanone: A pupal mortality factor in *Heliothis zea*: Sensitive larval stage and in plant activity in *Lycopersicon* sp. *Entomologia Experimentalis et Applicata* 47: 205–210.

Farrar, R.R. and Kennedy, G.G. (1990). Growth inhibitors in host plant resistance to insects: Examples from wild tomato with *Heliothis zea*. *Journal of Entomology Science* 25: 45–56.

Fatokun, C.A., Young, N.D. and Myers, G.O. (1997). Molecular markers and genome mapping in cowpea. In Singh, B.B., Mohan Raj, D.R., Dashiell, K.E. and Jackai, L.E.N. (Eds.), *Advances in Cowpea Research*, Devon, UK: Sayce, 352–360.

Fatokun, C.A., Menancio-Hautea, D.I., Danesh, D. and Young, N.D. (1992). Evidence for orthologus seed weight genes in cowpea and mung bean based on RFLP mapping. *Genetics* 132: 841–846.

Felton, G.W., Broaduray, R.M. and Duffey, S.S. (1989). Inactivation of protease inhibitor activity by plant derived quinines, compilations for host-plant resistance against noctuid herbivore. *Journal of Insect Physiology* 35: 981–990.

Feltus, F.A., Singh, H.P., Lohithaswa, H.C., Schulze, S.R., Silva, T.D. and Paterson, A.H. (2006). A comparative genomics strategy for targeted discovery of single-nucleotide polymorphisms and conserved-noncoding sequences in orphan crops. *Plant Physiology* 140:1183–1191.

Ferry, R.L. and Cuthbert, F.P. Jr. (1973). Factors affecting evaluation of fruitworm resistance in the tomato. *Journal of American Society of Horticulture Science* 98: 457–459.

Ferry, R.L. and Cuthbert, F.P. Jr. (1975). Inheritance of pod resistance to cowpea curculio infestation in southern peas. *Journal of Heredity* 66: 43–44.

Flint-Garcia, S.A., McMullen, M.D. and Darrah, L.L. (2003). Genetic relationship of stalk strength and ear height in maize. *Crop Science* 43: 23–31.

Flint-Garcia, S.A., Darrah, L.L., McMullen, M.D. and Hibbard, B.E. (2003). Phenotypic versus marker assisted selection for stalk strength and second generation European corn borer resistance in maize. *Theoretical and Applied Genetics* 107: 1331–1336.

Folkertsma, R.T., Sajjanar, G.M., Reddy, B.V.S., Sharma, H.C. and Hash, C.T. (2003). Genetic mapping of QTL associated with sorghum shoot fly (*Atherigona soccata*) resistance in sorghum (*Sorghum bicolor*). In *Final Abstracts Guide, Plant and Animal Genome XI*, Jan 11–15, 2003, San Diego, CA, 42. http://www.intl-pag.org/11/abstracts/P5d_P462_XI.html.

Frei, A., Blair, M.W., Cardona, C., Beebe, S.E., Gu, H. and Dorn, S. (2005). QTL mapping of resistance to *Thrips palmi* Karny in common bean. *Crop Science* 45: 379–387.

Fujita, D., Yoshimura, A. and Yashi, H. (2003). Detection of QTLs associated with antibiosis to green rice leafhopper, *Nephotettix cincticeps* Uhler, in four Indica rice varieties. *Rice Genetics Newsletter* 19: 38–39.

Fujita, D., Doi, K., Yoshimura, A. and Yashi, H. (2006). Molecular mapping of a novel gene, *Grh5*, conferring resistance to green rice leafhopper (*Nephotettix cincticeps* Uhler) in rice, *Oryza sativa* L. *Theoretical and Applied Genetics* 113: 567–573.

Fukuta, Y., Tamura, K., Hirac, M. and Oya, S. (1998). Genetic analysis of resistance to green rice leafhopper (*Nephotettix cincticeps* Uhler) in rice parental line Norin-PL6, using RFLP markers. *Breeding Science* 48: 243–249.

Fulton, T., Beck-Bunn, T., Emmatty, D., Eshed, Y., Lopez, J., Petiard, V., Uhlig, J., Zamir, D. and Tanksley, S.D. (1997). QTL analysis of an advanced backcross of *Lycopersicon peruvianum* to the cultivated tomato and comparisons with QTLs found in other wild species. *Theoretical and Applied Genetics* 95: 881–894.

Gaikwad, B.P., Darekar, K.S. and Chavan, U.D. (1991). Varietal reaction of eggplant against jassid. *Journal of Research, Maharashtra Agricultural Universities* 16(3): 354–356.

Gale, M.D. and Devos, K.M. (1998). Comparative genetics in the grasses. *Proceedings National Academy of Sciences USA* 95: 1971–1974.

Gannon, A.J. and Bach, C.E. (1996). Effects of soybean trichome density on Mexican bean beetle (Coleoptera: Coccinellidae) development and feeding preference. *Environmental Entomology* 25(5): 1077–1082.

Garcia, G.M., Stalker, H.T., Shroeder, E. and Kochert, G. (1996). Identification of RAPD, SCAR, and RFLP markers tightly linked to nematode resistance genes introgressed from *Arachis cardenasii* into *Arachis hypogaea*. *Genome* 39: 836–845.

Gaur, P.M. and Slinkard, A.E. (1990a). Inheritance and linkage of isozyme coding genes in chickpea. *Journal of Heredity* 81: 455–461.

Gaur, P.M. and Slinkard, A.E. (1990b). Genetic control and linkage relations of additional isozyme markers in chickpea. *Theoretical and Applied Genetics* 80: 648–656.

Gibson, D.M., Gallo, L.G., Krasnoff, S.B. and Ketchum, R.E.B. (1995). Increased efficiency of *Bacillus thuringiensis* subsp. *kurstaki* in combination with tannic acid. *Journal of Economic Entomology* 88: 270–277.

Gill, B.S., Hatchett, J.H. and Raupp, W.J. (1987). Chromosomal mapping of Hessian fly-resistance gene *H13* in the D genome of wheat. *Journal of Heredity* 78: 97–100.

Giri, A.P. and Kachole, M.S. (1998). Amylase inhibitors of pigeonpea (*Cajanus cajan*) seeds. *Phytochemistry* 47: 197–202.

Githiri, S.M., Kimani, P.M. and Pathak, R.S. (1996). Genetic linkage of the aphid resistance gene, *Rac*, in cowpea. *African Crop Science Journal* 4: 145–150.

Green, P.W.C., Stevenson, P.C., Simmonds, M.S.J. and Sharma, H.C. (2002). Can larvae of the pod-borer, *Helicoverpa armigera* (Lepidoptera: Noctuidae), select between wild and cultivated pigeonpea [*Cajanus* sp. (Fabaceae)]? *Bulletin of Entomological Research* 92: 45–51.

Green, P.W.C., Stevenson, P.C., Simmonds, M.S.J. and Sharma, H.C. (2003). Phenolic compounds on the pod-surface of pigeonpea, *Cajanus cajan*, mediate feeding behavior of *Helicoverpa armigera* larvae. *Journal of Chemical Ecology* 29: 811–821.

Gregory, P., Tingey, W.M., Ave, D.A. and Bouthyette, P.Y. (1986). Potato glandular trichomes: A physio-chemical defense mechanism against insects. In Green, M.B. and Hedin, P.A. (Eds.), *Natural Resistance of Plants to Pests: Role of Allelochemicals*. ACS Symposium Series 296. Washington, D.C., USA: American Chemical Society, 160–167.

Groh, S, Gonzalez-de Leon, D., Khairallah, M.M., Jiang, C., Bergvinson, D., Bohr, M., Hoisington, D.A. and Melchinger, A.E. (1998a). QTL mapping in tropical maize. III. Genomic regions for resistance to *Diatraea* spp. and associated traits in two RIL populations. *Crop Science* 38: 1062–1072.

Groh, S., Khairallah, M.M., Gonzalez de Leon, D., Willcox, M., Jiang, C., Hoisington, D.A. and Melchinger, A.E. (1998b). Comparison of QTLs mapped in RILs and their test-cross progenies of tropical maize for insect resistance and agronomic traits. *Plant Breeding* 117: 193–202.

Gualtieri, G., Kulikova, O., Limpens, E., Kim, D.J., Cook, D.R., Bisseling, T. and Geurts, R. (2002). Microsynteny between pea and *Medicago truncatula* in the *SYM2* region. *Plant Molecular Biology* 50: 225–235.

Guo, B.Z., Zhang, Z.J., Li, R.G., Widstrom, N.W., Snook, M.E., Lynch, R.E. and Plaisted, D. (2001). Restriction fragment length polymorphism markers associated with silk maysin, antibiosis to corn earworm (Lepidoptera: Noctuidae) larvae in a dent and sweet corn cross. *Journal of Economic Entomology* 94: 564–571.

Hadi, M.Z., McMullen, M.D. and Finer, J.J. (1996). Transformation of 12 different plasmids into soybean via particle bombardment. *Plant Cell Reports* 15: 500–505.

Haldane, J.B.S. (1919). The combination of linkage values, and the calculation of distances between the loci of linked factors. *Journal of Genetics* 8: 299–309.

Haley, S.D., Afanador, L. and Kelly, J.D. (1994). Selection for monogenic pest resistance traits with coupling- and repulsion-phase RAPD markers. *Crop Science* 34: 1061–1066.

Halward, T., Stalker, H.T. and Kochert, G. (1993). Development of an RFLP linkage map in diploid peanut species. *Theoretical and Applied Genetics* 87: 379–384.

Harsulkar, A.M., Giri, A.P., Patankar, A.G., Gupta, V.S., Sainani, M.N., Ranjekar, P.K. and Deshpande, V.V. (1999). Successive use of non-host plant proteinase inhibitors required for effective inhibition of *Helicoverpa armigera* gut proteinases and larval growth. *Plant Physiology* 121: 497–506.

Hash, C.T., Folkerstma, R.T., Ramu, P., Reddy, B.V.S., Mahalakshmi, V., Sharma, H.C., Rattunde, H.F.W., Weltzein, E.R., Haussmann, B.I.G, Ferguson, M.E. and Crouch, J.H. (2003). Marker assisted breeding across ICRISAT for terminal drought tolerance and resistance to shoot fly and *Striga* in sorghum. In Abstracts, *In the Wake of the Double Helix: From the Green Revolution to the Gene Revolution*, 27–31 May, 2003. Bologna, Italy: University of Bologna, 82. http://www.doublehelix.too.it.

Hedin, P.A., Jenkins, J.N., Collum, D.H., White, W.H. and Parrott, W.L. (1983). Multiple factors in cotton contributing to resistance to the tobacco budworm, *Heliothis virescens* (F.). In Hedin, P.A. (Ed.), *Plant Resistance to Insects*. ACS Symposium Series 206. Washington, D.C., USA: American Chemical Society, 347–365.

Heller, W. and Forkman, G. (1993). Biosynthesis of flavonoids. In Harborne, J.B. (Ed.), *The Flavonoids: Advances in Research Since 1986*. London, UK: Chapman and Hall, 499–535.

Hernandez, P., Martin, A. and Dorado, G. (1999). Development of SCARs by direct sequencing of RAPD products: A practical tool for the introgression and marker-assisted selection of wheat. *Molecular Breeding* 5: 245–253.

Hernandez, P., Dorado, G., Prieto, P., Gimenez, M.J., Ramirez, M.C., Laurie, D.A., Snape, J.W. and Martin, A. (2001). A core genetic map of *Hordeum chilense* and comparisons with maps of barley (*Hordeum vulgare*) and wheat (*Triticum aestivum*). *Theoretical and Applied Genetics* 102: 1259–1264.

Herselman, L., Thwaites, R., Kimmins, F.M., Courtois, B., van der Merwe, P.J. and Seal, S.E. (2004). Identification and mapping of AFLP markers linked to peanut (*Arachis hypogaea* L.) resistance to the aphid vector of groundnut rosette disease. *Theoretical and Applied Genetics* 109: 1426–1433.

Hinds, W.E. (1906). *Proliferation as a Factor in the Natural Control of the Mexican Cotton Boll Weevil.* Bureau of Entomology Bulletin 59. Washington, D.C., USA: U.S. Department of Agriculture.

Hirabayashi, H. and Ogawa, T. (1995). RFLP mapping of *BphI* (brown planthopper resistance gene) in rice. *Breeding Science* 45: 369–371.

Hirabayashi, H. and Ogawa, T. (1999). Identification and utilization of DNA markers linked to genes for resistance to brown planthopper (BPH) in rice. *Advances in Breeding Science* 41: 71–74.

Huang, N., Parco, A., Mew, T., Magpantay, G., McCouch, S., Guiderdoni, E., Xu, J.C., Subudhi, P., Angeles, E.R. and Khush, G.S. (1997). RFLP mapping of isozymes, RAPD and QTLs for grain shape, and brown planthopper resistance in a doubled haploid rice population. *Molecular Breeding* 3: 105–113.

Huang, Z., He, G., Shu, L., Li, X. and Zhang, Q. (2001). Identification and mapping of two brown planthopper resistance genes in rice. *Theoretical and Applied Genetics* 102: 929–934.

Hulburt, D.J., Boerma, H.R. and All, J.N. (2004). Effect of pubescence tip on soybean resistance to lepidopteran insects. *Journal of Economic Entomology* 97: 621–627.

Humphreys, M.W., Pasakinskiene, I., James, A.R. and Thomas, H. (1998). Physically mapping quantitative traits for stress resistance in forage grasses. *Journal of Experimental Botany* 49: 1611–1618.

Hussain, M.A. and Trehan, K.N. (1940). Final report on the scheme of investigation on the whitefly of cotton in the Punjab. *Indian Journal of Agricultural Sciences* 10: 101–109.

Huttel, B., Winter, P., Weising, K., Choumane, W., Weigand, F. and Kahl, G. (1999). Sequence-tagged microsatellite site markers for chickpea (*Cicer arietinum* L.). *Genome* 42: 210–217.

Huyen, T., Phan, T., Ellwood, S.R., Ford, R., Thomas, S. and Oliver, R. (2006). Differences in syntenic complexity between *Medicago truncatula* with *Lens culinaris* and *Lupinus albus*. *Functional Plant Biology* 33: 775–782.

Hyne, V., Kearsey, M.J., Pike, D.J. and Snape, J.W. (1995). QTL analysis: Unreliability and bias in estimation procedures. *Molecular Breeding* 1: 273–283.

Ilyas, M., Puri, S.N. and Rote, N.B. (1991). Effects of some morphophysiological characters of leaf on incidence of cotton whitefly. *Journal of Research, Maharashtra Agricultural Universities* 16(3): 386–388.

Ishaaya, I., Hirashima, A., Yablonski, S., Taurata, S. and Eto, M. (1991). Mimosine, a non-protein amino acid inhibits growth and enzyme systems in *Tribolium castaneum*. *Pesticide Biochemistry and Physiology* 39: 35–42.

Ishii, T., Brar, D.S., Multani, D.S. and Khush, G.S. (1994). Molecular tagging of genes for brown planthopper resistance and earliness introgressed from *Oryza australiensis* into cultivated rice, *O. sativa*. *Genome* 37: 217–221.

Jackai, L.E.N. and Oghiakhe, S. (1989). Podwall trichomes and resistance of two wild cowpea, *Vigna vexillata*, accessions to *Maruca testulalis* (Geyer) (Lepidoptera: Pyralidae) and *Clavigralla tomentosicollis* Stal (Hemiptera: Coreidae). *Bulletin of Entomological Research* 79: 595–605.

Jackson, D.M., Johnson, A.W. and Stephenson, M.G. (2002). Survival and development of *Heliothis virescens* (Lepidoptera: Noctuidae) larvae on isogenic tobacco lines with different levels of alkaloids. *Journal of Economic Entomology* 95: 1294–1302.

Jain, A., Ariyadasa, R., Kumar, A., Srivastava, M.N., Mohan, M. and Nair, S. (2004). Tagging and mapping of a rice gall midge resistance gene, *Gm8*, and development of SCARs for use in marker-aided selection and gene pyramiding. *Theoretical and Applied Genetics* 109: 1377–1384.

Jampatong, C., McMullen, M.D., Barry, D.B., Darrah, L.L., Byrne, P.F. and Kross, H. (2002). Quantitative trait loci for first- and second-generation European corn borer resistance derived from the maize inbred M047. *Crop Science* 41: 584–593.

Jeffree, C.E. (1986). The cuticle, epicuticular waxes and trichomes of plants, with reference to their structure, functions and evolution. In Juniper, B.E. and Southwood, T.R.E. (Eds.), *Insects and the Plant Surface*. London, UK: Edward Arnold Publishers, 23–64.

Jena, K.K., Jeung, J.U., Lee, J.H., Choi, H.C. and Brar, D.S. (2006). High resolution mapping of a new brown planthopper (BPH) resistance gene, *Bph18(t)*, and marker-assisted selection for BPH resistance in rice (*Oryza sativa* L.). *Theoretical and Applied Genetics* 112: 288–297.

Jena, K.K., Pasalu, I.C., Rao, Y.K., Varalaxmi, Y., Krishnaiah, K., Khush, G.S. and Kochert, G. (2003). Molecular tagging of a gene for resistance to brown planthopper in rice (*Oryza sativa* L.). *Euphytica* 129: 81–88.

Jiang, C., Wright, R.J., Woo, S.S., DelMonte, T.A. and Paterson, A.H. (2000). QTL analysis of leaf morphology in tetraploid *Gossypium* (cotton). *Theoretical and Applied Genetics* 100: 409–418.

Jiang, J. and Gill, B.S.D. (1994). New 18S.26S ribosomal RNA gene loci: chromosomal landmarks for the evolution of polyploid wheats. *Chromosoma* 103: 179–185.

Jing, S.R., Xing, C.Z., Yuan, Y.L. and Liu, S.L. (1997). Research on the breeding and utilization of a bollworm-resistant hybrid cotton variety. *China Cottons* 24(7): 15–17.

Johnson, B. (1953). The injurious effects of the hooked epidermal hairs of the French beans (*Phaseolus vulgaris* L.) on *Aphis craccivora* Koch. *Bulletin of Entomological Research* 44: 779–788.

Juvik, J.A., Shapiro, J.A., Young, T.E. and Mutschler, M.A. (1994). Acylglucoses from wild tomatoes alter behavior and reduce growth and survival of *Helicoverpa zea* and *Spodoptera exigua* (Lepidoptera: Noctuidae). *Journal of Economic Entomology* 87: 482–492.

Kaga, A. and Ishimoto, M. (1998). Genetic localization of a bruchid resistance gene and its relationship to insecticidal cyclopeptide alkaloids, the vignatic acids in mungbean (*Vigna radiata* L. Wilczek). *Molecular and General Genetics* 258: 378–384.

Karim, S., Riazuddin, S. and Dean, D.H. (1999). Interaction of *Bacillus thuringiensis* delta-endotoxins with midgut brush border membrane vesicles of *Helicoverpa armigera*. *Journal of Asia Pacific Entomology* 2: 153–162.

Karp, A., Edwards, K.J., Bruford, M., Funk, S., Vosman, B., Morgante, M., Seberg, O., Kremer, A., Boursot, P., Arctander, P., Tautz, D. and Hewitt, G.M. (1997). Molecular technologies for biodiversity evaluation: opportunities and challenges. *Nature Biotechnology* 15: 625–628.

Katsar, C.S., Paterson, A.H., Teetes, G.L. and Peterson, G.C. (2002). Molecular analysis of sorghum resistance to the greenbug (Homoptera: Aphididae). *Journal of Economic Entomology* 95: 448–457.

Kauffman, W.C. and Kennedy, G.G. (1989). Inhibition of *Campoletis sonorensis* parasitism to *Heliothis zea* and of parasitoid development by 2-tridecanone-mediated insect resistance of wild tomato. *Journal of Chemical Ecology* 15: 1919–1930.

Kawaguchi, M., Murata, K., Ishii, T., Takumi, S., Mori, N. and Nakamura, C. (2001). Assignment of a brown planthopper (*Nilaparvata lugens* Stal) resistance gene *bph4* to the rice chromosome 6. *Breeding Science* 51: 13–18.

Kazan, K., Muehlbauer, F.J., Weeden, N.F. and Ladizinsky, G. (1993). Inheritance and linkage relationships of morphological and isozyme loci in chickpea (*Cicer arietinum* L.). *Theoretical and Applied Genetics* 86: 417–426.

Kennedy, G.G. and Yamamoto, R.T. (1979). A toxic factor causing resistance in a wild tomato to the tobacco hornworm and some other insects. *Entomologia Experimentalis et Applicata* 26: 121–126.

Kennedy, G.G., Gould, F., de Ponting, O.M.B. and Stinner, R.E. (1987). Ecological, agricultural, genetic and commercial considerations in the development of insect resistant germplasms. *Environmental Entomology* 16: 327–338.

Khairallah, M.M., Bohn, M., Jiang. C., Deutsch, I.A., Jewell, D.C., Mihm, J.A., Melchinger, A.E., Gonzalez-deLeon, D. and Hoisington, D.A. (1998). Molecular mapping of QTL for southwestern corn borer resistance, plant height and flowering in tropical maize. *Plant Breeding* 117: 309–318.

Khairallah, M., Hoisington, D., Gonzalez de Leon, D., Bohn, M., Melchinger, A., Jewell, D.C., Deutsch, and Mihm, J.A. (1997). Location and effect of quantitative trait loci for southwestern corn borer and sugarcane borer resistance in tropical maize. In *Insect Resistant Maize: Recent Advances and Utilization. Proceedings of an International Symposium*, 27 November–3 December, 1994. Mexico City, Mexico: International Maize and Wheat Improvement Center, 148–154.

Khalifa, H. (1979). Breeding for bollworm resistance in cotton. *Cotton et Fibres Tropicales* 24: 309–314.

Khan, Z.R., Ciepla, A. and Norris, D.M. (1987). Behavioral and physiological responses of cabbage looper, *Trichoplusia ni* (Hubner), to steam distillates from resistant versus susceptible soybean plants. *Journal of Chemical Ecology* 13: 1903–1915.

Khan, Z.R., Ward, J.T. and Norris, D.M. (1986). Role of trichomes in soybean resistance to cabbage looper, *Trichoplusia ni*. *Entomologia Experimentalis et Applicata* 42: 109–117.

Kim, D.J., Kim, T.K., Choi, E.J., Park, W.C., Kim, T.H., Ahn, D.H., Yuan, Z., Blackall, L. and Keller, J. (2004). Fluorescence in situ hybridization analysis of nitrifiers in piggery wastewater treatment reactors. *Water Science and Technology* 49: 333–340.

Kisha, J.S.A. (1984). Whitefly, *Bemisia tabaci* infestations on tomato varieties and a wild *Lycopersicon* species. *Annals of Applied Biology* 104: 124–125.

Kisha, T.J., Sneller, C.H. and Diers, B.W. (1997). Relationship between genetic distance among parents and genetic variance in populations of soybean. *Crop Science* 37: 1317–1325.

Klun, J.A., Tipton, C.C. and Brindley, T.A. (1967). 2-4 dihydoxy-7-methodoxy-1,4-benoxazin-3-zone (DIMBOA) as active agent in the resistance of maize to the European corn borer. *Journal of Economic Entomology* 60: 1529–1533.

Knott, S.A. and Haley, C.S. (1992). Aspects of maximum-likelihood methods for the mapping of quantitative traits in line crosses. *Genetic Research* 60: 139–151.

Kobayashi, A., Kaneda, C., Ikeda, R. and Ikehashi, H. (1980). Inheritance of resistance to green rice leafhopper, *Nephotettix cincticeps* in rice. *Japanese Journal of Breeding* 30 (supplement 1): 56–57.

Kogan, M. (1982). Plant resistance in pest management. In Metcalf, R.H. and Luckmann, W.H. (Eds.), *Introduction to Insect Pest Management*. New York, USA: John Wiley & Sons, 93–134.

Kong, L., Ohm, H.W., Cambron, S.E. and Williams, C.E. (2005). Molecular mapping determines that Hessian fly resistance gene *H9* is located on chromosome 1A of wheat. *Plant Breeding* 124: 525–531.

Korzun, V., Malyshev, S., Voylokov, A.V. and Bomer, A. (2001). A genetic map of rye (*Secale cereale* L.) combining RFLP, isozyme, protein, microsatellite and gene loci. *Theoretical and Applied Genetics* 102: 709–717.

Kosambi, D.D. (1944). The estimation of map distances from recombination values. *Annals of Eugenics* 12: 172–175.

Krakowsky, M.D., Lee, M. and Holland, J.B. (2007). Genotypic correlation and multivariate QTL analyses for cell wall components and resistance to stalk tunneling by the European corn borer in maize. *Crop Science* 47: 485–488.

Krakowsky, M.D., Brinkman, M.J., Woodman-Clikeman, W.L. and Lee, M. (2002). Genetic components of resistance to stalk tunneling by the European corn borer in maize. *Crop Science* 42: 1309–1315.

Kretschmer, J.M., Chalmers, K.J., Manning, S., Karakousis, A., Bait, A.R., Islam, M.R., Logue, S.J., Chac, Y.W., Barker, S.J., Lance, R.C.M. and Langridge, P. (1997). RFLP mapping of the *Ha2* cereal cyst nematode resistance gene in barley. *Theoretical and Applied Genetics* 94: 1060–1064.

Kumar, H. (1992). Inhibition of ovipositional responses of *Chilo partellus* (Lepidoptera: Pyralidae) by the trichomes on the lower leaf surface of a maize. *Journal of Economic Entomology* 85: 1736–1739.

Kunzel, G., Korzum, L. and Meister, A. (2000). Cytologically integrated physical restriction fragment length polymorphism maps for the barley genome based on translocation breakpoints. *Genetics* 154: 397–412.

Kurata, N., Nagamura, Y., Yamamoto, K., Harushima, Y., Sue, N., Wu, J., Antonio, B.A., Shomura, A., Shimizu, T., Kuboki, Y., Toyama, T., Miyamoto, M., Kirihara, T., Hayasaka, K., Miyao, A., Monna, L., Zhong, H.S., Tamura, Y., Wang, Z.X., Momma, T., Yano, M., Sasaki, T. and Minobe, Y. (1994). A 300 kilobase interval map of rice including 883 expressed sequences. *Nature Genetics* 8: 365–372.

Kurata, N., Umehara, Y., Tanoue, H. and Sasaki, T. (1997). Physical mapping of the rice genome with yeast artificial chromosome clones. Special issue: *Oryza*: from molecular to plant. *Plant Molecular Biology* 35: 101–113.

Lal, M.N., Singh, S.S. and Singh, V.P. (1999). Reaction of mustard aphid, *Lipaphis erysimi* (Kalt.) to morphological characters of mustard, *Brassica juncea* L. *Journal of Entomological Research* 23(3): 221–223.

Lampert, E.P., Haynes, D.L., Sawyer, A.J., Jokinen, D.P., Wellso, S.G., Gulln, R.L. and Roberts, J.J. (1983). Effects of regional releases of resistant wheats on the population dynamics of the cereal leaf beetle (Coleoptera: Chrysomelidae). *Annals of Entomological Society of America* 76: 972–980.

Lander, E.S., Green, P., Abrahamson, J., Barlow, A., Daly, M.J., Lincoln, S.E. and Newburg, L. (1987). MAPMAKER: An interactive computer package for constructing primary genetic maps of experimental and natural populations. *Genomics* 1: 174–181.

Lapointe, S.L. and Tingey, W.M. (1984). Feeding response of the green peach aphid (Homoptera: Aphididae) to potato glandular trichomes. *Journal of Economic Entomology* 79: 1264–1268.

Lawlor, H.J., Siddique, K.H.M., Sedgley, R.H. and Thurling, N. (1998). Improving cold tolerance and insect resistance in chickpea and the use of AFLPs for the identification of molecular markers for these traits. *Acta Horticulturae* 461: 185–192.

Lee, E.A., Byrne, P.F., McMullen, M.D., Snook, M.E., Wiseman, B.R., Widstrom, N.W. and Coe, E.H. (1998). Genetic mechanisms underlying apimaysin and maysin synthesis and corn earworm antibiosis in maize (*Zea mays* L.) *Genetics* 149: 1997–2006.

Lin, B.Y., Peng, S.F., Chen, Y.J., Chen, H.S. and Kao, C.F. (1997). Physical mapping of RFLP markers on four chromosome arms in maize using terminal deficiencies. *Molecular and General Genetics* 256: 509–516.

Lin, J.J., Kuo, J., Ma, J., Saunders, J.A., Beard, H.S., MacDonald, M.H., Kenworthy, W., Ude, G.N. and Matthews, B.L. (1996). Identification of molecular markers in soybean: comparing RFLP, RAPD, and AFLP DNA mapping techniques. *Plant Molecular Biology Reporter* 14: 156–169.

Lincoln, S.E., Daly, M.J. and Lander, E.S. (1993). *Mapping Genes Controlling Quantitative Traits Using MAPMAKER/QTL Version 1.1: A Tutorial and Reference Manual.* 2nd ed. Cambridge, UK: Whitehead Institute for Biometrical Research, 920–927.

Liu, X.M., Gill, B.S. and Chen, M.S. (2005). Hessian fly resistance gene *H13* is mapped to a distal cluster of resistance genes in chromosome 6DS of wheat. *Theoretical and Applied Genetics* 111: 243–249.

Liu, X.M., Smith, C.M. and Gill, B.S. (2002). Mapping of microsatellite markers linked to the *Dn4* and *Dn6* genes expressing Russian wheat aphid resistance in wheat. *Theoretical and Applied Genetics* 104: 1042–1048.

Liu, X.M., Smith, C.M. and Gill, B.S. (2005). Allelic relationships among Russian wheat aphid resistance genes. *Crop Science* 45: 2273–2280.

Liu, X.M., Fritz, A.K., Reese, J.C., Wilde, G.E., Gill, B.S. and Chen, M.S. (2005). *H9, H10,* and *H11* compose a cluster of Hessian fly-resistance genes in the distal gene-rich region of wheat chromosome 1AS. *Theoretical and Applied Genetics* 110: 1473–1480.

Liu, X.M., Smith, C.M., Gill, B.S. and Tolmay, V. (2001). Microsatellite markers linked to six Russian wheat aphid resistance genes in wheat. *Theoretical and Applied Genetics* 102: 504–510.

Lukefahr, M.J., Haughtaling, J.E. and Graham, H.M. (1971). Suppression of *Heliothis* populations with glabrous cotton strains. *Journal of Economic Entomology* 64: 486–488.

Lukefahr, M.J., Houghtaling, J.E. and Graham, H.M. (1975). Suppression of *Heliothis* spp. with cottons containing combinations of resistant characters. *Journal of Economic Entomology* 68: 743–746.

Ma, Z.Q., Gill, B.S., Sorrells, M.E. and Tanksley, S.D. (1993). RFLP markers linked to 2 Hessian fly-resistance genes in wheat (*Triticum aestivum* L.) from *Triticum tauschii* (Coss) Schmal. *Theoretical and Applied Genetics* 85: 750–754.

Ma, Z.Q., Saidi, A., Quick, J.S. and Lapitan, N.L.V. (1998). Genetic mapping of Russian wheat aphid resistance genes *Dn2* and *Dn4* in wheat. *Genome* 41: 303–306.

MacIntosh, S.C., Kishore, G.M., Perlak, F.J., Marrone, P.G., Stone, T.B., Sims, S.R. and Fuchs, R.L. (1990). Potentiation of *Bacillus thuringiensis* insecticidal activity by serine protease inhibitors. *Journal of Agriculture and Food Chemistry* 38: 1145–1152.

Mahal, M.S., Dhawan, A.K. and Singh, B. (1980). Relative susceptibility of okra varieties to the leafroller, *Sylepta derogata* F. *Indian Journal of Ecology* 7(1): 155–158.

Malik, R., Smith, C.M., Harvey, T.L. and Brown-Guedira, G.L. (2003). Genetic mapping of wheat curl mite resistance genes *Cmc3* and *Cmc4* in common wheat. *Crop Science* 43: 644–650.

Martin-Sanchez, J.A., Gomez-Colmenarejo, M., Del Moral, J., Sin, E., Montes, M.J., Gonzalez Belinchon, C., Lopez-Brana, J. and Delibes, A. (2003). A new Hessian fly resistance gene (*H30*) transferred from the wild grass *Aegilops triuncialis* to hexaploid wheat. *Theoretical and Applied Genetics* 106: 1248–1255.

Masuzaki, S., Shigyo, M. and Yamauchi, N. (2006). Direct comparison between genomic constitution and flavonoid contents in *Allium* multiple alien addition lines reveals chromosomal locations of genes related to biosynthesis from dihydrokaempferol to quercetin glycosides in scaly leaf of shallot (*Allium cepa* L.). *Theoretical and Applied Genetics* 112: 607–617.

McCouch, S.R., Khush, G.S. and Tanksley, S.D. (1991). Tagging genes for disease and insect resistance via linkage to RFLP markers. In *Rice Genetics. II: Proceedings, Second International Rice Genetics Symposium*, 14–18 May, 1990. Manila, Philippines: International Rice Research Institute, 443–449.

McMullen, M.D. and Simcox, K.D. (1995). Genomic organization of disease and insect resistance genes in maize. *Molecular Plant Microbe Interactions* 8: 811–815.

McNally, K.L. and Mutschler, M.A. (1997). Use of introgression lines and zonal mapping to identify RAPD markers linked to QTL. *Molecular Breeding* 3: 203–212.

Menendez, C.M., Hall, A.E. and Gepts, P. (1997). A genetic linkage map of cowpea (*Vigna unguiculata*) developed from a cross between two inbred, domesticated lines. *Theoretical and Applied Genetics* 95: 1210–1217.

Miflin, B. (2000). Crop improvement in the 21st century. *Journal of Experimental Botany* 51: 1–8.

Miller, C.A., Altinkut, A. and Lapitan, N.L.V. (2001). A microsatellite marker for tagging *Dnl*, a wheat gene conferring resistance to the Russian wheat aphid. *Crop Science* 41: 1584–1589.

Minney, B.H.P., Gatehouse, A.M.R., Dobir, P., Dendy, J., Cardona, C. and Gatehouse, J.A. (1990). Biochemical bases of seed resistance to *Zabrotes subfasciatus* (bean weevil) in *Phaseolus vulgaris* (common bean): A mechanism for arcelin toxicity. *Journal of Insect Physiology* 36: 757–767.

Moericke, V. (1969). Host plant specific color behavior by *Hyalopterus pruni* (Aphididae). *Entomologia Experimentalis et Applicata* 12: 524–534.

Mohan, M., Nair, S., Bentur, J.S., Rao, U.P. and Bennett, J. (1994). RFLP and RAPD mapping of the rice *Gm2* gene confers resistance to biotype I of gall midge *(Orseolia oryzae)*. *Theoretical and Applied Genetics* 87: 782–788.

Mohan, M., Nair, S., Bhagwat, A., Krishna, T.G., Yano, M., Bhatia, C.R. and Sasaki, T. (1997). Genome mapping, molecular markers and marker-assisted selection in crop plants. *Molecular Breeding* 3: 87–103.

Moharramipour, S., Tsumuki, H., Sato, K. and Yoshida, H. (1997). Mapping resistance to cereal aphids in barley. *Theoretical and Applied Genetics* 94: 592–596.

Moreira, L.A., Mollema, C. and van Heusden, S. (1999). Search for molecular markers linked to *Liriomyza trifolii* resistance in tomato. *Euphytica* 109: 149–156.

Mudge, J., Cregan, P.B., Kenworthy, J.P., Kenworthy, W.J., Orf, J.H. and Young, N.D. (1997). Two microsatellite markers that flank the major soybean cyst nematode resistance locus. *Crop Science* 37: 1611–1615.

Mullis, K. (1990). The unusual origin of the polymerase chain reaction. *Scientific American* 262(4): 56–65.

Murata, K., Fujiwara, M., Mumi, H., Takumi, S., Mori, C. and Nakamura, C. (2000). A dominant brown planthopper resistance gene, *Bph9* locates on the long arm of rice chromosome 12. *Rice Genetics Newsletter* 17: 84–86.

Murray, J.D., Michaels, T.E., Cardona, C., Schaafsma, A.W. and Pauls, K.P. (2004). Quantitative trait loci for leafhopper (*Empoasca fabae* and *Empoasca kraemeri*) resistance and seed weight in the common bean. *Plant Breeding* 123: 474–479.

Musetti, L. and Neal, J.J. (1997). Resistance to the pink potato aphid, *Macrosiphum euphorbiae*, in two accessions of *Lycopersicon hirsutum* f. *glabratum*. *Entomologia Experimentalis et Applicata* 84: 137–146.

Myers, G.O., Fatokun, C.A. and Young, N.D. (1996). RFLP mapping of an aphid resistance gene in cowpea (*Vigna unguiculata* (L.) Walp. *Euphytica* 91: 181–187.

Nadimpalli, R.G., Jarret, R.L., Phatak, S.C. and Kochert, X.G. (1993). Phylogenetic relationships of the pigeonpea (*Cajanus cajan*) based on nuclear restriction fragment length polymorphism. *Genome* 36: 216–223.

Nagaraj, N., Reese, J.C., Tuinstra, M.R., Smith, C.M., Amand, P., Kirkham, M.B., Kofoid, K.D., Campbell, L.R. and Wilde, G.E. (2005). Molecular mapping of sorghum genes expressing tolerance to damage by greenbug (Homoptera: Aphididae). *Journal of Economic Entomology* 98: 595–602.

Nair, S., Bentur, J.S., Rao, U.P. and Mohan, M. (1995). DNA markers tightly linked to a gall midge resistance gene (*Gm2*) are potentially useful for marker-aided selection in rice breeding. *Theoretical and Applied Genetics* 91: 68–73.

Nair, S., Kumar, A., Srivastava, M.N. and Mohan, M. (1996). PCR-based DNA markers linked to a gall midge resistance gene, *Gm4t* has potential for marker-aided selection in rice. *Theoretical and Applied Genetics* 92: 660–665.

Nanda, U.K., Sasmal, A. and Mohanty, S.K. (1996). Varietal reaction of pigeonpea to pod borer *Helicoverpa armigera* (Hubner) and modalities of resistance. *Current Agricultural Research* 9: 107–111.

Narvel, J.M., Chu, W., Fehr, W., Cregan, P.B. and Shoemaker, R.C. (2000). Development of multiplex sets of simple sequence repeat DNA markers covering the soybean genome. *Molecular Breeding* 6: 175–183.

Narvel, J.M., Walker, D.R., Rector, B.G., All, J.N., Parrott, W.A. and Boerma, H.R. (2001). A retrospective DNA marker assessment of the development of insect resistant soybean. *Crop Science* 41: 1931–1939.

Natarajan, K. (1990). Investigations on the ovipositional preference of *Heliothis armigera* as affected by certain genotypes. In Jayaraj, S., Uthamasamy, S., Gopalan, M. and Rabindra, R.J. (Eds.), Heliothis *Management: Proceedings of National Workshop*. Coimbatore, Tamil Nadu, India: Tamil Nadu Agricultural University, 80–86.

Neal, J.J., Steffens, J.C. and Tingey, W.M. (1989). Glandular trichomes of *Solanum berthaultii* and its resistance to the Colorado potato beetle. *Entomologia Experimentalis et Applicata* 51: 133–140.

Nienhuis, J., Helentjaris, T., Slocum, M., Ruggero, B. and Schaefer, A. (1987). Restriction fragment length polymorphism analysis of loci associated with insect resistance in tomato. *Crop Science* 27: 797–803.

Nieto-Lopez, R.M. and Blake, T.K. (1994). Russian wheat aphid resistance in barley: Inheritance and linked molecular markers. *Crop Science* 34: 655–659.

Ohm, H.W., Sharma, H.C., Patterson, F.L., Ratcliffe, R.H. and Obani, M. (1995). Linkage relationships among genes on wheat chromosome 5A that condition resistance to Hessian fly. *Crop Science* 35: 1603–1607.

Oksanen, H., Pertlunen, V. and Kangas, E. (1970). Studies on chemical factors involved in the olfactory orientation of *Blastophagus piniperda* (Coleoptera: Scolytidae). *Contributions of Boyce Thompson Institute* 24: 275–282.

Palevitz, B.A. (2000). Rice genome gets a boost. Private sequencing yields a rough draft for public. http://www.the-scientist.com/yr2000/may/palevitz-pl-000501.html.

Panda, N. and Daugherty, D.M. (1975). Note on the antibiosis factor of resistance to corn earworm in pubescent genotypes of soybean. *Indian Journal of Agricultural Sciences* 45: 68–72.

Panda, N. and Khush, G.S. (1995). *Host Plant Resistance to Insects*. Wallingford, Oxon, UK: CAB International.

Pandian A., Ford, R. and Taylor, P.W.J. (2000). Transferability of sequence tagged microsatellite site (STMS) primers across four major pulses. *Plant Molecular Biology Reporter* 18(4): 395a–395h (www.uga.edu/ispmb).

Panguluri, S., Janaiah, K., Govil, J., Kumar, P. and Sharma, P. (2006). AFLP fingerprinting in pigeonpea (*Cajanus cajan* (L.) Millsp.) and its wild relatives. *Genetic Resources and Crop Evolution* 53: 523–531.

Parameswaran, S. (1983). *Ecology, Host Resistance and Management of the Cotton Stem Weevil*, Pempherulus affinis *Faust*, Ph.D. thesis, Tamil Nadu Agricultural University, Coimbatore, Tamil Nadu, India.

Parani, M., Lakshmi, M., Kumar, P.S. and Parida, A. (2000). Ribosomal DNA variation and phylogenetic relationships among *Cajanus cajan* (L.) Millsp. and its wild relatives. *Current Science* 78: 1235–1238.

Parmar, B.S. and Walia, S. (2001). Prospects and problems of phytochemical biopesticides. In Koul, O. and Dhaliwal, G.S. (Eds.), *Phytochemical Biopesticides.* Amsterdam, The Netherlands: Harvard Academic Publishers, 133–210.

Patanakamjorn, S. and Pathak, M.D. (1967). Varietal resistance of rice to the Asiatic rice borer, *Chilo suppressalis* (Lepidoptera: Crambidae) and its association with various plant characters. *Annals of Entomological Society of America* 60: 287–292.

Patankar, A.G., Harsulkar, A.M., Giri, A.P., Gupta, V.S., Sainani, M.N., Ranjekar, P.K. and Deshpande, V.V. (1999). Diversity in inhibitors of trypsin and *Helicoverpa armigera* gut proteinases in chickpea (*Cicer arietinum*) and its wild relatives. *Theoretical and Applied Genetics* 99: 719–726.

Paterson, A.H., Tanksley, S.D. and Sorrells, M.E. (1991). DNA markers in plant improvement. *Advances in Agronomy* 46: 39–90.

Pathak, M.D., Andres, F., Galacgae, N. and Raros, R. (1971). *Resistance of Rice Varieties to Striped Rice Borers.* Technical Bulletin 11. Manila, Philippines: International Rice Research Institute.

Penny, L.H., Scott, G.E. and Guthrie, W.D. (1967). Recurrent selection for European corn borer resistance in maize. *Crop Science* 7: 407–409.

Percy, R.G. (1999). Inheritance of cytoplasmic-virescent *cyt-V* and dense-glanding *dg* mutants in American pima cotton. *Crop Science* 39: 372–374.

Pereira, A. (2000). Plant genomics is revolutionising agricultural research. *Biotechnology Development Monitor* 40: 2–7.

Peter, A.J., Shanower, T.G. and Romeis, J. (1995). The role of plant trichomes in insect resistance: A selective review. *Phytophaga* 7: 41–64.

Pillemer, E.A. and Tingey, W.M. (1978). Hooked trichomes and resistance of *Phaseolus vulgaris* to *Empoasca fabae* (Harris). *Entomologia Experimentalis et Applicata* 24: 83–94.

Piskur, J. and Langkjaer, R.B. (2004). Yeast genome sequencing: the power of comparative genomics. *Molecular Microbiology* 53: 381–389.

Ponnaiya, B.W.X. (1951). Studies in the genus *Sorghum*: The cause of resistance in sorghum to the insect pest *Atherigona indica* M. *Madras University Journal, Section B* 21: 203–217.

Powell, W., Morgante, M., Andre, C., Hanafey, M., Vogel, J., Tingey, S. and Rafalski, A. (1996). The comparison of RFLP, RAPD, AFLP and SSR (microsatellite) markers for germplasm analysis. *Molecular Breeding* 2: 225–238.

Primrose, S.B. (1998). *Principles of Genome Analysis. A Guide to Mapping and Sequencing of DNA from Different Organisms.* Oxford, UK: Blackwell Science.

Prokopy, R.J. and Owens, E.D. (1983). Visual detection of plants by herbivorous insects. *Annual Review of Entomology* 28: 337–364.

Rajarajeswari, V. and Subbarao, I.V. (1997). Gossypol glands in relation to resistance to bollworm (*Helicoverpa armigera*) in upland cotton (*Gossypium hirsutum*). *Indian Journal of Agricultural Sciences* 67(7): 293–295.

Rao, D.V. (1967). Hardness of sugarcane varieties in relation to shoot borer (*Chilotraea infuscatellus* Snell.) incidence. *Andhra Agricultural Journal* 14: 99–105.

Ratnaparkhe, M.B., Gupta, V.S., Ven Murthy, M.R. and Ranjekar, P.K. (1995). Genetic fingerprinting of pigeonpea [*Cajanus cajan* (L.) Millsp.] and its wild relatives using RAPD markers. *Theoretical and Applied Genetics* 91: 893–898.

Rector, B.G., All, J.N., Parrott, W.A. and Boerma, H.R. (1998). Identification of molecular markers linked to quantitative trait loci for soybean resistance to corn earworm. *Theoretical and Applied Genetics* 96: 786–790.

Rector, B.G., All, J.N., Parrott, W.A. and Boerma, H.R. (1999). Quantitative trait loci for antixenosis resistance to corn earworm in soybean. *Crop Science* 39: 531–538.

Rector, B.G., All, J.N., Parrott, W.A. and Boerma, H.R. (2000). Quantitative trait loci for antibiosis resistance to corn earworm in soybean. *Crop Science* 40: 233–238.

Rembold, H. (1981). Malic acid in chickpea exudates: A marker for *Heliothis* resistance. *International Chickpea Newsletter* 4: 18–19.

Rembold, H., Wallner, P., Kohne, A., Lateef, S.S., Grune, M. and Weigner, Ch. (1990). Mechanism of host plant resistance with special emphasis on biochemical factors. In *Chickpea in the Nineties: Proceedings of the Second International Workshop on Chickpea Improvement*, 4–8 December, 1989. Patancheru, Andhra Pradesh, India: International Crops Research Institute for the Semi-Arid Tropics/International Center for Agricultural Research in the Dry Areas, 191–194.

Renganayaki, K., Fritz, A.K., Sadasivam, S., Pammi, S., Harrington, S.E., McCouch, S.R., Kumar, S.M. and Reddy, A.S. (2002). Mapping and progress toward map-based cloning of brown planthopper biotype-4 resistance gene introgressed from *Oryza officinalis* into cultivated rice, *O. sativa. Crop Science* 42: 2112–2117.

Renwick, J.A.A. and Radke, C.D. (1988). Sensory cues in host selection for oviposition by the cabbage butterfly, *Pieris rapae. Journal of Insect Physiology* 34: 251–257.

Roberts, J.J., Gallun, R.L., Patterson, F.L. and Foster, J.E. (1979). Effects of wheat leaf pubescens on Hessian fly. *Journal of Economic Entomology* 72: 211–214.

Robinson, S.H., Wolfonbarger, D.A. and Doday, R.H. (1980). Antixenosis of smooth leaf cotton to the ovipositional response of tobacco budworm. *Crop Science* 20: 646–649.

Roche, P.A., Alston, F.H., Maliepaard, C.A., Evans, K.M., Vrielink, R., Dunemann, F., Markussen, T., Tartarini, S., Brown, L.M., Ryder, C. and King, G.J. (1997). RFLP and RAPD markers linked to the rosy leaf curling aphid resistance gene (*Sd1*) in apple. *Theoretical and Applied Genetics* 94: 528–533.

Romeis, J., Shanower, T.G. and Peter, A.J. (1999). Trichomes of pigeonpea and two wild *Cajanus* species. *Crop Science* 39: 1–5.

Rupakala, A., Rao, D., Reddy, L., Upadhyaya, H.D. and Sharma, H.C. (2005). Inheritance of trichomes and resistance to pod borer (*Helicoverpa armigera*) and their association in interspecific crosses between cultivated pigeonpea (*Cajanus cajan*) and its wild relative *C. scarabaeoides. Euphytica* 145: 247–257.

Saiki, R.K., Gelfand, D.H., Stoffel, S., Scharf, S.J., Higuchi, R., Horn, G.T., Mullis, K.B. and Erlich, H.A. (1988). Primer-directed enzymatic amplification of DNA with a thermostable DNA polymerase. *Nature* 239: 487–497.

Saiki, R.K., Scharf, S., Faloona, F., Mullis, K.B., Horn, G.T., Erlich, H.A. and Arnheim, N. (1985). Enzymatic amplification of beta-globin genomic sequences and restriction site analysis for diagnosis of sickle cell anemia. *Science* 230: 1350–1354.

Saka, N., Toyama, T., Tuji, T., Nakamae, H. and Izawa, T. (1997). Fine mapping of green ricehopper resistant gene *Grh-3(t)* and screening of *Grh-3(t)* among green ricehopper resistant and green leafhopper resistant cultivars in rice. *Breeding Science* 47 (supplement): 1–55.

Sanford, L.L., Deahl, K.L., Sidden, S.L. and Ladd, T.L. Jr. (1992). Glykoalkaloid contents in tubers from *Solanum tuberosum* populations selected for potato leafhopper resistance. *American Potato Journal* 69: 693–703.

Santra, D.K., Tekeoglu, M., Ratnaparkhe, M., Kaiser, W.J. and Muehlbauer, F.J. (2000). Identification and mapping of QTLs conferring resistance to ascochyta blight in chickpea. *Crop Science* 40: 1606–1612.

Sardesai, N., Kumar, A., Rajyashri, K.R., Nair, S. and Mohan, M. (2002). Identification and mapping of an AFLP marker linked to *Gm7*, a gall midge resistance gene and its conversion to a SCAR marker for its utility in marker-aided selection in rice. *Theoretical and Applied Genetics* 105(5): 691–698.

Saxena, R.C. and Okech, S.H. (1985). Role of plant volatiles in resistance of selected rice varieties to brown planthopper, *Nilaparvata lugens* (Stal) (Homoptera: Delphacidae). *Journal of Chemical Ecology* 11: 1601–1616.

Schneider, K.A., Brothers, M.E. and Kelly, J.D. (1997). Marker-assisted selection to improve drought resistance in common bean. *Crop Science* 37: 51–60.

Schreiber, K. (1958). Uber unigt inhallsstoffe dar solannaceon und ihre Bedeutung far die kartoffel Kferristinz. *Entomologia Experimentalis et Applicata* 1: 28–37.

Selvi, A., Shanmugasundaram, P., Mohan Kumar, S. and Raja, J.A.J. (2002). Molecular markers for yellow stem borer, *Scirpophaga incertulas* (Walker) resistance in rice. *Euphytica* 124: 371–377.

Shanower, T.G., Yoshida, M. and Peter, A.J. (1997). Survival, growth, fecundity and behavior of *Helicoverpa armigera* (Lepidoptera: Noctuidae) on pigeonpea and two wild *Cajanus* species. *Journal of Economic Entomology* 90: 837–841.

Shapiro, A.M. and DeVay, J.E. (1987). Hypersensitivity reaction of *Brassica nigra* L. (Cruciferae) kills eggs of *Pieris* butterflies (Lepidoptera: Pieridae). *Oekologia* 71: 631–632.

Shappley, Z.W., Jenkins, J.N., Meredith, W.R. and McCarty, J.R. Jr. (1996). An RFLP linkage map of Upland cotton, *Gossypium hirsutum* L. *Theoretical and Applied Genetics* 97: 756–761.

Sharma, H.C. (1982). A note on extraction and purification of gossypol from pigment glands of cottonseeds. *Cotton Development* 12: 71–72.

Sharma, H.C. and Agarwal, R.A. (1981). Behavioral responses of larvae of spotted bollworm, *Earias vittella* Fabr. towards cotton genotypes. *Indian Journal of Ecology* 8: 223–228.

Sharma, H.C. and Agarwal, R.A. (1982a). Consumption and utilization of bolls of different cotton genotypes by larvae of *Earias vittella* F. and effect of gossypol and tannins on food utilization. *Zietschrift fur Angewandte Zoologie* 68: 13–38.

Sharma, H.C. and Agarwal, R.A. (1982b). Effect of some antibiotic compounds in *Gossypium* on the post-embryonic development of spotted bollworm (*Earias vittella* F.). *Entomologia Experimentalis et Applicata* 31: 225–228.

Sharma, H.C. and Agarwal, R.A. (1983a). Role of some chemical components and leaf hairs in varietal resistance in cotton to jassid, *Amrasca biguttula biguttula* Ishida. *Journal of Entomological Research* 7: 145–149.

Sharma, H.C., and Agarwal, R.A. (1983b). Factors affecting genotypic susceptibility to spotted bollworm (*Earias vittella* Fab.) in cotton. *Insect Science and Its Application* 4: 363–372.

Sharma, H.C., Agarwal, R.A. and Singh, M. (1982). Effect of some antibiotic compounds in cotton on post embryonic development of spotted bollworm (*Earias vittella* F.) and the mechanism of resistance in *Gossypium arboreum*. *Proceedings of Indian Academy of Sciences (B)* 91: 67–77.

Sharma, H.C., Bhagwat, V.R. and Saxena, K.B. (1997). *Biology and Management of Spotted Pod Borer, Maruca vitrata (Geyer)*. Patancheru, Andhra Pradesh, India: International Crops Research Institute for the Semi-Arid Tropics (CRISAT).

Sharma, H.C. and Crouch, J.H. (2004). Molecular marker assisted selection: A novel approach for host plant resistance to insects in grain legumes. In Ali, M., Singh, B.B., Kumar, S. and Dhar, V. (Eds.), *Pulses in New Perspective. Proceedings of the National Symposium on Crop Diversification and Natural Resource Management*, 20–22 December, 2003. Kanpur, Uttar Pradesh, India: Indian Society of Pulses Research and Development, Indian Institute for Pulses Research, 147–174.

Sharma, H.C. and Franzmann, B.A. (2002). Orientation of sorghum midge, *Stenodiplosis sorghicola*, females (Diptera: Cecidomyiidae) to color and host-odor stimuli. *Journal of Agricultural and Urban Entomology* 18: 237–248.

Sharma, H.C., Gaur, P.M. and Hoisington, D.A. (2005). Physico-chemical and molecular markers for host plant resistance to *Helicoverpa armigera*. In Saxena, H., Rai, A.B., Ahmad, R. and Gupta, S. (Eds.), *Recent Advances in* Helicoverpa *Management*. Kanpur, Uttar Pradesh, India: Indian Society of Pulses Research and Development, Indian Institute for Pulses Research, 84–121.

Sharma, H.C., Leuschner, K. and Vidyasagar, P. (1990). Factors influencing oviposition behavior of sorghum midge, *Contarinia sorghicola* Coq. *Annals of Applied Biology* 116: 431–439.

Sharma, H.C., Lopez, V.F. and Vidyasagar, P. (1994). Effect of panicle compactness and host plant resistance in sequential plantings on population increase of panicle-feeding insects in *Sorghum bicolor* (L.) Moench. *International Journal of Pest Management* 40: 216–221.

Sharma, H.C. and Norris, D.M. (1991). Chemical basis of resistance in soybean to cabbage looper, *Trichoplusia ni*. *Journal of Science of Food and Agriculture* 55: 353–364.

Sharma, H.C. and Norris, D.M. (1994a). Phagostimulant activity of sucrose, sterols and soybean leaf extractables to the cabbage looper, *Trichoplusia ni* (Lepidoptera: Noctuidae). *Insect Science and Its Application* 15: 281–288.

Sharma, H.C. and Norris, D.M. (1994b). Biochemical mechanisms of resistance to insects in soybean: Extraction and fractionation of antifeedants. *Insect Science and Its Application* 15: 31–38.

Sharma, H.C. and Nwanze, K.F. (1997). *Mechanisms of Resistance to Insects in Sorghum and Their Usefulness in Crop Improvement.* Information Bulletin no. 45. Patancheru, Andhra Pradesh, India: International Crops research Institute for the Semi-Arid tropics (ICRISAT).

Sharma, H.C. and Ortiz, R. (2002). Host plant resistance to insects: An eco-friendly approach for pest management and environment conservation. *Journal of Environmental Biology* 23: 11–35.

Sharma, H.C., Pampapathy, G. and Sullivan, D.J. (2003). Influence of host plant resistance on activity and abundance of natural enemies. In Ignacimuthu, S. and Jayaraj, S. (Eds.), *Biological Control of Insect Pests.* New Delhi, India: Phoenix Publishing House, 282–296.

Sharma, H.C., Sharma, K.K. and Crouch, J.H. (2004). Genetic transformation of crops for insect resistance: Potential and limitations. *CRC Critical Reviews in Plant Sciences* 23: 47–72.

Sharma, H.C., Vidyasagar, P. and Leuschner, K. (1990). Components of resistance to the sorghum midge, *Contarinia sorghicola. Annals of Applied Biology* 116: 327–333.

Sharma, H.C. and Youm, O. (1999). Host plant resistance in integrated pest management. In Khairwal, I.S., Rai, K.N., Andrews, D.J. and Harinarayana, H. (Eds.), *Pearl Millet Improvement.* New Delhi, India: Oxford and IBH Publishing, 381–415.

Sharma, H.C., Ahmad, R., Ujagir, R., Yadav, R.P., Singh, R. and Ridsdill-Smith, T.J. (2005). Host plant resistance to cotton bollworm/legume pod borer, *Helicoverpa armigera.* In Sharma, H.C. (Ed.), Heliothis/Helicoverpa *Management: Emerging Trends and Strategies for Future Research.* New Delhi, India: Oxford and IBH Publishing, 167–208.

Sharma, H.C., Crouch, J.H., Sharma, K.K., Seetharama, N. and Hash, C.T. (2002). Applications of biotechnology for crop improvement: Prospects and constraints. *Plant Science* 163: 381–395.

Sharma, H.C., Doumbia, Y.O., Scheuring, J.F., Ramaiah, K.V., Beninati, N.F. and Haidra, M. (1994). Genotypic resistance and mechanisms of resistance to the sorghum head bug, *Eurystylus immaculatus* Odh. in West Africa. *Insect Science and Its Application* 15: 39–48.

Sharma, H.C., Pampapathy, G., Dwivedi, S.L. and Reddy, L.J. (2003). Mechanisms and diversity of resistance to insect pests in wild relatives of groundnut. *Journal of Economic Entomology* 96: 1886–1897.

Sharma, H.C., Stevenson, P.C., Simmonds, M.S.J. and Green, P.W.C. (2001). *Identification of* Helicoverpa armigera *(Hübner) Feeding Stimulants and the Location of Their Production on the Pod-Surface of Pigeonpea* [Cajanus cajan (L.) *Millsp.*]. Final Technical Report. DfiD Competitive Research Facility Project [R 7029 (C)]. Patancheru, Andhra Pradesh, India: International Crops Research Institute for the Semi-Arid Tropics (ICRISAT).

Sharma, P.N., Ketipearachchi, Y., Murata, K., Torii, A., Takumi, S., Mori, N. and Nakamura, C. (2003). RFLP/AFLP mapping of a brown planthopper (*Nilaparvata lugens* Stal) resistance gene *Bph1* in rice. *Euphytica.* 129: 109–117.

Sharopova, N., McMullen, M.D., Schultz, L., Schroeder, S., Sanchez, H., Gardiner, J., Bergstrom, D., Houchins, K., Polacco, M., Edwards, K.J., Ruf, T., Register, J.C., Brower, C., Thompson, R., Chin, E., Lee, M., Liscum, I.I.I., Cone, E., Davis, G. and Coe, E.H. Jr. (2002). Development and mapping of SSR markers for maize. *Plant Molecular Biology* 48: 463–481.

Sherman, F. (1998). *An Introduction to the Genetics and Molecular Biology of the Yeast* Saccharomcyes erevisiae. http://dbb.urmc.rochester.edu/labs/sherman_f/yeast/Index.html.

Shoemaker, D.D., Schadt, E.E., Armour, Y.D., He, P., Garrett-Engel, P.D. and McDonayl, P.M. (2001). Experimental annotation of the human genome using microarray technology. *Nature* 409: 922–927.

Shon, C.C., Lee, M., Melchinger, A.E., Guthrie, W.D. and Woodman, W.I. (1993). Mapping and characterization of quantitative trait loci affecting resistance against second generation European corn borer in maize with the aid of RFLPs. *Heredity* 70: 648–659.

Simmonds, M.S.J. and Stevenson, P.C. (2001). Effects of isoflavonoids from *Cicer* on larvae of *Helicoverpa armigera. Journal of Chemical Ecology* 27: 965–977.

Simon, C.J. and Muehlbauer, F.J. (1997). Construction of a chickpea linkage map and its comparison with maps of pea and lentil. *Journal of Heredity* 88: 115–119.

Singh, R. and Agarwal, R.A. (1988). Influence of leaf-veins on ovipositional behaviour of jassid, *Amrasca biguttula biguttula* (Ishida). *Journal of Cotton Research and Development* 2(1): 41–48.

Singh, R. and Ellis, P.R. (1993). Sources, mechanisms and basis of resistance in crucifereae to cabbage aphid, *Brevicoryne brassicae*. *Bulletin of OILB/SROP* 16: 21–35.

Singh, S.P. and Jotwani, M.G. (1980). Mechanisms of resistance to shoot fly. III. Biochemical basis of resistance. *Indian Journal of Entomology* 42: 551–556.

Smith, C.M. (1989). *Plant Resistance to Insects.* New York, USA: John Wiley & Sons.

Smith, C.M. (2005). *Plant Resistance to Arthropods: Molecular and Conventional Approaches.* Dordrecht, The Netherlands: Springer Verlag.

Sogawa, K. and Pathak, M.D. (1970). Mechanisms of brown planthopper resistance of Mudgo variety of rice (Hemiptera: Delphacidae). *Applied Entomology and Zoology* 5: 145–158.

Somers, D.J., Isaac, P. and Edwards, K. (2004). A high-density microsatellite consensus map for bread wheat (*Triticum aestivum* L.). *Theoretical and Applied Genetics* 109: 1105–1114.

Song, Q.J., Marek, I.F., Shoemaker, R.C., Lark, K.G., Concibido, V.C., Delannay, X., Specht, J.E. and Cregan, P.B. (2004). A new integrated genetic linkage map of the soybean. *Theoretical and Applied Genetics* 109: 122–128.

Srivastava, C.P. and Srivastava, R.P. (1989). Screening for resistance to gram pod borer, *Heliothis armigera* (Hubner) in chickpea (*Cicer arietinum* L.) genotypes and observations on its mechanism of resistance in India. *Insect Science and Its Application* 10: 255–258.

Srivastava, S. and Narula, A. (2004). *Plant Biotechnology and Molecular Markers.* Berlin, Germany: Springer Verlag.

Stadler, E. (1986). Oviposition and feeding stimuli in leaf surface waxes. In Juniper, B.E. and Southwood, T.R.E. (Eds.), *Inects and Plant Surface.* London, UK: Edward Arnold, 105–121.

Stadler, E. and Buser, H.R. (1984). Defense chemicals in the leaf surface wax synergistically stimulate oviposition by phytophagous insects. *Experientia* 40: 1157–1159.

Starks, K.J., McMilan, W.W., Sekul, A.A. and Cox, H.C. (1965). Corn earworm larval feeding response to corn silk and kernel extracts. *Annals of Entomological Society of America* 58: 74–76.

Staub, J.E., Serquen, F.C. and Gupta, M. (1996). Genetic markers, map construction, and their application in plant breeding. *Hortscience* 31: 729–741.

Stevenson, P.C., Green, P.W.C., Simmonds, M.S.J. and Sharma, H.C. (2005). Physical and chemical mechanisms of plant resistance to *Helicoverpa*: Recent research on chickpea and pigeonpea. In Sharma, H.C. (Ed.), Helicoverpa/Heliothis *Management: Emerging Trends and Strategies for the Future Research.* New Delhi, India: Oxford & IBH Publishers, 215–228.

Stromberg, L.D., Dudley, J.W. and Rufener, G.K. (1994). Comparing conventional early generation selection with molecular marker-assisted selection in maize. *Crop Science* 34: 1221–1225.

Sundramurthy, V.T. and Chitra, K.L. (1992). Integrated pest management in cotton. *Indian Journal of Plant Protection* 20: 1–17.

Sunnucks, P., Wilson, A.C.C., Beheregaray, L.B., Zenger, K., French, J. and Taylor, A.C. (2000). SSCP is not so difficult: The application and utility of single-stranded conformation polymorphism in evolutionary biology and molecular ecology. *Molecular Ecology* 9: 1699–1710.

Tamura, K., Fukuta, Y., Hirae, M., Oya, S., Ashikawa, I. and Vagi, T. (1999). Mapping of the Grh/locus for green rice leafhopper resistance in rice using RFLP markers. *Breeding Science* 49: 11–14.

Tang, D.L. and Wang, W.G. (1996). Influence of contents of secondary metabolism substances in cotton varieties on the growth and development of cotton bollworm. *Plant Protection* 22: 6–9.

Tao, Y.Z., Hardy, A., Drenth, J., Henzell, R.G., Franzmann, B.A., Jordan, D.G., Butler, D.R. and McIntyre, C.L. (2003). Identifications of two different mechanisms for sorghum midge resistance through QTL mapping. *Theoretical and Applied Genetics* 107: 116–122.

Tar'an, B., Michaels, T.E. and Pauls, K.P. (2002). Genetic mapping of agronomic traits in common bean (*Phaseolus vulgaris* L.). *Crop Science* 42: 544–546.

Tar'an, B., Thomas, E., Michaels, T.E. and Pauls, K.P. (2003). Marker assisted selection for complex trait in common bean (*Phaseolus vulgaris* L.) using QTL-based index. *Euphytica* 130: 423–433.

Tekeoglu, M., Tullu, A., Kaiser, W.J. and Muehlbauer, F.J. (2000). Inheritance and linkage of two genes that confer resistance to fusarium wilt in chickpea. *Crop Science* 40: 1247–1251.

Terry, L.I., Chase, K., Jarvik, T., Orf, J., Mansur, L. and Lark, K.G. (2000). Soybean quantitative trait loci for resistance to insects. *Crop Science* 40: 375–382.

Terry, L.I., Chase, K., Orf, J., Jarvik, T., Mansur, L. and Lark, K.G. (1999). Insect resistance in recombinant inbred soybean lines derived from non-resistant parents. *Entomologia Experimentalis et Applicata* 91: 465–476.

Thayumanavan, B., Velusamy, R., Sadasivam, S. and Saxena, R.C. (1990). Phenolic compounds, reducing sugars and free amino acids in rice leaves of varieties resistant to rice thrips. *International Rice Research Newsletter* 15: 14–15.

Tsuji, T., Saka, N. and Izawa, T. (2003). Genetic analysis of the early maturity linked with the green rice leaf hopper-resistance gene, *Grh3(t)*. *Research Bulletin of the Aichi ken Agricultural Research Center* 35: 17–22.

Ujagir, R. and Khare, B.P. (1988). Susceptibility of chickpea cultivars to gram pod borer, *Heliothis armigera* (Hubner). *Indian Journal of Plant Protection* 16(1): 45–49.

Ulloa, M., Meredith, W., Shappley, Z. and Kahler, A. (2002). RFLP genetic linkage maps from four $F_{2.3}$ populations and a JoinMap of *Gossypium hirsutum* L. *Theoretical and Applied Genetics* 104: 200–208.

Uthamasamy, S. (1985). Effect of leafhopper, *Amrasca devastans* feeding on the organic acid content of okra leaves. *Madras Agricultural Journal* 72: 75.

Venter, E. and Botha, A.M. (2000). Development of markers linked to *Diuraphis noxia* resistance in wheat using a novel PCR-RFLP approach. *Theoretical and Applied Genetics* 100: 965–970.

Vidyachandra, B., Roy, J.K. and Bhaskar, D. (1981). Chemical differences in rice varieties susceptible or resistant to gall midges and stem borers. *International Rice Research Newsletter* 6(2): 7–8.

Villareal, J.M., Hautea, D.M. and Carpena, A.L. (1998). Molecular mapping of the bruchid resistance gene in mungbean *Vigna radiata* L. *Philippine Journal of Crop Science* 23 (supplement 1): 1–9.

Vos, P., Hogers, R., Bleeker, M., Rijans, M., Van de Lee, T., Homes, M., Frijters, A., Pot, J., Kuiper, M. and Zabeau, M. (1995). AFLP: A new technique for DNA fingerprinting. *Nucleic Acids Research* 23: 4407–4414.

Vroh Bi, I., Maquet, A., Baudoin, J.P., du Jardin, P., Jacquemin, J.M. and Mergeai, G. (1999). Breeding for "low-gossypol seed and high-gossypol plants" in upland cotton. Analysis of tri-species hybrids and backcross progenies using AFLPs and mapped RFLPs. *Theoretical and Applied Genetics* 99: 1233–1244.

Waiss, A.C. Jr., Chan, B.G., Elliger, C.A., Wiseman, B.R., McMillian, W.W., Widstrom, N.W., Zuber, M.S. and Keaster, A.J. (1979). Maysin, a flavone glycoside from corn silks with antibiotic activity toward corn earworm. *Journal of Economic Entomology* 72: 256–258.

Walker, D., Boerma, H.R., All, J. and Parrott, W.L. (2002). Combining cry1Ac with QTL alleles from PI 229358 to improve soybean resistance to lepidopteran pests. *Molecular Breeding* 9: 43–51.

Wallace, L.E., McNeal, F.H. and Berg, M.A. (1974). Resistance to both *Oulema melanopus* and *Cephus cinctus* in pubescent-leaved and solid-stemmed wheat selections. *Journal of Economic Entomology* 67: 105–107.

Wang, C., Yasui, H., Yoshimura, A., Su, C., Zhai, H. and Wan, J. (2003). Green rice leafhopper resistance gene transferred through backcrossing and CAPs marker assisted selection. *Chinese Agriculture Science* 2: 13–18.

Wang, C.H., Yasui, A., Yoshimura H. and Wan, J. (2004). Inheritance and QTL mapping of antibiosis to green leafhopper in rice. *Crop Science* 44: 1451–1457.

Wang, T., Xu, S.S., Harris, M.O., Hu, J. and Cai, L.L.X. (2006). Genetic characterization and molecular mapping of Hessian fly resistance genes derived from *Aegilops tauschii* in synthetic wheat. *Theoretical and Applied Genetics* 113: 611–618.

Weeden, N.F., Muehlbauer, F.J. and Ladizinsky, G. (1992). Extensive conservation of linkage relationships between pea and lentil genetic maps. *Journal of Heredity* 83: 123–129.

Weeden, N.F., Ellis, T.H.N., Timmerman Vaughan, G.M., Simon, C.J., Torres, A.M., Wolko, B. and Knight, R. (2000). How similar are the genomes of the cool season food legumes? In *Linking Research and Marketing Opportunities for Pulses in the 21st Century: Proceedings of the Third International Food Legumes Research Conference*, 22–26 September, 1997, Adelaide, Australia. Dordrecht, The Netherlands: Kluwer Academic Publishers, pp. 397–410.

Weng, Y. and Lazar, M.D. (2002). Amplified fragment length polymorphism- and simple sequence repeat-based molecular tagging and mapping of greenbug resistance gene *Gb3* in wheat. *Plant Breeding* 121: 218–223.

Weng, Y., Li, W., Devkota, R.N. and Rudd, J.C. (2005). Microsatellite markers associated with two *Aegilops tauschii*-derived greenbug resistance loci in wheat. *Theoretical and Applied Genetics* 110: 462–469.

Widstrom, N.W., Butron, A., Guo, B.Z., Wilson, D.M., Snook, M.E., Cleveland, T.E. and Lynch, R.E. (2003a). Control of preharvest aflatoxin contamination in maize by pyramiding QTL involved in resistance to ear-feeding insects and invasion by *Aspergillus* spp. *European Journal of Agronomy* 19: 563–572.

Widstrom, N.W., Wiseman, B.R., Snook, M.E., Nuessly, G.S. and Scully, B.T. (2003b). Registration of the maize population Zapalote Chico 2451F. *Crop Science* 43: 444–445.

Willcox, M.C., Khairallah, M.M., Bergvinson, D., Crossa, J., Deutsch, J.A., Edmeades, G.O., Gonzalez-de-Leon, D., Jiang, C., Jewell, D.C., Mihm, J.A., Williams, W.P. and Hoisington, D. (2002). Selection for resistance to southwestern corn borer using marker-assisted and conventional backcrossing. *Crop Science* 42: 1516–1528.

Williams, C.E., Collier, C.C., Sardesai, N., Ohm, H.W. and Cambron, S.E. (2003). Phenotypic assessment and mapped markers for *H31*, a new wheat gene conferring resistance to Hessian fly (Diptera: Cecidomyiidae). *Theoretical and Applied Genetics* 107: 1516–1523.

Williams, L.H. (1954). The feeding habits and food preferences of Acrididae and the factors which determine them. *Transactions of the Royal Entomological Society (London) (B)* 105: 423–454.

Williams, M.N.V., Pande, N., Nair, S., Mohan, M. and Bennett, J. (1991). Restriction fragment length polymorphism analysis of polymerase chain reaction products amplified from mapped loci of rice (*Oryza sativa* L.) genomic DNA. *Theoretical and Applied Genetics* 82: 489–498.

Winter, P., Benko-Iseppon, A.M., Huttel, B., Ratnaparkhe, M., Tullu, A., Sonnante, G., Pfaff, T., Tekeoglu, M., Santra, D., Sant, V.J., Rajesh, P.N., Kahl, G. and Muehlbauer, F.J. (2000). A linkage map of chickpea (*Cicer arietinum* L.) genome based on recombinant inbred lines from a *C. arietinum* x *C. reticulatum* cross: Localization of resistance genes for fusarium wilt races 4 and 5. *Theoretical and Applied Genetics* 101: 1155–1163.

Winter, P., Pfaff, T., Udupa, S.M., Huttel, B., Sharma, P.C., Sahi, S., Arreguin-Espinoza, R., Weigand, F., Muehlbauer, F.J. and Kahl, G. (1999). Characterization and mapping of sequence-tagged microsatellite sites in the chickpea (*Cicer arietinum* L.). *Genome* 262: 90–101.

Wiseman, B.R. and Isenhour, D.J. (1994). Resistance in sweet corn to corn earworm larvae. *Journal of Agricultural Entomology* 11: 157–163.

Wittenberg, A.H.J., van der Lee, T., Cayla, C., Kilian, A., Visser, R.G.F. and Schouten, H.J. (2005). Validation of the high-throughput marker technology DArT using the model plant *Arabidopsis thaliana*. *Molecular Genetics and Genomics* 274: 30–39.

Wu, L.F., Cai, Q.N. and Zhang, Q.W. (1997). The resistance of cotton lines with different morphological characteristics and their F_1 hybrids to cotton bollworm. *Acta Entomologica Sinica* 40: 103–109.

Xiao, J., Grandillo, S., Ahn, S.N.K., McCouch, S.R., Tanksley, S.D., Li, J. and Yuan, L. (1996). Genes from wild rice improve yield. *Nature* 384: 223–224.

Xu, X.F., Mei, H.W., Luo, L.J., Cheng, X.M. and Li, Z.K. (2002). RFLP-facilitated investigation of the quantitative resistance of rice to brown planthopper (*Nilaparvata lugens*). *Theoretical and Applied Genetics* 104: 248–253.

Yang, T.J., Kim, D.H., Kuo, G.C., Kumar, L., Young, N.D. and Park, H.G. (1998). RFLP marker-assisted selection in backcross breeding for introgression of the bruchid resistance gene in mungbean. *Korean Journal of Breeding* 30: 8–15.

Yang, S., Pang, W., Harper, J., Carling, J., Wenzl, P., Huttner, E., Zong, X. and Kilian, A. (2006). Low level of genetic diversity in cultivated pigeonpea compared to its wild relatives is revealed by diversity arrays technology (DArT). *Theoretical and Applied Genetics* 113: 585–595.

Yazawa, S., Yasui, H., Yoshimura, A. and Iwata, N. (1998). RFLP mapping of genes for resistance to green rice leafhopper (*Nephotettix cincticeps* Uhler) in rice cultivar DV85 using near isogenic lines. *Scientific Bulletin of the Faculty of Agriculture, Kyushu University* 52: 169–175.

Yencho, G.C., Bonierbale, M.W., Tingey, W.M., Plaisted, R.L. and Tanksley, S.D. (1996). Molecular markers locate genes for resistance to the Colorado potato beetle, *Leptinotarsa decemlineata*, in hybrid *Solanum tuberosum* X *S. berthaultii* potato progenies. *Entomologia Experimentalis et Applicata* 81: 141–154.

Yencho, G.C., Cohen, M.B. and Byrne, P.F. (2000). Applications of tagging and mapping insect resistance loci in plants. *Annual Review of Entomology* 45: 393–422.

Yoshida, M., Cowgill, S.E. and Wightman, J.A. (1995). Mechanisms of resistance to *Helicoverpa armigera* (Lepidoptera: Noctuidae) in chickpea: Role of oxalic acid in leaf exudates as an antibiotic factor. *Journal of Economic Entomology* 88: 1783–1786.

Yoshida, M., Cowgill, S.E. and Wightman, J.A. (1997). Roles of oxalic and malic acids in chickpea trichome exudates in host-plant resistance to *Helicoverpa armigera*. *Journal of Chemical Ecology* 23: 1195–1210.

Young, N.D., Fatokun, C.A., Menancio-Hautea, D. and Danesh, D. (1992a). RFLP mapping in cowpea. In Thottappilly G., Monti, G.L., Mohan Raj, D.R. and Moore, A.W. (Eds.), *Biotechnology, Enhancing Research on Tropical Crops in Africa*. Ibadan, Nigeria: International Institute of Tropical Agriculture (IITA), 237–246.

Young, N.D., Kumar, L., Menancio-Hautea, D., Danesh, D., Talekar, N.S., Shanmugasundarum, S. and Kim, D.H. (1992b). RFLP mapping of a major bruchid resistance gene in mungbean (*Vigna radiata*, L. Wilczek). *Theoretical and Applied Genetics* 84: 839–844.

Zhang, Z.S., Xiao, Y.H., Luo, M., Li, X.B., Luo, X.Y., Hou, L., Li, D.M. and Pei, Y. (2005). Construction of a genetic linkage map and QTL analysis of fiber-related traits in upland cotton (*Gossypium hirsutum* L.). *Euphytica* 144: 91–99.

Zhao, J.Z., Fan, X.L., Shi, X.P., Zhao, R.M. and Fan, Y.L. (1997). Gene pyramiding: An effective strategy of resistance management for *Helicoverpa armigera* and *Bacillus thuringiensis*. *Resistant Pest Management* 9(2): 19–21.

Zhu, J.L. (1981). Preliminary observations on the resistance of varieties of spring wheat to *Hydrellia griseola* (Fallen). *Insect Knowledge* 18: 213–214.

Zhu, L.C., Smith, C.M., Fritz, A., Boyko, E. and Flinn, M.B. (2004). Genetic analysis and molecular mapping of a wheat gene conferring tolerance to the greenbug (*Schizaphis graminum* Rondani). *Theoretical and Applied Genetics* 109: 289–293.

Zhu, L.C., Smith, C.M., Fritz, A., Boyko, E., Voothluru, P. and Gill, B.S. (2005). Inheritance and molecular mapping of new greenbug resistance genes in wheat germplasms derived from *Aegilops tauschii*. *Theoretical and Applied Genetics* 111: 831–837.

Zummo, G.R., Segers, J.C. and Benedict, J.H. (1984). Seasonal phenology of allelochemicals in cotton and resistance to bollworm (Lepidoptera: Noctuidae). *Environmental Entomology* 13: 1287–1290.

7

Genetic Transformation of Crops for Resistance to Insect Pests

Introduction

There is a continuing need to increase food production to meet food requirements in the future, particularly in the developing countries of Asia, Africa, and Latin America. Several biotic and abiotic constraints are a limiting factor in increasing the production and productivity of food crops. Losses due to insect pests, which represent one of the largest constraints in crop production, have been estimated at 14% of the total agricultural production (Oerke et al., 1994). In addition, insects also act as vectors of various plant pathogens, and the annual cost for controlling pest damage has been estimated to be US$10 billion. Large-scale application of insecticides for insect control results in toxic residues in food and food products, in addition to adverse effects on the nontarget organisms in the environment. Furthermore, the cost-benefit ratio of such practices can easily become negative in marginal cropping systems, particularly when other factors such as diseases or drought become a limiting factor in crop production. The losses due to insect pests can be minimized effectively through host plant resistance to insects through conventional plant breeding and use of biotechnological approaches.

The ability to isolate and manipulate single genes through recombinant DNA technology (Schnepf and Whiteley, 1981; Watson et al., 1987), together with the ability to insert specific genes into cultivars with desirable agronomic traits, and adaptation to environmental conditions in a particular region have opened up new vistas in crop improvement (Chilton, 1983). Significant progress has been made over the past two decades in introducing foreign genes into plants, and this has provided opportunities to modify crops to increase yields, impart resistance to insect pests and diseases, and improve nutritional quality and yield (H.C. Sharma et al., 2002, 2003). In this chapter, we focus on candidate genes with potential to impart resistance to insect pests, review the progress in developing transgenic plants with resistance to insects, and assess the future potential of this technology for making host plant resistance an effective weapon in pest management.

Genetic Transformation: The Protocols

The efficiency of tissue culture and transformation protocols is one of the most important components for successful generation of transgenic crops (Birch, 1997). The major components for developing transgenic plants are: (1) reliable tissue culture and regeneration systems, (2) preparation of gene constructs and transformation with suitable vectors, (3) efficient transformation techniques to introduce genes into the crop plants, (4) recovery and multiplication of transgenic plants, (5) molecular and genetic characterization of transgenic plants, (6) transferring genes of interest into elite cultivars, and (7) evaluation of transgenic plants for their effectiveness in controlling the target pests without being a hazard to the environment. Although several approaches have been tried successfully for genetic transformation of crop plants, only the following three approaches have been used widely to introduce genes into a wide range of crop plants (Potrykus, 1991; Dale, Irwin, and Scheffer, 1993; Seetharam et al., 2002).

- *Agrobacterium*-mediated gene transfer
- Microprojectile bombardment with DNA or biolistics
- Direct DNA transfer into isolated protoplasts

Agrobacterium-Mediated Gene Transfer

Agrobacterium tumefaciens (Smith and Townsend) has been used widely to transfer novel genes into crop plants. It is a soil-inhabiting bacterium that results in gall formation at the wound site in many dicotyledonous plants. This tumor-inducing capability is due to the presence of a large *Ti* (tumor-inducing) plasmid in virulent strains of *Agrobacterium*. Likewise, *Ri* (root-inducing) megaplasmids are found in virulent strains of *Agrobacterium rhizogenes* (Ricker et al.) Conn., the causative agent of "hairy root" disease. The *Ti* and *Ri* plasmids, and the molecular biology of crown gall and hairy root induction, have been studied in considerable detail (Zambryski et al., 1983; Zambryski, 1992). *Agrobacterium*-mediated transformation is brought about by incorporation of genes of interest from an independently replicating *Ti* plasmid within the *A. tumefaciens* cell, which then infects the plant cell and transfers the T-DNA containing the gene of interest into the chromosomes of the actively dividing cells of the host plant.

Microprojectile Bombardment with DNA or Biolistics

In the particle bombardment (biolistics) method, tungsten or gold microprojectiles are coated with the DNA to be inserted, and bombarded into cells or tissues capable of subsequent plant regeneration. Acceleration of heavy microprojectiles (0.5 to 5.0 μm diameter tungsten or gold particles) coated with DNA carry the genes into the cell and tissue (Klein et al., 1987; Sanford, 1990). The DNA-coated particles enter the plant cells and the DNA is incorporated in a small proportion of the treated cells. The transformed cells are then selected for plant regeneration.

Direct DNA Transfer into Isolated Protoplasts

Genetically engineered DNA can also be directly injected into nuclei of embryogenic single cells, which can be induced to regenerate plants in cell culture (Neuhaus et al., 1987).

This requires micromanipulation of single cells or small colonies of cells under the micro-scope, and precise injection of small amounts of DNA with a thin glass micropippette. Injected cells or clumps of cells are subsequently raised in *in vitro* culture, and regenerated into plants. In the protoplast transformation, the cell wall of the target cells is removed by enzymatic treatment, and the cells are covered by a plasma membrane (Zhang and Wu, 1988). The DNA can be added into cell suspension, which can be introduced by affecting the plasma membrane by polyethylene glycol or by passing an electric current through the protoplast suspension. The DNA gets incorporated into the genome of a few cells. A suit-able marker may be inserted to select the transformed protoplasts and the cell colonies that develop from them (Shimamoto et al., 1989).

DNA transfer into the protoplast has resulted in high levels of expression of the trans-gene. The *Bacillus thuringiensis* (*Bt*) (Berliner) toxin gene *cry2Aa2* operon expressed in tobacco chloroplasts resulted in *Bt* protein content of up to 45.3% of the total protein in mature leaves, which resulted in 100% mortality of cotton bollworm, *Heliothis virescens* (F.), and the beet armyworm, *Spodoptera exigua* (Hubner) (Cosa et al., 2002). Under rice chloro-plast transcription elements, the CryIIa5 *Bt* toxin accumulated up to 3% of total soluble protein in leaf tissue, which was 300 times more compared to expression of the same protein in nuclear transformed plants (Leelavathi et al., 2002). In another study, chloroplast transformed tobacco plants with the *cryIIa5* gene under the control of rice *psbA* transcrip-tional elements showed high levels of expression of *Bt* toxin without imposing a yield penalty. The transgenic plants with up to 3% toxin protein of total soluble protein with high levels of resistance to the noctuid, *Helicvoerpa armigera* (Hubner), have also been devel-oped (Reddy et al., 2002). Analysis of T_0, T_1, and T_2 generation plants revealed site-specific integration, maternal inheritance, and uniform expression of the transgenes. The chloro-plast transformation vector *pNRAB* carrying two expression cassettes for the spectinomy-cin resistance gene *aadA* and the insect resistance gene *cry1Aa10* (sited between the *rps7* and *ndhB*) resulted in chloroplast transformants at a frequency of four in 1,000 bombarded cotyledon petioles (Hou et al., 2003). Chloroplast transformation of oilseed rape with the *cry1Aa10* gene resulted in 47% mortality of the second instar larvae of the diamond back moth, *Plutella xylostella* (L.). Chloroplasts of transplastomic tobacco expressing the *cry1C* gene have also shown high levels of *Bt* protein (about 1% of total protein) and the plastid transgenes were not transmitted through pollen (C.H. Lin et al., 2003). This will facilitate not only improvement in breeding for insect-resistant plants, but also the prevention of contamination of transgenes among crop plants. The results suggested that overexpres-sion of insecticidal toxin coding genes in chloroplasts would be an effective strategy to produce transgenic plants with high efficiency for controlling the target pests, and delay the emergence of resistance among phytophagous pests. This also prevents the inadver-tent movement of the transgene into closely related wild relatives of the crops due to maternal inheritance of the chloroplasts (Bansal and Sharma, 2003).

Gene Expression

Efficient genetic engineering relies on being able to generate a specific gene product at the desired level of expression, in appropriate tissues, at the right time. This can be accom-plished by creating gene constructs that include promoters and/or transcription regula-tion elements that control the level, location, and timing of gene expression. A major constraint in developing effective transgenic plants has been the lack of promoters that offer a high level of site-specific gene expression in the crop species of interest. Generally, transgene expression is driven by constitutive promoters such as cauliflower mosaic virus 35S (*CaMV35S*) (Benfey and Chua, 1989, 1990) and *Actin1* (McElory, Rothenberg, and Wu, 1990). Although the *CaMV35S* promotor has been widely used for transformation of dicotyledons, it has low activity in monocotyledons. Moreover, the pattern of *CaMV35S* promoter activity in different tissues of transgenic plants is difficult to predict (Benfey and Chua, 1990). In general, it has been found that monocot promoters are more active in monocot tissues than in dicot tissues (Wilmink, van de Ven, and Dons, 1995). However, tissue-specific promoters have now been successfully employed for driving transgene expression in pith tissue. Phosphoenolpyruvate carboxylase (*PEPC*) from maize can be used for gene expression in green tissue (Hudspeth and Grula, 1989). From a crop yield potential perspective, insect-resistant transgenes should be expressed only in those organs likely to be attacked by the insects. Otherwise the plants may be highly resistant, but the metabolic cost may substantially reduce the crop yield. Constitutive promoters such as *CaMV35S* are effective in providing high levels of gene expression, but may have unanticipated consequences towards nontarget organisms because of expression of the transgene in all plant parts. Therefore, a more targeted expression of insecticidal genes by using tissue- and organ-specific promoters can form an important component for developing transgenic plants with resistance to insects (Wong, Hironaka, and Fischhoff, 1992; Svab and Maliga, 1993; McBride et al., 1995).

Transposon-mediated repositioning of transgenes is another strategy to generate plants that are free of selectable markers and T-DNA inserts (Cotsaftis et al., 2002). By using a minimal number of transformation events, a large number of transgene insertions in the genome can be obtained so as to benefit from position effects in the genome that contribute to higher levels of expression. The *cry1B* gene expressed under the control of maize *ubiquitin* promoter between minimal terminal inverted repeats of the maize *Ac-Ds* transposon system has been cloned, and the 5′ untranslated sequence of a *gfp* gene used as an excision marker. The results indicated that transposon-mediated relocation of the gene of interest is a powerful method for generating transgenic plants, and exploiting favorable position effects in the plant genome.

Genetic transformation typically involves a marker gene for resistance to antibiotics (kanamycin – *npt* gene) or herbicides (phosphoinothricin – *bar* gene) (Table 7.1, Figure 7.1), a replication site, and a multiple cloning site (MCS) with several restriction sites for DNA insertion. Foreign DNA can be inserted into the vector using restriction enzymes that recognize a specific DNA sequence. Insertion of foreign DNA interrupts gene expression of

TABLE 7.1

Selectable Markers Used in Genetic Transformation of Crops

Gene	Markers
Antibiotic	*Dhfr*: Dihydrofolate reductase—Methotrexate/trimethoprin
	hpt: Hygromycin phosphotransferase—Hygromycin B
	*npt*II: Neomycin phosphotransferase—Kanamycin, neomycin, G418
Herbicide	*als*: acetolactate synthase—Chlorsulfuron, imidazolinones
	aroA: 5-Enolpyruvylshikimate-3-phosphate synthase—Glyphosate
	bar: Phosphoinothricin acetyltransferase—Phosphoinothricin
	bar: Glufocinate—Basta

FIGURE 7.1 Selection of putative transgenic plants of sorghum with the herbicide glufocinate (BASTA) as a selection marker.

an identifiable protein product to indicate DNA incorporation. Construction of DNA sequence for incorporation into vectors consists of several components, for example, in toxin genes from *Bt*, the gene should be first converted from AT-rich (typical of bacteria) to GC-rich (typical of higher plants) to increase toxin expression. Most changes are made to the third codon, thereby minimizing changes in the amino acid sequence, and increasing the expression of *Bt* toxin by 10- to 100-fold. For expression of the *Bt* gene in the higher plants, a recognizable promoter and a terminator sequence must bracket the *Bt* gene. Popular constitutive promoters as described above include *CaMV35S*, *ubiquitin*, *PEPC*, and maize pollen specific promoter (Koziel et al., 1993). The size of the vectors ranges from 5,000 to 11,000bp, depending on the *Bt* gene and the promoter incorporated into the vector. Delivery of the vectors into the nucleus can be achieved by using *Agrobacterium*-mediated transformation and biolistic methods (Koziel et al., 1993).

Premature polyadenylation at times may lead to poor expression of the transgene (Haffani et al., 2000). Transformation of potato with the *cry3Ca1* gene resulted in transformants with poor expression of the transgene (Kuvshinov et al., 2001). No full-length transcript (2,300 nucleotides) was detected, but short transcripts of approximately 1,100 nucleotides were observed. The sites at which premature polyadenylation took place were not those that showed the highest degree of identity to the canonical AAUAAA motif, suggesting that premature polyadenylation may contribute to poor expression of transgenes in a foreign host. Expression of the transgene may also depend on the plant species in which the gene has been deployed. Expression level of the synthetic gene *cry9Aa* under the control of double *35S* promotor has been observed to be three to ten times lower in potato, cauliflower, and turnip as compared to that in tobacco (Kuvshinov et al., 2001). In tobacco plants transformed with a truncated native *cry9Aa* gene and with a translational fusion construct of the truncated native *cry9Aa* and *uidA* (GUS) gene, the expression level of the native *cry9Aa* gene ranged from 0.03 to 1 pg of *cry9Aa* mRNA per 1 µg of total RNA, while the expression level of the synthetic *cry9Aa* gene was five to ten times higher at the mRNA level, and at least 50 times higher at the translational level. Therefore, considerable care should be exercised while selecting a gene construct, the promotor, and the marker genes for genetic engineering of plants for resistance to insects.

Genetic Transformation of Crop Plants for Resistance to Insects

Genes from bacteria such as *B. thuringiensis, B. subtilis* (Ehrenberg) Gohn, and *B. sphaericus* Meyer and Neide, protease inhibitors, plant lectins, ribosome inactivating proteins, secondary plant metabolites, and small RNA viruses have been used alone or in combination with conventional host plant resistance to develop crop cultivars that suffer less damage from insect pests (Table 7.2). Genes conferring resistance to insects have been inserted into crop plants such as maize, *Zea mays* L., rice, *Oryza sativa* L., wheat, *Triticum aestivum* L., sorghum, *Sorghum bicolor* (L.) Moench, sugarcane, *Saccharam officinarum* L., cotton, *Gossypium hirsutum* L., potato, *Solanum tuberosum* L., tobacco, *Nicotiana tabacum* L., broccoli, *Brassica oleracea* var *italica* L., cabbage, *Brassica oleracea* var *capitata* L., chickpea, *Cicer arietinum* L., pigeonpea, *Cajanus cajan* (L.) Millsp., cowpea, *Vigna unguiculata* (L.) Walp., groundnut, *Arachis hypogea* L., tomato, *Lycopersicon esculentum* Mill., brinjal, *Solanum melongena* L., and soybean, *Glycine max* (L.) Merr. (Hilder and Boulter, 1999; H.C. Sharma et al., 2000, 2004; H.C. Sharma, Sharma, and Crouch, 2004). Genetically transformed crops with *Bt* genes have been deployed for cultivation in Argentina, Australia, Bulgaria, Canada, China, France, Germany, India, Indonesia, Mexico, Portugal, Romania, South Africa, Spain, the United States, Ukraine, and Uruguay (James, 2007). Although several transgenic crops with insecticidal genes have been introduced in the temperate regions, very little has been done to use this technology for improving productivity of crops that are important for food security in the developing countries, where the need for increasing food production is most urgent. Transgenic *Bt* cotton and maize have been grown largely on a commercial scale under high input, temperate, or subtropical cropping systems. The most urgent need to use this technology is in the tropical regions, where soil fertility, water availability, insect pests, and diseases severely constrain crop production. For transgenic plants to be useful as an effective weapon in pest management, they have to substitute, completely or partially, for the use of insecticides in crop production, and result in increased crop production and environment conservation. The bioefficacy of different toxin genes expressed in transgenic plants in different crops is discussed below.

Toxin Proteins from *Bacillus thuringiensis*

Ishiwata discovered this bacterium in 1901 from diseased silkworm, *Bombyx mori* L. larvae. Berliner (1915) isolated it from diseased larvae of *Ephestia kuhniella* Keller, and designated it as *Bacillus thuringiensis*. It is a Gram-positive bacterium, which produces proteinaceuos crystalline inclusion bodies during sporulation. Further research on *Bt* by Steinhaus (1951) led to renewed interest in it as a biopesticide. There are several subspecies of this bacterium, which are effective against lepidopteran, dipteran, and coleopteran insects. The identification of the *kurstaki* strain provided a boost for commercialization of *Bt*. The HD 1 strain identified by Dulmage (1981) is the most important *Bt* strain. There are over 50 registered *Bt* products with more than 450 formulations (Shewry and Gutteridge, 1992). *Bacillus thuringiensis* var. *israeliensis* has been used extensively for the control of mosquitoes (de Barjac and Sotherland, 1990). *Bacillus thuringiensis* var. *morrisoni* and *B. thuringiensis* var. *israeliensis* carry four genes that encode mosquito and black fly toxins Cry IVA, Cry IVB, Cry IVC, and Cry IVD (Bechtel and Bulla, 1976). Because of the crystalline nature of *Bt*-specific toxin proteins, the term Cry is used in gene and protein nomenculature. The toxin genes earlier were classified into four types, based on insect specificity and sequence homology (Hofte and Whiteley, 1989). Cry I type genes encode proteins of 130 kDa, and

TABLE 7.2

Insect-Resistant Transgenic Crops Based on δ-Endotoxins from *Bacillus thuringiensis*

Crop	Transgene	Remarks	References
Cotton	*cryIAc*	Effective against *Pectinophora gossypiella*. High levels of resistance in Coker 312 to *Trichoplusia ni*, *Spodoptera exigua*, and *Helicoverpa zea/Heliothis virescens*.	Wilson et al. (1992), Benedict et al. (1996)
	cryIAc	High mortality of *Helicoverpa armigera* in Shiyuan 321 and Zhongmiansuo 19, 3517, and 541.	S.D. Guo et al. (1999)
	cryIAc	S 545, S 591, S 636, S 1001 from Simian 3, and 1109 from Zhongmiansuo 12 exhibit high levels of resistance to *Helicoverpa armigera*.	Ni et al. (1996)
	cry1Ac	MECH 12, MECH 162 and MECH 184.	Mohan and Manjunath (2002)
Maize	*cry1Ab*	Resistant to European corn borer, *Ostrinia nubilalis*.	Kozeil et al. (1993), Armstrong et al. (1995)
	cry9C	Resistant to European corn borer, *Ostrinia nubilalis*.	Jansens et al. (1997)
	cryIAb	Effective against spotted stem borer, *Chilo partellus* and maize stalk borer, *Busseola fusca*.	van Rensburg (1999)
	cryIAb	Resistant to stem borers, *Diatraea grandiosella* and *Diatraea saccharalis*.	Bergvinson, Willcox, and Hoisington (1997)
	cry1Ab	Resistant to fall armyworm, *Spodoptera frugiperda* and Southwestern corn borer, *Diabrotica undecimpuncta howardi*.	Williams et al. (1997)
	cryIAb, cryIAc	Resistant to corn earworm, fall armyworm, Southwestern corn borer, and sugarcane borer.	Bohorova et al. (1999)
	cry34Ab1 and *cry 35Ab1*	Resistant to Western maize rootworm, *Diabrotica virgifera virgifera* and southern maize rootworm, *Diabrotica undecimpunctata howardi*.	Herman et al. (2002)
	cry1Ab	Line GH 0937 resistant to *Helicoverpa zea* and *Spodoptera frugiperda*.	Wiseman et al. (1999)
Rice	*cryIAb*	Basmati 370 and M 7 resistant to striped stem borer, *Chilo suppressalis* and leaf folder, *Cnaphalocrosis medinalis*.	Fujimoto et al. (1993)
		Resistant to yellow stem borer, *Scirpophaga incertulas* and striped stem borer, *Chilo suppressalis*.	Ghareyazie et al. (1997)
		Resistant to yellow stem borer, *Scirpophaga incertulas*.	Datta et al. (1998)
		Maintainer line R 68899B resistant to yellow rice stem borer, *Scirpophaga incertulas* and leaf folder, *Cnaphalocrosis medinalis*.	Alam et al. (1999)
		Transgenic plants resistant to yellow stem borer.	Nayak et al. (1997)
	cryIAc	Lines IR 64, Pusa Basmati 1, and Karnal Local caused complete mortality of yellow stem borer larvae within 4 days of infestation.	Khanna and Raina (2002)
	cryIIa	Resistant to leaf folder, *Cnaphalocrosis medinalis* and yellow stem borer, *Scirpophaga incertulas*.	Mqbool et al. (1998)
Sorghum	*cryIAc*	Moderately resistant to spotted stem borer, *Chilo partellus*.	Girijashankar et al. (2006)
Sugarcane	*cry1Ab*	Resistant to sugarcane borer, *Diatraea saccharalis*.	Arencibia et al. (1997)
Groundnut	*cry1Ac*	Complete mortality or up to 66% reduction in larval weight of lesser corn stalk borer, *Elasmopalpus lignosellus*.	Singsit et al. (1997)

continued

TABLE 7.2   **(continued)**

Crop	Transgene	Remarks	References
Chickpea	cry1Ac	Transgenic cultivars ICCV 1 and ICCV 6 inhibited development and feeding of *Helicoverpa armigera*.	Kar et al. (1997)
		Transgenic plants caused mortality of neonate larvae of *Helicoverpa armigera*.	Sanyal et al. (2005)
Tobacco	crylII	Resistant to Colorado potato beetle, *Leptinotarsa decemlineata*.	Perlak et al. (1993)
	cryIIa5	Resistant to *Helicoverpa armigera*.	Selvapandian et al. (1998)
	crylAa	Resistant to tobacco hornworm, *Manduca sexta*.	Barton, Whiteley, and Yang (1987)
	crylAb	Resistant to tobacco hornworm, *Manduca sexta* and budworm, *Heliothis virescens*.	Vaeck et al. (1987)
	cry1Ac	Resistant to tobacco budworm, *Heliothis virescens*.	McBride et al. (1995)
Potato	crylII	Resistant to Colorado potato beetle, *Leptinotarsa decemlineata*.	Jansens et al. (1995)
		Tolerance to Colorado potato beetle, *Leptinotarsa decemlineata*.	Perlak et al. (1993)
	crylIIA	Resistant to Colorado potato beetle, *Leptinotarsa decemlineata*.	Arpaia et al. (1997)
	crylIIB cry1Ab	Less damage to leaves by potato tuber moth, *Phthorimaea opercullela*.	Arpaia et al. (2000)
	cry1Ac9 cry5	Resistant to *Phthorimaea opercullela*.	Davidson et al. (2002)
	cry1Ab	Caused complete mortality of *Phthorimaea operculella* in Sangema, Cruza 148, and LT 8 cultivars.	Mohammed et al. (2000) Canedo et al. (1999)
	cry1Ab	Resistant to potato tuber moth, *Phthorimaea operculella*.	
	cryV		Douches et al. (2002)
Tomato	cry1Ab, cry1Ac	Resistant to lepidopteran pests.	Delannay et al. (1989); Van der Salm et al. (1994)
	cry1Ac	Resistant to *Helicoverpa armigera*.	Mandaokar et al. (2000)
		Resistant to tobacco hornworm, *Manduca sexta*.	Fischhoff et al. (1987)
Brinjal	crylIIb	Resistant to Colorado potato beetle, *Leptinotarsa decemlineata*.	Arencibia et al. (1997)
		Resistant to fruit borer, *Leucinodes orbonalis*.	Kumar et al. (1998)
Broccoli	cry1C	Resistant to diamondback moth, *Plutella xylostella*, cabbage looper, *Trichoplusia ni*, and cabbage butterfly, *Pieris rapae*.	Cao et al. (1999)
Chinese cabbage	cry1Ab, cry1Ac	Resistant to diamondback moth, *Plutella xylostella*.	Xiang et al. (2000)

are usually specific to lepidopteran larvae (Figures 7.2 and 7.3), type II genes encode for 70 kDa proteins that are specific to lepidopteran and dipteran larvae, and type III genes encode for 70 kDa proteins specific to coleopteran larvae. Type IV genes are specific to the dipteran larvae. The system was further extended to include type V genes that encode for proteins effective against lepidopteran and coleopteran larvae (Tailor et al., 1992). The *Bt* δ-endotoxins are now known to constitute a family of related proteins for which 140 genes have been described (Crickmore et al., 1998), with specificities for Lepidoptera, Coleoptera,

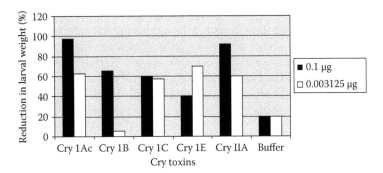

FIGURE 7.2 Reduction in weight of *Helicoverpa armigera* larvae fed on artificial diet treated with 0.1 and 0.003125 µg of different Cry toxins of *Bacillus thuringiensis*.

and Diptera. Several of these *Bt* toxin genes have been used for the genetic transformation of crop plants.

Cotton

Considerable progress has been made in developing cotton cultivars with *Bt* genes for resistance to bollworms, and there is a clear advantage of growing transgenic cotton in reducing the damage by bollworms and increasing cottonseed yield. The cotton cultivar Coker 312, transformed with the *cry1Ac* gene (having 0.1% toxin protein), has shown high levels of resistance to cabbage looper, *Trichoplusia ni* (Hubner), tobacco caterpillar, *S. exigua*, corn earworm, *Helicoverpa zea* (Boddie), and tobacco budworm, *H. virescens*. Cotton bollworm damage was reduced to 2.3% in flowers and 1.1% in bolls in the transgenic cotton, compared to 23% damage in flowers and 12% damage in bolls in the nontransgenic commercial cultivar, Coker 312 (Benedict et al., 1996). The cottonseed yield was 1,050 kg ha^{-1} in Coker 312 compared to 1,460 kg in *Bt* cotton. Cotton plants with *Bt* genes are effective against pink bollworm, *Pectinophora gossypiella* (Saunders) (Wilson et al., 1992). Survival of cotton bollworms (*H. zea* and *H. virescens*) has been found to be greater on squares and flower anthers than on other floral structures in Deltapine 5415 conventional cotton and transgenic NuCOTN 33B (Cotsaftis et al., 2002). Enzyme-linked immunosorbent assay (ELISA) indicated that *cry1Ac* expression varied in different plant parts, but bollworm survival did not correlate with the protein expression (Greenplate, 1999). Trends in Bollgard II were similar to Bollgard I (Gore, Leonard, and Adamczyk, 2001).

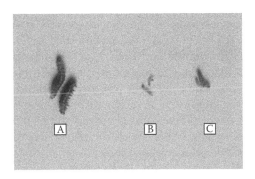

FIGURE 7.3 Five-day-old *Helicoverpa armigera* larvae reared on untreated artificial diet (A), and the diet treated with 0.1 µg mL^{-1} of Cry1Ac (B), and 0.1% commercial formulation (C) of *Bacillus thuringiensis*.

In China, the transgenic cotton lines S 545, S 591, S 636, and S 1001 from Simian 3, and Zh 1109 from Zhongmiansuo 12 with *Bt* genes have shown adverse affects on survival and development of *H. armigera* (Ni et al., 1996). Cotton cultivars Shiyuan 321, Zhongmiansuo 19, 3517, and 541 (transformed with *Bt* genes) have resulted in up to 96% mortality of cotton bollworm, *H. armigera* (S.D. Guo et al., 1999). A chimeric gene, *Bt29K* (coding sequences of activated Cry1Ac insecticidal protein) and an endoplasm reticulum-retarding signal peptide (*API-B*) gene have been transferred into two cotton varieties in China (H.N. Guo et al., 2003). Nine homozygous transgenic lines showed 90.0 to 99.7% mortality of *H. armigera* larvae, good agronomic traits, and expression of *Bt* proteins at a level of 0.09 to 0.17%. Insect resistance of homozygous lines expressing the activated chimeric *cry1Ac* and *API-B* was better than the lines expressing *cry1Ac* only (H.N. Guo et al., 2003). First- to fourth-instar larvae of *H. armigera* died on transgenic *Bt* cotton, the pupation decreased by 48.2 and 87.5%, and adult emergence by 66.7 and 100%, respectively. Egg laying decreased by 50.1 to 69.7%, and egg hatching by 80.6 to 87.8% (Cui and Xia, 1999). Feeding anther dust of transgenic cotton to the adults decreased the number of eggs and egg hatching by 59.8 and 72.1%, respectively. Neonates of *H. armigera* have the ability to detect and avoid transgenic *Bt*-cotton Zhong 30 and transgenic *CpTi-Bt* cotton, SGK 321 as compared to the nontransgenic cotton Shiyuan 321 (J.H. Zhang et al., 2004). The larvae consumed more food on *CpTi-Bt* transgenic cotton than on *Bt* transgenic cotton. Fourth instars were in equal numbers on transgenic and nontransgenic cottons, but food consumption on transgenic cotton was lower than that on the nontransgenic cotton. In no-choice tests involving fifth instars, significantly less time was spent on feeding on the two transgenic cottons. The neonates selectively feed on the nontransgenic cotton or the preferred plant parts.

In India, cotton hybrids (Mech 12, Mech 162, and Mech 184) carrying the *cry1Ac* gene were approved for cultivation in 2001, targeting bollworm, *H. armigera*, pink bollworm, *P. gossypiella*, spotted bollworm, *Earias vittella* (F.), and the spiny bollworm, *E. insulana* (Boisd.) (Mohan and Manjunath, 2002). Under integrated pest management, the transgenic cotton hybrid Mech 162 suffered lower damage by the bollworms *H. armigera* and *P. gossypiella* (11.5%) as compared to the conventional cotton (29.4%) (Bambawale et al., 2004). Only three sprays were needed for pest control on the transgenic cotton compared to seven sprays on the conventional cotton. The seed cotton yield was 12.4 q ha^{-1} in Mech 162 compared to 7.1 q ha^{-1} in the conventional cotton. There is considerable variation in the production of Cry1Ac toxin protein in different plants of the same and different hybrids, and over the crop-growing season, and *Cry1Ac* production declines during the fruiting stage (Kranthi et al., 2005). Different *Bt* transgenic hybrids vary in their susceptibility to *H. armigera*, and the mortality of *H. armigera* larvae declines substantially during the latter part of the crop growth. The larval numbers were significantly lower on the transgenic hybrids under high infestation, but the differences in larval density between the transgenic and nontransgenic hybrids under low levels of infestation were quite small (H.C. Sharma and Pampapathy, 2006). Bollworm damage in squares and bolls was significantly lower in the transgenic hybrids than in the nontransgenic ones, although there were a few exceptions. Differences in seed cotton yield between the transgenic and the nontransgenic hybrids were not significant under unprotected conditions at moderate levels of infestation. However, significant differences in seed cotton yield were observed under heavy bollworm infestation. Transgenic hybrids also suffered low shoot damage by spotted bollworm, *E. vittella*. Bollworm damage and seed cotton yield of the *Gossypium arboreum* L. varieties Aravinda and MDL 2450 have been found to be similar to the transgenic hybrids, suggesting that it would be useful to combine transgenic resistance to *H. armigera* with plant characteristics conferring resistance to this pest or nontarget insect pests in the region to realize the full potential of transgenic plants for sustainable crop production (H.C. Sharma and Pampapathy, 2006).

Maize

Transgenic maize expressing *Bt* toxin genes is quite effective against the European corn borer (ECB), *Ostrinia nubilalis* (Hubner) (Koziel et al., 1993; Armstrong et al., 1995; Archer et al., 2000). *Bt*-maize is quite effective in preventing ECB damage, and generally produces higher grain yields than the nontransgenic crop (Clark et al., 2000). First-generation ECB damage is reduced or eliminated with the use of the *Bt* hybrids. In the absence of ECB pressure, the performance of transgenic hybrids is similar to their nontransgenic counterparts. Yield of isoline hybrids is 10% lower than the standard and *Bt* hybrids regardless of ECB infestation (Lauer and Wedberg, 1999), but *Bt* hybrids generally yield 4 to 8% more than the standard hybrids under severe ECB pressure. Transgenic maize expressing *cry9C* (from *B. thuringiensis* subsp. *tolworthi*) is also effective against ECB (Jansens et al., 1997). Maize plants transformed with *Bt* genes have also been found to be effective against the spotted stem borer, *Chilo partellus* (Swinhoe) and the maize stalk borer, *Busseola fusca* (Fuller) in Southern Africa (van Rensburg, 1999). The spotted stem borer is more susceptible than the maize stalk borer to transgenic maize with *Bt* genes. Maize plants with *Cry1Ab* gene are also resistant to the Southwestern corn borer, *Diatraea grandiosella* (Dyar) and sugarcane borer, *Diatraea saccharalis* (F.) (Bergvinson, Willcox, and Hoisington, 1997).

The *Bt*-transformed plants exhibit greater resistance to *D. grandiosella* than those derived from conventional host plant resistance (Bergvinson, Willcox, and Hoisington, 1997). Williams et al. (1997) developed transgenic corn hybrids, which sustained significantly less leaf feeding damage by fall armyworm, *Spodoptera frugiperda* (J.E. Smith) and Southwestern corn borer, *Diabrotica undecimpuncta howardi* (Barber) than the resistant cultivars derived through conventional breeding. Resistance to fall armyworm and near immunity to Southwestern maize borer observed in these transgenic maize hybrids is the highest level of resistance documented for these insect pests. Transgenic tropical maize inbred lines with *cry1Ab* or *cry1Ac* genes with resistance to corn earworm, fall armyworm, Southwestern corn borer, and sugarcane borer have also been developed (Bohorova et al., 1999). A binary insecticidal crystal protein (*bICP*) from *B. thuringiensis* strain PS149B1 [composed of a 14 kDa protein (*cry34Ab1*) and a 44 kDa protein (*cry35Ab1*)] have been co-expressed in transgenic maize plants, which provide effective control of western maize rootworm, *Diabrotica virgifera virgifera* Leconte under field conditions (Herman et al., 2002). The 14 kDa protein is also active alone against the southern maize rootworm, *D. undecimpunctata howardi*, and was synergized by a 44 kDa protein. Transgenic crops have also been observed to have beneficial effects on nontarget pests, for example, maize hybrids with *cry1Ab* also suffer less *Fusarium* ear rot than their nontransgenic counterparts (Munkvold, Hellmich, and Rice, 1999). Novartis Sweetcorn and GH 0937 hybrids containing the *Bt* gene are highly resistant to *H. zea* and *S. frugiperda* (Lynch et al., 1999a,b; Wiseman et al., 1999). The resistance of maize to insect pests increased with an increase in backcross generations (Z.H. Liu et al., 2000). The transfer ratio of the extrinsic *Bt* gene is greater through macrogametes than through microgametes.

Rice

Rice plants having 0.05% toxin of the total soluble leaf protein have shown high levels of resistance to the striped stem borer, *Chilo suppressalis* (Walker) and rice leaf folder,

Cnaphalocrosis medinalis (Guen.) (Fujimoto et al., 1993). Scented varieties of rice (Basmati 370 and M 7) have also been transformed with *cryIIa*, confering resistance to yellow rice stem borer, *Scirpophaga incertulas* (Walker) and the rice leaf folder, *C. medinalis* (Mqbool et al., 1998). Truncated *cry1Ab* gene has been introduced into several *indica* and *japonica* rice cultivars by microprojectile bombardment and protoplast systems (Datta et al., 1998). Rice lines transformed with the synthetic *cry1Ac* gene are highly resistant to yellow stem borer, *S. incertulas* (Nayak et al., 1997), and those with the *cry1Ab* gene are resistant to the striped stem borer, *C. suppressalis* and the yellow stem borer, *S. incertulas* (Ghareyazie et al., 1997). The *cry1Ab* gene has also been inserted into the maintainer line, R 68899B with enhanced resistance to yellow stem borer, *C. suppressalis* (Alam et al., 1999). Khanna and Raina (2002) developed *Bt* transgenics of elite *indica* rice breeding lines (IR 64, Pusa Basmati 1, and Karnal Local) with synthetic *cryAc* gene. Selected *Bt* lines of IR 64 and Pusa Basmati 1, having *Bt* titers of 0.1% (of total soluble protein) showed 100% mortality of yellow stem borer larvae within four days of infestation in cut stems as well as at the vegetative stage in whole plant assays. Husnain et al. (2002) expressed *cry1Ab* in Basmati rice under the control of three promoters (*PEPC*, *ubiquitin*, and pollen specific promotor derived from the *Bp10* gene of *Brassica napus* L. in *pGEM 4Z*). Toxin protein expression was 0.05% of the total protein in stems under the control of the *PEPC* promotor alone or in combination with the pollen specific promotor, but was nearly 0.15% of the total protein under the control of the *ubiquitin* promotor, suggesting that a specific promotor can be used to limit the expression of *cry1Ab* gene in desired plant parts. There were no significant differences in the progenies of either the *indica* × *japonica* or *japonica* × *japonica* crosses, indicating that transgenic *Bt* rice can be used in crossbreeding elite cultivars with resistance to insects (Cui et al., 2001). Zhu et al. (2003) observed that a dominant locus, and the linkage between the *cry1Ab* and *bar* gene, followed 3:1 segregation in T_1, was stably homozygotic in T_2, and followed 1:1 and 3:1 segregation in BC_1-BC_4 and BCF_2 (self-cross of BC), respectively.

Sorghum

Toxins from *B. thuringiensis* var *morrisoni* have shown biological activity against the sorghum shoot fly, *Atherigona soccata* (Rondani). Cry1Ac, Cry1C, Cry1E, and Cry2A are moderately effective against the spotted stem borer, *C. partellus*, while Cry1Ac is effective against *H. armigera* (H.C. Sharma et al., 2004). Sorghum plants having the *cry1Ac* gene have been developed for resistance to spotted stem borer, *C. partellus* (Harshavardhan et al., 2002; Seetharama et al., 2002; Girijashankar et al., 2005). Feeding by the neonate larvae of *C. partellus* on the leaf discs from transgenic plants was 60% lower compared to that on the nontransgenic control plants, and the weight gain by the larvae was reduced by 36%. The larval mortality was 40% more in the larvae fed on leaf discs from transgenic plants compared to that on the nontransgenic control plants.

Sugarcane

The truncated *cry1Ab* gene in sugarcane has shown significant activity against the sugarcane borer, *D. saccharalis* despite low expression of the *Bt* protein (Arencibia et al., 1997). Transgenic cane tissue incorporated into artificial diet caused low body weights, prolonged development time, and high mortality in Mexican rice borer, *Eoreuma loftini* (Dyar) (Samad and Leyva, 1998). Larvae that survived to adulthood laid fewer eggs compared to larvae fed on artificial diet with nontransgenic sugarcane or artificial diet alone. Egg hatch was also poor in the transgenic treatment. A population of 42 transgenic sugarcane (Ja60-5) clones expressing a truncated *cry1Ab* gene has been evaluated in the field. Five elite transgenic clones showed agronomic and industrial traits similar to the nontransgenic plants of Ja60-5, but a small number of qualitative traits were different. A total of 51 polymorphic DNA bands (out of the 1,237 analyzed) were identified by extensive AFLP and RAMP analysis, showing rare but consistent genomic changes in the transgenic plants as compared to C_1 and C_2 control plants (Arencibia et al., 1999).

Oilseed Crops

A codon-modified *cry1Ac* gene has been introduced into groundnut by using microprojectile bombardment (Singsit et al., 1997). The immunoassay of plants selected with hygromycin has shown the expression of Cry1Ac protein up to 0.16% of the total soluble protein. Complete mortality or up to 66% reduction in larval weight has been recorded in the lesser corn stalk borer, *Elasmopalpus lignosellus* (Zeller). There was a negative correlation between larval survival and larval weight of the lesser corn stalk borer with the amount of *Bt* protein.

Grain Legumes

A tissue culture and regeneration protocol has been developed for chickpea, which has been found to be useful for genetic transformation of this crop (Jayanand, Sudarasanam, and Sharma, 2003). Chickpea cultivars ICCV 1 and ICCV 6, transformed with the *cry1Ac* gene, have been found to inhibit the development of and feeding by *H. armigera* (Kar et al., 1997). Sanyal et al. (2005) developed transgenic chickpea plants with *cry1Ac* gene under the control of *CaMV35S* promotor and *nptII* as a selection marker. The transgenic plants showed protein expression levels of 14.5 to 23.5 ng mg^{-1} of extractable protein. Plants with Cry1Ac expression levels of >10 ng mg^{-1} showed >80% mortality of the neonate larvae of *H. armigera*. Transgenic plants with *cry1Ac* gene are also being tested for resistance to *H. armigera* (Ramakrishna Babu et al., 2005). Pigeonpea plants transformed with *cry1Ab, cry1Ac*, and soybean trypsin inhibitor (*SBTI*) genes have been developed at ICRISAT, and are being tested against *H. armigera* (Gopalaswamy et al., 2003; Sree Latha et al., 2005; K.K. Sharma, Ananda Kumar, and Sharma, 2005; K.K. Sharma, Lavanya, and Anjaiah, 2006). Pigeonpea plants transformed with *cry1E-C* gene from *Bt* under the control of *CaMV35S* promotor and *nptII* as a selection marker have shown resistance to the larvae of tobacco caterpillar, *Spodoptera litura* (F.) (Surekha et al., 2005).

Tobacco

The first transgenic tobacco plants with *Bt* toxin genes were produced in 1987 (Barton, Whiteley, and Yang, 1987; Fischhoff et al., 1987; Vaeck et al., 1987). These plants expressed full length or truncated *Bt* toxin genes *cry1A* under the control of constitutive promoters. The expression was quite low in tobacco plants, resulting in only 20% mortality of tobacco hornworm, *Manduca sexta* L. larvae. Truncated *cry1A* genes encoding for the toxic N-terminal fragment provided better protection to tobacco and tomato plants. Plants transformed with truncated gene expressed about 0.02% of total leaf soluble protein. Gene truncation, use of different promoters, enhancer sequences, and fusion proteins resulted in only a marginal improvement in gene expression (Barton, Whiteley, and Yang, 1987; Vaeck et al., 1987).

The *cry1IIA* gene from *B. thuringiensis* subsp. *tenebrionis* has also been expressed in transgenic tobacco (Sutton, Havstad, and Kemp, 1992), and five classes of sequences that mimic eukaryotic processing signals (which may be responsible for the low levels of transcription and translation) were identified and eliminated. The GC content of the gene was raised from 36% to 49%, and the codon usage was changed to be more plant-like. When the synthetic gene was driven by the *CaMV35S* promoter and the alfalfa mosaic rymovirus translational enhancer, up to 0.6% of CryIIIA toxin of the total protein was observed in transgenic tobacco plants. The transgenic plants showed resistance to the Colorado potato beetle, *Leptinotarsa decemlineata* (Say). Another study by Perlak et al. (1993) also confirmed that the synthetic *cryIII* gene in tobacco is effective for the control of Colorado potato beetle, *L. decemlineata*. Tobacco plants containing the *cryIIa5* gene are also resistant to *H. armigera* (Selvapandian et al., 1998), and the effectiveness of this toxin is comparable to Cry1Ab or Cry1Ac.

Potato

Second-generation transgenic potatoes grown from tubers of transgenic plants expressing the *cryIAc* gene showed 10% mortality of first-instar larvae of *Phthorimaea operculella* (Zeller) after 48 hours of feeding on leaf discs (Ebora, Ebora, and Sticklen, 1994). The second-instar larvae were slightly less capable of surviving on leaf discs from transgenic plants than those fed on the untransformed plants after 24 hours of feeding. Synthetic *cryIII* gene has also been expressed in potato plants with resistance to Colorado potato beetle, *L. decemlineata* (Jansens et al., 1995). Transgenic potato plants containing the *cry1Ab* gene (*Bt* 884) and a truncated gene *cry1Ab6* resulted in less damage to the leaves by the potato tuber moth, *P. operculella*. However, the size of the leaf tunnels increased over time in plants containing only the *Bt* 884 gene, while no increase was observed in tunnel length in plants containing the *cry1Ab6* gene (Arpaia et al., 2000). Transgenic LT 8 and Sangema tubers remained uninfested by *P. operculella* for 6 months. However, no significant effects were observed on the nontarget species such as *Liriomyza huidobrensis* (Blanch.), *Russelliana solanicola* Tuthill, and *Myzus persicae* (Sulzer). Damage to the fourth terminal leaf by *Epitrix cucumeris* (Harris) was 20 to 31% lower than in nontransgenic plants (Stoger et al., 1999). Davidson et al. (2002) developed transgenic lines of Ilam Hardy and Iwa with *cry1Ac9* gene. A transgenic line from each cultivar inhibited larval growth of *P. operculella* by over 40%, and the line derived from Ilam Hardy prevented pupation of all larvae. A modified gene of *B. thuringiensis* var *tolworthi* (*cryIIIB*) has shown insecticidal activity toward neonate larvae of Colorado potato beetle (Arpaia et al., 1997). Picentia and the wild species, *Solanum integrifolium* (Poir.), have also been transformed with a wild type (wt) and four mutagenized versions of *Bt43* belonging to the *cryIII* class (Innacone, Grieco, and Cellini, 1997). Adult males feeding on high-level *Bt*-expressing transgenic potatoes were able to mate and produce mobile sperm, but the females were impaired in their reproductive ability because their ovaries were not fully developed (Stewart, Feldman, and LeBlanc, 1999).

New Leaf *Bt*-transgenic potatoes provide substantial ecological and economic benefits to potato growers (Hoy, 1999). The *cry5*-Lemhi Russet and *cry5*-Atlantic potato lines have shown up to 100% mortality of first instar larvae of the potato tuber moth (Mohammed et al., 2000). The insertion and expression of the *cry1Ab* into potato cultivars Sangema, Cruza 148, and LT 8 have resulted in up to 100% larval mortality of *P. operculella* (Canedo et al., 1999). The codon-modified *Bt-cry5* gene (revised nomenclature *cryIIaI*) has been inserted into the cultivar Spunta to provide resistance to Colorado potato beetle, *L. decemlineata*, and potato tuber moth, *P. operculella* (Douches et al., 2002). Two transgenic "Spunta" clones, G2 and G3, produced high levels of mortality in first instars of potato tuber moth in detached-leaf bioassays (80 to 83% mortality), laboratory tuber tests

(100% mortality), and field trials in Egypt (99 to 100% undamaged tubers). Reduced feeding by Colorado potato beetle first instars was also observed in detached-leaf bioassays (80 to 90% reduction). Field trials in the United States demonstrated that the performance of these transgenic lines was comparable to "Spunta." These *Bt-cry5* transgenic potato plants with high potato tuber moth resistance can be used in integrated pest management programs.

Vegetables

Expression of *Bt* genes in tomato was one of the first examples of genetically modified plants with resistance to insects (Fischhoff et al., 1987). Tomato plants expressing *cry1Ab* and *cry1Ac* genes are effective against the lepidopteran insects (Delannay et al., 1989; Van der Salm et al., 1994). Expression of *cry1Ac* gene in tomato is highly effective against *H. armigera* (Mandaokar et al., 2000). The transgenic tomato plants expressing *cry1Ab* suffered significantly lower damage by *H. armigera* than the nontransgenic control plants in the laboratory, greenhouse, and field. The *Bt*-transgenic plants caused 100% mortality of *H. armigera* larvae (Kumar and Kumar, 2004). Transformed brinjal plants have also shown insecticidal activity against the fruit borer, *Leucinodes orbonalis* Guen. (Kumar et al., 1998).

Synthetic *cry1C* gene introduced into broccoli, *B. oleracea* subsp. *italica*, provides protection not only from the susceptible diamondback moth, *P. xylostella* larvae, but also from diamondback moth selected for moderate levels of resistance to Cry1C (Cao et al., 1999). Synthetic *cry1Ab* gene inserted into broccoli cultivar cv. Pusa Broccoli KTS-1 (Viswakarma et al., 2004) and *cry1Ab* in cabbage (Bhattacharya et al., 2002), and *cry1Ab* or *cry1Ac* genes in *Brassica campestris* L. subsp. *parachinensis* (Xiang et al., 2000) have shown resistance to *P. xylostella*. Transgenic cauliflower plants transformed with synthetic *cry9Aa* gene have also shown high levels of activity against *P. xylostella* (Kuvshinov et al., 2001). Transgenic broccoli containing *cry1C* is resistant to the cabbage looper, *T. ni*, and cabbage butterfly, *Pieris rapae* (L.). L.B. Lin et al. (2001) observed a mortality of 94.4% in *P. rapae* larvae reared in the laboratory on transgenic rapeseed leaves as compared with 15.0% mortality on the nontransgenic cultivar, Xiangyou 16. Under field conditions, the larval mortality was 46.5% on the transgenic rapeseed compared to 21.3% mortality on the control plots.

Fruits and Forest Trees

Walnut, *Juglans regia* L., trees expressing a modified *cry1Ac* gene have been found to be effective against the first instar codling moth, *Cydia pomonella* (L.) larvae (Leslie et al., 2001). Production of Cry1Ac protein was confirmed by Western analysis. There was a good correspondence between GUS activity, protein expression, and insect mortality. Alteration of the wild-type nucleic acid sequence was important in increasing the bioefficacy. Dandekar et al. (2002) reported that apple and walnut transformed with *Bt* genes conferred enhanced resistance to the codling moth, *C. pomonella* and other lepidopteran insects.

Ornamentals

Shinoyama et al. (2003) introduced a modified *cry1Ab* gene into chrysanthemum, *Chrysanthemum morifolium* syn. *D. morifolium* (Ramat). The level of accumulation of Cry1Ab protein ranged from 10.5 to 80 ng per 50 µg of total soluble protein (0.021 to 0.16% of the total protein). All the larvae of *H. armigera* died during the first instar on the leaves of transformed lines with expression level of >47.6 ng per 50 µg of total protein.

Vegetative Insecticidal Proteins

A supernatant of vegetative *Bacillus cereus* (Frankland and Frankland) culture has two compounds; *VIP 1* and *VIP 2*, which have been shown to possess toxic effects toward insects (Estruch et al., 1997). *VIP 3* is highly toxic to *Agrotis* and *Spodoptera* (Estruch et al., 1996). The activity of these proteins is similar to δ-endotoxins of *Bt*. The acute toxicity of vegetative insecticidal proteins is in the same range as that of the δ-endotoxins from *Bt*. They induce gut paralysis, followed by complete lysis of the gut epithelium cells, resulting in larval mortality. The *vip3Aa14* gene from *B. thuringiensis tolworthi* was expressed in *Escherichia coli* Escherich using expression vector *pET29a* (Bhalla et al., 2005). The expressed Vip3Aa14 protein was found in cytosolic supernatant as well as pellet fraction, but the protein was more abundant in the cytosolic supernatant fraction. Both full-length and truncated (devoid of signal sequence) Vips were highly toxic to the larvae of *S. litura* and *P. xylostella*. Rang et al. (2005) produced the *Vip3Ba1* protein in *E. coli* and tested against the European corn borer, *O. nubilalis* and the diamondback moth *P. xylostella*. The expressed protein resulted in significant growth delays, but had no larvicidal effect, indicating that its host range might be different than that of Vip3A proteins. Several transgenic events of cotton and corn are under field testing to be deployed for pest management in the near future (Christou et al., 2006).

Toxin Proteins from *Photorhabdus luminescens*

The *tcdA* gene of *Photorhabdus luminescens* (Thomas and Poinar) Boemare et al. encoding a 283 kDa protein (toxin A), which is highly toxic to insects, has been expressed in *Arabidopsis thaliana* (L.) Heyn. The transgenic plants expressed the protein at 700 ng mg^{-1} of extractable protein, and were highly toxic to tobacco hornworm, *M. sexta* (D. Liu et al., 2003). Toxin A isolated from the transgenic plants also inhibited the growth of the Southern corn rootworm, *D. undecimpunctata howardi*. Addition of 5'- and 3'-untranslated regions of a tobacco osmotin gene (*osm*) increased toxin A production 10-fold and recovery of insect-resistant lines 12-fold. In the best line, high expression of toxin A and insect resistance were maintained for at least five generations. The intact *tcdA* mRNA represented the largest effective transgenic transcript produced in plants to date.

Secondary Plant Metabolites

Many secondary plant metabolites, such as alkaloids, steroids, foliar phenolic esters, terpenoids, saponins, flavonoids, and nonprotein amino acids, act as potent protective chemicals. Some of the secondary plant metabolites are produced in response to insect feeding (H.C. Sharma and Agarwal, 1983; Ebel, 1986; H.C. Sharma and Norris, 1991). Systemically induced responses are modified through synthesis and action of jasmonic acid via its lipid precursor, for example, linoleic acid in tomato. Exogenous application of jasmonate induces the production of proteinase inhibitors. Xu et al. (1993) observed enhanced resistance in rice by wounding methyl jasmonate and abscisic acid in transgenic plants. Effective manipulation of secondary metabolites by introduction (or elimination by antisense RNA technology) of enzyme encoding sequences is quite difficult (Hallahan et al., 1992; McCaskill and Croteau, 1998), and increased production of many of these chemicals may impose a measurable cost in productivity potential of crop plants (Vrieling, van Wijk, and Swa, 1991). Such cost is not involved in natural protection mechanisms based on protective proteins (Brown, 1988). Expression of relatively large amounts of a foreign protein such as cowpea proteinase inhibitor (*CpTi*) does not impose a cost in yield in transgenic plants (Hilder and Gatehouse, 1991).

Monoterpenes, the C10 isoprenoids, are a large family of natural products that are best known as constituents of the essential oils and defensive oleoresins of aromatic plants (Mahmoud and Croteau, 2002). In addition to ecological roles in pollinator attraction, allelopathy, and plant defense, monoterpenes are also used extensively in the food, cosmetic, and pharmaceutical industries. The importance of these plant products has prompted studies on the monoterpene biosynthetic pathway, cloning of the relevant genes, and development of genetic transformation techniques for agronomically significant monoterpene-producing plants. Metabolic engineering of monoterpene biosynthesis in the model plant peppermint, *Mentha piperita* L., has resulted in yield increase and compositional improvement of the essential oil, and also provided strategies for manipulating flavor and fragrance production, and plant defense (Mahmoud and Croteau, 2002).

Expression of a bacterial cytokinin biosynthesis gene (*PI-II-ipt*) in *Nicotiana plumbaginifolia* (L.) plants has been correlated with enhanced resistance to *M. sexta* and *M. persicae* (Smigocki, Heu, and Buta, 2000). The *PI-II-ipt* gene has also been expressed in *N. tabacum* and *L. esculentum*, and similar antifeedant effects were observed with the transgenic tobacco, but not in tomato. A 30 to 50% reduction in larval weight gain was observed with some of the tomato plants, but these results could not be repeated consistently. Leaf surface extracts from transgenic *N. plumbaginifolia* leaves killed 100% of *M. sexta* second instars at concentrations of 0.05%, whereas the *N. tabacum* extracts were at least 20 times less active. Extract suspensions were stable for up to two days at ambient temperatures below 42°C, and for at least three months at 4°C when stored in the dark. High performance liquid chromatography (HPLC) analysis of the *N. plumbaginifolia* extracts yielded an active fraction that reduced hatching of *M. sexta* eggs by 30% and killed first, second, and third instars within 24, 48, and 72 hours of exposure, respectively. The activity appears to be associated with oxygen-containing aliphatic compounds, possibly diterpenes. Based on partial characterization of activity, the production, secretion, or accumulation of secondary metabolites in leaves of cytokinin-producing *PI-II-ipt N. plumbaginifolia* plants appears to be responsible for the observed insect resistance.

Protease Inhibitors

The enzyme inhibitors act on key insect gut digestive enzymes such as α-amylase and proteinases. Several kinds of α-amylase and proteinase inhibitors present in seeds and vegetative organs in plants influence food utilization by the phytophagous insects (Ryan, 1990; Konarev, 1996; Chrispeels, Grossi-de-Sa, and Higgins, 1998; Gatehouse and Gatehouse, 1998). Protease inhibitors (PIs) of plants are involved in a number of functions, including the control of endogenous proteolytic enzymes (Richardson, 1977), the reserve of ammonia and sulfur amino acids within the storage organs (Pusztai, 1972; Tan-Wilson et al., 1985), and the plant defense against insect and nematode attack (Sijmons, 1993; Thomas et al., 1994; Urwin et al., 1995; Lawrence and Koundal, 2002). In tomato and tobacco plants, protease inhibitors have been found to accumulate in response to infection by pathogenic microorganisms (Peng and Black, 1976; Rickauer, Fournier, and Esquerre-Tugaye, 1989). Since protease inhibitors are primary gene products, they are excellent candidates for engineering insect resistance into plants. Disruption of amino acid metabolism by inhibition of protein digestion has been one of the targets for use in insect control (Johnson et al., 1989). Genes encoding inhibitors specific for serine-proteases are the main digestive proteases in most lepidopteran insects (Boulter, 1993). Deployment of protease inhibitors for insect control requires a detailed analysis of particular insect-plant interaction. The ability of some insect species to compensate for protease inhibition by switching to an alternative proteolytic activity or overproducing the existing proteases may limit the application of protease inhibitors in such species (Jongsma et al., 1995). Adaptive mechanisms elevate the levels of other classes of proteinases to compensate for the trypsin activity inhibited by dietary proteinase inhibitors.

Soybean Kunitz type trypsin inhibitor (*SBTI*) and soybean Bowman-Birk type trypsin-chymotrypsin inhibitor (*SBBI*) reduced the larval weight of *H. armigera*, and such effects were greater for *SBTI* than *SBBI* (Johnston, Gatehouse, and Anstee, 1993; Shukla, Arora, and Sharma, 2005). Larvae feeding on diet containing 0.234 mM *SBTI* also reduced the trypsin-like enzyme activity in the gut of *H. armigera*. There is considerable diversity in protease inhibitors of cultivated chickpea and its wild relatives (Patankar et al., 1999). The diversity in proteinase activity in *H. armigera* gut and the flexibility in their expression during developmental stages depending on the diet provides a basis for selection of proper PIs for use in transgenic plants (Patankar et al., 2001). *Helicoverpa armigera* fed on chickpea showed more than 2.5- to 3-fold proteinase activity than those fed on pigeonpea, and cotton. Over 90% of the gut proteinase activity of the fifth-instar larvae is of the serine proteinase type, while the second-instar larvae showed the presence of metalloproteases, aspartic-, cysteine-, and serine-proteinases. Proteinase inhibitors with multi-inhibitory activities were generated by replacement of phytocystatin domains in sunflower multi-cystatin (*SMC*) by the serine proteinase inhibitor *BGIT* from bitter gourd, *Momordica charantia* L. seeds. Inanaga et al. (2001) compared the *chimaeric* inhibitors and the recombinant *SMC* (*r-SMC*) in relation to their effects on the growth of *S. exigua* larvae. When the second instar larvae were reared on a diet containing *rSMC*, *SMC-T3*, or SMC-T23 for ten days, a significant reduction in weight gain was observed (Inanaga et al., 2001). Mean weights of larvae with *rSMC*, *SMC-T3*, and *SMC-T23* diets were 43, 32, and 43 mg, respectively, compared to 60 mg in the control diet. In contrast, *BGIT* had little effect on the growth of the *S. exigua* larvae. The results suggested that *SMC-T3* with two phytocystatin domains and one serine proteinase inhibitor domain is an efficient inhibitor of proteinases in *S. exigua* larvae (Inanaga et al., 2001). Considerable progress has been made in deploying proteinase inhibitors in transgenic plants for controlling various insect pests (Table 7.3).

Tobacco

Transgenic tobacco plants expressing cowpea trypsin inhibitor (*CpTi*) have shown resistance to *H. armigera* (F.P. Zhang et al., 1998). Transgenic tobacco plants expressing high levels of *SBTI* have shown greater resistance than the tobacco plants expressing *CpTi* against *H. virescens*. The *SBTI* is also more effective than *CpTi* in reducing the proteolytic activity of gut extracts obtained from full-grown larvae of *H. armigera*. Proteolysis by gut extracts showed 40-fold more inhibition by *SBTI* than *CpTi* (Gatehouse et al., 1993). However, *CpTi* is considered to be more useful for genetic transformation, because unlike many serine protease inhibitors (*SPIs*), it is not deleterious to mammals (Pusztai et al., 1992). In another study, *H. armigera* larvae fed on transgenic tobacco plants expressing *SBTI* gene showed normal growth and development (Nandi et al., 1999). In another study, transgenic tobacco plants expressing *SBTI* resulted in increased insect mortality, reduced insect growth, and reduced plant damage by *H. virescens* (Hilder et al., 1987), *H. zea* (Hoffman et al., 1992), *Spodoptera littoralis* (Boisd.), *M. sexta* (Yeh et al., 1997; McManus et al., 1999), and *H. armigera* (Charity et al., 1999). *Helicoverpa armigera* larvae fed on plants expressing the giant taro, *Alocasia macrorrhiza* (L.) G. Don proteinase inhibitor

TABLE 7.3

Insect-Resistant Transgenic Crops Expressing Protease Inhibitors from Plants

Crop	Transgene	Remarks	References
Tobacco	CpTi	Resistant to *Helicoverpa armigera*.	Zhao et al. (1998); F.O. Zhang et al. (1998); Gatehouse et al. (1993)
	SBTI	Reduces the proteolytic activity of gut extracts obtained from *Helicoverpa armigera* larvae.	
		Increased insect mortality, reduced insect growth, and reduced plant damage by *Heliothis virescens, Heliothis zea, Spodoptera littoralis, Manduca sexta*, and *Helicoverpa armigera*.	Hilder et al. (1987); Hoffman et al. (1992); Yeh et al. (1997); McManus et al. (1999); Charity et al. (1999)
	GTPI	Growth inhibition of 22 to 40% of *Helicoverpa armigera*.	Wu et al. (1997)
Potato	Kti_3, C-II, and PU-IV from soybean	Reduction in larval weight of *Spodoptera littoralis* by 50%.	Marchetti et al. (2000)
Cotton	CpTi	Resistant to *Helicoverpa armigera*.	Li et al. (1998)
Maize	Oryzacystatin	Resistant to Southern maize rootworm, *Diabrotica undecimpunctata*	Edmonds et al. (1996)
Rice	Kunitz type SBTI	Resistant to Brown planthopper, *Nilaparvata lugens*.	Lee et al. (1999)
		Resistant to insect pests.	Duan et al. (1996)
	CpTi	Resistant to *Chilo suppressalis* and *Sesamia inferens*.	Xu et al. (1996)
	WTI-1B	Retarded larval growth of *Chilo suppressalis*.	Mochizuki et al. (1999)
Sugarcane	Proteinase inhibitor II	Increased antibiosis to larvae of sugarcane grubs, *Antitrogus consanguineus*.	Nutt et al. (1999)
Cabbage	CpTi	Resistant to *Pieris rapae*.	Fang et al. (1997)
Potato	OCI	Caused 53% mortality of *Leptinotarsa decemlineata* larvae.	Lecardonnel et al. (1999)
Wheat	TI from barley	Significant reduction in survival rate of the Angoumois grain moth, *Sitotroga cerealella*.	Altpeter et al. (1999)
Chinese cabbage	Modified CpTi (sck)	Resistant to *Pieris rapae* in lab and field.	G.D. Yang et al. (2002)
Tobacco	*Brassica juncea*, BjTi	Resistant to tobacco cut worm, *Spodoptera litura*.	Mandal et al. (2002)
Tobacco	PI 2 (MTI-2)	Caused larval mortality and decreased mean larval weight of *Spodoptera littoralis*.	De Leo et al. (1998)

(*GTPI*) have shown growth inhibition of 22 to 40%, but no larval mortality was observed (Wu et al., 1997).

Serine PI cDNA from cabbage, *B. oleracea* var. *capitata* (*BoPI*), has been subcloned and expressed in transgenic tobacco to test the ability of *BoPI* to enhance resistance to insects in a heterologous system (Pulliam et al., 2001). The transgenic plants containing *BoPI* gene were better than the other transgenic plants with different PI genes, and compared favorably with *Bt cry1Ac* transgenic plants in a bioassay with tobacco budworm, *H. virescens*. High levels of mustard trypsin PI 2 (*MTI-2*) expressed in tobacco and *Arabidopsis* have shown deleterious effects against *S. littoralis* larvae, causing larval mortality and reduction in feeding and larval weight (De Leo et al., 1998). However, larvae fed on leaves from plants expressing *MTI-2* at low expression levels did not show increased mortality, but a net gain in weight and faster development compared with the control larvae. These observations were correlated with the differential expression of digestive proteases in the larval gut, overexpression of existing proteases on low *MTI-2* expression level plants, and induction of new proteases on high *MTI-2* expression level plants. The results emphasized the need for development of a proteinase inhibitor-based defense strategy for plants with optimum level of proteinase inhibitor activity relative to the pest's sensitivity threshold to the proteinase inhibitor being expressed. The cDNA for bovine spleen trypsin inhibitor (*SI*), a homologue of bovine pancreatic trypsin inhibitor (*BPTI*), including the natural mammalian presequence, has also been expressed in tobacco (Christeller et al., 2002). Stability of *SI* was shown by the presence of protein at high levels in completely senescent leaves. Modifications to the cDNA (3′ and 5′ changes and minor codon changes) resulted in a 20-fold variation in expression. Expression of modified *SI* in transgenic tobacco leaves at 0.5% total soluble protein reduced both survival and growth of *H. armigera* larvae feeding on leaves from the late first instar. In larvae surviving for eight days, midgut trypsin activity was reduced in *SI* tobacco-fed larvae, chymotrypsin activity increased, while the activities of leucine aminopeptidase (cytosol aminopeptidase) and elastase-like chymotrypsin remained unaltered (Christeller et al., 2002). *Manduca sexta* encoded protease inhibitors expressed in *N. tabacum* also provide protection against insects (Thomas et al., 1995a).

Potato

Transgenic potato expressing the cysteine proteinase inhibitor, oryzacystatin (*OCI*) gene resulted in up to 53% mortality of *L. decemlineata* (Lecardonnel et al., 1999). Feeding young females of *L. decemlineata* on foliage from "Kennebec" potato (K 52) transformed with *OCI* did not affect survival, incidence of diapause, relative growth rate, and reproductive fitness (Cloutier et al., 2000). However, efficiency of conversion of ingested foliage during postemergence, growth, and adaptation of the digestive proteolytic system to the inhibitory effect of *OCI* were reduced. Three soybean protease inhibitor genes (*KTi₃*, *C-II*, and *PI-IV*), when transferred into potato, showed variable expression among different plants (Archetti et al., 2000). The level of resistance to *S. littoralis* was particularly high in tobacco, where many plants caused complete mortality of the larvae, while in potato, the larval mortality was much less frequently achieved, but resulted in a reduction of larval weight gain by 50%. A highly significant correlation was observed between inhibitor content and larval weight. Larval weight gain has been found to be dependent on the mid-gut proteolytic activity.

Cotton

Transgenic cotton lines expressing *CpTi* are resistant to *H. armigera* (Li et al., 1998). Proteinase inhibitors of *M. sexta* have also been expressed in transgenic cotton to provide protection against insects (Thomas et al., 1995b). Protease inhibitors accumulated to approximately 0.1% of total protein, depending on the tissue analyzed. Fecundity of cotton white fly, *Bemisia tabaci* (Genn.), was reduced significantly compared to that on the controls.

Maize

The cysteine protease inhibitor oryzacystatin is effective for controlling the Southern maize rootworm, *D. undecimpunctata* (Edmonds et al., 1996).

Rice

Expression of the potato trypsin inhibitor gene confers resistance to insect pests in rice (Duan et al., 1996). Constitutive expression of *CpTi* increases resistance to *C. suppressalis* and *Sesamia inferens* (Walker) in rice (Xu et al., 1996). A synthetic gene (*mwti11b*) coding for a winged bean trypsin inhibitor (*WTI-11B*) significantly retarded the larval growth of *C. suppressalis* in rice (Mochizuki et al., 1999). Accumulation of Kunitz type *SBTI* in rice also confers resistance to the brown planthopper, *Nilaparvata lugens* (Stal) (Lee et al., 1999).

Wheat

The Angoumois grain moth, *Sitotroga cerealella* (Oliver), is one of the major storage pests of cereals, and no antibiotic resistance in wheat against this insect has been identified to date. Characterization of the midgut proteases of *S. cerealella* larvae revealed that the major digestive proteases were trypsin-like and alpha-chymotrypsin-like serine proteases (Shukle and Wu, 2003). The potential value of naturally occurring plant protease inhibitors as resistance factors for *S. cerealella* has been assessed in bioassays using artificial seeds prepared by freeze-drying a flour paste in Teflon moulds and then coating the seeds with gelatin. Soybean trypsin inhibitor (Kunitz inhibitor) had an adverse effect on the development of the insect and suggested that protease inhibitors might serve as a transgenic resistance factor (Shukle and Wu, 2003). A significant reduction in survival rate of the Angoumois grain moth, *S. cerealella* has been observed on transgenic wheat seeds expressing the trypsin inhibitor from barley (Altpeter et al., 1999). However, it did not have a significant protective effect against leaf-feeding insects.

Sugarcane

Transgenic sugarcane plants expressing potato proteinase inhibitor II have shown increased antibiosis to larvae of sugarcane grubs, *Antitrogus consanguineus* (Blackburn) (Nutt et al., 1999). *Antitrogus consanguineus* larvae feeding for six weeks on transgenic plants carrying the potato proteinase inhibitor II gene gained only 4.2% of the weight of grubs feeding on nontransgenic control plants (Nutt et al., 2001).

Vegetables

The *CpTi* gene in cabbage, *B. oleracea* var *capitata* (cultivars Yingchun and Jingfeng), has shown resistance to *P. rapae* (Fang et al., 1997). However, adults of *Psylliodes chrysocephala* (L.) feed identically on leaf discs from control or transformed plants of oilseed rape expressing constitutively the cysteine proteinase inhibitor, oryzacystatin I (*OCI*) (Western et al., 1998). Trypsin inhibitor gene from sweet potato expressed in Taiwan cauliflower, *B. oleracea* var. *botrytis*, showed resistance to insects (L.C. Ding, Yeh, and Wang, 1998). *CpTI* gene and *nptII* gene were introduced into *Brassica campestris* L. ssp. *chinensis*. The insect resistance of the transgenic plants was maintained in the progeny (She et al., 2000). Modified *CpTi* gene (*sck*) has been successfully transferred into Chinese cabbage, *B. campestris* subsp. *pekinensis* cultivars (GP 11 and Zhongbai 4) (G.D. Yang et al., 2002). The resistance of the transgenic plants to *P. rapae* was observed in laboratory and field conditions. A trypsin inhibitor from Indian mustard, *Brassica juncea* (L.) Czern. (*BjTi*, a precursor of a 2S seed storage protein), showed a soybean trypsin inhibitor active site-like motif (*GPFRI*) at the expected processing site (Mandal et al., 2002). The *BjTi* is a thermostable Kunitz-type trypsin inhibitor that inhibits trypsin at a molar ratio of 1:1. The 20 kDa *BjTi* was purified from mid-mature seeds. Third-generation transgenics expressing *BjTi* at 0.28 to 0.83% of soluble leaf protein showed remarkable resistance against the tobacco cutworm, *S. litura*. This novel trypsin inhibitor can be used in transforming seed crops for protection to their vegetative parts and early seed stages, when insect damage is maximal. As the seeds mature, the trypsin inhibitor will be naturally processed to the inactive storage protein that is safe for consumption.

Chrysanthemum

Cysteine proteinase inhibitor from potato (multicystatin, *PMC*) has been expressed in chrysanthemum, *Dendranthema grandiflorum* (Ramat.) at a level of 0.13% of total protein (Seetharam et al., 2002). No correlation between reduction in oviposition rate by the western flower thrips, *Frankliniella occidentalis* (Perg.) and *PMC* expression could be established, which may be due to the relatively low expression level of *PMC* in chrysanthemum.

Alpha Amylase Inhibitors

Six different types of α-amylase inhibitors: lectin-like, knottin-like, cereal-type, Kunitz-like, gamma-purothionin-like, and thaumatin-like, can be used in pest control (Franco et al., 2002). These inhibitors show remarkable structural diversity leading to different modes of inhibition and different specificity profiles against diverse α-amylases. Specificity of inhibition is an important issue as the introduced inhibitor should affect neither the plant's own α-amylases nor the nutritional value of the crop. Of particular interest are some bifunctional inhibitors with additional favorable properties such as proteinase inhibitory activity or chitinase activity. Alpha-amylase inhibitors are attractive candidates for the control of seed weevils as these are highly dependent on starch as an energy source. Insect α-amylases (α-1, 4-glucan-4-glucanohydrolases, EC 3.2.1.1) constitute a family of endoamylases that catalyze the hydrolysis of α-1,4-glycosidic linkages in starch components, glycogen, and other carbohydrates. The enzyme plays a key role in carbohydrate metabolism of microorganisms, plants, and animals. Several insects, especially those similar to weevils that feed on starchy seeds during larval and/or adult stages, depend on their α-amylases for survival. The α-amylase inhibitors from *Phaseolus vulgaris* L. seeds are detrimental to the development of Mexican bean weevil, *Callosobruchus maculatus* (Fab.) (Ishimoto and Chrispeels, 1996). Amylase inhibitors from pigeonpea inhibit 22% amylase activity in *H. armigera* (Giri and Kachole, 1998). Amylase and protease inhibitors in artificial diet increased insect mortality and showed adverse effects on growth and development of larvae.

Transgenic tobacco plants expressing amylase inhibitors from wheat (*WAAI*) increase the mortality of the lepidopteran larvae by 30 to 40% (Carbonero et al., 1993), and those from bean (*BAAI*) to *Callosobruchus* spp. (Shade et al., 1994; Schroeder et al., 1995) (Table 7.4). Enhanced levels of resistance to the bruchids have also been obtained in seeds of transgenic adzuki beans with α-amylase gene (Ishimoto et al., 1996). Alpha-amylase inhibitors (*alpha Al-1* and *alpha Al-2*) are effective in protecting peas from the weevil damage under field conditions (Morton et al., 2000). *Alpha Al-1* inhibits pea bruchid, *Bruchus pisorum* L. α-amylase by 80%, while *alpha Al-2* inhibits the enzyme by 40%. *Alpha Al-2* delays the maturation of the larvae, while alpha *Al-1* results in larval mortality.

TABLE 7.4

Insect-Resistant Transgenic Crops Expressing α-Amylase Inhibitors

Crop	Inserted Gene	Useful Trait	References
Pea	α-amylase inhibitors from *Phaseolus vulgaris*	Resistant to bruchids.	Shade et al. (1994)
Tobacco	α-amylase inhibitors from wheat (WAAI)	Increased the mortality of lepidopteran larvae by 30 to 40%.	Carbonero et al. (1993)
	α-amylase inhibitors from bean (*BAAI*)	Resistant to *Callosobruchus* sp.	Shade et al. (1994); Schroeder et al. (1995)
Adzuki beans	α-amylase	Enhanced levels of resistance to bruchids.	Ishimoto et al. (1996)
Pea	α-amylase inhibitors (*alpha Al-1*, and *alpha Al-2*)	*alpha Al-1* inhibits pea bruchid α-amylase by 80%.	Morton et al. (2000)

Enzymes

Several enzymes expressed in transgenic plants have shown resistance to lepidopteran insects (Purcell et al., 1993; Smigocki et al., 1993; Corbin et al., 1994; X. Ding et al., 1998). Cholesterol oxidase from *Streptomyces* is highly toxic to cotton boll weevil, *Anthonomus grandis* (Boh.) (Cho et al., 1995), while polyphenol oxidases and peroxidases increase the inhibitory effect of 5CQA (5-caffeoyl quinic acid) and cholorogenic acid by oxidizing the dihydroxy groups to ubiquinones that covalently bind to nucleophilic (–SH2 and –NH2) groups of proteins, peptides, and amino acids. Behle et al. (2002) produced transgenic plants of tomato, *L. esculentum*, and South American tobacco, *Nicotiana sylvestris* Speg. & Comes. and *N. tabacum* expressing 5 to 400 times higher peroxidase activity than corresponding tissues of wild-type plants. The larvae of *H. zea* consumed 1.5 times less food on leaf discs from transgenic plants compared with the leaf discs from the corresponding nontransgenic plants. Lipoxigenase from soybean has also been shown to exhibit toxic effects towards insects (Shukle and Murdock, 1983), and has been expressed in transgenic plants, but resistance to insects has not been demonstrated.

The insect moulting enzyme chitinase, which degrades chitin to low-molecular-weight soluble and insoluble oligosaccharides, is a metabolic target of selective pest control (Chit Kramer and Muthukrishnan, 1997). The cDNA and genomic clones for the chitinase from the hornworm, *M. sexta* have been isolated and characterized. Transgenic plants that express hornworm chitinase constitutively exhibit resistance to insects. Transgenic tobacco plants expressing the chitinase gene have shown resistance to several lepidopteran insects (X. Ding et al., 1998). Accumulation of bean chitinase (*BCH*) increased as the potato plant developed, with maximum levels recorded in mature plants (Down et al., 2001). A transformed entomopathogenic virus that produces the enzyme chitinase displayed enhanced insecticidal activity. Chitinase also potentiates the efficacy of the toxin from *B. thuringiensis*. Field studies involving alfalfa, *Medicago sativa* L., plants (parental, transgenic alpha-amylase-producing and transgenic lignin peroxidase-producing plants) indicated that the lignin peroxidase transgenic plants had significantly lower shoot weight, and higher nitrogen and phosphorus content, than the parental or transgenic amylase plants (Donegan et al., 1999). Significantly higher population levels of culturable, aerobic spore forming, and cellulose-utilizing bacteria, lower activity of the soil enzymes dehydrogenase and alkaline phosphatase, and higher soil pH levels were also associated with the lignin peroxidase transgenic plants.

Plant Lectins

Plant lectins are a heterogeneous group of sugar binding proteins, which have a protective function against a range of organisms. Plant lectins are particularly effective against the sap sucking Hemiptera (Shukle and Murdock, 1983; Czapla and Lang, 1990; Powell et al., 1993, 1995; Hilder et al., 1995) (Table 7.5). Lectins from *Canavalia ensiformis* (L.) DC (*Concanavalin A*), *Amaranthus caudatus* L., *Lens culinaris* Medik., and *Galanthus nivalis* L. induced significant mortality in *Acyrthosiphon pisum* (Harris) (Rahbe et al., 1995). Snowdrop, garlic, and chickpea lectins have shown adverse effects on survival, growth, and development of *H. armigera* (Shukla, Arora, and Sharma, 2005; H.C. Sharma et al., 2005). Many mannose-binding lectins were toxic to *A. pisum*. Concanavalin A has also been tested against *Aphis gossypii* Glover, *Aulacorthum solani* Kalten., *Macrosiphum euphorbiae* (Thomas), *Macrosiphum albifrons* Essig, and *M. persicae* at concentrations between 10 and 1500 μg mL^{-1}. Mortality was quite variable from one species to another. No part of the digestive tract contained detectable amounts of endoprotease activity (Rahbe et al., 1995). Plant lectins have shown biological activity against a range of insects. However, consideration should be given with regard to their deployment in transgenic plants because of their known toxicity to mammals and humans.

Tobacco

The gene for snowdrop lectin (*GNA*) expressed in tobacco plants has shown enhanced resistance to peach potato aphid, *M. persicae*, and pea lectin in tobacco to *H. virescens* (Boulter et al., 1990). Greater insecticidal activity has been observed in chitin binding lectins and the lectin gene in wheat germ and common bean. Fall armyworm, *S. frugiperda*

TABLE 7.5

Insect-Resistant Transgenic Crops Expressing Plant Lectin Genes

Crop	Inserted Gene	Useful Trait	References
Tobacco	GNA from *Galanthus nivalis*	Resistant to peach potato aphid, *Myzus persicae*.	Boulter et al. (1990)
	Pea lectin	Resistant to *Heliothis virescens*.	Boulter et al. (1990)
Cotton	GNA	Reduced larval weight of cotton bud worm, *Heliothis virescens*.	Satyendra, Stewart, and Wilkins (1998)
Rice	GNA	Resistant to brown planthopper, *Nilaparvata lugens* and green leafhopper, *Nephotettix virescens*.	C.D. Yang et al. (1998)
		Resistant to potato leafhopper, *Empoasca fabae*.	Habibi, Backus, and Czapla (1992)
Potato	GNA and ConA	Less susceptible to peach potato aphid, *Myzus persicae*.	Gatehouse et al. (1999)
Tomato	GNA	Reduction in larval biomass of tomato moth, *Leacanobia oleracea*.	Fitches, Gatehouse, and Gatehouse (1997)
Maize	Wheat agglutinin (WGA)	Moderate activity against *Ostrinia nubilalis* and *Diabrotica* sp.	Maddock et al. (1991)
Mustard	WGA	Antibiotic, antifeedant, and insecticidal to mustard aphid, *Lipaphis erysimi*.	Kanrar et al. (2002)
Sugarcane	GNA	Increased antibiosis to larvae of sugarcane grubs, *Antitrogus consanguineus*.	Nutt et al. (1999)

larvae fed on transformed plants of tobacco expressing the protein *Tarin 1* from *Colocasia esculenta* (L.) Schott. exhibited retarded growth, and reduced pupation and biomass, and increased larval mortality (Leal Bertioli et al., 2003). *Tarin 1* also inhibited the growth of *Pseudomonas syringae* van Hall (pv. tomato) under *in vitro* conditions. Root damage by *Meloidogyne javanica* (Treub.) was greater in the control plants than in the transformed plants, but the results were not statistically significant (Leal Bertioli et al., 2003).

Potato

Transgenic potatoes expressing *GNA* and concanavalin A (*ConA*) were less susceptible to peach-potato aphid, *M. persicae* (Gatehouse et al., 1995, 1996, 1999). Larval biomass of the tomato moth, *Lecanobia oleracea* (L.) is reduced in artificial diet containing *GNA*, and on excised leaves of transgenic tomato (Fitches, Gatehouse, and Gatehouse, 1997), which may result in lower fecundity of the female moths. Accumulation of the transgene product in transformed potato plants expressing GNA increased as the potato plant developed, with maximum levels found in mature plants (Down et al., 2001). The variation in accumulation of GNA in transgenic plants within a line of clonal replicates was correlated with expression of resistance to the larvae of tomato moth, *L. oleracea*. Growing conditions affected the levels of GNA expression in transgenic plants and the expression of resistance to *L. oleracea* (Down et al., 2001).

Cotton

Larvae of cotton budworm, *H. virescens*, fed on transformed cotton with the lectin gene have a reduced weight, but there was no effect on larval survival (Satyendra, Stewart, and Wilkins, 1998).

Cereals

Transgenic maize expressing wheat agglutinin has shown moderate activity against *O. nubilalis* and *Diabrotica* sp. (Maddock et al., 1991). Snowdrop lectin at levels greater than 0.04% decreases the fecundity but not the survival of the grain aphid, *Sitobion avenae* (F.) (Stoger et al., 1999). Transgenic haploid rice shoots with *GNA* have shown resistance to brown planthopper, *N. lugens*, green planthopper, *Nephotettix virescens* (Distant) (C.D. Yang et al., 1998; Tinjuangjun et al., 2000; Tang et al., 2001), rice small brown planthopper, *Laodelphax striatellus* (Fallen) (Sun, Wu, and Tang, 2002), and potato leafhoppers, *Empoasca fabae* (Harris) (Habibi, Backus, and Czapla, 1992). In plants where *GNA* expression is tissue-specific (phloem and epidermal layer), or constitutive, the green planthopper survival has been reduced by 23 and 53%, respectively (Foissac et al., 2000). The brown planthopper nymphs tended to avoid plants expressing *GNA*, and avoidance was less pronounced and took a longer time to develop on plants where *GNA* expression was tissue specific. In contrast to brown planthopper, the green planthopper nymphs were attracted to plants expressing *GNA*.

Sugarcane

Transgenic sugarcane plants engineered to express *GNA* have shown increased antibiosis to larvae of sugarcane grubs, *A. consanguineus* (Nutt et al., 1999). Larvae feeding on the roots of transgenic sugarcane plants gained only 20.6% of the weight of controls (Nutt et al., 2001).

Addition of transgenic sugarcane tissue with *GNA* into artificial diet enhanced larval growth in sugarcane borer, *D. saccharalis*, resulting in higher larval and pupal weight compared with a diet with nontransgenic sugarcane, but this effect was not observed in the second generation (Setamou et al., 2002). In contrast, larval survival, percentage adult emergence, and female fecundity of *Eoreuma loftini* (Dyar) were significantly reduced when fed a transgenic sugarcane diet compared with a nontransgenic sugarcane diet.

Vegetables

Wheat germ agglutinin (*WGA*) is antimetabolic, antifeedant, and insecticidal to the mustard aphid, *Lipaphis erysimi* (Kalt.) (Kanrar et al., 2002). Bioassays using leaf discs showed that feeding on transgenics plants induced high mortality and significantly reduced fecundity of aphids. However, mammalian toxicity of this lectin is high, and it may not be a good candidate for use in genetic transformation. The *gna* expressed in transgenic Chinese cabbage (plants 282-3 and 282-9) was inherited in simple Mendelian fashion, and showed resistance to *M. persicae* (G.D. Yang et al., 2003).

Viruses, Neurotoxins, and Insect Hormones

Neurotoxins

Spiders and scorpions produce powerful neurotoxins that have been expressed in transgenic organisms (Barton and Miller, 1991). Jiang et al. (1995) constructed the gene expressing the venom of an Australian spider, *Androctonus australis* Hector, by annealing partially complementary single-stranded oligonucleotide using codes preferred in plants. This gene was then introduced into a plasmid for expression in plants. Transgenic plants of tobacco have been obtained containing an insecticidal spider peptide gene, and some of these plants have shown resistance to *H. armigera* (Jiang, Zhu, and Chen, 1996). The role of neurotoxins from insects and spiders needs to be studied in greater detail before they are deployed in other organisms and plants because of their possible toxicity to mammals.

Viruses

Enhancin genes from *T. ni* or *H. armigera* baculoviruses introduced into tobacco plants resulted in slower development and increased larval mortality on some transgenic lines (Cao et al., 2002). The majority of the transgenic plants had little or no inhibitory effect. A recombinant *Autographa californica* (Speyer) nucleopolyhedrovirus (AcNPV) with the enhancin gene from *T. ni* granulovirus, *AcEnh26*, expressed in tobacco plants showed a 10-fold increase in AcNPV infection (Hayakawa et al., 2000). Transgenic tobacco transformed with the *T. ni* granulovirus enhancin gene has also been shown to enhance baculovirus infection in the larvae (Hayakawa et al., 2004). Long-term feeding of lyophilized transgenic tobacco material with the enhancin gene resulted in adverse effects on the larvae of *Mythimna separata* (Walk.) and *S. exigua*. The results suggested that baculovirus enhancin gene products have potential for use in pest management.

Neuropeptides and Peptidic Hormones

Neuropeptides and small peptidic hormones are another interesting class of molecules that can be deployed as possible insecticides through transgenic plants (Tortiglione et al.,

1999; Altstein et al., 2000). These molecules regulate several physiological processes in insects and are active at very low concentrations. An alteration of their titer in the insect could cause severe functional modifications. There are several examples of neuropeptides encoded by a single gene coding for multiple copies of one or more peptides. Backbone cyclic (*BBC*) neuropeptide-based antagonists (*NBA*) has been applied to the insect pyrokinin/ pheromone biosynthesis activating neuropeptide (*PBAN*) family. It has led to the discovery of potent antagonists and metabolically stable peptidomimetitic antagonists devoid of agonistic activity, which inhibited *PBAN*-mediated activities in moths *in vivo* (Altstein et al., 2000). There are possibilities for deploying these molecules through transgenic plants to disrupt physiological processes of insects. Transgenic tobacco plants have been produced expressing a precursor of a regulatory peptide from *Aedes aegypti* (L.) (Trypsin Modulating and Oostatic Factor, *Aea-TMOF*), which interferes with the development of tobacco budworm, *M. sexta* larvae, spaced by dibasic residues, *Arg-Arg*, as potential post-translational cleavage sites (Tortiglione et al., 2002). Peptide extracts from transgenic plants had *TMOF* activity and inhibited *in vitro* the biosynthesis of serine proteases. This activity was consistently present in T_1 plants and absent in control plants. Tobacco budworm *H. virescens* larvae fed on leaves from the transgenic plants showed a reduced growth rate compared to those fed with control plants.

The potential for *GNA* to act as a carrier protein to deliver an insect neuropeptide from *M. sexta* allatostatin (*Manse-AS*) to the hemolymph of lepidopteran larvae has been examined by expressing a *GNA/Manse-AS* fusion protein (*FP*) in *E. coli* and feeding purified *FP* to larvae of the tomato moth, *L. oleracea* (Fitches et al., 2002). The fusion protein administered at 1.5 or 0.5% of dietary proteins strongly inhibited feeding and prevented the growth of fifth-instar larvae, whereas neither *GNA* nor *Manse-AS* alone, nor a mixture of *GNA* and *Manse-AS* in control treatments, had deleterious effects at similar levels. Elevated levels of material reacting with anti-*Manse-AS* antibodies were detected in the hemolymph of insects fed with diets containing *FP*, suggesting that transport of the peptide occurred (Fitches et al., 2002). Evidence for the delivery of intact *FP* to the hemolymph was provided by the co-elution of *Manse-AS*-like immunoreactivity with standard *FP* after size exclusion chromatography of hemolymph from *FP*-fed larvae. *GNA/Manse-AS* and similar fusion proteins offer a novel and effective strategy for delivering insect neuropeptides by oral administration, which could be used in conjunction with expression in transgenic plants for pest management (Fitches et al., 2002).

Gamma-aminobutyrate (*GABA*), which accumulates in plants in response to biotic and abiotic stresses via activation of glutamate decarboxylase, acts as an inhibitory neurotransmitter in insect pests (Shelp, van Cauwenberghe, and van Bown, 2003). Ingested *GABA* disrupts nerve functioning and causes damage to oblique-banded leaf roller, *Choristoneura rosaceana* (Harris) and tobacco budworm, *H. virescens*. Feeding by tobacco budworm and oblique-banded leaf roller larvae stimulated *GABA* accumulation in soybean and tobacco. Elevated levels of endogenous *GABA* in genetically engineered tobacco deterred feeding by tobacco budworm larvae and infestation by the northern root-knot nematode, *Meloidogyne hapla* Chitwood. Genetically engineered plants overexpressing glutamate decarboxylase and having high *GABA*-producing potential have potential for use in integrated pest management (Shelp, van Cauwenberghe, and van Bown, 2003).

Biotin-Binding Proteins

Biotin-binding proteins (*BBPs*), such as avidin and streptavidin, represent potent insect control compounds, which could be delivered via transgenic plants (Malone et al., 2002).

Black field cricket nymphs, *Teleogryllus commodus* (Walker) had significantly reduced growth and survival when fed on lettuce leaves painted with purified avidin. Clover root weevils, *Sitona lepidus* Gyllen., adults were unharmed when fed clover foliage painted with avidin. In contrast, neonate or one-week-old *S. lepidus* larvae had poor survival when fed on artificial diets containing avidin or streptavidin. Neonate larvae of Argentine stem weevils, *Listronotus bonariensis* (Kuschel) had significantly reduced survival when fed with artificial diet containing streptavidin or avidin. Slugs, *Deroceras reticulatum* (Muller), and snails, *Cantareus aspersus* (Muller), were not harmed when fed with avidin-painted lettuce. Similar numbers of eggs were laid and galls produced by the root-knot nematodes, *M. javanica* (Treub) Chitwood, *M. hapla*, and *M. incognita* (Kofoid and White) Chitwood inoculated onto transgenic tobacco plants expressing avidin and nontransgenic controls (Malone et al., 2002). Transgenic tobacco expressing avidin from 3.1 to 4.6 mM, and from 1.9 to 11.2 mM in apple; and streptavidin from 11.4 to 24.5 mM in tobacco and from 0.4 to 14.6 mM in apple (biotin-binding proteins) conferred a high level of insect resistance in transformed tobacco plants to potato tuber moth, *P. operculella*, and in apple plants to light brown apple moth, *Epiphyas postvittana* (Walker) (Markwick et al., 2003). More than 90% of potato tuber moth larvae died on tobacco plants expressing either avidin or streptavidin genes within nine days of inoculation. Mortality of light brown apple moth larvae was greater on three avidin-expressing (89.6 to 80.1%) and two streptavidin-expressing (90 to 82.5%) apple plant lines than on nontransformed control plants (14.1%) after 21 days. Weight of light brown apple moth larvae was also significantly reduced by feeding on apple shoots expressing avidin and streptavidin at >3.8 mM (Markwick et al., 2003).

Antibodies

Genes that are based on antibody technology can also be exploited for genetic transformation of crop plants (Hilder and Boulter, 1999). Single chain antibodies (ScFvs) can be used to block the function of essential insect proteins, which serve as control agents against nematodes, pathogens, and viruses (Van Engelen et al., 1994; Rosso et al., 1996). This approach of controlling insects would offer the advantage of allowing some degree of selection for specificity effects, so that insect pests, but not the beneficial organisms are targeted. The development of a delivery system from transgenic plants to the insect hemolymph will remove a key constraint in the transgenic approach to crop protection. Mi-mediated resistance is developmentally regulated, and protects mature plants but not seedlings against the aphid, *Macrosiphum euphorbiae* (Thomas) infestation (Goggin et al., 2004). Mi-1.2 is transcribed in the leaves prior to the onset of aphid resistance, and the transcript levels are comparable in seedling and flowering stages. Constitutive overexpression of Mi-1.2 in transgenic plants did not hasten the onset of aphid resistance in seedlings or boost the level of resistance observed in flowering plants. The results suggested that Mi-1.2 transcription levels did not modulate aphid resistance in tomato leaves (Goggin et al., 2004).

Inducible Resistance

Mechanical wounding and insect damage result in transient increase in activity of polyphenol oxidase (Dhankher and Gatehouse, 2003). However, there is no systemic induction of this enzyme following wounding, insect damage, or application of methyl jasmonate.

The activation of jasmonic acid biosynthesis by cell wall elicitors, the peptide systemin and other compounds is shown to relate to the function of jasmonates in plants (Creelman and Mullet, 1997). Jasmonate can modulate gene expression at the level of translation, RNA processing, and transcription (Creelman and Mullet, 1997). Prosystemin, a compound biologically active as systemin (Ryan and Pearce, 1998), when assayed for proteinase inhibitor induction in young tomato plants, has been found to be active in the alkalinization response in cultured cells (Dombrowski, Pearce, and Ryan, 1999). Prosystemin or large fragments of prosystemin can be an active inducer of defense responses in tomato leaves. However, *M. sexta* larvae feeding on tomato plants constitutively expressing a prosystemin antisense gene had approximately three times higher growth rates than larvae feeding on nontransformed control plants (Orozco-Cardenas, McGurl, and Ryan, 1993). Prosystemin mRNA levels in antisense and control plants were correlated with levels of proteinase inhibitor I and II protein levels after 6 and 12 days of larval feeding, indicating that plant resistance to insects can be modulated by genetically engineering a gene encoding a component of the inducible systemic signaling system regulating a plant defensive response. Despite several reports on successful protection of plants and trees against phytophagous insects, defense strategies based on protease inhibitor expression in plants have not resulted in any commercial applications so far. This could be due to the insects' capacity to react to protease inhibitors, and the protease inhibitor expression levels in transgenic plants. Use of the bacterial isopentenyl transferase (*ipt*) gene, involved in cytokinin biosynthesis (fused proteinase inhibitor II [*PI-IIK* gene]) in *N. plumbaginifolia* reduces *M. sexta* feeding by 70% (Smigocki et al., 1993), and retards the development of peach potato aphid, *M. persicae*. Zeatin and zeatin-riboside levels in leaves remaining on *PI-II-ipt* plants after hornworm feeding are elevated by about 70-fold. Exogenous application of zeatin to the *PI-II-ipt* leaves enhanced the level of resistance to the tobacco hornworm and completely inhibited the normal development of the green peach aphid. Jin et al. (2000) developed transgenic cabbage with *cry1Ab3* gene under the control of the soybean wound-inducible *vspB* promoter. Cabbage plants with *cry1Ab3* gene were resistant to larvae of the diamondback moth, *P. xylostella*, whereas plants with wild-type *Bt* gene were susceptible.

Gene Pyramiding

Combining conventional host plant resistance with *Bt* genes can provide a germplasm base to achieve durable resistance to insect pests (Table 7.6). Westedt et al. (1998) examined Lemhi Russet and two lines with resistance to *P. operculella*. USDA8380-1 (leaf leptines)

TABLE 7.6

Gene Pyramiding to Increase the Effectiveness of Transgenic Plants for Pest Management

Transgene(s)	Remarks	References
Serine PIs + *Bt*	Enhances activity of *Bt* genes.	MacIntosh et al. (1990)
Tannic acid + *Bt*	Increases activity of *Bt* genes.	Gibson et al. (1995)
cry1Ac + *CpTi*	More effective than plants expressing *Bt* gene alone.	Zhao et al. (1998)
cry1Ac + *cry1C*	Plants producing both toxins caused rapid mortality.	Cao et al. (2002)
cry1Ac + *cry1Ab*	Showed greater insecticidal activity.	Stewart et al. (2001)

and L235-4 (glandular trichomes), along with the *cryV-Bt*-transgenic lines of each genotype, were tested for resistance to potato tuber moth, *P. operculella*. Nearly 54% mortality was observed when first-instar larvae were fed on the leaves of USDA8380-1. High levels of expression occurred in *cryV-Bt* transgenic lines, with up to 96% mortality. Walker et al. (2002) combined a QTL conditioning corn earworm resistance in soybean PI 229358, and *cry1Ac* transgene from the recurrent parent Jack-*Bt* into BC_2F_3 plants by marker-assisted selection. The segregating individuals were genotyped and SSR markers linked to an antibiosis/antixenosis QTL on linkage group M, and were tested for the presence of Cry1Ac. Few larvae of corn earworm and soybean looper survived on leaves expressing the Cry1Ac protein. Though not as great as the effect of Cry1Ac, the PI 229358-derived LG M QTL also had a detrimental effect on larval weights of both species, and on defoliation by corn earworm, but did not reduce defoliation by soybean looper, indicating that combining transgene- and QTL-mediated resistance to lepidopteran pests may be a viable strategy for insect control.

Transgenic plants of cotton with *Bt* + *GNA* conferred resistance to bollworm, *H. armigera* and the cotton aphid, *A. gossypii* (Z. Liu et al., 2003). Segregation of resistant and susceptible plants in F_2 and BC_1 populations fitted the 3:1 and 1:1 ratios, respectively, indicating that resistance to bollworm was controlled by one pair of dominant genes and inherited in Mendelian manner with no cytoplasmic effects. Allelic tests showed that the resistance gene in transgenic line TBG and that of insect-resistant transgenic cotton strains Zhongxin 94, R 19, Shanxi 94-24, Shuangkang 1, and Xinmian 33B might be inserted in different chromosomes. The *Bt* gene and *OC* gene have been co-transformed to tobacco chloroplast with the particle bombardment method. Transgenic tobacco containing both genes had enhanced toxicity to the larvae of corn earworm, *H. zea*, as compared to plants containing only the *Bt* or *OC* gene alone. The *Bt* and *OC* genes were inherited maternally (Su et al., 2002). Sugarcane plants transformed with both the GNA and a proteinase inhibitor from *Nicotiana alata* Link and Oho. caused a significant negative effect on the weight gain of *Dermolepida albohirtum* (Waterhouse) larvae (Nutt et al., 2001). Transformed entomopathogenic virus that produces chitinase displayed enhanced insecticidal activity (Chit Kramer and Muthukrishnan, 1997). Chitinase also potentiates the efficacy of the toxin from *B. thuringiensis*.

Gahakwa et al. (2000) studied the stability of transgene expression in 40 independent rice plants representing 11 diverse cultivated varieties. Each line contained three or four different transgenes delivered by particle bombardment, either by co-transformation, or in the form of a co-integrate vector. Approximately 75% of the lines showed Mendelian inheritance of all transgenes, suggesting integration at a single locus. The levels of transgene expression varied among different lines, but primary transformants showing high-level expression of the *GNA*, *gusA*, *hpt*, and *bar* genes were transmitted to the progeny. Six transgenes (three markers and three insect resistance genes) were stably expressed over four generations of transgenic rice plants. Transgene expression was stable, and this represented a step toward genetic engineering from model varieties to elite breeding lines grown in different parts of the world.

Conclusions

The ideal transgenic technology should be commercially viable, environmentally benign (biodegradable), easy to use in diverse agroecosystems, and have a wide spectrum of activity against the target insect pests. It should also be harmless to the natural enemies and nontarget organisms, target the sites in insects that have developed resistance to the conventional insecticides, be flexible enough to allow ready deployment of alternatives (if and when the resistance is developed in insect populations), and preferably produce acute rather than chronic effects on the target insects. Some of the criteria can be achieved by exploiting genes that are based on antibody technology. Single chain antibodies can be used to block the function of essential insect proteins. The development of a delivery system for toxins from the transgenic plants to the insect hemolymph will remove a key constraint in the transgenic approach to crop protection. Incorporation of insecticidal genes in crop plants will have a tremendous effect on pest management. We need to pursue a management strategy that reflects the insect biology, insect-plant interactions, and their influence on the natural enemies to prolong the life span of transgenic crops. Emphasis should also be placed on combining exotic genes with conventional host plant resistance, and also with traits conferring resistance to other insect pests and diseases of importance in a crop in the target region. Several genes conferring resistance to insects can also be deployed as multilines or synthetics. While several crops with commercial viability have been transformed in the developed world, very little has been done to use this technology to increase food production in the harsh environments of the tropics. There is a need to use these tools for providing resistance to insects in cereals, legumes, and oil seed crops that are a source of sustenance for poorer sections of the society. Equally important is the need to follow the biosafety regulations and make this technology available to farmers who cannot afford the high cost of seeds and chemical pesticides. International research centers, advanced research institutions, and the national agricultural research systems need to play a major role in promoting biotechnology for food security of poor people in the developing countries.

References

Alam, M.F., Datta, K., Abrigo, E., Oliva, N., Tu, J., Virmani, S.S. and Datta, S.K. (1999). Transgenic insect-resistant maintainer line (IR68899B) for improvement of hybrid rice. *Plant Cell Reports* 18: 572–575.

Altpeter, F., Diaz, I., McAuslane, H., Gaddour, K., Carbonero, P. and Vasil, I.K. (1999). Increased insect resistance in transgenic wheat stably expressing trypsin inhibitor *CMe*. *Molecular Breeding* 5: 53–63.

Altstein, M., Ben-Aziz, O., Schefler, I., Zeltser, I. and Gilon, C. (2000). Advances in the application of neuropeptides in insect control. *Crop Protection* 19: 547–555.

Archer, T.L., Schuster, G., Patrick, C., Cronholm, G., Bynum, E.D. Jr. and Morrison, W.P. (2000). Whorl and stalk damage by European and Southwestern corn borers to four events of *Bacillus thuringiensis* transgenic maize. *Crop Protection* 19: 181–190.

Archetti, S., Delledonne, M., Fogher, C., Chiaba, C., Chiesa, F., Savazzini, F. and Giordano, A. (2000). Soybean Kunitz, C-II and PI-IV inhibitor genes confer different levels of insect resistance to tobacco and potato transgenic plants. *Theoretical and Applied Genetics* 101: 519–526.

Arencibia, A., Vazquez, R.I., Prieto, D., Tellez, P., Carmona, E.R., Coego, A., Hernandez, L., de la Riva, G.A. and Selman-Housein, G. (1997). Transgenic sugarcane plants resistant to stem borer attack. *Molecular Breeding* 3: 247–255.

Arencibia, A.D., Carmona, E.R., Cornide, M.T., Castiglione, S., O'Relly, J., Chinea, A., Oramas, P. and Sala, F. (1999). Somaclonal variation in insect-resistant transgenic sugarcane (*Saccharum* hybrid) plants produced by cell electroporation. *Transgenic Research* 8: 349–360.

Armstrong, C.L., Parker, G.B., Pershing, J.C., Brown, S.M., Sanders, P.R., Duncan, D.R., Stone, T., Dean, D.A., DeBoer, D.L., Hart, J., Howe, A.R., Morrish, F.M., Pajeau, M.E., Peterse, W.L., Reich, B.J., Rodriguez, R., Santino, C.G., Sato, S.J., Schuler, W., Sims, S.R., Stehling, S., Tarochione, L.J. and Fromm, M.E. (1995). Field evaluation of European corn borer control in progeny of 173 transgenic corn events expressing an insecticidal protein from *Bacillus thuringiensis. Crop Science* 35: 550–557.

Arpaia, S., De Marzo, L., Di Leo, G.M., Santoro, M.E., Mennella, G. and Vanloon, J.J.A. (2000). Feeding behaviour and reproductive biology of Colorado potato beetle adults fed transgenic potatoes expressing the *Bacillus thuringiensis* Cry3B endotoxin. *Entomologia Experimentalis et Applicata* 95: 31–37.

Arpaia, S., Mennella, G., Onofaro, V., Perri, E., Sunseri, F. and Rotino, G.L. (1997). Production of transgenic eggplant (*Solanum melongena* L.) resistant to Colorado potato beetle (*Leptinotarsa decemlineata* Say). *Theoretical and Applied Genetics* 95: 329–334.

Bambawale, O.M., Singh, A., Sharma, O.P., Bhosle, B.B., Lavekar, R.C., Dhandapani, A., Kanwar, V., Tanwar, R.K., Rathod, K.S., Patnge, N.R. and Pawar, V.M. (2004). Performance of *Bt* cotton (MECH 162) under integrated pest management in farmers' participatory field trial in Nanded district, Central India. *Current Science* 86: 1628–1633.

Bansal, K.C. and Sharma, R.K. (2003). Chloroplast transformation as a tool for prevention of gene flow from GM crops to weedy or wild relatives. *Current Science* 84: 1286–1287.

Barton, K.A. and Miller, M.J. (1991). Insecticidal toxins in plants. *European Patent* 0431829.

Barton, K., Whiteley, H. and Yang, N.S. (1987). *Bacillus thuringiensis* δ-endotoxin in transgenic *Nicotiana tabacum* provides resistance to lepidopteran insects. *Plant Physiology* 85: 1103–1109.

Bechtel, D.B. and Bulla, L.A. Jr. (1976). Electron microscope study of sporulation and parasporal crystal formation in *Bacillus thuringiensis. Journal of Bacteriology* 127: 1472–1483.

Behle, R.W., Dowd, P.F., Tamez Guerra, P. and Lagrimini, L.M. (2002). Effect of transgenic plants expressing high levels of a tobacco anionic peroxidase on the toxicity of *Anagrapha falcifera* nucleopolyhedrovirus to *Helicoverpa zea* (Lepidoptera: Noctuidae). *Journal of Economic Entomology* 95: 81–88.

Benedict, J.H., Sachs, E.S., Altman, D.W., Deaton, D.R., Kohel, R.J., Ring, D.R. and Berberich, B.A. (1996). Field performance of cotton expressing CryIA insecticidal crystal protein for resistance to *Heliothis virescens* and *Helicoverpa zea* (Lepidoptera: Noctiudae). *Journal of Economic Entomology* 89: 230–238.

Benfey, P.N. and Chua, N.H. (1989). Regulated genes in transformed plants. *Science* 244: 174–181.

Benfey, P.N. and Chua, N.H. (1990). The cauliflower mosaic virus 35S promotor combinatorial regulation of transcription in plants. *Science* 250: 959–966.

Bergvinson, D., Willcox, M.N. and Hoisington, D. (1997). Efficacy and deployment of transgenic plants for stem borer management. *Insect Science and Its Application* 17: 157–167.

Berliner, E. (1915). Uber die Schalffsuchi der Mehlmottenraupo (*Ephestia kuhniella* Zell.) und thren Erreger, *Bacillus thuringiensis* n. sp. *Zietschrift fur Angewabdte Entomolgie* 2: 29–56.

Bhalla, R., Dalal, M., Panguluri, S.K., Jagadish, B., Mandaokar, A.D., Singh, A.K. and Kumar, P.A. (2005). Isolation, characterization and expression of a novel vegetative insecticidal protein gene of *Bacillus thuringiensis. FEMS Microbiology Letters* 243: 467–472.

Bhattacharya, R.C., Viswakarma, N., Bhat, S.R., Kirti, P.B. and Chopra, V.L. (2002). Development of insect-resistant transgenic cabbage plants expressing a synthetic cryIA(b) gene from *Bacillus thuringiensis. Current Science* 83: 146–150.

Birch, R.G. (1997). Plant transformation: Problems and strategies for practical application. *Annual Review of Plant Physiology and Molecular Breeding* 48: 297–326.

Bohorova, N., Zhang, W., Julstrum, P., McLean, S., Luna, B., Brito, R.M., Diaz, L., Ramos, M.E., Estanol, P. and Pacheco, M. (1999). Production of transgenic tropical maize with *cryIAb* and *cryIAc* genes via microprojectile bombardment of immature embryos. *Theoretical and Applied Genetics* 99: 437–444.

Boulter, D. (1993). Insect pest control by copying nature using genetically engineered crops. *Phytochemistry* 34: 1453–1466.

Boulter, D., Edwards, G.A., Gatehouse, A.M.R., Gatehouse, J.A. and Hilder, V.A. (1990). Additive protective effects of incorporating two different higher plant derived insect resistance genes in transgenic tobacco plants. *Crop Protection* 9: 351–354.

Brown, D.G. (1988). The cost of plant defense: An experimental analysis with inducible proteinase inhibitors in tomato. *Oecologia* 76: 467–470.

Canedo, V., Benavides, J., Golmirzaie, A., Cisneros, F., Ghislain, M. and Lagnaoui, A. (1999). Assessing Bt-transformed potatoes for potato tuber moth, *Phthorimaea operculella* (Zeller), management. In *Impact on a Changing World. International Potato Center Program Report 1997–1998*. Lima, Peru: International Potato Center (CIP), 161–169.

Cao, J., Ibrahim, H., Garcia, J.J., Mason, H., Granados, R.R. and Earle, E.D. (2002). Transgenic tobacco plants carrying a baculovirus enhancin gene slow the development and increase the mortality of *Trichoplusia ni* larvae. *Plant Cell Reports* 21: 244–250.

Cao, J., Tang, J.D., Strizhov, N., Shelton, A.M. and Earle, E.D. (1999). Transgenic broccoli with high levels of *Bacillus thuringiensis* Cry1C protein control diamondback moth larvae resistant to Cry1A or Cry1C. *Molecular Breeding* 5: 131–141.

Carbonero, P., Royo, J., Diaz, I., Garcia-Maroto, F., Gonzalez-Hidalgo, E., Gutierez, C. and Casanera, P. (1993). Cereal inhibitors of insect hydrolases (α-amylases and trypsin): Genetic control, transgenic expression and insect pests. In Bruening, G.J., Garcia-Olmedo, F. and Ponz, F.J. (Eds.), *Workshop on Engineering Plants Against Pests and Pathogens*, 1–13 January, 1993. Madrid, Spain: Instituto Juan March de Estudios Investigacions.

Charity, J.A., Anderson, M.A., Bittisnich, D.J., Whitecross, M. and Higgins, T.J.V. (1999). Transgenic tobacco and peas expressing a proteinase inhibitor from *Nicotiana alata* have increased resistance. *Molecular Breeding* 5: 357–365.

Chilton, M.D. (1983). A vector for introducing new genes into plants. *Scientific American* 248: 50–59.

Chit Kramer, K.J. and Muthukrishnan, S. (1997). Insect chitinases: Molecular biology and potential use as biopesticides. *Insect Biochemistry and Molecular Biology* 27: 887–900.

Cho, H., Choi, K., Yamashita, M., Morikawa, H. and Murooka, Y. (1995). Introduction and expression of the *Streptomyces* cholesterol oxidase gene (*choA*), a potent insecticidal protein active against boll weevil larvae, into tobacco cells. *Applied Microbiology and Biotechnology* 44: 133–138.

Chrispeels, M.J., Grossi-de-Sa, M.F. and Higgins, T.J.V. (1998). Genetic engineering with α-amylase inhibitors seeds resistant to bruchids. *Seed Science Research* 8: 257–263.

Christou, P., Capell, T., Kohli, A., Gatehouse, J.A. and Gatehouse, A.M.R. (2006). Recent developments and future prospects in insect pest control in transgenic crops. *Trends in Plant Science* 11: 302–308.

Christeller, J.T., Burgess, E.P.J., Mett, V., Gatehouse, H.S., Markwick, N.P., Murray, C., Malone, L.A., Wright, M.A., Philip, B.A., Watt, D., Gatehouse, L.N., Lovei, G.L., Shannon, A.L., Phung, M.M., Watson, L.M. and Laing, W.A. (2002). The expression of a mammalian proteinase inhibitor, bovine spleen trypsin inhibitor in tobacco and its effects on *Helicoverpa armigera* larvae. *Transgenic Research* 11: 161–173.

Clark, T.L., Foster, J.E., Kamble, S.T. and Heinrichs, E.A. (2000). Comparison of Bt (*Bacillus thuringiensis* Berliner) maize and conventional measures for control of the European corn borer (Lepidoptera: Crambidae). *Journal of Entomology Science* 35: 118–128.

Cloutier, C., Jean, C., Fournier, M., Yelle, S. and Michaud, D. (2000). Adult Colorado potato beetles, *Leptinotarsa decemlineata* compensate for nutritional stress on oryzacystatin I-transgenic potato plants by hypertrophic behavior and over-production of insensitive proteases. *Archives of Insect Biochemistry and Physiology* 44: 69–81.

Corbin, D.R., Greenplate, J.T., Wong, E.Y. and Purcell, J.P. (1994). Cloning of an insecticidal cholesterol oxidase gene and its expression in bacteria and plant protoplasts. *Applied Environmental Microbiology* 60: 4239–4244.

Cosa De, B., Moar, W., Lee, S.B., Miller, M. and Daniell, H. (2002). Overexpression of the Bt cry2Aa2 operon in chloroplasts leads to formation of insecticidal crystals. *Nature Biotechnology* 19: 71–74.

Cotsaftis, O., Sallaud, C., Breitler, J.C., Meynard, D., Greco, R., Pereira, A. and Guiderdoni, E. (2002). Transposon-mediated generation of T-DNA- and marker-free rice plants expressing a Bt endo-toxin gene. *Molecular Breeding* 10: 165–180.

Creelman, R.A. and Mullet, J.E. (1997). Biosynthesis and action of jasmonates in plants. *Annual Review of Plant Physiology and Plant Molecular Biology* 48: 355–381.

Crickmore, N., Ziegler, D.R., Fietelson, J., Schnepf, E., Van Rie, J., Lereclus, D., Baum, J. and Dean, D.H. (1998). Revision of the nomenclature for *Bacillus thuringiensis* pesticidal crystal proteins. *Microbiology and Molecular Biology Reviews* 62: 807–813.

Cui, J.J. and Xia, J.Y. (1999). Effects of transgenic Bt cotton on development and reproduction of cotton bollworm. *Acta Agricultura University, Henan* 33: 20–24.

Cui, H.R., Wang, Z.H., Shu, Q.Y., Wu, D.X., Xia, Y.W. and Gao, M.W. (2001). Agronomic traits of hybrid progenies between *Bt* transgenic rice and conventional rice varieties. *Chinese Journal of Rice Science* 15: 101–106.

Czapla, T.H. and Lang, B.A. (1990). Effects of plant lectins on the larval development of European corn borer (Lepidoptera: Pyralidae) and Southern corn rootworm (Coleoptera: Chysomelidae). *Journal of Economic Entomology* 83: 2480–2485.

Dale, P.J., Irwin, J.A. and Scheffler, J.A. (1993). The experimental and commercial release of transgenic crop plants. *Plant Breeding* 111: 1–22.

Dandekar, A.M., Fisk, H.J., McGranahan, G.H., Uratsu, S.L., Bains, H., Leslie, C.A., Tamura, M., Escobar, M., Labavitch, J., Grieve, C., Gradziel, T., Vail, P.V., Tebbets, S.J., Sassa, H., Tao, R., Viss, W., Driver, J., James, D. and Passey, A. (2002). Teo-G different genes for different folks in tree crops: What works and what does not. *Proceedings of the Colloquium—Genetic Stability of Transgenes Under Field Conditions.* 96th ASHS Annual Conference, 29 July, 1999, Minneapolis, Minnesota, USA: *HortScience* 37: 281–286.

Datta, K., Vasquez, A., Tu, J., Torrizo, L., Alam, M.F., Oliva, N., Abrigo, E., Khush, G.S. and Datta, S.K. (1998). Constitutive and tissue-specific differential expression of the cryIA(b) gene in trans-genic rice plants conferring resistance to rice insect pests. *Theoretical and Applied Genetics* 97: 20–30.

Davidson, M.M., Jacobs, J.M.E., Reader, J.K., Butler, R.C., Frater, C.M., Markwick, N.P., Wratten, S.D. and Conner, A.J. (2002). Development and evaluation of potatoes transgenic for a cry1Ac9 gene conferring resistance to potato tuber moth. *Journal of American Society of Horticulture Science* 127: 590–596.

de Barjac, H. and Sotherland, D.J. (1990). *Bacterial Control of Mosquitoes and Blackflies.* New Brunswick, New Jersey, USA: Rutgers University Press.

Delannay, X., LaVallee, B.J., Proksch, R.K., Fuchs, R.L., Sims, S.K., Greenplate, J.T., Marrone, P.G., Dodson, R.B., Augustine, J.J., Layton, J.G. and Fischhoff, D.A. (1989). Field performance of transgenic tomato plants expressing *Bacillus thuringiensis* var *kurstaki* insect control protein. *BioTechnology* 7: 1265–1269.

De Leo, F., Bonade-Bottino, M.A., Ceci, L.R., Gallerani, R. and Jouanin, L. (1998). Opposite effects on *Spodoptera littoralis* larvae of high expression level of a trypsin inhibitor in transgenic plants. *Plant Physiology* 118: 997–1004.

Dhankher, O.P. and Gatehouse, J.A. (2003). Non-systemic induction of polyphenol oxidase in pea and chickpea after wounding. *Physiology and Molecular Biology of Plants* 9: 125–129.

Ding, L.C., Hu, C.Y., Yeh, K.W. and Wang, P.J. (1998). Development of insect-resistant transgenic cauliflower plants expressing the trypsin inhibitor gene isolated from local sweet potato. *Plant Cell Reports* 17: 854–860.

Ding, X., Gopalakrishnan, B., Johnson, L.B., White, F.F., Wang, X., Morgan, T.D., Kramer, K.J. and Muthukrishnan, S. (1998). Insect resistance of transgenic tobacco expressing an insect chitinase gene. *Transgenic Research* 7: 77–84.

Dombrowski, J.E., Pearce, G. and Ryan, C.A. (1999). Proteinase inhibitor-inducing activity of the prohormone prosystemin resides exclusively in the C-terminal systemin domain. *Proceedings National Academy of Sciences USA* 96: 12947–12952.

Donegan, K.K., Seidler, R.J., Doyle, J.D., Porteous, L.A., Digiovanni, G., Widmer, F. and Watrud, L.S. (1999). A field study with genetically engineered alfalfa inoculated with recombinant *Sinorhizobium meliloti*: Effects on the soil ecosystem. *Journal of Applied Ecology* 36: 92–936.

Douches, D.S., Li, W., Zarka, K., Coombs, J., Pett, W., Grafius, E. and El Nasr, T. (2002). Development of Bt-cry5 insect-resistant potato lines 'Spunta-G2' and 'Spunta-G3.' *HortScience* 37: 1103–1107.

Down, R.E., Ford, L., Bedford, S.J., Gatehouse, L.N., Newell, C., Gatehouse, J.A. and Gatehouse, A.M.R. (2001). Influence of plant development and environment on transgene expression in potato and consequences for insect resistance. *Transgenic Research* 10: 223–236.

Duan, X., Li, X., Xue, Q., Abo el Saad, M., Xu, D. and Wu, R. (1996). Transgenic rice plants harboring an introduced potato proteinase inhibitor II gene are insect resistant. *Nature Biotechnology* 14: 494–498.

Dulmage, H.T. (1981). Insecticidal activity of isolates of *Bacillus thuringiensis* and their potential for pest control. In Burgess, H.D. (Ed.), *Microbial Control of Pests and Plant Diseases*. London, UK: Academic Press, 129–141.

Ebel, J. (1986). Phytoalexin synthesis: The biochemical analysis of the induction process. *Annual Review of Phytopatholgy* 24: 235–264.

Ebora, R.V., Ebora, M.M. and Sticklen, M.B. (1994). Transgenic potato expressing the *Bacillus thuringiensis* CryIA(c) gene effects on the survival and food consumption of *Phthorimaea operculella* (Lepidoptera: Gelechiidae) and *Ostrinia nubilalis* (Lepidoptera: Noctuidae). *Journal of Economic Entomology* 87: 1122–1127.

Edmonds, H.S., Gatehouse, L.N., Hilder, V.A. and Gatehouse, J.A. (1996). The inhibitory effects of the cysteine protease inhibitor, oryzacystatin, on digestive proteases and on larval survival and development of the southern corn rootworm (*Diabrotica undecimpunctata* Howard). *Entomologia Experimentalis et Applicata* 78: 83–94.

Estruch, J.J., Carrozzi, N.B., Desai, N., Duck, N.B., Warren, G.W. and Koziel, M.G. (1997). Transgenic plants: Emerging approach to pest control. *Nature Biotechnology* 15: 137–141.

Estruch, J.J., Warren, G.W., Mullins, M.A., Nye, G.J., Craig, J.A. and Koziel, M.G. (1996). Vip3A, a novel *Bacillus thuringiensis* vegetative insecticidal protein with a wide spectrum of activities against lepidopteran insects. *Proceedings National Academy of Sciences USA* 93: 5389–5394.

Fang, H.J., Li, D.L., Wang, G.L. and Li, Y.H. (1997). An insect-resistant transgenic cabbage plant with the cowpea trypsin inhibitor (CpTi) gene. *Acta Botanica Sinica* 39: 940–945.

Fischhoff, D.A., Bowdish, K.S., Perlak, F.J., Marrone, P.G., MvCormick, S.M., Niedermeyer, J.G., Dean, D.A., Kusano-Kretzmer, K., Mayer, E.J., Rochester, D.E., Rogers, S.G. and Fraley, R.T. (1987). Insect tolerant tomato plants. *Bio/Technology* 5: 807–812.

Fitches, E., Gatehouse, A.M.R. and Gatehouse, J.A. (1997). Effects of snowdrop lectin (GNA) delivered via artificial diet and transgenic plants on the development of tomato moth (*Lecanobia oleracea*) larvae in laboratory and glasshouse trials. *Journal of Insect Physiology* 43: 727–739.

Fitches, E., Audsley, N., Gatehouse, J.A. and Edwards, J.P. (2002). Fusion proteins containing neuropeptides as novel insect control agents: Snowdrop lectin delivers fused allatostatin to insect haemolymph following oral ingestion. *Insect Biochemistry and Molecular Biology* 32: 1653–1661.

Foissac, X., Nguyen, T.L., Christou, P., Gatehouse, A.M.R. and Gatehouse, J.A. (2000). Resistance to green leafhopper (*Nephotettix virescens*) and brown planthopper (*Nilaparvata lugens*) in transgenic rice expressing snowdrop lectin (*Galanthus nivalis* agglutinin; GNA). *Journal of Insect Physiology* 46: 573–583.

Franco, O.L., Rigden, D.J., Melo, F.R. and Grossi de Sa, M.F. (2002). Plant alpha-amylase inhibitors and their interaction with insect alpha-amylases: Structure, function and potential for crop protection. *European Journal of Biochemistry* 269: 397–412.

Fujimoto, H., Itoh, K., Yamamoto, M., Kayozuka, J. and Shimamoto, K. (1993). Insect resistant rice generated by a modified delta endotoxin genes of *Bacillus thuringiensis*. *Bio/Technology* 11: 1151–1155.

Gahakwa, D., Maqbool, S.B., Fu, X., Sudhakar, D., Christou, P. and Kohli, A. (2000). Transgenic rice as a system to study the stability of transgene expression: Multiple heterologous transgenes show similar behaviour in diverse genetic backgrounds. *Theoretical and Applied Genetics* 101: 388–399.

Gatehouse, A.M.R. and Gatehouse, J.A. (1998). Identifying proteins with insecticidal activity: Use of encoding genes to produce insect-resistant transgenic crops. *Pesticide Science* 52: 165–175.

Gatehouse, A.M.R., Davison, G.M., Stewart, J.N., Gatehouse, L.N., Kumar, A., Geoghegan, I.E., Birch, A.N.E. and Gatehouse, J.A. (1999). Concanavalin A inhibits development of tomato moth (*Lecanobia oleracea*) and peach potato aphid (*Myzus persicae*) when expressed in transgenic potato plants. *Molecular Breeding* 5: 153–165.

Gatehouse, A.M.R., Down, R.E., Powell, K.S., Sauvion, N., Bahbe, Y., Newell, C.A., Merryweather, A., Hamilton, W.D.O. and Gatehouse, J.A. (1996). Transgenic potato plants with enhanced resistance to the peach-potato aphid *Myzus persicae. Entomologia Experimentalis et Applicata* 79: 295–307.

Gatehouse, A.M.R., Powell, K.S., Van Damme, E.J.M. and Gatehouse, J.A. (1995). Insecticidal properties of plant lectins. In Pusztai, A. and Bardocz, S. (Eds.), *Lectins, Biomedical Perspectives*. London, UK: Taylor and Francis.

Gatehouse, A.M.R., Shi, Y., Powell, K.S., Brough, C., Hilder, V.A., Hamilton, W.D.O., Newell, C., Merryweather, A., Boutler, D. and Gatehouse, J.A. (1993). Approaches to insect resistance using transgenic plants. *Philosophical Transactions of the Royal Society of London, Biological Sciences* (B) 342: 279–286.

Ghareyazie, B., Alinia, F., Menguito, C.A., Rubia, L.G., de Palma, J.M., Liwanag, E.A., Cohen, M.B., Khush, G.S. and Bennett, J. (1997). Enhanced resistance to two stem borers in an aromatic rice containing a synthetic CryIA(b) gene. *Molecular Breeding* 3: 401–414.

Gibson, D.M., Gallo, L.G., Krasnoff, S.B. and Ketchum, R.E.B. (1995). Increased efficiency of *Bacillus thuringiensis* subsp. *kurstaki* in combination with tannic acid. *Journal of Economic Entomology* 88: 270–277.

Giri, A.P. and Kachole, M.S. (1998). Amylase inhibitors of pigeonpea (*Cajanus cajan*) seeds. *Phytochemistry* 47: 197–202.

Girijashankar, V., Sharma, H.C., Sharma, K.K., Swathisree, V., Sivarama Prasad, L., Bhat, B.V., Royer, M., Secundo, B.S., Narasu, L.M., Altosaar, I. and Seetharama, N. (2005). Development of transgenic sorghum for insect resistance against the spotted stem borer (*Chilo partellus*). *Plant Cell Reports* 24: 513–522.

Goggin, F.L., Shah, G., Williamson, V.M. and Ullman, D.E. (2004). Developmental regulation of Mi-mediated aphid resistance is independent of Mi-1.2 transcript levels. *Molecular Plant Microbe Interactions* 17: 532–536.

Gopalaswamy, S.V.S., Kumar, S., Subaratnam, G.V., Sharma, H.C. and Sharma, K.K. (2003). Transgenic pigeonpea: A new tool to manage *Helicoverpa armigera*. In *National Symposium on Bioresources, Biotechnology and Bioenterprise*, 19–20 November 2003, Osmania University, Hyderabad, Andhra Pradesh, India.

Gore, J., Leonard, B.T. and Adamczyk, J.J. (2001). Bollworm (Lepidoptera: Noctuidae) survival on Bollgard and Bollgard II cotton flower bud and flower components. *Journal of Economic Entomology* 94: 1445–1451.

Greenplate, J.T. (1999). Quantification of *Bacillus thuringiensis* insect control protein Cry1A(c) over time in Bollgard cotton fruit and terminals. *Journal of Economic Entomology* 92: 1377–1383.

Guo, H.N., Wu, J.H., Chen, X.Y., Luo, X.L., Lu R., Shi, Y.J., Qin, H.M., Xiao, J.L. and Tian, Y.C. (2003). Cotton plants transformed with the activated chimeric *cry1Ac* and *API-B* genes. *Acta Botanica Sinica* 45: 108–113.

Guo, S.D., Cui, H.Z., Xia, L.Q., Wu, D.L., Ni, W.C., Zhang, Z.L., Zhang, B.L. and Xu, Y.J. (1999). Development of bivalent insect-resistant transgenic cotton plants. *Scientia Agricultura Sinica* 32: 1–7.

Habibi, J., Backus, E.A. and Czapla, T.H. (1992). Effect of plant lectins on survival of potato leafhopper. *Proceedings, XIX International Congress of Entomology*, Beijing, China, 373.

Haffani, Y.Z., Overney, S., Yelle, S., Bellemare, G. and Belzile, F.J. (2000). Premature polyadenylation contributes to the poor expression of the *Bacillus thuringiensis cry3CA1* gene in transgenic potato plants. *Molecular and General Genetics* 264: 82–88.

Hallahan, D.L., Pickett, J.A., Wadham, L.J., Wallsgrove, R.M. and Woodcock, C.M. (1992). Potential of secondary metabolites in genetic engineering of crops for resistance. In Gatehouse, A.M.R., Hilder, V.A. and Boulter, D. (Eds.), *Plant Genetic Manipulation for Crop Protection*. Wallingford, UK: CAB International, 215–248.

Harshavardhan, D., Rani, T.S., Sharma, H.C., Richa, A. and Seetharama, N. (2002). Development and testing of *Bt* transgenic sorghum. In *International Symposium on Molecular Approaches to Improve Crop Productivity and Quality*, 22–24 May, 2002. Coimbatore, Tamil Nadu, India: Tamil Nadu Agricultural University.

Hayakawa, T., Hashimoto, Y., Mori, M., Kaido, M., Shimojo, E., Furusawa, I. and Granados, R.R. (2004). Transgenic tobacco transformed with the *Trichoplusia ni* granulovirus enhancin gene affects insect development. *BioControl Science and Technology* 14: 211–214.

Hayakawa, T., Shimojo, E.I., Mori, M., Kaido, M., Furusawa, I., Miyata, S., Sano, Y., Matsumoto, T., Hashimoto, Y. and Granados, R.R. (2000). Enhancement of baculovirus infection in *Spodoptera exigua* (Lepidoptera: Noctuidae) larvae with *Autographa californica* nucleopolyhedrovirus or *Nicotiana tabacum* engineered with a granulovirus enhancin gene. *Applied Entomology and Zoology* 35: 163–170.

Herman, R.A., Scherer, P.N., Young, D.L., Mihaliak, C.A., Meade, T., Woodsworth, A.T., Stockhoff, B.A. and Narva, K.E. (2002). Binary insecticidal crystal protein from *Bacillus thuringiensis*, strain PS149B1: Effects of individual protein components and mixtures in laboratory bioassays. *Journal of Economic Entomology* 95: 635–639.

Hilder, V.A. and Boulter, D. (1999). Genetic engineering of crop plants for insect resistance: A critical review. *Crop Protection* 18: 177–191.

Hilder, V.A. and Gatehouse, A.M.R. (1991). Phenotypic costs to plants of an extra gene. *Transgenic Research* 1: 54–60.

Hilder, V.A., Gatehouse, A.M.R., Sheerman, S.E., Baker, R.F. and Boulter, D. (1987). A novel mechanism of insect resistance engineered into tobacco. *Nature* 330: 160–163.

Hilder, V.A., Powell, K.S., Gatehouse, A.M.R., Gatehouse, J.A., Gatehouse, L.N., Shi, Y., Hamilton, W.D.O., Merryweather, A., Newell, C.A., Timans, J.C., Peumans, W.J., Van Damme, E. and Boulter, D. (1995). Expression of snowdrop lectin in transgenic tobacco plants results in added protection against aphids. *Transgenic Research* 4: 18–25.

Hoffmann, M.P., Zalom, F.G., Wilson, L.T., Smilanick, J.M., Malyj, L.D., Kisen, J., Hilder, V.A. and Barnes, W.M. (1992). Field evaluation of transgenic tobacco containing genes encoding *Bacillus thuringiensis* δ-endotoxin or cowpea trypsin inhibitor: Efficacy against *Helicoverpa zea* (Lepidoptera: Noctuidae). *Journal of Economic Entomology* 85: 2516–2522.

Hofte, H. and Whiteley, H.R. (1989). Insecticidal crystal proteins of *Bacillus thuringiensis*. *Microbiology Review* 53: 242–255.

Hou, B.K., Zhou, Y.H., Wan, L.H., Zhang, Z.L., Shen, G.F., Chen, Z.H. and Hu, Z.M. (2003). Chloroplast transformation in oilseed rape. *Transgenic Research* 12: 111–114.

Hoy, C.W. (1999). Colorado potato beetle resistance management strategies for transgenic potatoes. *American Journal of Potato Research* 76: 215–219.

Hudspeth, R.L. and Grula, J.W. (1989). Structure and expression of maize gene encoding the phosphoenolpyruvate carboxylase isozyme involved in C4 photosynthesis. *Plant Molecular Biology* 12: 579–589.

Husnain, T., Asad, J., Mqbool, S.B., Datta, S.K. and Raizuddin, S. (2002). Variability in expression of *cry1Ab* gene in indica basmati rice. *Euphytica* 128: 121–128.

Inanaga, H., Kobayasi, D., Kouzuma, Y., Aoki Yasunaga, C., Iiyama, K. and Kimura, M. (2001). Protein engineering of novel proteinase inhibitors and their effects on the growth of *Spodoptera exigua* larvae. *Bioscience, Biotechnology and Biochemistry* 65: 2259–2264.

Innacone, R., Grieco, P.D. and Cellini, F. (1997). Specific sequence modifications of a Cry3B endotoxin gene result in high levels of expression and insect resistance. *Plant Molecular Biology* 34: 485–496.

Ishimoto, M. and Chrispeels, M.J. (1996). Protective mechanism of the Mexican bean weevil against high levels of alpha-amylase inhibitor in the common bean. *Plant Physiology* 111: 393–401.

Ishimoto, M., Sato, T., Chrispeels, M.J. and Kitamura, K. (1996). Bruchid resistance of transgenic azuki bean expressing seed alpha amylase inhibitor of common bean. *Entomologia Experimentalis et Applicata* 79: 309–315.

James, C. (2007). *Global Status of Commercialized Biotech/GM Crops: 2006*. ISAAA Briefs no. 35. Ithaca, New York, USA: International Service for Acquisition on Agri-Biotech Applications (ISAAA). http://www.isaaa.org/resources/publications/briefs/35.

Jansens, S., Cornelissen, M., Clercq, R. de, Reynaerts, A. and Peferoen, M. (1995). *Phthorimaea opercullella* (Lepidoptera: Gelechiidae) resistance in potato by expression of *Bacillus thuringiensis* Cry IA(b) insecticidal crystal protein. *Journal of Economic Entomology* 88: 1469–1476.

Jansens, S., Vliet, A. van, Dickburt, C., Buysse, L., Piens, C., Saey, B., de Wulf, A., Gossele, V., Paez, A. and Gobel, E. (1997). Transgenic corn expressing a Cry9C insecticidal protein from *Bacillus thuringiensis* protected from European corn borer damage. *Crop Science* 37: 1616–1624.

Jayanand, B., Sudarsanam, G. and Sharma, K.K. (2003). An efficient protocol for regeneration of whole plant of chickpea (*Cicer arietinum* L.) by using auxiliary meristem explant derived from *in vitro* germinated seedlings. *In Vitro Cellular and Development Biology—Plant* 39: 171–179.

Jiang, H., Zhu, Y.X. and Chen Z.L. (1996). Insect resistance of transformed tobacco plants with a gene of a spider insecticidal peptide. *Acta Botanica Sinica* 38: 95–99.

Jiang, H., Zhu, Y.X., Wang, Y.P., Wang, Z.P. and Zhang, Z.L. (1995). Synthesis of the spider insecticidal gene and construction of a plasmid expressing in plants. *Acta Botanica Sinica* 37: 321–325.

Jin, R.G., Liu, Y.B., Tabashnik, B.E. and Borthakur, D. (2000). Development of transgenic cabbage (*Brassica oleracea* var. *capitata*) for insect resistance by *Agrobacterium tumefaciens*-mediated transformation. *In Vitro Cellular and Developmental Biology—Plant* 36: 231–237.

Johnson, R., Narvaez, J., An, G. and Ryan, C. (1989). Expression of proteinase inhibitors I and II in transgenic tobacco plants: Effects on natural defense against *Manduca sexta* larvae. *Proceedings National Academy of Sciences USA* 86: 9871–9875.

Johnston, K.A., Gatehouse, J.A. and Anstee, J.H. (1993). Effect of soybean protease inhibitors on growth and development of larval *Helicoverpa armigera*. *Journal of Insect Physiology* 39: 657–664.

Jongsma, M.A., Bakker, P.L., Peters, J., Bosch, D. and Stiekema, W.J. (1995). Adaptation of *Spodoptera exigua* larvae to plant proteinase inhibitors by induction of proteinase activity insensitive to inhibition. *Proceedings National Academy of Sciences USA* 92: 8041–8045.

Kanrar, S., Venkateswari, J., Kirti, P.B. and Chopra, V.L. (2002). Transgenic Indian mustard (*Brassica juncea*) with resistance to the mustard aphid (*Lipaphis erysimi* Kalt.). *Plant Cell Reports* 20: 976–981.

Kar, S., Basu, D., Das, S., Ramkrishnan, N.A., Mukherjee, P., Nayak, P. and Sen, S.K. (1997). Expression of CryIA(c) gene of *Bacillus thuringiensis* in transgenic chickpea plants inhibits development of pod borer (*Heliothis armigera*) larvae. *Transgenic Research* 6: 177–185.

Khanna, H.K. and Raina, S.K. (2002). Elite indica transgenic rice plants expressing modified Cry1Ac endotoxin of *Bacillus thuringiensis* show enhanced resistance to yellow stem borer (*Scirpophaga incertulas*). *Transgenic Research* 11: 411–423.

Klein, T.M., Wolf, E.D., Wu, R. and Sanford, J.C. (1987). High-velocity microprojectiles for delivering nucleic acids into living cells. *Nature* 327: 70–73.

Konarev, A.V. (1996). Interaction of insect digestive enzymes with plant protein inhibitors and host-parasite co-evolution. *Euphytica* 92: 89–94.

Koziel, M.G., Beland, G.L., Bowman, C., Carozzi, N.B., Crenshaw, R., Crossland, L., Dawson, J., Desai, N., Hill, M., Kadwell, S., Launis, K., Lewis, K., Maddox, D., McPherson, K., Meghji, M.R., Merlin, E., Rhodes, R., Warren, G.W., Wright, M. and Evola, S.V. (1993). Field performance of elite transgenic maize plants expressing an insecticidal protein derived from *Bacillus thuringiensis*. *Bio/Technology* 11: 194–200.

Kranthi, K.R., Naidu, S., Dhawad, C.S., Tatwawadi, A., Mate, K., Patil, E., Bharose, A.A., Behere, G.T., Wadaskar, R.M. and Kranthi, S. (2005). Temporal and intra-plant variability in Cry1Ac expression in *Bt*-cotton and its influence on the survival of the cotton bollworm, *Helicoverpa armigera* (Hubner) (Noctuidae: Lepidoptera). *Current Science* 89: 291–298.

Kumar, H. and Kumar, V. (2004). Tomato expressing Cry1A(b) protein from *Bacillus thuringiensis* protected against tomato fruit borer, *Helicoverpa armigera* (Hubner) (Lepidoptera: Noctuidae) damage in the laboratory, greenhouse and field. *Crop Protection* 23: 135–139.

Kumar, P.A., Mandaokar, A., Sreenivasu, K., Chakrabarti, S.K., Bisaria, S., Sharma, S.R., Kaur, S. and Sharma, R.P. (1998). Insect-resistant transgenic brinjal plants. *Molecular Breeding* 4: 33–37.

Kuvshinov, V., Koivu, K., Kanerva, A. and Pehu, E. (2001). Transgenic crop plants expressing synthetic cry9Aa gene are protected against insect damage. *Plant Science* 160: 341–353.

Lauer, J. and Wedberg, J. (1999). Grain yield of initial Bt corn hybrid introductions to farmers in the Northern Corn Belt. *Journal of Production Agriculture* 12: 373–376.

Lawrence, P.K. and Koundal, K.R. (2002). Plant protease inhibitors in control of phytophagous insects. *Electronic Journal of Biotechnology.* http://www.ejb.org/content/vol5/issue1/full/3/htm.

Leal Bertioli, S.C.M., Pascoal, A.V., Guimaraes, P.M., de Sa, M.F.G., Guimaraes, R.L., Monte, D.C. and Bertioli, D.J. (2003). Transgenic tobacco plants expressing Tarin 1 inhibit the growth of *Pseudomonas syringae* pv. *tomato* and the development of *Spodoptera frugiperda. Annals of Applied Biology* 143: 349–357.

Lecardonnel, A., Chauvin, L., Jouanin, L., Beaujean, A., Prevost, G. and Sangwan Norreel, B. (1999). Effects of rice cystatin I expression in transgenic potato on Colorado potato beetle larvae. *Plant Science* 140: 71–79.

Lee, S.I., Lee, S.H., Koo, J.C., Chun, H.J., Lim, C.O., Mun, J.H., Song, Y.H. and Cho, M.J. (1999). Soybean Kunitz trypsin inhibitor (SKTI) confers resistance to the brown planthopper (*Nilaparvata lugens* Stal) in transgenic rice. *Molecular Breeding* 5: 1–9.

Leelavathi, S., Selvapandiyan, A., Raman, R., Giovanni, F. and Shukla, V. (2002). Analysis of chloroplast transformed tobacco plants with cry1Ia5 under rice psbA transcriptional elements reveal high level expression of Bt toxin without imposing yield penalty and stable inheritance of transplastome. *Molecular Breeding* 9: 259–269.

Leslie, C.A., McGranahan, G.H., Dandekar, A.M., Uratsu, S.L., Vail, P.V., Tebbets, J.S. and Germain, E. (2001). Development and field-testing of walnuts expressing the Cry1A(c) gene for lepidopteran insect resistance. *Fourth International Walnut Symposium*, Bordeaux, France, 13–16 September, 1999. *Acta Horticulturae* 544: 195–199.

Li, Y.E., Zhu, Z., Chen, Z.X., Wu, X., Wang, W. and Li, S.J. (1998). Obtaining transgenic cotton plants with cowpea trypsin inhibitor. *Acta Gossypii Sinica* 10: 237–243.

Lin, C.H., Chen, Y.Y., Tzeng, C.C., Tsay, H.S. and Chen, L.J. (2003). Expression of a *Bacillus thuringiensis* cry1C gene in plastid confers high insecticidal efficacy against tobacco cutworm—a *Spodoptera* insect. *Botanical Bulletin of Academia Sinica* 44: 199–210.

Lin, L.B., Guan, C.Y., Wang, G.H., Chen, S.Y. and Li, X. (2001). Studies on resistance to cabbage butterfly (*Pieris rapae*) larvae in transgenic insect-resistant rapeseed (*Brassica napus*). *Journal of Yunnan Agricultural University* 16: 203–205.

Liu, D., Burton, S., Glancy, T., Li, Z.S., Hampton, R., Meade, T. and Merlo, D.J. (2003). Insect resistance conferred by 283-kDa *Photorhabdus luminescens* protein TcdA in *Arabidopsis thaliana. Nature Biotechnology* 21: 1222–1228.

Liu, Z., Guo, W.Z., Zhu, X.F., Zhu, Z. and Zhang, T.Z. (2003). Inheritance analysis of resistance of transgenic *Bt* + GNA cotton line to *Helicoverpa armigera. Journal of Agricultural Biotechnology* 11: 388–393.

Liu, Z.H., Tang, J.H., Hu, Y.M., Ji, H.Q., Ji, L.Y., Qiu, D.Y. and Wang, A.D. (2000). Identification of corn-borer resistant plant with transferred *Bt* gene in different backcross generations of corn. *Scientia Agricultura Sinica* 33: 152–155.

Lynch, R.E., Wiseman, B.R., Plaisted, D. and Warnick, D. (1999a). Evaluation of transgenic sweet corn hybrids expressing CryIA(b) toxin for resistance to corn earworm and fall armyworm (Lepidoptera: Noctuidae). *Journal of Economic Entomology* 92: 246–252.

Lynch, R.E., Wiseman, B.R., Sumner, H.R., Plaisted, D. and Warnick, D. (1999b). Management of corn earworm and fall armyworm (Lepidoptera: Noctuidae) injury on a sweet corn hybrid expressing a *CryIA (b)* gene. *Journal of Economic Entomology* 92: 1217–1222.

MacIntosh, S.C., Kishore, G.M., Perlak, F.J., Marrone, P.G., Stone, T.B., Sims, S.R. and Fuchs, R.L. (1990). Potentiation of *Bacillus thuringiensis* insecticidal activity by serine protease inhibitors. *Journal of Agriculture and Food Chemistry* 38: 1145–1152.

Maddock, S.E., Hufman, G., Isenhour, D.J., Roth, B.A., Raikhel, N.V., Howard, J.A. and Czapla, T.H. (1991). Expression in maize plants of wheatgerm agglutinin, a novel source of insect resistance. In *3rd International Congress of Plant Molecular Biology*. Tucson, Arizona, USA. Abstract 372.

Mahmoud, S.S. and Croteau, R.B. (2002). Strategies for transgenic manipulation of monoterpene biosynthesis in plants. *Trends in Plant Science* 7: 366–373.

Malone, L.A., Burgess, E.P.J., Mercer, C.F., Christeller, J.T., Lester, M.T., Murray, C., Phung, M.M., Philip, B.A., Tregidga, E.L., and Todd, J.H. (2002). Effects of biotin-binding proteins on eight species of pasture invertebrates. *New Zealand Plant Protection* 55: 411–415.

Mandal, S., Kundu, P., Roy, B. and Mandal, R.K. (2002). Precursor of the inactive 2S seed storage protein from the Indian mustard *Brassica juncea* is a novel trypsin inhibitor: Characterization, post-translational processing studies, and transgenic expression to develop insect-resistant plants. *Journal of Biological Chemistry* 277: 37161–37168.

Mandaokar, A.D., Goyal, R.K., Shukla, A., Bisaria, S., Bhalla, R., Reddy, V.S., Chaurasia, A., Sharma, R.P., Altosaar, I. and Kumar, P.A. (2000). Transgenic tomato plants resistant to fruit borer (*Helicoverpa armigera* Hubner). *Crop Protection* 19: 307–312.

Marchetti, S., Delledonne, M., Fogher, C., Chiaba, C., Chiesa, F., Savazzini, F. and Giordano, A. (2000). Soybean Kunitz, *C-II* and *PI-IV* inhibitor genes confer different levels of insect resistance to tobacco and potato transgenic plants. *Theoretical and Applied Genetics* 101: 519–526.

Markwick, N.P., Docherty, L.C., Phung, M.M., Lester, M.T., Murray, C., Yao, J.L., Mitra, D.S., Cohen, D., Beuning, L.L., Kutty Amma, S., Christeller, J.T. and Yao, J.L. (2003). Transgenic tobacco and apple plants expressing biotin-binding proteins are resistant to two cosmopolitan insect pests, potato tuber moth and light brown apple moth, respectively. *Transgenic Research* 12: 671–681.

McBride, K.E., Svab, Z., Schaaf, D.J., Hogan, P.S., Stalker, D.M. and Maliga, P. (1995). Application of a chimeric *Bacillus* gene in chloroplasts leads to extraordinary level of an insecticidal protein in tobacco. *Bio/Technology* 13: 362–365.

McCaskill, D. and Croteau, R. (1998). Some caveats for bioengineering terpenoid metabolism in plants. *Trends in Biotechnology* 16: 349–355.

McElory, D., Rothenberg, M. and Wu, R. (1990). Structural characterization of a rice actin gene. *Plant Molecular Biology* 14: 163–171.

McManus, M.T., Burgess, E.P.J., Philip, B., Watson, L.M., Laing, W.A., Voisey, C.R. and White, D.W.R. (1999). Expression of the soybean (Kunitz) trypsin inhibitor in transgenic tobacco: effects on larval development of *Spodoptera litura*. *Transgenic Research* 8: 383–395.

Mochizuki, A., Nishizawa, Y., Onodera, H., Tabei, Y., Toki, S., Habu, Y., Ugaki, M. and Ohashi, Y. (1999). Transgenic rice plants expressing a trypsin inhibitor are resistant against rice stem borers, *Chilo suppressalis*. *Entomologia Experimentalis et Applicata* 93: 173–178.

Mohammed, A., Douches, D.S., Pett, W., Grafius, E., Coombs, J., Liswidowati, W.L. and Madkour, M.A. (2000). Evaluation of potato tuber moth (Lepidoptera: Gelechiidae) resistance in tubers of Bt-Cry5 transgenic potato lines. *Journal of Economic Entomology* 93: 472–476.

Mohan, K.S. and Manjunath, T.M. (2002). *Bt* cotton: India's first transgenic crop. *Journal of Plant Biology* 29: 225–236.

Morton, R.L., Schroeder, H.E., Bateman K.S., Chrispeels, M.J., Armstrong, E. and Higgins, T.J.V. (2000). Bean alpha-amylase inhibitor 1 in transgenic peas (*Pisum sativum*) provides complete protection from pea weevil (*Bruchus pisorum*) under field conditions. *Proceedings National Academy of Sciences USA* 97: 3820–3825.

Mqbool, S.B., Husnain, T., Raizuddin, S. and Christou, P. (1998). Effective control of yellow rice stem borer and rice leaf folder in transgenic rice *indica* varieties Basmati 370 and M7 using novel δ-endotoxin Cry2A *Bacillus thuringiensis* gene. *Molecular Breeding* 4: 501–507.

Munkvold, G.P., Hellmich, R.L. and Rice, L.G. (1999). Comparison of fumonisin concentrations in kernels of transgenic *Bt* maize hybrids and nontransgenic hybrids. *Plant Disease* 83: 130–138.

Nandi, A.K., Basu, D., Das, S. and Sen, S.K. (1999). High level expression of soybean trypsin inhibitor gene in transgenic tobacco plants failed to confer resistance against damage caused by *Helicoverpa armigera*. *Journal of Bioscience* 24: 445–452.

Nayak, P., Basu, D., Das, S., Basu, A., Ghosh, D., Ramakrishnan, N.A., Ghosh, M. and Sen, S.K. (1997). Transgenic elite *indica* rice plants expressing CryIAc delta-endotoxin of *Bacillus thuringiensis* are resistant against yellow stem borer (*Scirpophaga incertulas*). *Proceedings National Academy of Sciences USA* 94: 2111–2116.

Neuhaus, G., Spangenberg, G., Scheid, O.M. and Schweiger, H.G. (1987). Transgenic rapeseed plants obtained by the injection of DNA into microscope derived embryoids. *Theoretical and Applied Genetics* 75: 30–36.

Ni, W.C., Huang, J.Q., Guo, S.D., Shu, C.G., Wu, J.Y., Wang, W.G., Zhang, Z.L., Chen, S., Mao, L.Q., Wang, Y., Xu, Y.J., Gu, L.M., Zhou, B.L., Shen, X.L. and Xiao, S.H. (1996). Transgenic bollworm-resistant cotton plants containing the synthetic gene coding *Bacillus thuringiensis* insecticidal protein. *Jiangsu Journal of Agricultural Sciences* 12: 1–6.

Nutt, K.A., Allsopp, P.G., Geijskes, R.J., McKeon, M.G., Smith, G.R. and Hogarth, D.M. (2001). Canegrub resistant sugarcane. *Proceedings of the XXIV Congress*, Brisbane, Australia, 17–21 September, 2001. *International Society of Sugarcane Technologists* 2: 582–584.

Nutt, K.A., Allsopp, P.G., McGhie, T.K., Shepherd, K.M., Joyce, P.A., Taylo, G.O., McQualter, R.B., Smith, G.R. and Hogarth, D.M. (1999). Transgenic sugarcane with increased resistance to cane-grubs. In *Proceedings of the 1999 Conference of the Australian Society of Sugarcane Technologists*, 27–30 April, 1999. Townsville, Brisbane, Australia, 171–176.

Oerke, E.C., Dehne, H.W., Schonbeck, F. and Weber, A. (1994). *Crop Production and Crop Protection: Estimated Losses in Major Food and Cash Crops*. Amsterdam, The Netherlands: Elsevier.

Orozco-Cardenas, M., McGurl, B. and Ryan, C.A. (1993). Expression of an antisense prosystemin gene in tomato plants reduces resistance toward *Manduca sexta* larvae. *Proceedings National Academy of Sciences USA* 90: 8273–8276.

Patankar, A.G., Giri, A.P., Harsulkar, A.M., Sainani, M.N., Deshpande, V.V., Ranjekar, P.K. and Gupta, V.S. (2001). Complexity in specificities and expression of *Helicoverpa armigera* gut proteinases explains polyphagous nature of the insect pest. *Insect Biochemistry and Molecular Biology* 31: 453–464.

Patankar, A.G., Harshulkar, A.M., Giri, A.P., Gupta, V.S., Sainani, A.M., Ranjekar, P.K. and Deshpande, V.V. (1999). Diversity in inhibitors of trypsin and *Helicoverpa armigera* gut proteinases in chick-pea (*Cicer arietinum*) and its wild relatives. *Theoretical and Applied Genetics* 99: 719–726.

Peng, J.H. and Black, L.L. (1976). Increased proteinase inhibitor activity in response to infection of resistant tomato plants by *Phytophthora infestans*. *Phytopathology* 66: 958–963.

Perlak, F.J., Stone, T.B., Muskopf, Y.N., Petersen, L.J., Parker, G.B., McPherson, S.A., Wyman, J., Love, S., Reed, G., Biever, D. and Fischhoff, D.A. (1993). Genetically improved potatoes: protection from damage by Colorado potato beetles. *Plant Molecular Biology* 22: 313–321.

Potrykus, I. (1991). Gene transfer to plants: Assessment of published approaches and results. *Annual Review of Plant Physiology and Molecular Biology* 42: 205–225.

Powell, K.S., Gatehouse, A.M.R., Hilder, V.A. and Gatehouse, J.A. (1993). Antimetabolic effects of plant lectins and plant and fungal enzymes on the nymphal stages of two important rice pests, *Nilaparvata lugens* and *Nephotettix cincticeps*. *Entomologia Experimentalis et Applicata* 66: 119–126.

Powell, K.S., Gatehouse, A.M.R., Hilder, V.A. and Gatehouse, J.A. (1995). Antifeedant effects of plant lectins and an enzyme on the adult stage of the rice brown planthopper, *Nilaparvata lugens*. *Entomologia Experimentalis et Applicata* 75: 51–59.

Pulliam, D.A., Williams, D.L., Broadway, R.M. and Stewart, C.N. (2001). Isolation and characteriza-tion of a serine proteinase inhibitor cDNA from cabbage and its antibiosis in transgenic tobacco plants. *Plant Cell Biotechnology and Molecular Biology* 2: 19–32.

Purcell, J.P., Greenplate, J.T., Jennings, M.G., Ryers, J.S., Pershing, J.C., Sims, S.R., Prinsen, M.J., Corbin, D.R., Tran, M., Sammons, R.D. and Stonard, R.J. (1993). Cholesterol oxidase: A potent insecticidal protein active against boll weevil larvae. *Biochemical and Biophysical Research Communications* 196: 1406–1413.

Pusztai, A. (1972). Metabolism of trypsin-inhibitory proteins in the germinating seeds of kidney bean (*Phaseolus vulgaris*). *Planta* 107: 121–129.

Pusztai, A., Grant, G., Brown, D.J., Stewart, J.C., Bardocz, S., Ewen, S.W.B., Gatehouse, A.M.R. and Hilder, V.A. (1992). Nutritional evaluation of the trypsin (EC3.4.21.4) inhibitor from cowpea (*Vigna unguiculata*). *British Journal of Nutrition* 68: 783–791.

Rahbe, Y., Sauvion, N., Febvay, G., Peumans, W.J. and Gatehouse, A.M.R. (1995). Toxicity of lectins and processing of ingested proteins in the pea aphid *Acyrthosiphon pisum*. *Entomologia Experimentalis et Applicata* 76: 143–155.

Ramakrishna Babu, A., Sharma, H.C., Subaratnam, G.V. and Sharma, K.K. (2005). Development of transgenic chickpea (*Cicer arietinum* L.) with *Bt cry1Ac* gene for resistance to pod borer, *Helicoverpa armigera*. In *IVth International Food Legumes Research Conference: Food Legumes for Nutritional Security and Sustainable Agriculture*, 18–22 October, 2005. New Delhi, India: Indian Agricultural Research Institute, 58.

Rang, C., Gil, P., Neisner, N., Van Rie, J. and Frutos, R. (2005). Novel Vip3-related protein from *Bacillus thuringiensis*. *Applied Environmental Microbiology* 71: 6276–6281.

Reddy, V.S., Sadhu, L., Selvapandiyan, A., Raman, R., Giovanni, G., Shukla, V. and Bhatnagar, R.K. (2002). Analysis of chloroplast transformed tobacco plants with cry1Ia5 under rice psbA transcriptional elements reveal high level expression of *Bt* toxin without imposing yield penalty and stable inheritance of transplastome. *Molecular Breeding* 9: 259–269.

Richardson, M. (1977). The proteinase inhibitors of plants and microorganisms. *Phytochemistry* 16: 159–169.

Rickauer, M., Fournier, J. and Esquerre-Tugaye, M.T. (1989). Induction of proteinase inhibitors in tobacco cell suspension culture by elicitors of *Phytopthora parasitica* var. *nocotianae*. *Plant Physiology* 90: 1065–1070.

Rosso, M.N., Schooten, A., Roosien, J., Borst Vrenssen, T., Hussey, R.S., Gommers, F.J., Bakker, J., Scho, A. and Abad, P. (1996). Expression and functional characterization of a single chain antibody directed against secretions involve a plant nematode infection process. *Biochemical and Biophysical Research Communications* 220: 255–263.

Ryan, C. (1990). Protease inhibitors in plants: Genes for improving defenses against insects and pathogens. *Annual Review of Phytopathology* 28: 425–449.

Ryan, C.A. and Pearce, G. (1998). Systemin: A polypeptide signal for plant defensive genes. *Annual Review of Cellular and Developmental Biology* 14: 1–17.

Samad, M.A. and Leyva, A. (1998). Integrated pest management and transgenic plants for insect control in Texas sugarcane. *Proceedings of the Inter American Sugarcane Seminar, Crop Production and Mechanization*, 9–11 September, 1998, Miami, Florida, USA: 127–131.

Sanford, J.C. (1990). Biolistic plant transformation. *Physiologia Plantarum* 79: 206–209.

Sanyal, I., Singh, A.K., Meetu, K. and Devindra, A.V. (2005). *Agrobacterium*-mediated transformation of chickpea (*Cicer arietinum* L.) with *Bacillus thuringiensis cry1Ac* gene for resistance against pod borer insect *Helicoverpa armigera*. *Plant Science* 168: 1135–1146.

Satyendra, R., Stewart, J.M. and Wilkins, T. (1998). Assessment of resistance of cotton transformed with lectin genes to tobacco budworm. In *Special Report*. Fayetteville: Arkansas Agricultural Experiment Station, University of Arkansas, 95–98.

Schnepf, H.E. and Whiteley, H.R. (1981). Cloning and expression of *Bacillus thuringiensis* crystal protein gene in *Escherichia coli*. *Proceedings National Academy of Sciences USA* 78: 2893–2897.

Schroeder, H.E., Gollasch, S., Moore, A., Tabe, L.M., Craig, S., Hardie, D.C., Chrispeels, M.J., Spencer, D. and Higgins, T.J.V. (1995). Bean alpha-amylase inhibitor confers resistance to pea weevil (*Bruchus pisorum*) in transgenic peas (*Pisum sativum* L.). *Plant Physiology* 107: 1233–1239.

Seetharam, A., Kuiper, G., Visser, P.B., de Kogel, W.J., Udayakumar, M. and Jongsma, M.A. (2002). Expression of potato multicystatin in florets of chrysanthemum and assessment of resistance to western flower thrips, *Frankliniella occidentalis*. *Acta Horticulturae* 572: 121–129.

Seetharama, N., Mythili, P.K., Rani, T.S., Harshavardhan, D., Ranjani, A. and Sharma, H.C. (2001). Tissue culture and alien gene transfer in sorghum. In Singh, R.P. and Jaiwal, P.K. (Eds.), *Plant Genetic Engineering. Vol. 2. Improvement of Food Crops.* Houstan, TX: Sci-Tech Publishing Company, 235–266.

Selvapandian, A., Reddy, V.S., Ananda Kumar, P., Tiwari, K.K. and Bhatnagar, R.K. (1998). Transformation of *Nicotiana tabaccum* with a native *cry1Ia5* gene confers complete protection against *Heliothis armigera*. *Molecular Breeding* 4: 473–478.

Setamou, M., Bernal, J.S., Legaspi, J.C., Mirkov, T.E. and Legaspi, B.C. Jr. (2002). Evaluation of lectin-expressing transgenic sugarcane against stalk borers (Lepidoptera: Pyralidae): Effects on life history parameters. *Journal of Economic Entomology* 95: 469–477.

Shade, R.E., Schroeder, H.E., Pueyo, J.J., Tabe, L.M., Murdock, L.L., Higgins, T.J.V. and Chrispeels, M.J. (1994). Transgenic pea seeds expressing α-amylase inhibitor of the common bean are resistant to bruchid beetles. *BioTechnology* 12: 793–796.

Sharma, H.C. and Agarwal, R.A. (1983). Role of some chemical components and leaf hairs in varietal resistance in cotton to jassid, *Amrasca biguttula buguttula* Ishida. *Journal of Entomological Research* 7: 145–149.

Sharma, K.K., Ananda Kumar, P. and Sharma, H.C. (2005). Insecticidal genes and their potential in developing transgenic crops for resistance to *Heliothis/Helicoverpa*. In Sharma, H.C. (Ed.), Heliothis/Helicoverpa *Management: Emerging Trends and Strategies for Future Research*. New Delhi, India: Oxford and IBH Publishers, 255–274.

Sharma, H.C. and Norris, D.M. (1991). Chemical basis of resistance in soybean to cabbage looper, *Trichoplusia ni*. *Journal of Science of Food and Agriculture* 55: 353–364.

Sharma, H.C. and Pampapathy, G. (2006). Influence of transgenic cotton on the relative abundance and damage by target and non-target insect pests under different protection regimes in India. *Crop Protection* 25: 800–813.

Sharma, H.C., Sharma, K.K. and Crouch, J.H. (2004). Genetic transformation of crops for insect resistance: Potential and limitations. *CRC Critical Reviews in Plant Sciences* 23: 47–72.

Sharma, H.C., Ananda Kumar, P., Seetharama, N., Hari Prasad, K.V. and Singh, B.U. (2004). Role of transgenic plants in pest management in sorghum. In Seetharama, N. and Godwin, I. (Eds.), *Sorghum Tissue Culture and Transformation*. New Delhi, India: Oxford and IBH Publishing, 117–130.

Sharma, H.C., Crouch, J.H., Sharma, K.K., Seetharama, N. and Hash, C.T. (2002). Applications of biotechnology for crop improvement: prospects and constraints. *Plant Science* 163: 381–395.

Sharma, H.C., Sharma, K.K., Seetharama, N. and Crouch, J.H. (2003). The utility and management of transgenic plants with *Bacillus thuringiensis* genes for protection from pests. *New Seeds Journal* 5: 53–76.

Sharma, H.C., Sharma, K.K., Seetharama, N. and Ortiz, R. (2000). Prospects for transgenic resistance to insects. *Electronic Journal of Biotechnology*. http//:www.scielo.cl/scielo.php?pid.

Sharma, H.C., Van Driesche, Arora, R. and Sharma, K.K. (2005). Biological activity of lectins from grain legumes and garlic against the legume pod borer, *Helicoverpa armigera*. *International Chickpea and Pigeonpea Newsletter* 12: 50–53.

Sharma, K.K., Lavanya, M. and Anjaiah, V. (2006). *Agrobacterium*-mediated production of transgenic pigeonpea (*Cajanus cajan* L. Millsp.) expressing the synthetic *Bt cry1Ab* gene. *In Vitro Cellular and Developmental Biology—Plant* 42: 165–173.

She, J.M., Cai, X.N., Zhu, Z., Zhu, W.M., Zhang, C.X. and Yuan, X.H. (2000). Acquisition of insect-resistant transgenic plants of *Brassica campestris* ssp. *chinensis* L. *Jiangsu Journal of Agricultural Sciences* 16: 79–82.

Shelp, B.J., van Cauwenberghe, O.R. and van Bown, A.W. (2003). Gamma aminobutyrate: From intellectual curiosity to practical pest control. *Canadian Journal of Botany* 81: 1045–1048.

Shewry, P.R. and Gutteridge, S. (Eds.). (1992). *Plant Protein Engineering*. Cambridge, UK: Cambridge University Press.

Shimamoto, K., Terada, R., Iwaza, T. and Fujimoto, H. (1989). Fertile transgenic rice plants regenerated from transformed protoplasts. *Nature* 338: 274–276.

Shinoyama, H., Mochizuki, A., Komano, M., Nomura, Y. and Nagai, T. (2003). Insect resistance in transgenic chrysanthemum [*Dendranthema grandiflorum* (Ramat.) Kitamura] by the introduction of a modified delta-endotoxin gene of *Bacillus thuringiensis*. *Breeding Science* 53: 359–367.

Shukla, S., Arora, R. and Sharma, H.C. (2005). Biological activity of soybean trypsin inhibitor and plant lectins against cotton bollworm/legume pod borer, *Helicoverpa armigera*. *Plant Biotechnology* 22: 1–6.

Shukle, R.H. and Murdock, L.L. (1983). Lipoxygenase, trypsin inhibitor, and lectin from soybeans: Effect on larval growth of *Manduca sexta* (Lepidoptera: Sphingidae). *Environmental Entomology* 12: 787–791.

Shukle, R.H. and Wu, L. (2003). The role of protease inhibitors and parasitoids on the population dynamics of *Sitotroga cerealella* (Lepidoptera: Gelechiidae). *Environmental Entomology* 32: 488–498.

Sijmons, P.C. (1993). Plant-nematode interactions. *Plant Molecular Biology* 23: 917–931.

Singsit, C., Adang, M.J., Lynch, R.E., Anderson, W.F., Aiming, W., Cardineau, G. and Ozias-Akins, P. (1997). Expression of a *Bacillus thuringiensis cryIA(c)* gene in transgenic peanut plants and its efficacy against lesser cornstalk borer. *Transgenic Research* 6: 169–176.

Smigocki, A., Heu, S. and Buta, G. (2000). Analysis of insecticidal activity in transgenic plants carrying the ipt plant growth hormone gene. EUCARPIA TOMATO 2000. XIV meeting of the EUCARPIA Tomato Working Group, Warsaw, Poland, 20–24 August, 2000. *Acta Physiologia Plantarum* 22: 295–299.

Smigocki, A., Neal, J.W. Jr., McCanna, I. and Douglass, L. (1993). Cytokinin-mediated insect resistance in *Nicotiana* plants transformed with the *ipt* gene. *Plant Molecular Biology* 23: 325–335.

Sree Latha, G., Sharma, H.C., Manohar Rao, D., Royer, M. and Sharma, K.K. (2005). Genetic transformation of pigeonpea [*Cajanus cajan* (L.) Millsp.] with *Bt cry1Ac* gene and the evaluation of transgenic plants for resistance to *Helicoverpa armigera*. In *IVth International Food Legumes Research Conference: Food Legumes for Nutritional Security and Sustainable Agriculture*, 18–22 October, 2005. New Delhi, India: Indian Agricultural Research Institute, 58.

Steinhaus, E.A. (1951). Possible use of *Bacillus thuringiensis* Berliner as an aid in the biological control of the alfalfa caterpillar. *Hilgardia* 20: 350–381.

Stewart, S.D., Adamczyk, J.J. Jr., Knighten, K.S. and Davis, F.M. (2001). Impact of *Bt* cottons expressing one or two insecticidal proteins of *Bacillus thuringiensis* Berliner on growth and survival of noctuid (Lepidoptera) larvae. *Journal of Economic Entomology* 94: 752–760.

Stewart, J.G., Feldman, J. and LeBlanc, D.A. (1999). Resistance of transgenic potatoes to attack by *Epitrix cucumeris* (Coleoptera: Chrysomelidae). *Canadian Entomologist* 131: 423–431.

Stoger, E., Williams, S., Christou, P., Down, R.E. and Gatehouse, J.A. (1999). Expression of the insecticidal lectin from snowdrop (*Galanthus nivalis* agglutinin; GNA) in transgenic wheat plants: Effects on predation by the grain aphid *Sitobion avenae*. *Molecular Breeding* 5: 65–73.

Su, N., Sun, M., Yang, B., Meng, K., Liu, C.Y., Ni, P.C. and Shen, G.F. (2002). The insect resistance of OC and Bt transplastomic plants and the phenotype of their progenies. *Hereditas Beijing* 24: 288–292.

Sun, X., Wu, A. and Tang, K. (2002). Transgenic rice lines with enhanced resistance to the small brown planthopper. *Crop Protection* 21: 511–514.

Surekha, Ch., Beena, M.R., Arundhati, A., Singh, P.K., Tuli, R., Dutta-Gupta, A. and Kirti P.B. (2005). *Agrobacterium*-mediated transformation of pigeonpea [*Cajanus cajan* (L.) Millsp.] using embryonal segments and development of transgenic plants for resistance against *Spodoptera*. *Plant Science* 169: 1074–1080.

Sutton, D.W., Havstad, P.K. and Kemp, J.D. (1992). Synthetic cryIIIA gene from *Bacillus thuringiensis* improved for high expression in plants. *Transgenic Research* 1: 228–236.

Svab, Z. and Maliga, P. (1993). High frequency plastid transformation in tobacco by selection for a *acd* A gene. *Proceedings National Academy of Sciences USA* 90: 913–917.

Tailor, R., Tippett, J., Gibb, G., Pells, S., Pike, D., Jordan, L. and Ely, S. (1992). Identification and characterization of a novel *Bacillus thuringiensis*-endotoxin entomocidal to coleopteran and lepidopteran larvae. *Molecular Microbiology* 7: 1211–1217.

Tang, K., Hu, Q.A., Sun, X., Wan B.L., Qi, H. and Lu, X. (2001). Development of transgenic rice pure lines with enhanced resistance to rice brown planthopper. *In Vitro Cellular and Developmental Biology—Plant* 37: 334–340.

Tan-Wilson, A.L., Hartl, P.M., Delfel, N.E. and Wilson, K.A. (1985). Differential expression of Kunitz and Bowman-Birk soybean proteinase inhibitors in plant and callus tissues. *Plant Physiology* 78: 310–314.

Thomas, J.C., Adams, D.G., Keppenne, V.D., Wasmann, C.C., Brown, J.K., Kanost, M.R. and Bohnert, H.J. (1995a). *Manduca sexta* encoded protease inhibitors expressed in *Nicotiana tabacum* provide protection against insects. *Plant Physiology and Biochemistry* 33: 611–614.

Thomas, J.C., Adams, D.G., Keppenne, V.D., Wasmann, C.C., Brown, J.K., Kanosh, M.R. and Bohnert, H.J. (1995b). Proteinase inhibitors of *Manduca sexta* expressed in transgenic cotton. *Plant Cell Reports* 14: 758–762.

Thomas, J.C., Wasmann, C.C., Echt, C., Dunn, R.L., Bohnert, H.J. and McCoy, T.J. (1994). Introduction and expression of an insect proteinase inhibitor in alfalfa (*Medicago sativa* L.). *Plant Cell Reports* 14: 31–36.

Tinjuangjun, P., Loc, N.T., Gatehouse, A.M.R., Gatehouse, J.A. and Christou, P. (2000). Enhanced insect resistance in Thai rice varieties generated by particle bombardment. *Molecular Breeding* 6: 391–399.

Tortiglione, C., Fanti, P., Pennacchio, F., Malva, C., Breuer, M., Loof, A. De., Monti, L.M., Tremblay, E. and Rao, R. (2002). The expression in tobacco plants of *Aedes aegypti* trypsin modulating oostatic factor alters growth and development of the tobacco budworm, *Heliothis virescens*. *Molecular Breeding* 9: 159–169.

Tortiglione, C., Malva, C., Pennacchio, F., Rao, R. and Scarascia Mugnozza, G.T. (1999). New genes for pest control. In Porceddu, E. and Pagnotta, M.A. (Eds.), *Genetics and Breeding for Crop Quality and Resistance. Proceedings of the XV EUCARPIA Congress*, 20–25 September, 1998, Viterbo, Italy. Dordrecht, The Netherlands: Kluwer Academic Publishers, 159–163.

Urwin, P.E., Atkins, H.J., Waller, D.A. and McPherson, M.J. (1995). Engineered oryzacystatin-I expressed in transgenic hairy roots confers resistance to *Globodera pallida*. *Plant Journal* 8: 121–131.

Vaeck, M., Reynaerts, A., Hofte, H., Jansens, S., DeBeuckleer, M., Dean, C., Zabeau, M., Van Montagu, M. and Leemans, J. (1987). Transgenic plants protected from insect attack. *Nature* 327: 33–37.

Van der Salm, T., Bosch, D., Honee, G., Feng, I., Munsterman, E., Bakker, P., Stiekema, W.J. and Visser, B. (1994). Insect resistance of transgenic plants that express modified *cry1A(b)* and *cry1C* genes: A resistance management strategy. *Plant Molecular Biology* 26: 51–59.

Van Engelen, F.A., Schouen, A., Molthoff, J.W., Roosien, J., Salinas, J., Dirkse, W.G., Schots, A., Bakker, P., Gommers, F., Jongsma, M.A., Bosch, D. and Stiekma, W.J. (1994). Coordinate expression of antibody subunit genes yields high levels of functional antibodies in roots of transgenic tobacco. *Plant Molecular Biology* 26: 1701–1710.

van Rensburg, J.B.J. (1999). Evaluation of *Bt*-transgenic maize for resistance to the stem borers *Busseola fusca* (Fuller) and *Chilo partellus* (Swinhoe) in South Africa. *South African Journal of Plant and Soil* 16: 38–43.

Viswakarma, N., Bhattacharya, R.C., Chakrabarty, R., Dargan, S., Bhat, S.R., Kirti, P.B., Shastri, N.V. and Chopra, V.L. (2004). Insect resistance of transgenic broccoli ('Pusa Broccoli KTS-1') expressing a synthetic cryIA(b) gene. *Journal of Horticultural Science and Biotechnology* 79: 182–188.

Vrieling, K., van Wijk, C.A.M. and Swa, Y.J. (1991). Costs assessment of the production of pyrrolizidine alkaloids in ragwort (*Senecio jacobaca* L.). *Oecologia* 97: 541–546.

Walker, D., Boerma, H.R., All, J. and Parrott, W. (2002). Combining cry1Ac with QTL alleles from PI 229358 to improve soybean resistance to lepidopteran pests. *Molecular Breeding* 9: 43–51.

Watson, J.D., Hopkins, N.H., Roberts, J.W., Steitz, J.A. and Weiner, A.M. (1987). *Molecular Biology of the Gene*. San Francisco, California, USA: Benjamin/Cummings Publishing.

Westedt, A.L., Douches, D.S., Pett, W. and Grafius, E.J. (1998). Evaluation of natural and engineered resistance mechanisms in *Solanum tuberosum* for resistance to *Phthorimaea operculella* (Lepidoptera: Gelechiidae). *Journal of Economic Entomology* 91: 552–556.

Western, A.W., Bloschl, G., Grayson, R.B., Girard, C., LeMetayer, M., Zaccomer, B., Bartlet, E., Williams, I., Bonade-Bottino, M., Pham-Delegue, M.H. and Jouanin L. (1998). Growth stimulation of beetle larvae reared on a transgenic oilseed rape expressing a cysteine proteinase inhibitor. *Journal of Insect Physiology* 44: 263–270.

Williams, W.P., Sagers, J.B., Hanten, J.A., Davis, F.M. and Buckley, P.M. (1997). Transgenic corn evaluated for resistance to fall armyworm and Southwestern corn borer. *Crop Science* 37: 957–962.

Wilmink, A., van de Ven, B.C.E. and Dons, J.J.M. (1995). Activity of constitutive promoters in various species from the Liliaceae. *Plant Molecular Biology* 28: 949–955.

Wilson, W.D., Flint, H.M., Deaton, R.W., Fischhoff, D.A., Perlak, F.J., Armstrong, T.A., Fuchs, R.L., Berberich, S.A., Parks, N.J. and Stapp, B.R. (1992). Resistance of cotton lines containing a *Bacillus thuringiensis* toxin to pink bollworm (Lepidoptera: Gelechiidae) and other insects. *Journal of Economic Entomology* 85: 1516–1521.

Wiseman, B.R., Lynch, R.E., Plaisted, D. and Warnick, D. (1999). Evaluation of *Bt* transgenic sweet corn hybrids for resistance to corn earworm and fall armyworm (Lepidoptera: Noctuidae) using a meridic diet bioassay. *Journal of Entomological Sciences* 34: 415–425.

Wong, E.Y., Hironaka, C.M. and Fischhoff, D.A. (1992). *Arabidopsis thaliana* small subunit leader and transit peptide enhance expression of *Bacillus thuringiensis* proteins in transgenic plants. *Plant Molecular Biology* 20: 81–93.

Wu, Y.R., Llewellyn, D., Mathews, A. and Dennis, E.S. (1997). Adaptation of *Helicoverpa armigera* (Lepidoptera: Noctuidae) to a proteinase inhibitor expressed in transgenic tobacco. *Molecular Breeding* 3: 371–380.

Xiang, Y., Wong, W.K.R., Ma, M.C. and Wong, R.S.C. (2000). *Agrobacterium*-mediated transformation of *Brassica campestris* ssp. *parachinensis* with synthetic *Bacillus thuringiensis cry1A(b)* and *cry1A(c)* genes. *Plant Cell Reports* 19: 251–256.

Xu, D., McElroy, D., Thoraburg, R.W. and Wu, R. (1993). Systemic induction of a potato pin 2 promoter by wounding methyl jasmonate and abscisic acid in transgenic rice plants. *Plant Molecular Biology* 22: 573–588.

Xu, D.P., Xue, Q.Z., McElroy, D., Mawal, Y., Hilder, V.A. and Wu, R. (1996). Constitutive expression of a cowpea trypsin inhibitor gene, *CpTi*, in transgenic rice plants confers resistance to two major rice insect pests. *Molecular Breeding* 2: 167–173.

Yang, C.D., Tang, K.X., Wu, L.B., Li, Y., Zhao, C.Z., Liu, G.J. and Shen, D.L. (1998). Transformation of haploid rice shoots with snowdrop lectin gene (*GNA*) by *Agrobacterium*-mediated transformation. *Chinese Journal of Rice Science* 12: 129–133.

Yang, G.D., Zhu, Z., Li, Y., Zhu, Z.J., Shangguan, X.X. and Wu, X. (2003). Expression and inheritance of snowdrop lectin gene (gna) in Chinese cabbage. *Acta Horticulturae Sinica* 30: 341–342.

Yang, G.D., Zhu, Z., Li, Y., Zhu, Z.J., Xiao, G.F. and Wei, X.L. (2002). Obtaining transgenic plants of Chinese cabbage resistant to *Pieris rapae* L. with modified CpTI gene (*sck*). *Acta Horticulturae Sinica* 29: 224–228.

Yeh, K.W., Lin, M.L., Tuan, S.J., Chen, Y.M., Lin, C.Y. and Kao, S.S. (1997). Sweet potato (*Ipomoea batatas*) trypsin inhibitors expressed in transgenic tobacco plants confer resistance against *Spodoptera litura*. *Plant Cell Reports* 16: 696–699.

Zambryski, P.C. (1992). Chronicles from the *Agrobacterium*-plant cell DNA transfer story. *Annual Review of Plant Physiology and Molecular Biology* 43: 465–490.

Zambryski, P.C., Joss, H., Genetello, C., Leemans, J., VanMontagu, L.M. and Schell, J. (1983). Ti-plasmid vector for the introduction of DNA into plant cells without alteration of their normal regeneration capacity. *EMBO Journal* 2: 2143–2150.

Zhang, F.P., Yin, W., Yu, C., Lan, L.B., Zhen, Z. and Hui, L.X. (1998). On the resistance of CpTI transgenic tobacco plants to cotton bollworm. *Acta Phytopathologica Sinica* 24: 331–335.

Zhang, J.H., Wang, C.Z., Qin, J.D. and Guo, S.D. (2004). Feeding behaviour of *Helicoverpa armigera* larvae on insect-resistant transgenic cotton and non-transgenic cotton. *Journal of Applied Entomology* 128: 218–225.

Zhang, W. and Wu, R. (1988). Efficient regeneration of transgenic plants from rice protoplasts and correctly regulated expression of the foreign gene in plants. *Theoretical and Applied Genetics* 76: 835–840.

Zhao, J.Z., Shi, X.P., Fan, X.L., Zhang, C.Y., Zhao, R.M. and Fan, Y.L. (1998). Insecticidal activity of transgenic tobacco co-expressing *Bt* and *CpTI* genes on *Helicoverpa armigera* and its role in delaying the development of pest resistance. *Rice Biotechnology Quarterly* 34: 9–10.

Zhu, C.X., Yao, F.Y., Wen, F.J. and Song, Y.Z. (2003). Genetics of *cry1A(b)* gene and its mediated resistance in transgenic rice. *Acta Phytophylacica Sinica* 30: 1–7.

8

Genetic Engineering of Entomopathogenic Microbes for Pest Management

Introduction

Entomopathogenic bacteria, viruses, fungi, nematodes, and protozoa have a great potential as a component of integrated pest management. However, they still account for <3% of the total pesticide market, and formulations based on the bacterium, *Bacillus thuringiensis* (Berliner) account for 80% to 90% of the commercial microbial pesticides. The commercial value of microbial pesticides is estimated at US$100 million per year (Meadows, 1990; Neale, 1997). The major constraint to the use of biopesticides is the need for simultaneous management of three biological systems: the pathogen, the prey, and the crop. Their greatest application is in vegetables, gardens, and orchard crops. Despite several advantages of biological insecticides, many factors have hindered their commercial success and practical effectiveness. Some of the problems associated with the use of microbial pesticides for pest management include:

- Quality and effectiveness;
- Unstable formulations and delivery systems;
- Sensitivity to light, relative humidity, and heat;
- Short shelf-life, especially in hot and humid conditions; and
- Limited host range and specificity to a particular stage of the insect.

Effective use of microbial pesticides often requires a more complex infrastructure than is needed for conventional pesticides. These include control of environmental conditions in the production area, an efficient system to deliver the product to the farmers, and training the farmers on the efficient use of biopesticides. With the exception of fungi, microbial insecticides do not kill the target insect on contact, but need to be ingested. The insect must feed on some plant tissue carrying the biopesticide before the agent can cause the mortality. Insects that bore into the plant tissue or remain hidden inside the plant structures are much less susceptible than insects feeding on foliage. All developmental stages of the insect are not equally susceptible to the microbial agents. As a result, the user must time the application precisely to prevent crop damage from an insect population exceeding the economic threshold. Some aspects of insect behavior also influence the performance of biopesticides. If major plant growth occurs after application of the biopesticide, then the unprotected plant surfaces become prone to insect damage. While proper timing of application is important, it may also be necessary to formulate the biological insecticide with feeding attractants such as molasses, which lure the insect to plant surfaces carrying the biocontrol agent. Contact poisons are superior in that an insect traversing a sprayed plant surface enroute to its preferred feeding site will contact the toxic residues regardless of its feeding habits.

Host specificity, which makes biopesticides ecologically and environmentally attractive, also constitutes a serious drawback. The narrow host range of biopesticides prevents their successful use to control a multiplicity of insect pests that feed on crop plants during the course of the growing season. For example, growers of crops such as sorghum, maize, rice, sugarcane, pigeonpea, groundnut, and cotton must contend with over a dozen different pest species, each with a different behavior and numbers that change dramatically over the crop growing season. As no single biological insecticide has been able to cope with the diversity of pests in a crop or cropping system, the necessity of multiple applications presents an economic obstacle to large-scale application of biological insecticides. Potency is another factor that has limited the use of biological insecticides. For economic reasons, the user will not apply an insecticide prophylactically in anticipation of an outbreak that might not occur or might not reach significant proportions. The insecticide must therefore be capable of controlling an insect infestation once the evidence of a threat is clearly established. Some biological insecticides, though highly effective under laboratory conditions, have been found to be ineffective under field conditions. Lack of potency has been ascribed, in some cases, to poor storage capability, short residual toxicity, slow mortality, requirements in the field for optimum temperature, relative humidity, and sunlight. Application of DNA-based technology promises new strains of entomopathogenic microbes with greater potency, increased host range, and adaptation to the harsh environmental conditions, and thus render them as an effective weapon for pest management in the future.

Natural versus Engineered Microbes

Microbial pesticides based on natural strains have been used for a long time. Several formulations based on different strains of *B. thuringiensis* (*Bt*) have been used over the past five decades. Many of these formulations have been improved by selecting or identifying more potent strains of the bacterium. Improvement of naturally occurring strains has increased the usefulness of biocontrol agents, and has taken advantage of the natural recombination based on conjugation, transduction, and transformation (Tortora, Funke, and Case, 1989; Schnepf et al., 1998). Transformation involves the death and lysis of the bacterial cell, and release of the DNA, which under certain circumstances can be taken up by the surrounding bacteria. This allows the transformation of the bacteria under controlled conditions in the laboratory, and results in new strains suitable for certain purposes (Stewart, 1992). Transduction is the result of viral infection in bacteria, resulting in the production of viral DNA and a coat protein, followed by assembly of the virus particle, and release of such particles from the cell. Transducing particles are formed by packaging of the host DNA in place of the viral DNA. In conjugation, the transfer of DNA takes place through cell-to-cell contact. It is controlled by plasmids, but can also be controlled by transposones. Replacement with alternate plasmids results in a different spectrum of genetic activity. Some plasmids can be transferred easily, while others require the presence of transfer proficient plasmids before genetic transfer can take place. These changes occur in nature, where exchange of the genetic material is a common phenomenon (Walter and Siedler, 1992).

Recent advances have made it easier to insert foreign DNA into organisms at will. Through this process, bacteria capable of performing different functions have been produced, for example, those producing mammalian proteins, hormones, or toxins from other organisms. This may involve the transfer of genetic material to another microorganism or it may be expressed differently, posing a new challenge. The exchange of DNA between microbes may be influenced by conjugal plasmids, cloning vectors, transducing phases, or transposones used, and the method used. Release of a number of genes into the environment may affect the natural process of selection and evolution, and addition of a large amount of genetic material can also increase the probability of mutations. To limit the exchange of genetic material, plasmids have been produced by altering the nucleotide sequences that are unable to function in the exchange process. However, under certain circumstances, helper plasmids can enter the cell, resulting in the mobilization of the disarmed plasmid (Zylstra, Cuskey, and Olson, 1992).

Genetic Engineering of Microbes

Several applications of genetic engineering of microbial pesticides will emerge in future. Microorganisms producing insecticidal secondary metabolites can be genetically engineered to produce such chemicals more efficiently. Engineered microbes can also be used to produce the intermediates needed for synthesis of insecticides. However, such chemicals are not polypeptides, and thus are not a product of a single gene, but are chemicals produced by multistep biosynthetic pathways. And optimization of the end product would require characterization and cloning of all the genes involved in the process so that all the steps in the process are performed in a concerted manner without disturbing the cell metabolism.

Entomopathogenic Viruses

The family of viruses belonging to Baculoviridae has the greatest potential as biopesticides because of their specificity to arthropods (Table 8.1). Narrow host range and slow action limit the use of entomopathogenic viruses. Genetic engineering provides an effective means of improving virus activity, and to this end, a variety of recombinant "rapid action" nuclear polyhedrosis viruses (NPVs) have been developed (Figure 8.1). A major focus of research on insect viruses involves engineering the viruses to express genes from other organisms that code for insecticidal proteins, which either kill the insects or disrupt normal physiological functions. Recombinant baculoviruses expressing specific toxins can be generated rapidly, and applied on a range of crops. Genes encoding neurotoxins have

TABLE 8.1

Some Commonly Used/Commercial Baculovirus Products for Insect Control

Baculovirus	Trade Name	Target Host(s)	References
Mb-NPV	Mamestrin, Virin-EKS	*Mamestra brassicae*	Cunningham (1995), Lacey et al. (2001), Inceoglu et al. (2001)
Px-GV	—	*Plutella xylostella*	Moscardi (1999)
Ha-NPV	HaNPV	*Helicoverpa armigera*	Moscardi (1999)
H-NPV	NPV	*Helicoverpa zea*	Ignoffo and Couch (1981), Moscardi (1999)
H-NPV	NPV	*Heliothis virescens*	Igonffo and Couch (1981), Moscardi (1999)
Ac-NPV	VPN 80	*Autographa californica*	Vail, Anderson, and Jay (1973), Cunningham (1995)
Tn-NPV	—	*Trichoplusia ni*	Ignoffo (1964)
Ld-NPV	Dispovirus	*Lymantria dispar*	Cunningham (1995), Shapiro, Robertson, and Bell (1987), Lacey et al. (2001)
Cp-GV	CYD-X, Madex, Caprovirisine, Granusal	*Cydia pomonella*	Jackson et al. (1992), Moscardi (1999)
Ao-GV	Capex	*Adoxophyes orana*	Lacey et al. (2001), Inceoglu et al. (2001)
Af-MNPV	AfNPV	*Heliothis, Helicoverpa, Spodoptera*	Lacey et al. (2001)
Ac-MNPV	Gusano	Lepidoptera	Lacey et al. (2001)
Hz-SNPV	Gemstar	*Helicoverpa* spp.	Lacey et al. (2001)
Or NPV	—	*Oryctes rhinoceros*	Bedford (1980)
Se-MNPV	SPOD-X	*Spodoptera exigua*	Cherry et al. (1997)
Sl NPV	Spodopterin	*Spodoptera littoralis*	Cunningham (1995), Cherry et al. (1997)
Sl-NPV	—	*Spodoptera litura*	Moscardi (1999)
Ag-NPV	Polygen, Multigen, Nitral, Protege	*Anticarsia gemmatalis*	Moscardi (1999), Moscardi and Sosa-Gomez (2000)

also been expressed in baculoviruses or larvae infected with chimeric baculoviruses. Recombinant baculoviruses expressing scorpion neurotoxins have been developed (Carbonell et al., 1988; Tomalski and Miller, 1991; Stewart et al., 1991; Bonning and Hammock, 1996). Venomous toxins produced by scorpions and spiders (Van Rie et al., 1990;

FIGURE 8.1   Larva of *Helicoverpa armigera* infected with the nuclear polyhedrosis virus (HaNPV). The vilulence of HaNPV can be improved through genetic engineering.

Wigle et al., 1990) and hymenopteran insects (Piek et al., 1989; Piek, 1990) can be used for increasing the effectiveness of baculoviruses. There is a distinct possibility of expressing excitatory and depressant proteins, as well as *Bt* toxins, in baculoviruses. Such an expression would increase the effectiveness of the NPVs, and can also be exploited to broaden their host range. Signal consequences of all the genes expressed in baculoviruses are processed appropriately, producing toxin proteins with amino acid terminal sequences similar to those of native proteins (Luckow and Summers, 1988), but the ratio of expression varies, depending on the protein being expressed and the signal sequence.

Baculoviruses with Neurotoxins

One of the major focuses for genetic modification has been to increase their effectiveness through insertion of neurotoxin genes from arthropods, which act by blocking the synaptic transmission or by inhibiting the activity of ion channels. Because of their specificity, natural toxins have become quite useful in designing new biopesticides, including baculoviruses (Zlotkin et al., 1971a, 1971b, 1971c; Zlotkin, 1988). Genes that code for scorpion, *Androctonus australis* Hector (*AaIT*), and mite venom have been engineered into the baculovirus (AcNPV) from *Autographa californica* (Speyer) to increase the effectiveness of this virus (Tomalski and Miller, 1991; Stewart et al., 1991; Bonning and Hammock, 1996) (Table 8.2). The *AaIT* shows remarkable specificity to insects and has no apparent effects on mammals (Zlotkin et al., 1971a; Fontecilla-Camps, 1989; DeDianous, Hoarau, and Rochat, 1987). The toxin proteins bind to the sodium channel proteins and affect the neuronal membranes (Lester et al., 1982; Catterall, 1984; Zlotkin and Gordon, 1985; Zlotkin, 1988).

TABLE 8.2

Genetically Modified Baculoviruses for Pest Management

Insect Species/NPV	Gene	GMO/Promoter	References
Autographa californica NPV (Ac-NPV)	*Androctonus australis* toxin gene (*AaIT*)	Ac-AaIT	Stewart et al. (1991), Hoover et al. (1995)
Ac-MNPV	*AaIT*	Ac–AaIT / *P10*	McCutchen et al. (1991)
Bombyx mori NPV (Bm-NPV)	*AaIT*	Bm-AaIT	Maeda et al. (1991)
Ac-NPV	*AaIT*	VAc-LdPD	Du and Thiem (1997)
Ac-NPV	Enhancin gene from *Trichoplusia ni* (*AcENh26*)	AcENh26	Hayakawa et al. (2000)
Heliothis zea NPV (Hz-NPV)	*AaIT*	Hz-AaIT	Treacy, Rensner, and All (2000)
Helicoverpa armigera NPV (Ha-SNPV)	*Egt*	HaSNPV-egt	Sun et al. (2004)
Ac-MNPV	*Mu-Aga-IV—Agelenopsis aperta* *AsII—Anemonia sulcata* *Sh1—Stydactyla*	*Phsp70* promoter	Prikhod ko et al. (1998)
Baculovirus	*Pyemotes tritici, Tox34*	*VEV-Tox34*	Tomalski and Miller (1991)
Ac-MNPV	Juvenile esterase from *Heliothis virescens*	*p10* promotor	Hammock et al. (1990)
Baculovirus	Diuretic hormone	—	Maeda (1989)
Ac-MNPV	Viral gene (*egt*), which codes for UDP-glycosyl transferase	—	O'Reilly and Miller (1991)

High selectivity of scorpion neurotoxins makes them an ideal choice to improve the efficacy of baculoviruses. The *AaIT* neurotoxin, specific to insect larvae, is associated with a single protein, which is distinct from those associated with toxicity to mammals (Zlotkin et al., 1971b). The scorpion venom toxins are neuropeptides of 60 to 70 amino acids cross-linked by four disulfate bonds (Bernard, Courard, Rochat, 1979; Anderson, 1992). The *AaIT* causes fast and reversible paralysis in insects, similar to pyrethroids (Zlotkin et al., 1991). Maeda et al. (1991) expressed the cDNA of *AaIT* in NPV from silkworm, *Bombyx mori* L., and achieved a significant increase in biological activity. However, the larvae had to be infected by injecting the baculovirus instead of by feeding because the recombinant baculovirus was polyhedron-negative. A polyhedron-positive baculovirus (*AcMNPV*) has now been developed that expresses the *AaIT* from a synthetic gene under the control of the *p10* promoter (McCutchen et al., 1991; Stewart et al., 1991). Bioassay with the chimeric baculovirus *vAcUW2* (Carbonell et al., 1988) has shown a significant decrease (40%) in time to kill the larvae of Southern armyworm, *Spodoptera eridania* (Cramer), cabbage looper, *Trichoplusia ni* (Hubner), and tobacco budworm, *Heliothis virescens* (Fab.). The recombinant virus is active in lepidopteran larvae when fed orally, and is expected to be stable under field conditions (Kuzio, Jaques, and Faulkner, 1989).

Wild-type *A. californica* nuclear polyhedrosis virus (*WT AcNPV*) has been modified to encode for an insect-selective toxin derived from the venom of the scorpion, *A. australis* to produce the recombinant virus *AcAaIT* (Hoover et al., 1995). Larvae of tobacco budworm, *H. virescens*, infected with the baculovirus *AcAaIT* fell off the plant 5 to 11 hours before death, and were unable to climb back on to the plant to continue feeding. These larvae were capable of feeding if placed on diet or confined to a leaf with a clip cage. Larvae infected with *AcAaIT* consumed significantly less foliage than the larvae infected with *WT AcNPV* or uninfected controls. Absence of liquefied cadavers on plants following treatment with *AcAaIT* may be more desirable to growers and consumers relative to treatment with wild-type virus. Du and Thiem (1997) developed a new recombinant *A. californica* nuclear polyhedrosis virus, *vAcLdPD*, bearing only the gene for host range factor 1 (*hrf-1*) controlled by its own promoter. Recombinant *AcNPV* with the enhancin gene from *T. ni* granulovirus, *AcEnh26*, has been propagated in Sf9 cells (Hayakawa et al., 2000). The infected cultured cells were combined with either *AcNPV* occlusion bodies (OBs) or *Spodoptera exigua* (Hubner) NPV (SeNPV) OBs and fed to third-instar larvae of *S. exigua*. Feeding larvae with *AcEnh26*-infected cells resulted in a 21- and 10-fold enhancement of infection by *AcNPV* and *SeNPV*, respectively, as compared to the controls.

Insecticidal properties of *AcAaIT* and corn earworm, *Helicoverpa zea* (Boddie), NPV (*HzAaIT*) genetically altered with toxin from *A. australis* have been evaluated against selected Heliothine species by Treacy, Rensner, and All (2000). The LD_{50} based on diet-overlay bioassays showed *AcAaIT* and *HzAaIT* to be equally virulent against larval tobacco budworm, *H. virescens*, but *HzAaIT* showed 1,335-fold greater biological activity than *AcAaIT* against corn earworm, *H. zea*. The *HzAaIT* killed the larvae at a faster rate than the nontransformed HzNPV (LT_{50} 2.5 and 5.6 days, respectively). In the greenhouse, foliar sprays of *AcAaIT* and *HzAaIT* were found to be equally effective in controlling *H. virescens* on cotton. However, *HzAaIT* was superior to *AcAaIT* against *H. zea* on cotton. Cotton treated with *AcAaIT* or *HzAaIT* at 10×10^{11} occlusion bodies (OB) ha^{-1} averaged 2.5 and 16.2 nondamaged flower buds per plant, respectively. Quicker killing speed exhibited by *HzAaIT* led to more improved protection than with HzNPV. Field trials indicated that *HzAaIT* at 5 to 12×10^{11} OB ha^{-1} provided better control of Heliothine complex in cotton than *Bt*, and equal to spinosad, but slightly poorer than pyrethroid and carbamate insecticides.

HzNPV is better for designing recombinant clones as an insecticide targeted at the multispecies Heliothine complex. Pesticidal properties of HaSNPV have also been improved by deleting the ecdysteroid UDP-glucosyltransferase (*egt*) gene from its genome (recombinant *HaSNPV-EGTD*) and incorporating an insect-selective toxin gene from the scorpion, *A. australis* (recombinant *HaSNPV-AaIT*) (Sun et al., 2004). No differences were observed in inactivation rates of the two recombinant *HaSNPVs* and their parent wild type, *HaSNPV-WT*. The average half-life of *HaSNPV* was 0.39 to 0.90 days, and inactivation rates correlated well with solar radiation.

The effects of expressing three *PsynXIV*-promoted toxin genes, *mag4*, *sat2*, and *ssh1*, that encode secretable and potent neurotoxins: *mu-Aga-IV* from *Agelenopsis aperta* (Gertsch), *AsII* from *Anemonia sulcata* Pennant, and *ShI* from *Stichadactyla*, respectively, on the ability of the *A. californica* nuclear polyhedrosis virus (AcMNPV) to effectively kill its host have been studied (Prikhod ko et al., 1998). The toxin genes *mu-Aga-IV* and *AsII* act at distinct sites on voltage-sensitive sodium channels of insects and synergistically promote channel opening. These toxins also had a synergistic effect against *Lucilia sericata* Meig. and *Spodoptera frugiperda* (J.E. Smith). Viruses expressing toxin genes under the control of the *Phsp70* promoter were more effective as bioinsecticides than under the control of the *PsynXIV* promoter. The *S. frugiperda* and *T. ni* larvae were most affected by *mu-Aga-IV* and *AsII*, respectively. When *mag4* and *sat2* were simultaneously expressed under the control of *Phsp70*, the properties were similar to two viruses expressing each of the toxin genes individually, except that larval feeding time was reduced by 10%, indicating a small advantage to coproducing synergistic toxins. There is considerable scope to produce baculoviruses expressing insect depressant toxins. Depressant insect toxin *LqhIT2* from Israeli yellow scorpion, *Leiurus quinquestriatus* (Ehren.) Hebreaus and *BjIT2* from the black scorpion, *Buthotus judaicus* Simon are nontoxic to mammals, but toxic to insects (Lester et al., 1982). The depressant toxins are quite different from the exitant toxins, and 50 mg of *LqhIT* induces complete paralysis in 50 minutes. AcDNA fragment uncoding *LqhIT2* has been synthesized (Kopeyan et al., 1990; Zlotkin et al., 1991). Full-length DNA encoding *LqhIT2* has been sequenced, and contains a signal sequence of 63bp preceding the 183 base pair coding sequence (Zilberberg, Zlotkin, and Gurevitz, 1992).

The spider toxins block neuromuscular transmission mediated by L-glutamate receptors and the ion channels. Polypeptide neurotoxins from *Hololena curta* (McCook) have 36 to 38 amino acids, and result in paralysis and presynaptic blockade (Stapleton et al., 1990). The straw mite, *Pyemotes tritici* (Lagreze-Fossat and Montane), produces a 27 kDa protein, which is toxic to a number of insect species. The mite toxin *TxP-1* is as effective as *AaIT* in the insect nervous system (Tomalski et al., 1989). A recombinant baculovirus expressing insect specific neurotoxic gene from *P. tritici*, *lox34*, has been constructed, and found to be effective on insects (Tomalski and Miller, 1991). The baculoviruses expressing the mite toxin gene may be more readily accepted by the public under field conditions. There is a need to understand the structure-function relationships to design toxin genes with improved virulence, improved quantity of toxin protein, and stability during and after production. Venom proteins from the ant, *Paraponera clavata* (Fab.), have also been purified (Piek et al., 1991). Toxins secreted by the wasps are highly toxic to lepidopteran larvae (Piek, May, Spanger, 1980; Piek et al., 1989). Venoms from *Colpa interrupta* (F.) and *Ampulex compressa* (Fab.) show activity on mammals, as well as on the cockroach, *Periplaneta americana* (L.). Peptides of 9 to 18 amino acid residues of wasps and ant venoms are toxic, and cause irreversible blocking of the insects' central nervous system (Piek, 1990, 1991). These can also be considered as potential genes for genetic transformation of baculoviruses.

Baculoviruses Expressing Insect Diuretic Hormones

The genes for juvenile hormone (JHE) esterase and diuretic hormone have also been expressed in baculoviruses (Maeda, 1989; Hammock et al., 1990; Bonning and Hammock, 1996). Proteases that perforate the insect cell membranes and hasten systemic spread of infection have also been expressed in NPVs (Harrison and Bonning, 2001). Baculoviruses expressing insect diuretic hormone that affect insect morphogenesis have also been developed (Maeda, 1989). Hammock et al. (1990) developed a recombinant NPV expressing juvenile hormone esterase. Excessive levels of JHE reduced feeding and delayed the development of insect larvae. Baculoviruses expressing JHE with point mutations designed to increase stability have been found to be more effective than wild-type JHE. A more dramatic effect has been obtained by selecting the viral gene *egt* (which codes for UDP glycosyl transferase, needed for production of the insect molting hormone ecdysterone) (O'Reilly and Miller, 1991).

Expression of Entomopoxvirus in Bacteria

A virus-enhancing factor (EF) from *Mythimna separata* (Walk.), entomopoxvirus, has been successfully expressed using an *Escherichia coli* Escherich expression system, *pGEX-2T* (Hukuhara, Hayakawa, and Wijonarko, 2001). The EF enhanced the fusion between the *PsunMNPV* and the cultured insect cell line, SIE-MSH-805-H. The resulting lysates of transformed *E. coli* cells enhanced *PsunMNPV* infection when treated with either trypsin or thrombin. The EF enhanced *Pseudaletia unipuncta* (Walk.) (=*M. separata*) multi-nucleopolyhedrovirus (*PsunMNPV*) infection in larvae of the armyworm, *P. unipuncta*.

Effect of Insecticides on Biological Activity of Baculoviruses Expressing Neurotoxins

The effect of deltamethrin has been studied on the larvae of *L. sericata*, *H. virescens*, and *T. ni* infected with *A. californica* NPV expressing *mu-Aga-IV* from *A. aperta* and *AsII* from *A. sulcata* (Popham et al., 1998). Co-application of deltamethrin conferred no advantage in field applications of recombinant baculovirus. Application of pyrethroids probably will have no adverse effect on recombinant virus efficacy unless it deters larval feeding and virus acquisition.

Role of Baculoviruses in Pest Management

To prolong the insecticidal activity of the recombinant baculoviruses expressing neurotoxins, it is important to ensure continued production of the neurotoxin through appropriate regulatory elements or by increasing the copy number of the target gene. It is still too early to assess how such engineered viruses will fit into integrated pest management (IPM) programs. Insect pathogenic viruses usually attack a narrow taxonomic range of insects and must be ingested to have an effect. Even if the range of insects that a virus can affect is broadened by genetic engineering, the recombinant viruses may affect the predatory and parasitic insects if the virus is infective to them. Although most researchers would like to produce an engineered organism that cannot survive on its own in the wild, there is some concern that new traits in engineered viruses may be passed on to other more competent viruses or the "disarmed" engineered viruses might mutate to a more independent form. How much environmental damage such a virus might cause has to be determined. The most important challenge to biotechnology with regard to insect viruses is the development of methods for mass production that do not require live insect hosts. If right media could be developed for producing effective NPVs, this would greatly enhance their utility in IPM programs. Sequencing of entire genomes of NPVs will throw more light on the structure-function relationships of baculoviruses, which will facilitate their production with better safety. The promise of baculoviruses for pest control is yet to be realized, because of low speed of kill, long incubation period in the host, and failure to reduce insect damage. The availability of selective and fast-acting viral insecticides would certainly reduce overdependence on synthetic pesticides.

Entomopathogenic Bacteria

Several entomopathogenic bacteria belonging to Bacilliaceae, Micrococcaceae, Lactobacillaceae, and Pseudomonaceae have been identified, of which *Bacillus sphaericus* Meyer and Neide, *B. thuringiensis*, *B. thuringiensis* var. *israeliensis*, *B. thuringiensis* var. *kurstaki*, *B. subtilis*, (Ehrenberg) Gohn, *Paenibacillus* (*Bacillus*) *lentimorbus* Dutky, and *Paenibacillus* (*Bacillus*) *poppillae* Dutky have been studied in considerable detail as biological control agents (Miller, Lingg, and Bulla, 1983; Charles, Neilsen-Leroux, and Delécluse, 1996). *Bacillus poppillae* and *Serratia entomophila* Grimont et al. have been used for the control of white grubs, while *B. sphaericus* has been used for the control of mosquito larvae (Klein and Jackson, 1992; Klein and Kaya, 1995). *Bacillus sphaericus* is more persistent in polluted habitats, recycles under certain conditions, but has a narrow host range (Nicolas, Regis, and Rios, 1994; Charles, Nielsen-Leroux, and Delécluse, 1996). The bacterium *Photorhabdus luminescens* (Thomas and Poinar) Boemare et al., which lives in the gut of entomophagous nematodes, produces a toxin, which consists of four native complexes encoded by *tca*, *tcb*, *tcc*, and *tcd*. Both *tca* and *tcd* encode complexes with high oral toxicity to tobacco hornworm, *Manduca sexta* L., and therefore are potential alternatives to *Bt* (Bowen et al., 1998). Genes from *P. luminescens* that encode insecticidal toxins have recently been described from other insect-associated bacteria such as *S. entomophila*, an insect pathogen, and *Yersinia pestis* KIM, the causative agent of bubonic plague, which has a flea vector (Waterfield et al., 2001).

Bacillus thuringiensis

Bacillus thuringiensis, which has been exploited extensively for insect control, is an aerobic, Gram-positive endospore-forming bacterium, and is widespread in natural environments in the soil. Ishiwata discovered this bacterium in 1901 from diseased silkworm, *B. mori* larvae. Berliner (1915) isolated it from diseased larvae of *Ephestia kuhniella* (Zeller), and designated it as *B. thuringiensis*. Its genome size is 2.4 to 5.7 million bp (Carlson, Caugant, and Kolsto, 1994). *Bacillus thuringiensis* isolates have several extra chromosomal elements; some are circular and others linear. Proteins comprising the parasporal crystal are generally encoded by large plasmids (González, Dulmage, and Carlton, 1981). Sequences hybridizing to *cry* gene probes occur commonly among *B. thuringiensis* chromosomes (Carlson and Kolsto, 1993), but the degree to which these chromosomal homologs contribute to production of crystal proteins is not clear. *Bacillus thuringiensis* harbors a large number of transposable elements, including insertion sequences and transposons. Transposable elements are probably involved in amplification of *cry* genes in the bacterial cell. The *cry1A* genes are flanked by two sets of inverted repeat sequences (Kronstad and Whiteley, 1984), which have been designated as IS*231* and IS*232* (Lereclus et al., 1992; Lysenko, 1983), and belong to the IS*4* and IS*21* family of insertion sequences, respectively (Rezsohazy et al., 1993; Menou et al., 1990). In *B. thuringiensis* subsp. *israelensis*, IS*231W* is adjacent to the *cry11Aa* gene. IS*231*-related DNA sequences have also been found in strains of *B. cereus* Frankland and *B. mycoides* Flugge (Leonard, Chen, and Mahillon, 1997). IS*240* is invariably present in dipteran-active strains (Rezsohazy et al., 1993). Insertion sequences have also been found upstream of *cry1Ca* (Smith et al., 1994) and downstream of *cry2Ab* (Hodgman et al., 1993). A transposable element designated as Tn*5401* has been isolated from a coleopteran-active *B. thuringiensis* (Baum, 1994), which is located downstream of the *cry3Aa* gene. Tn*4430* mediates the transfer of nonconjugative plasmids by a conduction process (Green, Battisti, and Thorne, 1989).

Gene Expression and Cry Structure

A common characteristic of all *cry* genes is their expression during the stationary phase, and their products accumulate in the cell to form a crystal inclusion, which accounts for 20 to 30% of the dry weight of the sporulated cells. Crystal protein synthesis and accumulation in *B. thuringiensis* are controlled by a variety of mechanisms. Sporulation is controlled by successive activation of sigma factors, which bind the core RNA polymerase to direct the transcription from sporulation-specific promoters. The *cry1Aa* gene is a sporulation-dependent *cry* gene expression in the mother cell. Two transcription start sites have been mapped (*BtI* and *BtII*) (Wong, Schnepf, and Whiteley, 1983), of which *BtI* is active between T_2 and T_6 stages of sporulation, while *BtII* is active from T_5 onwards. *In vitro* transcription experiments have indicated that at least two other *cry* genes (*cry1Ba* and *cry2Aa*) contain either *BtI* alone or *BtI* with *BtII* (Brown and Whiteley, 1988).

The expression of *cry* genes is considered to be sporulation dependent (Poncet et al., 1997). However, low levels of transcription of *cry4Aa*, *cry4Ba*, and *cry11Aa* genes have been detected during the transition phase (Yoshisue et al., 1995), and may be controlled by RNA polymerase (Poncet et al., 1997). The *cry3Aa* gene from *B. thuringiensis* var. *tenebrionis* is expressed during the vegetative phase, although to a lesser extent than during the stationary phase (De Souza, Lecadet, and Lereclus, 1993; Malvar, Gawron Burke, and Baum, 1994; Sekar, 1988). The *cry3Aa* expression increases in mutant strains incapable of initiating sporulation (Agaisse and Lereclus, 1994; Malvar and Baum, 1994; Lereclus et al., 1995; Salamitou et al., 1996). The *cry3Aa* expression is activated by a non-sporulation-dependent mechanism during the transition from exponential growth to the stationary phase (Agaisse and Lereclus, 1994; Salamitou et al., 1996).

The toxin proteins generally form crystalline inclusions in the mother cell, and depending on protoxin composition, the crystals have various forms: bipyramidal (Cry1), cuboidal (Cry2), flat rectangular (Cry3A), irregular (Cry3B), spherical (Cry4A and Cry4B), and rhomboidal (Cry11A) (Schnepf et al., 1998). The ability of the protoxins to crystallize may decrease their susceptibility to premature proteolytic degradation. However, the crystals have to be solubilized rapidly in the insect gut to become biologically active. Structure and solubility characteristics of the crystal presumably depend on secondary structure of the protoxin, energy of the disulfide bonds, and presence of additional *B. thuringiensis*-specific components. The cysteine-rich C-terminal half of the Cry1 protoxins possibly contributes to crystal structure through the formation of disulfide bonds (Bietlot et al., 1990). The cysteine-rich C-terminal region is absent from the 73 kDa Cry3A protoxins and this protein forms a flat, rectangular crystal inclusion in which the polypeptides do not appear to be linked by disulfide bridges (Bernhard, 1986). Analysis of the three-dimensional structure of the Cry3A has revealed the presence of four intermolecular salt bridges, which might participate in the formation of the crystal inclusion (Li, Carroll, and Ellar, 1991). Crystallization of Cry2A (71 kDa) and Cyt1A (27 kDa) requires the presence of accessory proteins (Agaisse and Lereclus, 1995; Baum and Malvar, 1995). These proteins may act at the post-translational level to stabilize the nascent protoxin molecule and facilitate crystallization. Cry1Ia toxin has been found in supernatant of *B. thuringiensis* cultures as a processed polypeptide of 60 kDa (Kostichka et al., 1996).

Because of the crystalline nature of these proteins, the term Cry is used in gene and protein nomenclature. Several *cry* gene promoters have been identified, and their sequences determined (Brizzard, Schnepf, and Kronstad, 1991; Brown, 1993; Yoshisue et al., 1993; Dervyn et al., 1995). The toxin genes earlier were classified into four types, based on insect specificity and sequence homology (Hofte and Whiteley, 1989). CryI-type genes encode proteins of 130 kDa, and are usually specific to lepidopteran larvae. Type II genes encode for 70 kDa proteins that are specific to lepidopteran and dipteran larvae, while type III genes encode for 70 kDa proteins specific to coleopteran larvae. Type IV genes are specific to the dipteran larvae. The system was further extended to include type V genes that encode for proteins effective against lepidopteran and coleopteran larvae (Tailor et al., 1992). The *Bt* δ-endotoxins are now known to constitute a family of related proteins for which over 140 genes have been described (Crickmore et al., 1998), with specificities for Lepidoptera, Coleoptera, and Diptera. The crystalline proteins get solubilized in the insect midgut at high pH, releasing proteins called δ-endotoxins. The toxin portion is derived from the N-terminal half of the protoxin, while the C-terminal portion is involved in the formation of parasporal inclusion bodies and is usually hydrolyzed into small peptides (Choma et al., 1990).

Mode of Action

The mode of action of the *B. thuringiensis* Cry toxins involves solubilization of the crystal protein in the insect midgut, proteolytic processing of the protoxin by proteases to toxin, binding of the toxin to midgut receptors, and insertion of the toxin into the apical membrane to create ion channels or pores. For most lepidopterans, protoxins are solubilized under the alkaline conditions of the insect midgut (Hofmann et al., 1988). Differences in the extent of solubilization often are associated with differences in toxicity of Cry proteins to various insect species (Aronson et al., 1991; Du, Martin, and Nickerson, 1994; Meenakshisundaram and Gujar, 1998). Decreased solubility could be one potential mechanism for insect resistance to *Bt* proteins (McGaughey and Whalon, 1992). In cotton bollworm, *H. zea*, CryIIA is less soluble than Cry1Ac, and fails to bind to a saturable binding component in the midgut brush border membrane (English et al., 1994). The unique mode of action of CryIIA may provide a useful tool for management of resistance to *Bt* toxins. Although binding of the Cry toxins to the receptors determines the species sensitivity to various toxins, there are distinct exceptions, for example, Cry1Ac binds to the ligand bands of beet armyworm, *S. exigua* brush border membrane proteins, but there is very little toxicity to the insect (Garczynski, Crim, and Adang, 1991; Garczynski and Adang, 1995). Cry1Ab is more toxic to the gypsy moth, *Lymantria dispar* (L.), than Cry1Ac, but does not bind well with the receptors in the brush border membrane (Wolfersberger, 1990).

After solubilization, many protoxins need to be processed by midgut proteases to become activated toxins (Lecadet and Dedonder, 1967; Tojo and Aizawa, 1983). The major proteases of the lepidopteran insect midgut are of trypsin (Milne and Kaplan, 1993; Lecadet and Dedonder, 1966) or chymotrypsin type (Johnston et al., 1995; Peterson, Fernando, and

Wells, 1995; Novillo, Castanera, and Ortego, 1997). The Cry1A protoxins are digested to a 65 kDa protein starting at the C-terminus, and proceeds towards the 55 to 65 kDa toxic core (Chestukhina et al., 1982; Choma et al., 1990). The carboxy-terminal end of the protoxin is clipped off in 10 kDa sections (Choma, Surewicz, and Kaplan, 1991). The mature Cry1A toxin is cleaved at R28 at the amino-terminal end (Nagamatsu et al., 1984), while Cry1Ac is cleaved at K623 on the carboxy-terminal end (Bietlot et al., 1989). Activated toxin binds to specific receptors on the apical brush border membrane vesicles (BBMV) of the midgut (Hofmann and Luthy, 1986; Hofmann et al., 1988). Irreversible binding is associated with membrane insertion (Van Rie et al., 1989; Ihara et al., 1993; Rajamohan et al., 1995) and requires the insertion of domain I (Flores et al., 1997). Cry1Aa and Cry1Ab bind to purified *M. sexta* aminopeptidase N (APN) (Masson et al., 1995). Cry1Ac also binds irreversibly to purified *L. dispar* APN (Vadlamudi, Ji, and Bulla, 1993; Valaitis et al., 1997). Cry1Ab receptor is believed to be a cadherin-like 210 kDa membrane protein (Francis and Bulla, 1997; Keeton and Bulla, 1997), while the Cry1Ac and Cry1C receptors have been identified as APN proteins with molecular weights of 120 and 106 kDa, respectively (Knight, Crickmore, and Ellar, 1994; Sangadala et al., 1994; Luo, Lu, and Adang, 1996).

Incorporation of purified 120 kDa APN into planar lipid bilayers catalyzes channel formation by Cry1Aa, Cry1Ac, and Cry1C (Schwartz et al., 1997). There is some evidence that domain II from either Cry1Ab or Cry1Ac promotes binding to the larger protein, while domain III of Cry1Ac promotes binding to APN (de Maagd et al., 1996a, 1996b). Alkaline phosphatase has also been proposed to be a Cry1Ac receptor (Sangadala et al., 1994). In *H. virescens*, three aminopeptidases bind to Cry1Ac. The 170 kDa APN binds to Cry1Aa, Cry1Ab, and Cry1Ac, but not Cry1C or Cry1E. N-Acetylgalactosamine inhibits the binding of Cry1Ac, but not of Cry1Aa or Cry1Ab. In gypsy moth, *L. dispar*, the Cry1Ac receptor seems to be APN, while Cry1Aa and Cry1Ab bind to a 210 kDa BBMV protein (Valaitis et al., 1995, 1997). In *P. xylostella* (Luo, Tabashnik, and Adang, 1997) and *B. mori* (Yaoi et al., 1997), the APN appears to function as a Cry1Ac binding protein. A gene encoding a Cry1Ab-binding APN has been cloned in *M. sexta* and its homolog in *P. xylostella* (Denolf et al., 1997). Insertion into the apical membrane of the columnar epithelial cells follows the initial receptor-mediated binding, rendering the toxin insensitive to proteases and monoclonal antibodies (Wolfersberger, Hofmann, and Luthy, 1986), inducing ion channels or nonspecific pores in the target membrane.

Several studies have demonstrated the association of insect specificity of a toxin with its affinity for specific receptors on BBMV (Hofmann and Luthy, 1986; Hofmann et al., 1988; Van Rie et al., 1989). Insect specificity of Cry1Aa and Cry1Ac is localized in the central domain of the toxin in *B. mori* and *T. ni*, and the central and C-terminal domain for *H. virescens* (Ge, Shivarova, and Dean, 1989; Ge et al., 1991; Bosch and Honee, 1993). Specificity and binding domains were colinear for Cry1Aa against *B. mori* (Lee et al., 1992). However, Wolfersberger (1990) observed that Cry1Ab was more active than Cry1Ac against gypsy moth, *L. dispar*, larvae, despite exhibiting a relatively weaker binding affinity. A number of other examples indicating the lack of correlation between receptor binding affinity and insecticidal activity have also been reported (Van Rie et al., 1989; Garczynski, Crim, and Adang, 1991; Sanchis and Ellar, 1993). Liang, Patel, and Dean (1995) observed that affinity of Cry1Ab was not directly related to toxin activity in gypsy moth, but observed a direct correlation between the irreversible binding rate and toxicity. Minor changes in binding usually do not have a major effect on toxicity. Binding affinity as measured by competition binding or irreversible binding may effect toxicity. The same mutation in a toxin can have different effects in different insects. Different toxins may have the same amino acid sequence in the loops of domain II (e.g., Cry1Ab and Cry1Ac), and yet bind to different receptors. Specificity differences for Cry1C between *S. frugiperda* Sf9 cells and *Aedes aegypti* (L.) larvae can be changed radically by single point mutations in the loops (Smith and Ellar, 1994). Domain III has also been implicated in receptor binding of Cry1Ac in *H. virescens* (Ge, Shivarova, and Dean, 1989; Ge et al., 1991). Domain switching experiments have also suggested a role for Cry1Ab domain III in binding to BBMV in *S. exigua* (de Maagd et al., 1996b).

Biopesticides Based on Bacillus thuringiensis

Natural isolates of *B. thuringiensis* produce several crystal proteins, each of which exhibits different target specificity (Hofte and Whiteley, 1989; Lambert and Peferoen, 1992). Its toxins are known to kill insect species belonging to Lepidoptera, Coleoptera, and Diptera (Hofte and Whiteley, 1989), and nematodes (Feitelson, Payne, and Kim, 1992). The HD 1 strain identified by Dulmage (1981) is the most important *Bt* product in the market. The first commercial *Bt* product, "Sporeine," was marketed in France (Luthy, Cordier, and Fischer, 1982). The major breakthrough came with the development of two commercial *Bt* products, "Thuricide" and "Dipel" in the 1960s. Formulations based on *Bt* occupy the key position, accounting for nearly 90% of the total biopesticide sales worldwide (Neale, 1997), with annual sales of nearly US$90 million (Lambert and Peferoen, 1992). Certain combinations of Cry proteins have also been shown to exhibit synergistic effects (Chang et al., 1993; Crickmore et al., 1995; Lee et al., 1996; Poncet et al., 1995; Wu, Johnson, and Federici, 1994). There are over 50 registered *Bt* products with more than 450 formulations (Shewry and Gutteridge, 1992) (Table 8.3). *Bacillus israeliensis* has also been used extensively for the control of mosquitoes (de Barjac and Sotherland, 1990). *Bacillus thuringiensis* var. *morrisoni* and *B. israelensis* carry four genes that encode mosquito and black fly toxins CryIVA, CryIVB, CryIVC, and CryIVD (Bechtel and Bulla, 1976). *Bacillus thuringiensis* also produces cytotoxins that synergize the activity of Cry toxins. A conjugation-like system has been used to transfer Cry-encoding plasmids from one strain to another (González, Brown, and Carlton, 1982), but most *cry* genes are not readily transmissible by this process. A number of

TABLE 8.3

Natural *Bacillus thuringiensis* Products in Use for Pest Control

Bt Strain	Product	Target Insects	References
Natural *Kurstaki* HD-1	Biobit, Dipel, Foray	Lepidoptera	Shah and Goettel (1999), Baum et al. (1999)
Kurstaki HD-1	Javelin, Steward, Thuricide, Vault	Lepidoptera	Shah and Goettel (1999), Baum et al. (1999)
Kurstaki	Bactospeine, Futura	Lepidoptera	Navon (2000)
Kurstaki	Able, Costar	Lepidoptera	Shah and Goettel (1999), Navon (2000)
Kurstaki	Bio-TI	Lepidoptera	Navon (2000)
Aizawai	Florbac, Xentari	Armyworms	Shah and Goettel (1999), Navon (2000)
Tenebrionis	Novodor	Coleoptera	Shah and Goettel (1999), Navon (2000)
Tenebrionis	Trident	Coleoptera	Shah and Goettel (1999), Navon (2000)
Galleriae	Spicturin	Lepidoptera	Shah and Goettel (1999), Navon (2000)
YB 1520	Mainfeng, *Bt* 8010 Rijin	Lepidoptera	
CT 43	Shuangdu	Lepidoptera, Coleoptera, and Diptera	Navon (2000)

transconjugant and naturally occurring strains producing Cry proteins distinct from those of *B. thuringiensis* subsp. *kurstaki*, including strains of *B. thuringiensis* subsp. *aizawai* and *B. thuringiensis* subsp. *morrisoni*, have been developed for use in pest control. It has been reported that the encapsulated products persist on crops twice as long as conventional products (Feitelson, Payne, and Kim, 1992). However, some field data have shown that persistence is still not long enough to increase the risk to nontarget organisms. Two encapsulated products called MVP^R (lepidopteran active) and M-Trak^R (coleopteran-active) have also been developed (Gelemter and Schwab, 1993).

Genetic Engineering of Bacteria

The immediate challenge for genetic engineering of bacteria is to: (1) increase the potency of the toxin(s), (2) broaden the activity spectrum, (3) improve the persistence under field conditions, and (4) reduce the production costs. The problems associated with specificity, shelf life, potency, and presence of viable spores have been overcome by using modern tools in microbiology and genetic engineering. Attempts have also been made to develop high-temperature- and ultraviolet-resistant strains of *Bt* (Salama, Ali, and Sharaby, 1991). The ultraviolet-resistant strains show high toxicity, although it is known that UV causes plasmid curing. However, the heat-resistant strains have shown considerable reduction in toxicity, which may be attributed to either plasmid curing during incubation at 48°C or to the introduction of foreign DNA. The genetically engineered bacteria can be killed by heat or chemical treatment and processed into a form that can be sprayed on crops. Due to its rigidity, the dead bacterial cell walls form a protective microcapsule that prevents the toxin from degrading. However, there are too many unknowns to determine whether the new molecular approach for developing transgenic *Bt* toxin-based pesticides will be beneficial in IPM. It is possible that the approach could be used to tailor-make *Bt*-based pesticides that affect only a single insect species. This could be quite beneficial since there will be no disruption of ecological dynamics of the system. The products could also be tailored in a way that they do not cause selection for resistance in other insects that do not damage the treated crop. It may be difficult to develop an on-farm IPM program with so many specific pesticides. Therefore, the use of engineered strains might be limited to farming systems that have only a few major pest problems.

Electroporation technology has been widely used to transform vegetative cells with plasmid DNA (Bone and Ellar, 1989; Lereclus et al., 1989a, 1989b; Mahillon et al., 1989; Masson, Prefontaine, and Brousseau, 1989; Schurter, Geiser, and Mathe, 1989). Macaluso and Mettus (1991) reported that some *B. thuringiensis* strains restrict methylated DNA. Plasmid DNA isolated from *Bacillus megaterium* De Bary or *E. coli* transformed *B. thuringiensis* with much higher frequencies than did DNA isolated from *B. subtilis* or *E. coli*. A number of shuttle vectors have been used to introduce cloned *cry* genes into *B. thuringiensis* (Gawron-Burke and Baum, 1991). Integrational vectors have also been used to insert *cry* genes by homologous recombination into plasmids (Lereclus et al., 1992; Adams et al., 1994) or chromosomes (Kalman et al., 1995). Plasmid vector systems employing *B. thuringiensis* site-specific recombination have been used to construct recombinant *B. thuringiensis* strains for new bioinsecticide products (Baum, Kakefuda, and Gawron-Burke, 1996; Sanchis et al., 1996, 1997). Applications of this technique have included disruptions of *cry* and *cyt* genes to assess their contribution to biological activity (Delécluse et al., 1993; Poncet et al., 1993), and inactivation of protease production genes to increase crystal production and stability (Tan and Donovan, 1995; Donovan, Tan, and Slaney, 1997). Progress in understanding *cry* gene expression has allowed the construction of asporogenous *B. thuringiensis*

TABLE 8.4

Genetically Modified Commercial *Bacillus thuringiensis* Products for Pest Control

Bt Strain	Product	Target Insect	References
Aizawai recipient, *kurstaki* donor	Agree, Design (transconjugant)	Lepidoptera (*Bt*-resistant *Plutella xylostella*)	Baum et al. (1999); Navon (2000), Joung and Cote (2000)
Kurstaki recipient, *aizawai* donor	Condor, Cutlass (transconjugant)	Lepidoptera	Baum et al. (1999); Navon (2000); Joung and Cote (2000)
Kurstaki	CRYMAX, Leptinox	Lepidoptera	Baum et al. (1999); Navon (2000); Joung and Cote (2000)
Kurstaki recipient	Raven	Lepidoptera, Coleoptera	Baum et al. (1999); Navon (2000); Joung and Cote (2000)
δ-endotoxin encapsulated in *Pseudomonas fluorescens*	MVP, MATTCH, M-Trak (CellCap®)	Lepidoptera, Coleoptera	Joung and Cote (2000)

strains that produce crystals, but the crystals remain encapsulated in the mother cell (Lereclus et al., 1995; Bravo et al., 1996).

Several commercial products (Table 8.4) and novel strains of *Bt* have also been developed through genetic engineering (Table 8.5). The shuttle vector *pHT3101* and its derivative *pHT408* bearing a copy of a *cryIAa* delta-endotoxin gene have been transferred into several *Bt* subspecies through phage *CP-54Ber*-mediated transduction (Lecadet et al., 1992). In Cry− and Cry+ native recipients, the introduction of the *cryIAa* gene resulted in the formation of large bipyramidal crystals that were active against P. *xylostella*. Transductants displaying dual specificity have been constructed by using new isolates LM63 and LM79, which have larvicidal activity against *Phaedon cochleariae* (Fab.) and *Leptinotarsa decemlineata* (Say). It was not possible to introduce *pHT7911* into *B. thuringiensis* subsp. *entomocidus*, *aizawai*, or *israelensis* by transduction. However, electroporation was successful, and transformants expressing the toxin gene *cryIIIA* carried by *pHT7911* showed high levels of expression of the cloned gene. *CP-54Ber*-mediated transduction has been found to be useful for introducing cloned crystal protein genes into *Bt*, creating strains with new combinations of genes. Sudarsan et al. (1994) studied the colonizing ability of a transcipient strain of *B. megaterium* carrying a lepidopteran-specific *cryIAa* gene of *B. thuringiensis* var. *kurstaki* strain HD1 in the phyllospheres of various economically important plants. The transcipient strain remained on the leaves of cotton and okra for more than 28 days, but its survival in phyllospheres of mulberry, groundnut, chickpea, tomato, and rice was limited to about 3 to 5 days. The persistence of *B. thuringiensis* was extremely short (<4 days) on all the crop plants tested. *Bacillus polymyxa* (Prazmowski) colonizes the crop better than *Bt* (Sudha, Jayakumar, and Sekar, 1999). Inoculation with *B. polymyxa* in rice plants increased the shoot and the root growth. The *cry1Ac* gene expressed in *B. polymyxa* (BP113) showed toxic effects against first-instar larvae of yellow stem borer of rice, *Scirpophaga incertulas* (Walk.).

The Alternative Delivery Systems for *Bt* Toxins

In addition to traditional formulations, alternative means of delivery, including endophytic bacteria such as *Leifsonia* (*Clavibacter*) *xyli* Davis et al. (Lampel et al., 1994), have been used to deliver the toxins to the target insects (Gelemter and Schwab, 1993). Plant-colonizing bacteria such as *Pseudomonas fluorescens* Migula, *P. cepacia* Burkholder, *Rhizobium leguminosarum*

TABLE 8.5

Genetic Engineering of Bacteria to Improve Their Biological Activity for Insect Control

Bacteria	Gene/Promoter	Remarks	References
Bacillus thuringiensis	Transformation between strains—Condor®, Foil®, MVP®, and M-Trak®	Effective against Coleoptera and Lepidoptera	Carlton (1988)
	Temperature- and ultraviolet-resistant strains	Lepidoptera	Salama, Ali, and Sharaby (1991)
	cryAa/pHT3101 and pHT4.8, *cryIIIA/pHT7911*	Effective against *Plutella xylostella*, *Phaedon cochleariae*, and *Leptinotarsa decemlineata*	Lecadet et al. (1992)
Bacillus megaterium	*cry1Aa*	The transipient strain had longer survival on leaves of several crops	Sudarsan et al. (1994)
Bacillus polymyxa	*cry1Ac*	Increased plant growth, and effective against *Scirpophaga incertulas*	Sudha, Jayakumar, and Sekar (1999)
Leifsonia (Clavibacter) xyli	Deliver *Bt* toxins	—	Lampel et al. (1994)
Bacillus circus	Deliver *Bt* toxins	—	Mahaffe, Moar, and Kloepper (1994)
Pseudomonas fluorescens	*cry1Ac/Omegen-km*	Effective against *Eldana saccharina*	Herrera et al. (1997)
	cry1Ac/electroporation	Highly effective against *Helicoverpa armigera*	Duan, Zhang, and Xu (2002a)
Pseudomonas cepacia	*cry1Ac/Omegen-km*	Colonizes the sugarcane plants endophytically	Black, Huckett, and Botha (1955)
Rhizobium leguminosarum	65 kDa polypeptide Coleoptera toxin/*pKT230*	Effective against *Gastrophysa viridula* and *Sitona lepidus*	Skot et al. (1990)
Azospirillium lipoferum	*cryqAa/pRKC*	—	Udayasuriyan et al. (1995)
Enterobacter gergoviae	*cry1A*	Symbiont bacteria in the gut of *Pectinophora gossypiella*	Kuzina et al. (2002)

Jordan, and *Azospirillium* spp. have also been used to produce and deliver the *Bt* proteins (Obukowicz et al., 1986a, 1986b; Stock et al., 1990; Udayasuriyan et al., 1995).

The *B. thuringiensis* crystal genes have been introduced into *E. coli*, *B. subtilis*, *B. megaterium*, and *P. fluorescens* (Schnepf and Whiletey, 1981; Gawron-Burke and Baum, 1991). Fermentation of recombinant pseudomonads has also been used to produce biopesticide formulations consisting of Cry inclusions encapsulated in dead cells. These encapsulated forms of the Cry proteins have shown improved persistence in the environment. Production or activity of certain Cry proteins in *P. fluorescens* has been improved by the use of chimeric *cry* genes containing a substantial portion of the Cry1Ab carboxyl-terminal region (Thompson et al., 1995). The rationale for using live endophytic or epiphytic bacteria as hosts is to prolong the persistence of cry proteins in the field by using a host that can propagate itself at the site of feeding and continue to produce crystal protein. The *cry1Ac* gene has been introduced into the endophytic bacterium *L. xyli* on an integrative plasmid

(Lampel et al., 1994), and the resulting recombinant strain has been used to inoculate corn for the control of European corn borer, *Ostrinia nubilalis* (Hubner) infestation (Tomasino et al., 1995). Endophytic isolates of *B. cereus* have been used as hosts for the *cry2Aa* gene (Mahaffee, Moar, and Kloepper, 1994), and a *B. megaterium* isolate that persists in the phyllosphere (Bora et al., 1994) has been used as a host for *cry1A* genes. Similarly, *cry* genes have been transferred into other plant colonizers, including *Azospirillum* spp., *R. leguminosarum*, *P. cepacia*, and *P. fluorescens* (Obukowicz et al., 1986a, 1986b; Skot et al., 1990; Stock et al., 1990; Udayasurian et al., 1995). Nambiar, Ma, and Iyer (1990) expressed toxin gene from *B. thuringiensis* subsp. *israelensis* into *Bradyrhizobium* species that fix nitrogen in nodules of pigeonpea. The plasmid was transferred by conjugative mobilization into a *Bradyrhizobium* species that nodulates pigeonpea. Experiments in a greenhouse indicated that this provided protection against root nodule damage by larvae of *Rivellia angulata* (Hendel). Alternative delivery systems have also been sought for the dipteran-active toxins of *B. thuringiensis* subsp. *israelensis* to increase their persistence in aquatic systems. Such hosts include *B. sphaericus* (Bar et al., 1991; Poncet et al., 1994), *Caulobacter crescentus* Poindaxter (Thanabalu et al., 1992), and the cyanobacteria, *Agmenellum quadruplicatum* (Menegh.) Brebisson. (Stevens et al., 1994) and *Synechococcus* spp. (Soltes-Rak et al., 1993).

The gene encoding a 65 kDa polypeptide toxin from *B. thuringiensis* subsp. *tenebrionis*, which is lethal to coleopteran insects, has been cloned in *E. coli* in the broad host range vector *pKT230* and subsequently transferred to *R. leguminosarum* by conjugation, producing two major polypeptides of 73 and 68 kDa (Skot et al., 1990). Cell extracts from toxin-producing rhizobia were toxic to larvae of *Gastrophysa viridula* (Deg.). Bioassays also showed that the delta-endotoxin was toxic to larvae of *Sitona lepidus* Gyllen. Pea and white clover plants suffered lower root and nodule damage by larvae of *S. lepidus* when inoculated with *Rhizobium* strains containing the toxin gene.

Cyanobacteria, *Agmellenum* sp., *Snechococcus* sp., and *C. croscentus* have also been used to deliver the *Bt* toxins in aquatic environments (Thanabalu et al., 1992; Soltes-Rak et al., 1993; Stevens et al., 1994). Other species of bacteria have been used on a much smaller scale in pest control. The *cry1Ac* gene has been introduced into the chromosome of *P. fluorescens* isolate 14 using an artificial transposon-carrying vector, *Omegon-Km*. Bioassays on *Eldana saccharina* Walker larvae have shown that the strain carrying the gene was as toxic as the one carrying it on *pKT240* (Herrera, Snyman, and Thomson, 1994; Herrera et al., 1997). Sugarcane treated with *P. fluorescens* 14: *Omegon-Km*-cry suffered lower stem borer damage than the untreated sugarcane. A transgenic strain of *P. fluorescens* containing the *cry1Ac* gene from a wild-type strain of *Bt* has been evaluated for its ability to colonize the external surface of sugarcane (Black, Huckett, and Botha, 1995). Plants inoculated with nontransgenic and transgenic strains of *P. fluorescens* showed that viable cells were recoverable for three months, although *cry1Ac* gene was detectable by polymerase chain reaction (PCR) for up to eight months. Material treated with the combination innoculum showed that the nontransgenic *P. fluorescens* strain outnumbered the transgenic counterpart. When the *cry1Ac* gene was integrated into *P. fluorescens* P303-1 by electroporation (Duan, Zhang, and Xu, 2002; Duan et al., 2002), the engineered bacteria were highly insecticidal to cotton bollworm, *H. armigera*. The LC_{50} against the second instars after five days was 50.13 to 192.87 µg g^{-1}. Mortality of the treated larvae showed a positive correlation, while weight showed a negative correlation with the bacteria concentration. *In vitro* bioassays with second-instar *H. armigera* showed that the engineered strains PT45, PT51, PT61, and PT71 had greater insecticidal activity than *Bt* strain HD-73 (Duan, Zhang, and Xu, 2002). Larvae fed on leaves suffered 70.0, 60.0, 60.0, 66.7, and 33.3% mortality, respectively. Larval weight showed a negative correlation with concentration.

A fusion plasmid *pRKC* was constructed using *pACYC184, RSF1010,* and a kanamycin-resistance cartridge from *pUC4K* to introduce *cryIAa* gene into *Azospirillum* spp. (Udayasuriyan et al., 1995). With the pRKC plasmid, the number of putative transconjugants obtained in *Azospirillum lipoferum* (Beijerinck) Tarrand et al. was about 300-fold greater than in *A. brasilense* (Tarrand et al.). Expression of the *cryIAa* gene was not apparent in SDS-PAGE of the *A. lipoferum* transconjugants harboring *pBTF8*. However, *E. coli* transformants with the *pBTF8* from *A. lipoferum* transconjugants produced a 135 kDa Cry protein, indicating that the *cry* gene was intact in the transconjugants. Production of molecules with toxic activity in genetically transformed symbiotic bacteria of insect pests can be used as another tool for biological control. The symbiont *Enterobacter gergoviae* Brenner et al. isolated from the gut of the pink bollworm, *Pectinophora gossypiella* (Saunders) has been transformed to express *cyt1A* (Kuzina et al., 2002). The transgenic bacteria can be used to spread genes encoding insecticidal proteins to populations of agricultural insects or as replacement for chemical insecticides.

Entomopathogenic Fungi

More than 700 species of fungi have been reported to be pathogenic to insects, of which nearly 10 have been deployed for insect control (Hajek and St. Leger, 1994). Species pathogenic to insect pests are: *Metarhizium anisopliae* (Metsch.), *M. flavoviride* (Metsch.), *Nomuraea rileyi* (Farlow) Samson, *Beauveria bassiana* (Balsamo), *Verticillium lecanii* (Zimon) Viegas, *Aschersonia aleyrodis* Webber, and *Paecilomyces farinosus* (Holm ex Gray) Brown & Smith (Aima, 1975; Mohamed, Sikorowski, and Bell, 1977; Abbaiah et al., 1988; Gopalakrishnan and Narayanan, 1989, 1990; Ferron, Fargues, and Riba, 1991; Hajek and St. Leger, 1994; Saxena and Ahmed, 1997; Uma Devi et al., 2001). Species belonging to the Hyphomycetes are pathogenic to a range of insects, particularly against Homoptera. Commercial products based on *B. bassiana, M. anisopliae, M. flavoviride, N. rileyi,* and *A. aleyrodis* have a good potential for biological control of insects, and are currently in use or under development as pesticides. The endophytic nature of *B. bassiana* offers the potential for season-long control of insects (Anderson and Lewis, 1991). Sensitivity to solar radiation, microbial antagonists, host behavior, physiological condition and age, temperature, relative humidity, and pesticides influence the effectiveness of entomopathogenic fungi (Hajek and St. Leger, 1994). Successful use of entomopathogenic fungi will depend on the use of the right formulation, dosage, and timing.

Adhesion of fungal spores to host cuticle and their germination is a prerequisite for efficacy of fungal pathogens. It is widely accepted that 90% relative humidity (RH) is required for germination of fungal spores, and this is a big handicap in widespread use of mycopesticides. However, special formulations of fungi in oil can overcome this problem by creating high RH microclimates around the spores, enabling entomopathogenic fungi to function in low RH environments (Bateman et al., 1993). On germination, the fungus penetrates the cuticle (setae and intersegmental membrane) of the insect, and grows in the hemocoel of the insect leading to death of the insects. The fungus also grows saprophytically, and in due course, the hyphae re-emerge and sporulate. Mass production of different entomopathogenic fungi is quite easy. Glucose-yeast extract–basal salts, agar, and carrot medium can be used for multiplication of *M. anisopliae*. Zapek Dox Broth (containing 2% chitin and 3% molasses) is good for growth and sporulation of most entomopathogenic fungi (Srinivasan, 1997). For commercial-scale production, however, one would require a solid-state fermentation system. The advantages associated with the use of fungi for insect control is the fact that they can infect an insect through the cuticle, thus removing the need for ingestion. They are generally not hazardous to mammals, have no toxic residues, and give long-term control. Recent breakthroughs in formulation, strain selection, and production have provided new impetus for the inclusion of fungal-based pesticides in IPM programs. A number of facilities for mass production of *M. anisopliae* and *B. bassiana* have now been established in several countries. Several commercial fungal products are currently available. Bio 1020 has been registered in Germany for the control of black vine weevil, *Otiorhynchus sulcatus* (F.). This product is composed of dried mycelial granules, and has a shelf life of up to six months at low temperatures.

The major drawbacks associated with fungal pesticides include relative instability, requirement for moist conditions for spore germination, invasion, and growth, and slow rates of mortality. The problems associated with moisture requirement may be overcome by selecting isolates that are less humidity dependent. In order to improve the insect mortality, it is also possible to select naturally occurring potent fungi and use them to improve other strains through protoplast fusion or anastomosis. The prospects for genetic improvement of entomopathogenic fungi are quite good (Ferron, Fargues, and Riba, 1991; St. Leger and Roberts, 1997). Developments in molecular biology of entomopathogenic fungi will provide the basic understanding of mechanisms of pathogenesis and produce recombinant fungi with increased virulence (Charnley, Cobb, and Clarkson, 1997). There is considerable genetic variability in natural strains of *B. bassiana* and their virulence to the spotted stem borer, *Chilo partellus* (Swinhoe) (Uma Devi et al., 2001). Genetic manipulation of fungi still remains a difficult task considering the large size of the fungal genome and the fact that most fungal toxins are complex molecules encoded by several genes. Davila, Zambrano, and Castillo (2001) used 40 primers to gain an understanding of the mechanisms for producing recombinant fungi with increased virulence. Molecular markers can also be used for detecting gene sequences that confer on fungi the capacity to cause disease in insects (Zambrano, Davila, and Castillo, 2002).

Genetic manipulation can be used to improve tolerance of entomopathogenic fungi to fungicides (Figure 8.2), thereby promoting their utility in IPM programs (Table 8.6). A benlate-resistance gene from *Aspergillus* has been used to transform *M. anisopliae* (Goettel et al., 1989). The transformants grew at benomyl concentrations up to 10 times that inhibit wild type, and were mitotically stable on either selective or nonselective medium or insect tissue. The transformants were pathogenic to the sphingid, *M. sexta*, producing both appressoria and the cuticle-degrading enzyme chymoelastase in the presence of 50 μg mL^{-1} of benomyl. The gene encoding the cuticle-degrading protease (*Pr1*) has been inserted into the genome of the same fungus. Bernier et al. (1989) introduced benomyl resistance (beta-tubulin) gene from *Neurospora crassa* (Draft) (encoding resistance to benomyl) into *M. anisopliae*. The transformants were mitotically stable when subcultured on nonselective agar and retained the ability to infect and kill larvae of *M. sexta*. The benomyl-resistant phenotype persisted in re-isolates from insect cadavers. St. Leger et al. (1995) cotransformed *M. anisopliae* with two plasmids (*pNOM102* and *pBENA3*) containing the beta-glucuronidase and benomyl resistance genes, using electroporation and biolistic delivery systems. The cotransformants showed pathogenicity to *B. mori*. The *bar* gene from *Streptomyces hygroscopicus* (Jensen) Waksman and Henrici under the control of *Aspergillus nidulans* (Eidam) Winter *trpC* promoter and terminator sequences has been inserted into *Paecilomyces fumosoroseus* (Wise) Brawn & Smith (Cantone and Vandenberg, 1999). Evaluation of selected transformants revealed two mutant strains with altered sporulation

FIGURE 8.2   Larva of *Helicoverpa armigera* infected with the entomopathogenic fungus, *Beauveria bassiana*, which can be improved for resistance to fungicides for use in integrated pest management.

capacity and virulence to Russian wheat aphid, *Diuraphis noxia* (Kurdj.). Using electroporation, *B. bassiana* GK2016 protoplasts have been transformed to methyl 1,2-benzimidazole carbamate (MBC) resistance with the *N. crassa* MBC-resistant *beta-tubulin* gene (Pfeifer and Khachatourians, 1992). The MBC phenotype was stable and transformants grew in the presence of 5 μg MBC mL^{-1}. Sandhu et al. (2001) obtained *B. bassiana* transformants with conventional protoplasting and eletroporation, and polyethylene glycol (PEG) treatment. Strains transformed with *pSV50* harboring the *beta-tubulin* gene of *N. crassa* grew well on benomyl concentrations at 10 μg mL^{-1} unlike the recipient strain. The transformants were

TABLE 8.6

Genetic Engineering of Entomopathogenic Fungi to Increase Their Effectiveness for Pest Management

Fungus	Gene/Promoter	Remarks	References
Metarhizium anisopliae	Benlate resistance gene from *Aspergillus*	Grew better in benomyl, and pathogenic to *Manduca sexata*	Goettel et al. (1989)
	Benomyl resistance gene (*beta tubulin*) from *Neurospora crassa*	Transformants effective against *Manduca sexta*	Bernier et al. (1989)
	Beta glucorinidase and *beta tubulin* gene/PBENA3	Transformants were pathogenic to *Bombyx mori*	St. Leger et al. (1995)
Paecilomyces fumoseroseus	Bar gene/*trpC*	Transformants virulent to *Diuraphis noxia*	Cantone and Vandenberg (1999)
Beauveria bassiana	Beta tubulin/*pSV50*	Transformants pathogenic to *Helicoverpa armigera*	Sudha, Jayakumar, and Sekar (1999)
	1,2-benzimidazol resistance with *beta tubulin* gene	Stable transformants	Pfeifer and Khachatourians (1992)
Erwinia herbicola	*cry1Aa1/pUNG*	Transformants toxic to *Plutella xylostella*	Lin et al. (2002)

mitotically stable on either selective or nonselective medium. Southern blot and hybridization of undigested fungal DNA of wild type and four transformants, probed with beta-tubulin sequence of *pSV50*, showed hybridization at the high *Mr* region of genomic DNA in four transformants. Virulence tests of the transformants showed that there was no significant loss in the pathogenicity toward third-instar larvae of *H. armigera*.

The *cry1Aa1* gene encoding insecticidal crystal protein (ICP) from the *pES1* has been cloned into *E. coli* plasmid vector *pTZ19U* to form *pUN4* (Lin et al., 2002). The *pUN4* was transferred into three isolates of epiphytic fungus, *Erwinia herbicola* (Löhnis) Dye by electroporation. The transformed *E. herbicola* strains *Eh4*, *Eh5*, and *Eh6* expressed the toxin protein and conferred insecticidal activity. Protein extracts from the *E. herbicola* transformants (2.5 mg mL^{-1}) resulted in 94 to 100% mortality of diamondback moth, *P. xylostella*. At 0.312 mg of protein extract mL^{-1}, *Eh4* and *Eh5* showed 26% and 42% mortality, respectively. *Eh4* or *Eh5* resulted in up to 84% mortality after 48 hours at 1.25 mg protein extract mL^{-1}. Insecticidal activity in the transformed *E. herbicola* strains makes it a good candidate as an environmentally friendly biopesticide.

Entomopathogenic Protozoa

Protozoa play an important role in population regulation of insect pests (Maddox, 1987; Brooks, 1988). They are host specific, slow acting, and produce chronic infections. They develop in a living insect, and require an intermediate host. Microsporidia are the most common protozoa infecting insects. They are persistent in nature, recycle, and affect reproduction and overall fitness of insects. Only a few species have been used in inundative releases (Solter and Becnel, 2000). The grasshopper pathogen, *Nosema locustae* Canning has been registered and used commercially (Henry and Oma, 1981). Their major drawback is low level of mortality and the need to produce them on living hosts. Genetic engineering can be used to overcome some of these drawbacks in the future.

Entomopathogenic Nematodes

Entomopathogenic nematodes of the genera *Steinernema* and *Heterorhabditis* have emerged as excellent candidates for biological control of insect pests (Atkinson, 1993). Entomopathogenic nematodes are associated with the bacterium, *Xenorhabdus* and are quite effective against a wide range of soil-inhabiting insects. The relationship between nematodes and the bacterium is symbiotic because the nematodes cannot reproduce inside the insects without the bacterium, and the bacterium cannot enter the insect hemocoel without the nematode and cause the infection (Poinar, 1990). Broad host range, high virulence, safety to nontarget organisms, and their effectiveness under certain conditions have made them ideal biological control agents. Liquid formulations and better application strategies have allowed nematode-based products to be quite competitive for pest management in high-value crops. Nematodes are generally more expensive to produce than insecticides, and their effectiveness is limited to certain niches and insect species. There is a need to improve culturing techniques, formulations, quality, and the application technology.

Genetic improvements in entomopathogenic nematodes may expand their potential as biocontrol agents by increasing search capacity, virulence, and resistance to environmental factors (Burnell and Dowds, 1996; Gaugler and Hashmi, 1996). Heat-shock-resistant protein has been inserted in *Heterorhabditis bacteriophora* Poinar, which is 18 times better than the wild types in surviving at high temperatures (Gaugler, Lewis, and Stuart, 1997; Gaugler, Wilson, and Shearer, 1997).

The approach of improving the efficacy of existing biological control agents has recently been extended to nematodes that live in a symbiotic relationship with bacteria. When the nematodes invade a specific insect host, the bacterium *Xenorhabdus* is released in the insect, thereby killing the insect within two days (Poinar and Thomas, 1966). The bacterium is believed to kill the insect by producing a specific toxin and attempts have been made to clone the gene responsible for producing the toxin. The aim is to engineer a more effective toxin that can be delivered by the nematode. Although nematodes must still be produced commercially using live insect hosts, this has not negated their potential utility in IPM systems. Research currently underway is focused on determining the best techniques for application of nematodes in crop and orchard systems. The major problem is the sensitivity of nematodes to environmental conditions. A detailed knowledge of the ecology and behavior of the nematodes and their hosts is likely to provide solutions to this problem.

Biosafety Considerations for Using Genetically Engineered Microbes

Biopesticides are often developed from species that are ubiquitous in natural environments. Residual effect of the biopesticides is thus of less importance. Therefore, the regulatory criteria used for their release as commercial products may not be as stringent as for synthetic chemicals. However, their effects on beneficial insects, allergenicity, and pathogenicity to humans must be evaluated. Risk assessment for genetically engineered microorganisms includes the probability of dissemination of the genetic material to the indigenous microbes and to other organisms, the risk of being expressed differently, and change in the level of gene expression due to mobilization of transfer deficient plasmids or self-transfer. To study this aspect, self-transmissible and non-self-transmissible plasmids have been created, which carry resistance to trimethoprin and sulfonamide. These plasmids are of immense value in assessing the survival and transfer of genetic material under natural conditions (Cuskey, 1992; Levin, 1995; Tzotzos, 1995). Addition of large amounts of genetic material into organisms increases the likelihood of mutations, and if the genetic material is exchanged with the indigenous microbes, the chances of mutation will also increase. Genetic engineering results in many more copies of the genetic material in significant amounts in different species and increases the probability of gene transfer and mutation. To limit the exchange of genetic material, plasmids have been produced by changing the nucleotide sequences that are unable to function in the exchange process and prevent the transfer of genetic material into select hosts. However, under certain circumstances, helper plasmids can enter the cell and mobilize the disarmed plasmid (Zylstra, Cuskey, and Olson, 1992). Risk associated with the use of genetically engineered biopesticides requires the identification of the hazard involved and the level of exposure, stability, persistence in nature, transferability, effective dosage, virulence, potency, and the host range.

Biopesticides are employed with the sole purpose of inhibiting the growth or reproduction of insect pests or causing immediate death. They are modified either to increase their efficiency to control the pest populations or increase their ability to survive under natural conditions. Chemical agents do not have the ability to survive and reproduce under natural conditions, while the biological control agents have. Therefore, the risk assessment for biological control agents has to be different from that of synthetic pesticides (WHO, 1973; FAO, 1986). The biological control agents have the ability to disperse on a large scale, can be subjected to drift, subsequent regrowth, and secondary spread (Levin, 1995). Potential for interaction between the released organisms and the environment have been discussed by several scientists (Kalmakoff and Miles, 1980; Stozky and Babich, 1984; Fry and Day, 1990). Attributes of major concern in risk assessment are:

- Nature of genetic alteration;
- Phenotype of the wild-type organism;
- Phenotype of the genetically engineered organism; and
- Interaction with the environment.

Each of these attributes needs to be properly defined as determined by the knowledge available and the inherent characteristics of the attribute. Genetic alteration includes stability and characteristics of the DNA added, nature of alteration (addition, deletion, etc.), the vector, and the RNA or DNA, which remains in the altered genome. Phenotype includes the level of domestication, ease of control, pest status, range, and prevalence of gene exchange for the wild-type organism and infectivity, changes in the utilization of substrate, diseases and natural enemies, susceptibility to antibiotics, environmental adaptation, and similarity to the previously released altered organisms. Change in the limits of adaptation to the environmental conditions requires closer scrutiny. Detailed information about the product, toxicology, anticipated and known effects on nontarget organisms, infectivity to vertebrates, and hazard assessment are important components for risk assessment. It should provide information on all aspects of the organism, possible interactions with the environment at the test site, long-term effects, genetic exchange, containment, and monitoring. Information about cultural and economic impact will play a major role in decision making about the testing and release of genetically modified microbes for pest management.

In most countries, the requirements for registration of genetically modified microbial pesticides include product analysis (product identity, manufacturing process, physical and chemical properties, and the adjuvants), toxicology (acute toxicity, chronic effects, hypersensitivity, oncogenicity, and reproductive effects), and effects on nontarget organisms (nontarget plants and insects, aquatic and wildlife organisms, and human beings) (WHO, 1973; FAO, 1986). Data for hazard assessment includes testing procedures (dosage-response curve, and maximum dosage used if positive results are obtained), and effects on different nontarget organisms (birds, dietary effects on avians, acute toxicity to fish, nontarget plants and insects, honeybees, and freshwater and estuarine invertebrates). For assessing the exposure risks, the data should include biological fate of the gene (habitat, survival, replication, gene flow, gene construct, probability of gene transfer and expression, and expression level), and chemical fate of the gene (fate of the gene or gene product in soil and water). Protocols should be established to generate data, and assure quality and reliability of the data. Bioassays about the product, its toxicology and effects on nontarget organisms, environmental conditions, and performance (including limitations) need to be properly described.

There are serious impediments to the commercial deployment of recombinant NPVs. Cost-effective means of mass production and *in vivo* production for rapid-action NPVs is unlikely to be commercially viable. Another significant barrier is deregulation by the biosafety regulatory agencies and registration. Currently, regulatory oversight of genetically modified organisms (GMOs) is extremely cautious, and this has undoubtedly extended development time, increased costs, and impacted commercial confidence in an early return on investment. To date, several prototype products, including rapid-action NPVs, have been tested in the field although none have been registered for commercial use. The entire genome of *H. armigera* NPV has been sequenced, and this information can now be used for matching gene(s) with virulence (Chen et al., 2002). It may be possible to use molecular markers linked to virulence to study a range of NPV strains for use in biopesticide formulations.

There is a need for consideration of biosafety aspects before releasing recombinant baculoviruses into the environment, taking into consideration the nature of the genetic material and the environment into which the recombinant baculoviruses are to be released. Genes expressing insect-specific neurotoxic peptides augment the insecticidal activity of baculoviruses and acquire altered host range. The data so far suggests that host range determination is confined to a specific locus on the baculovirus genome. Therefore, probability of horizontal mobilization of the toxin gene to nontarget insect populations or in wild relatives in the field may not be greater than occurs in nature. Baculoviruses sprayed on fruits and vegetables are safe to human beings (Carter, 1984). Long-term exposure to NPVs is safer for workers, and no antibodies have been observed in workers exposed to products of *H. zea* SNPV (Ignoffo and Couch, 1981). Therefore, there is a possibility that the recombinant baculovirus can be deployed for pest management in the future.

Hartig et al. (1989) evaluated the NPV of *A. californica* (*AcNPV*) by using *in vitro* test systems for toxicity and transforming potential in mammalian cells. Mass cultures of CV-1 and W138 cells were unaffected by *AcNPV*. Human foreskin cells grew more slowly after inoculation, but eventually produced healthy monolayers. The sensitivities of the inhibition of reproductive survivability (IRS) assays indicated slight *AcNPV* toxicity to CV-1, W138, and human fore-skin cells. Toxicity was not ameliorated when gradient-purified or inactivated virus was used, suggesting that the toxic component of the preparation is part of the virion or copurifies with it. *AcNPV* was not toxic to and did not transform BALB/c 3T3 cells or primary cell cultures derived from Syrian hamster embryo cells (SHE). Unlike the BALB/c 3T3 transformation assay, the SHE assay detected no spontaneous transformants. The SHE transformation assay can employ simian adenovirus 7 as a positive control. The results suggested that *in vitro* assessment of viral pesticide toxicity should employ the IRS assay and that transformation assessment is best done with the SHE-simian adenovirus 7 procedure.

Conclusions

Greater awareness of the problems associated with development of insect resistance to insecticides, environmental hazards associated with synthetic pesticides, and pesticide residues in food and food products has necessitated greater emphasis on use of biopesticides for pest management in the future. There is a need to develop strategies for using microorganisms alone or in combination with synthetic pesticides for pest management. Considerable progress has been made in developing more virulent or effective strains of entomopathogenic bacteria, viruses, fungi, and nematodes through genetic engineering.

However, biosafety concerns have been raised regarding the deployment of the genetically engineered biocontrol agents for pest management. There is a need for a thorough investigation regarding the fate of genetically modified organisms in the environment, and their interaction with wild relatives and nontarget organisms. Products based on *Bt* at times may not be compatible with some pesticides. For example, some insecticides have a significant antifeeding effect, while a *Bt* preparation has to be ingested for a desired effect. Genetically engineered microbial pesticides can play an important role in pest management and reduce the amounts of pesticides applied for pest control, leading to reduction of pesticide residues in food and food products, and resulting in conservation of the environment. A fast-track process can be adopted for testing and release of genetically engineered microbes, which are known to give satisfactory results and have a relatively safer track record in different environments.

References

Abbaiah, K., Satyanarayanan, A., Rao, K.T. and Rao, N.V. (1988). Incidence of fungal disease on *Heliothis armigera* larvae in Andhra Pradesh, India. *International Pigeonpea Newsletter* 8: 11.

Adams, L.F., Mathewes, S., O'Hara, P., Petersen, A. and Gürtler, H. (1994). Elucidation of the mechanism of CryIIIA overproduction in a mutagenized strain of *Bacillus thuringiensis* var. *tenebrionis*. *Molecular Microbiology* 14: 381–389.

Agaisse, H. and Lereclus, D. (1994). Expression in *Bacillus subtilis* of the *Bacillus thuringiensis cryIIIA* toxin gene is not dependent on a sporulation-specific sigma factor and is increased in a *spo0A* mutant. *Journal of Bacteriology* 176: 4734–4741.

Agaisse, H. and Lereclus, D. (1995). How does *Bacillus thuringiensis* produce so much insecticidal crystal protein? *Journal of Bacteriology* 177: 6027–6032.

Aima, P.J. (1975). Infection of pupae of *Heliothis armigera* by *Paecilomyces farinosus*. *New Zealand Journal of Forestry Science* 5: 42–44.

Anderson, C. (1992). Researchers ask for help to save key pesticide. *Nature* 355: 661.

Anderson, L. and Lewis, L.C. (1991). Suppression of *Ostrinia nubilalis* (Hubner) (Lepidoptera: Pyralidae) by endophytic *Beauveria bassiana* (Balsamo) Vuillemin. *Environmental Entomology* 20: 1207–1211.

Aronson, A.I., Han, E.S., McGaughey, W. and Johnson. D. (1991). The solubility of inclusion proteins from *Bacillus thuringiensis* is dependent upon protoxin composition and is a factor in toxicity to insects. *Applied Environmental Microbiology* 57: 981–986.

Atkinson, H. (1993). Opportunities for improved control of plant parasitic nematodes via plant biotechnology. In Beadle, D.J., Copping, D.H.L., Dixon, G.K. and Holloman, D.W. (Eds.), *Opportunities for Molecular Biology in Crop Production*. Farnham, UK: British Crop Protection Conference, 257–266.

Bar, E., Lieman Hurwitz, J., Rahamim, E., Keynan, A. and Sandler, N. (1991). Cloning and expression of *Bacillus thuringiensis israelensis* δ-endotoxin in *B. sphaericus. Journal of Invertebrate Pathology* 57: 149–158.

Bateman, R., Carey, M., Moore, D. and Prior, C. (1993). The enhanced infectivity of *Metarhizium flavoviride* in oil formulations to desert locusts at low humidities. *Annals of Applied Biology* 122: 145–152.

Baum, J.A. (1994). Tn5401, a new class II transposable element from *Bacillus thuringiensis. Journal of Bacteriology* 176: 2835–2845.

Baum, J.A., Kakefuda, M. and Gawron Burke, C. (1996). Engineering *Bacillus thuringiensis* bioinsecticides with an indigenous site-specific recombination system. *Applied Environmental Microbiology* 62: 4367–4373.

Baum, J.A. and Malvar, T. (1995). Regulation of insecticidal crystal protein production in *Bacillus thuringiensis. Molecular Microbiology* 18: 1–12.

Baum, J.A., Timothy, B.J. and Carlton, B.C. (1999). *Bacillus thuringiensis.* In Hall, F.R. and Monn, J.J. (Eds.), *Biopesticides: Use and Delivery.* Totowa, New Jersey, USA: Humana Press, 189–209.

Bechtel, D.B. and Bulla, L.A. Jr. (1976). Electron microscope study of sporulation and parasporal crystal formation in *Bacillus thuringiensis. Journal of Bacteriology* 127: 1472–1483.

Bedford, G.O. (1980). Biology, ecology, and control of palm rhinoceros beetles. *Annual Review of Entomology* 25: 309–339.

Berliner, E. (1915). Uber die Schalffsuchi der Mehlmottenraupo (*Ephestia kuhniella* Zell.) und thren Erreger, *Bacillus thuringiensis* n. sp. *Zietschfit fur Angewandte Entomologie* 2: 29–56.

Bernhard, K. (1986). Studies on the delta-endotoxin of *Bacillus thuringiensis* var. *tenebrionis. FEMS Microbiology Letters* 33: 261–265.

Bernier, L., Cooper, R.M., Charnley, A.K. and Clarkson, J.M. (1989). Transformation of the entomo-pathogenic fungus *Metarhizium anisopliae* to benomyl resistance. *FEMS Microbiology Letters* 60: 261–266.

Bietlot, H.P., Carey, P.R., Pozsgay, M. and Kaplan, H. (1989). Isolation of carboxyl-terminal peptides from proteins by diagonal electrophoresis: Application to the entomocidal toxin from *Bacillus thuringiensis. Annals of Biochemistry* 181: 212–215.

Bietlot, H.P., Vishnubhatta, I., Carey, P.R., Pozsgay, M. and Kaplan, H. (1990). Characterization of the cysteine residues and disulfide linkages in the protein crystal of *Bacillus thuringiensis. Biochemistry Journal* 267: 309–316.

Black, K.G., Huckett, B.I. and Botha, F.C. (1995). Ability of *Pseudomonas fluorescens*, engineered for insecticidal activity against sugarcane stalk borer, to colonise the surface of sugarcane plants. *Proceedings of the Annual Congress, South African Sugar Technologists' Association* 69: 21–24.

Bone, E.J. and Ellar, D.J. (1989). Transformation of *Bacillus thuringiensis* by electroporation. *FEMS Microbiology Letters* 58: 171–178.

Bonning, B.C. and Hammock, B.D. (1996). Development of recombinant baculovirus for insect control. *Annual Review of Entomology* 41:191–210.

Bora, R.S., Murty, M.G., Shenbagarathai, R. and Sekar, V. (1994). Introduction of a lepidopteran-specific insecticidal crystal protein gene of *Bacillus thuringiensis* subsp. *kurstaki* by conjugal transfer into a *Bacillus megaterium* strain that persists in the cotton phyllosphere. *Applied Environmental Microbiology* 60: 214–222.

Bowen, D., Rocheleau, T.A., Blackburn, M., Andreev, O., Golubeva, E., Bhartia, R. and Ffrench-Constant, R.H. (1998). Insecticidal toxins from the bacterium *Photorhabdus luminescens. Science* 280: 2129–2132.

Bravo, A., Agaisse, H., Salamitou, S. and Lereclus, D. (1996). Analysis of *crylAa* expression in *sigE* and *sigK* mutants of *Bacillus thuringiensis. Molecular and General Genetics* 250: 734–741.

Brizzard, B.L., Schnepf, H.E. and Kronstad, J.W. (1991). Expression of the *crylB* crystal protein gene of *Bacillus thuringiensis. Molecular and General Genetics* 231: 59–64.

Brooks, W.M. (1988). Entomogenous protozoa. In Ignoffo, C.M. and Mandava, N.B. (Eds.), *Handbook of Natural Pesticides, Vol. V. Microbial Insecticides, Part A: Entomogenous Protozoa and Fungi.* Boca Raton, Florida, USA: CRC Press, 1–149.

Brown, K.L. (1993). Transcriptional regulation of the *Bacillus thuringiensis* subsp. *thompsoni* crystal protein gene operon. *Journal of Bacteriology* 175: 7951–7957.

Brown, K.L. and Whiteley, H.R. (1988). Isolation of a *Bacillus thuringiensis* RNA polymerase capable of transcribing crystal protein genes. *Proceedings National Academy of Sciences USA* 85: 4166–4170.

Burnell, A.M. and Dowds, B.C.A. (1996). The genetic improvement of entomopathogenic nematodes and their symbiont bacteria: Phenotypic targets, genetic limitations and an assessment of possible hazards. *Biocontrol Science and Technology* 6: 435–447.

Cantone, F.A. and Vandenberg, J.D. (1999). Genetic transformation and mutagenesis of the entomopathogenic fungus *Paecilomyces fumosoroseus. Journal of Invertebrate Pathology* 74: 281–288.

Carbonell, L.F., Hodge, M.R., Tomalski, M.D. and Miller, L.K. (1988). Synthesis of a gene coding for an insect specific scorpion neurotoxin and attempts to express it using baculovirus vectors. *Gene* 73: 409–418.

Carlson, C.R. and Kolstø, A.B. (1993). A complete physical map of a *Bacillus thuringiensis* chromosome. *Journal of Bacteriology* 175: 1053–1060.

Carlson, C.R., Caugant, D.A. and Kolstø, A.B. (1994). Genotypic diversity among *Bacillus cereus* and *Bacillus thuringiensis* strains. *Applied Environmental Microbiology* 60: 1719–1725.

Carlton, B.C. (1988). Genetic improvement of *Bacillus thuringiensis* as a biopesticide. In Roberts, D.W. and Granados, R.R. (Eds.), *Biological Pesticides and Novel Plant Resistance for Insect Pest Management*. Ithaca, New York, USA: Boyce Thompson Institute, 38–43.

Carter, J.B. (1984). Viruses as pest control agents. *Biotechnology and Genetic Engineering Reviews* 1: 365–419.

Catterall, W.A. (1984). The molecular basis of neuronal excitability. *Science* 223: 653–659.

Chang, C., Yu, Y.M., Dai, S.M., Law, S.K. and Gill, S.S. (1993). High-level *cryIVD* and *cytA* gene expression in *Bacillus thuringiensis* does not require the 20-kilodalton protein, and the coexpressed gene products are synergistic in their toxicity to mosquitoes. *Applied Environmental Microbiology* 59: 815–821.

Charles, J.F., Nielsen-Leroux, C. and Delécluse, C. (1996). *Bacillus sphaericus* toxins: Molecular biology and mode of action. *Annual Review of Entomology* 41: 451–472.

Charnley, A.K., Cobb, B. and Clarkson, J.M. (1997). Towards the improvement of fungal insecticides. In Evans, H.F. (Eds.), *Microbial Insecticides: Novelty or Necessity? Proceedings, British Crop Protection Council Symposium* 68: 115–126.

Chen, X., Zhang, W.J., Wong, J., Chun, G., Lu, A., McCutchen, B.F., Presnail, J.K., Herrmann, R., Dolan, M., Tingey, S., Hu, Z.H. and Vlak, J.M. (2002). Comparative analysis of the complete genome sequences of *Helicoverpa zea* and *Helicoverpa armigera* single-nucleocapsid nucleopolyhedroviruses. *Journal of General Virology* 83: 673–684.

Cherry, A.J., Parnell, M.A., Garzywacz, D. and Jones, K.A. (1997). The optimization of *in vivo* nuclear polyhedrosis virus production in *Spodoptera exempta* (Walker) and *Spodoptera exigua* (Hubner). *Journal of Invertebrate Pathology* 70: 50–58.

Chestukhina, G.G., Kostina, L.I., Mikhailova, A.L., Tyurin, S.A., Klepikova, F.S. and Stepanov, V.M. (1982). The main features of *Bacillus thuringiensis* delta-endotoxin molecular structure. *Archives of Microbiology* 132: 159–162.

Choma, C.T., Surewicz, W.K. and Kaplan, H. (1991). The toxic moiety of the *Bacillus thuringiensis* protoxin undergoes a conformational change upon activation. *Biochemical and Biophysical Research Communications* 179: 933–938.

Choma, C.T., Surewicz, W.K., Carey, P.R., Pozsgay, M. and Raynor, T. (1990). Unusual proteolysis of the protoxin and toxin from *Bacillus thuringiensis*: Structural implications. *European Journal of Biochemistry* 189: 523–527.

Crickmore, N., Bone, E.J., Williams, J.A. and Ellar, D.J. (1995). Contribution of the individual components of the δ-endotoxin crystal to the mosquitocidal activity of *Bacillus thuringiensis* subsp. *israelensis*. *FEMS Microbiology Letters* 131: 249–254.

Crickmore, N., Zeigler, D.R., Feitelson, J., Schnepf, E., Van Rie, J., Lereclus, D., Baum, J. and Dean, D.H. (1998). Revision of the nomenclature for the *Bacillus thuringiensis* pesticidal crystal proteins. *Microbiology and Molecular Biology Reviews* 62: 807–813.

Cunningham, J.C. (1995). *Baculoviruses* as microbial insecticides. In Renveni, R. (Ed.), *Novel Approaches to Integrated Pest Management*. Boca Raton, Florida, USA: Lewis Press, 261–292.

Cuskey, S.M. (1992). Biological containment of genetically engineered microorganisms. In Levin, M.A., Siedler, R.J. and Rogul, N. (Eds.), *Microbial Ecology: Principles, Methods, and Applications*. New York, USA: McGraw-Hill, 911–918.

Davila, M., Zambrano, K. and Castillo, M.A. (2001). Use of RAPD technique for the identification of DNA fragments likely related to virulence in entomopathogenic fungus. *Bioagro* 13: 93–98.

de Barjac, H. and Sotherland, D.J. (1990). *Bacterial Control of Mosquitoes and Blackflies*. New Brunswick, New Jersey, USA: Rutgers University Press.

DeDianous, S., Hoarau, R. and Rochat, H. (1987). Reexamination of the specificity of the scorpion *Androctonus australis* Hector insect toxin towards arthropods. *Toxicon* 25: 411–417.

Delécluse, A., Poncet, S., Klier, A. and Rapoport, G. (1993). Expression of *cryIVA* and *cryIVB* genes, independently or in combination, in a crystal-negative strain of *Bacillus thuringiensis* subsp. *israelensis*. *Applied Environmental Microbiology* 59: 3922–3927.

de Maagd, R.A., Kwa, M.S.G., van der Klei, H., Yamamoto, T., Schipper, B., Vlak, J.M., Stiekema, W.J. and Bosch, D. (1996a). Domain III substitution in *Bacillus thuringiensis* CryIA(b) results in superior toxicity for *Spodoptera exigua* and altered membrane protein recognition. *Applied Environmental Microbiology* 62: 1537–1543.

de Maagd, R.A., van der Klei, H., Bakker, P.L., Stiekema, W.J. and Bosch, D. (1996b). Different domains of *Bacillus thuringiensis* δ-endotoxins can bind to insect midgut membrane proteins on ligand blots. *Applied Environmental Microbiology* 62: 2753–2757.

Denolf, P., Hendrickx, K., Van Damme, J., Jansens, S., Peferoen, M., Degheele, D. and Van Rie, J. (1997). Cloning and characterization of *Manduca sexta* and *Plutella xylostella* midgut aminopeptidase N enzymes related to *Bacillus thuringiensis* toxin-binding proteins. *European Journal of Biochemistry* 248: 748–761.

Dervyn, E., Poncet, S., Klier, A. and Rapoport, G. (1995). Transcriptional regulation of the *cryIVD* gene operon from *Bacillus thuringiensis* subsp. *israelensis*. *Journal of Bacteriology* 177: 2283–2291.

De Souza, M.T., Lecadet, M.M. and Lereclus, D. (1993). Full expression of the *cryIIIA* toxin gene of *Bacillus thuringiensis* requires a distant upstream DNA sequence affecting transcription. *Journal of Bacteriology* 175: 2952–2960.

Donovan, W.P., Tan Y. and Slaney, A.C. (1997). Cloning of the *nprA* gene for neutral protease A of *Bacillus thuringiensis* and effect of *in vivo* deletion of *nprA* on insecticidal crystal protein. *Applied Environmental Microbiology* 63: 2311–2317.

Du, C., Martin, P.A.W. and Nickerson, K.W. (1994). Comparison of disulfide contents and solubility at alkaline pH of insecticidal and noninsecticidal *Bacillus thuringiensis* protein crystals. *Applied Environmental Microbiology* 60: 3847–3853.

Du, X. and Thiem, S.M. (1997). Characterization of host range factor 1 (*hrf-1*) expression in *Lymantria dispar* L. nucleopolyhedrovirus and recombinant *Autographa californica* M nucleopolyhedrovirus-infected IPLB-Ld652Y cells. *Virology* 227: 420–430.

Duan, C.X., Zhang, Q.W. and Xu, J. (2002). *In vitro* and plant bioassays of insecticidal engineered *Pseudomonas fluorescens* in cotton with cotton bollworms. *Chinese Journal of Biological Control* 18: 67–70.

Duan, C.X., Zhang, Q.W., Xu, J., Zhang, Z.W., Xiong, Y.K. and Yang, Q.Q. (2002). Construction of insecticidal engineered strains of *Pseudomonas fluorescens* and their insecticidal activity. *Acta Entomologica Sinica* 45: 419–424.

Dulmage, H.T. (1981). Insecticidal activity of isolates of *Bacillus thuringiensis* and their potential for pest control. In Burges, H.D. (Ed.), *Microbial Control of Pests and Plant Diseases*. London, UK: Academic Press, 129–141.

English, L., Robbins, H.L., Von Tersch, M.A., Kulesza, C.A., Ave, D., Coyle, D., Jany, C.S. and Slatin, S.L. (1994). Mode of action of CryIIA: A *Bacillus thuringiensis* delta-endotoxin. *Insect Biochemistry and Molecular Biology* 24: 1025–1035.

FAO (Food and Agriculture Organization). (1986). *International Code of Conduct on the Distribution and Use of Pesticides*. Rome, Italy: Food and Agriculture Organization.

Feitelson, J.S., Payne, J. and Kim, L. (1992). *Bacillus thuringiensis*: Insects and beyond. *Biotechnology* 10: 271–275.

Ferron, P., Fargues, J. and Riba, G. (1991). Fungi as microbial insecticides against pests. In Arora, D.K., Ajelio, L. and Mukherji, K.G. (Eds.), *Handbook of Applied Mycology*. New York, USA: Dekker, 665–706.

Flores, H., Soberon, X., Sanchez, J. and Bravo, A. (1997). Isolated domain II and III from the *Bacillus thuringiensis* Cry1Ab delta-endotoxin binds to lepidopteran midgut membranes. *FEBS Letters* 414: 313–318.

Fontecilla-Camps, J.C. (1989). Three dimensional model of the insect directed scorpion toxin from *Androctonus australis* Hector and its implication for the evolution of scorpion toxins in general. *Journal of Molecular Evolution* 29: 63–67.

Francis, B.R. and Bulla, L.A. Jr. (1997). Further characterization of BT-R₁, the cadherin-like receptor for Cry1Ab toxin in tobacco hornworm (*Manduca sexta*) midguts. *Insect Biochemistry and Molecular Biology* 27: 541–550.

Fry, J.C. and Day, M.J. (1990). Release of genetically engineered and other microorganisms. In Fry, J.C. and Day, M.J. (Eds.), *Release of Genetically Engineered and Other Microorganisms*. Cambridge, UK: Cambridge University Press, 120–136.

Garczynski, S.F. and Adang, M.J. (1995). *Bacillus thuringiensis* Cry1A(c) δ-endotoxin binding aminopeptidase in the *Manduca sexta* midgut has a glycosyl-phosphatidylinositol anchor. *Insect Biochemistry and Molecular Biology* 25: 409–415.

Garczynski, S.F., Crim, J.W. and Adang, M.J. (1991). Identification of putative insect brush border membrane-binding molecules specific to *Bacillus thuringiensis* δ-endotoxin by protein blot analysis. *Applied Environmental Microbiology* 57: 2816–2820.

Gaugler, R. and Hashmi, S. (1996). Genetic engineering of an insect parasite. In Stelow, J. (Ed.), *Genetic Engineering Principles and Methods*. New York, USA: Plenum Press, 135–137.

Gaugler, R., Lewis, E. and Stuart, R.J. (1997). Ecology in the service of biological control: The case of entomopathogenic nematodes. *Oecologia* 109: 483–489.

Gaugler, R., Wilson, M. and Shearer, P. (1997). Field release and environmental fate of a transgenic entomopathogenic nematode. *Biological Control* 9: 75–80.

Gawron Burke, C. and Baum, J.A. (1991). Genetic manipulation of *Bacillus thuringiensis* insecticidal crystal protein genes in bacteria. In Setlow, J.K. (Ed.), *Genetic Engineering: Principles and Methods*, Vol. 13. New York, USA: Plenum Press, 237–263.

Ge, A.Z., Shivarova, N.I. and Dean, D.H. (1989). Location of the *Bombyx mori* specificity domain on a *Bacillus thuringiensis* δ-endotoxin protein. *Proceedings National Academy of Sciences USA* 86: 4037–4041.

Ge, A.Z., Rivers, D., Milne, R. and Dean, D.H. (1991). Functional domains of *Bacillus thuringiensis* insecticidal crystal proteins: Refinement of *Heliothis virescens* and *Trichoplusia ni* specificity domains on CryIA(c). *Journal of Biological Chemistry* 266: 17954–17958.

Gelemter, W. and Schwab, G.E. (1993). Transgenic bacteria, virus, algae and other microorganisms as *Bacillus thuringiensis* toxin delivery systems. In Entwistle, P.F., Cory, J.S., Bailey, M. and Higgs, S. (Eds.), Bacillus thuringiensis, *an Environmental Biopesticide: Theory and Practice*. Chichester, UK: John Wiley & Sons, 89–124.

Goettel, M.S., St. Leger, R.J., Bhairi, S., Jung, M.K., Oakley, B.R. and Staples, R.C. (1989). Transformation of the entomopathogenic fungus, *Metarhizium anisopliae*, using the *Aspergillus nidulans benA3* gene. *Current Genetics* 17: 129–132.

González, J.M. Jr., Brown, B.J. and Carlton, B.C. (1982). Transfer of *Bacillus thuringiensis* plasmids coding for δ-endotoxin among strains of *B. thuringiensis* and *B. cereus*. *Proceedings National Academy of Sciences USA* 79: 6951–6955.

González, J.M. Jr., Dulmage, H.T. and Carlton, B.C. (1981). Correlation between specific plasmids and δ-endotoxin production in *Bacillus thuringiensis*. *Plasmid* 5: 351–365.

Gopalakrishnan, C. and Narayanan, K. (1989). Epizootiology of *Nomuraea rileyi* (Farlow) Samson in field populations of *Helicoverpa* (*Heliothis*) *armigera* (Hübner) in relation to three host plants. *Journal of Biological Control* 3: 50–52.

Gopalakrishnan, C. and Narayanan, K. (1990). Studies on the dose-mortality relationship between the entomofungal pathogen *Beauveria bassiana* (Bals.) Vuillemin and *Heliothis armigera* (Hub.) (Lepidoptera: Noctuidae). *Journal of Biological Control* 4(2): 112–115.

Green, B.D., Battisti, L. and Thorne, C.B. (1989). Involvement of Tn4430 in transfer of *Bacillus anthracis* plasmids mediated by *Bacillus thuringiensis* plasmid pXO12. *Journal of Bacteriology* 171: 104–113.

Hajek, A.E. and St. Leger, R.J. (1994). Interactions between fungal pathogens and insect hosts. *Annual Review of Entomology* 39: 293–322.

Hammock, B.D., Bonning, B.B., Possee, R.D., Hanzlik, T.N. and Maeda, S. (1990). Expression and effects of the juvenile hormone esterase in a baculovirus vector. *Nature* 344: 458–461.

Harrison, R.L. and Bonning, B.C. (2001). Use of proteases to improve the insecticidal activity of baculoviruses. *Biological Control* 20: 199–209.

Hartig, P.C., Chapman, M.A., Hatch, G.G. and Kawanishi, C.Y. (1989). Insect virus: Assays for toxic effects and transformation potential in mammalian cells. *Applied and Environmental Microbiology* 55: 1916–1920.

Hayakawa, T., Shimojo, E.I., Mori, M., Kaido, M., Furusawa, I., Miyata, S., Sano, Y., Matsumoto, T., Hashimoto, Y. and Granados, R.R. (2000). Enhancement of baculovirus infection in *Spodoptera exigua* (Lepidoptera: Noctuidae) larvae with *Autographa californica* nucleopolyhedrovirus or *Nicotiana tabacum* engineered with a granulovirus enhancin gene. *Applied Entomology and Zoology* 35: 163–170.

Henry, J.E. and Oma, E.A. (1981). Pest control by *Nosema locustae*, a pathogen of grasshoppers and crickets. In Burges, H.D. (Ed.), *Microbial Control of Pests and Plant Diseases 1970–1980*. London, UK: Academic Press, 573–586.

Herrera, G., Snyman, S.J. and Thomson, J.A. (1994). Construction of bioinsecticidal strain of *Pseudomonas fluorescens* active against the sugarcane borer, *Eldana saccharina*. *Applied Environmental Microbiology* 60: 682–690.

Herrera, G., Snyman, S.J., Thomson, J.A. and Mihm, J.A. (1997). Construction of a bioinsecticidal strain of *Pseudomonas fluorescens* active against sugarcane borer. In *Insect Resistant Maize: Recent Advances and Utilization. Proceedings of an International Symposium*, 27 November–3 December, 1994. El Batan, Mexico: International Maize and Wheat Improvement Center, 159–162.

Hodgman, T.C., Ziniu, Y., Shen, J. and Ellar, D.J. (1993). Identification of a cryptic gene associated with an insertion sequence not previously identified in *Bacillus thuringiensis*. *FEMS Microbiology Letters* 114: 23–29.

Hofmann, C. and Lüthy, P. (1986). Binding and activity of *Bacillus thuringiensis* delta-endotoxin to invertebrate cells. *Archives of Microbiology* 146: 7–11.

Hofmann, C., Lüthy, P., Hütter, R. and Pliska, V. (1988). Binding of the delta-endotoxin from *Bacillus thuringiensis* to brush-border membrane vesicles of the cabbage butterfly (*Pieris brassicae*). *European Journal of Biochemistry* 173: 85–91.

Hofte, H. and Whiteley, H.R. (1989). Insecticidal crystal proteins of *Bacillus thuringiensis*. *Microbiology Reviews* 53: 242–255.

Hoover, K., Schultz, C.M., Lane, S.S., Bonning, B.C., Duffey, S.S., McCutchen, B.F. and Hammock, B.D. (1995). Reduction in damage to cotton plants by a recombinant baculovirus that knocks moribund larvae of *Heliothis virescens* off the plant. *Biological Control: Theory and Applications in Pest Management (USA)* 5: 419–426.

Hukuhara, T., Hayakawa, T. and Wijonarko, A. (2001). A bacterially produced virus enhancing factor from entomopoxvirus enhances nucleopolyhedrovirus infection in armyworm larvae. *Journal of Invertebrate Pathology* 78: 25–30.

Ignoffo, C.M. (1964). Production and virulence of a nuclear polyhedrosis virus from larvae of *Trichoplusia ni* (Hubner) reared on a semi-synthetic diet. *Journal of Invertebrate Pathology* 6: 318–326.

Ignoffo, C.M. and Couch, T.L. (1981). The neucleopolyhedrosis virus of *Heliothis* species as a microbial insecticide. In Burges, H.D. (Ed.), *Microbial Control of Pests and Plant Diseases 1970–1990*. London, UK: Academic Press, 329–362.

Ihara, H., Kuroda, E., Wadano, A. and Himeno, M. (1993). Specific toxicity of δ-endotoxins from *Bacillus thuringiensis* to *Bombyx mori*. *Bioscience, Biotechnology and Biochemistry* 57: 200–204.

Inceoglu, A.B., Kamita, S.G., Hinton, A.C., Huang, Q., Severson, T.F., Kang, K. and Hammock, B.D. (2001). Recombinant baculoviruses for insect control. *Pest Management Science* 57: 981–987.

Jackson, D.M., Drown, G.C., Nardin, G.C. and Johnson, D.W. (1992). Autodissemination of baculoviruses for management of tobacco bud worms. *Journal of Economic Entomology* 85: 710–719.

Johnston, K.A., Lee, M.J., Brough, C., Hilder, V.A., Gatehouse, A.M.R. and Gatehouse, J.A. (1995). Protease activities in the larval midgut of *Heliothis virescens*: Evidence for trypsin and chymotrypsin-like enzymes. *Insect Biochemistry and Molecular Biology* 25: 375–383.

Joung, K.B. and Cote, J.C. (2000). A review of the environmental impact of the microbial insecticide *Bacillus thuringiensis*. *Technical Bulletin No. 29*. Montreal, Canada: Horticulture Research and Development Centre, Quebec University. http://res2.agrca/stjean/crdh.htm.

Kalmakoff, J. and Miles, J.A.R. (1980). Ecological approaches to the use of microbial pathogens in insect control. *Bioscience* 30: 344–347.

Kalman, S., Kiehne, K.L., Cooper, N., Reynoso, M.S. and Yamamoto, T. (1995). Enhanced production of insecticidal proteins in *Bacillus thuringiensis* strains carrying an additional crystal protein gene in their chromosomes. *Applied Environmental Microbiology* 61: 3063–3068.

Keeton, T.P. and Bulla, L.A. Jr. (1997). Ligand specificity and affinity of BT-R$_1$, the *Bacillus thuringiensis* Cry1A toxin receptor from *Manduca sexta*, expressed in mammalian and insect cell cultures. *Applied Environmental Microbiology* 63: 3419–3425.

Klein, M.G. and Jackson, T.A. (1992). Bacterial diseases of scarabs. In Jackson, T.A. and Glare, T.R. (Eds.), *Use of Pathogens in Scarab Pest Management*. Andover, Maryland, USA: Intercept Ltd., 43–61.

Klein, M.G. and Kaya, H.K. (1995). *Bacillus* and *Serratia* species for scarab control. *Memoirs of the Institute of Oswaldo Cruz* 90: 87–95.

Knight, P.J., Crickmore, N. and Ellar, D.J. (1994). The receptor for *Bacillus thuringiensis* Cry1A(c) δ-endotoxin in the brush border membrane of the lepidopteran *Manduca sexta* is aminopeptidase-N. *Molecular Microbiology* 11: 429–436.

Kopeyan, C., Mansuelle, P., Sampieri, F.T., Bahraoui, E.M., Rochat, H. and Granier, C. (1990). Primary structure of scorpion anti-insect toxins isolated from the venom of *Leiurus quinquestriatus quinquestriatus*. *FEBS Letters* 261: 423–426.

Kostichka, K., Warren, G.W., Mullins, M., Mullins, A.D., Craig, J.A., Koziel, M.G. and Estruch, J.J. (1996). Cloning of a *cryV*-type insecticidal protein gene from *Bacillus thuringiensis*: The *cryV*-encoded protein is expressed early in stationary phase. *Journal of Bacteriology* 178: 2141–2144.

Kronstad, J.W. and Whiteley, H.R. (1984). Inverted repeat sequences flank a *Bacillus thuringiensis* crystal protein gene. *Journal of Bacteriology* 160: 95–102.

Kuzina, L.V., Miller, E.D., Ge, B.X. and Miller, T.A. (2002). Transformation of *Enterobacter gergoviae* isolated from pink bollworm (Lepidoptera: Gelechiidae) gut with *Bacillus thuringiensis* toxin. *Current Microbiology* 44: 1–4.

Kuzio, J., Jaques, R. and Faulkner, P. (1989). Identification of p74, a gene essential for virulence of baculovirus occlusion bodies. *Virology* 173: 759–763.

Lacey, L.A., Frutos, R., Kaya, H.K. and Vail, P. (2001). Insect pathogens as biological control agents. Do they have a future? *Biological Control* 21: 230–238.

Lambert, B. and Peferoen, M. (1992). Insecticidal promise of *Bacillus thuringiensis*. Facts and mysteries about a successful biopesticide. *BioScience* 42: 112–122.

Lampel, J.S., Canter, G.L., Diimock, M.B., Kelly, J.L., Anderson, J.J., Uratani, B.B., Foulke, J.S., Jr. and Turner, J.T. (1994). Integrative cloning, expression, and stability of the *cry1A(c)* gene from *Bacillus thuringiensis* subsp. *kurstaki* in a recombinant strain of *Clavibacter xyli* subsp. *cynodontis*. *Applied Environmental Microbiology* 60: 501–508.

Lecadet, M.M. and Dedonder, R. (1966). Les protéases de *Pieris brassicae*. I. Purification et propriétés. *Bulletin of Society of Chemical Biology* 48: 631–660.

Lecadet, M.M. and Dedonder, R. (1967). Enzymatic hydrolysis of the crystals of *Bacillus thuringiensis* by the proteases of *Pieris brassicae*. I. Preparation and fractionation of the lysates. *Journal of Invertebrate Pathology* 9: 310–321.

Lecadet, M.M., Chaufaux, J., Ribier, J. and Lereclus, D. (1992). Construction of novel *Bacillus thuringiensis* strains with different insecticidal activities by transduction and transformation. *Applied and Environmental Microbiology* 58: 840–849.

Lee, M.K., Curtiss, A., Alcantara, E. and Dean, D.H. (1996). Synergistic effect of the *Bacillus thuringiensis* toxins CryIAa and CryIAc on the gypsy moth, *Lymantria dispar*. *Applied Environmental Microbiology* 62: 583–586.

Lee, M.K., Milne, R.E., Ge, A.Z. and Dean, D.H. (1992). Location of a *Bombyx mori* receptor binding region on a *Bacillus thuringiensis* δ-endotoxin. *Journal of Biological Chemistry* 267: 3115–3121.

Léonard, C., Chen, Y. and Mahillon, J. (1997). Diversity and different distribution of IS*231*, IS*232* and IS*240* among *Bacillus cereus*, *Bacillus thuringiensis* and *Bacillus mycoides*. *Microbiology* 143: 2537–2547.

Lereclus, D., Agaisse, H., Gominet, M. and Chaufaux, J. (1995). Overproduction of encapsulated insecticidal crystal proteins in a *Bacillus thuringiensis spoOA* mutant. *Bio/Technology* 13: 67–71.

Lereclus, D., Arantès, O., Chaufaux, J. and Lecadet, M.M. (1989a). Transformation and expression of a cloned δ-endotoxin gene in *Bacillus thuringiensis*. *FEMS Microbiology Letters* 60: 211–218.

Lereclus, D., Bourgouin, C., Lecadet, M.M., Klier, A. and Rapoport, G. (1989b). Role, structure, and molecular organization of the genes coding for parasporal δ-endotoxins of *Bacillus thuringiensis*. In Smith, I., Slepecky, R.A. and Setlow, P. (Eds.), *Regulation of Procaryotic Development Structural and Functional Analysis of Bacterial Sporulation and Germination*. Washington, D.C., USA: American Society for Microbiology, 255–276.

Lereclus, D., Vallade, M., Chaufaux, J., Arantes, O. and Rambaud, S. (1992). Expansion of the insecticidal host range of *Bacillus thuringiensis* by *in vivo* genetic recombination. *Bio/Technology* 10: 418–421.

Lester, D., Lazarovici, P., Pelhate, M. and Zlotkin, E. (1982). Two insect toxins from the venom of the scorpion *Buthotus judaicus*. Purification, characterization and action. *Biochimistry Biophysics Acta* 701: 370–381.

Levin, S.A. (1995). Microbial pesticides: Safety considerations. In Tzotzos, T. (Ed.), *Genetically Modified Microorganisms: A Guide to Biosafety*. London, UK: Commonwealth Agricultural Bureau, International, 93–109.

Li, J., Carroll, J. and Ellar, D.J. (1991). Crystal structure of insecticidal δ-endotoxin from *Bacillus thuringiensis* at 2.5 Å resolution. *Nature* 353: 815–821.

Liang, Y., Patel, S.S. and Dean, D.H. (1995). Irreversible binding kinetics of *Bacillus thuringiensis* CryIA δ-endotoxins to gypsy moth brush border membrane vesicles is directly correlated to toxicity. *Journal of Biological Chemistry* 270: 24719–24724.

Lin, C.H., Huang, W.C., Tzeng, C.C. and Chen, L.J. (2002). Insecticidal activity and plasmid retention of the epiphytic *Erwinia herbicola* strains transformed with the *Bacillus thuringiensis cry1Aa1* gene. *Plant Protection Bulletin, Taipei* 44: 21–36.

Luckow, V.A. and Summers, M.D. (1988). Trends in development of baculovirus vectors. *Bio/Technology* 6: 47–75.

Luo, K., Lu, Y.J. and Adang, M.J. (1996). A 106 kDa form of aminopeptidase is a receptor for *Bacillus thuringiensis* CryIC delta-endotoxin in the brush border membrane of *Manduca sexta*. *Insect Biochemistry and Molecular Biology* 26: 783–791.

Luo, K., Tabashnik, B.E. and Adang, M.J. (1997). Binding of *Bacillus thuringiensis* Cry1Ac toxin to aminopeptidase in susceptible and resistant diamondback moths (*Plutella xylostella*). *Applied Environmental Microbiology* 63: 1024–1027.

Luthy, P., Cordier, J. and Fischer, H. (1982). *Bt* as a bacterial insecticide: Basic considerations and applications. In Kurstak, E. (Ed.), *Microbial and Viral Pesticides*. New York, USA: Marcel Dekker, 35–74.

Lysenko, O. (1983). *Bacillus thuringiensis*: evolution of a taxonomic conception. *Journal of Invertebrate Pathology* 42: 295–298.

Macaluso, A. and Mettus, A.M. (1991). Efficient transformation of *Bacillus thuringiensis* requires non-methylated plasmid DNA. *Journal of Bacteriology* 173: 1353–1356.

Maddox, J.V. (1987). Protozoan diseases. In Fuxa, J.R. and Tanada, Y. (Eds.), *Epizootiology of Insect Diseases*. New York, USA: John Wiley & Sons, 417–452.

Maeda, S. (1989). Increased insecticidal effect by a recombinant baculovirus carrying a synthetic diuretic hormone gene. *Biochemical and Biophysical Research Communications* 165: 1177–1183.

Maeda, S. Volrath, S.L., Hanzlik, T.N., Harper, S.A., Maddox, D.W., Hammock, B.D. and Fowler, E. (1991). Insecticidal effects of an insect specific neurotoxin expressed by a recombinant baculovirus. *Virology* 184: 777–780.

Mahaffee, W.F., Moar, W.J. and Kloepper, J.W. (1994). Bacterial endophytes genetically engineered to express the CryIIA δ-endotoxin from *Bacillus thuringiensis* subsp. *kurstaki*. In Ryder, M.H., Stevens, P.M. and Bowen, G.D. (Eds.), *Improving Plant Productivity with Rhizosphere Bacteria*. East Melbourne, Victoria, Australia: CSIRO Publications, 245–246.

Mahillon, J., Chungjatupornchai, W., Decock, J., Dierickx, S., Michiels, F., Peferoen, M. and Joos, H. (1989). Transformation of *Bacillus thuringiensis* by electroporation. *FEMS Microbiology Letters* 60: 205–210.

Malvar, T. and Baum, J.A. (1994). Tn*5401* disruption of the *spo0F* gene, identified by direct chromosomal sequencing, results in CryIIIA overproduction in *Bacillus thuringiensis*. *Journal of Bacteriology* 176: 4750–4753.

Malvar, T., Gawron Burke, C. and Baum, J.A. (1994). Overexpression of *Bacillus thuringiensis* HknA, a histidine protein kinase homolog, bypasses early Spo-mutations that result in CryIIIA overproduction. *Journal of Bacteriology* 176: 4742–4749.

Masson, L., Prefontaine, G. and Brousseau, R. (1989). Transformation of *Bacillus thuringiensis* vegetative cells by electroporation. *FEMS Microbiology Letters* 60: 273–278.

Masson, L., Lu, Y.J., Mazza, A., Brousseau, R. and Adang, M.J. (1995). The CryIA(c) receptor purified from *Manduca sexta* displays multiple specificities. *Journal of Biological Chemistry* 270: 20309–20315.

McCutchen, B.F., Choudary, P.V., Crenshaw R., Maddox, D., Kamita, S.G., Palekar, N., Volrath, S., Fowler, E. and Hammock, D. (1991). Development of a recombinant baculovirus expressing an insect-selective neurotoxin: potential for pest control. *Bio/Technology* 9: 848–852.

McGaughey, W.H. and Whalon, M.E. (1992). Managing insect resistance to *Bacillus thuringiensis* toxins. *Science* 258: 1451–1455.

Meadows, M. (1990). Environmental release of *Bacillus thuringiensis*. In Fry, J.C. and Day, N. (Eds.), *Release of Genetically Engineered and Other Microorganisms*. New York, USA: McGraw-Hill, 120–136.

Meenakshisundaram, K.S. and Gujar, G.T. (1998). Proteolysis of *Bacillus thuringiensis* subspecies *kurstaki* endotoxin with midgut proteases of some important lepidopterous species. *Indian Journal of Experimental Biology* 36: 593–598.

Menou, G., Mahillon, J., Lecadet, M.M. and Lereclus, D. (1990). Structural and genetic organization of IS*232*, a new insertion sequence of *Bacillus thuringiensis*. *Journal of Bacteriology* 172: 6689–6696.

Miller, L.K., Lingg, A.J. and Bulla, L.A. Jr. (1983). Bacterial, viral, and fungal insecticides. *Science* 219: 715–721.

Milne, R. and Kaplan, (1993). Purification and characterization of a trypsin-like digestive enzyme from spruce budworm (*Choristoneura fumiferana*) responsible for the activation of δ-endotoxin from *Bacillus thuringiensis*. *Insect Biochemistry and Molecular Biology* 23: 663–673.

Mohamed, A.K.A., Sikorowski, P.P. and Bell, J.V. (1977). Susceptibility of *Heliothis zea* larvae to *Nomuraea rileyi* at various temperatures. *Journal of Invertebrate Pathology* 30: 414–417.

Moscardi, F. (1999). Assessment of the application of baculoviruses for control of Lepidoptera. *Annual Review of Entomology* 44: 257–289.

Moscardi, F. and Sosa-Gomez, D.R. (2000). Microbial control of insect pests of soybean. In Lacey, L.A. and Kaya, H.K. (Eds.), *Field Manual of Techniques in Invertebrate Pathology. Application and Evaluation of Pathogens for Control of Insects and Other Invertebrate Pests*. Dordrecht, The Netherlands: Kluwer Academic Press, 447–460.

Nagamatsu, Y., Itai, Y., Hatanaka, C., Funatsu, G. and Hayashi, K. (1984). A toxic fragment from the entomocidal crystal protein of *Bacillus thuringiensis*. *Agricultural and Biological Chemistry* 48: 611–619.

Nambiar, P.T.C., Ma, S.W. and Iyer, V.N. (1990). Limiting an insect infestation of nitrogen-fixing root nodules of the pigeonpea (*Cajanus cajan*) by engineering the expression of an entomocidal gene in its root nodules. *Applied and Environmental Microbiology* 56: 2866–2869.

Navon, A. (2000). *B. thuringiensis* insecticides in crop protection: Reality and prospects. *Crop Protection* 19: 669–676.

Neale, M.C. (1997). Bio-pesticides–harmonization of registration requirements within EU directive 91–414. An industry view. *OEPP Bulletin* 27: 89–93.

Nicolas, L., Regis, L.N. and Rios, E.M. (1994). Role of the exosporium in the stability of the *Bacillus sphaericus* binary toxin. *FEMS Microbiology Letters* 124: 271–276.

Novillo, C., Castañera, P. and Ortego, F. (1997). Characterization and distribution of chymotrypsin-like and other digestive proteases in Colorado potato beetle larvae. *Archives of Insect Biochemistry and Physiology* 36: 181–201.

Obukowicz, M.G., Perlak, F.J., Kusano Kretzmer, K., Mayer, E.J. and Watrud, L.S. (1986a). Integration of the delta endotoxin gene of *Bacillus thuringiensis* into the chromosome of root-colonizing pseudomonads using Tn5. *Gene* 45: 327–331.

Obukowicz, M.G., Perlak, F.J., Kusano Kretzmer, K., Mayer, E.J., Bolten, S.L. and Watrud, L.S. (1986b). Tn5-mediated integration of the delta-endotoxin gene from *Bacillus thuringiensis* into the chromosome of root-colonizing pseudomonads. *Journal of Bacteriology* 168: 982–989.

O'Reilly, D.R. and Miller, L.K. (1991). Improvement of a baculovirus pesticide by deletion of EGT gene. *BioTechnology* 9: 1086–1089.

Peterson, A.M., Fernando, G.J.P. and Wells, M.A. (1995). Purification, characterization and cDNA sequence of an alkaline chymotrypsin from the midgut of *Manduca sexta*. *Insect Biochemistry and Molecular Biology* 25: 765–774.

Pfeifer, T.A. and Khachatourians, G.G. (1992). *Beauveria bassiana* protoplast regeneration and transformation using electroporation. *Applied Microbiology and Biotechnology* 38: 376–381.

Piek, T. (1990). Neurotoxins from venoms of the Hymenoptera: Twenty-five years of research in Amsterdam. *Comparative Biochemistry and Physiology. C. Comparative Pharmacology and Toxicology* 96: 223–233.

Piek, T. (1991). Neurotoxic kinins from wasp and ant venoms. *Toxicon* 29: 139–149.

Piek, T., May, T.E. and Spanger, W. (1980). Paralysis of locomotion in insects by the venom of the digger wasp *Philanthus triangulum*. In *Insect Neurobiology and Pesticide Action* (*Neurotox '79*). London, UK: Society of Chemical Industry, 219–226.

Piek, T., Duval, A., Hue, B., Karst, H., Lapied, B., Mantel, P., Nakajima, T., Pelhate, M. and Schmidt, J.O. (1991). Poneratoxin, a novel peptide neurotoxin from the venom of the ant, *Paraponera clavata*. *Comparative Biochemistry and Physiology. C. Comparative Pharmacology and Toxicology* 99: 487–495.

Piek, T., Hue, B., Lind, A., Mantel, P., van Marle, J. and Visser, J.H. (1989). The venom of *Ampulex compressa* effects on behaviour and synaptic transmission of cockroaches. *Comparative Biochemistry and Physiology. C. Comparative Pharmacology and Toxicology* 92: 175–183.

Poinar, G.O. and Thomas, G.M. (1966). Significance of *Achromobacter nematophillus* Poinar and Thomas (Achromobacteraceae: Eubacteriales) in the development of the nematode, DD136 (*Neoaplectana* sp: Steinernematidae). *Parasitology* 56: 385–390.

Poinar, G.O. Jr. (1990). Taxonomy and biology of Steinernematidae and Heterorhabditidae. In Gaugler, R. and Kaya, H.K. (Eds.), *Entomopathogenic Nematodes in Biological Control*. Boca Raton, Florida, USA: CRC Press, 23–61.

Poncet, S., Anello, G., Delécluse, A., Klier, A. and Rapoport, G. (1993). Role of the CryIVD polypeptide in the overall toxicity of *Bacillus thuringiensis* subsp. *israelensis*. *Applied Environmental Microbiology* 59: 3928–3930.

Poncet, S., Delécluse, A., Anello, G., Klier, A. and Rapoport, G. (1994). Transfer and expression of the *cryIVB* and *cryIVD* genes of *Bacillus thuringiensis* subsp. *israelensis* in *Bacillus sphaericus* 2297. *FEMS Microbiology Letters* 117: 91–96.

Poncet, S., Delécluse, A., Klier, A. and Rapoport, G. (1995). Evaluation of synergistic interactions between the CryIVA, CryIVB and CryIVD toxic components of *B. thuringiensis* subsp. *israelensis* crystals. *Journal of Invertebrate Pathology* 66: 131–135.

Poncet, S., Dervyn, E., Klier, A. and Rapoport, G. (1997). Spo0A represses transcription of the cry toxin genes in *Bacillus thuringiensis*. *Microbiology* 143: 2743–2751.

Popham, H.J.R., Prikhod ko, G.G., Felcetto, T.J., Ostlind, D.A., Warmke, J.W., Cohen, C.J. and Miller, L.K. (1998). Effect of deltamethrin treatment on lepidopteran larvae infected with baculoviruses expressing insect-selective toxins μ-Aga-IV, as II, or sh 1. *Biological Control* 12: 79–87.

Prikhod ko, G.G., Popham, H.J.R., Felcetto, T.J., Ostlind, D.A., Warren, V.A., Smith, M.M., Garsky, V.M., Warmkeo, J.W. Cohen, C.J. and Miller, L.K. (1998). Effects of simultaneous expression of two sodium channel toxin genes on the properties of baculoviruses as biopesticides. *Biological Control* 12: 66–78.

Rajamohan, F., Alcantara, E., Lee, M.K., Chen, X.J., Curtiss, A. and Dean, D.H. (1995). Single amino acid changes in domain II of *Bacillus thuringiensis* CryIAb δ-endotoxin affect irreversible binding to *Manduca sexta* midgut membrane vesicles. *Journal of Bacteriology* 177: 2276–2282.

Rezsöhazy, R., Hallet, B., Delcour, J. and Mahillon, J. (1993). The IS4 family of insertion sequences: Evidence for a conserved transposase motif. *Molecular Microbiology* 9: 1283–1295.

Rochat, H., Bernard, P. and Courard, F. (1979). Scorpion toxins: Chemistry and mode of action. In Ceccarelli, B. and Clementi, F. (Eds.), *Advances in Cytopharmacology*, vol. 3. New York, USA: Raven Press, 325–334.

Salama, H.S., Ali, A.M.M. and Sharaby, A. (1991). *Bacillus thuringiensis* Berliner resistant to high temperature and ultraviolet radiation. *Journal of Applied Entomology* 112: 520–524.

Salamitou, S., Agaisse, H., Bravo, A. and Lereclus, D. (1996). Genetic analysis of *cryIIIA* gene expression in *Bacillus thuringiensis*. *Microbiology* 142: 2049–2055.

Sanchis, V. and Ellar, D.J. (1993). Identification and partial purification of a *Bacillus thuringiensis* CryIC δ-endotoxin binding protein from *Spodoptera littoralis* gut membranes. *FEBS Letters* 316: 264–268.

Sanchis, V., Agaisse, H., Chaufaux, J. and Lereclus, D. (1996). Construction of new insecticidal *Bacillus thuringiensis* recombinant strains by using the sporulation non-dependent expression system of *cryIIIA* and a site specific recombination vector. *Journal of Biotechnology* 48: 81–96.

Sanchis, V., Agaisse, H., Chaufaux, J. and Lereclus, D. (1997). A recombinase-mediated system for elimination of antibiotic resistance gene markers from genetically engineered *Bacillus thuringiensis* strains. *Applied Environmental Microbiology* 63: 779–784.

Sandhu, S.S., Unkles, S.E., Rajak, R.C. and Kinghorn, J.R. (2001). Generation of benomyl resistant *Beauveria bassiana* strains and their infectivity against *Helicoverpa armigera*. *Biocontrol Science and Technology* 11: 245–250.

Sangadala, S., Walters, F.S., English, L.H. and Adang, M.J. (1994). A mixture of *Manduca sexta* aminopeptidase and phosphatase enhances *Bacillus thuringiensis* insecticidal CryIA(c) toxin binding and ^{86}Rb$^+$-K$^+$ efflux *in vitro*. *Journal of Biological Chemistry* 269: 10088–10092.

Saxena, H. and Ahmed, R. (1997). Field evaluation of *Beauveria bassiana* against *Helicoverpa armigera* (Hübner) infecting chickpea. *Journal of Biological Control* 11: 93–96.

Schnepf, H.E. and Whitley, H.R. (1981). Cloning and expression of *Bacillus thuringiensis* crystal protein gene in *Escherichia coli*. *Proceedings National Academy of Sciences USA* 78: 2893–2897.

Schnepf, E., Crickmore, N., Van Rie, J., Lereclus, D., Baum, J., Feitelson, J., Zeigler, D.R. and Dean, D.H. (1998). *Bacillus thuringiensis* and its pesticidal crystal proteins. *Microbiology and Molecular Biology Reviews* 62: 775–806.

Schurter, W., Geiser, M. and Mathé, D. (1989). Efficient transformation of *Bacillus thuringiensis* and *B. cereus* via electroporation: Transformation of acrystalliferous strains with a cloned delta-endotoxin gene. *Molecular and General Genetics* 218: 177–181.

Schwartz, J.L., Lu, Y.J., Sohnlein, P., Brousseau, R., Laprade, R., Masson, L. and Adang, M.J. (1997). Ion channels formed in planar lipid bilayers by *Bacillus thuringiensis* toxins in the presence of *Manduca sexta* midgut receptors. *FEBS Letters* 412: 270–276.

Sekar, V. (1988). The insecticidal crystal protein gene is expressed in vegetative cells of *Bacillus thuringiensis* var. *tenebrionis*. *Current Microbiology* 17: 347–349.

Shah, P.A. and Goettel, M.S. (1999). Directory of microbial control products. Society for Invertebrate Pathology. Division of Microbial Control. http://www.sipweb.org/directory.htm.

Shapiro, M., Robertson, J.R. and Bell, R.A. (1987). Quantitative and qualitative differences in gypsy moth (Lepidoptera: Lymantriidae) nucleopolyhedrosis virus produced in different-aged larvae. *Journal of Economic Entomology* 79: 1174–1177.

Shewry, P.R. and Gutteridge, S. (Eds.). (1992). *Plant Protein Engineering*. Cambridge, UK: Cambridge University Press.

Skot, L., Harrison, S.P., Nath, A., Mytton, L.R. and Clifford, B.C. (1990). Expression of insecticidal activity in *Rhizobium* containing the δ-endotoxin gene cloned from *Bacillus thuringiensis* subsp. *tenebrionis. Plant and Soil* 127: 285–295.

Smith, G.P. and Ellar, D.J. (1994). Mutagenesis of two surface-exposed loops of the *Bacillus thuringiensis* CryIC δ-endotoxin affects insecticidal specificity. *Biochemistry Journal* 302: 611–616.

Smith, G.P., Ellar, D.J., Keeler, S.J. and Seip, C.E. (1994). Nucleotide sequence and analysis of an insertion sequence from *Bacillus thuringiensis* related to IS150. *Plasmid* 32: 10–18.

Solter, L.F. and Becnel, J.J. (2000). Entomopathogenic microsporida. In Lacey, L.A. and Kaya, H.K. (Eds.), *Field Manual of Techniques in Invertebrate Pathology: Application and Evaluation of Pathogens for Control of Insects and Other Invertebrate Pests.* Dordrecht, The Netherlands: Kluwer Academic Press, 231–254.

Soltes-Rak, E., Kushner, D.J., Williams, D.D. and Coleman, J.R. (1993). Effect of promoter modification on mosquitocidal *cryIVB* gene expression in *Synechococcus* sp. strain 7942. *Applied Environmental Microbiology* 59: 2404–2410.

Srinivasan, T.R. (1997). Studies on Pathogenicity and Virulence of *Metarhizium anisopliae* (Metsch.), *Metarhizium flavoviride* (Metsch.) and *Nomuraea rileyi* (Farlow) Samson and Management of *Helicoverpa armigera* (Hübner) and *Spodoptera litura* (F.) in *Lycopersicon esculentum* (L.). Ph.D. thesis, Tamil Nadu Agricultural University, Coimbatore, Tamil Nadu, India.

St. Leger, R.J. and Roberts, D.W. (1997). Engineering improved mycoinsecticides. *Trends in Biotechnology* 15: 83–85.

St. Leger, R.J, Shimizu, S., Joshi, L., Bidochka, M.J. and Roberts, D.W. (1995). Co-transformation of *Metarhizium anisopliae* by electroporation or using the gene gun to produce stable GUS transformants. *FEMS Microbiology Letters* 131: 289–294.

Stapleton, A., Blankenship, D.T., Ackermann, B.L., Chenn, T.M., Goroder, G.W., Manley, G.D., Plafreyman, M.G., Coutant, J.E. and Cardi, A.D. (1990). Curtatoxins, neurotoxic insecticidal polypeptides isolated from the funnel web spider *Hololena curta. Journal of Biology and Chemistry* 265: 2054–2059.

Stevens, S.E. Jr., McMurphy, R.C., Lamoreaux, W.J. and Coons, L.B. (1994). A genetically engineered mosquitocidal cyanobacterium. *Journal of Applied Physiology* 6: 187–197.

Stewart, C.J. (1992). Natural transformation and its potential for gene transfer in the environment. In Levin, M.A., Siedler, R.J. and Rogul, N. (Eds.), *Microbial Ecology, Principles, Methods and Applications.* New York, USA: McGraw-Hill, 283–311.

Stewart, L.M.D., Hirst, M., Ferber, M.L., Merryweather, A.T., Cayley, P.J. and Possee, R.D. (1991). Construction of an improved baculovirus insecticide containing an insect specific toxin gene. *Nature* 352: 85–88.

Stock, C.A., McLoughlin, T.J., Klein, J.A. and Adang, M.J. (1990). Expression of a *Bacillus thuringiensis* crystal protein gene in *Pseudomonas cepacia* 526. *Canadian Journal of Microbiology* 36: 879–884.

Stozky, G. and Babich, H. (1984). Fate of genetically engineered microbes in the environment. *rDNA Technical Bulletin* 7: 163–189.

Sudarsan, N., Suma, N.R., Vennison, S.J. and Sekar, V. (1994). Survival of a strain of *Bacillus megaterium* carrying a lepidopteran-specific gene of *Bacillus thuringiensis* in the phyllospheres of various economically important plants. *Plant and Soil* 167: 321–324.

Sudha, S.N., Jayakumar, R. and Sekar, V. (1999). Introduction and expression of the cry1Ac gene of *Bacillus thuringiensis* in a cereal-associated bacterium, *Bacillus polymyxa. Current Microbiology* 38:163–167.

Sun, X.L., Sun, X.C., van-der Werf, W. and Hu, Z.H. (2004). Field inactivation of wild-type and genetically modified *Helicoverpa armigera* single nucleocapsid nucleopolyhedrovirus in cotton. *Biocontrol Science and Technology* 14: 185–192.

Tailor, R., Tippett, J., Gibb, G., Pells, S., Pike, D., Jordan, L. and Ely, S. (1992). Identification and characterization of a novel *Bacillus thuringiensis*-endotoxin entomocidal to coleopteran and lepidopteran larvae. *Molecular Microbiology* 7: 1211–1217.

Tan, Y. and Donovan, W. (1995). Cloning and characterization of the alkaline protease gene of *Bacillus thuringiensis.* Abstract Q-40. In *95th General Meeting of the American Society for Microbiology.* Washington, D.C., USA: American Society for Microbiology, 406.

Thanabalu, T., Hindley, J., Brenner, S., Oei, C. and Berry, C. (1992). Expression of the mosquitocidal toxins of *Bacillus sphaericus* and *Bacillus thuringiensis* subsp. *israelensis* by recombinant *Caulobacter crescentus*, a vehicle for biological control of aquatic insect larvae. *Applied Environmental Microbiology* 58: 905–910.

Thompson, M.A., Schnepf, H.E. and Feitelson, J.S. (1995). Structure, function, and engineering of *Bacillus thuringiensis* toxins. In Setlow, J.K. (Ed.), *Genetic Engineering: Principles and Methods*, vol. 17. New York, USA: Plenum Press, 99–117.

Tojo, A. and Aizawa, K. (1983). Dissolution and degradation of δ-endotoxin by gut juice protease of silkworm, *Bombyx mori*. *Applied and Environmental Microbiology* 45: 576–580.

Tomalski, M. and Miller, L.K. (1991). Insect paralysis by baculovirus mediated expression of a mite neurotoxin gene. *Nature* 252: 82–85.

Tomalski, M.D., Kutney, R., Bruce, W.A., Brown, M.R., Blum, M.S. and Travis, J. (1989). Purification and characterization of insect toxins derived from the mite, *Pyemotes tritici*. *Toxicon* 27: 1151–1167.

Tomasino, S.F., Leister, R.T., Dimock, M.B., Beach, R.M. and Kelly, J.L. (1995). Field performance of *Clavibacter xyli* subsp. *cynodontis* expressing the insecticidal crystal protein gene *cry1A(c)* of *Bacillus thuringiensis* against European corn borer in field corn. *Biological Control* 5: 442–448.

Tortora, G.J., Funke, B.R. and Case, C.L. (1989). *Microbiology*. New York, USA: Cummings, 201–225.

Treacy, M.F., Rensner, P.E. and All, J.N. (2000). Comparative insecticidal properties of two nucleopolyhedrovirus vectors encoding a similar toxin gene chimer. *Journal of Economic Entomology* 93: 1096–1104.

Tzotzos, T. (Ed.). (1995). *Genetically Modified Microorganisms: A Guide to Biosafety*. Wallingford, Oxon, UK: Commonwealth Agricultural Bureau, International.

Udayasuriyan, V., Nakamura, A., Masaki, H. and Uozumi, T. (1995). Transfer of an insecticidal protein gene of *Bacillus thuringiensis* into plant-colonizing *Azospirillum*. *World Journal of Microbiology and Biotechnology* 11: 163–167.

Uma Devi, K., Padmavathi, J., Sharma, H.C. and Seetharama, N. (2001). Laboratory evaluation of the virulence of *Beauveria bassiana* isolates to the sorghum shoot borer, *Chilo partellus* Swinhoe (Lepidoptera: Pyralidae) and their characterization by RAPD-PCR. *World Journal of Microbiology and Biotechnology* 17: 131–137.

Vadlamudi, R.K., Ji, T.H. and Bulla, L.A. Jr. (1993). A specific binding protein from *Manduca sexta* for the insecticidal toxin of *Bacillus thuringiensis* subsp. Berliner. *Journal of Biological Chemistry* 268: 12334–12340.

Vail, P.V., Anderson, S.J. and Jay, D.L. (1973). New procedures for rearing cabbage loopers and other lepidopterous larvae for propagation of nuclear polyhedrosis virus. *Environmental Entomology* 2: 339–344.

Valaitis, A.P., Lee, M.K., Rajamohan, F. and Dean, D.H. (1995). Brush border membrane aminopeptidase-N in the midgut of the gypsy moth serves as the receptor for the CryIA(c) δ-endotoxin of *Bacillus thuringiensis*. *Insect Biochemistry and Molecular Biology* 25: 1143–1151.

Valaitis, A.P., Mazza, A., Brousseau, R. and Masson, L. (1997). Interaction analyses of *Bacillus thuringiensis* Cry1A toxins with two aminopeptidases from gypsy moth midgut brush border membranes. *Insect Biochemistry and Molecular Biology* 27: 529–539.

Van Rie, J., Jansens, S., Hofte, H., Degheele, D. and Van Mellaert, H. (1989). Specificity of *Bacillus thuringiensis* endotoxins. Importance of specific receptors on the brush border membrane of the mid-gut of target insects. *European Journal of Biochemistry* 186: 239–247.

Van Rie, J., McGhaughey, W.H., Johnson, D.E., Barnett, B.D. and Van Mellaert, H. (1990). Mechanism of insect resistance to the microbial insecticide *Bacillus thuringiensis*. *Science* 247: 72–74.

Visser, B., Bosch, D. and Honée, G. (1993). Domain-function studies of *Bacillus thuringiensis* crystal proteins: A genetic approach. In Entwistle, P.F., Cory, J.S., Bailey, M.J. and Higgs S. (Eds.), Bacillus thuringiensis, *An Environmental Biopesticide: Theory and Practice*. Chichester, UK: John Wiley & Sons, 71–88.

Walter, M.V. and Seidler, R.J. (1992). Measurement of conjugal gene transfer in terrestrial ecosystems. In Levin, M.A., Siedler, R.J. and Rogul, N. (Eds.), *Microbial Ecology: Principles, Methods, and Applications*. New York, USA: McGraw Hill, 311–326.

Waterfield, N.R., Bowen, D.J., Fetherston, J.D., Perry, R.D. and Ffrench-Constant, R.H. (2001). The *tc* genes of *Photorhabdus*: A growing family. *Trends in Microbiology* 9: 185–191.

WHO (World Health Organization). (1973). *Uses of Viruses for the Control of Insect Pests and Disease Vectors*. WHO Technical Report Series no 531. Geneva, Switzerland: World Health Organization.

Wigle, D.T., Samenic, R.M., Wilkins, K., Riedel, D., Ritter, L., Morrison, H.I. and Mao, Y. (1990). Mortality study of Canadian male farm operators: Non-Hodgkin's lymphoma mortality and agricultural practices in Saskatchewan. *Journal of National Cancer Institute* 82: 575–582.

Wolfersberger, M.G. (1990). The toxicity of two *Bacillus thuringiensis* δ-endotoxins to gypsy moth larvae is inversely related to the affinity of binding sites on midgut brush border membranes for the toxins. *Experientia* 46: 475–477.

Wolfersberger, M.G., Hofmann, C. and Lüthy, P. (1986). Interaction of *Bacillus thuringiensis* delta-endotoxin with membrane vesicles isolated from lepidopteran larval midgut. In Falmagne, P., Alouf, J.E., Fehrenbach, F.J., Jeljaszewicz, J. and Thelestam, M. (Eds.), *Bacterial Protein Toxins*. Stuttgart, Germany: Gustav Fischer Verlag, 237–238.

Wong, H.C., Schnepf, H.E. and Whiteley, H.R. (1983). Transcriptional and translational start sites for the *Bacillus thuringiensis* crystal protein gene. *Journal of Biological Chemistry* 258: 1960–1967.

Wu, D., Johnson, J.J. and Federici, B.A. (1994). Synergism of mosquitocidal toxicity between CytA and CryIVD proteins using inclusions produced from cloned genes of *Bacillus thuringiensis*. *Molecular Microbiology* 13: 965–972.

Yaoi, K., Kadotani, T., Kuwana, H., Shinkawa, A., Takahashi, T., Iwahana, H. and Isato, R. (1997). Aminopeptidase N from *Bombyx mori* as a candidate for the receptor of *Bacillus thuringiensis* Cry1Aa toxin. *European Journal of Biochemistry* 246: 652–657.

Yoshisue, H., Ihara, K., Nishimoto, T., Sakai, H. and Komano, T. (1995). Expression of the genes for insecticidal crystal proteins in *Bacillus thuringiensis*: *cryIVA*, not *cryIVB*, is transcribed by RNA polymerase containing Sigma H and that containing Sigma E. *FEMS Microbiology Letters* 127: 65–72.

Yoshisue, H., Nishimoto, T., Sakai, H. and Komano, T. (1993). Identification of a promoter for the crystal protein-encoding gene *cryIVB* from *Bacillus thuringiensis* subsp. *israelensis*. *Gene* 137: 247–251.

Zambrano, B.K., Davila, M. and Castillo, M.A. (2002). Detection of fungi DNA fragments and their possible relation with entomopathogenic protein synthesis activity. *Revista de la Facultad de Agronomia* 19: 185–193.

Zilberberg, N., Zlotkin, E. and Gurevitz, M. (1992). Molecular analysis of cDNA and the transcript encoding the depressant insect selective neurotoxin of the scorpion *Leiurus quinquestriatus* (Ehren.). *Insect Biochemistry and Molecular Biology* 22: 199–203.

Zlotkin, E. (1988). Neurotoxins. In Lunt, G.G. and Olsen, R.W. (Eds.), *Comparative Invertebrate Neurochemistry*. London, UK: Croom Helm, 256–324.

Zlotkin, E. and Gordon, D. (1985). Detection, purification and receptor binding assays of insect selective neurotoxins derived from scorpion venom. In Breer, H. and Miller, T.A. (Eds.), *Neurochemical Techniques in Insect Research*. New York, USA: Springer Verlag, 243–295.

Zlotkin, E., Eitan, M., Bindokas, V.P., Adams, M.E., Moyer, M., Burkhart, W. and Fowler, E. (1991). Functional duality and structural uniqueness of depressant insect selective neurotoxins. *Biochemistry* 30: 4814–4821.

Zlotkin, E., Frankel, G., Miranda, F. and Lissitzky, S. (1971a). The effect of scorpion venom potency. *Toxicon* 9: 1–8.

Zlotkin, E., Miranda, F., Kupeyan, G. and Lissitzky, S. (1971b). A new toxic protein in the venom of scorpion *Androctonus australis* Hector. *Toxicon* 9: 9–13.

Zlotkin, E., Rochat, H., Kupeyan, C., Miranda, F. and Lissitzky, S. (1971c). Purification and properties of the "insect" toxin from the venom of the scorpion *Androctonus australis* Hector. *Bioichimie* 53: 1073–1078.

Zylstra, G.J., Cuskey, S.M. and Olson, R.H. (1992). Construction of plasmids for use in survival and gene transfer research. In Levin, M.A., Siedler, R.J. and Rogul, N. (Eds.), *Microbial Ecology: Principles, Methods, and Applications*. New York, USA: McGraw-Hill, 363–371.

9

Genetic Engineering of Natural Enemies for Integrated Pest Management

Introduction

Indiscriminate use of pesticides has led to serious concerns relating to their adverse effects on nontarget organisms, pesticide residues in food and food products, pest resurgence, development of resistance to insecticides, toxic effects on human beings, and environmental pollution (Metcalf and Luckmann, 1982). Widespread pesticide use has also led to elimination of natural enemies and, as a result, it has become practically difficult to control several insect species through the currently available chemical pesticides. Therefore, it is important to adopt pest control strategies that are ecologically sound, economically practical, and socially acceptable. In this context, development and deployment of natural enemies that are adapted to extremes of climatic conditions or are capable of tolerating sublethal doses of pesticides can play an important role in pest management. Recent advances in molecular biology have broadened the available techniques for genetic manipulation of arthropods for a variety of traits in species of interest (Atkinson, Pinkerton, and O'Brochta, 2001; Kramer, 2004). Release of genetically improved arthropods for suppression of pest populations has been undertaken in the past, and the possible applications of genetically modified arthropods have expanded considerably (Braig and Yan, 2002). Incorporation of foreign DNA into the genome has expanded the possibilities for genetic transformation of insects, although it has also raised a few questions. The first genetically modified insects were produced 20 years ago with the restoration of wild-type eye color in a mutant strain of *Drosophila melanogaster* Meigen (Rubin and Spradling, 1982), followed by transformation of the Mediterranean fruit fly, *Ceratitis capitata* Weid. (Loukeris et al., 1995).

Beneficial arthropods can be transformed for a variety of traits, and deployed more effectively as biocontrol agents. Some of the major constraints in using natural enemies in pest control are the difficulties involved in mass rearing and their ability to withstand adverse conditions. The molecular techniques can also be used to understand the genetics and physiology of reproduction and control of sex ratio, and this information can be used

TABLE 9.1

Applications of Biotechnology to Improve Arthropod Natural Enemies

Characteristics	References
Modify the genome of natural enemies	Handler and O'Brochta (1991)
Change sex ratio of parasitoids	Bownes (1992)
Cryopreservation	Mazur et al. (1992)
Develop genetic linkage maps	Sobral and Honeycutt (1993)
Identify biotypes	Ballinger-Crabtree, Black, and Miller (1992)
Improve artificial diets	Thompson (1990)
Monitor establishment and dispersal	Edwards and Hoy (1993)
Parentage analysis and genetic changes	Scott and Williams (1993)

to improve rearing of natural enemies for biological control (Table 9.1). Selection for resistance to pesticides, lack of diapause, and enhanced temperature tolerance has been successful. However, most research has been focused on selection for resistance to pesticides (Hoy, 1990a). Genetic improvement has proved to be practical and cost effective when the trait(s) limiting the efficacy can be identified, and the improved strain retains the fitness (Headley and Hoy, 1987). Biotechnological interventions can also be used to broaden the host range of natural enemies or enable their production on artificial diet or nonhost insect species that are easy to multiply under laboratory conditions. Some of the desirable characteristics for transgenic insects include pathogen resistance, environmental hardiness, increased fecundity, and improved host-seeking ability (O'Brochta and Atkinson, 1998).

There is tremendous scope for developing natural enemies with genes for resistance to pesticides and ability to withstand adverse weather conditions (Hoy, 1992a). Genetically modified natural enemies can be used more effectively in pest management programs in combination with the conventional insecticides. Pesticide-resistant predators and parasitoids have been evaluated in the field and are being used in several integrated pest management (IPM) programs (Hoy, 1990a). The transgenic predatory mite, *Metaseiulus occidentalis* Nesbitt has been used for the control of spider mite, *Tetranychus uirticae* Koch (Presnail et al., 1997). If the natural populations of predatory mites can be replaced with the one that is resistant to pesticides, then there will be considerable reduction in pesticide use as the pesticide application would only affect the crop pests, but not the pesticide-resistant predatory mites.

Another potential application of biotechnology could be the use of insects as a delivery vehicle for vaccines (Crampton et al., 1999). A blood-sucking insect can be transformed to express an antigen and deliver it through saliva that the insect would inject into the vertebrate host during feeding (Coates et al., 1999; Crampton et al., 1999). Genetic transformation can also be used to modify commercial production of materials such as silk, honey, lac, and biomaterials (Mori and Tsukada, 2000) or production of pharmaceuticals and biomolecules (Yang et al., 2002). However, release of genetically modified insects might have a potential risk to the environment (Spielman, Beier, and Kiszewski, 2002). This is of particular concern when the same vector transmits several disease-causing pathogens, as it might be quite difficult to develop transgenic individuals incapable of transmitting different pathogens. Genetic improvement can be useful when:

- The natural enemy is known to be a potentially effective biological control agent, except for one limiting factor.
- The limiting trait is primarily influenced by a single major gene.

- The gene can be obtained by selection, mutagenesis, or cloning.
- The manipulated strain is fit and effective.
- The released strain can be maintained in some form of reproductive isolation.

Some of the biotechnological interventions that can help improve the efficiency of biological control of insect pests are discussed below.

Techniques for Genetic Engineering of Arthropods

Genes for use in genetic engineering of arthropods can be isolated from either closely or distantly related organisms for insertion into the target insect species. It may also be possible to isolate a gene from the species being manipulated, alter it, and reinsert it into a germ line. The ability to genetically transform nondrosophilid insects offers new approaches to analyzing molecular genetic systems in insects. A consequence of increased knowledge that will arise from these studies will lead to the development of new strategies for pest control that will be specific for the target insect. Most of the research on genetic engineering of insects has been aimed at gaining an understanding of gene regulation or developmental processes. Cloned genes could also be modified by *in vitro* mutation to achieve a desired phenotype. Various techniques employed for genetic engineering of insects (Table 9.2) are described below.

DNA Injection into Eggs

Microinjection into eggs has been employed widely to deliver DNA into the embryos of the target insect species. Different injection methods are required for different insect species (Miller et al., 1987; McGrane et al., 1988; Milne Phillips and Krell, 1988; Morris, Eggleston, and Crampton, 1989). The eggs of the phytoseiid mite, *M. occidentalis* are difficult to dechlorinate and dehydrate, and the needle tip needs to be modified (Presnail and Hoy, 1992). The DNA can be injected into early preblastoderm eggs present within the adult females

TABLE 9.2

Methods for Genetic Transformation of Arthropods

Technique	Example	References
Microprojectiles	*Drosohila* embryos	Baldarelli and Lengyel (1990)
Electroporation	*Drosophila*	Kamdar, von Allmen, and Finnerty (1992)
Maternal microinjection	*Metaseiulus occidentalis*	Presnail and Hoy (1992)
Microinjection of eggs		Miller et al. (1987), McGrane et al. (1988), Morris, Eggleston, and Crampton (1989)
Soaking dechlorinated eggs in DNA	*Drosophila*	Walker (1989)
Sperm as vectors of DNA	*Lucilia cuprina* and *Apis mellifera* L. DNA bound externally	Atkinson et al. (1991), Milne, Phillips, and Krell (1988)
Transformation of cultured cells	*Aedes albopictus* (Skuse)	Fallon (1991)
Transplant nuclei and cells	*Drosophila*	Zalokar (1981)

of *M. occidentalis* by inserting the needle through the cuticle of the gravid female. This results in relatively high survival and stable transformation (Presnail and Hoy, 1992).

Maternal Microinjection

Maternal microinjection involves injection of DNA into an adult female to deliver DNA to the eggs. The adaptation of this method to the parasitoid, *Cardiochiles diaphaniae* Marsh has been reported by Presnail and Hoy (1996). The results of preliminary dissections and injections with the plasmid *pJKP2* suggested that this method could be used to deliver injected DNA to the ovaries of the wasp, *C. diaphaniae*. Of the 16 females that were injected and allowed to produce offspring, several individuals contained the plasmid sequences. In another experiment, the plasmid *phsopd*, containing the parathion hydrolase gene (*opd*) of *Pseudomonas diminuta* Leifson and Hugh, was injected. The plasmid probe hybridized to high molecular weight DNA from three of 38 G1 adults. A fourth adult produced a hybridization pattern consistent with integration of plasmid into the nuclear genome. The results indicated that maternal microinjection can result in transformation of the parasitoid. The persistence of DNA injected into two species of the adult female phytoseiid, *M. occidentalis* and its transmission to serial eggs deposited by them has been assessed by polymerase chain reaction (PCR) (Presnail and Hoy, 1994). The effect of DNA concentration on persistence and transmission was examined in females of *M. occidentalis* microinjected with plasmid DNA at three different concentrations (250, 500, 750 ng μL^{-1}) and allowed to deposit one to five eggs. The plasmid DNA was found in 82% of the females assayed and in 70% of all the eggs, including the fifth egg produced after microinjection. Transmission of DNA to multiple eggs was also examined in *Amblyseius finlandicus* (Oud.). The persistence and presence of plasmid DNA in both eggs and females suggested that maternal microinjection was a more efficient method for DNA delivery than traditional egg microinjection, and that it could be useful as a DNA delivery system in other phytoseiids.

Presnail et al. (1997) transformed four lines of *M. occidentalis* by maternal microinjection. Putatively transformed lines were identified by standard PCR in the G1 generation. After 30 generations, the lines were examined by Southern blot hybridization, and the plasmid probe was hybridized to uncut high molecular weight DNA from all four lines, indicating that the transgene was associated with high molecular weight DNA. In restriction-digested Southern blots, one of the four lines displayed a hybridization pattern consistent with integration of two plasmids into the chromosomal DNA. All four colonies were confirmed to be positive by PCR after 150 generations. Colonies examined for mRNA expression after 100 generations displayed PCR products consistent with transcription of the introduced genetic sequences. Microinjection of early preblastoderm eggs in gravid females of *M. occidentalis* resulted in relatively high levels of survival and transformation (Presnail and Hoy, 1992). Transformation was achieved without the aid of any transposase-producing helper plasmid. The predatory mite was transformed with a plasmid containing the *Escherichia coli* Escherich beta-galactosidase gene regulated by the *Drosophila hsp70* heat-shock promoter. Putatively transformed lines were isolated based on beta-galactosidase activity in first-generation larvae. Transformation was confirmed in the sixth generation by PCR amplification of a region spanning the *Drosophila/E. coli* sequences. This method can also be adapted to other beneficial arthropods, particularly the phytoseiids.

Sperm-Mediated Transfer

DNA has been introduced successfully into individual honeybees by inseminating virgin queens with linearized DNA and semen (Robinson et al., 2000), and the trait was inherited as extrachromosomal arrays. This technique, along with transposable elements (P-elements), can be used for stable integration of exogenous DNA.

Paratransgenesis

Foreign genes can also be expressed in endosymbionts in insect gut or other organs rather than by the insect. This system has been used in the case of tsetse flies, *Glossina morsitans* Westwood and kissing bugs, *Rhodnius prolixus* (Stal). Genetic engineering of endosymbionts in tsetse fly to express antitrypanosome agents could prevent transmission of sleeping sickness (Aksoy, 2000). Flies with recombinant *Sodalis* species expressing antimicrobial peptide attacin reduced the infection rate to 3,000 (Aksoy, 2003). The symbiont of kissing bug, *R. prolixus*, the vector of Chagas disease, has been transformed with a marker gene and gene for cecropin A, a trypanocidal peptide found in insect hemolymph (Durvasula et al., 1997). Cecropin was stably expressed and killed *Trypanosoma cruzi* Chaggas without affecting the kissing bugs.

Vectors for Genetic Engineering of Arthropods

Transgenic expression requires an appropriate promoter and other regulatory elements. Four different transposable elements from four separate families of eukaryotic transposable elements can be used to transform nondrosophilid insects (Atkinson and O'Brochta, 1999). In addition, viral and bacterial endosymbiont gene delivery systems have also been developed for some insect species. Many genes have been cloned and inserted into *Drosophila* by P-element-mediated transformation, but very few cloned genes are of value for genetic manipulation of beneficial arthropods.

Transposable Elements

Transposable elements, called P-elements, are mobile pieces of DNA that have the ability to move from one site on a chromosome to another (Liao, Rehm, and Rubin, 2000). They move vertically to the progeny, but they can also move horizontally between individuals of different species (Kidwell, 1992). The P-elements can be used to transform insects by macroinjecting the embryos with two circular DNA plasmids (a vector and a helper). The vector plasmid contains the transposable element, exogenous DNA to be incorporated under the control of promoter, and a marker gene. Since the vector lacks the transposase, the transposase is provided by the helper plasmid. The helper plasmid catalyzes the excision of exogenous DNA and the marker gene from the vector plasmid, and helps in their integration/insertion into the host genome (Rubin and Spradling, 1982). The P-elements are effective only in the case of *Drosophila*, but type II transposable elements, unrelated to P-elements, have been used in other insects. *Minos*, a member of the *Tc1* family, has been used for transformation of *C. capitata* (Loukeris et al., 1995) and mosquito, *Anopheles stephensi* Lister (Catteruccia et al., 2000). *Mariner*, consisting of several subfamilies, has been used in the case of *Lucilia cuprina* Weid. (Coates et al., 1997). The *hAT* elements (*mariner* and *hobo*) have a broad host range. Interplasmid transposition assays have shown that *hobo* is mobile in *Musca domestica* (L.) and *Helicoverpa armigera* (Hubner) (O'Brochta et al., 1994; Pinkerton, O'Brochta, and Atkinson, 1996). *PiggyBack* elements isolated from cabbage looper, *Trichoplusia ni* (Hubner), have also been used to transform nondrosophilid insects (Kramer, 2004).

Viral Vectors

Densoviruses have been used for genetic transformation of insects, particularly mosquitoes. They have a narrow host range, are relatively stable in the environment, and no microinjection is necessary (Barreau, Jousset, and Cornet, 1994). Alphavirus, *Sindbis* (*SIN*), serves as an mRNA template for translation (Strauss, Rice, and Strauss, 1984). This has been used for stable cytoplasmic expression of foreign DNA. Foreign DNA can be inserted downstream of this promoter (*dsSIN*). Alpine viruses are infectious to humans, and this may be inappropriate for genetic transformation of insects. Alphavirus, *Sindbis*, can be used to deliver high levels of gene expression *in vivo* in a number of nonhost arthropod species without causing cytopathic effects in infected cells or impairing development (Lewis et al., 1999). Recombinant *Sindbis* virus has a great potential for analyzing the change in the function of developmental genes during diversification. Densoviruses from *Aedes aegypti* (L.) (*AcDNV*) and *Junonia coenia* (Hubner) (*JCDNV*) have also been used to create transducing vectors to express the reporter gene in insect cells (Giraud, Devauchelle, and Bergoin, 1992; Afanasiev et al., 1994). Using recombinant *Sindbis* virus, the function of the homeotic gene *Ultrabithorax* in the development of butterfly, *Precis coenia* (Hubner), wings, and beetle, *Tribolium castaneum* (Herbst.), embryos has been investigated. Ectopic *Ultrabithorax* expression in butterfly forewing imaginal discs was sufficient to cause transformation of forewings in adults, including scale morphology and pigmentation. Expression of *Ultrabithorax* in beetle embryos outside of its endogenous expression domain affects normal development of the body wall cuticle and appendages. Homeotic genes probably play an important role in diversification of arthropod appendages. *Ultrabithorax* is sufficient to confer hindwing identity in butterflies and alters normal development of anterior structures in beetles.

Baculoviruses

Corsaro, DiRenzo, and Fraser (1989) optimized transfection conditions for cloning a *UND-K* derivative of the IPLB-HZ 1075 cell line of the corn earworm, *Helicoverpa zea* (Boddie), using a calcium phosphate coprecipitation technique and the DNA genome of the *H. zea* S-type nuclear polyhedrosis virus (NPV). The technique permitted relatively efficient *in vitro* manipulation of the virus genome. Human 5-lipoxygenase (EC 1.13.11.34), the key enzyme involved in the transformation of arachidonic acid to the potent biologically active leukotrienes, has been overexpressed in cells of *Spodoptera frugiperda* (J.E. Smith) using the *Autographa californica* Speyer NPV strain E2 expression system (Funk et al., 1989). Although infected cells were able to express mutant 5-lipoxygenase protein, enzyme activity was not substantially altered, suggesting the nonessential nature of certain histidines in binding iron at the putative ferric catalytic site. The major antibacterial proteins attacin, cecropin, and lysozyme are secreted into the hemolymph of the saturniid, *Hyalophora cecropia* (L.) upon challenge with bacteria. Attacin has been isolated and the corresponding amino acid sequence suggested that basic attacin is synthesized as a 233-residue preproprotein confirmed by cloning the cDNA fragment encoding the basic attacin in the *A. californica* NPV downstream of the polyhedrin promoter and expressing the protein in *S. frugiperda* cells. Attacin has also been produced in last-instar larvae of *T. ni* after injection of recombinant virus (Gunne, Hellers, and Steiner, 1990). Based on protein processing pattern, it has been suggested that a protease produced by the *S. frugiperda* cells cleaves Arg45-Arg46, producing mature attacin. The instability of the attacin proteins is rationalized in terms of their random-coil structure, which was deduced from circular dichroism measurements. The *A. californica* NPV has also been used as a vector for expression of foreign genes in *B. mori* (Mori et al., 1995). Homologous recombination of the host gene and the virus allows specific alteration of gene sequences (Yamao et al., 2002). The baculovirus expression system has been used to modify silk proteins or production of hormones in silkworm (Kadonookuda et al., 1995; Yang et al., 2002).

Genetic Improvement of Beneficial Arthropods

Genetic transformation can be used to improve the efficacy of natural enemies and their adaptation to the environment, resistance to insecticides, control of disease vectors, cryopreservation, and improve the rearing systems. A number of genes have been cloned and expressed in different insect species (Table 9.3). Various aspects of genetic improvement of natural enemies through genetic engineering have been discussed below.

Cryopreservation

One of the potential applications of biotechnology is the development of cryobiological methods for preserving embryos of arthropod biological control agents. Currently, arthropod biological control agents can be maintained only by continuous rearing or by holding

TABLE 9.3

Novel Genes Cloned for Genetic Manipulation of Insects

Gene of Interest	Species	References
Acetylcholinesterase (*Ace*)—pesticide resistance	*Drosophila*	Hall and Spierer (1986) Fournier et al. (1989)
	Anopheles stephensi	Hall and Malcolm (1991) Hoffmann, Fournier, and Spierer (1992)
Resistance to parathion	*Cardiochiles diaphaniae*	Presnail and Hoy (1996)
Parathion hydrolase (*opd*)—parathion resistance	*Spodoptera frugiperda*	Dumas, Wild, and Rauschel (1990)
	Drosophila melanogaster	Philips et al. (1990)
Esterase β1 amplification core—organophosphate resistance	*Culex* species	Mouches et al. (1986, 1990)
Glutathione *S*-transferase (*MdGST1*)—organohosphate resistance	*Musca domestica*	Wang, McCommas, and Syvanen (1991)
γ-Aminobutyric acid A (*GABA$_A$*)—dieldrin resistance	*Drosophila*	Ffrench-Constant et al. (1991)
Cytochrome (*P450-Ba*)—DDT resistance	*Drosophila*	Waters et al. (1992)
Glutathione *S*-transferase (*DmGST1-1*)—DDT resistance	*Drosophila*	Toung, Hsieh, and Tu (1990)
Metallothionein genes (*Mtn*)—copper resistance	*Drosophila*	Theodore, Ho, and Marni (1991)

continued

TABLE 9.3 (continued)

Gene of Interest	Species	References
Multidrug resistance (*Mdr49* and *Mdr65*)—colchicine resistance	*Drosophila*	Wu et al. (1991)
Neomycin phosphotransferase (*neo*)—resistance to kanamycin and neomycin (*G418*)	Transposon *Tn5*	Beck et al. (1982)
Insect microbiocide (Defensin A)	*Aedes aegypti*	Kokoza et al. (2000)
Trypanocidal protein (Cecropin A) antibody fragment	*Rhodnius prolixus*	Durvasula et al. (1997)
Resistance to yellow fever virus antiparasitic genes	*Rhodnius prolixus* *Aedes aegypti* *Anopheles stephensi*	Durvasula et al. (1999) Higgs et al. (1998) Ito et al. (2002)
Catalase—hydrogen peroxide resistance	*Drosophila*	Orr and Sohal (1992)
Wild-type eye color	*Bactrocera dorsalis* (Hendel) *Ceratitis capitata*	Handler and McCombs (2000)
	Aedes aegypti	Loukeris et al. (1995)
		Handler et al. (1998), Michel et al. (2001)
Biomolecule production	*Bombyx mori* L.	Jasinskiene et al. (1998), Kadonookuda et al. (1995), Ho et al. (1998), Yang et al. (2002)
Luciferase	*Aedes aegypti*	Ward et al. (2001), Jordan et al. (1998), Johnson et al. (1999)
Infection in antennal cells	*Manduca sexata* L.	Franco et al. (1998)
Green fluorescent protein (GFP)	*Bombyx mori*	Toshiki et al. (2000), Olson et al. (2000), Yamao et al. (2002)
	Aedes aegypti	Pinkerton et al. (2000), Afanasiev et al. (1999)
	Anopheles stephensi	Catteruccia et al. (2000)
	Anopheles gambiae Giles	Afanasiev et al. (1999)
	Tribolium castaneum	Berghammer et al. (1999)
	Stomoxys calcitrans L.	O'Brochta, Atkinson, and Lehane (2000)
	Pectinophora gossypiella	Peloquin et al. (2000)
	Apis mellifera	Robinson et al. (2000)
	Musca domestica	Hediger et al. (2001)
	Anastrepha suspensa Loew.	Handler and Harrell (2001)
	Lucilia cuprina	Heinrich et al. (2002)
	Athalia rosae L.	Sumitani et al. (2003)

specimens in diapause (for those species that undergo diapause). This is expensive and can lead to loss of colonies, as well as to genetic drift or contamination of the colonies. Mazur et al. (1992) demonstrated that embryos of the fruit fly, *D. melanogaster*, can be preserved in liquid nitrogen and then thawed to develop into viable and fertile adults. If cryopreservation can be adapted to other arthropods, a significant saving in rearing costs can be achieved. More importantly, valuable collections of arthropod natural enemies could also be maintained indefinitely.

Altering Biological Attributes

Altering longevity of certain arthropods might be beneficial, and research on mechanisms of ageing may provide useful genes in the future. A cloned catalase gene inserted into *D. melanogaster* by P-element-mediated transformation has been shown to provide resistance to hydrogen peroxide, although it did not prolong the lifespan of flies (Orr and Sohal, 1992). Shortening developmental rate, enhancing progeny production, altering sex ratio, extending temperature and relative humidity tolerances, and altering host or habitat preferences could enhance the effectiveness of biological control agents (Hoy, 1976). O'Brochta and Atkinson (1998) suggested the use of transgenic insects for the control of insect-borne diseases by making the insects incapable of transmitting the disease.

Quality Control of Insect Cultures and Mass Production

Maintaining quality in laboratory-reared arthropods is difficult due to possible genetic changes caused by inadvertent selection, inbreeding, genetic drift, and founder effects (Stouthammer, Luck, and Werren 1992; Hopper, Roush, and Powell, 1993). Biotechnological approaches can be used for mass production and quality control of *Trichogramma* spp. (de Almeida, da Silva, and de Medeiros, 1998). If high-quality and inexpensive artificial diets for predators and parasitoids are available, biological control programs would no longer be restricted by inefficient mass production methods. Beckendorf and Hoy (1985) suggested that recombinant DNA techniques could make arthropod natural enemies more efficient and less expensive. Once a gene has been cloned, it can be inserted into a number of beneficial species. One of the significant benefits of recombinant DNA techniques may be that it will be easier to maintain quality in transgenic arthropods. The ability to manipulate and insert genetic material into the genome of *Drosophila* has been used to develop a fundamental understanding of genetics, biochemical interactions, development, and behavior of insects (Lawrence, 1992). Genetic engineering of arthropods other than *Drosophila* has been attempted, but with limited success (Walker, 1989; Handler and O'Brochta, 1991; Hoy, 1994; Kramer, 2004).

Adaptation to Extreme Environmental Conditions

Increased tolerance to cold in frost-susceptible arthropods or tolerance to heat and dry conditions, for example, in *Trichogramma* species, can be achieved by genetic transformation (Figure 9.1). Antifreeze protein genes cloned from the wolf-fish, *Anarhichas lupus* (L.) have been expressed in transgenic *Drosophila* (Rancourt et al., 1990; Rancourt, Davies, and Walker, 1992), using the *hsp70* promoter and yolk polypeptide promoters of *Drosophila*. Although additional work is required, the results suggest that subtropical or tropical species of arthropod natural enemies could become adapted to a much broader range of climates.

Improving Resistance to Insecticides

Several insect parasitoids can be improved for resistance to insecticides for use in integrated pest management (Figures 9.2 and 9.3). Potentially useful resistance genes that have been cloned include parathion hydrolase gene (*opd*) from *Pseudomonas diminuta* Leifson and Hugh and *Flavobacterium*, cyclodiene resistance gene (γ-aminobutryic acid A, $GABA_A$) from *Drosophila*, β-tubulin genes from *Neurospora crassa* (Draft) and *Septoria nodorum* (Berk.) that confer resistance to benomyl, acetylcholinesterase gene (*Ace*) from *D. melanogaster* and

FIGURE 9.1 The egg parasitoid, *Trichogramma chilonis*, which can be improved for adaptation to harsh environments.

FIGURE 9.2 *Helicoverpa* larval parasitoid, *Cotesia* sp. (cocoons), which can be improved for resistance to insecticides.

FIGURE 9.3 The lacewing, *Chrysoperla carnea*, which can be improved for resistance to insecticides.

the mosquito *A. stephensi*, glutathione *S*-transferase gene (*GST1*) from *M. domestica*, a cytochrome P450-B1 gene (*CYP6A2*) associated with dichlorodiphenyltrichloroethane (DDT) resistance in *Drosophila*, and amplification core and esterase *B1* gene from *Culex* responsible for resistance to organophosphates (Atkinson, Pinkerton, and O'Brochta, 2001). Metallothionein genes have been cloned from *Drosophila* and other organisms that appear to function in homeostasis of copper and cadmium (Theodore, Ho, and Marni, 1991). These genes may provide resistance to fungicides containing copper in arthropod natural enemies. In many crops, fungicides may have a negative impact on beneficial arthropods such as phytoseiid predators. The organophoshate-degrading enzymes that confer resistance to paraoxon and parathion have been expressed in *D. melanogaster* (Benedict, Scott, and Cockburn, 1994; Benedict, Salazar, and Collins, 1995), and *C. diaphaniae* (Presnail and Hoy, 1996).

Females of *M. occidentalis* resistant to carbaryl-organophosphate-sulfur lived longer (25.3 days versus 19.7 days), exhibited a high intrinsic rate of increase (0.243 versus 0.182 individuals per female per day), and shorter generation time (13.9 days versus 17.0 days), when reared on a diet of 0 to 48-hour-old eggs rather than a diet of mixed actives of *Tetranychus pacificus* McGregor on bean, *Phaseolus vulgaris* L., leaf discs (Bruce, Oliver, and Hoy, 1990). The female to male ratio of the progeny was 2.1:1 when reared on eggs, and 2.0:1 when reared on mixed actives, suggesting that diet influences sex ratio. There were no significant differences in oviposition rates for repeatedly mated and females mated once reared on a diet of younger eggs (0 to 24 hours old) compared to a diet of older eggs (72- to 96-fold) of *T. pacificus*. The COS strain of *M. occidentalis* exhibited life-table parameters comparable to those of other strains, suggesting that the reproductive attributes of this predator were not altered as a result of artificial laboratory selection. Microbial genes conferring resistance to pesticides can function in arthropods. The *opd* gene, isolated from *Pseudomonas* and conferring resistance to organophosphates, has been inserted, using a baculovirus expression vector into cultured fall armyworm, *S. frugiperda*, cells and larvae (Dumas, Wild, and Rauschel, 1990). Philips et al. (1990) also transferred the *opd* gene into *D. melanogaster*. The *opd* gene was put under the control of the *Drosophila* heat-shock promoter, *hsp70*, and stable enzyme was produced and accumulated with repeated induction. It is likely that this gene could be used to confer resistance to organophosphates in beneficial arthropod species, as well as serving as a selectable marker for detecting transformation of insect pests.

Dominant Repressible Lethal Genetic System

Traditionally, sterile insect technique (SIT) has been employed to control several insects (Knipling, 1955). However, this system depends on large-scale production of the target insect, and use of irradiation, or chemical sterilization. The SIT has been used against several insect species, with considerable success. Experience has shown that it works effectively when sufficient resources are available or it is applied to geographically isolated pest populations. To achieve effective control, a high ratio of sterile to normal males has to be maintained for several generations. The process of rearing the insects under laboratory conditions and sterilization results in reduced fitness (Holbrook and Fujimoto, 1970). Reduced fitness of the sterile insects requires a high ratio between sterile and wild-type insect populations. Because the sterile insects die after one generation, the releases have to be made repeatedly. Because of these problems, a number of alternatives have been proposed, including hybrid sterility and cytoplasmic incompatibility (Whitten, 1985), chromosomal translocations (Serebrovsky, 1940), and conditional lethal traits. The conditional lethal system was first proposed by LaChance and Knipling (1962). As the conditional lethal genes become effective only under certain conditions, there is sufficient chance for the gene to spread into the wild populations. The conditional lethal gene could be temperature dependent or results in failure to diapause. The female killing system using a y-linked translocation can achieve better control than the SIT (Foster, 1991). Release of insects carrying a dominant lethal (RIDL) gene has been proposed as an alternative to the conventional techniques used for insect sterilization (Thomas et al., 2000; Alphey and Andreasen, 2001; Schliekelman and Gould, 2000a, 2000b). It is based on the use of a dominant, repressible, female-specific gene for insect control. A sex-specific promoter or enhancer gene is used to drive the expression of a repressible transcription factor, which in turn controls the production of a toxic gene product. A non-sex-specific expression of the repressible transcription factor can also be used to regulate a selectively lethal gene product. Insects produced through genetic transformation using this approach do not require sterilization through irradiation, and can be released in the ecosystem to mate with the wild population to produce the sterile insects, which will be self-perpetuating. *Notch60g11* is a sex-linked mutation in *D. melanogaster* that causes dominant, cold-sensitive lethality in heterozygous embryos (Fryxell and Miller, 1995). A population of normal *D. melanogaster* has been driven to extinction by adding an equal number of homozygous *Notch60g11* mutants to each of three successive generations at 18°C. *Notch60g11* homozygotes reared at 26°C showed normal viability and mating success, even in competition with a wild-type insect population, presumably because of the developmental stage-specificity of the *Notch60g11* mutation. Because *Notch60g11* is a frame shift mutation in a gene that is highly conserved in arthropods and vertebrates, this autocidal biological control strategy could be used in any insect species that reproduces sexually and lives in a temperate climate.

Markers and Promoters

Some genes are useful for identifying transformants, including microbial genes such as neomycin or *G418* resistance, chloramphenicol acetyltransferase (*CAT*), and β-galactosidase. Relatively few genes cloned from *Drosophila* can be used directly for transforming beneficial arthropods, but they could serve as probes for homologous sequences in other arthropods. Expression of transgenes can be regulated by using different promoters. Many people have used the heat shock protein *hsp70* from *D. melanogaster* to allow temporal control of gene expression with a change in temperature. Tissue-specific expression has been achieved with *Apyrase* in mosquito salivary glands. The *ug* promoter has been used to drive the expression of an agent with anti-*Plasmodium* activity, a blood meal triggered expression of antipathogen factors in mosquito fat-bodies (Kokoza et al., 2000). Carboxypeptidase promoters can be used to drive blood inducible gene expression in *A. aegypti* (Moreira et al., 2000).

Genetic manipulation through DNA-mediated genetic transformation also requires a reliable genetic marker. The use of the polymerase chain reaction (PCR) to identify DNA markers, particularly markers identified by a random sample of the genome, such as random amplified polymorphic DNA (RAPD), offers a highly efficient method for detecting genetic changes in arthropod populations (Arnheim, White, and Rainey, 1990; Williams et al., 1990; Hadrys, Balick, and Schierwater, 1992). The RAPD-PCR method is also of potential value for monitoring, establishment, and dispersal of specific biotypes of arthropod natural enemies (Chapco et al., 1992; Edwards and Hoy, 1993). Green fluorescent protein (GFP) has been expressed in insects, and can be used as a marker to identify the transformants (Berghammer, Klingler, and Wimmer, 1999). The GFP can be linked with different promoters that allow expression in different tissues and in different insects. Peloquin et al. (2000) used the GFP in pink bollworm, *Pectinophora gossypiella* (Saunders), pre-blastoderm embryos and analyzed *in vivo* the expression of DNA that encodes the enhanced GFP. PCR has also been also used to amplify the expected 579bp enhanced GFP DNA fragment from protein-positive pink bollworm.

Identification of transformed individuals could also be achieved by using a pesticide resistance gene, such as the *opd* gene as a selectable marker. Another option is to use the neomycin (*neo*) antibiotic resistance gene, which functions in both *Drosophila* and mosquitoes, and is less likely to provoke concern about risks of releasing transgenic arthropods into the environment. Use of insecticide resistance genes as selection markers might lead to unintended genetic selection for resistance to insecticides, and horizontal transfer of the transgene may lead to some complications in the environment. Pesticide resistance genes have been used in arthropods when the goal is to develop insecticide-resistant strains of insects for biological control. Another marker is the β-galactosidase gene (*lacZ*) isolated from *E. coli* and regulated by the *Drosophila hsp70* promoter, which has been expressed in both *Drosophila* and the phytoseiid predator, *M. occidentalis* (Presnail and Hoy, 1992). If an appropriate selectable marker is not available, identifying transformed lines can be accomplished with PCR and subsequent analysis by Southern blot hybridization or an immunological procedure.

Environmental Release and Potential Risks

Experience indicates that the probability that a new organism will become established is small. Risk assessment should include questions about survival, reproduction, and dispersal of transgenic species, and their effects on other species. Questions may also be raised about the inserted DNA, its stability, and possible effect on other species should the genetic material move from one species to another. Historical examples of biological invasions or classical biological control have demonstrated lack of predictability, low level of successful establishment, and the importance of scale, specificity, and the speed of evolution (Ehler, 1990). The greater the genetic novelty, the greater is the possibility of surprising results. Molecular markers can be used to understand dispersal and interactions between the species (Williamson, 1992).

Transgenic insects may not pose a greater threat than the ones selected through conventional approaches. Release of insects modified through conventional approaches has occurred for many years (Braig and Yan, 2002). Hoy (2000) has given a list of assays that may be considered for risk assessment of genetically modified organisms. The major concerns include: (1) attributes of the species (potential for gene flow to other species, role in ecosystem, mobility, longevity, and fecundity), (2) the nature of the transgene (function and stability), (3) the environment (alternate hosts, presence of species in the environment, and possibilities for indirect dissemination), and (4) comparison of the transgenic with nontransgenic phenotype. Very little attention has been paid to evaluate intentional or accidental release of transgenic insects in the environment. Extreme mobility, small size, and large numbers of insects that exist in laboratories the world over have a high probability of accidental escape of transgenic insects into the environment. Scant attention has been paid to monitoring and actual impact of accidental escapes.

In general, transgenic insects have a reduced fitness, but it may not be true in all cases. There are no established procedures in place to conduct risk assessment for transgenic insects, as the goal is to spread them as far as possible (Hoy, 2000). Insects with insecticide resistance traits might pose a threat to public health (Hoy et al., 1997). Each transgenic species should be evaluated on a case-by-case basis depending on whether an insect species is a pest or a beneficial arthropod. Host/prey specificity is usually important to ensure that arthropod biological control agents control the target pests. Changes in host specificity of a vector insect or agricultural pest should be carefully evaluated. Host range or preference should be documented through laboratory or greenhouse no-choice tests. Temperature and relative humidity tolerances and diapause attributes often restrict geographical distribution of arthropods. Changes in responses to abiotic factors should be demonstrable with growth chamber and laboratory tests by comparing the responses of transgenic with those of the unmanipulated strains. Laboratory tests should also determine whether the trait is transmitted to the progeny. Likewise, genetic evaluations should determine whether the inserted gene(s) are maintained in their original insertion site. For this reason, it may be a good idea to avoid the use of transformation methods, such as transposable-element vectors, that could result in movement of the inserted DNA. Even a transposable-element vector lacking the transposase gene could possibly move if a transposase were supplied by a helper transposable element that was native to the engineered species. Likewise, it may be important to determine that the inserted gene cannot be transmitted to the pest species, particularly if it confers resistance to pesticides (Tiedje et al., 1989; Hoy, 1992b).

Concerns have also been raised about the safety of classical biological control agents. The environmentalists are particularly concerned about the preservation of native flora and fauna (Howarth, 1991). The era of accepting classic biological control as environmentally risk free appears to have passed (Harris, 1985; Ehler, 1990; Hoy, 1992b). Protocols for evaluating the risks associated with releasing parasitoids and predators that have been manipulated with rDNA techniques do not currently exist, but will probably include, as a minimum, the questions or principles governing the release of genetically modified living organisms (Tiedje et al., 1989; Hoy, 1990b, 1992b, 1995). Some of the concerns that need to be addressed are discussed below.

Genetic Exchange with Natural Populations

Genetic variation, in general, is much greater in natural populations than in transgenic insects. Mating between transgenic and wild populations would expose the transgenic insects to a range of genetic backgrounds that would allow for broad nonadditive effects. These interactions may alter vector competence and host range, and influence life history parameters by affecting survival and reproduction (Gubler, 1993). Novel traits may spread to a large geographical area in a short span of time (Daborn et al., 2002). The risks associated with accidental release of a transgene are largely associated with the nature of the transgene, the marker gene, and their expression patterns. Interspecific transfer of the transposable element *mariner* has been suggested for *Drosophila* and *Zaprionus* (Maruyama and Hartl, 1991), and the lepidopteran, *H. cecropia* (Lidholm, Gudmundsson, and Boman, 1992). While the interspecific transfer of *mariner* is suspected only on the basis of DNA sequence similarities and no specific vector has been identified, the data are consistent with the hypothesis that transposable elements can move between different species and orders of arthropods (Kidwell, 1992; Robertson, 1993).

Horizontal Gene Flow

If horizontal transmission of DNA (or microorganisms) between arthropods occurs, even if exceedingly rarely, there is no guarantee that genes inserted into one species will be completely stable. Naturally occurring horizontal transmission of DNA between species may have provided some of the variability upon which evolution has been based, but the extent and nature of this naturally occurring gene transfer are just being determined. The release of transgenic arthropods will have to be evaluated on the basis of probable benefits and potential risks. Horizontal gene transfer should be taken seriously while considering the release of transgenic arthropods, and such a risk is more associated with the use of transposable elements. The *hobo* transposable element spreads through *D. melanogaster* (Pascal and Periquet, 2003). Laboratory experiments have demonstrated the ability of P-elements to spread rapidly to populations in which they were absent (Good et al., 1989). The transposable element *mariner* probably spreads horizontally between insects (Robertson and Lampe, 1995). Little is known about the transfer of gut symbionts horizontally, but it occurs over a short span of time in insects sharing common food sources (Huigens et al., 2000). Baculoviruses are capable of incorporating genes from the host genome (Fraser et al., 1995), but such a route has not been demonstrated convincingly. Risk assessment for transgenic insects is complicated by our inability to judge the possibilities for horizontal gene transfer, and the level to which the harm may occur.

Transgene Instability

Stability of the introduced gene is necessary to ensure that the transgenic population will carry the gene with adequate expression, and that the released population will not pose a hazard to the environment. Transposable elements have the ability to move within the insect genome, and this may influence the stability of the introduced gene if the transposable element is used from the same family as the vector (O'Brochta and Atkinson, 1996). *Hermes* and *hobo* transposable elements belong to the same *hAT* family, although their transposase elements are only 55% identical (Sundararajan, Atkinson, and O'Brochta, 1999).

There were no differences between a wild type and genetically modified population of *M. occidentalis* in daily egg production, hatchability at three temperatures (28.5, 33.5, and 38.2°C) and four relative humidities (32.5, 75.5, 93, and 100% RH), diapause incidence, or proportion of female progeny produced. Eggs did not hatch at 38°C with any of the relative humidities tested, or at 33.5°C with 100% RH (Li and Hoy, 1996). *Metaseilus occidentalis* did not survive on leaves of *Phaseolus vulgaris* L., *Citrus jambhiri* Lush., *Zea mays* L., *Lycopersicon esculentum* Mill., *Capsicum annuum* L., or *Hedera helix* L. without prey or on a diet of pollen alone. Adult females did not survive on eggs and larvae of *Papilio cresphontes* Cramer and *Eurema daira* (Godart). As no significant differences between the transgenic and wild-type strains were found for the traits tested, the transgenic strain is not expected to exhibit any new biological attributes in a short-term field release. Genetic manipulation of phytoseiid species has yielded transgenic strains, but none have been released into the environment (McDermott and Hoy, 1997). Previous data suggested that *M. occidentalis* could not survive the wet humid summer season. A nontransgenic strain of *M. occidentalis* has been released on soybean plants infested with *T. urticae*. CLIMEX, a population growth model that uses climatic factors to determine whether a given poikilothermic species can colonize and persist in new geographic areas, indicated that *M. occidentalis* cannot persist through the wet season, although it may be able to establish and persist through the autumn, winter, and spring months.

Conclusions

Molecular approaches will play an important role in exploring insect phylogeny and diversity, while genetic engineering will enable the production of insects with known functions. Biotechnological tools offer the potential of genetically modifying the insects for disease control, production of vaccines, more robust natural enemies with resistance to pesticides, or better adaptation to extremes of climatic conditions, increased host range, and efficient rearing systems. Although considerable progress has been made in the laboratory, these advances have not been accompanied by product development, delivery, and risk assessment. Experience with plant biotechnology may serve as a guideline, and also as a warning, if the public does not understand completely the benefits of the development of genetically modified insects for health, agriculture, and industrial purposes. There is a need to develop mechanisms to drive the gene of interest through natural populations, testing, deployment, and risk assessment.

References

Afanasiev, B.N., Kozlov, Y.V., Carlson, J.O. and Beaty, B.J. (1994). Densovirus of *Aedes aegypti* as an expression vector in mosquito cells. *Experimental Parasitology* 79: 322–339.

Afanasiev, B.N., Ward, T.W., Beaty, B.J. and Carlson, J.C. (1999). Transduction of *Aedes aegypti* mosquitoes with vectors derived from *Aedes* densovirus. *Virology* 257: 62–72.

Aksoy, S. (2000). Tsetse: A haven for microorganisms. *Parasitology Today* 16: 114–118.

Aksoy, S. (2003). Control of tsetse flies and trypanosomes using molecular genetics. *Veterinary Parasitology* 115: 125–145.

Alphey, L. and Andreasen, M. (2001). Dominant lethality and insect population control. *Molecular and Biochemical Parasitology* 121: 173–178.

Arnheim, N., White, T. and Rainey, W.E. (1990). Application of PCR: Organismal and population biology. *BioScience* 40: 174–182.

Atkinson, P.W. and O'Brochta, D.A. (1999). Genetic transformation of non-drosophilid insects by transposable elements. *Annals of the Entomological Society of America* 92: 930–936.

Atkinson, P.W., Pinkerton, A.C. and O'Brochta, D.A. (2001). Genetic transformation systems in insects. *Annual Review of Entomology* 46: 317–346.

Atkinson, P.W., Hines, E.R., Beaton, S., Matthaei, K.I., Reed, K.C. and Bradley, M.P. (1991). Association of exogenous DNA with cattle and insect spermatozoa *in vitro*. *Molecular Reproduction and Development* 23: 1–5.

Baldarelli, R.M. and Lengyel, J.A. (1990). Transient expression of DNA after ballistic introduction into *Drosophila* embryos. *Nucleic Acids Research* 18: 5903–5904.

Ballinger-Crabtree, M.E., Black, W.C. and Miller, B.R. (1992). Use of genetic polymorphisms detected by the random amplified polymorphic DNA polymerase chain reaction (RAPD-PCR) for differentiation and identification of *Aedes aegypti* subspecies and populations. *American Journal of Tropical Medicine and Hygiene* 47: 893–901.

Barreau, C., Jousset, F.X. and Cornet, M. (1994). An efficient and easy method of infection of mosquito larvae from virus-contaminated cell cultures. *Journal of Virological Methods* 49: 153–156.

Beck, E., Ludwig, G., Auerswald, E.A., Reiss, B. and Schaller, H. (1982). Noctuidae sequence and exact localization of the *Neomycin phosphotransferase* gene from transposon Tn5. *Gene* 19: 327–336.

Beckendorf, S.K. and Hoy, M.A. (1985). Genetic improvement of arthropod natural enemies through selection, hybridization or genetic engineering techniques. In Hoy, M.A. and Herzog, D.C. (Eds.), *Biological Control in Agricultural IPM Systems*. Orlando, Florida, USA: Academic Press, 167–187.

Benedict, M.Q., Salazar, C.E. and Collins, F.H. (1995). A new dominant selectable marker for genetic transformation; *Hsp70-opd*. *Insect Biochemistry and Molecular Biology* 25: 1061–1065.

Benedict, M.Q., Scott, M.J. and Cockburn, A.F. (1994). High-level expression of the bacterial *opd* gene in *Drosophila melanogaster*: Improved inducible insecticide resistance. *Insect Molecular Biology* 3: 247–252.

Berghammer, A.J., Klingler, M. and Wimmer, E.A. (1999). A universal marker for transgenic insects. *Nature* 402: 370.

Bownes, M. (1992). Molecular aspects of sex determination in insects. In Crampton, J.M. and Eggleston, P. (Eds.), *Insect Molecular Science*. London, UK: Academic Press, 76–100.

Braig, H.R. and Yan, G. (2002). The spread of genetic constructs in natural insect populations. In Letourneau, D.K. and Burrows, B.E. (Eds.), *Genetically Engineered Organisms: Assessing Environmental and Human Health Effects*. Boca Raton, Florida, USA: CRC Press, 251–314.

Bruce Oliver, S.J. and Hoy, M.A. (1990). Effect of prey stage on life-table attributes of a genetically manipulated strain of *Metaseiulus occidentalis* (Acari: Phytoseiidae). *Experimental and Applied Acarology* 9: 201–217.

Catteruccia, F., Nolan, T., Loukeris, T.G., Blass, C., Savakis, C., Kafatos, F.C. and Crisanti, A. (2000). Stable germline transformation of the malaria mosquito *Anopheles stephensi*. *Nature* 405: 959–962.

Chapco, W., Ashton, N.W., Martel, R.K.B. and Antonishyn, N. (1992). A feasibility study of the use of random amplified polymorphic DNA in the population genetics and systematics of grasshoppers. *Genome* 35: 569–574.

Coates, C.J., Jasinskiene, N., Pott, G.B. and James, A.A. (1999). Promoter-directed expression of recombinant fire-fly luciferase in the salivary glands of *Hermes*-transformed *Aedes aegypti*. *Gene* 226: 317–325.

Coates, C.J., Turney, C.L., Frommer, M., O'Brochta, D.A. and Atkinson, P.W. (1997). Interplasmid transposition of the *mariner* transposable element in non-drosophilid insects. *Molecular and General Genetics* 253: 728–733.

Corsaro, B.G., DiRenzo, J. and Fraser, M.J. (1989). Transfection of cloned *Heliothis zea* cell lines with the DNA genome of the *Heliothis zea* nuclear polyhedrosis virus. *Journal of Virological Methods* 25: 283–292.

Crampton, J.M., Stowell, S.L., Karras, M. and Sinden, R.E. (1999). Model systems to evaluate the use of transgenic haematophagous insects to deliver protective vaccines. *Parasitologia* 41: 473–477.

Daborn, P.J., Yen, J.L., Bogwitz, M.R., Le Goff, G., Feil, E., Jeffers, S., Tijet, N., Perry, T., Heckel, D., Batterham, P., Feyereisen, R., Wilson, T.G. and Ffrench-Constant, R.H. (2002). A single P450 allele associated with insecticide resistance in *Drosophila*. *Science* 297: 2253–2256.

de Almeida, R.P., da Silva, C.A.D. and de Medeiros, M.B. (1998). Biotechnology for mass production and management of *Trichogramma* for use in biological control. *Documentos Centro Nacional de Pesquisa do Algodao* 60: 1–61.

Dumas, D.P., Wild, J.R. and Rauschel, F.M. (1990). Expression of *Pseudomonas* phosphotriesterase activity in the fall armyworm confers resistance to insecticides. *Experientia* 46: 729–734.

Durvasula, R.V., Gumbs, A., Panackal, A., Kruglov, O., Aksoy, S., Merrifield, R.B., Richards, F.F. and Beard, C.B. (1997). Prevention of insect-borne disease: An approach using transgenic symbiotic bacteria. *Proceedings National Academy of Sciences USA* 94: 3274–3278.

Durvasula, R.V., Gumbs, A., Panackal, A., Kruglov, O., Taneja, J., Kang, A.S., Cordon Rosales, C., Richards, F.F., Whitham, R.G. and Bread, C.B. (1999). Expression of a functional antibody fragment in the gut of *Rhodnius prolixus* via transgenic bacterial symboint *Rhodococcus rhodnii*. *Medical and Veterinary Entomology* 13: 115–119.

Edwards, O.R. and Hoy, M.A. (1993). Polymorphism in two parasitoids detected using random amplified polymorphic DNA (RAPD) PCR. *Biological Control: Theory and Application in Pest Management* 3: 243–257.

Ehler, L.E. (1990). Environmental impact of introduced biological control agents: Implications for agricultural biotechnology. In Marois, J.J. and Bruyening, G. (Eds.), *Risk Assessment in Agricultural Biotechnology*. Publication No. 1928. Berkeley, CA: Division of Agriculture and Natural Resources, University of California, USA, 85–96.

Fallon, A. (1991). DNA-mediated gene transfer: applications to mosquitoes. *Nature* 352: 828–829.

Ffrench-Constant, R.H., Mortlock, D.P., Shaffer, C.D., MacIntyre, R.J. and Roush, R.T. (1991). Molecular cloning and transformation of cyclodiene resistance in *Drosophila*: An invertebrate γ-aminobutyric acid subtype A receptor locus. *Proceedings National Academy of Sciences USA* 88: 7209–7213.

Foster, G.G. (1991). Simulation of genetic control. Homozygous-viable pericentric inversions in field female killing system. *Theoretical and Applied Genetics* 82: 368–378.

Fournier, D., Karch, F., Bride, J., Hall, L.M.C., Berge, J.B. and Spierer, P. (1989). *Drosophila melanogaster* acetylcholinesterase gene structure, evolution and mutations. *Journal of Molecular Biology* 210: 15–22.

Franco, M., Rogers, M.E., Shimizu, C., Shike, H., Vogt, R.G. and Burns, J.C. (1998). Infection of Lepidoptera with a pseudotyped retroviral vector. *Insect Biochemistry and Molecular Biology* 28: 819–825.

Fraser, M.J., Cary, L., Boonvisudhi, K. and Wang, H.G. (1995). Assay for movement of lepidopteran transposon IFP2 in insect cells using a baculovirus genome as a target DNA. *Virology* 211: 397–407.

Fryxell, K.J. and Miller, T.A. (1995). Autocidal biological control: A general strategy for insect control based on genetic transformation with a highly conserved gene. *Journal of Economic Entomology* 88: 1221–1232.

Funk, C.D., Gunne, H., Steinar, H., Izumi, T. and Samuelsson, B. (1989). Native and mutant 5-lipoxygenase expression in a baculovirus/insect cell system. *Proceedings National Academy of Sciences USA* 86: 2592–2596.

Giraud, C., Devauchelle, G. and Bergoin, M. (1992). The densovirus of *Junonia coenia* (Jc DNV) as an insect cell expression vector. *Virology* 186: 207–218.

Good, A.G., Meister, G.A., Brock, H.W., Grigliatti, T.A. and Hickey, D.A. (1989). Rapid spread of transposable P-elements in experimental populations of *Drosophila melanogaster*. *Genetics* 122: 387–396.

Gubler, D.J. (1993). Release of exotic genomes. *Journal of the American Mosquito Control Association* 9: 104.

Gunne, H., Hellers, M. and Steiner, H. (1990). Structure of preproattacin and its processing in insect cells infected with a recombinant baculovirus. *European Journal of Biochemistry* 187: 699–703.

Hadrys, H., Balick, M. and Schierwater, B. (1992). Applications of random amplified polymorphic DNA (RAPD) in molecular ecology. *Molecular Ecology* 1: 55–63.

Holbrook, F.R. and Fujimoto, M.S. (1970). Mating competitiveness of unirradiated and irradiated Mediterranean fruit fly. *Proceedings National Academy of Sciences USA* 94: 7520–7525.

Hall, L.M.C. and Malcolm, C.A. (1991). The acetycholinesterase gene of *Anopheles stephensi*. *Cellular and Molecular Neurobiology* 11: 131–141.

Hall, L.M.C. and Spierer, P. (1986). The *ACE* locus on *Drosophila melanogaster* structural gene for acetylcholinesterase with an unusual 5' leader. *European Molecular Biology Organization Journal* 5: 2949–2954.

Handler, A.M. and Harrell, R.A.I. (2001). Transformation of the Caribbean fruit fly, *Anastrepha suspensa*, with a *piggyBac* vector marked with polyubiquitin regulated GFP. *Insect Biochemistry and Molecular Biology* 31: 199–205.

Handler, A.M. and McCombs, S.D. (2000). The *piggyBac* transposon mediates germ-line transformation in the Oriental fruit fly and closely related elements in its genome. *Insect Molecular Biology* 9: 605–612.

Handler, A.M. and O'Brochta, D.A. (1991). Prospects for gene transformation in insects. *Annual Review of Entomology* 36, 159–183.

Handler, A.M., McCombs, S.D., Fraser, M.J. and Saul, S.H. (1998). The lepidopteran transposon vector, *piggyBac*, mediates germ-line transformation in the Mediterranean fruit fly. *Proceedings National Academy of Sciences USA* 95: 7520–7525.

Harris, P. (1985). Biocontrol and the law. *Bulletin of the Entomological Society of Canada* 17(1): 1–2.

Headley, J.C. and Hoy, M.A. (1987). Benefit/cost analysis of an integrated mite management program for almonds. *Journal of Economic Entomology* 80: 555–559.

Hediger, M., Niessen, M., Wimmer, E.A., Dubendorfer, A. and Bopp, D. (2001). Genetic transformation of the housefly *Musca domestica* with the lepidopteran derived transposon *piggyBac*. *Insect Molecular Biology* 10: 113–119.

Heinrich, J.C., Li, X., Henry, R.A., Haack, N., Stringfellow, L., Heath, A.C.G. and Scott, M.J. (2002). Germ-line transformation of the Australian sheep blowfly *Lucilia cuprina*. *Insect Molecular Biology* 11: 1–10.

Higgs, S., Rayner, J.O., Olson, K.E., Davis, B.S., Beaty, B.J. and Blair, C.D. (1998). Engineered resistance in *Aedes aegypti* to a West African and a South American strain of yellow fever virus. *American Journal of Tropical Medicine and Hygiene* 58: 663–670.

Ho, W.K., Meng, Z.Q., Lin, H.R., Poon, C.T., Leung, Y.K., Yan, K.T., Dias, N., Che, A.P., Liu, J., Zheng, W.M., Sun, Y. and Wong, A.O. (1998). Expression of grass carp growth hormone by baculovirus in silkworm larvae. *Biochim Biophys Acta* 1381: 331–339.

Hoffmann, F., Fournier, D. and Spierer, P. (1992). Minigene rescue acetylcholinesterase lethal mutations in *Drosophila melanogaster*. *Journal of Molecular Biology* 223: 17–22.

Hopper, K.R., Roush, R.T. and Powell, W. (1993). Management of genetics of biological control introductions. *Annual Review of Entomology* 38: 27–51.

Howarth, F.G. (1991). Environmental effects of classical biological control. *Annual Review of Entomology* 36: 485–509.

Hoy, M.A. (1976). Genetic improvement of insects: Fact or fantasy. *Environmental Entomology* 5: 833–839.

Hoy, M.A. (1990a). Pesticide resistance in arthropod natural enemies: Variability and selection response. In Roush, R.T. and Tabashnik, B.E. (Eds.), *Pesticide Resistance in Arthropods*. New York, USA: Chapman and Hall, 203–236.

Hoy, M.A. (1990b). Genetic improvement of arthropod natural enemies: Becoming a conventional tactic? In Baker, R.R. and Dunn, P.E. (Eds.), *New Directions in Biological Control: Alternatives for Suppressing Agricultural Pests and Diseases. UCLA Symposium on Molecular and Cellular Biology.* New Series Vol. 112. New York, USA: Alan R. Liss, 405–417.

Hoy, M.A. (1992a). Biological control of arthropods: Genetic engineering and environmental risks. *Biological Control* 2: 166–170.

Hoy, M.A. (1992b). Criteria for release of genetically-improved phytoseiids: An examination of the risks associated with release of biological control agents. *Environmental and Applied Acarology* 14: 393–416.

Hoy, M.A. (1994). *Insect Molecular Genetics*. San Diego, California, USA: Academic Press.

Hoy, M.A. (1995). Impact of risk analyses on pest-management programs employing transgenic arthropods. *Parasitology Today* 11: 229–232.

Hoy, M.A. (2000). Transgenic arthropods for pest management programs: risks and realities. *Experimental and Applied Acarology* 24: 463–495.

Hoy, M.A., Gaskalla, R.D., Capinera, J.L. and Keierleber, C.N. (1997). Laboratory containment of transgenic arthropods. *American Entomologist* 43: 206–256.

Huigens, M.E., Luck, R.F., Klaassen, R.H.G., Maas, M.F.P.M., Timmermans, M.J.T.N. and Stouthamer, R. (2000). Infectious parthenogenesis. *Nature* 405: 178–179.

Ito, J., Ghosh, A.K., Moreira, L.A., Wimmer, E.A. and Jacobs Lorena, M. (2002). Transgenic anopheline mosquitoes impaired in transmission of a malaria parasite. *Nature* 417: 452–455.

Jasinskiene, N., Coates, C.J., Benedict, M.Q., Cornel, A.J., Salazar Rafferety, C., James, A.A. and Collins, F.H. (1998). Stable transformation of the yellow fever mosquito, *Aedes aegypti*, with the Hermes element from the housefly. *Proceedings National Academy of Sciences USA* 95: 3743–3747.

Johnson, B.W., Olson, K.E., Allen Miura, T., Rayms Keller, A., Carlson, J.O., Coates, C.J., Jasinskiene, N., James, A.A., Beaty, B.J. and Higgs, S. (1999). Inhibition of luciferase expression in transgenic *Aedes aegypti* mosquitoes by Sindbis virus expression of antisense luciferase RNA. *Proceedings National Academy of Sciences USA* 96: 13399–13403.

Jordan, T.V., Shike, H., Boulo, V., Cedeno, V., Fang, Q., Davis, B.S., Jacobs Lorena, M., Higgs, S., Fryxell, K.J. and Burns, J.C. (1998). Pantropic retroviral vectors mediate somatic cell transformation and expression of foreign genes in dipteran insects. *Insect Molecular Biology* 7: 215–222.

Kadonookuda, K., Yamamoto, M., Higashino, Y., Taniai, K., Kato, Y., Choudhury, S., Xu, J.H., Choi, S., Sugiyama, M., Nakashima, K., Maeda, S. and Yamakawa, M. (1995). Baculovirus-mediated production of the human growth hormone in larvae of the silkworm, *Bombyx mori*. *Biochemical and Biophysical Research Communications* 213: 389–396.

Kamdar, P., von Allmen, G. and Finnerty, V. (1992). Transient expression of DNA in *Drosophila* via electroporation. *Nucleic Acids Research* 20: 3526.

Kidwell, M.G. (1992). Horizontal transfer of P-elements and other short inverted repeat transposons. *Genetica* 86: 275–286.

Knipling, E.F. (1955). Possibilities of insect control or eradication through the use of sexually sterile males. *Journal of Economic Entomology* 45: 459–462.

Kokoza, V., Ahmed, A., Cho, W., Jasinskiene, N., James, A.A. and Raikhel, A. (2000). Engineering blood meal-activated systemic immunity in the yellow fever mosquito, *Aedes aegypti*. *Proceedings National Academy of Sciences USA* 97: 9144–9149.

Kramer, M.G. (2004). Recent advances in transgenic arthropod technology. *Bulletin of Entomological Research* 94: 95–110.

LaChance, L. and Knipling, E.F. (1962). Control of insects through genetic manipulations. *Annals of Entomological Society of America* 55: 515–520.

Lawrence, P.A. (1992). *The Making of a Fly. The Genetics of Animal Design.* London, UK: Blackwell Scientific Publications.

Lewis, D.L., DeCamillis, M.A., Brunetti, C.R., Halder, G., Kassner, V.A., Selegue, J.E., Higgs, S. and Carroll, S.B. (1999). Ectopic gene expression and homeotic transformations in arthropods using recombinant Sindbis viruses. *Current Biology* 9: 1279–1287.

Li, J.B. and Hoy, M.A. (1996). Adaptability and efficacy of transgenic and wild-type *Metaseiulus occidentalis* (Acari: Phytoseiidae) compared as part of a risk assessment. *Experimental and Applied Acarology* 20: 563–573.

Liao, G.C., Rehm, E.J. and Rubin, G.M. (2000). Insertion site preferences of the P-transposable element in *Drosophila melanogaster*. *Proceedings National Academy of Sciences USA* 97: 3347–3351.

Lidholm, D.A., Gudmundsson, G.H. and Boman, H.G. (1992). A highly receptive, mariner-like element in the genome of *Hyalophora cecropia*. *Journal of Biological Chemistry* 266: 11518–11521.

Loukeris, T.G., Livadaras, I., Arca, B., Zabalkou, S. and Savakis, C. (1995). Gene transfer into the Medfly, *Ceratitis capitata*, with a *Drosophila hydei* transposable element. *Science* 270: 2002–2005.

Maruyama, K. and Hartl, D.L. (1991). Evidence for interspecific transfer of the transposable element *mariner* between *Drosophila* and *Zaprionus*. *Journal of Molecular Evolution* 33: 514–524.

Mazur, P., Cole, K.W., Hall, J.W., Schreuders, P.D. and Mahowald, A.P. (1992). Cryobiological preservation of *Drosophila* embryos. *Science* 258: 1932–1935.

McDermott, G.J. and Hoy, M. A. (1997). Persistence and containment of *Metaseiulus occidentalis* (Acari: Phytoseiidae) in Florida: Risk assessment for possible releases of transgenic strains. *Florida Entomologist* 80: 42–53.

McGrane, V., Carlson, J.O., Miller, B.R. and Beaty, B.J. (1988). Microinjection of DNA into *Aedes triseriatus* ova and detection of integration. *American Journal of Tropical Medicine and Hygiene* 39: 502–510.

Metcalf, R.L. and Luckmann, W.H. (Eds.). (1982). *An Introduction to Insect Pest Management.* New York, USA: John Wiley & Sons.

Miller, L.H., Sakai, R.K., Romans, P., Gwadz, W., Kantoff, P. and Coon, H.G. (1987). Stable integration and expression of a bacterial gene in the mosquito *Anopheles gambiae*. *Science* 237: 779–781.

Milne, C.P. Jr, Phillips, J.P. and Krell, P.J. (1988). Microinjection of early honeybee embryos. *Journal of Agricultural Research* 27: 84–89.

Moreira, L.A., Edwards, M.J., Adhami, F., Jasinskiene, N., James, A.A. and Jacobs-Lorena, M. (2000). Robust gut-specific gene expression in transgenic *Aedes aegypti* mosquitoes. *Proceedings National Academy of Sciences USA* 97: 10895–10898.

Mori, H. and Tsukada, M. (2000). New silk protein: Modification of silk protein by gene engineering for production of biomaterials. *Reviews in Molecular Biotechnology* 74: 95–103.

Mori, H., Yamao, M., Nakazawa, H., Sugahara, H., Shirai, N., Matsubara, F., Sumida, M. and Imamura, T. (1995). Transovarian transmission of a foreign gene in the silkworm, *Bombyx mori*, by *Autographa californica* nuclear polyhedrosis virus. *Bio/Technology* 13: 1005–1007.

Morris, A.C., Eggleston, P. and Crampton, J.M. (1989). Genetic transformation of the mosquito *Aedes aegypti* by micro-injection of DNA. *Medical and Veterinary Entomology* 3: 1–7.

Mouches, C., Pasteur, N., Berge, J.B., Hyrien, O., Raymond, M., de Saint Vincent, B.R., de Silvestri, M. and Georghiou, G.P. (1986). Amplification of an esterase gene is responsible for insecticide resistance in California *Culex* mosquitoe. *Science* 233: 778–780.

Mouches, C., Pauplin, Y., Agarwal, M., Lemieux, L., Herzog, M., Abadon, M., Boyssat Arnaouty, V., Hyrien, O., de Saint Vincent, B.R., Georghiou, G.P. and Pasteur, N. (1990). Characterization of amplification core and esterase B1 gene responsible for insecticide resistance in *Culex*. *Proceedings National Academy of Sciences USA* 98: 2574–2578.

O'Brochta, D.A. and Atkinson, P.W. (1996). Transposable elements and gene transformation in non-drosophilid insects. *Insect Biochemistry and Molecular Biology* 26: 739–753.

O'Brochta, D.A. and Atkinson, P.W. (1998). Building a better bug: Inserting new genes into a few specific insect species could stop some infectious diseases, benefit agriculture and produce innovative materials. *Scientific American* 279: 60–65.

O'Brochta, D.A., Atkinson, P.W. and Lehane, M.J. (2000). Transformation of *Stomoxys calcitrans* with a Hermes gene vector. *Insect Molecular Biology* 9: 531–538.

O'Brochta, D.A., Warren, W.D., Saville, K.J. and Atkinson, P.W. (1994). Interplasmid transposition of *Drosophila hobo* elements in non-drosophilid insects. *Molecular and General Genetics* 244: 9–14.

Olson, K.E., Myles, K.M., Seabaugh, R.C., Higgs, S., Carlson, J.O. and Beaty, B.J. (2000). Development of a Sindbis virus expression system that efficiently expresses green fluorescent protein in midgets of *Aedes aegypti* following *per os* infection. *Insect Molecular Biology* 9: 57–65.

Orr, W.C. and Sohal, R.S. (1992). The effects of catalase gene overexpression on life span and resistance to oxidative stress in transgenic *Drosophila melanogaster*. *Archives of Biotechnology and Biophysics* 297: 35–41.

Pascual, L. and Periquet, G. (2003). Distribution of *hobo* transposable elements in natural populations of *Drosophila melanogaster*. *Molecular Biology and Evolution* 8: 282–296.

Peloquin, J.J., Thibault, S.T., Staten, R. and Miller, T.A. (2000). Germ-line transformation of pink bollworm (Lepidoptera: Gelechiidae) mediated by the *piggyBac* transposable element. *Insect Molecular Biology* 9: 323–333.

Phillips, J.P., Xin, J.H., Kirby, K., Milne, C.P. Jr., Krell, P. and Wild, J.R. (1990). Transfer and expression of an organophosphate insecticide-degrading gene from *Pseudomonas* in *Drosophila melanogaster*. *Proceedings National Academy of Sciences USA* 87: 8155–8159.

Pinkerton, A.C., O'Brochta, D.A. and Atkinson, P.W. (1996). Mobility of *hAT* transposable elements in the Old World bollworm, *Helicoverpa armigera*. *Insect Molecular Biology* 5: 223–227.

Pinkerton, A.C., Michel, K., O'Brochta, D.A. and Atkinson, P.W. (2000). Green fluorescent protein as a genetic marker in transgenic *Aedes aegypti*. *Insect Molecular Biology* 9: 1–10.

Presanil, J.K. and Hoy, M.A. (1992). Stable genetic transformation of a beneficial arthropod by microinjection. *Proceedings National Academy of Sciences USA* 89: 7732–7736.

Presnail, J.K. and Hoy, M.A. (1994). Transmission of injected DNA sequences to multiple eggs of *Metaseiulus occidentalis* and *Amblyseius finlandicus* (Acari: Phytoseiidae) following maternal microinjection. *Experimental and Applied Acarology* 18: 319–330.

Presnail, J.K. and Hoy, M.A. (1996). Maternal microinjection of the endoparasitoid *Cardiochiles diaphaniae* (Hymenoptera: Braconidae). *Annals of the Entomological Society of America* 89: 576–580.

Presnail, J.K., Jeyaprakash, A., Li, J. and Hoy, M.A. (1997). Genetic analysis of four lines of *Metaseiulus occidentalis* (Acari: Phytoseiidae) transformed by maternal microinjection. *Annals of the Entomological Society of America* 90: 237–245.

Rancourt, D.E., Davies, P.L. and Walker, V.K. (1992). Differential translatability of antifreeze protein mRNAs in a transgenic host. *Biochimica et Biophysica Acta* 1129: 188–194.

Rancourt, D.E., Peters, I.D., Walker, V.K. and Davies, P.L. (1990). Wolffish antifreeze protein from transgenic *Drosophila*. *Bio/Technology* 8: 453–457.

Robertson, H.M. (1993). The mariner transposable element is widespread in insects. *Nature* 362: 241–245.

Robertson, H.M. and Lampe, D.J. (1995). Recent horizontal transfer of a *mariner* transposable element among and between Diptera and Neuroptera. *Molecular Biology and Evolution* 12: 850–862.

Robinson, K.O., Ferguson, H.J., Cobey, S., Vaessin, H. and Smith, B.H. (2000). Sperm-mediated transformation of the honey bee, *Apis mellifera*. *Insect Molecular Biology* 9: 625–634.

Rubin, G.M. and Spradling, A.C. (1982). Genetic transformation of *Drosophila* with transposable element vectors. *Science* 218: 348–353.

Schliekelman, P. and Gould, F. (2000a). Pest control by introduction of a conditional lethal trait on multiple loci. Potential, limitations, and optical strategies. *Journal of Economic Entomology* 93: 1543–1565.

Schliekelman, P. and Gould, F. (2000b). Pest control by release of insects carrying a female-killing allele on multiple loci. *Journal of Economic Entomology* 93: 1566–1579.

Scott, M.D. and Williams, S.M. (1993). Comparative reproductive success of communally breeding burying beetles as assessed by PCR with randomly amplified polymorphic DNA. *Proceedings National Academy of Sciences USA* 90: 2242–2245.

Serebrovsky, A.S. (1940). On the possibility of a new method for the control of insect pests. *Zoologicheskii Zhurnal* 19: 618–630.

Sobral, B.W.S. and Honeycutt, R.J. (1993). High output genetic mapping of polyploids using PCR-generated markers. *Theoretical and Applied Genetics* 86: 105–112.

Spielman, A., Beier, J.C. and Kiszewski, A.E. (2002). Ecological and community considerations in engineering arthropods to suppress vector-borne disease. In Letourneau, D.K. and Burrows, B.E. (Eds.), *Genetically Engineered Organisms: Assessing Environmental and Human Health Effects.* Boca Raton, Florida, USA: CRC Press, 315–329.

Stouthammer, R., Luck, R.F. and Werren, J.H. (1992). Genetics of sex determination and the improvement of biological control using parasitoids. *Environmental Entomology* 21: 427–535.

Strauss, E.G., Rice, C.M. and Strauss, J.H. (1984). Complete nucleotide sequence of the genomic RNA of Sindbis virus. *Virology* 133: 92–110.

Sumitani, M., Yamamoto, D.S., Oishi, K., Lee, J.M. and Hatakeyama, M. (2003). Germline transformation of the sawfly, *Athalia rosae* (Hymenoptera: Symphyta), mediated by a *PiggyBac*-derived vector. *Insect Biochemistry and Molecular Biology* 33: 449–458.

Sundararajan, P., Atkinson, P.W. and O'Brochta, D.A. (1999). Transposable element interactions in insects: Crossmobilization of *hobo* and *Hermes*. *Insect Molecular Biology* 8: 359–368.

Theodore, L., Ho, A. and Marni, G. (1991). Recent evolutionary history of the metallothionein gene *mtn* in *Drosophila*. *Genetical Research* 58: 203–210.

Thomas, D.D., Donnelly, C.A., Wood, R.J. and Alphey, L.S. (2000). Insect population control using a dominant repressible, lethal genetic system. *Science* 287: 2474–2476.

Thompson, S.N. (1990). Nutritional considerations in propagation of entomophagous species. In Baker, R.R. and Dunn, P.E. (Eds.), *New Directions in Biological Control: Alternatives for Suppressing Agricultural Pests and Diseases. UCLA Symposia on Molecular and Cellular Biology.* New Series, Vol. 112. New York, USA: Alan R. Liss, 389–404.

Tiedje, J.M., Clowell, R.K., Grossman, Y.L., Hodson, R.E., Lenski, R.E., Mack, R.M. and Regal, P.J. (1989). The planned introduction of genetically engineered organisms: Ecological considerations and recommendations. *Ecology* 70: 298–315.

Toshiki, T., Chantal, T., Corinne, R., Roshio, K., Eappen, A., Mari, K., Natuo, K., Jean-Luc, T., Bernard, M., Gerard, C., Paul, S., Malcolm, F., Jean-Claude, P. and Pierre, C. (2000). Germline transformation of the silkworm *Bombyx mori* L. using a *piggyBac* transposon-derived vector. *Nature Biotechnology* 18: 81–84.

Toung, Y.P.S., Hsieh, T.S. and Tu, C.P.D. (1990). *Drosophila* glutathione S-transferase 1-1 shares a region of sequence homology with the maize glutathione S-transferase III. *Proceedings National Academy of Sciences USA* 87: 31–35.

Walker, V.K. (1989). Gene transfer in insects. *Advances in Cell Culture* 7: 87–124.

Wang, J.Y., McCommas, S. and Syvanen, M. (1991). Molecular cloning of a glutathione S-transferase overproduced in an insecticide-resistant strain of the housefly (*Musca domestica*). *Molecular and General Genetics* 227: 260–266.

Ward, T.W., Jenkins, M.S., Afanasiev, B.N., Edwards, M.J., Duda, B.A., Suchman, E., Jacobs-Lorena, M., Beaty, B.J. and Carlson, J.O. (2001). *Aedes aegypti* transducing densovirus pathogenesis and expression in *Aedes aegypti* and *Anopheles gambiae* larvae. *Insect Molecular Biology* 10: 397–405.

Waters, L.C., Zelhof, A.C., Shaw, B.J. and Chang, L.Y. (1992). Possible involvement of the long terminal repeat of transposable element 17.6 in regulating expression of an insecticide resistance-associated P450 gene in *Drosophila*. *Proceedings National Academy of Sciences USA* 89: 4855–4859.

Whitten, M.J. (1985). The conceptual basis for genetic control. In Kerkut, G.A. and Gilbert, L.I. (Eds.), *Comprehensive Insect Physiology, Biochemistry, and Pharmacology. Vol. 12. Insect Control.* Oxford, UK: Pergamon Press, 465–528.

Williams, J.G.K., Kubelik, A.R., Livak, K.J., Rafalski, J.A. and Tingey, S.V. (1990). DNA polymorphisms amplified by arbitrary primers are useful as genetic markers. *Nucleic Acids Research* 18(22): 6531–6535.

Williamson, M. (1992). Environmental risks from the release of genetically modified organisms (GMOs): The need for molecular ecology. *Molecular Ecology* 1: 3–8.

Wu, C.T., Budding, M., Griffin, M.S. and Croop, J.M. (1991). Isolation and characterization of *Drosophila* multidrug resistance gene homologs. *Molecular and Cellular Biology* 11: 3940–3948.

Yamao, M., Katayama, N., Nakazawa, H., Yamakawa, M., Hayashi, Y., Hara, S., Kamei, K. and Mori, H. (2002). Gene targeting in the silkworm by use of a baculovirus. *Genes and Development* 13: 511–516.

Yang, G., Chen, Z., Cui, D., Li, B. and Wu, X. (2002). Production of recombinant human calcitonin from silkworm (*B. mori*) larvae infected by baculovirus. *Current Pharmaceutical Design* 9: 323–329.

Zalokar, M. (1981). A method for injection and transplantation of nuclei and cells in *Drosophila* eggs. *Experientia* 37: 1354–1356.

10

Deployment of Insect-Resistant Transgenic Crops for Pest Management: Potential and Limitations

Introduction

There is a continuing need to increase food production, particularly in the developing countries of Asia, Africa, and Latin America. This increase in food production has to come from increased productivity of major crops grown on existing cultivable lands. One practical means of achieving greater yields is to minimize the pest-associated losses, which are estimated at 14% of the total agricultural production (Oerke et al., 1994). Insect pests, diseases, and weeds cause an estimated loss of US$243.4 billion in eight major field crops (42%), out of total attainable production of $568.7 billion worldwide. Among these, insects cause an estimated loss of 90.4 billion, diseases 76.8 billion, and weeds 64.0 billion. Current crop protection costs are valued at $31 billion annually (James, 2007). Insects not only cause direct loss to the agricultural produce, but also act as vectors of various plant pathogens. In addition, there are extra costs in the form of insecticides applied for pest control, currently valued at US$10 billion annually.

Massive application of pesticides to minimize losses due to insect pests, diseases, and weeds results in adverse effects on the beneficial organisms, leaves pesticide residues in the food chain, and causes environmental pollution. A large number of insects have also developed resistance to insecticides, and over 650 cases of insect resistance have been documented. *Helicoverpa armigera* (Hubner) has shown resistance to several groups of insecticides, resulting in widespread failure of pest control operations. The cotton whitefly, *Bemisia tabaci* (Genn.), tobacco caterpillar, *Spodoptera litura* (Fab.), green peach/potato aphid, *Myzus persicae* (Sulzer), cotton aphid, *Aphis gossypii* Glover, and diamondback moth, *Plutella xylotella* (L.) have developed high levels of resistance to insecticides. Development of resistance to insecticides has necessitated the application of higher dosages of the same pesticide or more number of pesticide applications. It is in this context that insect-resistant transgenic plants can play a major role in integrated pest management (IPM) in the future (Sharma et al., 2002a; Sharma, Sharma, and Crouch, 2004).

Progress in Development of Insect-Resistant Transgenic Crops

The first insect-resistant transgenic crop was field-tested in 1994, and large-scale commercial cultivation started in 1996 in the United States (McLaren, 1998). Since then, the area planted to transgenic crops has increased from 1.7 million hectare in 1996 to over 100 million hectare in 2006, and 21 countries grew transgenic crops (James, 2007) (Figure 10.1). Most of the area under transgenic crops is in the United States, followed by Argentina, Canada, Brazil, India, and China. The transgenic crops include soybean, maize, cotton, and canola. Among the traits deployed, most area has been planted to herbicide-resistant soybean, followed by *Bt* maize and cotton, and herbicide-tolerant maize, cotton, and canola. Herbicide-tolerant soybean, maize, canola, and cotton occupied 71% or 63.7 million hectare, followed by 18% or 16.2 million hectare of *Bt* transgenic crops. Nearly one-quarter of the crop area planted to transgenic crops is grown in developing countries, mainly in Argentina, China, South Africa, and India, and over 75% of the farmers are small resource-poor farmers in developing countries.

Much of the success in developing insect-resistant transgenic plants with genes from *Bacillus thuringiensis* (Berliner) (*Bt*) has come from in-depth genetic and biochemical information generated on this bacterium. Genes encoding for δ-endotoxins from *B. thuringiensis* have been cloned since the 1980s (Schnef and Whiteley, 1981). Genetically modified plants with resistance to insects were developed in the mid-1980s (Barton, Whiteley, and Yang, 1987; Fischhoff et al., 1987; Hilder et al., 1987; Vaeck et al., 1987), and since then, there has been rapid progress in developing transgenic plants with insect resistance in several crops (Hilder and Boulter, 1999; Sharma, Sharma, and Crouch, 2004). The progress made in development and deployment of insect-resistant transgenic plants with different toxin genes for pest management in different crops is discussed below.

Effectiveness of Transgenic Crops for Controlling Insect Pests

Cotton

There is a clear advantage of growing transgenic cotton in reducing bollworm damage and increasing cottonseed yield (Figure 10.2). Cotton plants with *Bt* genes are effective against

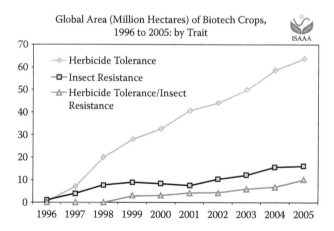

FIGURE 10.1   Area under transgenic crops worldwide. (From Clive James, 2005.)

FIGURE 10.2   Performance of *Bt* transgenic and nontransgenic cotton hybrids with and without insecticide protection.

pink bollworm, *Pectinophora gossypiella* (Saunders) (Wilson et al., 1992), cabbage looper, *Trichoplusia ni* (Hubner), tobacco caterpillar, *Spodoptera exigua* (Hubner), and cotton bollworms, *Helicoverpa zea* Boddie/*Heliothis virescens* (F.) (Benedict et al., 1996), and *H. armigera* (Hubner) (Guo et al., 1999). Field trials in the Andalusian region of Spain have also confirmed the effectiveness of *Bt* transgenic cotton in providing protection against *H. armigera*, *P. gossypiella*, and *Earias insulana* (Boisd). In addition to reduction in the number of insecticidal sprays, the transgenic cotton was also relatively safe to beneficial arthropods (Novillo, Soto, and Costa, 1999).

In China, transgenic cotton cultivars Shiyuan 321, Zhongmiansuo 19, 3517, and 541 resulted in up to 96% mortality of cotton bollworm, *H. armigera* (Guo et al., 1999). Sukang 102 exceeded the nontransgenic control and Simian 3 in boll weight (Xu et al., 2000). New cotton hybrids derived from cross-pollination between *Bt*-transgenic cotton as a donor parent and traditionally bred cotton cultivars possessing high yield and resistance to cotton diseases and insect pests have shown competitive and transgressive dominances, and the lint output increased by 11.7% to 54.6% (Z.B. Wu et al., 2003). Dong et al. (2004) reported that yield increase in *Bt*-transgenic cotton varieties under good agronomic management and pesticide application was only marginal. However, the hybrids resulted in 20% increase in productivity over the conventional cultivars. Rapid increases in adoption of *Bt* cotton have been attributed to hybrid seed production, seedling transplanting, and low planting densities.

In India, the cotton hybrids (Mech 12, Mech 162, Mech 184, RCH 2) carrying the *cry1Ac* gene have been approved for cultivation (Mohan and Manjunath, 2002). Since then, many hybrids have been released for cultivation in different regions in the country (Table 10.1).

TABLE 10.1

Transgenic Events with Resistance to Insects Used for Crop Improvement in Different Crops

Event	Institution/ Company	Description
		Cotton, *Gossypium hirsutum* L.
COT102	Syngenta Seeds Inc.	Insect-resistant cotton produced by inserting gene from *Bacillus thuringiensis* AB88. The *APH4* gene from *E. coli* was used as a selectable marker.
MON531/757/1076	Monsanto	Insect-resistant cotton produced by inserting gene from *Bacillus thuringiensis* subsp. *kurstaki* (B.t.k.).
		Tomato, *Lycopersicon esculentum* L.
5345	Monsanto	Resistance to lepidopteran pests through the *cry1Ac* gene from *Bacillus thuringiensis* spp. *kurstaki*.
		Potato, *Solanum tuberosum* L.
RBMT15-101, -02, -15.	Monsanto	Colorado potato beetle and potato virus Y produced by inserting the *cry3A* protein and gene encoding for PVY.
RBMT21-129, -350, BMT22-082.	Monsanto	Colorado potato beetle and potato leafroll virus (PLRV) resistant potatoes produced by inserting the *cry3A* and replicase encoding gene from PLRV.
		Maize, *Zea mays* L.
176	Syngenta Seeds Inc.	Insect-resistant maize produced by inserting *cry1Ab* for resistance to European corn borer.
BT11 (X4334CBR, X4734CBR)	Syngenta Seeds Inc.	Insect-resistant and herbicide-tolerant maize inserting the *cry1Ab* gene from *Bacillus thuringiensis* var. *kurstaki*, and the phosphoinothricin acelytransferase (PAT) encoding gene from *S. viridochromogen*.
MON802	Monsanto	Insect-resistant and glyphosate herbicide-tolerant line produced by inserting the genes encoding the protein from *Bacillus thuringiensis* and the 5-5′-enolpyruvylshikimate-3-phosphate synthase (*A. tumefaciens* strain CP4).
MON809	Pioneer HiBred International Inc.	Resistant to European corn borer through introduction of a synthetic *cry1Ab* gene.
MON810	Monsanto	Insect-resistant maize produced by inserting the *cry1Ab* gene from *Bacillus thuringiensis* kurstaki (HD-1). The genetic modification affords resistance to attack by the European corn borer.

Source: http://www.agbios.com/dbase.php

Transgenic hybrids suffered 20 to 40% less bollworm damage than their nontransgenic counterparts under unprotected conditions (Figure 10.3). Among these, Mech 184 suffered the least boll damage (<20%) (Sharma and Pampapathy, 2006). There were no differences in cottonseed yield of transgenic and nontransgenic versions of Mech 12 and Mech 162 under protected or unprotected conditions (Figure 10.4), under moderate levels of bollworm damage. However, their yield under unprotected conditions was very low as both of these hybrids were severely damaged by cotton leafhopper, *Amrasca biguttula biguttula* Ishida. Transgenic hybrid Mech 184 yielded more than the nontransgenic counterpart, both under protected and unprotected conditions. This hybrid was also less susceptible to cotton leafhopper, *A. biguttula biguttula* (<4 jassids per plant) as compared to Mech 12 and

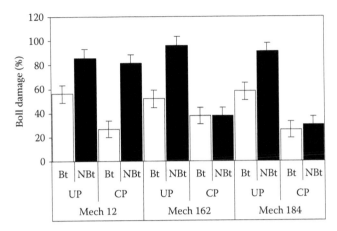

FIGURE 10.3 Cotton bollworm/legume pod borer, *Helicoverpa armigera*, damage in three cotton *cry1Ac* transgenic hybrids under protected and unprotected conditions. UP, unprotected; CP, completely protected.

Mech 162 (6 to 14 jassids per plant) (Figure 10.5). Under farmers' field conditions, the transgenic hybrid Mech 162 suffered less damage by the bollworms, *H. armigera* and *P. gossypiella* (11.5%) as compared to the conventional cotton (29.4%) in plots under integrated pest management (Bambawale et al., 2004). Only three sprays were needed for pest control on the transgenic cotton compared to seven sprays on the conventional cotton variety. Qaim and Zilberman (2003) reported that pesticide use was reduced by 70% on the *Bt*-transgenic hybrids, while the yields increased by 60%.

Cereals

Maize plants with *cry1Ab* (Koziel et al., 1993; Armstrong et al., 1995; Archer et al., 2000) and *cry9C* (Jansens et al., 1997) are highly effective against the European corn borer (ECB), *Ostrinia nubilalis* (Hubner). Transgenic maize with *Bt* genes has also been found to be effective against the spotted stem borer, *Chilo partellus* (Swinhoe) and maize stalk borer, *Busseola*

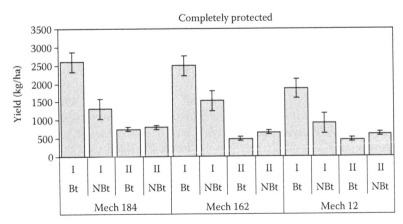

FIGURE 10.4 Cottonseed yield in three cotton *cry1Ac* transgenic hybrids under protected and unprotected conditions. I and II, First and second picking, respectively.

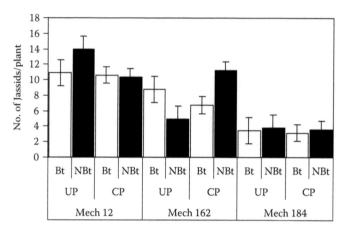

FIGURE 10.5   Cotton leafhopper, *Amrasca biguttulla biguttulla*, incidence in three cotton *cryIAc* transgenic hybrids under protected and unprotected conditions. UP, unprotected; CP, completely protected.

fusca Fuller in southern Africa (van Rensburg, 1999), sugarcane borers, *Diatraea grandiosella* (Dyar) and *Diatraea saccharalis* F. (Bergvinson, Willcox, and Hoisington, 1997), fall armyworm, *Spodoptera frugiperda* (J.E. Smith), and Southwestern corn rootworm, *Diabrotica undecim-punctata howardi* Barber (Williams et al., 1997). Insecticidal crystal protein from B. *thuringiensis* strain PS149B1 provides effective control of western maize rootworm, *Diabrotica virgifera virgifera* Leconte (Herman et al., 2002). Field studies indicated significantly greater resistance to insect feeding in Tex6 plants as compared with nontransgenic B 73 plants, primarily to flea beetles, *Chaetocnema* spp. (Dowd and White, 2002) and corn earworm, *H. zea* (approximately three- fold). At harvest, similar trends in reduction of numbers of damaged kernels as well as infection by *Fusarium* spp. were noted. Transgenic *Bt* maize is now grown in five countries in Europe. Transgenic hybrids with *cryIAb* gene showed less insect feeding and *Fusarium* ear rot than their nontransgenic counterparts (Munkvold, Hellmich, and Rice, 1999). Good control of major stem borer species, *C. partellus*, *Eldana saccharina* Walker, and *Sesamia calamistis* Hampson has been observed in controlled field trials in Kenya (Mugo et al., 2005). In South Africa, genetically modified white maize with the *cryIAb* gene gave greater yields and resulted in less damage by the stalk borers, *B. fusca* and *C. partellus* than the nontransgenic cultivars (Keetch et al., 2005). The total gain from borer-resistant maize was estimated to be over 15.5 million Euros.

Transgenic rice cultivars with resistance to insects have not been released for cultivation, although in several events *Bt* genes have been developed for resistance to the striped stem borer, *Chilo suppressalis* (Walker) (Fujimoto et al., 1993), yellow rice stem borer, *Scirpophaga incertulas* (Walker) (Mqbool et al., 1998), and rice leaf folder, *Cnaphalocrosis medinalis* (Guenee). Rice varieties with *cryIAc* gene have shown better resistance to yellow stem borer (Nayak et al., 1997), while those with the *cryIAb* are resistant to striped stem borer and the yellow stem borer (Ghareyazie et al., 1997). Plants of the *cryIAb*-transformed line 827 were more resistant to young larvae of *S. incertulas*, *C. suppressalis*, and *C. medinalis* than the control plants at the vegetative stage, but not at the flowering stage (Alinia et al., 2000). Sorghum plants with the *cryIAc* gene with resistance to spotted stem borer, *C. partellus*, have been identified (Girijashankar et al., 2005). The truncated *cryIAb* gene in sugarcane has shown significant activity against the sugarcane borer, *D. saccharalis*, despite low expression of the *Bt* protein (Arencibia et al., 1997). None of these transgenic plants have been deregulated for cultivation by farmers. Transgenic *Bt* rice released in Iran in 2004 was grown on over 4,000 ha in 2005 (James, 2007).

Grain Legumes

Groundnut plants with the *cry1Ac* gene have shown resistance to lesser corn stalk borer, *Elasmopalpus lignosellus* (Zeller) (Singsit et al., 1997). Chickpea cultivars transformed with the *cry1Ac* gene have been found to inhibit feeding by *H. armigera* (Kar et al., 1997; Sanyal et al., 2005). Transgenic pigeonpea plants with *cry1Ab* (Gopalswamy et al., 2003; Sharma, Lavanya, and Anjaiah, 2006) and *cry1E* genes have also been developed (Surekha et al., 2005).

Potato

Transgenic potato plants containing *cy1Ab* and a truncated gene *cry1Ab6* resulted in less damage by the potato tuber moth, *Phthorimaea opercullela* (Zeller) (Arpaia et al., 2000). Picentia and the wild species *Solanum integrifolium* Poir. have also been transformed with four mutagenized versions of *Bt* 43 belonging to *cryIII* (Innacone, Grieco, and Cellini, 1997). The *cry5*-Lemhi Russet and *cry5*-Atlantic potato lines have shown up to 100% mortality of first-instar larvae of the potato tuber moth (Mohammed et al., 2000). Transgenic potato expressing the oryzacystatin (*OCI*) gene resulted in up to 53% mortality of *L. decemlineata* (Lecardonnel et al., 1999). Some of the potato cultivars have been tested on farmers' fields and released for cultivation.

Vegetables

Tomato plants expressing *cry1Ab* and *cry1Ac* genes are effective against the lepidopteran insects (Fischhoff et al., 1987; Delannay et al., 1989; Van der Salm et al., 1994). Tomato plants with the *cry1Ac* gene are highly effective against *H. armigera* (Mandaokar et al., 2000). Transformed brinjal plants have also shown significant insecticidal activity against the fruit borer, *Leucinodes orbonalis* (Guen.) (Kumar et al., 1998). A synthetic *cry1C* gene in broccoli provides protection from cabbage looper, *T. ni*, cabbage butterfly, *Pieris rapae* L., and diamondback moth, *P. xylostella* (Cao et al., 1999). Chinese cabbage, *Brassica campestris* subsp. *parachinensis* with *cry1Ab* or *cry1Ac* genes is resistant to *P. xylostella* (Xiang et al., 2000).

Deployment of Transgenic Plants for Pest Management

Transgenic crops will have similar effects on insect damage and population dynamics as the insect-resistant varieties derived through conventional breeding (Sharma and Ortiz, 2002). The effects of transgenic resistance are expected to be much higher than those of conventional varieties, and such effects should be cumulative over time. Models predicting the effect of insect-resistant cultivars on insect abundance have been developed for several insect pests (Luginbill and Knipling, 1969; Knipling, 1979; Sharma, 1993; Sharma and Ortiz, 2002). As the insects are continuously exposed to the transgene product in genetically modified plants, there are distinct possibilities of a different type of effect on the population dynamics of the target and nontarget pests. There is a need to develop a better understanding of the field performance of insect-resistant transgenic cultivars under diverse environmental conditions to assess their role in IPM programs. Long-term suppression of pest populations would be governed by insect-host plant interactions, reproductive rate, dispersal propensity, and regional abundance of the transgenic crops (Carriere et al., 2003). Pink bollworm, *P. gossypiella*, population density has declined considerably in regions where *Bt* cotton was abundant. Such long-term suppression has not been observed with *Bt* sprays, indicating that transgenic crops have opened up new avenues for pest management.

Deployment of insect-resistant transgenic crops will lead either to an increase in the economic threshold level (ETL) or delay the time required by the insects to reach the ETL, depending on the nature of resistance and the stage of the insect on which the ETL is based. Crop growth and pest incidence should be monitored carefully so that appropriate control measures are initiated in time based on ETLs. Care should be taken to use alternate control options such as natural enemies, biopesticides [nucleopolyhedrosis viruses (NPV), entomopathogenic fungi, and nematodes], and natural plant products (neem, custard apple, and *Pongamia*), which do not disturb the natural control agents. Use of pesticide formulations such as seed treatment with systemic insecticides or soil application of granules of systemic insecticides to control sucking pests early in the season, and spraying selective insecticides may be considered to suppress insect populations in the beginning of the season to minimize adverse effects on the natural enemies (Sharma and Ortiz, 2000; Sharma et al., 2002b). Broad-spectrum and toxic insecticides should be used only during the peak activity periods of the target or nontarget pests. Efforts should be made to rotate pesticides with different modes of action, and avoid repetition of insecticides belonging to the same group or insecticides that fail to give effective control of the target insects. Deployment of transgenic plants should be based on the overall philosophy of IPM, taking into account alternate mortality factors, reduction of selection pressure, and monitoring populations for resistance development to design more effective management strategies. To increase the effectiveness and usefulness of transgenic plants, it is important to develop a strategy to prolong the life and effectiveness of transgenic crops for pest management (Sharma et al., 2002b).

It is important to implement IPM strategies from the very beginning to increase the effectiveness and prolong the life of the transgenic crops. However, in the absence of effective alternatives, farmers continue to rely heavily on insecticides. Once commercialized, the insect-resistant transgenic crops should be able to fulfill this need. While *Bt* cotton can largely control the bollworm complex, sucking pests and other nonlepidopteran insects will have to be tackled by combining host plant resistance (HPR) and biological, cultural, and chemical control methods. Transgenic crops cannot be effective as a stand-alone technology to manage insect pests, but can serve as one of the options in IPM. Combining various IPM strategies may be effective in delaying development of resistance to the transgene in pest populations. The IPM systems should diversify the mortality factors so that resistant strains are not selected for a single mortality mechanism, and thus, reduce the selection pressure on the pest populations for any single mortality factor. The following approaches may be followed for increasing the effectiveness of transgenic crops for pest management.

Transgenic Crops and Chemical Control

Transgenic crops can make a critical contribution in reducing the dosage and frequency of insecticide application. For rational pest management, transgenic cultivars have to be deployed in combination with low dosages of insecticides (Schell, 1997). Transgenic cotton

(NuCotn 33) acts additively in combination with polyhedrosis virus AcNPV-AaIT in reducing bollworm damage (All and Treacy, 1997). Insects such as *H. virescens*, *H. zea*, *T. ni*, and *S. exigua* are many times more sensitive to lamda cyhalothrin (Karate) sprays when they have a prior exposure to *Bt* (Harris et al., 1998). Insecticides are highly effective for bollworm control on transgenic cotton even at lower rates of application (Brickle et al., 1999). Susceptibility of *S. exigua* to alpha-cypermethrin, methomyl, profenofos, and chlorfluazuron was 1- to 1.6-fold lower in larvae fed with *Bt* cotton than in larvae fed with common cotton (Xue, Dong, and Zhang, 2002). However, the activities of acetylcholinesterase and carboxylesterase were greater in the larvae fed with *Bt* cotton than those fed with common cotton. The enhanced insecticidal activity enables a more practical resistance management strategy for transgenic crops.

The *Bt* sweet corn hybrids appear to be ideal candidates for use in IPM programs for both the fresh and processing sweet corn markets, and their use should drastically reduce the quantity of insecticides currently used to control these pests in sweet corn. First-generation European corn borer, *O. nubilalis* damage was reduced or eliminated with the use of the *Bt* hybrids (Archer et al., 2000). However, yield of *Bt* hybrids was 8% less than standard hybrids when an insecticide was applied. The value of protection offered by *Bt* maize is generally lower than the seed premiums (Hyde et al., 1999, 2001). Kernel damage was reduced to $1.7\,cm^2$ with five applications of methomyl compared to 172 and $50\,cm^2$ on Silver Queen and Bonus cultivars, respectively (Lynch et al., 1999). Corn earworm and European corn borer control in transgenic sweet corn has been found to be superior to that achieved in nontransgenic varieties sprayed with insecticides (Doohan et al., 2002). Damaged ears of transgenic varieties ranged from 0 to 6% and most damage was restricted to the tip of the silk end. In contrast, 40% of nontransgenic varieties had damaged ears, despite regular spraying of insecticides. Ears of nontransgenic varieties were damaged more severely and many late-instar larvae were recovered. The percentage decrease in insecticide use doubled from 13.2% in 1996 to 26.0% in 1998 (Pilcher et al., 2002), and scouting for European corn borers decreased from 91 to 75%. After having planted *Bt* maize and obtained excellent control of European corn borer, most farmers believed that this insect had been causing more yield loss than previously suspected. In Germany, grain yield of *Bt*-maize was approximately 14 to 15% higher than for untreated maize, and insecticide and *Bt* treatments showed profits of 18 to 55, and 84 to 93 Euros ha^{-1}, respectively (Degenhardt, Horstmann, and Mulleder, 2003).

Newleaf potato plants have been found to be highly effective in suppressing populations of Colorado potato beetle, *Leptinotarsa decemlineata* (Say), and provided better control than weekly sprays of a microbial *Bt*-based formulation containing Cry3Aa, biweekly applications of permethrin, or early- and mid-season applications of systemic insecticides (phorate and disulfoton) (Reed et al., 2001). Biweekly applications of permethrin significantly reduced the abundance of several major generalist predators, and resulted in significant increases in the abundance of green peach potato aphid, *M. persicae*, a vector of viral diseases. Transgenic *Bt* potato, *Bt*-based microbial formulations, and systemic insecticides appeared to be compatible for IPM in potato. In contrast, the broad-spectrum pyrethroid insecticide permethrin was less compatible with the IPM programs. Thus, transgenic crops can be used in conjunction with chemical control without any detrimental or antagonistic effects.

Transgenic Crops and Biological Control

Recombinant DNA technology has allowed development of plants that are well suited for use along with biological control. Unfortunately, plant breeders have continued to attempt to breed for total resistance, and biocontrol specialists have ignored the role of the plant in ensuring successful foraging behavior by the natural enemies (Poppy and Sutherland, 2004). Although some scientists have highlighted the need to consider both the bottom-up (plant defense) and top-down (biocontrol) control of insect pests, there have been few serious attempts to combine these approaches. As more is understood about the proximate and ultimate causes of direct and indirect defenses, the potential exists for engineering plants that combine both strategies. This new possibility for controlling insect pests, which will combine natures' own defenses with man's ingenuity, may stack the odds in our favor in the continual struggle against insect pests.

A reduction in numbers of eggs and larvae of the insect host may also affect the activity of natural enemies. The significance of such effects would depend on the importance of the immature stages of the target insect pest for maintaining the populations of the natural enemies. Transgenic crops may reduce the numbers of certain natural enemies in some areas, but their populations will be maintained on other crops that serve as an alternate host to the target pests. A few of the known predators are specialists on one insect, and hence, the populations of predators and parasites with a wide host range would be maintained on other insect species (Fitt, Mares, and Llewellyn, 1994). Within-field impact may be greater for parasitoids that are monophagous (Sharma, Arora, and Pampapathy, 2007). The populations of such natural enemies can only be maintained on alternate insect hosts on nontransgenic crops or other hosts of the target pest (Dhillon and Sharma, 2007).

Manipulation of crop environment based on detailed understanding of tritrophic interactions can contribute to improvements in biological control of insect pests in transgenic crops. Particular attention should be given to compatibility of plant resistance and biological control (Verkerk, Leather, and Wright, 1998). Allelochemicals, refugia, intercropping, crop backgrounds, fertilization regimes, parasitoid conditioning (by host plants), and transgenic crops affect the activity of natural enemies. *Trichogramma* increased the yields by 2% to 3%, while insecticides increased the yields by 7% to 10%. Financial losses for the *Trichogramma* treatment amounted to 52–57 Euros ha^{-1}, while insecticide and *Bt* treatments showed profits of 18 to 55 and 84 to 93 Euros ha^{-1}, respectively. Larval mortality increased with an increase in density of natural enemies (Mascarenhas and Luttrell, 1997). The genotype on which *H. zea* larvae were predisposed significantly affected larval survival in environments with low, medium, and high density of natural enemies at 24, 96, and 48 hours post infestation. Larvae that were conditioned on BTK cotton tended to survive at lower rates than the larvae conditioned on non-BTK cotton. Length of exposure was a significant factor affecting larval survival at all post infestation time intervals in a cage environment with no natural enemies. In the absence of natural enemies, larval survival was correlated positively with length of exposure. Larvae that were conditioned for 4 days before field infestation survived at significantly higher rates than the larvae that were conditioned for 1 and 2 days.

Transgenic Crops and Cultural Control

A negative linear relationship has been observed between yield and percentage of susceptible plants in mixtures of transgenic and nontransgenic plants across planting densities (Nault et al., 1995). In diamondback moth-infested plots, transgenic plants had low levels of damage both as a pure stand and in mixtures. Nontransgenic plants in diamondback moth-infested plots suffered high levels of defoliation and produced less biomass and seed yield than the transgenic plants (Ramachandran et al., 2000). Relative crowding coefficient (RCC), a measure of competition between two plant types, ranged from 0.6 to 1.1 in plots where there was no diamondback moth infestation, and 1.1 to 12.8 in plots where there was diamondback moth infestation. No competitive advantage was observed for either plant type in seed mixtures when there was no diamondback moth infestation. Moth densities in all plots (mixed and nonmixed) declined after two generations (Riggin Bucci and Gould, 1997). Percentage of parasitism by the parasitoid, *Diadegma insulare* (Cresson) was not significantly different between mixed and nonmixed plots, and abundance of the four most common diamondback moth, *P. xylostella*, predators did not differ between mixed and nonmixed plots. There were no differences in numbers of predators on transgenic versus nontransgenic plants. Intrafield mixtures may decrease the density of target pests such as the diamondback moth, while not adversely affecting the natural enemies. Therefore, there is considerable potential for integration of transgenic plants and cultural practices for IPM.

Advantages and Limitations of Insect-Resistant Transgenic Crops

There will be tremendous potential benefits to the environment through the deployment of genetically modified crops. Deployment of insect-resistant crops has been estimated to result in a one million kilogram reduction in pesticides applied for pest control in the United States in 1999 as compared to 1998 (NRC, 2000). Papaya with transgenic resistance to ringspot virus (Gonsalves, 1998) has been grown in Hawaii since 1996. Rice yellow mottle virus (RYMV), which is difficult to control with conventional approaches, can now be controlled through transgenic rice, which will provide insurance against total crop failure. Globally, herbicide-resistant soybean, insect-resistant maize, and genetically improved cotton account for 85% of the total area under transgenic crops. Transgenic plants with insecticidal genes are set to feature prominently in pest management in both developed and the developing world in future. Such an effort will play a major role in minimizing insect-associated losses, increase crop production, and improve the environment. Brookes and Barfoot (2005) have estimated that use of genetically modified crops has led to a decrease in pesticide use by 172,000 metric tons, reduction in greenhouse gas emissions by 10 million metric tons, diesel fuel savings by 1.8 billion liters, increase in net income for farmers by $27 billion ($4.16 billion in China and $124.2 million in India), reduced environmental index quotient (EIQ) by 14%, and generated global market of biotech crops by $5.5 billion in 2006. Major advantages of growing transgenic crops are:

- A major reduction in insecticide sprays;
- Increased activity of natural enemies;
- Reduced exposure of nontarget organisms to pesticides; and
- Reduction in pesticide residues in food and food products.

The benefits of growing transgenic crops to growers have been higher yield, lower input costs, and easier agronomic management. These factors are likely to have substantial impact on the livelihoods of farmers in both developed and developing countries. However, there is considerable public debate on the issue, and several claims to the contrary have also been published. In many developing countries, small-scale farmers suffer pest-related yield losses because of technical and economic constraints. Pest-resistant genetically modified crops can contribute to increased yields and agricultural growth in such situations. Adopting transgenic crops offers the additional advantage of controlling insect pests that have become resistant to commonly used insecticides (Sharma et al., 2003). For rational pest management, transgenic cultivars have to be deployed in combination with low dosages of insecticides (Schell, 1997), and other methods of pest control.

Developments in plant biotechnology offer both promises and challenges. Close proximity of transgenic crops to the sprayed fields of nontransgenic crops may result in greater damage in transgenic crops because of insect migration from sprayed fields to the transgenic crops, and the resultant increase in pest pressure may reduce the benefits of transgenic crops. Toxins from *Bt* have been widely used as "natural" insecticides for many years. However, an increase in the presence of *Bt* toxins in the ecosystem (via transgenic crops) may significantly increase the pressure on insect populations to evolve into resistant biotypes. The evidence on these issues is still inconclusive, and there is a need for careful monitoring before the transgenic crops are deployed on a large scale (Sharma and Ortiz, 2000; Sharma et al., 2002b). One of the approaches to overcome these problems is to develop a new generation of transgenic crops with better genes, and use combinations of genes to delay the development of resistance in insect populations. Transgenic crops at times fail to provide adequate control of insect pests due to environmental influences on gene expression or development of resistance, ineffectiveness against secondary pests, and adverse effects on nontarget organisms. The major limitations of using transgenic plants are:

- Secondary pests are not controlled in the absence of sprays for the major pests.
- The need to control the secondary pests through chemical sprays will kill the natural enemies and thus offset one of the advantages of transgenics.
- Proximity to sprayed fields and insect migration may reduce the benefits of transgenics.
- Development of resistance in insect populations may limit the usefulness of transgenics.
- Effects on nontarget organisms.
- Gene escape into the environment.
- Social and ethical issues.

Stability of Transgene Expression

Stable expression of the transgene is an important component for commercial deployment and success of transgenic crops. Impairment of transgene expression may occur due to mutations, deletions, and DNA rearrangement. In general, transgenes are not predisposed for mutations and recombinations, and are known to occur at a frequency of 10^{-4} to 10^{-6} as in endogenous genes (Peterhans et al., 1990). Reduced expression can occur via *cis*- or *trans*-acting regulatory elements present in the environment of the transgene. Changes in transgene expression occur at the transcriptional and posttranscriptional level that affect the stability, processing, or transport of m-RNA, and the resultant proteins. Homology in DNA sequence can occur due to multiple homologous copies (multiple direct or inverted repeats) and endogenous genes sharing sequences with the transgene. Inactivation can be effected in the *cis*- (same locus) or *trans*- (allelic positions) modes. Transgenes containing multiple, complete, or partial repeats at one locus may be silenced. In many cases, silencing does not occur in the primary transformants, but in the subsequent generations (Kilby, Leyser, and

Furner, 1992; Assaad, Tucker, and Signer, 1993). Homology-induced silencing plays a role in regulation of endogenous genes whereby the endogenous repeats are frequently hypermethylated and packaged into transcriptionally inactive heterochromatin. These inactivations are also *trans* acting (Mayer, Heidmann, and Niedenhof, 1993). Common promoter sequences between two genes occupying nonallelic locations may result in *trans*-inactivation of the primary transgene (Matzke, Neuhuber, and Matzke, 1993). A mere 90 bp of homology within a *35S* promoter fragment between a second inserted gene and the hypermethylated copies of a primary transgene with *35S* promoter sequences result in silencing of the inserted second gene (Vaucheret, 1993). Homology-based gene silencing (co-suppression) has been observed to occur only in a fraction of transformants that carry a sense copy of the endogenous gene. At the molecular level, co-suppression involves transcriptional inhibition (Ingelbrecht et al., 1994) and posttranscriptional RNA degradation, while transcription is unaffected (Mueller et al., 1995). Even single copy transgenes get inactivated due to DNA methylation (Antequera, Boyes, and Bird, 1990). Condensation of chromatin structures is mediated by proteins that specifically recognize methylated DNA (Lewis et al., 1992). The methylation pattern of the integration region also has a significant influence on the methylation state of the transgene (Prols and Meyer, 1992).

Performance Limitations

The *Bt* toxins cannot produce the same dramatic effects on insect mortality as the synthetic insecticides. The farmers need to be educated about the efficacy and mode of action of transgenic crops. The expectations have to be real, and remedial measures should be taken as the situation warrants. The effects of the transgenic crops on insects will be relatively slower, but cumulative over time. Transgenic crops may not be able to withstand the pest density in some seasons. Therefore, careful monitoring of pest populations is an essential component of pest management involving the transgenic crops. The value of the transgenic crops can be best realized when deployed as a component of pest management for sustainable crop production (Sharma and Ortiz, 2000). However, enough information has not been generated involving transgenic crops in a genuine IPM system to demonstrate long-term benefits of the transgenic crops, especially if environmental and human health hazards are taken into account. Currently deployed transgenic crops produce only one *Bt* toxin protein, while the *Bt* strains used for commercial formulations produce several toxins in addition to other factors that increase insect mortality. The current *cryIAb* construct employs PEP-carboxylase promoter, which enables expression in green tissue and, as a result, the expression is greater in young plants. Insects that migrate into the plant whorl or stem tissue with incomplete chlorophyll formation may escape the toxin protein. If the toxin is expressed in insufficient amounts in such tissues, the insects can develop mechanisms to withstand low levels of toxins in the transgenic plants. Behavioral avoidance of the tissue expressing the toxin gene can be another component in insect resistance to the transgenic plants. Therefore, care should be taken to express the toxins in sufficient amounts at the site of damage or feeding by the insects.

Secondary Pest Problems

Most crops are attacked by a large number of insect pests. In the absence of competition from the major insect pests, the secondary pests tend to assume a major pest status (Hilder and Boulter, 1999). The *Bt* toxins may be ineffective against some insect pests, for example, leaf hoppers, mirid bugs, root feeders, mites, etc. This will offset some of the advantages expected of the cultivation of transgenic crops. Management of stinkbugs is necessary in transgenic *Bt* cotton if more than 20% of the bolls are damaged in mid- to late-season (Greene, Turnipseed, and Sullivan, 1997). There are no differences between transgenic and nontransgenic cultivars in boll weevil or aphid damage, beneficial arthropods, or fiber characteristics (Parker and Huffman, 1997). Effective and timely control measures should be adopted for the control of secondary pests on transgenic crops. Although there is a trend to develop target-specific compounds for chemical control, it will be desirable to have genes with a broad-spectrum of activity for use in genetic transformation of crops, provided this does not influence the beneficial organisms.

Insect Sensitivity

There are many species of insects that are not susceptible to the currently available *Bt* proteins. There is a need to broaden the pool of genes that can be effective against insects that are not sensitive to the currently available genes. Since first-generation transgenic crops have only one *Bt* toxin gene, lack of control of less sensitive species may present another problem in pest management. This is not the same as development of resistance (which is a progressive decrease in sensitivity to a chemical by a population in response to the use of a product to kill the insects). If there is low or no sensitivity to a chemical in an insect species, it is not resistance. *Helicoverpa virescens* is less sensitive to Cry1Aa, Cry1C and Cry1E, while *Spodoptera littoralis* (Boisd.) is insensitive to most of the *Bt* toxins (Gill, Cowles, and Pietrantonio, 1992). *Spodoptera litura* (F.) is less sensitive to toxins from *B. thuringiensis* var *kurstaki* than *H. armigera*, *Achoea janata* L., *P. xylostella*, and *Spilosoma obliqua* Walker (Meenakshisundaram and Gujar, 1998). *Bt* toxins CryIC and Cry1E, which are active against *H. virescens* (MacIntosh et al., 1991), are ineffective against *H. armigera* (Chakrabarti et al., 1998). Cry1B is less effective against *H. armigera* (Chakrabarti et al., 1998) and *H. virescens* (Hofte and Whiteley, 1989). Thus, there are considerable differences in the sensitivity of different insect species to various *Bt* toxins, and due care has to be taken to deploy *Bt* toxins in different crops or cropping systems.

Evolution of Insect Biotypes

Experience from conventional breeding has shown that there is a direct relationship between the deployment of insect-resistant cultivars and the evolution of new insect biotypes. In the case of greenbug, *Schizaphis graminum* (Rondani), the breeding programs continue to struggle to keep pace with the evolution of new biotypes (Wood, 1971; Daniels, 1977). However, there is no relationship between the deployment of greenbug-resistant wheat cultivars and the development of new greenbug biotypes (Porter et al., 1997). For sorghum, only three of the 11 biotypes of greenbug have shown a correlation between the use of resistant hybrids and the development of new biotypes. Even within the three biotypes, no clear cause-and-effect relationship has been established. Based on analysis of these specific insect-plant interactions, future plant resistance efforts should focus on the use of the most effective resistance genes despite past predictions of what effect these genes may have on insect population genetics.

Environmental Influence on Gene Expression

There have been some failures in insect control through the transgenic crops. Cotton bollworm, *H. virescens* destroyed *Bt* cottons due to high tolerance to *Bt* toxin, CryIAc in Texas

(Kaiser, 1996). Similarly, *H. armigera* and *H. punctigera* destroyed the cotton crop in the second half of the crop-growing season in Australia because of reduced production of *Bt* toxins in the transgenic crops (Hilder and Boulter, 1999). Possible causes for the failure of insect control may be due to: (1) inadequate production of the toxin, (2) effect of the environment on expression of the transgene, (3) locally resistant insect populations, and (4) development of resistance due to inadequate management.

A cotton crop flooded with 3 to 4 cm deep water for 12 days lost resistance to insects significantly compared with the control plants irrigated normally (Y.R. Wu et al., 1997). A similar reaction has been observed in *Bt* cotton, which grew in overcast and rainy weather continuously for 21 days. When the water logging was over, the cotton plants recovered gradually and their insect resistance increased to some extent. Under flooded conditions, the activity of superoxide dismutase increased considerably in *Bt* cotton plants at first, and then dropped continuously. Epistatic and environmental effects on foreign gene expression could influence the stability, efficacy, and durability (Sachs et al., 1996). CryIA gene expression is variable and is influenced by genetic and environmental factors. The CryIA phenotype segregates as a simple, dominant Mendelian trait. However, non-Mendelian segregation occurred in some lines derived from MON 249. Expression of transgene is influenced by: (1) site of gene insertion, (2) gene construct, (3) epistasis, (4) somaclonal mutations, and (5) physical environment. Appropriate evaluation and selection procedures should be used in a breeding program to develop crop varieties with stable expression of the transgene.

Conclusions

Incorporation of insecticidal genes in crop plants will have a tremendous effect on pest management. Emphasis should be placed on combining exotic genes with conventional host plant resistance, and also with traits conferring resistance to other insect pests and diseases of importance in the target region. Although several crops with commercial viability have been transformed in the developed world, very little has been done to use this technology to develop insect-resistant cultivars to increase food production in the harsh environments of the tropics. There is a need to follow integrated pest management (IPM) practices to make transgenic crops a viable technology for pest control. Emphasis should be given to commercial and regulatory requirements for the deployment of transgenic crops, including spatial and temporal expression of the transgene, and food and feed safety of the product.

References

Alinia, F., Ghareyazie, B., Rubia, L., Bennett, J. and Cohen, M.B. (2000). Effect of plant age, larval age, and fertilizer treatment on resistance of a cry1Ab-transformed aromatic rice to lepidopterous stem borers and foliage feeders. *Journal of Economic Entomology* 93: 484–493.

All, J.N. and Treacy, M.F. (1997). Improved control of *Heliothis virescens* and *Helicoverpa zea* with a recombinant form of *Autographa californica* nuclear polyhedrosis virus and interaction with Bollgard R cotton. In *Proceedings, Beltwide Cotton Conference*, 6–10 January, 1997, New Orleans, vol. 2. Memphis, Tennessee, USA: National Cotton Council, 1294–1296.

Antequera, F., Boyes, J. and Bird, A. (1990). High levels of de novo methylation and altered chromatin structure at CpG islands in cell lines. *Cell* 62: 503–514.

Archer, T.L., Schuster, G., Patrick, C., Cronholm, G., Bynum, E.D. Jr. and Morrison, W.P. (2000). Whorl and stalk damage by European and Southwestern corn borers to four events of *Bacillus thuringiensis* transgenic maize. *Crop Protection* 19: 181–190.

Arencibia, A., Vazquez, R.I., Prieto, D., Tellez, P., Carmona, E.R., Coego, A., Hernandez, L., de-la Riva, G.A. and Selman-Housein, G. (1997). Transgenic sugarcane plants resistant to stem borer attack. *Molecular Breeding* 3: 247–255.

Armstrong, C.L., Parker, G.B., Pershing, J.C., Brown, S.M., Sanders, P.R., Duncan, D.R., Stone, T., Dean, D.A., DeBoer, D.L., Hart, J., Howe, A.R., Morrish, F.M., Pajeau, M.E., Peterse, W.L., Reich, B.J., Rodriguez, R., Santino, C.G., Sato, S.J., Schuler, W., Sims, S.R., Stehling, S., Tarochione, L.J. and Fromm, M.E. (1995). Field evaluation of European corn borer control in progeny of 173 transgenic corn events expressing an insecticidal protein from *Bacillus thuringiensis*. *Crop Science* 35: 550–557.

Arpaia, S., De Marzo, L., Di Leo, G.M., Santoro, M.E., Mennella, G. and Vanloon, J.J.A. (2000). Feeding behaviour and reproductive biology of Colorado potato beetle adults fed transgenic potatoes expressing the *Bacillus thuringiensis* Cry3B endotoxin. *Entomologia Experimentalis et Applicata* 95: 31–37.

Assaad, F.F., Tucker, K.L. and Signer, E.R. (1993). Epigenetic repeat: Induced gene silencing (RIGS) in *Arabidopsis*. *Plant Molecular Biology* 22: 1067–1085.

Bambawale, O.M., Singh, A., Sharma, O.P., Bhosle, B.B., Lavekar, R.C., Dhandapani, A., Kanwar, V., Tanwar, R.K., Rathod, K.S., Patnge, N.R. and Pawar, V.M. (2004). Performance of *Bt* cotton (MECH 162) under integrated pest management in farmers' participatory field trial in Nanded district, Central India. *Current Science* 86: 1628–1633.

Barton, K., Whiteley, H. and Yang, N.S. (1987). *Bacillus thuringiensis* δ-endotoxin in transgenic *Nicotiana tabacum* provides resistance to lepidopteran insects. *Plant Physiology* 85: 1103–1109.

Benedict, J.H., Sachs, E.S., Altman, D.W., Deaton, D.R., Kohel, R.J., Ring, D.R. and Berberich, B.A. (1996). Field performance of cotton expressing CryIA insecticidal crystal protein for resistance to *Heliothis virescens* and *Helicoverpa zea* (Lepidoptera: Noctuidae). *Journal of Economic Entomology* 89: 230–238.

Bergvinson, D., Willcox, M.N. and Hoisington, D. (1997). Efficacy and deployment of transgenic plants for stem borer management. *Insect Science and Its Application* 17: 157–167.

Brickle, D.S., Turnipseed, S.G., Sullivan, M.J. and Dugger, P. (1999). The efficacy of different insecticides and rates against bollworms (Lepidoptera: Noctuidae) in B.T. and conventional cotton. In Richter, D. (Ed.), *Proceedings, Beltwide Cotton Production and Research Conference*, 3–7 January, 1999, Orlando, Florida, vol. 2. Memphis, Tennessee, USA: National Cotton Council, 934–936.

Brookes, G. and Barfoot, P. (2005). GM crops: The global economic and environmental impact—The first nine years. *AgBio Forum* 8: 187–196.

Cao, J., Tang, J.D., Strizhov, N., Shelton, A.M. and Earle, E.D. (1999). Transgenic broccoli with high levels of *Bacillus thuringiensis* Cry1C protein control diamondback moth larvae resistant to Cry1A or Cry1C. *Molecular Breeding* 5: 131–141.

Carriere, Y., Ellers Kirk, C., Sisterson, M., Antilla, L., Whitlow, M., Dennehy, T.J. and Tabashnik, B.E. (2003). Long-term regional suppression of pink bollworm by *Bacillus thuringiensis* cotton. *Proceedings National Academy of Sciences USA* 100: 1519–1523.

Chakrabarti, S.K., Mandaokar, A.D., Kumar, P.A. and Sharma, R.P. (1998). Synergistic effect of Cry1A(c) and Cry1F delta-endotoxins of *Bacillus thuringiensis* on cotton bollworm, *Helicoverpa armigera*. *Current Science* 75: 663–664.

Daniels, N.E. (1977). Seasonal populations and migration of the greenbug in the Texas panhandle. *Southwestern Entomologist* 2: 20–26.

Degenhardt, H., Horstmann, F. and Mulleder, N. (2003). Bt-maize in Germany: Experiences with cultivation from 1998 to 2002. *Mais* 31: 75–77.

Delannay, X., LaVallee, B.J., Proksch, R.K., Fuchs, R.L., Sims, S.K., Greenplate, J.T., Marrone, P.G., Dodson, R.B., Augustine, J.J., Layton, J.G. and Fischhoff, D.A. (1989). Field performance of transgenic tomato plants expressing *Bacillus thuringiensis* var *kurstaki* insect control protein. *Bio/Technology* 7: 1265–1269.

Dhillon, M.K. and Sharma, H.C. (2007). Survival and development of *Campoletis chlorideae* on various insect and crop hosts: Implications for *Bt*-transgenic crops. *Journal of Applied Entomology* 131: 179–185.

Dong, H., Li, W., Tang, W. and Zhang, D. (2004). Development of hybrid *Bt* cotton in China: A successful integration of transgenic technology and conventional techniques. *Current Science* 86: 778–782.

Doohan, D.J., Felix, J., Jasinski, J., Welty, C. and Kleinhenz, M.D. (2002). Insect management and herbicide tolerance in near-isogenic sister lines of transgenic and non-transgenic sweet corn. *Crop Protection* 21: 375–381.

Dowd, P.F. and White, D.G. (2002). Corn earworm, *Helicoverpa zea* (Lepidoptera: Noctuidae) and other insect associated resistance in the maize inbred Tex6. *Journal of Economic Entomology* 95: 628–634.

Fischhoff, D.A., Bowdish, K.S., Perlak, F.J., Marrone, P.G., McCormick, S.M., Niedermeyer, J.G., Dean, D.A., Kusano-Kretzmer, K., Mayer, E.J., Rochester, D.E., Rogers, S.G. and Fraley, R.T. (1987). Insect tolerant tomato plants. *Bio/Technology* 5: 807–812.

Fitt, G., Mares, C.L. and Llewellyn, D.J. (1994). Field evaluation and potential ecological impact of transgenic cottons (*Gossypium hirsutum*) in Australia. *Biocontrol Science and Technology* 4: 535–548.

Fujimoto, H., Itoh, K., Yamamoto, M., Kayozuka, J. and Shimamoto, K. (1993). Insect resistant rice generated by a modified delta endotoxin genes of *Bacillus thuringiensis*. *Bio/Technology* 11: 1151–1155.

Ghareyazie, B., Alinia, F., Menguito, C.A., Rubia, L.G., Palma, J.M. de Liwanag, E.A., Cohen, M.B., Khush, G.S. and Bennett, J. (1997). Enhanced resistance to two stem borers in an aromatic rice containing a synthetic CryIA(b) gene. *Molecular Breeding* 3: 401–414.

Gill, S.S., Cowles, E.A. and Pietrantonio, F.V. (1992). The mode of action of *Bacillus thuringiensis* endotoxins. *Annual Review of Entomology* 37: 615–636.

Girijashankar, V., Sharma, H.C., Sharma, K.K., Swathisree, V., Sivarama Prasad, L., Bhat, B.V., Royer, M., Secundo, B.S., Narasu, L.M., Altosaar, I. and Seetharama, N. (2005). Development of transgenic sorghum for insect resistance against the spotted stem borer (*Chilo partellus*). *Plant Cell Reports* 24: 513–522.

Gonsalves, D. (1998). Control of papaya ringspot virus in papaya: A case study. *Annual Review of Phytopathology* 36: 415–437.

Gopalswamy, S.V.S., Kumar, S., Subaratnam, G.V., Sharma, H.C. and Sharma, K.K. (2003). Transgenic pigeonpea: A new tool to manage *Helicoverpa armigera*. In *National Symposium on Bioresources, Biotechnology and Bioenterprise*, 19–20 November, 2003. Hyderabad, Andhra Pradesh, India: Osmania University.

Greene, J.K., Turnipseed, S.G. and Sullivan, M.J. (1997). Treatment thresholds for stink bugs in transgenic *Bt* cotton. In *Proceedings, Beltwide Cotton Conference*, 6–10 January, 1997. New Orleans, Lousiana, vol. 2. Memphis, Tennessee, USA: National Cotton Council, 895–898.

Guo, S.D., Cui, H.Z., Xia, L.Q., Wu, D.L., Ni, W.C., Zhang, Z.L., Zhang, B.L. and Xu, Y.J. (1999). Development of bivalent insect-resistant transgenic cotton plants. *Scientia Agricultura Sinica* 32: 1–7.

Harris, J.G., Hershey, C.N., Watkins, M.J. and Dugger, P. (1998). The usage of Karate (lambda-cyhalothrin) oversprays in combination with refugia, as a viable and sustainable resistance management strategy for B.T. cotton. *Proceedings, Beltwide Cotton Conference*, 5–9 January, 1998, San Diego, California, vol. 2. Memphis, Tennessee, USA: National Cotton Council, 1217–1220.

Herman, R.A., Scherer, P.N., Young, D.L., Mihaliak, C.A., Meade, T., Woodsworth, A.T., Stockhoff, B.A. and Narva, K.E. (2002). Binary insecticidal crystal protein from *Bacillus thuringiensis*, strain PS149B1: Effects of individual protein components and mixtures in laboratory bioassays. *Journal of Economic Entomology* 95: 635–639.

Hilder, V.A. and Boulter, D. (1999). Genetic engineering of crop plants for insect resistance: A critical review. *Crop Protection* 18: 177–191.

Hilder, V.A., Gatehouse, A.M.R., Sheerman, S.E., Baker, R.F. and Boulter, D. (1987). A novel mechanism of insect resistance engineered into tobacco. *Nature* 330: 160–163.

Hofte, H. and Whiteley, H.R. (1989). Insecticidal crystal proteins of *Bacillus thuringiensis. Microbiology Review* 53: 242–255.

Hyde, J., Martin, M.A., Preckel, P.V. and Edwards, C.R. (1999). The economics of *Bt* corn: Valuing protection from the European corn borer. *Review of Agricultural Economics* 21: 442–454.

Hyde, J., Martin, M.A., Preckel, P.V., Dobbins, C.L. and Edwards, C.R. (2001). An economic analysis of non-Bt corn refuges. *Crop Protection* 20: 167–171.

Ingelbrecht, I., Van Houdt, H., Van Montague, M. and Depicker, A. (1994). Post-transcriptional silencing of reporter transgenes in tobacco correlates with DNA methylation. *Proceedings National Academy of Sciences USA* 91: 10502–10506.

Innacone, R., Grieco, P.D. and Cellini, F. (1997). Specific sequence modifications of a Cry3B endotoxin gene result in high levels of expression and insect resistance. *Plant Molecular Biology* 34: 485–496.

James, C. (2007). *Global Status of Commercialized Biotech/GM Crops: 2006.* ISAAA Briefs no. 35. Ithaca, New York, USA: International Service for Acquisition on Agri-Biotech Applications (ISAAA). http://www.isaaa.org/resources/publications/briefs/35.

Jansens, S., van Vliet, A., Dickburt, C., Buysse, L., Piens, C., Saey, B., de Wulf, A., Gossele, V., Paez, A. and Gobel, E. (1997). Transgenic corn expressing a Cry9C insecticidal protein from *Bacillus thuringiensis* protected from European corn borer damage. *Crop Science* 37: 1616–1624.

Kaiser, J. (1996). Pests overwhelm Bt cotton crop. *Nature* 273: 423.

Kar, S., Basu, D., Das, S., Ramkrishnan, N.A., Mukherjee, P., Nayak, P. and Sen, S.K. (1997). Expression of CryIA(c) gene of *Bacillus thuringiensis* in transgenic chickpea plants inhibits development of pod borer (*Heliothis armigera*) larvae. *Transgenic Research* 6: 177–185.

Keetch, D.P., Webster, J.W., Ngqaka, A., Akanbi, R. and Mahlanga, P. (2005). Bt maize for small scale farmers: A case study. *African Journal of Biotechnology* 4: 1505–1509.

Kilby, N.J., Leyser, H.M.O. and Furner, I.J. (1992). Promoter methylation and progressive transgene inactivation in *Arabidopsis. Plant Molecular Biology* 20: 103–112.

Knipling, E.F. (1979). The basic principles of insect population suppression. *Bulletin of Entomological Society of America* 12: 7–15.

Koziel, M.G., Beland, G.L., Bowman, C., Carozzi, N.B., Crenshaw, R., Crossland, L., Dawson, J., Desai, N., Hill, M., Kadwell, S., Launis, K., Lewis, K., Maddox, D., McPherson, K., Meghji, M.R., Merlin, E., Rhodes, R., Warren, G.W., Wright, M. and Evola, S.V. (1993). Field performance of elite transgenic maize plants expressing an insecticidal protein derived from *Bacillus thuringiensis. Bio/Technology* 11: 194–200.

Kumar, P.A., Mandaokar, A., Sreenivasu, K., Chakrabarti, S.K., Bisaria, S., Sharma, S.R., Kaur, S. and Sharma, R.P. (1998). Insect-resistant transgenic brinjal plants. *Molecular Breeding* 4: 33–37.

Lecardonnel, A., Chauvin, L., Jouanin, L., Beaujean, A., Prevost, G. and Sangwan-Norreel, B. (1999). Effects of rice cystatin I expression in transgenic potato on Colorado potato beetle larvae. *Plant Science* 140: 71–79.

Lewis, J.D., Meehan, R.R., Henzel, W.J., Maurer-Fogy, I., Jeppesen, P., Klein, F. and Bird, A. (1992). Purification, sequence and cellular localization of a novel chromosomal protein that binds to methylated DNA. *Cell* 69: 905–914.

Luginbill, P. Jr. and Knipling, E.F. (1969). Suppression of wheat stem fly with resistant wheat. United States Department of Agriculture/Agriculture Research Service (USDA/ARS). *Production Research Reports* 107: 1–9.

Lynch, R.E., Wiseman, B.R., Plaisted, D. and Warnick, D. (1999). Evaluation of transgenic sweet corn hybrids expressing CryIA(b) toxin for resistance to corn earworm and fall armyworm (Lepidoptera: Noctuidae). *Journal of Economic Entomology* 92: 246–252.

MacIntosh, S.C., Stone, T.B., Jokerst, R.S. and Fuchs, R.L. (1991). Binding of *Bacillus thuringiensis* proteins to a laboratory selected strain of *Heliothis virescens. Proceedings National Academy of Sciences USA* 188: 8930–8933.

Mandaokar, A.D., Goyal, R.K., Shukla, A., Bisaria, S., Bhalla, R., Reddy, V.S., Chaurasia, A., Sharma, R.P., Altosaar, I. and Kumar, P.A. (2000). Transgenic tomato plants resistant to fruit borer (*Helicoverpa armigera* Hubner). *Crop Protection* 19: 307–312.

Mascarenhas, V.J. and Luttrell, R.G. (1997). Combined effect of sublethal exposure to cotton expressing the endotoxin protein of *Bacillus thuringiensis* and natural enemies of survival of bollworm (Lepidoptera: Noctuidae) larvae. *Environmental Entomology* 26: 939–945.

Matzke, M.A., Neuhuber, F. and Matzke, A.J.M. (1993). A variety of epistatic interactions can occur between partially homologous transgene loci brought together by sexual crossing. *Molecular and General Genetics* 236: 379–386.

Mayer, P., Heidmann, I. and Niedenhof, I. (1993). Differences in DNA methylation are associated with a paramutation phenomenon in transgenic *Petunia*. *The Plant Journal* 4: 86–100.

McLaren, J.S. (1998). The success of transgenic crops in the USA. *Pesticide Outlook* 9: 36–41.

Meenakshisundaram, K.S. and Gujar, G.T. (1998). Proteolysis of *Bacillus thuringiensis* subspecies *kurstaki* endotoxin with midgut proteases of some important lepidopterous species. *Indian Journal of Experimental Biology* 36: 593–598.

Mohammed, A., Douches, D.S., Pett, W., Grafius, E., Coombs, J., Liswidowati, W.L. and Madkour, M.A. (2000). Evaluation of potato tuber moth (Lepidoptera: Gelechiidae) resistance in tubers of Bt-Cry5 transgenic potato lines. *Journal of Economic Entomology* 93: 472–476.

Mohan, K.S. and Manjunath, T.M. (2002). Bt cotton: India's first transgenic crop. *Journal of Plant Biology* 29: 225–236.

Mqbool, S.B., Husnain, T., Raizuddin, S. and Christou, P. (1998). Effective control of yellow rice stem borer and rice leaf folder in transgenic rice *indica* varieties Basmati 370 and M 7 using novel δ-endotoxin Cry2A *Bacillus thuringiensis* gene. *Molecular Breeding* 4: 501–507.

Mueller, E., Gilbert, J., Davenport, G., Brigneti, G. and Baulcombe, D.C. (1995). Homology-dependent resistance: Transgenic virus resistance in plants related to homology-dependent gene silencing. *The Plant Journal* 7: 1001–1013.

Mugo, S., De Groote, H., Bergvinson, D., Mulaa, M., Songa, J. and Gichuki, S. (2005). Developing maize for resource poor farmers: Recent advances in IRMA project. *African Journal of Biotechnology* 41: 1490–1504.

Munkvold, G.P., Hellmich, R.L. and Rice, L.G. (1999). Comparison of fumonisin concentrations in kernels of transgenic Bt maize hybrids and nontransgenic hybrids. *Plant Disease* 83: 130–138.

NRC (National Research Council). (2000). *Genetically Modified Pest-Protected Plants: Science and Regulation*. Washington, D.C., USA: National Research Council.

Nault, B.A., Follett, P.A., Gould, F. and Kennedy, G.G. (1995). Assessing compensation for insect damage in mixed plantings of resistant and susceptible potatoes. *American Potato Journal* 72: 157–176.

Nayak, P., Basu, D., Das, S., Basu, A., Ghosh, D., Ramakrishnan, N.A., Ghosh, M. and Sen, S.K. (1997). Transgenic elite *indica* rice plants expressing CryIAc delta-endotoxin of *Bacillus thuringiensis* are resistant against yellow stem borer (*Scirpophaga incertulas*). *Proceedings National Academy of Sciences USA* 94: 2111–2116.

Novillo, C., Soto, J. and Costa, J. (1999). Results in Spain of varieties of cotton, genetically modified against bollworms. *Boletin de Sanidad Vegetal, Plagas* 25: 383–393.

Oerke, E.C., Dehne, H.W., Schonbeck, F. and Weber, A. (1994). *Crop Production and Crop Protection: Estimated Losses in Major Food and Cash Crops*. Amesterdam, The Netherlands: Elsevier.

Parker, R.D. and Huffman, R.L. (1997). Evaluation of insecticides for boll weevil control and impact on non-target arthropods on non-transgenic and transgenic B.t. cotton cultivars. *Proceedings, Beltwide Cotton Production and Research Conference*, 6–10 January, 1997, New Orleans, Louisiana, vol. 2. Memphis, Tennessee, USA: National Cotton Council, 1216–1221.

Peterhans, A., Schlupmann, H., Basse, C. and Pazkowski, J. (1990). Intrachromosomal recombination in plants. *EMBO Journal* 9: 3437–3445.

Pilcher, C.D., Rice, M.E., Higgins, R.A., Steffey, K.L., Hellmich, R.L., Witkowski, J., Calvin, D., Ostlie, K.R. and Gray, M. (2002). Biotechnology and the European corn borer: Measuring historical farmer perceptions and adoption of transgenic Bt corn as a pest management strategy. *Journal of Economic Entomology* 95: 878–892.

Poppy, G.M. and Sutherland, J.P. (2004). Can biological control benefit from genetically-modified crops? Tritrophic interactions on insect-resistant transgenic plants. *Physiological Entomology* 29: 257–268.

Porter, D.R., Burd, J.D., Shufran, K.A., Webster, J.A. and Teetes, G.L. (1997). Greenbug (Homoptera: Aphididae) biotypes: Selected by resistant cultivars or preadapted opportunists? *Journal of Economic Entomology* 90: 1055–1065.

Prols, F. and Meyer, P. (1992). The methylation patterns of chromosomal integration regions influence gene activity of transferred DNA in *Petunia hybrida*. *The Plant Journal* 2: 465–475.

Qaim, M. and Zilberman, D. (2003). Yield effects of genetically modified crops in developing countries. *Science* 299: 900–902.

Ramachandran, S., Buntin, G.D., All, J.N., Raymer, P.L. and Stewart, C.N. Jr. (2000). Intraspecific competition of an insect-resistant transgenic canola in seed mixtures. *Agronomy Journal* 92: 368–374.

Reed, G.L., Jensen, A.S., Riebe, J., Head, G. and Duan, J.J. (2001). Transgenic Bt potato and conventional insecticides for Colorado potato beetle management: Comparative efficacy and non-target impacts. *Entomologia Experimentalis et Applicata* 100: 89–100.

Riggin Bucci, T.M. and Gould, F. (1997). Impact of intraplot mixtures of toxic and nontoxic plants on population dynamics of diamondback moth (Lepidoptera: Plutellidae) and its natural enemies. *Journal of Economic Entomology* 90: 241–251.

Sachs, E.S., Benedict, J.H., Taylor, J.F., Stelly, D.M., Davis, S.K. and Altman, D.W. (1996). Pyramiding CryIA(b) insecticidal protein and terpenoids in cotton to resist tobacco budworm (Lepidoptera: Noctuidae). *Environmental Entomology* 25: 1257–1266.

Sanyal, I., Singh, A.K., Meetu, K. and Devindra, A.V. (2005). *Agrobacterium*-mediated transformation of chickpea (*Cicer arietinum* L.) with *Bacillus thuringiensis cry1Ac* gene for resistance against pod borer insect *Helicoverpa armigera*. *Plant Science* 168: 1135–1146.

Schell, J. (1997). Cotton carrying the recombinant insect poison Bt toxin: No case to doubt the benefits of plant biotechnology. *Current Opinion in Biotechnology* 8: 235–236.

Schnef, H.E. and Whiteley, H.R. (1981). Cloning and expression of *Bacillus thuringiensis* crystal protein gene in *Escherichia coli*. *Proceedings National Academy of Sciences USA* 78: 2893–2897.

Sharma, H.C. (1993). Host plant resistance to insects in sorghum and its role in integrated pest management. *Crop Protection* 12: 11–34.

Sharma, H.C., Arora, R. and Pampapathy, G. (2007). Influence of transgenic cottons with *Bacillus thuringiensis cry1Ac* gene on the natural enemies of *Helicoverpa armigera*. *BioControl* 52: 469–489.

Sharma, K.K., Lavanya, M. and Anjaiah, V. (2006). *Agrobacterium*-mediated production of transgenic pigeonpea (*Cajanus cajan* L. Millsp.) expressing the synthetic bt cry1Ab gene. *In Vitro Cellular & Developmental Biology – Plant* 42: 165–173.

Sharma, H.C. and Ortiz, R. (2000). Transgenics, pest management, and the environment. *Current Science* 79: 421–437.

Sharma, H.C. and Ortiz, R. (2002). Host plant resistance to insects: An eco-friendly approach for pest management and environment conservation. *Journal of Environmental Biology* 23: 111–135.

Sharma, H.C. and Pampapathy, G. (2006). Influence of transgenic cotton on the relative abundance and damage by target and non-target insect pests under different protection regimes in India. *Crop Protection* 25: 800–813.

Sharma, H.C., Sharma, K.K. and Crouch, J.H. (2004). Genetic engineering of crops for insect control: Effectiveness and strategies for gene deployment. *CRC Critical Reviews in Plant Sciences* 23: 47–72.

Sharma, H.C., Crouch, J.H., Sharma, K.K., Seetharama, N. and Hash, C.T. (2002a). Applications of biotechnology for crop improvement. *Plant Science* 163: 381–395.

Sharma, H.C., Sharma, K.K., Seetharama, N. and Crouch, J.H. (2002b). Development and deployment of transgenic plants with *Bacillus thuringiensis* genes for pest management. In Sashidhar Rao, Maruthi Mohan, P. and Subramanyam, C. (Eds.), *Developments in Microbial Biochemistry and Its Impact on Biotechnology*. Hyderabad, Andhra Pradesh, India: Department of Biochemistry, University College of Science, Osmania University, 25–47.

Sharma, H.C., Sharma, K.K., Seetharama, N. and Crouch, J.H. (2003). The utility and management of transgenic plants with *Bacillus thuringiensis* genes for protection from pests. *New Seeds Journal* 5: 53–76.

Singsit, C., Adang, M.J., Lynch, R.E., Anderson, W.F., Aiming, W., Cardineau, G. and Ozias-Akins, P. (1997). Expression of a *Bacillus thuringiensis* CryIA(c) gene in transgenic peanut plants and its efficacy against lesser cornstalk borer. *Transgenic Research* 6: 169–176.

Surekha, Ch., Beena, M.R., Arundhati, A., Singh, P.K., Tuli, R., Dutta-Gupta, A. and Kirti, P.B. (2005). *Agrobacterium*-mediated transformation of pigeonpea [*Cajanus cajan* (L.) Millsp.] using embryonal segments and development of transgenic plants for resistance against *Spodoptera*. *Plant Science* 169: 1074–1080.

Vaeck, M., Reynaerts, A., Hofte, H., Jansens, S., DeBeuckleer, M., Dean, C., Zabeau, M., Van Montagu, M. and Leemans, J. (1987). Transgenic plants protected from insect attack. *Nature* 327: 33–37.

Van der Salm, T., Bosch, D., Honee, G., Feng, I., Munsterman, E., Bakker, P., Stiekema, W.J. and Visser, B. (1994). Insect resistance of transgenic plants that express modified Cry1A(b) and Cry 1C genes: A resistance management strategy. *Plant Molecular Biology* 26: 51–59.

van Rensburg, J.B.J. (1999). Evaluation of Bt-transgenic maize for resistance to the stem borers *Busseola fusca* (Fuller) and *Chilo partellus* (Swinhoe) in South Africa. *South African Journal of Plant and Soil* 16: 38–43.

Vaucheret, H. (1993). Identification of a general silencer for 19S and 35S promoters in a transgenic tobacco plant: 90 bp of homology in the promoter sequence are sufficient for trans-inactivation. *C.R. Academic des Sciences de la vie Paris* 316: 1471–1483.

Verkerk, R.H.J., Leather, S.R. and Wright, D.J. (1998). The potential for manipulating crop-pest-natural enemy interactions for improved insect pest management. *Bulletin of Entomological Research* 88: 493–501.

Williams, W.P., Sagers, J.B., Hanten, J.A., Davis, F.M. and Buckley, P.M. (1997). Transgenic corn evaluated for resistance to fall armyworm and southwestern corn borer. *Crop Science* 37: 957–962.

Wilson, W.D., Flint, H.M., Deaton, R.W., Fischhoff, D.A., Perlak, F.J., Armstrong, T.A., Fuchs, R.L., Berberich, S.A., Parks, N.J. and Stapp, B.R. (1992). Resistance of cotton lines containing a *Bacillus thuringiensis* toxin to pink bollworm (Lepidoptera: Gelechiidae) and other insects. *Journal of Economic Entomology* 85: 1516–1521.

Wood, E.A. Jr. (1971). Designation and reaction of three biotypes of the greenbug cultured on resistant and susceptible cultivars of sorghum. *Journal of Economic Entomology* 64: 183–185.

Wu, Y.R., Llewellyn, D., Mathews, A. and Dennis, E.S. (1997). Adaptation of *Helicoverpa armigera* (Lepidoptera: Noctuidae) to a proteinase inhibitor expressed in transgenic tobacco. *Molecular Breeding* 3: 371–380.

Wu, Z.B., Yang, Y.H., Liu, X.L., Shi, Z.G. and Chen, P. (2003). Screening of new combinations of resistant hybrid cotton to insect pests and evaluation of heterosis. *Acta Phytophylacica Sinica* 30: 243–249.

Xiang, Y., Wong, W.K.R., Ma, M.C. and Wong, R.S.C. (2000). *Agrobacterium*-mediated transformation of *Brassica campestris* ssp. *parachinensis* with synthetic *Bacillus thuringiensis* Cry1A(b) and Cry1A(c) genes. *Plant Cell Reporter* 19: 251–256.

Xu, L.H., He, X.H., Li, G.F. and Yang, D.Y. (2000). Bolling regularity of transgenic insect-resistant cotton lines "Sukang 102" and "Sukang 310." *Jiangsu Journal of Agricultural Sciences* 16: 208–211.

Xue, M., Dong, J. and Zhang, C.S. (2002). Effect of feeding Bt cotton and other plants on the changes of development and insecticide susceptibilities of lesser armyworm *Spodoptera exigua* (Hubner). *Acta Phytopathologica Sinica* 29: 13–18.

11

Transgenic Resistance to Insects: Interactions with Nontarget Organisms

Introduction

Plant resistance to insects has evolved over millions of years, and considerable progress has been made in development and deployment of crop plants with resistance to insects. Counter adaptation by the herbivores to the resistance mechanisms, and the response of the natural enemies to the insect-resistant plants and the insect host determine the outcome of such a complex interaction (Sharma and Waliyar, 2003). Such an interaction is also influenced by the abiotic environmental factors. The intentional selection and breeding for resistance to insects has led to the development of insect-resistant cultivars in several crops and as a result the need to apply insecticides for insect control has been reduced considerably. Insect-resistant cultivars are in general compatible with the natural enemies, although plant resistance at times can have detrimental affects on the activity and abundance of natural enemies. The latter has not received any adverse publicity, as the adverse effects of insect-resistant cultivars on the nontarget organisms are still far less than those of broad-spectrum insecticides used for pest management, which may result in complete elimination of the natural enemies of crop pests.

Significant progress has been made over the past two decades in handling and introduction of novel genes into crop plants, and has provided opportunities to increase yields, impart resistance to biotic and abiotic stress factors, and improve nutrition. In addition to widening the pool of useful genes, genetic engineering has also allowed the use of several desirable genes in a single event, thus reducing the time required to introgress novel genes into the elite background. Genes from bacteria such as *Bacillus thuringiensis* (Berliner) (*Bt*) have been used successfully for pest control through transgenic crops on a commercial scale (Hilder and Boulter, 1999; Sharma et al., 2000; Shelton, Zhao, and Roush, 2002). The *Bt* toxins are environmentally benign (Lambert and Peferoen, 1992), and their use in transgenic plants will avoid numerous hazards associated with the use of synthetic insecticides in insect control. Protease inhibitors, plant lectins, ribosome inactivating proteins, secondary plant metabolites, vegetative insecticidal proteins, and small RNA viruses can also be deployed through the transgenic plants alone or in combination with *Bt* genes for crop protection (Sharma et al., 2000; Sharma and Pampapathy, 2006).

Introduction of transgenic crops will lead to an increase in crop productivity, and result in preservation of natural habitats because less land may be used for agriculture. A number of ecological and economic issues need to be addressed while considering the production and deployment of transgenic crops for insect control (Sharma and Ortiz, 2000). The most important consideration is the immediate reduction in the amount of pesticides applied for pest control. The number of pesticide applications on a crop such as cotton varies from 10 to 25, and most of the sprays are directed against key pests such as *Heliothis virescens* F., *Helicoverpa zea* (Boddie), and *H. armigera* (Hubner). In situations where transgenic crops are introduced as a component of pest management, the number of pesticide sprays is likely to be reduced by two third to half. Reduction in number of pesticide sprays would lead to increased activity of natural enemies, while some of the minor pests may tend to attain higher pest densities in the absence of sprays applied for the control of major pests, for example, plant hoppers, *Amrasca biguttula biguttula* Ishida, and stink bug, *Nezara viridula* (L.), on *Bt*-cotton (Bundy, McPherson, and Herzog, 2000; Sharma and Pampapathy, 2006).

The deployment of insect-resistant transgenic plants for increasing the production and productivity of crops for sustainable crop production has raised some concerns about the real or conjectural effects of transgenic plants on nontarget organisms, including human beings (Miller and Flamm, 1993; Sharma et al., 2002), and evolution of resistant strains of insects (Williamson, Perrins, and Fitter, 1990). As a result, the caution has given rise to doubt because of lack of adequate information. However, genetically modified organisms have a better predictability of gene expression than conventional breeding methods, and transgenes are not conceptually different than the use of native genes or organisms modified by conventional technologies. Historically, crop plants have not been subjected to risk or safety analysis or risk management, and are improved by cross-pollination between plants with desirable traits, or with species that are sexually compatible.

Efficacy of transgenic crops for controlling the target and the nontarget pests, and their effects on natural enemies need to be determined on a regional basis. The significance of such effects would depend on the importance of the immature stages of the target insect for maintaining the populations of the natural enemies. Transgenic plants may reduce the numbers of certain natural enemies in areas planted with transgenic crops, but their populations may be maintained on the other crops that serve as a host to the target pests. A few of the known predators are specialists on one insect and hence, the populations of the predators and parasitoids with a wide host range would be maintained on other insect species (Fitt, Mares, and Llewellyn, 1994; Jasinski et al., 2001; Dhillon and Sharma, 2007). Within-field impact may be greater for parasitoids that are monophagous, and populations of such natural enemies can only be maintained on the nontransgenic crops or other hosts of the target pest. Finally, the effects of transgenic crops on the abundance of natural enemies should be compared with the nontransgenic fields of the same crop where the natural enemies may be virtually absent because of heavy pesticide application.

Transgenic plants released for commercial use must contain information on environmental safety assessment. To generate such information, studies need to be conducted under laboratory and field conditions, following standard ecotoxicological procedures used for insecticides. Ecotoxicological studies on insecticides are designed based on their mode of action, mode of application (foliar sprays drift to neighboring plants and also affect nontarget organisms), and their presence (residues) in the environment (for a limited period of time due to rapid degradation). However, such studies may be insufficient to measure the ecological effects of transgenic plants on nontarget organisms. Endogenous production of *Bt* toxins in all parts of the plant throughout the crop-growing season may lead to a different mode of exposure of herbivores, and thus their parasites and predators. Also, altered metabolism of plants (by virtue of production of toxin protein or insertion of the novel gene) may also affect the interaction between the plants, insects, and their natural enemies. Therefore, there is a need for long-term studies involving insect-resistant transgenic plants, including lethal, sublethal, and chronic effects in nontarget organisms.

Bt Sprays, Transgenic Plants, and Nontarget Organisms

The *Bt* spores and toxins are degraded quickly. Under UV light, the toxins decay within a few hours to a few days, while the spores need soil for germination (Schutte and Riede, 1998). In comparison, the toxin proteins in transgenic plants are produced throughout the crop cycle in all parts of the plant. Formulations based on *Bt* have shown toxic effects against some nontarget insects (Wagner et al., 1996), but only a few of the parasitoids and predators may be directly harmed by *Bt* sprays (Flexner, Lighthart, and Croft, 1986; Deml and Dettner, 1998). Application of *Bt* as foliar sprays does not cause any harmful effects to microorganisms or to vertebrates (Kreutzweiser et al., 1996). *Bt* sprays have shown negative effects against 24 species each of parasites and predators, but no adverse effects have been observed against 78 predators and 87 parasitoids (Deml and Dettner, 1998). Indirect effects through nutritional quality of the prey have been observed on 12 predators and eight parasitoids. High levels of *Bt* toxicity have been observed only against two species, but low levels of toxicity have been observed against 29 species. Toxins from *B. thuringiensis* var. *tenebrionis* have shown no effects against 9 of the 11 species examined, but high mortality and reduced fecundity have been observed in two species. Under field conditions, Newleaf (Cry3Aa) potato plants have been found to be highly effective in suppressing populations of Colorado potato beetle, *Leptinotarsa decemlineata* (Say) and provided better control than weekly sprays of a microbial *Bt*-based formulation containing Cry3Aa, without reducing the number of the generalist predators, and thus is compatible with integrated pest management (IPM) programs (Reed et al., 2001).

Transgenic plants express semiactivated proteins, which are ingested by the insects. In this process, no crystal solubilization and protoxin-toxin conversion is necessary in the insect midgut. Increased activity has been observed in some herbivorous insects when fed with toxin instead of the protoxin, suggesting that such insects lack the appropriate enzyme composition or pH that converts the protoxin to toxin (Moar et al., 1995). The truncated and activated toxins expressed in transgenic plants constitutively in high doses kill the insects much faster than the *Bt* sprays, which need to be activated by the insect enzymes. Arthropod species that have the binding sites but cannot activate the toxins could be harmed by the toxins from the transgenic plants. However, *Bt* toxins can be activated in different ways, leading to a change in their specificity (Haider, Knowles, and Ellar, 1986). While *Bt* sprays might affect all the insects in an ecosystem wherever they are sprayed, the toxins from the transgenic plants affect only the insects that ingest the toxins from the plants through feeding or indirectly through the insect host or prey (Figure 11.1). Therefore, toxicity assays from *Bt* sprays cannot be substituted for the toxins expressed in transgenic plants. The *Bt* protein molecule is quite complex, and its mode of action on nontarget organisms is not understood properly. Little information is available on mode of action in

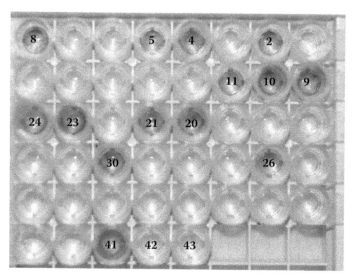

FIGURE 11.1   Detection of *Bt* toxins in arthropods in transgenic cotton under field conditions using ELISA. High levels of *Bt* toxin were detected in ash weevil (cell numbers 10 and 24) and grasshoppers (23 and 30); while low levels of *Bt* toxins were detected in spiders (2 and 20), leafhoppers (4 and 26), green bug (5), red cotton bug (8), larvae of tobacco caterpillar (11), and dusky cotton bug (21). 41 = Positive control. 42 = Negative control. 43 = Blank.

insensitive insect species, and higher trophic level insects such as natural enemies. Such information is important in the ecological context, particularly with the deployment of transgenic plants expressing high levels of *Bt*-toxins constitutively.

Interaction of Transgenic Crops with Nontarget Organisms: Protocols for Ecotoxicological Evaluation of Transgenic Plants

Ecologically, risk is the result of exposure hazard, which is a function of concentration, exposure time, and distribution. In the case of *Bt*-insecticides, the toxins are applied topically, whereas the toxins are expressed constitutively in all plant parts in the transgenic plants, albeit at different concentrations. Therefore, in addition to the chewing type of insects, the sap sucking and tissue boring insects are also exposed to the toxin proteins. The *Bt*-insecticides degrade rapidly in the field due to UV light, and their persistence in the environment is low (Behle, McCurie, and Shasha, 1997). However, constitutive expression of *Bt* toxins in transgenic plants exposes the insects to the toxin proteins throughout the life cycle, and across generations. Extended temporal and spatial expression results in greater exposure, and possibly may present a greater risk. As a result, all insects colonizing a crop during the growing season may ingest some amount of toxin proteins, a proportion of which may pass on to the natural enemies in the processed form.

Therefore, there is a need to standardize the protocols for ecotoxicological testing of transgenic plants on nontarget organisms, including higher trophic level organisms. Such studies need to take into account the novel and extended route of exposure caused by constitutive expression of the toxin protein in transgenic plants. The protocols should include

TABLE 11.1

Assessment of Biosafety of Transgenic Crops to Nontarget Organisms

Assessment of Biosafety of Transgenic Crops	Organisms/Tests
Which nontarget organism should be included in the assessment?	Coccinellids, lacewings, hymenopteran parasitoids, pollinators, bacteria, fungi, and earthworms in the soil
What kind of nontarget tests should be carried out?	Acute and chronic toxicity effects
How should the biosafety tests be carried out?	Laboratory, contained field trials, and long-term field trials

testing for chronic, lethal, sublethal, and acute toxicity effects (Table 11.1). Basing ecological studies solely on bitrophic studies that provide the highly processed toxin protein directly to the nontarget organisms may not be sufficient as such studies may not provide information on synergistic interaction between plants, insects, and the natural enemies. Routes of exposure should include the natural situation, where the toxin is actually entering the food, and then delivered to the higher trophic level. Test organisms should include ecologically relevant species, and consider all the multitrophic interactions involved (Table 11.2). There is a need for detailed information on Hymenoptera, which also includes honeybees, bumble bees, and wild bees because of their role in pollination. Diptera and Coleoptera also contain some important parasites and predators, and it is crucial to have information on the effects of transgenic plants on these insects. It is important to differentiate between different life stages as only the larval stages act as parasites or predators in several species, while the adults feed on pollen and nectar. In some species, such as coccinellids, both adults and larvae feed on the herbivore insects. There is a need to generate information on predator-prey ratio, and population dynamics of the insect host and its prey. Such information is essential for learning about existing biological control mechanisms, their effectiveness, ecological effects, and potential changes in the future. First, the attention should be focused on important species. If important species have been identified, a tiered testing procedure from laboratory to field should be followed to assess the risk that the deployment of a specific transgenic crop would pose to key nontarget organisms, similar to the testing procedure used in ecotoxicology (Dutton, Romeis, and Bigler, 2003a) (Table 11.3).

Use of insecticide treatments often leads to significant differences in treatment effects. However, the potential adverse effects of transgenic plants on beneficial insects are likely to be more sublethal and chronic in nature, and the long-term intergenerational effects on population dynamics in the field may persist over many years. Quite often, on becoming adults, the insects migrate away from the field in search of food, mating, and oviposition. Different habitats provide different sets of conditions for the insects. One-year small plot studies may not provide adequate information on such complex interactions. Therefore, long-term monitoring should be carried out wherever transgenic crops are deployed.

TABLE 11.2

Nontarget Organisms Likely to Get Exposed to Transgenic Crops under Natural Conditions

Category of Organisms	Test Organisms
Species of ecological importance in the ecosystem	Natural enemies of crop pests
Species of economic importance in the ecosystem	Pollinators
Species likely to get exposed to insecticidal proteins	Nontarget herbivores and natural enemies
Species that are rare or endangered	Protected Lepidoptera, Collembola, etc.

TABLE 11.3

Protocols for Assessment of Biosafety of Transgenic Plants to Nontarget Organisms

Test	Effect	Action
Tier I: Host/prey ingests the toxin	**No**	No need for further tests
Yes		
Tier I: Host/prey sensitive to toxin	**No**	No need for further tests
Yes		
Tier II: Parasitoid/predator ingest the toxin	**No**	Host prey-mediated effects expected; need to be assessed in transgenic plants
Yes		
Tier II: Parasitoid/predator sensitive to toxin	**No**	Host prey-mediated effects expected; need to be assessed in transgenic plants
Yes		
Tier III: Direct and prey-mediated effects expected, but not separable		Direct and prey-mediated effects need to be assessed in transgenic plants under contained and open field trials for a long period, and outcomes compared with current pest control practices

The risk posed by transgenic plants to entomophagous arthropods also depends on exposure and their sensitivity to the insecticidal protein. It is therefore essential to determine if and at what level the organisms are exposed to the transgene (Dutton, Romeis, and Bigler, 2003b). Exposure is associated with the feeding behavior of phytophagous and entomophagous arthropods and temporal and spatial expression of the insecticidal protein. For organisms that are potentially exposed to the insecticidal protein, sensitivity tests should be performed to assess the toxicity. The route of exposure (fed directly or via herbivorous prey) and the origin of the *Bt* (from transgenic plants or incorporated into artificial diet) strongly influence the degree of mortality (Hilbeck, 2001). In choice feeding trials where *Chrysoperla carnea* (Stephan) could choose between nontransgenic maize- and transgenic *Bt*-maize-fed *Spodoptera littoralis* (Boisd.), the later instars showed a significant preference for nontransgenic maize-fed *S. littoralis*, while this was not the case when the choice was between *Bt*- and isogenic maize-fed aphids. Partially or moderately resistant plants expressing quantitative rather than single gene traits and affecting the target pest sublethally may provide a more meaningful assessment of the effects of transgenic plants on nontarget organisms. Transgenic plants with less severe nontarget effects may allow for better incorporation of such plants in integrated pest management or biological control programs using multiple control strategies, and thus reducing the selection pressure for development of resistance to the transgene in pest populations (Hilbeck, 2001). Cowgill and Atkinson (2003) suggested a sequential approach for testing the effects of protease inhibitor-expressing crops on nontarget herbivorous insects. For the first four tiers of the approach, potatoes expressing cystatins (cysteine proteinase inhibitors) were used as an example. Although the plants had high levels of resistance to potato cyst nematodes, *Globodera pallida* (Stone) and *Globodera rostochiensis* (Wollen.), the results showed that they have negligible impact on the nontarget herbivorous insect, *Eupteryx aurata* (L.) Skar. The approach consisted of five tiers: (1) conduct field surveys to characterize nontarget invertebrate fauna of a crop, (2) perform histochemical assays to identify a subset of herbivores with a particular type of digestive proteolytic enzymes, (3) conduct controlled environment or small-scale field trials, (4) evaluate the impact of the protease inhibitor-expressing plants on the selected nontarget species, and (5) conduct field trials to compare the relative effect of transgenic plants and current management practices such as pesticide use on selected species.

Influence of Transgenic Crops on Diversity of Nontarget Organisms

One of the major concerns of transgenic crops is their effects on nontarget organisms, about which little is known at the moment. Herbivores, detritivores, and many of their predators and parasitoids in arable systems are sensitive to changes in the environment, including weed communities that result from the introduction of genetically modified herbicide-tolerant crops (Sharma et al., 2002; Hawes et al., 2003). The information that use of genetically modified corn may have toxic effects on larvae of the monarch butterfly, *Danaus plexippus* (L.) (Losey, Raynor, and Carter, 1999; Obrycki et al., 2001), has generated a huge amount of public debate. However, several studies have subsequently revealed that the impact of the current *Bt* maize varieties on monarch butterfly populations is negligible (Sears et al., 2001; Gatehouse, Ferry, and Raemaekers, 2002). Wraight et al. (2000) concluded that there is no relationship between mortality of *Papilio polyxenes* Fab., and pollen deposition from transgenic maize on its host plants. Pollen from the transgenic plants failed to cause any mortality under laboratory conditions. An assessment of the impact of *Bt* cotton pollen on two important economic insects, the Chinese tussah silkworm, *Antheraea pernyi* (Guer.) and the mulbery silkworm, *Bombyx mori* L., has also been conducted, and it was concluded that the adverse effect is negligible (Wu, Peng, and Jia, 2003). Transgenic pollen prolonged duration of first-instar larvae of silkworm, *B. mori*, compared with the nonpollen treatment, but were not significantly different from those fed with pollen from nontransgenic cotton or maize (Li et al., 2002). Body weight of the third-instars fed on transgenic pollen was significantly more than that of the controls. Consequently, it was considered that pollen from transgenic insect-resistant cotton and maize has negligible adverse effects on the growth and development of the silkworm.

There were no differences in abundance of aerial fauna, including Aphididae, Cicadellidae, Araneae, and Coleoptera (Chrysomelidae, Coccinellidae, and Staphylinidae), and other insects living in close contact with these, or with the European corn borer, *Ostrinia nubilalis* (Hubner), such as hymenopteran parasitoids, and Syrphidae in isogenic transgenic maize (Lozzia, 1999). The biodiversity of the carabid communities was quite low for both types of maize and no significant differences were found in any of the indices analyzed. *Bt*-maize has no adverse effects on nontarget arthropod communities (Bourguet et al., 2002). Numbers of nontarget arthropods such as aphids [*Metopolophium dirhodum* (Walker.), *Rhopalosiphum padi* (L.), and *Sitobion avenae* (F.)], bug, *Orius insidiosus* (Say), syrphid, *Syrphus corollae* (Fab.), ladybird beetle, *Coccinella septempunctata* L., lacewing, *C. carnea*, thrips, and hymenopteran parasitoids did not differ between the *Bt* and non-*Bt* maize. However, thrips were more abundant on *Bt* maize at one site than in non-*Bt* maize (Bourguet et al., 2002). A trend towards a community effect on flying arthropods has been observed, with lower abundance of adult Lepidoptera, flies in the families Lonchopteridae, Mycetophilidae, and Syrphidae; and the hymenopteran parasitoids belonging to Ceraphronidae (Candolfi et al., 2004). The effects were weak and restricted to two sampling dates corresponding to anthesis. However, a significant reduction was observed on the community of plant dwellers with insecticides, such as Karate, Xpress, and Delfin. Karate and Xpress showed a prolonged effect on soil dwellers (Candolfi et al., 2004). Larvae of

Coleomagilla maculata (DeGeer), *O. insidiosus,* and *C. carnea* (except those of *C. maculata*) did not complete development on pollen diet, although pollen was provided to larvae for a limited period of time (Pilcher et al., 1997). However, pollen contains relatively low amounts of *Bt* toxin, and the larvae generally do not feed on pollen under natural conditions. The concentration of Cry1Ab in pollen is very low (2.57 to 2.94 µg per gram dry weight) as compared with that of the leaves (Koziel et al., 1993; Fearing et al., 1997). Only the adults feed on pollen to supplement their diet for ovary maturation.

Diversity of the arthropod community in transgenic cotton has been found to be similar to that in conventional cotton without spraying, but the Shannon's index for total arthropod community and neutral insect subcommunity in *Bt* cotton fields was significantly higher than in sprayed plots at the mid- and late-growing stages of cotton (Li et al., 2003; Men, Ge, and Liu, 2003). *Bt* cotton may increase the numbers and stability of arthropods in a cotton ecosystem and improve the management of insect pests. The diversity of the arthropod community in *Bt* cotton fields is greater than that in conventional cotton, suggesting that *Bt* cotton is highly favorable for integrated management of cotton pests (Wu, Peng, and Jia, 2003). However, Cui and Xia (2000a, 2000b) observed that *Bt* transgenic cottons resulted in a decrease in species abundance and numbers of individual species. In the absence of cotton bollworm, *H. armigera* on transgenic cotton (the dominant pest in cotton), the red spider mite and thrips become the dominant pests. *Bt* cotton was effective in controlling *H. armigera*, but the populations of *Aphis gossypii* Glover, *Thrips tabaci* Lind., and *Lygus lucorum* (Mayer-Dur) increased in *Bt* cotton fields under natural or chemical control compared to that in the normal cotton fields (Sun et al., 2002). The population of *Chrysopa formosa* Brauer increased in transgenic cotton fields, but that of predators such as *Orius minutus* (L.), *Deraeocoris punctulatus* (Fallen), and several parasitoids decreased in transgenic cotton fields. Seed treatment of transgenic cotton with imidacloprid resulted in an increase in numbers of coccinellid beetles, *C. septempunctata* and *Cheilomenes sexmaculatus* (Fab.); green lace wing, *C. carnea*; Lynx spider, *Oxyopes javanus* Thorell; orb spider, *Argiope minuta* (Karsh); wolf spider, *Lycosa pseudoannulata* (Boes. and Str.); long-jawed spider, *Tetragnatha javana* (Thorell); *Neoscona theisi* (Walch.); and *Peucetia viridana* (Stal) (Kannan, Uthamasamy, and Mohan, 2004). The populations of nontarget insect pests were significantly greater in transgenic cotton than those in normal cotton plots. Species diversity and numbers varied between bivalent (*cry1Ac + CpTi*) and univalent (*cry1Ac*) transgenic plots (Sun et al., 2003). The population of *A. gossypii, A. biguttula biguttula,* and *Bemisia tabaci* (Genn.) was lower by 33.0%, 50.6%, and 22.7%, in bivalent cotton than in univalent cotton, respectively. However, the population of *T. tabaci* and *L. lucorum* was higher by 208.9 and 18.4%, respectively. The numbers of leafhoppers, *A. biguttula biguttula*; grey weevil, *Myllocerus* sp.; red cotton bug, *Dysdercus keonigii* F.; and aphids, *A. gossypii* have been found to be similar between the *Bt* transgenic and nontransgenic cotton (Sharma and Pampapathy, 2006). However, damage by the bollworms, *H. armigera* and *Earias vittella* (F.) was significantly lower in transgenic cotton compared to nontransgenic cotton. There were no apparent effects of transgenic cotton on the relative abundance of predatory spiders, *Clubiona* sp. and *Neoscona* sp., coccinellid, *C. sexmaculatus* and the chrysopid, *C. carnea* (Figure 11.2). However, the abundance of spiders, coccinellids, and chrysopids was very low in insecticide-protected plots toward the end of the cropping season (Sharma, Pampapathy, and Arora, 2007).

There were no differences in trap captures of major ground-dwelling coleopteran predators such as carabids and staphylinids between *Bt*-and non-*Bt* potato fields treated with weekly sprays of a microbial *Bt*-based formulation containing Cry3Aa, biweekly applications of permethrin, early and mid-season in-furrow applications of systemic insecticides

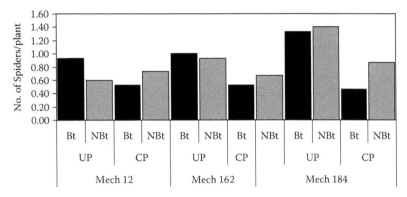

FIGURE 11.2 Abundance of predatory spiders and coccinellids on *Bt* transgenic and nontransgenic cottons under protected and unprotected conditions in the field. UP, unprotected and CP, completely protected.

(phorate and disulfoton), and untreated control (Duan et al., 2004). However, weekly sprays of permethrin significantly reduced the trap capture of ground-dwelling spiders (Araneae), while there were no significant differences in the capture of spiders between *Bt* and non-*Bt* potato fields treated with *Bt* sprays, systemic insecticides, and untreated control. Significantly more springtails were captured in potato fields treated with permethrin than in any other treatment regime.

Influence of Transgenic Crops on Activity and Abundance of Pollinators

The occurrence of the transgene product in nectar and pollen, as well as the foraging behavior on transgenic plants, should be considered while studying the effects of transgene or its products under laboratory and field conditions (Malone and Pham-Delègue, 2001). In addition, indirect effects related to secondary changes in plant signals should also be examined, where the aim is to assess the risk of transgene dispersal in the environment (Pierre and Pham-Delègue, 2000). There are no significant effects of *Bt*-transgenic crops on the honeybee, *Apis mellifera* L., and bumblebees, *Bombus* sp. (Malone and Pham-Delègue, 2001). Activated *Bt* toxin mixed with water or honey did not show any adverse effects on larvae and adults. Arpaia (1996) supplied CryIIIB toxin in supplemental syrup to *A. mellifera* colonies at two concentrations (400 and 2000 times higher than the expected protein content in pollen from *Bt*-transgenic plants), and did not observe any toxic effect on the larvae. There was no effect on pupae, indicating that transgenic crops producing CryIIIB toxin poses no risk to the pollinators.

However, a number of studies have revealed direct toxic effects of purified serine-type protease inhibitors (PIs) on adult honeybees when fed at high concentrations under laboratory conditions. These PI's include BBI (Bowman-Birk inhibitor) (Pham-Delègue et al., 2000), BPTI (basic pancreatic inhibitor) (Malone et al., 1995), Pot-I and Pot-II (Malone et al., 1998), and SBTI (soybean trypsin inhibitor) (Malone et al., 1995; Burgess, Malone, and Christeller, 1996; Pham-Delègue et al., 2000). In contrast, ingestion of high doses of purified Cystatin or OC-I from rice, belonging to the cystein-type inhibitors, did not cause any effect on bee survival (Girard et al., 1998). Trypsin inhibitor and WGA do not exhibit acute toxicity towards *A. mellifera* (Belzunces et al., 1994). *In vivo*, trypsin inhibitor and WGA caused a decrease in the amount of trypsin activity, but did not have a significant effect on esterase activity. *In vitro*, trypsin inhibitor inhibited about 80% of nonspecific protease activity and 100% of trypsin activity. Serine proteases, but not the cystein proteases, play a major role during protein degradation in honeybee (Jimenez and Gilliam, 1989). In addition to direct toxic effects, ingestion of PIs can affect the learning performance of the honeybee. Such a detrimental effect has been reported for the serine-type PIs such as BBI (Pham-Delègue et al., 2000) and CpTI (Picard-Nizou et al., 1997). However, cystein-type PIs have no effect on the insects' learning performance (Girard et al., 1998; Pham-Delègue et al., 2000).

Genetically modified oilseed rape, *Brassica napus* L., expressing heterologous chitinase in somatic tissue for enhanced disease resistance did not cause any adverse effects on honeybee, *A. mellifera* foraging behavior (Picard-Nizou et al., 1995). Insects do not discriminate between conventional and transgenic oilseed rape resistant to glufosinate (Pierre et al., 2003). The diversity and density of the foraging insects has been found to be similar on the transgenic and nontransgenic cultivars, as was the foraging behavior. Honeybees flew indifferently across these genotypes and no differences have been observed in nectar and pollen between the transgenic and nontransgenic plants. Thus, we may assume that the transgenic crops do not pose a major threat to the activity and abundance of pollinators.

Interaction of Transgenic Crops with Predators

Plant breeders have continued to breed for high levels of resistance to insects, and biocontrol specialists have ignored the role of the host plant in ensuring successful foraging by natural enemies. The new possibilities for controlling insect pests, which will combine both "nature's" own defenses with man's ingenuity, may stack the odds in favor of natural enemies and against insect pests (Poppy and Sutherland, 2004). Recombinant DNA technology will allow plants to be designed that are well suited for use along with biological control. However, transgenic crops have to be assessed for their effects on the environment, including the possible impact on nontarget arthropods, many of which are important for biological control of insect pests (Romeis et al., 2004). A framework for risk assessment on transgenic crops to nontarget arthropods has been proposed by Romeis, Dutton, and Bigler (2004). As a first step, it is important to determine which entomophagous arthropods play a major role in regulating the pest populations in an agroecosystem, and which ones may be at risk. Because the risk that transgenic plants pose to entomophagous arthropods depends on both their exposure and sensitivity to the insecticidal proteins, it is essential to determine if, and at what level, organisms are exposed to the transgene product. Exposure will be associated with the feeding behavior of phytophagous and entomophagous arthropods and the tissue- and cell-specific temporal and spatial expression of the insecticidal proteins. For organisms that are potentially being exposed to the insecticidal protein, sensitivity tests should be performed to assess the toxicity. The testing procedure adopted should depend on the feeding behavior under natural conditions.

It is difficult to assess the effects of transgenic plants on the abundance of generalist predators, as their populations fluctuate in repeat cycles of several generations. The test plots have to be quite large (up to 1 ha or more) to measure such effects on mobile insects such as lacewings, coccinellids, wasps, etc. Of the nontarget insects, the generalist predators are less exposed to the transgene product as it is likely that not all of the prey will be contaminated. Therefore, it is necessary to differentiate them from the host-specific parasitoids, which are more likely to be affected by the toxins if the insect host acquires the toxins from the plants.

Synergistic/Neutral Interactions

Bt *Toxins*

Predator levels in *Bt* cotton fields have been found to be significantly greater than those in conventional cotton fields where insecticides were used for controlling the bollworm, *H. armigera* (Wu, Peng, and Jia, 2003). As the predator population increases, the outbreak of cotton aphid, *A. gossypii* in mid-season is controlled effectively, while the mirids become a major pest in *Bt* cotton because of reduced number of insecticide applications against *H. armigera. Campylomma diversicornis* Reuter, a dominant predator on eggs and newly hatched larvae of *H. armigera*, plays an important role in controlling insect pests in transgenic *Bt* cotton fields (W.X. Liu et al., 2000).

No major differences have been observed in the abundance of generalist predators in fields with transgenic and nontransgenic crops (Hoffmann et al., 1992; Flint et al., 1995; Luttrell et al., 1995; Sims, 1995; Orr and Landis, 1997; Wang and Xia, 1997). Cui and Xia (1999) did not observe a significant increase in populations of predatory arthropods in the *Bt* cotton field, except that of *Propylea japonica* (Thun.). There was no apparent effect of transgenic cotton on the relative abundance of predatory spiders, *Clubiona* sp. and *Neoscona* sp., coccinellid, *C. sexmaculatus*, and the chrysopid, *C. carnea*. (Sharma, Pampapathy, and Arora, 2007). In *Bt* transgenic maize, temporal abundance of nontarget arthropods (bug, *O. insidiosus*; the syrphid, *S. corollae*; the ladybird, *C. septempunctata*; and the lacewing, *C. carnea*) varied greatly over the season, but did not differ between *Bt* and non-*Bt* maize (Bourguet et al., 2002). None of the coccinellids (three taxa), carnivorous carabids (three taxa), and ants were affected by *cry3A*-transgenic potatoes resistant to the Colorado potato beetle, *L. decemlineata. Orius insidiosus* was more abundant in transgenic fields than in nontransgenic fields in one year, while spiders were abundant in another year. The predator activity is not affected by pure transgenic and mixed seed potato crop (Riddick, Dively, and Barbosa, 1998). However, significant effects on predator numbers were observed because of decline in the numbers of the insect hosts.

There are no adverse effects of *Bt* maize on pre-imaginal development or mortality of *C. carnea* when reared on *R. padi* that had fed on *Bt* maize (Lozzia et al., 1998). Since there is no *Bt* toxin in the phloem as established by Raps et al. (2001), the aphids fed to the predators may have no *Bt* toxin, and thus no adverse effects of the transgenic plants could be measured. Dutton et al. (2002) observed that *C. carnea* larvae were not affected when fed spider mites, *Tetranychus urticae* Koch, that had been reared on *Bt* maize, even though the spider mites were found to contain more *Cry1Ab* toxin as compared to the lepidopteran larvae. Third instars of *C. carnea* showed a significant preference for *S. littoralis* fed nontransgenic maize. However, no preference was displayed when *C. carnea* had the choice between *R. padi* fed on transgenic or nontransgenic maize. This lack of preference for *R. padi* fed on either transgenic or nontransgenic maize has been ascribed to the absence of the *Bt* protein in the phloem. In prey combinations with *S. littoralis* and *R. padi*, all three larval stages of *C. carnea* showed a preference for *R. padi* regardless of whether they had fed on transgenic or nontransgenic maize (Meier and Hilbeck, 2001). Romeis, Dutton, and Bigler (2004) suggested that *C. carnea* larvae are not sensitive to Cry1Ab, and that earlier reported negative effects of *Bt* maize were prey-quality mediated rather than direct toxic effects. No statistically significant effects were observed on survival, aphid consumption, development, or reproduction in *Hippodamia convergens* (Guerin-Meneville) fed on *Myzus persicae* (Sulzer) reared on potatoes expressing δ-endotoxin of *B. thuringiensis* subsp. *tenebrionis* (Dogan et al., 1996). Presence of the *cry1Ab* gene showed no marked effects on predation by the wolf spider, *Pirata subpiraticus* (Boes. and Str.) on the rice leaf folder, *Cnaphalocrocis medinalis* (Gn.) and the brown planthopper, *Nilaparvata lugens* (Stal) (Z.C. Liu et al., 2003). No significant differences were observed in insect mortality from egg hatch to adult eclosion, and total developmental period in *Orius majusculus* (Reuter) nymphs fed on *Anaphothrips obscurus* (Muller), a thysanopteran pest of maize, reared on isogenic and transgenic maize plants expressing *cry1Ab* (Zwahlen et al., 2000).

Lectins

Aphids fed on artificial diet containing GNA showed no deleterious effects on adult longevity, but resulted in a consistent trend for improved fecundity of the ladybird beetle, *H. convergens*. Egg production increased by 70%, suggesting that GNA is not deleterious to ladybirds (Down et al., 2003). About 50% of the potential activity of Oryzacystatin I (OCI) has been recovered in extracts of *Perillus bioculatus* Fab. feeding on *L. decemlineata* reared on OCI-potato foliage, indicating that the predator was sensitive to OCI (Bouchard, Michaud, and Cloutier, 2003). However, *P. bioculatus* survived on OCI prey and developed normally, indicating its ability to compensate prey-mediated exposure to the OCI inhibitor. Confinement of *P. bioculatus* to potato foliage provided no evidence that potato plant-derived nutrition is a viable alternative to predation. Restriction to potato foliage was inferior to free water for short-term survival of nonfeeding first-instar larvae, suggesting that OCI, an effective inhibitor of a fraction of digestive enzymatic potential in *P. bioculatus*, may not interfere with its predation potential when expressed in potato plants fed to its prey at a maximum level of 0.8% of total soluble proteins in mature foliage (Bouchard, Michaud, and Cloutier, 2003). No significant effects have been observed on development, survival, and progeny production of ladybird larvae fed on aphids from transgenic plants (Down et al., 2003).

Antagonistic Interactions

Bt *Toxins*

Sun et al. (2003) reported that the population density of the majority of natural enemies (including predatory and parasitic enemies) was significantly lower in transgenic cotton than in normal cotton. The populations of *P. japonica, Lysiphlebia japonica* (Ashm.), and *Allothrombium ovatum* (Zhang and Xin) in bivalent transgenic cotton were lower than in univalent cotton by 30.4, 42.8, and 46.8%, respectively, whereas the density of the eggs of *Chrysopa sinica* Tjeder and Araneida was lower by 20.0% and 27.4% (not significant), and that of *C. sinica* and *O. minutus* was higher by 27.0% and 8.9%, respectively. Significantly lower numbers of *C. maculata* larvae have been recorded in open as well as caged plots of transgenic sweet corn with *cry1Ab* gene as compared with non-*Bt* treatments. No significant effects were observed within a year on overall density of the beneficial insect populations between *Bt* and non-*Bt* sweet corn. However, long-term field studies with larger sample size should be conducted to characterize the effects of transgenic crops on natural enemies in the field (Wold et al., 2001). Abundance of *Lebia grandis* Dubrony has been found to be lower in pure and mixed crops of transgenic potatoes than in pure nontransgenic potato crop (Riddick, Dively, and Barbosa, 1998). Feeding larvae of the coccinellid, *C. sex-maculatus*, on *H. armigera* larvae reared on *Bt*-intoxicated diet results in slow growth of the coccinellid larvae (Figure 11.3). However, the coccinellid grubs and adults feed on several insects under field conditions, and therefore, may not be exposed to high levels of *Bt* toxins or only to prey of poor nutritional quality.

Larvae of *C. carnea* fed on pollen from transgenic maize did not complete development on the pollen diet, although pollen was provided to the larvae for a limited period of time (Pilcher et al., 1997). Larvae of *C. carnea* fed lepidopteran larvae (such as *S. littoralis*) reared on maize expressing *cry1Ab* are affected adversely (Hilbeck et al., 1998a, 1998b, 1999; Dutton et al., 2002). Predator larvae were also affected when fed pure activated *Bt* toxin in artificial diet (Hilbeck et al., 1998b). Prey-mediated effects of transgenic plants have been studied through intoxicated *L. decemlineata* eaten by *C. maculata*, with some reduction in consumption rate and survival of *C. maculata* (Riddick and Barbosa, 1998). These studies not only showed the direct toxicity of *Bt* toxins to the predators, but also revealed plant-herbivore-natural enemy interactions that contributed to increased mortality mediated by insect-resistant transgenic plant fed prey that were not lethally affected by the transgenic plants. However, further studies may be necessary to ascertain whether the adverse effects were from the intoxicated prey or from the *Bt* toxin directly. Transgenic *Bt* sweet corn and the foliar insecticides indoxacarb and spinosad are all less toxic to the most abundant predators in sweet corn (*C. maculata, Hippodamia axyridis* Pallas, and *O. insidiosus*) than the

FIGURE 11.3 Effect of feeding *Helicoverpa armigera* larvae reared on *Bt*-intoxicated diet to the grubs of the coccinellid predator *Cheilomenes sexmaculatus* (left) as compared to those fed on untreated artificial diet (right).

pyrethroid lambda-cyhalothrin (Musser and Shelton, 2003). Control of *O. nubilalis*, *Spodoptera frugiperda* (J.E. Smith), and *H. zea* by *Bt* sweet corn and spinosad was better than or equal to application of lambda-cyhalothrin.

Protease Inhibitors

Pentatomids are affected adversely when fed on proteinase inhibitor contaminated prey (Ashouri et al., 1998; Walker et al., 1998). Prey reared on transgenic oilseed rape expressing oryzacystatin (OC-1) inhibited ladybird, *H. axyridis* digestive enzymes *in vitro*, but had no effect *in vivo*. Ladybird was able to upregulate digestive proteases in response to the inhibitor and had no adverse effects on survival and development of the coccinellid when fed on *P. xylostella* larvae, which had accumulated the transgene product in the larval tissues at levels of up to 3 ng per gut (Ferry et al., 2003). The predatory stinkbug, *Podisus maculiventris* Say may suffer some indirect adverse effects as a result of foraging on prey in crops expressing either *GNA* or *CpTI* due to prey being of inferior quality rather than to direct toxicity of the transgene products (Bell et al., 2003). A significant reduction in growth and female adult weight was observed when *P. maculiventris* was provided with tomato moth, *Lacanobia oleracea* (L.) larvae injected with either GNA or CpTI at 10 µg day^{-1}. Weight reduction was 11.3% and 16.6% in GNA-fed and CpTI-fed *P. maculiventris*, respectively. The males, however, were not significantly affected. The female bugs that had not been exposed to the transgene proteins as nymphs showed no reduction in fecundity when provided with prey injected with either GNA or CpTI at the same dose. When provided with prey that had been reared on transgenic plants expressing either GNA or CpTI, no effects on the survival of nymphs were observed and only small, largely nonsignificant, reductions in weights were recorded throughout pre-adult development. Male nymphs fed on prey reared on diet with GNA prolonged the pre-adult development by 0.8 days. The next generation adults showed a significant reduction in egg production when fed on GNA-fed prey.

Lectins

Two-spotted ladybirds, *Adalia bipunctata* L., fed for 12 days on peach potato aphid, *M. persicae* colonizing transgenic potatoes expressing GNA showed a decrease in fecundity, egg viability, and longevity (Birch et al., 1999). However, subsequent studies by Down et al. (2000) showed that the observed effects were prey-quality mediated rather than the direct effects of GNA. Adverse effects on ladybird reproduction caused by eating peach-potato aphids from transgenic potatoes were reversed after switching the ladybirds to pea aphids from nontransgenic bean plants. GNA showed high levels of toxicity to grubs of *C. septempunctata* (Dhillon et al., 2008).

Interaction of Transgenic Crops with Parasitoids

The effects of transgenic crops on the parasitoids vary across crops and the cropping systems. Some of the variation may be due to differences in pest abundance between the transgenic and the nontransgenic crops.

Synergistic/Neutral Interactions

Bt *Toxins*

Cotton bollworm parasitoid, *Campoletis sonorensis* (Cameron), and transgenic plants act synergistically, decreasing the survival of *H. virescens* larvae beyond the level expected for an additive interaction. Synergistic increases in mortality and parasitism have also been detected when development rates on toxic plants and control plants were equal (Johnson and Gould, 1992). The parasitoid, *Cardiochiles nigriceps* (Viereck) does not reduce the survival of *H. virescens* larvae significantly, and its activity is not influenced by the transgenic plants (Johnson, 1997; Johnson, Gould, and Kennedy, 1997). Riggin Bucci and Gould (1997) reported that percentage parasitism, by the parasitoid of diamond back moths, *Diadegma insulare* (Cresson) was not significantly different between the mixed and nonmixed plots of transgenic plants. There are no adverse effects of transgenic maize on parasitization of the European corn borer, *O. nubilalis* by *Eriborus tenebrans* (Graven) and *Macrocentrus grandii* (Goid.) (Orr and Landis, 1997). Similarly, Volkmar and Freier (2003) did not observe any clear effect of *Bt* transgenic maize on the activity and abundance of parasitoids. The parasitoids, *Oedothorax apicatus* (Black.) and *Erigone atra* (Black.) were dominant species in all fields, while *E. dentipalpis* (Wider) was dominant at Halle. *Meioneta rurestris* (Koch.), *Porrhomma microphthalmum* (Cambridge), and *Pardosa agrestis* (Westring) were dominant at Oderbruch. Parasitism of *H. armigera* eggs has been found to be greater on transgenic *Bt* cotton (probably because of healthy crop and more egg laying) than on the nontransgenic one (Figure 11.4), but there were no differences in larval parasitism. However, parasitism by the larval-pupal parasitoids was quite low on the transgenic cotton as very few larvae survived long enough to ensure complete development of larval-pupal parasitoids.

Lectins

Transgenic sugarcane expressing GNA and ingested via *Eoreuma loftini* (Dyar) was not acutely toxic to the parasitoid, *Parallorhogas pyralophagus* (Marsh) (Tomov and Bernal, 2003). In the first generation, adult longevity increased by two days; and cocoon to adult and egg to adult developmental times were prolonged by one day in parasitoids exposed to

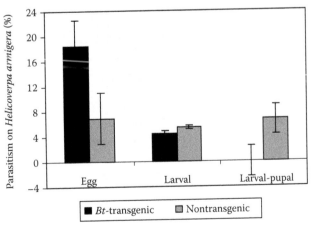

FIGURE 11.4 Parasitization of *Helicoverpa armigera* eggs and larvae on transgenic and nontransgenic cottons under field conditions.

transgenic sugarcane expressing *GNA*. In the second generation, adult longevity was reduced by three days, adult size by 5%, and egg load by 24%. Neither the fitness of females of *Aphidius ervi* (Haliday) nor the sex ratio of the progeny was affected when reared on aphids fed on transgenic plants expressing *OCIDELTAD86* for nematode resistance (Cowgill, Danks, and Atkinson, 2004). There are no detrimental effects of the transgenic oilseed rape (expressing *cry1Ac* and *OC-I*) on the ability of the parasitoid, *Diaeretiella rapa* (McIntosh) to control aphid, *M. persicae* populations. Adult parasitoid emergence and sex ratio were not altered on the transgenic oilseed rape as compared with the nontransgenic plants (Schuler et al., 2001). *Bt* oilseed rape caused 100% mortality of a *Bt*-susceptible strain but no mortality of the *Bt*-resistant *P. xylostella* strain NO-QA (Schuler et al., 2004). *Cotesia plutellae* (Kurdjumov) eggs hatched, but premature host mortality did not allow *C. plutellae* larvae to complete their development in *Bt*-susceptible *P. xylostella* feeding on *Bt* leaves. In contrast, *C. plutellae* developed to maturity in *Bt*-resistant *P. xylostella* fed on *Bt* oilseed rape leaves, and there was no effect of *Bt* plants on percentage parasitism, time to emergence from hosts, time to adult emergence, and adult emergence from cocoons (Schuler et al., 2004). Parasitoids that had attacked *Bt*-resistant *P. xylostella* larvae on transgenic plants suffered no measurable adverse effects of *Bt* toxins on the behavior of adults or on larval survival. Continued ability of *C. plutellae* to locate and parasitize *Bt*-resistant *P. xylostella* on transgenic crops might help to constrain the spread of genes for *Bt* resistance (Schuler et al., 1999). Adult *C. plutellae* females are more attracted to *Bt* plants damaged by *Bt*-resistant *P. xylostella* larvae than by susceptible hosts (Schuler et al., 2003). Mixtures of *Bt* and wild-type plants demonstrated that the parasitoid is as effective in controlling *Bt*-resistant *P. xylostella* larvae on *Bt* plants as on wild-type plants (Schuler et al., 2003).

Antagonistic Interactions

Bt *Toxins*

Egg parasitism of third-generation noctuids in *Bt*-transgenic cotton is lower than in the conventional cottons (Wang and Xia, 1997). In transgenic cotton, abundance of the parasitoids, *C. chlorideae* and *Microplitis* sp. decreased by 79.2% and 87.5%, and 88.9% and 90.7%, respectively, in natural and integrated control plots (Cui and Xia, 1997, 1998). Ren et al. (2004) observed that transgenic cotton (*cry 1Ac* and *CpTi* cotton, Zhongkang 310) suppressed the growth and development of *H. armigera* parasitized by *Microplitis mediator* (Haliday) and *C. chlorideae* or nonparasitized larvae. Cocoon formation and cocoon weight decreased by 26.1% and 1 mg, respectively, while for *C. chlorideae*, the reductions were 17.9% and 5.1 mg, respectively. Under laboratory conditions, rearing of the parasitoid *C. chlorideae* on *H. armigera* larvae fed on the Cry1Ac-intoxicated diet results in reduced cocoon formation (Figure 11.5) (Sharma, Dhillon, and Arora, 2008). There is a significant reduction in survival of *C. chlorideae* reared on *H. armigera* larvae fed on *Bt*-transgenic cotton leaves (Figure 11.6). However, the reduced survival of the parasitoid is not due to the direct toxicity of *Bt* to the parasitoid, but early mortality of the host larvae (Sharma, Pampapathy, and Arora, 2007). The larvae exhibited delayed development on transgenic cotton and, in some cases, abnormal development was also observed. The total hemolymph protein content of the larvae fed on transgenic cotton was lower than that on the nontransgenic cotton, which might result in delayed and abnormal development of the parasitoid (Ren et al., 2004). Decreased parasitism has been attributed to lower density and poor nutritional quality of the bollworm, *H. armigera* larvae (Cui and Xia, 1999; Wu, Peng, and Jia, 2003). Parasitism of *H. virescens* larvae by *C. sonorensis* was lower on *Bt* transgenic tobacco than on the nontransgenic tobacco, and was lower for *Bt*-susceptible than the *Bt*-resistant larvae

FIGURE 11.5 The hymenopteran parasitoid, *Campoletis chlorideae* female laying eggs in the second-instar larva of *Helicoverpa armigera* reared on *Bt*-intoxicated artificial diet/transgenic plants. Right: egg, larva, and pupa of *C. chlorideae*.

(Johnson, Gould, and Kennedy, 1997). Decreased feeding by *H. virescens* larvae on transgenic plants may be responsible for the observed differences in parasitism, because *C. sonorensis* locates host larvae using cues from damaged plants. When mean larval survival was used to estimate fitness of the nonadapted insects relative to the *Bt* toxin-adapted insects, it was found that *C. sonorensis* might delay adaptation to *Bt*-transgenic plants (Johnson, Gould, and Kennedy, 1997). Bourguet et al. (2002) observed a significant variation in parasitism by the tachinids, *Lydella thompsoni* Herting and *Pseudoperichaeta nigrolineata* Walker across locations, and greater parasitism was recorded in normal than in *Bt* transgenic maize.

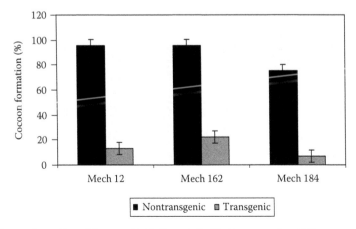

FIGURE 11.6 Cocoon formation of the parasitoid, *Campoletis chlorideae* reared on *Helicoverpa armigera* larvae fed on the leaves of *Bt* transgenic and nontransgenic cottons.

Protease Inhibitors

Soybean Kunitz inhibitor (SKTI) fed to *L. oleracea* larvae in an artificial diet has been detected in hemolymph of the parasitoid, *Eulophus pennicornis* (Nees) larvae, indicating that proteinase inhibitors in the host diet can be delivered to a parasitoid via the host hemolymph (Down et al., 1999). If transgenic plants expressing proteinase inhibitors for protection against insect pests are to form a component of integrated pest management systems, possible adverse effects, whether direct or indirect, of transgene expression on parasitoids should be taken into consideration. The success of the parasitic wasp, *E. pennicornis* was reduced by the presence of CpTI in host diet and on transgenic potato expressing the transgene for resistance to *L. oleracea* (Bell et al., 2001a). Parasitoid progeny that developed on *L. oleracea* reared on CpTI-containing diets, however, were not adversely affected.

Lectins

The success of the parasitic wasp, *E. pennicornis* was not reduced by the presence of GNA in any of the diets or by the length of feeding of the host prior to parasitism (Bell et al., 1999). However, the mean numbers of *E. pennicornis* wasps that developed on *L. oleracea* reared from the third-instar on the GNA-containing maize diet was significantly higher than on the controls, although the differences were not significant. Progeny of *E. pennicornis* that developed on *L. oleracea* reared on GNA-containing diets showed little or no alteration in size, longevity, egg load, and fecundity when compared with wasps that had developed on hosts fed on respective control diets. The results suggested that expression of *GNA* in transgenic crops will not adversely affect the ability of the ectoparasitoid, *E. pennicornis* to utilize *L. oleracea* as a host (Bell et al., 1999). GNA-containing artificial diet has a detrimental effect on aphid parasitoids. Effects of GNA appeared to be host-quality mediated rather than constituting direct toxic effects of GNA (Couty et al., 2001a; Couty, Clark, and Poppy, 2001). GNA delivered via artificial diet to the aphid, *M. persicae* can be transferred through the trophic levels and has a dose-dependent effect on the parasitoid, *A. ervi* (Couty et al., 2001b). Parasitoid larvae excreted most of the ingested GNA in the meconium, but some of it was detected in the pupae. Although *A. ervi* development was not affected when developing within hosts feeding on transgenic potato leaves, this probably reflected suboptimum expression of the toxin in the transgenic potato line used. Bell et al. (2001b) suggested that GNA-expressing potatoes and *E. pennicornis* were compatible for controlling *L. oleracea*. GNA expressed in tomato enhanced the parasitism by *E. pennicornis* on tomato moth in the greenhouse.

Interactions of Transgenic Plants with Fauna and Flora in the Rhizosphere

Potential effects of genetically transformed crops on nontarget species are not restricted only to the environment above ground, but also on those inhabiting the soil rhizosphere (Jepson, Croft, and Pratt, 1994). Some genetically engineered crops affect soil ecosystems (Griffiths, Geoghegan, and Robertson, 2000), but the long-term significance of any of these changes is unclear. Effects of transgene products on nontarget organisms may decrease the rate of plant decomposition, and of carbon and nitrogen levels, thus affecting soil fertility. Similarly, declining species diversity of soil microorganisms in some cases can cause lower community diversity and reduce productivity above ground (Van der Heijden et al., 1998). Organisms in the rhizosphere such as Collembola, nematodes, protozoa, fungi, bacteria, and earthworms should be included in risk assessment studies, but have received little attention (Groot and Dicke, 2002).

Toxins from the transgenic plants are introduced into the soil primarily through incorporation of the crop residues into the soil after crop harvest (Tapp and Stotzky, 1998a) or through the root exudates (Saxena, Flores, and Stotzky, 1999). Insecticidal proteins produced by *Bt* bind rapidly and tightly on clays, both pure mined clay minerals and soil clays, on humic acids extracted from soil, and on complexes of clay and humic acids. Binding reduces susceptibility of the proteins to microbial degradation (Stotzky, 2004). Vertical movement of Cry1Ab protein, either purified or in root exudates, or biomass of *Bt* corn, decreased as the concentration of the clay minerals kaolinite or montmorillonite in soil increased. The toxins produced in *Bt* plants retain their biological activity when bound to the soil, so accumulation of these toxins is likely to occur in the ecosystem. *Bt* toxins are absorbed and bound around the soil particles in <30 minutes, suggesting that toxins from root exudates or crop residue remain free in the soil for a short time (Venkateswerulu and Stotzky, 1990, 1992; Crecchio and Stotzky, 1998), which reduces its availability to the microbes. This may lead to accumulation of the toxins in the environment. The persistence of toxins in the soil could improve insect control or lead to development of resistance in insects inhabiting the soil. *Bt* toxins released from the root exudates bind to the soil particles and remain active up to 180 days (Saxena and Stotzky, 2001), an association that interferes with biodegradation. Toxins are present in the soil rhizosphere throughout the crop growth, and for several months after crop maturity (Saxena and Stotzky, 2000). Purified Cry1Ab protein and the protein released from *Bt* corn exhibited binding and persistence in soil. Insecticidal protein was also released in root exudates of *Bt* potato (Cry3A protein) and rice (Cry1Ab protein), but not in root exudates of *Bt* canola, cotton, and tobacco (Cry1Ac).

Earthworms ingest the bound toxins, but are not affected by them. However, earthworms may function as intermediaries through which the toxins are passed on to other trophic levels. When purified toxin from *B. thuringiensis* ssp. *kurstaki* is added into the soil, the pesticidal activity against *Manduca sexta* L. has been recorded up to 234 days (Tapp and Stotzky, 1998b). These estimates are quite longer than the persistence reported earlier (8 to 17 days for the purified toxins, and 2 to 41 days from biomass of transgenic corn, cotton, and potato) (Palm et al., 1994; Sims and Holdan, 1996; Sims and Ream, 1997). The level of Cry1Ac protein in samples collected three months after the last season's tillage has been evaluated using both enzyme-linked immunosorbent assays (ELISA) and bioassays with a susceptible insect species, *H. virescens*. Both methods revealed that no detectable Cry1Ac protein was present in any of the soil samples collected from within or outside the Bollgard fields (Head et al., 2002). Based on the results from reference standards, the limit of detection for the ELISA was 3.68 ng of extractable protein per gram of soil, and that of the bioassay (measured by EC_{50}) was 8 ng of biologically active protein per gram of soil. Cry1Ac protein accumulated as a result of continuous use of transgenic *Bt* cotton, and subsequent incorporation of plant residues into the soil by postharvest tillage is extremely low and does not result in detectable biological activity. Transgenic *Bt* corn decomposed at a slower rate in soil than nontransgenic corn, possibly because *Bt* corn had more lignins than the nontransgenic corn. Biomass of *Bt* canola, cotton, potato, rice, and tobacco also decomposed at a lower rate than the biomass of respective near-isogenic nontransgenic plants. However, the lignin content of these *Bt* plants, which was significantly less than that of *Bt* corn, was not significantly different from that of their near-isogenic nontransgenic counterparts, although it was consistently higher.

Bt transgenic corn cultivation appears to have no significant influence on the nemato-fauna, neither at the level of genus composition nor with regard to biodiversity (Manachini and Lozzia, 2002). Nevertheless, one region did have a change in trophic group composition: fungal feeding nematodes were more numerous in the soil with *Bt* corn, while in the field with the isogenic hybrid, the bacterial feeders were higher than the *Bt* corn. The Cry1Ab protein had no consistent effects on nontarget organisms (earthworms, nematodes, proto-zoa, bacteria, fungi) in soil or *in vitro*. The Cry1Ab protein was not taken up from soil by non-*Bt* corn, carrot, radish, or turnip grown in soil in which *Bt* corn had been grown or into which biomass of *Bt* corn had been incorporated (Stotzky, 2004). No effects have been detected in culturable bacteria, fungi, protozoa, and nematodes from the *Bt*-maize fields (Saxena and Stotzky, 2001). Under field conditions, the microflora of *Bt* transgenic potato plants has been observed to be minimally different from that of chemically and microbi-ally treated commercial potato plants (Donegan et al., 1996). It is unlikely that expression of *Bt* and any other genes in transgenic plants would have an adverse effect on the soil microflora. There are no significant differences in mortality or weight of the earthworm, *Lumbricus terrestris* L., after 40 days in soil planted with *Bt* maize or after 45 days in soil amended with *Bt* maize (Saxena and Stotzky, 2001). Toxin has been detected in the gut and casts of earthworms, but is cleared in two to three days after being placed in fresh soil.

A 200-day study has been carried out to investigate the impact of *Bt* transgenic corn on immature and adult *L. terrestris* in the field and in the laboratory. No lethal effects of trans-genic *Bt* corn on immature and adult earthworms were observed (Zwahlen et al., 2003). Immature *L. terrestris* in the field had a very similar growth pattern when fed either trans-genic (*Bt*) or nontransgenic (non-*Bt*) corn litter. No significant differences were observed in relative weights of *L. terrestris* adults fed on transgenic or nontransgenic corn during the first 160 days in a laboratory trial, but after 200 days, the adult *L. terrestris* exhibited a weight loss of 18% when fed transgenic corn litter compared to a weight gain of 4% in those fed on nontransgenic corn. Further studies are necessary to see whether or not this difference in relative weight was due to the degradation of Cry1Ab toxin in corn residues (Zwahlen et al., 2003). Percentage infection of *H. virescens* larvae by *Nomuraea rileyi* (Farlon) was greater on *Bt* transgenic plants, and higher in the normal larvae than in *Bt*-resistant larvae, and might accelerate adaptation to *Bt* transgenic plants (Johnson, Gould, and Kennedy, 1997).

The proteinase inhibitor concentration in the transgenic plant litter after 57 days was 0.05% of the sample at day 0, and was not detectable in subsequent sampling (Donegan et al., 1997). Although the carbon content of the transgenic plant litter was comparable to that of the parental plant litter on sample day 0, it became significantly lower over time. Nematode populations in the soil surrounding the transgenic plant litter bags were greater than those in the soil surrounding parental plant litter bags and had a different trophic group composition, including a significantly higher ratio of fungal feeding nematodes to bacterial feeding nematodes on sample day 57. In contrast, Collembola populations in the soil surrounding the transgenic plant litter bags were significantly lower than in the soil surrounding parental plant litter bags, indicating that under field conditions, the protei-nase inhibitor remained immunologically active in buried transgenic plant litter for at least 57 days and that decomposing parental and transgenic plant litter differed in quality (carbon content) and in the response of exposed soil organisms (Collembola and nematodes) (Donegan et al., 1997). Susceptibility of *Nosema pyrausta* (Paillot)-infected European corn borer, *O. nubilalis*, to *Bt* formulation Dipel ES is greater than to the uninfected larvae. Dipel ES reduced *N. pyrausta* spore production in larvae feeding on diet containing Dipel ES, suggesting that *Bt* maize might have some adverse effects on the survival and continual impact of *N. pyrausta* as a regulating factor on European corn borer populations (Pierce, Solter, and Weinzierl, 2001).

Conclusions

The development and deployment of transgenic plants with insect resistance will continue to expand in the future. This approach of controlling insects would offer the advantage of allowing some degree of selection for specificity effects, so that pests but not the beneficial organisms are targeted. The effects of transgenic plants on the activity and abundance of natural enemies vary across crops and the insect species involved. The adverse effects of transgenic crops on natural enemies are largely through poor quality of the host or early mortality of the insect host, rather than through direct toxicity to the natural enemies. Wherever the transgenic crops have shown adverse effects on natural enemies, these effects may still be far lower than those of the broad-spectrum pesticides. Deployment of transgenic plants would also lead to a major reduction in the number of insecticide sprays, resulting in increased activity of natural enemies and a safer environment in which to live.

References

Arpaia, S. (1996). Ecological impact of Bt-transgenic plants. I. Assessing possible effects of CryIIIB toxin on honey bee (*Aphis mellifera* L.) colonies. *Journal of Genetics and Breeding* 50: 315–319.

Ashouri, A.S., Overney, S., Michaud, D. and Cloutier, C. (1998). Fitness and feeding are affected in the two spotted stinkbug, by the cysteine proteinase inhibitor, Oryzacystatin I. *Archives of Insect Biochemistry and Physiology* 38: 74–83.

Aulrich, K., Flachowsky, G., Daenicke, R. and Halle, I. (2001). Bt-corn in animal nutrition. *Ernahrungsforschung* 46(1): 13–20.

Behle, R.W., McCurie, M.R. and Shasha, B.S. (1997). Effects of sunlight and simulated rain on residual toxicity of *Bacillus thuringiensis* formulations. *Journal of Economic Entomology* 90: 1560–1566.

Bell, H.A., Down, R.E., Fitches, E.C., Edwards, J.P. and Gatehouse, A.M.R. (2003). Impact of genetically modified potato expressing plant-derived insect resistance genes on the predatory bug *Podisus maculiventris* (Heteroptera: Pentatomidae). *Biocontrol Science and Technology* 13: 729–741.

Bell, H.A., Fitches, E.C., Down, R.E., Marris, G.C., Edwards, J.P., Gatehouse, J.A. and Gatehouse, A.M.R. (1999). The effect of snowdrop lectin (GNA) delivered via artificial diet and transgenic plants on *Eulophus pennicornis* (Hymenoptera: Eulophidae), a parasitoid of the tomato moth *Lacanobia oleracea* (Lepidoptera: Noctuidae). *Journal of Insect Physiology* 45: 983–991.

Bell, H.A., Fitches, E.C., Marris, G.C., Bell, J., Edwards, J.P., Gatehouse, J.A. and Gatehouse, A.M.R. (2001a). Transgenic GNA expressing potato plants augment the beneficial biocontrol of *Lacanobia oleracea* (Lepidoptera; Noctuidae) by the parasitoid *Eulophus pennicornis* (Hymenoptera; Eulophidae). *Transgenic Research* 10: 35–42.

Bell, H.A., Fitches, E.C., Down, R.E., Ford, L., Marris, G.C., Edwards, J.P., Gatehouse, J.A. and Gatehouse, A.M.R. (2001b). Effect of dietary cowpea trypsin inhibitor (CpTI) on the growth and development of the tomato moth *Lacanobia oleracea* (Lepidoptera: Noctuidae) and on the success of the gregarious ectoparasitoid *Eulophus pennicornis* (Hymenoptera: Eulophidae). *Pest Management Science* 57: 57–65.

Belzunces, L.P., Lenfant, C., Di Pasquale, S. and Colin, M.E. (1994). *In vivo* and *in vitro* effects of wheat germ agglutinin and Bowman-Birk soybean trypsin inhibitor, two potential transgene products, on midgut esterase and protease activities from *Apis mellifera*. *Comparative Biochemistry and Physiology* 109B: 63–69.

Birch, A.N.E., Geoghegan, I.E., Majerus, M.E.N., McNicol, J.W., Hackett, C.A., Gatehouse, A.M.R. and Gatehouse, J.A. (1999). Tri-trophic interactions involving pest aphids, predatory 2-spot ladybirds and transgenic potatoes expressing snowdrop lectin for aphid resistance. *Molecular Breeding* 5: 75–83.

Bouchard, E., Michaud, D. and Cloutier, C. (2003). Molecular interactions between an insect predator and its herbivore prey on transgenic potato expressing a cysteine proteinase inhibitor from rice. *Molecular Ecology* 12: 2429–2437.

Bourguet, D., Chaufaux, J., Micoud, A., Delos, M., Naibo, B., Bombarde, F., Marque, G., Eychenne, N. and Pagliari, C. (2002). *Ostrinia nubilalis* parasitism and the field abundance of non-target insects in transgenic *Bacillus thuringiensis* corn (*Zea mays*). *Environmental Biosafety Research* 1: 49–60.

Bundy, C.S., McPherson, R.M. and Herzog, G.A. (2000). An examination of the external and internal signs of cotton boll damage by stink bugs (Heteroptera: Pentatomidae). *Journal of Entomological Science* 35: 402–410.

Burgess, E.P.J., Malone, L.A. and Christeller, J.T. (1996). Effects of two proteinase inhibitors on the digestive enzymes and survival of honey bees (*Apis mellifera*). *Journal of Insect Physiology* 42: 823–828.

Candolfi, M.P., Brown, K., Grimm, C., Reber, B. and Schmidli, H. (2004). A faunistic approach to assess potential side-effects of genetically modified Bt-corn on non-target arthropods under field conditions. *Biocontrol Science and Technology* 14: 129–170.

Couty, A., Clark, S.J. and Poppy, G.M. (2001). Are fecundity and longevity of female *Aphelinus abdominalis* affected by development in GNA-dosed *Macrosiphum euphorbiae*? *Physiological Entomology* 26: 285–286.

Couty, A., de la Viña, G., Clark, S.J., Kaiser, L., Pham-Delègue, M.H. and Poppy, G.M. (2001a). Direct and indirect sub-lethal effects of *Galanthus nivalis* agglutinin (GNA) on the development of a potato-aphid parasitoid, *Aphelinus abdominalis* (Hymenoptera: Aphelinidae). *Journal of Insect Physiology* 47: 553–561.

Couty, A, Down, R.E., Gatehouse, A.M.R., Kaiser, L., Pham-Delègue, M.H. and Poppy, G.M. (2001b). Effects of artificial diet containing GNA and GNA-expressing potatoes on the development of the aphid parasitoid *Aphidius ervi* Haliday (Hymenoptera: Aphidiidae). *Journal of Insect Physiology* 47: 1357–1366.

Cowgill, S.E. and Atkinson, H.J. (2003). A sequential approach to risk assessment of transgenic plants expressing protease inhibitors: Effects on nontarget herbivorous insects. *Transgenic Research* 12: 439–449.

Cowgill, S.E., Danks, C. and Atkinson, H.J. (2004). Multitrophic interactions involving genetically modified potatoes, nontarget aphids, natural enemies and hyperparasitoids. *Molecular Ecology* 13: 639–647.

Crecchio, C. and Stotzky, G. (1998). Insecticidal activity and biodegradation of the toxin from *Bacillus thuringiensis* subsp. *kurstaki* bound to humic acids from soil. *Soil Biology and Biochemistry* 30(4): 463–470.

Cui, J.J. and Xia, J.Y. (1997). The effect of Bt transgenic cotton on the feeding function of major predators. *China Cottons* 24: 19.

Cui, J.J. and Xia, J.Y. (1998). Effects of early seasonal strain of Bt transgenic cotton on population dynamics of main pests and their natural enemies. *Acta Gossypii Sinica* 10: 255–262.

Cui, J.J. and Xia, J.Y. (1999). Effect of transgenic Bt cotton on the population dynamics of natural enemies. *Acta Gossypii Sinica* 11: 84–91.

Cui, J.J. and Xia, J.Y. (2000a). Effects of Bt transgenic cotton on the structures and composition of insect community. *Journal of Yunnan Agricultural University* 15(1): 342–345.

Cui, J.J. and Xia, J.Y. (2000b). Effects of transgenic Bt cotton R93-6 on the insect community. *Acta Entomologica Sinica* 43: 43–51.

Deml, R. and Dettner, K. (1998). Wirkungen *Bacillus thuringiensis*-toxin – produzierender Pflanzen auf Ziel- und Nichtzielorganismen – eine Standortbestimmung. *Umweltbundesamt Texte* 36: 120 S.

Dhillon, M.K. and Sharma, H.C. (2007). Survival and development of *Campoletis chlorideae* on various insect and crop hosts: Implications for *Bt*-transgenic crops. *Journal of Applied Entomology* 131: 179–185.

Dhillon, M.K., Lawo, N., Sharma, H.C. and Romeis, J. (2008). Direct effects of *Galanthus nivalis* agglutinin (GNA) and avidin on the ladybird beetle, *Coccinella septempunctata*. *IOBC/WPRS/SROP Bulletin* 33: 43–50.

Dogan, E.B., Berry, R.E., Reed, G.L. and Rossignol, P.A. (1996). Biological parameters of convergent lady beetle (Coleoptera: Coccinellidae) feeding on aphids (Homoptera: Aphididae) on transgenic potato. *Journal of Economic Entomology* 89: 1105–1108.

Donegan, K.K., Schaller, D.L., Stone, J.K., Ganio, L.M., Reed, G., Hamm, P.B. and Seidler, R.J. (1996). Microbial populations, fungal species diversity and plant pathogen levels in field plots of potato plants expressing the *Bacillus thuringiensis* var. *tenebrionis* endotoxin. *Transgenic Research* 5: 25–35.

Donegan, K.K., Seidler, R.J., Fieland, V.J., Schaller, D.L., Palm, C.J., Ganio, L.M., Cardwell, D.M. and Steinberger, Y. (1997). Decomposition of genetically engineered tobacco under field conditions: Persistence of the proteinase inhibitor I product and effects on soil microbial respiration and protozoa, nematode and microarthropod populations. *Journal of Applied Ecology* 34: 767–777.

Down, R.E., Ford, L., Mosson, H.J., Fitches, E., Gatehouse, J.A. and Gatehouse, A.M.R. (1999). Protease activity in the larval stage of the parasitoid wasp, *Eulophus pennicornis* (Nees) (Hymenoptera: Eulophidae); effects of protease inhibitors. *Parasitology* 119: 157–166.

Down, R.E., Ford, L., Woodhouse, S.D., Davison, G.M., Majerus, M.E.N., Gatehouse, J.A. and Gatehouse, A.M.R. (2003). Tritrophic interactions between transgenic potato expressing snowdrop lectin (GNA), an aphid pest [peach-potato aphid; *Myzus persicae* (Sulz.)] and a beneficial predator (2-spot ladybird; *Adalia bipunctata* L.). *Transgenic Research* 12: 229–241.

Down, R.E., Ford, L., Woodhouse, S.D., Raemaekers, R.J.M., Leitch, B., Gatehouse, J.A. and Gatehouse, A.M.R. (2000). Snowdrop lectin (GNA) has no acute toxic effect on a beneficial insect predator, the 2-spot ladybird (*Adalia bipunctata* L.). *Journal of Insect Physiology* 46: 379–391.

Duan, J.J., Head, G., Jensen, A. and Reed, G. (2004). Effects of transgenic *Bacillus thuringiensis* potato and conventional insecticides for Colorado potato beetle (Coleoptera: Chrysomelidae) management on the abundance of ground-dwelling arthropods in Oregon potato ecosystems. *Environmental Entomology* 33: 275–281.

Dutton, A., Romeis, J. and Bigler, F. (2003a). Test procedure to evaluate the risk that insect-resistant transgenic plants pose to entomophagous arthropods. In Van Driesche, R.G. (Ed.), *Proceedings, First International Symposium on Biological Control of Arthropods*, Honolulu. FHTET-2003-05. Morgantown, USA: U.S. Department of Agriculture, Forest Service, 466–472.

Dutton, A., Romeis, J. and Bigler, F. (2003b). Assessing the risks of insect resistant transgenic plants on entomophagous arthropods: Bt-maize expressing Cry1Ab as a case study. *BioControl* 48: 611–636.

Dutton, A., Klein, H., Romeis, J. and Bigler, F. (2002). Uptake of Bt-toxin by herbivores feeding on transgenic maize and consequences for the predator, *Chrysoperla carnea*. *Ecological Entomology* 27: 441–447.

Fearing, P.L., Brown, D., Vlachos, D., Meghji, M. and Privalle, L. (1997). Quantitative analysis of CryIA(b) expression in Bt maize plants, tissues, and silage and stability of expression over successive generations. *Molecular Breeding* 3: 169–176.

Ferry, N., Raemaekers, R.J.M., Majerus, M.E.N., Jouanin, L., Port, G., Gatehouse, J.A. and Gatehouse, A.M.R. (2003). Impact of oilseed rape expressing the insecticidal cysteine protease inhibitor oryzacystatin on the beneficial predator *Harmonia axyridis* (multicoloured Asian ladybeetle). *Molecular Ecology* 12: 493–504.

Fitt, G., Mares, C.L. and Llewellyn, D.J. (1994). Field evaluation and potential ecological impact of transgenic cottons (*Gossypium hirsutum*) in Australia. *Biocontrol Science and Technology* 4: 535–548.

Flexner, J.L., Lighthart, B. and Croft, B.A. (1986). The effects of microbial pesticides on non-target, beneficial arthropods. *Agriculture, Ecosystems and Environment* 16: 203–254.

Flint, H.M., Henneberry, T.J., Wilson, F.D., Holguin, E., Parks, N. and Buehler, R.E. (1995). The effects of transgenic cotton, *Gossypium hirsutum* L.; containing *Bacillus thuringiensis* toxin genes for the control of the pink bollworm, *Pectinophora gossypiella* (Saunders) (Lepidoptera, Gelechiidae) and other arthropods. *Southwestern Entomologist* 20: 281–292.

Gatehouse, A.M.R., Ferry, N. and Raemaekers, R.J.M. (2002). The case of the monarch butterfly: A verdict is returned. *Trends in Genetics* 18: 249–251.

Girard, C., Picard Nizou, A.L., Grallien, E., Zaccomer, B., Jouanin, L. and Pham-Delègue, M.H. (1998). Effects of proteinase inhibitor ingestion on survival, learning abilities and digestive proteinases of the honeybees. *Transgenic Research* 7: 239–246.

Griffiths, B.S., Geoghegan, I.E. and Robertson, W.M. (2000). Testing genetically engineered potato, producing the lectins GNA and ConA, on nontarget soil organisms and processes. *Journal of Applied Ecology* 37: 159–170.

Groot, A.T. and Dicke, M. (2002). Insect-resistant transgenic plants in a multi-trophic context. *Plant Journal* 31: 387–406.

Haider, M.Z., Knowles, B.H. and Ellar, D.J. (1986). Specificity of *Bacillus thuringiensis* var. *colemani* insecticidal delta-endotoxin is determined by differential proteolytic processing of the protoxin by larval gut proteases. *European Journal of Biochemistry* 156: 531–540.

Hawes, C., Haughton, A.J., Osborne, J.L., Roy, D.B., Clark, S.J., Perry, J.N., Rothery, P., Bohan, D.A., Brooks, D.R., Champion, G.T., Dewar, A.M., Heard, M.S., Woiwod, I.P., Daniels, R.E., Young, M.W., Parish, A.M., Scott, R.J., Firbank, L.G. and Squire, G.R. (2003). Responses of plants and invertebrate trophic groups to contrasting herbicide regimes in the farm scale evaluations of genetically modified herbicide-tolerant crops. *Philosophical Transactions of the Royal Society of London. Biological Sciences (Series B)* 358: 1899–1913.

Head, G., Surber, J.B., Watson, J.A., Martin, J.W. and Duan, J.J. (2002). No detection of Cry1Ac protein in soil after multiple years of transgenic Bt cotton (Bollgard) use. *Environmental Entomology* 31: 30–36.

Hilbeck, A. (2001). Implications of transgenic, insecticidal plants for insect and plant biodiversity. *Perspectives in Plant Ecology, Evolution and Systematics* 4: 43–61.

Hilbeck, A., Baumgartner, M., Fried, P.M. and Bigler, F. (1998a). Effects of transgenic *Bacillus thuringiensis* corn-fed prey on mortality and development time of immature *Chrysoperla carnea* (Neuroptera: Chrysopidae). *Environmental Entomology* 27: 480–487.

Hilbeck, A., Moar, W.J., Pusztai, C.M., Filippini, A. and Bigler, F. (1998b). Toxicity of *Bacillus thuringiensis* Cry1Ab toxin to the predator, *Chrysoperla cornea* (Neuroptera: Chrysopidae). *Environmental Entomology* 27: 1255–1263.

Hilbeck, A., Moar, W.J., Pusztai, C.M., Filippini, A. and Bigler, F. (1999). Prey mediated effects of Cry1Ab toxin and protoxin on the predator, *Chrysoperla carnea*. *Entomologia Experimentalis et Applicata* 91: 305–316.

Hilder, V.A. and Boulter, D. (1999). Genetic engineering of crop plants for insect resistance: A critical review. *Crop Protection* 18: 177–191.

Hoffmann, M.P., Zalom, F.G., Wilson, L.T., Smilanick, J.M., Malyj, L.D., Kiser, J., Hilder V.A. and Barnes, W.M. (1992). Field evaluation of transgenic tobacco containing genes encoding *Bacillus thuringiensis* delta-endotoxin or cowpea trypsin inhibitor: Efficacy against *Helicoverpa zea* (Lepidoptera: Noctuidae). *Journal of Economic Entomology* 85: 2516–2522.

Jasinski, J., Eisley, B., Young, C., Willson, H. and Kovach, J. (2001). Beneficial arthropod survey in transgenic and non-transgenic field crops in Ohio. *Special Circular, Ohio Agricultural Research and Development Center* 179: 99–102.

Jepson, P.C., Croft, B.A. and Pratt, G.E. (1994). Test systems to determine the ecological risks posed by toxin release from *Bacillus thuringiensis* genes in crop plants. *Molecular Ecology* 3: 81–89.

Jimenez, D.R. and Gilliam, M. (1989). Age-related changes in midgut ultrastructure and trypsin activity in the honey bee, *Apis mellifera. Apidologie* 20: 287–303.

Johnson, M.T. (1997). Interaction of resistant plants and wasp parasitoids of tobacco budworm (Lepidoptera: Noctuidae). *Environmental Entomology* 26: 207–214.

Johnson, M.T. and Gould, R. (1992). Interaction of genetically engineered host plant resistance and natural enemies of *Heliothis virescens* (Lepidoptera: Noctuidae) in tobacco. *Environmental Entomology* 21: 586–597.

Johnson, M.T., Gould, F. and Kennedy, G.G. (1997). Effects of natural enemies on relative fitness of *Heliothis virescens* genotypes adapted and not adapted to resistant host plants. *Entomologia Experimentalis et Applicata* 82: 219–230.

Kannan, M., Uthamasamy, S. and Mohan, S. (2004). Impact of insecticides on sucking pests and natural enemy complex of transgenic cotton. *Current Science* 86: 726–729.

Koziel, M.G., Beland, G.L., Bowman, C., Carozzi, N.B., Crenshaw, R., Crossland, L., Dawson, J., Desai, N., Hill, M., Kadwell, S., Launis, K., Lewis, K., Maddox, D., McPherson, K., Meghji, M.R., Merlin, E., Rhodes, R., Warren, G.W., Wright, M. and Evola, S.V. (1993). Field performance of elite transgenic maize plants expressing an insecticidal protein derived from *Bacillus thuringiensis*. *Bio/Technology* 11: 194–200.

Kreutzweiser, D.P., Gringorten, J.L., Thomas, D.R. and Butcher, J.T. (1996). Functional effects of the bacterial insecticide *Bacillus thuringiensis* var. *kurstaki* on aquatic communities. *Ecotoxicology and Environment Safety* 33: 271–280.

Lambert, B. and Peferoen, M. (1992). Insecticidal promise of *Bacillus thuringiensis*. Facts and mysteries about a successful biopesticide. *BioScience* 42: 112–122.

Li, W.D., Wu, K.M., Chen, X.X., Feng, H.Q., Xu, G.A. and Guo, Y.Y. (2003). Effects of transgenic cotton carrying Cry1A + CpTI and Cry1Ac genes on the diversity of arthropod community in cotton fields in northern area of North China. *Journal of Agricultural Biotechnology* 11: 383–387.

Li, W.D., Ye, G.Y., Wu, K.M., Wang, X.Q. and Guo, Y.Y. (2002). Evaluation of impact of pollen grains from Bt, Bt/CpTI transgenic cotton and Bt corn plants on the growth and development of the mulberry silkworm, *Bombyx mori* Linnaeus (Lepidoptera: Bombycidae). *Agricultural Sciences in China* 1: 1334–1343.

Liu, W.X., Wan, F.H., Zhang, F. and Meng, Z.J. (2000). Bionomics of the cotton bollworm predator, *Campylomma diversicornis*. *Chinese Journal of Biological Control* 16: 148–151.

Liu, Z.C., Ye, G.Y., Fu, Q., Zhang, Z.T. and Hu, C. (2003). Indirect impact assessment of transgenic rice with cry1Ab gene on predations by the wolf spider, *Pirata subpiraticus*. *Chinese Journal of Rice Science* 17: 175–178.

Losey, J.E., Raynor, L.S. and Carter, M.E. (1999). Transgenic pollen harms monarch larvae. *Nature* 399: 214.

Lozzia, G.C. (1999). Biodiversity and structure of ground beetle assemblages (Coleoptera Carabidae) in Bt corn and its effects on non target insects. *Bollettino di Zoologia Agraria e di Bachicoltura* 31: 37–50.

Lozzia, G.C., Furlanis, C., Manachini, B. and Rigamonti, I.E. (1998). Effects of Bt corn on *Rhopalosiphum padi* L. (Rhynchota: Aphididae) and on its predator *Chrysoperla carnea* Stephen (Neuroptera: Chrysopidae). *Bollettino di Zoologia Agrariae di Bachicoltura* 30: 153–164.

Luttrell, R.G., Mascarenhas, V.J., Schneider, J.C., Parker, C.D. and Bullock, P.D. (1995). Effect of transgenic cotton expressing endotoxin protein on arthropod population in Mississippi cotton. In *Proceedings, Beltwide Cotton Production Research Conference*. Memphis, Tennessee, USA: National Cotton Council of America, 760–763.

Malone, L.A. and Pham-Delègue, M.H. (2001). Effects of transgene products on honey bees (*Apis mellifera*) and bumblebees (*Bombus* sp.). *Apidologie* 32: 287–304.

Malone, L.A., Burgess, E.P.J., Christeller, J.T. and Gatehouse, H.S. (1998). *In vivo* responses of honey bee midgut proteases to two protease inhibitors from potato. *Journal of Insect Physiology* 44: 141–147.

Malone, L.A., Giacon, H.A., Burgess, E.P.J., Maxwell, J.Z., Christeller, J.T. and Laing, W.A. (1995). Toxicity of trypsin endopeptidase inhibitors to honey bees (Hymenoptera: Apidae). *Journal of Economic Entomology* 88: 46–50.

Manachini, B. and Lozzia, G.C. (2002). First investigations into the effects of Bt corn crop on Nematofauna. *Bollettino di Zoologia Agraria e di Bachicoltura* 34: 85–96.

Meier, M.S. and Hilbeck, A. (2001). Influence of transgenic *Bacillus thuringiensis* corn-fed prey on prey preference of immature *Chrysoperla carnea* (Neuroptera: Chrysopidae). *Basic and Applied Ecology* 2(1): 35–44.

Men, X.Y., Ge, F. and Liu, X.H. (2003). Diversity of arthropod communities in transgenic Bt cotton and nontransgenic cotton agroecosystems. *Environmental Entomology* 32: 270–275.

Miller, H.I. and Flamm, E.L. (1993). Biotechnology and food regulation. *Current Opinion in Biotechnology* 4: 265–268.

Moar, W.J., Pusztai-Carey, M., van Faassen, H., Bosch, D., Frutos, R., Rang, C., Luo, K. and Adang, M.J. (1995). Development of *Bacillus thuringiensis* CryIC resistance by *Spodoptera exigua* (Hubner) (Lepidoptera: Noctuidae). *Applied Environmental Microbiology* 61: 2086–2092.

Musser, F.R. and Shelton, A.M. (2003). Bt sweet corn and selective insecticides: Impacts on pests and predators. *Journal of Economic Entomology* 96: 71–80.

Obrycki, J.J., Losey, J.E., Taylor, O.R. and Jesse, L.C.H. (2001). Transgenic insecticidal corn: Beyond insecticidal toxicity to ecological complexity. *BioScience* 51: 353–361.

Orr, D.B. and Landis, D.A. (1997). Oviposition of European corn borer (Lepidoptera: Pyralidae) and impact of natural enemy populations in transgenic versus isogenic corn. *Journal of Economic Entomology* 90: 905–909.

Palm, C.J., Donegan, K.K., Harris, D. and Siedler, C.J. (1994). Quantification in soil of *Bacillus thuringiensis* var. *kurstaki* endotoxin from transgenic plants. *Molecular Ecology* 3: 145–151.

Pham-Delègue, M.H., Girard, C., Le Métayer, M., Picard-Nizou, A.L., Hennequet, C., Pons, O. and Jouanin, L. (2000). Long-term effects of soybean protease inhibitors on digestive enzymes, survival and learning abilities of honeybees. *Entomologia Experimentalis et Applicata* 95: 21–29.

Picard Nizou, A.L., Pham-Delègue, M.H., Kerguelen, V., Douault, P., Marilleau, R., Olsen, L., Grison, R., Toppan, A. and Masson, C. (1995). Foraging behaviour of honeybees (*Apis mellifera* L.) on transgenic oilseed rape (*Brassica napus* L. var. *oleifera*). *Transgenic Research* 4: 270–276.

Picard Nizou, A.L., Grison, R., Olsen, L., Pioche, C., Arnold, G. and Pham-Delègue, M.H. (1997). Impact on proteins used in plant genetic engineering toxicity and behavioral study in the honeybee. *Journal of Economic Entomology* 90: 1710–1716.

Pierce, C.M.F., Solter, L.F. and Weinzierl, R.A. (2001). Interactions between *Nosema pyrausta* (Microsporidia: Nosematidae) and *Bacillus thuringiensis* subsp. *kurstaki* in the European corn borer (Lepidoptera: Pyralidae). *Journal of Economic Entomology* 94: 1361–1368.

Pierre, J. and Pham-Delègue, M.H. (2000). How to study the impact of genetically modified rape on bees? *OCL Oleagineux, Crops Gras, Lipides* 7(4): 341–344.

Pierre, J., Marsault, D., Genecque, E., Renard, M., Champolivier, J. and Pham-Delègue, M.H. (2003). Effects of herbicide-tolerant transgenic oilseed rape genotypes on honey bees and other pollinating insects under field conditions. *Entomologia Experimentalis et Applicata* 108: 159–168.

Pilcher, C.D., Obrycki, J.J., Rice, M.E. and Lewis, L.C. (1997). Pre-imaginal development, survival and field abundance of insect predators on transgenic *Bacillus thuringiensis* corn. *Environmental Entomology* 26: 446–454.

Poppy, G.M. and Sutherland, J.P. (2004). Can biological control benefit from genetically-modified crops? Tritrophic interactions on insect-resistant transgenic plants. *Physiological Entomology* 29: 257–268.

Raps, A., Kehr, J., Gugerli, P., Moar, W.J., Bigler, F. and Hilbeck, A. (2001). Immunological analysis of phloem sap of *Bacillus thuringiensis* corn and of the non-target herbivore *Rhopalosiphum padi* (Homoptera: Aphididae) for the presence of Cry1Ab. *Molecular Ecology* 10(2): 525–533.

Reed, G.L., Jensen, A.S., Riebe, J., Head, G. and Duan, J.J. (2001). Transgenic Bt potato and insecticides for Colorado potato beetle management: Comparative efficacy and non-target impacts. *Entomologia Experimentalis et Applicata* 100(1): 89–100.

Ren, L., Yang, Y.Z., Li, X., Miao, L., Yu, Y.S. and Qin, Q.L. (2004). Impact of transgenic Cry1A plus CpTI cotton on *Helicoverpa armigera* (Lepidoptera: Noctuidae) and its two endoparasitoid wasps *Microplitis mediator* (Hymenoptera: Braconidae) and *Campoletis chlorideae* (Hymenoptera: Ichneumonidae). *Acta Entomologica Sinica* 47: 1–7.

Riddick, E.W. and Barbosa, P. (1998). Impact of Cry3A-intoxicated *Leptinotarsa decemlineata* (Coleoptera: Chrysomelidae) and pollen on consumption, development, and fecundity of *Coleomegilla maculata* (Coleoptera: Coccinellidae). *Annals of the Entomological Society of America* 91: 303–307.

Riddick, E.W., Dively, G. and Barbosa, P. (1998). Effect of a seed-mix deployment of Cry3A-transgenic and nontransgenic potato on the abundance of *Lebia grandis* (Coleoptera: Carabidae) and *Coleomegilla maculata* (Coleoptera: Coccinellidae). *Annals of the Entomological Society of America* 91: 647–653.

Riddick, E.W., Dively, G. and Barbosa, P. (2000). Season-long abundance of generalist predators in transgenic versus nontransgenic potato fields. *Journal of Entomological Science* 35: 349–359.

Riggin Bucci, T.M. and Gould, F. (1997). Impact of intraplot mixtures of toxic and nontoxic plants on population dynamics of diamondback moth (Lepidoptera: Plutellidae) and its natural enemies. *Journal of Economic Entomology* 90: 241–251.

Romeis, J., Dutton, A. and Bigler, F. (2004). *Bacillus thuringiensis* toxin (Cry1Ab) has no direct effect on larvae of the green lacewing *Chrysoperla carnea* (Stephens) (Neuroptera: Chrysopidae). *Journal of Insect Physiology* 50: 175–183.

Romeis, J., Sharma, H.C., Sharma, K.K., Das, S. and Sarmah, B.K. (2004). The potential of transgenic chickpeas for pest control and possible effects on non-target arthropods. *Crop Protection* 23: 923–938.

Saxena, D., Flores, S. and Stotzky, G. (1999). Insecticidal toxin in root exudates from Bt corn. *Nature* 402: 480.

Saxena, D. and Stotzky, G. (2000). Insecticidal toxin from *Bacillus thuringiensis* is released from roots of transgenic Bt corn *in vitro* and *in situ*. *FEMS Microbial Ecology* 33: 35–39.

Saxena, D. and Stotzky, G. (2001). *Bacillus thuringiensis* (Bt) toxin released from root exudates and biomass of Bt corn has no apparent effect on earthworms, nematodes, protozoa, bacteria, and fungi in soil. *Soil Biology and Biochemistry* 33: 1225–1230.

Schuler, T.H., Denholm, I., Clark, S.J., Stewart, C.N. and Poppy, G.M. (2004). Effects of Bt plants on the development and survival of the parasitoid *Cotesia plutellae* (Hymenoptera: Braconidae) in susceptible and Bt-resistant larvae of the diamondback moth, *Plutella xylostella* (Lepidoptera: Plutellidae). *Journal of Insect Physiology* 50: 435–443.

Schuler, T.H., Denholm, I., Jouanin, L., Clark, S.J., Clark, A.J. and Poppy, G.M. (2001). Population-scale laboratory studies of the effect of transgenic plants on nontarget insects. *Molecular Ecology* 10: 1845–1853.

Schuler, T.H., Potting, R.P.J., Denholm, I. and Poppy, G.M. (1999). Parasitoid behaviour and Bt plants. *Nature* 401: 825–826.

Schuler, T.H., Potting, R.P.J., Denholm, I., Clark, S.J., Clark, A.J., Stewart, C.N. and Poppy, G.M. (2003). Tritrophic choice experiments with Bt plants, the diamondback moth (*Plutella xylostella*) and the parasitoid *Cotesia plutellae*. *Transgenic Research* 12: 351–361.

Schütte, G. and Riede, M. (1998). *Bacillus thuringiensis*-Toxine in Kulturpflanzen. In Nutzung der Gentechnik im Agrarsektor der USA – Die Diskussion von Versuchsergebnissen und Szenarien zur Biosicherheit, Umweltbundesamt Texte Nr. 47/98. Berlin, Germany.

Sears, M.K., Hellmich, R.L., Stanley-Horn, D.E., Oberhauser, K.S., Pleasants, J.M., Mattila, H.R., Siegfried, B.D. and Dively, G.P. (2001). Impact of Bt corn pollen on monarch butterfly populations: A risk assessment. *Proceedings National Academy of Sciences USA* 98: 11937–11942.

Sharma, H.C. and Ortiz, R. (2000). Transgenics, pest management, and the environment. *Current Science* 79: 421–437.

Sharma, H.C. and Pampapathy, G. (2006). Influence of transgenic cotton on the relative abundance and damage by target and non-target insect pests under different protection regimes in India. *Crop Protection* 25: 800–813.

Sharma, H.C., Dhillon, M.K. and Arora, R. (2008). Effects of *Bacillus thuringiensis* ∂-endotoxin-fed *Helicoverpa armigera* on the survival and development of the parasitoid, *Campoletis chlorideae*. *Entomologia Experimentalis et Applicata* 126: 1–8.

Sharma, H.C., Pampapathy, G. and Arora, R. (2007). Influence of transgenic cottons with *Bacillus thuringiensis cry1Ac* gene on the natural enemies of *Helicoverpa armigera*. *BioControl* 52: 469–489.

Sharma, H.C. and Waliyar, F. (2003). Genetic diversity, arthropod response, and pest management. In Waliyar, F., Collette, L. and Kenmore, P.E. (Eds.), *Beyond the Gene Horizon: Sustaining Agricultural Productivity and Enhancing Livelihoods Through Optimization of Crop and Crop-Associated Diversity with Emphasis on Semi-Arid Tropical Agro-ecosystems*, 23–25 September, 2002. Patancheru, Andhra Pradesh, India: International Crops Research Institute for the Semi-Arid Tropics (ICRISAT), and Rome, Italy: Food and Agriculture Organization, 66–88.

Sharma, H.C., Seetharama, N., Sharma, K.K. and Ortiz, R. (2002). Transgenic plants: Environmental concerns. In Singh, R.P. and Jaiwal, P.K. (Eds.), *Plant Genetic Engineering. Vol. 1. Applications and Limitations*. Houston, Texas, USA: Sci-Tech Publishing, 387–428.

Sharma, H.C., Sharma, K.K., Seetharama, N. and Ortiz, R. (2000). Prospects for using transgenic resistance to insects in crop improvement. *Electronic Journal of Biotechnology* 3, no 2. http://www.ejb.org/content/vol3/issue2/full/20.

Shelton, A.M., Zhao, J.Z. and Roush, R.T. (2002). Economic, ecological, food safety, and social consequences of the deployment of Bt transgenic plants. *Annual Review of Entomology* 47: 845–881.

Sims, S.R. (1995). *Bacillus thuringiensis* var. *kurstaki* (CryIA (c)) protein expressed in transgenic cotton: Effects on beneficial and other non-target insects. *Southwestern Entomologist* 20: 493–500.

Sims, S.R. and Holden, L.R. (1996). Insect bioassay for determining soil degradation of *Bacillus thuringiensis* subsp. *kurstaki* CryIA(b) protein in corn tissue. *Environmental Entomology* 25: 659–664.

Sims, S.R. and Ream, J.E. (1997). Soil inactivation of *Bacillus thuringiensis* sub sp. *kurstaki* CryIIA insecticidal protein within transgenic cotton tissue: Laboratory microcosm and field study. *Journal of Agricultural and Food Chemistry* 45: 1502–1505.

Stotzky, G. (2004). Persistence and biological activity in soil of the insecticidal proteins from *Bacillus thuringiensis*, especially from transgenic plants. *Plant and Soil* 266: 77–90.

Sun, C.G., Xu, J., Zhang, Q.W., Feng, H.B., Wang, F. and Song, R. (2002). Effect of transgenic Bt cotton on population of cotton pests and their natural enemies in Xinjiang. *Chinese Journal of Biological Control* 18: 106–110.

Sun, C.G., Zhang, Q.W., Xu, J., Wang, Y.X. and Liu, J.L. (2003). Effects of transgenic Bt cotton and transgenic Bt + CpTI cotton on the population dynamics of main cotton pests and their natural enemies. *Acta Entomologica Sinica* 46: 705–712.

Tapp, H. and Stotzky, G. (1998a). Absorption and binding of the insecticidal toxin proteins from *Bacillus thuringiensis*. sub sp. *kurstaki* and sub sp. *tenebrionis* on clay minerals. *Soil Biology and Biochemistry* 26: 663–679.

Tapp, H. and Stotzky, G. (1998b). Persistence of the insecticidal toxin from *Bacillus thuringiensis* subsp. *kurstaki* in soil. *Soil Biochemistry* 30(4): 471–476.

Tomov, B.W. and Bernal, J.S. (2003). Effects of GNA transgenic sugarcane on life history parameters of *Parallorhogas pyralophagus* (Marsh) (Hymenoptera: Braconidae), a parasitoid of Mexican rice borer. *Journal of Economic Entomology* 96: 570–576.

Van der Heijden, M.G.A., Klironomos, J.N., Ursic, M., Moutoglis, P., Streitwolf-Engel, R., Boller, T., Weimken, A. and Sanders, I.R. (1998). Mycorrhizal fungal diversity determines plant biodiversity, ecosystem variability and productivity. *Nature* 396: 69–71.

Venkateswerulu, G. and Stotzky, G. (1990). A simple method for isolation of the antilepidopteran toxin from *Bacillus thuringiensis* subsp. *kurstaki*. *Biotechnology and Applied Biochemistry* 12: 245–251.

Venkateswerulu, G. and Stotzky, G. (1992). Binding of the toxin and protoxin protein of *Bacillus thuringiensis* subsp. *kurstaki* on clay minerals. *Current Microbiology* 25: 1–9.

Volkmar, C. and Freier, B. (2003). Spider communities in Bt maize and not genetically modified maize fields. *Zeitschrift fur Pflanzenkrankheiten und Pflanzenschutz* 110: 572–582.

Wagner, D.L., Peacock, J.W., Carter, J.L. and Talley, S.E. (1996). Field assessment of *Bacillus thuringiensis* on nontarget Lepidoptera. *Environmental Entomology* 25(6): 1444–1454.

Walker, A.J., Ford, L., Majerus, M.E.N., Geoghegan, I.E., Birch, N., Gatehouse, J.A. and Gatehouse, A.M.R. (1998). Characterization of the mid-gut digestive proteinase activity of the two-spot ladybird, *Adalia bipunctata*, and its sensitivity to proteinase inhibitors. *Insect Biochemistry and Molecular Biology* 28: 173–180.

Wang, C.Y. and Xia, J.Y. (1997). Differences of population dynamics of bollworms and of population dynamics of major natural enemies between Bt transgenic cotton and conventional cotton. *China Cottons* 24: 13–15.

Williamson, M., Perrins, J. and Fitter, A. (1990). Releasing genetically engineered plants. Present proposal and possible hazards. *Trends in Ecology and Evolution* 5: 417–419.

Wold, S.J., Burkness, E.C., Hutchison, W.D. and Venette, R.C. (2001). In-field monitoring of beneficial insect populations in transgenic corn expressing a *Bacillus thuringiensis* toxin. *Journal of Entomological Science* 36(2): 177–187.

Wraight, C.L., Zangerl, A.R., Carroll, M.J. and Berenbaum, M.R. (2000). Absence of toxicity of *Bacillus thuringiensis* pollen to black swallowtails under field conditions. *Proceedings National Academy of Sciences USA* 97: 7700–7703.

Wu, K.M., Peng, Y.F. and Jia, S.R. (2003). What we have learnt on impacts of Bt cotton on non-target organisms in China. *AgBiotechNet* 5: 1–4.

Zwahlen, C., Hilbeck, A., Howald, R. and Nentwig, W. (2003). Effects of transgenic Bt corn litter on the earthworm *Lumbricus terrestris*. *Molecular Ecology* 12: 1077–1086.

Zwahlen, C., Nentwig, W., Bigler, F. and Hilbeck, A. (2000). Tritrophic interactions of transgenic *Bacillus thuringiensis* corn, *Anaphothrips obscurus* (Thysanoptera: Thripidae), and the predator *Orius majusculus* (Heteroptera: Anthocoridae). *Environmental Entomology* 29(4): 846–850.

12

Development of Resistance to Transgenic Plants and Strategies for Resistance Management

Introduction

Biopesticides, particularly those based on *Bacillus thuringiensis* (*Bt*) Berliner, are valuable alternatives to synthetic pesticides. Because of limited stability under field conditions and narrow activity spectrum, their use in pest management represents only a small fraction of the total pesticide usage in agriculture. Deployment of *Bt* Cry proteins in transgenic plants has overcome some of these drawbacks, and several cultivars of cotton, maize, potato, tomato, and tobacco have been released for cultivation. Use of *Bt* transgenic crops is expected to reduce pesticide use and provide a capability to control some of the insect pests that have developed high levels of resistance to conventional pesticides. However, concerns have been raised that deployment of insect-resistant transgenic plants will lead to development of resistance in insect populations to the transgene, and to herbicide and antibiotic genes used as selective markers. While some of these concerns may be real, others seem to be highly exaggerated. Therefore, careful thought should be given to the production and release of transgenic crops in different agroecosystems. Insect pest populations have a remarkable capacity to develop resistance, and over 500 species of insects have shown resistance to synthetic insecticides (Georghiou, 1990; Moberg, 1990; Rajmohan, 1998). Therefore, there is a need to take a critical view of resistance to *Bt* and other novel genes that are being deployed for pest management.

The spores and toxins of *Bt* are degraded quickly, and have a half-life of one week to a few months. Under UV light, the toxins are degraded within a few hours to a few days, while the spores need soil for germination. *Bt* toxins can persist in the soil for a longer period when absorbed on to humic acids. In such cases, the insecticidal activity is retained, and the toxins are more resistant to biodegradation. Toxin accumulation increases with an increase in the amount of humic acids (Crecchio and Stotzky, 1998). In comparison, transgenic plants produce toxin proteins throughout the crop cycle in all parts of the plant. Transgenic plants express semiactivated proteins, which are ingested by the insects.

In this process, no crystal solubilization and protoxin-toxin conversion is necessary in the insect midgut. Increased toxicity has been observed in some herbivorous insects when fed with the toxin instead of the pro-toxin, suggesting that such insects lack the appropriate enzyme composition or pH necessary to convert pro-toxin to toxin (Moar et al., 1995). The truncated and activated toxins expressed in transgenic plants in high doses kill the insects much faster than *Bt* toxins applied as sprays, which need to be activated by enzymes in the insect midgut. While *Bt* sprays might affect all the insects in an ecosystem wherever they are sprayed, the toxins from the transgenic plants affect only the insects that ingest the toxin from the plants through feeding or indirectly through the insect host or prey. Therefore, toxicity results and development of resistance to *Bt* formulations cannot be substituted for the toxins expressed in transgenic plants. Development of resistance and insect sensitivity to *Bt* toxins and other novel genes is not properly understood, and such information is important in the context of deployment of transgenic plants for pest management (Sharma and Ortiz, 2000). Since plants expressing only the *Bt* cry proteins have been released for cultivation, this chapter will largely focus on issues related to development of resistance to *Bt* Cry proteins and strategies for resistance management for sustainable crop production.

Tremendous progress had been made in developing insect-resistant transgenic plants with genes from *Bt*. Genetically modified plants with resistance to insects were developed in the mid-1980s (Barton, Whiteley, and Yang, 1987; Vaeck et al., 1987), and since then, there has been rapid progress in developing transgenic plants with insect resistance in several crops (Hilder and Boulter, 1999; Sharma, Sharma, and Crouch, 2004). Transgenic crops are now being grown on over 100 million ha worldwide (James, 2007). The success of *Bt* transgenic crops has far exceeded expectations, and does not preclude development of resistance to *Bt* toxins expressed in transgenic plants in the future. Key factors delaying development of resistance in insect populations to *Bt* crops include refuges of non-*Bt* host plants that enable survival of susceptible individuals, low resistance allele frequencies, recessive inheritance of resistance to *Bt* toxins, fitness costs associated with resistance, and the performance of resistant populations on non-*Bt* hosts. The relative importance of these factors may vary among different crops and cropping systems. Widespread and prolonged cultivation of *Bt* transgenic crops represents the largest adoption of insect-resistant crops over time and space (Ferre and Van Rie, 2002). Before large-scale deployment of transgenic crops for pest management was undertaken, many scientists predicted development of resistance to transgenic crops at a fast rate, a view re-inforced by selection for resistance to *Bt* Cry proteins under laboratory conditions. However, monitoring of insects in regions with high adoption of *Bt* crops has not yet led to detection of resistance in field populations of European corn borer, *Ostrinia nubilalis* (Hubner), pink bollworm, *Pectinophora gossypiella* (Saunders), corn earworm, *Helicoverpa zea* (Boddie), and tobacco budworm, *Heliothis virescens* (Fab.) in the United States; and *Helicoverpa armigera* (Hubner) in China, India, and Australia. Despite expectations that insect pests would rapidly evolve resistance to transgenic crops, increases in the frequency of resistance alleles caused by exposure to *Bt* crops in field have not yet been documented (Tabashnik et al., 2003). In laboratory and greenhouse tests, several strains with resistance to *Bt* toxins have been selected. However, strains with 70- to 10,000-fold resistance to *Bt* toxins in the artificial diet failed to complete development on *Bt* transgenic crops. Resistance to *Bt* appears to be costly and there is a rapid decline of resistance in populations collected from greenhouses and then maintained in the laboratory without selection.

With the development of resistance to *Bt* toxins, the value of transgenic crops will diminish greatly due to lower sensitivity of the target pest to the *Bt* toxins. One of the consequences of such a development would be that the farmers would resort to large-scale application of broad-spectrum insecticides, which will lead to serious environmental hazards associated with the use of synthetic insecticides. The potential for development of resistance to *Bt* proteins is not only of concern to farmers, but to scientists, extension agencies, and the transgenic plant industry. The investments made in the past would be turned useless unless this issue is addressed on an urgent basis. Most of the transgenic plants produced so far have *Bt* genes under the control of cauliflower mosaic virus (*CaMV35S*) constitutive promoter, and this system may lead to development of resistance in the target and nontarget insect species as the toxins are expressed in all parts of the plant (Harris, 1991). Toxin production also decreases over the crop-growing season, which may lead to development of resistance to the toxin used because of sublethal concentrations, and to other related *Bt* toxins to which the insect populations may initially be quite sensitive. Low doses of the toxins eliminate the most sensitive individuals of a population, leaving a population in which resistance can develop much faster. Since most *Bt* toxins have a similar mode of action, resistance developed against one toxin may also lead to development of cross-resistance to other toxins. Therefore, there is a need to take a critical look at the potential for development of resistance to *Bt* transgenic crops and develop strategies to deploy different *Bt* toxins alone or in combination with other novel genes and plant traits associated with resistance to insect pests in different crops.

Factors Influencing Insect Susceptibility to *Bt* Toxins

Variation in Insect Populations for Susceptibility to *Bt* Toxins

Considerable variation has been observed in the toxicity of Dipel® and purified Cry1Ac protein in 15 geographically diverse populations of *H. zea* and *H. virescens* collected from several locations in the United States (Stone and Sims, 1993). The variability in susceptibility of different populations of *H. zea* (16- to 52-fold) and *H. virescens* (17- to 71-fold) to *B. thuringiensis* formulations such as Javelin WG®, Dipel ES®, and Condor OF® was in contrast to Cry1A toxins. The LC_{50} ranges of *Bt* formulations and Cry toxins for field-collected populations were similar to those for laboratory colonies of *H. virescens*, but differed widely for *H. zea* (Luttrell, Wan, and Knighten, 1999).

Wide variation in susceptibility of *H. armigera* populations collected from different locations in India was first reported by Gujar et al. (2000) by using discriminating concentrations of *B. thuringiensis* var. *kurstaki* HD-1 strain (0.54 µg endotoxin g^{-1} diet) and HD-73 (1.4 µg endotoxin g^{-1} diet) (Figure 12.1). Chandrashekar et al. (2005) described a 16-fold variation in susceptibility of neonates of *H. armigera* (LC_{50} 0.023 µg to 0.372 µg^{-1} mL). Studies on susceptibility of *H. armigera* populations to *Bt* toxins from different parts of India indicated that Cry1Ac was most toxic, followed by Cry1Aa, and Cry1Ab (Kranthi, Kranthi, and Wanjari, 2001). The LC_{50} values ranged from 0.07 to 0.99 µg mL^{-1} (14-fold) for Cry1Aa, 0.69 to 9.94 µg mL^{-1} (14-fold) for Cry1Ab, and 0.01 to 0.67 µg mL^{-1} (67-fold) of diet for Cry1Ac. Across locations, the LC_{50} values were 0.62 µg mL^{-1} for Cry1Aa, 4.43 µg mL^{-1} for Cry1Ab, and 0.100 µg mL^{-1} of diet for Cry1Ac, which can be used as baseline susceptibility indices for resistance monitoring in the future. The LC_{99} values derived from cumulative data were 515 µg mL^{-1} for Cry1Aa, 13,385 µg mL^{-1} for Cry1Ab, and 75 µg mL^{-1} of diet for Cry1Ac, and these values represent the diagnostic doses for monitoring of resistance in *H. armigera* to *Bt* toxins. The range of LC_{50} values for Cry1Ac in *H. armigera* populations in Pakistan

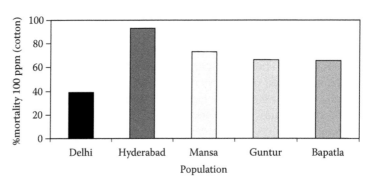

FIGURE 12.1   Variation in susceptibility of *Helicoverpa armigera* to *Bacillus thuringiensis* at different locations in India. (From Gujar, G.T. et al., 2000. *Current Science* 78: 995–1001. With permission.)

have shown more than 16-fold resistance to *Bt* toxins (Karim and Riazuddin, 1999). Pre-adaptation of insects to various temperature regimes, differences in genetic make-up of populations, and use of xenobiotics might be responsible for variations in susceptibility of insects to *B. thuringiensis* (Broza, 1986; Chandrashekar et al., 2005).

Extensive investigations on baseline susceptibility of *H. armigera* to *Bt* have been carried out in China (Zhao et al., 1996). Shen et al. (1998) reported that populations of *H. armigera* from Yanggu (Shadong), Handan (Hebei), Xinxian (Henan), Xiaoxian (Anhui), and Fengxian (Jiangsu) were resistant to *Bt* as compared to the susceptible strain. Populations of *H. armigera* from Yanggu and Xinxian were also resistant to transgenic cotton. This variation in insect susceptibility was attributed to large-scale use of *B. thuringiensis* var. *kurstaki*, and significant cultivation of transgenic cotton in China. Wu, Guo, and Nan (1999) reported 100-fold variation in susceptibility of *H. armigera* from five regions in China to Cry1Ac. Further studies on monitoring of insect susceptibility to Cry1Ac were carried out for the populations sampled from *B. thuringiensis* transgenic cotton fields by Wu, Guo, and Gao (2002), who found five-fold variation in IC_{50} (concentration producing 50% inhibition of larval development of third-instars). The IC_{50} ranged from 0.020 to 0.105 μg^{-1} mL, 0.016 to 0.099 μg^{-1} mL, and 0.016 to 0.080 μg^{-1} mL for 1998, 1999, and 2000 insect populations, respectively.

Holloway and Dang (2000) did not observe any differences in susceptibility of *H. armigera* and *Helicoverpa punctigera* (Wallengren) in cotton, in contrast to sweetcorn, using a discriminating concentration (10.02%) of MVP II formulation of Cry1Ac in Australia. *Helicoverpa armigera* is more tolerant to *Bt* toxins than *H. punctigera* (Liao, Heckel, and Akhurst, 2002). Only Cry1Ab, Cry1Ac, Cry2Aa, Cry2Ab, and Vip 3 killed *H. armigera* at dosages that could be considered to be effective for controlling this pest. There were no differences in the relative toxicity of Cry1Fa and Cry1Ac for *H. punctigera*, but Cry1Fa showed little toxicity to *H. armigera*. The differences in susceptibility to Cry1Ac and Cry2Aa were significant.

The LC_{50} of Cry1Ac to *Earias vittella* (F.) ranges from 0.006 to 0.105 μg g^{-1} of diet, representing a 17.5-fold variation in susceptibility from 27 locations in India (Kranthi et al., 2004). Variation in a majority of sites was around 10%. The strains collected from South India showed greater variation in susceptibility to *Bt* toxins. Reed and Halliday (2001) established Cry9C baseline susceptibility for field-collected populations of European corn borer, *O. nubilalis* and Southwestern corn borer, *Diatraea grandiosella* (Dyar). For the European corn borer, LC_{50} values ranged from 13.2 to 65.1 ng cm^{-2}, and the LC_{90} values from 46.5 to 214 ng cm^{-2}. The LC_{50} values for neonate larvae of Southwestern corn borer were 16.9 to 39.9 ng cm^{-2}, and the LC$_{90}$ values were 40.3 to 157 ng cm^{-2}. The most sensitive Southwestern corn borer colony was collected from the Mississippi delta exhibiting an LC$_{50}$ value of 22.6 ng cm^{-2}, which displayed the widest LC$_{90}$ confidence limits of 40.3 to 94.8 ng cm^{-2}. This baseline data can be used for monitoring of resistance to Cry9C in the future. Baseline susceptibility studies have also been carried out for Cry1Ab to fall armyworm, *Spodoptera frugiperda* (J.E. Smith) (Lynch et al., 2003). Comparison of LC$_{50}$ values for various colonies did not indicate an appreciable change in susceptibility of this insect from 1998 to 2000.

The LC$_{50}$ for diamondback moth, *Plutella xylostella* (L.), larvae from different locations in India varied from 1.0 to 10.97 mg per liter. Nine generations of selection for resistance resulted in an increase in LC$_{50}$ from 2.76 to 5.28 mg per liter (using a selection pressure of 6.4 mg per liter), indicating that there is a possibility of developing resistance to *Bt* in this insect in India (Mohan and Gujar, 2000). The susceptibility to different toxin proteins of *Bt* ranged from 0.14 to 3.74 µg mL^{-1} for Cry1Aa, 0.007 to 1.25 for Cry1Ab, 0.18 to 2.47 for Cry1Ac, and 0.12 to 3.0 µg for Cry1C (Mohan and Gujar, 2002). There were small differences in susceptibility of *Sesamia nonagrioides* (Lefebre) populations, but no differences were observed in two populations of *O. nubilalis* in Spain. In another study, differences in five populations of *O. nubilalis* from the United States were five-fold in susceptibility to Dipel, but the populations responded quite rapidly for resistance to *Bt*, indicating that low levels of susceptibility do not necessarily imply low potential to respond to selection pressure. Very high differences in susceptibility to *Bt* products and Cry1Ab and Cry1Ac have been found in the case of *H. virescens* and *H. zea* (Stone and Sims, 1993; Zhao et al., 1996; Wu, Gou, and Nan, 1999).

Effect of Host Plant on Insect Susceptibility to Cry Proteins

The susceptibility of neonates of *H. armigera* to *B. thuringiensis* ssp. *kurstaki* (HD-73) ranged from twofold (LC$_{50}$ 84.5 to 164.2 µg per liter) for insects collected from chickpea to about five-fold (LC$_{50}$ 51.1 to 247.7 µg per liter) for insects collected from cotton (Gujar et al., 2004) (Figure 12.2). The insects collected from pearl millet were twice as tolerant as those collected from cotton and sunflower at Sirsa, Haryana, India. Host-specific colonies raised from insects collected from cotton in Bhatinda, Punjab, India (by incorporating cabbage, cauliflower, chickpea, green pea, pearl millet, and pigeonpea into the artificial diet) indicated that insects reared on cabbage-, cauliflower-, and pearl millet-based diets

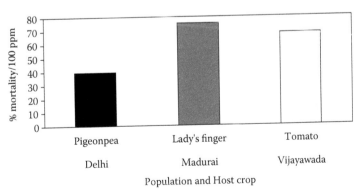

FIGURE 12.2 Variation in susceptibility of *Helicoverpa armigera* to *Bacillus thuringiensis* across host plants. (From Gujar, G.T. et al., 2004. *Entomologia Experimentalis et Applicata* 113: 165–172. With permission.)

were more susceptible than those reared on chickpea-, green pea-, and pigeonpea-based diets, suggesting that the host plant exercises a considerable influence on insect susceptibility to *Bt* toxins.

Detection, Development, and Monitoring of Resistance

Detection of Resistance

The probability of detecting one or more rare resistant larvae depends on sample size, the density of larvae on nontransformed plants, and assumed frequency of resistant phenotypes in a given population. Probability of detection increases with an increase in sample size, background density, or the frequency of resistant individuals (Venette, Hutchison, and Andow, 2000). Following binomial probability theory, if a frequency of 10^{-3} to 10^{-4} is expected for the resistance alleles, 10^3 to 10^4 samples must be collected to have 95% probability of locating one or more resistant larvae. Estimates of the phenotypic frequency of resistance from an in-field screen can be useful for estimating initial frequency of resistant alleles.

Deeba et al. (2003) developed an insect growth inhibition assay to monitor development of resistance to *Bt* toxins in spotted bollworm, *E. vittella*. A discrimination-dose-assay based on 160 g of seeds of *Bt* cotton with the *cryAc* gene in 1.3 liters of artificial diet has been proposed to monitor *H. armigera* resistance to *Bt* (Kranthi et al., 2005a). Cotton leaf feeding assay has been also used to assess the toxicity of Cry1Aa, Cry1Ab, and Cry1Ac to the semilooper, *Anomis flava* (F.) (Kranthi and Kranthi, 2000). The LC_{50} values of Cry1Ac, Cry1Ab, and CryA1a for neonate larvae were 0.79 to 1.11, 3.48 to 4.12, and 4.98 to 6.08 ng cm^{-2} of leaf, respectively; and of Cry1Ac for the fourth-instars ranged from 12.91 to 21.14 ng cm^{-2}, and for Cry1Aa and Cry1Ab, the values ranged from 53.0 to 138 ng cm^{-2}. Cry1Ab at a concentration corresponding to the upper limit of LC_{99} produced >99% mortality, while in the case of Cry1Ac, further adjustments and validation may be necessary.

Liu and Tabashnik (1998) devised a procedure to eliminate a recessive allele conferring resistance to *B. thuringiensis* in *P. xylostella* populations composed of resistant and susceptible individuals. The susceptible homozygous and heterozygous individuals were killed with a diagnostic concentration of Cry1Ab and Cry1Aa. The LC_{50} of Cry1Ab at five days was sevenfold lower for the susceptible strain than for the heterogeneous strain. Diagnostic concentrations of Cry1Ab and Cry1Ac for *O. nubilalis* based on LC_{99} and ED_{99} (based on earlier baseline data) did not detect any resistance development in populations collected from different locations in the United States (Marcon et al., 2000).

Development of Resistance under Laboratory Conditions

The ability of arthropods to develop resistance depends on the genetic variability of insects used in the selection, presence of resistance alleles, selection pressure, and duration of selection over the generations. The resistant larvae are able to feed and complete larval development and produce normal pupae from which fertile adults emerge. Resistance to *B. thuringiensis* var *kurstaki* was first reported in Indian meal moth, *Plodia interpunctella* (Hubner) (McGaughey, 1985; Van Rie et al., 1990). A 100-fold resistance was recorded after

15 generations, and reached 250-fold after 36 generations of selection with Dipel® (McGaughey and Beeman, 1988). However, the resistant colony was still susceptible to a strain of *Bt* expressing toxin proteins different than Cry1A. The resistant colony was 800-fold resistant to Cry1Ab, but fourfold more susceptible to Cry1Ca (McGaughey and Johnson, 1987). However, selection for resistance to Dipel in another colony increased resistance to Cry1Ab, without increasing susceptibility to Cry1Ca (McGaughey and Johnson, 1992, 1994; Herrero, Oppert, and Ferre, 2001). Colonies of *P. interpunctella* selected for resistance to other strains of *Bt* showed low levels of resistance to Cry1Ca probably because of low frequencey of resistance alleles to this toxin (Ferre and Van Rie, 2002).

Almond moth, *Cadra cautella* (Walker) (McGaughey and Beeman, 1988), cotton leaf beetle, *Chrysomela scripta* (F.) (Bauer et al., 1994), tobacco budworm, *H. virescens* (Stone, Sims, and Marrone, 1989; MacIntosh et al., 1991; Gould et al., 1992), European corn borer, *O. nubilalis* (Bolin, Hutchinson, and Andow, 1995; Bolin, Hutchinson, and Davis, 1996), cotton leafworm, *Spodoptera littoralis* (Boisd.) (Moar et al., 1995), and cabbage looper, *Trichoplusia ni* (Hubner) (Estada and Ferré, 1994) have also been selected for resistance *Bt* toxins under laboratory conditions.

Diamondback moth, *P. xylostella*, is the only insect known to have developed resistance to *Bt* under field conditions in regions heavily sprayed with *Bt* formulations. The highest levels of resistance recorded have been 30-fold (Tabashnik et al., 1990). Selection under laboratory conditions has resulted in up to 1,000-fold resistance (Tabashnik et al., 1993), and the resistant colony displayed high levels of resistance to Cry1 toxins, while the levels of resistance to Cry2a were not significant (Tabashnik et al., 1993, 1994a, 1996). Over 90% of the larvae from the resistant colony survived on transgenic canola expressing *cry1Ac*, while the unselected ones died on the transgenic plants (Ramachandran et al., 1998b). A resistant strain from the Philippines expressed 200-fold resistance to Cry1Ac (Ferre et al., 1991), but was susceptible to Cry1Ba and Cry1Ca. Following selection with Cry1Ab and later with hybrid Cry1A protoxin, partial resistance was detected against Cry1A toxins, but not against Cry1Ca, Cry1Fa, and Cry1Ja (Tabashnik et al., 1997b). In a colony from Florida, 1500-fold resistance was observed in the second generation, which fell to 300-fold in subsequent generations in the absence of selection pressure, and remained stable thereafter (Tang et al., 1996). These insects were resistant to Cry1A toxins, but were susceptible to toxin proteins such as Cry1Ba, Cry1Ca, Cry1Da, and Cry9Ca (Lambert, Bradley, and van Duyn, 1996; Tang et al., 1996). Selection under laboratory conditions with Cry1Ca, and later on transgenic broccoli expressing *cry1Ca*, increased resistance levels to 12,400-fold (Zhao et al., 2000a), suggesting that *P. xylostella* has the capability to develop resistance against several *Bt* proteins in a short span of time. Several colonies of *P. xylostella* have been selected for resistance to *Bt* proteins, and resistance is retained in the absence of selection pressure for several generations, but the expression of resistance to different toxins is quite variable (Sayyed et al., 2000).

High levels of resistance (10,000-fold) have been obtained in a colony of *H. virscens* to Cry1Ac (Gould et al., 1995), which also exhibited resistance to Cry1Ab and Cry1Fa, but no resistance to Cry1Ba and Cry1Ca. Following continued selection for resistance to Cry1Ac or Cry1Ab, moderate levels of resistance were recorded against Cry2a (Stone, Sims, and Marrone, 1989; Sims and Stone, 1991; Kota et al., 1999). Further selection for resistance to Cry2Aa resulted in development of resistance to Cry1Ac and Cry2a. However, the larvae from the resistant colony suffered 100% mortality on tobacco plants expressing *cry2Aa* (Kota et al., 1999). Larvae of *H. armigera* have shown potential to develop resistance to Cry1Ac under selection pressure; 76-fold resistance was observed after nine generations (Kranthi et al., 2000) and 31.4-fold after six generations (Chandrashekar and Gujar, 2004).

Studies on the stability of resistance have shown relatively high sustenance of resistance (about 50- to 100-fold over the susceptible laboratory strain) (Akhurst, James, and Bird, 2000). Resistance to Cry1Ac in *H. armigera* has been detected after 16 generations of continuous selection (Akhurst et al., 2003). The resistance ratio (RR) peaked approximately 300-fold at generation 21, after which it declined and oscillated between 57- and 111-fold. First-instar *H. armigera* from generation 25 (RR = 63) were able to complete larval development on transgenic cotton expressing *cry1Ac* and produce fertile adults. There is a fitness cost associated with resistance to *Bt* on cotton and on artificial diet. The BX strain is susceptible to commercial *Bt* spray formulations Dipel and XenTari (which contain multiple insecticidal crystal proteins), but is resistant to the MVP formulation, which contains only Cry1Ac. This strain is also resistant to Cry1Ab, but not to Cry2Aa or Cry2Ab. Liang, Tan, and Guo (2000) reported about 43-fold resistance in *H. armigera* to *Bt* cotton over 16 generations of selection on the basis of mean concentration for 50% weight loss of larvae. Meng et al. (2004) selected *H. armigera* on transgenic cotton using a leaf-feeding method for over 45 generations. Resistance to transgenic *Bt* cotton was detected after 12 generations. The survival rates of F_{12} neonates feeding on leaves (for 4 days) and boll-opening (five days) of R19 line were 34 and 72%, respectively, compared to 0% and 40% for the nonselected strain. Resistance to *B. thuringiensis* HD-1 (Dipel) in neonates (F_{12}) was six-fold. After 42 generations, the strain developed very high levels of resistance to Cry1Ac protoxin, 210 g kg^{-1} MVPII wettable powder, and 200 g liter^{-1} MVPII liquid formulation—the resistance ratios being 540-, 580-, and 510-fold, respectively. Selection of *P. gossypiella* larvae for resistance to Cry1Ac resulted in 300-fold resistance, and cross resistance to Cry1Aa and Cry1Ab, low levels of resistance to Cry1Bb, but no resistance to Cry1Ca, Cry1Da, Cry1Fa, Cry2Aa, and Cry9Ca. Larvae from the resistant colony showed 40% survival on the *Bt* transgenic cotton expressing *cry1Ac* (Tabashnik et al., 2000a). A resistant colony has also been selected on artificial diet containing leaf powder from the *Bt* transgenic cotton. A strain of recombinant *Pseudomonas fluorescens* Migula engineered to express the 130 kDa delta-endotoxin protein of *B. thuringiensis* subsp. *kurstaki* strain HD-1 has been used to select a resistant strain of *H. virescens* for 14 generations (Sims and Stone, 1991). After selection, the resistant strain was 12-fold more resistant to the endotoxin than the susceptible strain.

Decreased susceptibility to Dipel has been observed in European corn borer, *O. nubilalis*, colonies selected for resistance to this product under laboratory conditions (Huang et al., 1999). Selection for resistance to Cry1Ac resulted in 160-fold resistance to this toxin. However, continued selection resulted in a decrease in resistance (Bolin, Hutchinson, and Andow, 1999). The highest level of resistance in *O. nubilalis* occurred between generations 7 (14-fold) and 9 (32-fold) for three different strains (Chaufaux et al., 2001). For each strain, the level of resistance fluctuated from generation to generation, although there were consistently significant decreases in toxin susceptibility across generations. These results suggested that low levels of resistance are common among widely distributed *O. nubilalis* populations.

Selection for resistance in beet armyworm, *Spodoptera exigua* (Hubner), to Cry1Ca resulted in 850-fold resistance, and cross resistance to Cry1Ab, Cry2Aa, and Cry9Ca (Moar et al., 1995). The larvae of beet armyworm pupated earliest when fed a Cry1Ac diet (first 20% to pupate) and produced offspring that developed significantly faster on a Cry1Ac diet than the parental strain (Sumerford, 2002). After two generations of selection, the selected population pupated two days faster than the parental population. The selected group also developed faster on transgenic *Bt* cotton (NuCOTN 33B) leaves than the parental strain. Individuals selected on media containing Cry1Ac developed no more rapidly on artificial diet containing Cry2Aa than the parental control colony of *S. exigua*. Selection for resistance in *S. littorallis* to *Bt* strain expressing Cry1Ca resulted in 500-fold resistance, and showed low levels of resistance to Cry1Ab and Cry1Da, but no resistance to Cry1Fa (Muller-Cohn et al., 1996).

A colony of *Leptinotarsa decemlineata* (Say) from fields sprayed with *Bt* formulation containing Cry3Aa, and selection for resistance to this toxin for 29 generations resulted in 293-fold resistance (Whalon et al., 1993; Rahardja and Whalon, 1995). However, foliage from transgenic potato lines expressing *cry3Aa* inhibited feeding by the grubs (Altre, Grafius, and Whalon, 1996). High levels of resistance have also been observed in *C. scripta* under selection pressure (Ballester et al., 1999). The selected strain displayed resistance to Cry1Ba, but not to Cyt1Aa (Federici and Bauer, 1998).

Development of Resistance under Field Conditions

The ability of insects to overcome plant resistance is always a grave risk, and ways to delay the onset of resistance in insect populations will be an ongoing debate as transgenic crops are deployed for cultivation on a large scale. Extensive and intensive exposure of insect pests to *Bt* toxins through transgenic crop plants or other tactics may lead to development of resistance to *Bt*. Development of resistance to *Bt* under field conditions may not be a serious issue since the *Bt* and the insect pests have co-evolved for millions of years under natural conditions (Bauer, 1995; Tabashnik, 1994). Because of limited exposure and several toxins produced by *Bt*, the rate of development of resistance under natural conditions may not be high. In transgenic plants, the insects are continuously exposed to the exotic genes, and there are possibilities of resistance development in the target insects.

Soybean looper, *T. ni*, collected from soybean and *Bt* cotton has been found to be less susceptible to *Bt* formulation, Condor XL in dosage-mortality and discriminating dosage bioassays than the reference strain (Mascarenhas et al., 1998). In some insect species, the probability of development of resistance may be very low, for example, *O. nubilalis* has been observed to develop tolerance to low levels of Cry1Ab in the artificial diet, but it has not been possible to initiate or sustain the insect colonies at concentrations in the diet closer to the actual levels expressed in the transgenic maize plants (Lang et al., 1996).

Kranthi and Kranthi (2004) estimated that when 40% of the total area under cotton in India would be covered with *Bt* cotton, it would take 11 years for resistance gene frequency to reach 0.5 in *H. armigera* populations if no pest control measures are adopted. If control operations cause 90% mortality, then it would take 45 years for resistance allele frequency to reach 0.5. Using a single locus simulation model involving ecological, biological, and operational factors, Zhao et al. (2000b) showed that resistance allele frequency of *H. armigera* to Cry1A toxin will increase from 0.001 to 0.5 after 38 generations (nine years) in a typical cropping system in China, where corn fields act as a natural refuge for *Bt* cotton. If the whole area in a region is cropped with *Bt* cotton, the expected time for field failure of *Bt* cotton will be only 26 generations (seven years). Similar views have been expressed for development of resistance to *H. zea*, which is less susceptible to *Bt* cotton (Han and Caprio, 2002). Baseline susceptibility studies and the ability of *H. armigera* to develop resistance to *Bt* have to some extent justified the fears of likelihood of development of resistance under field conditions.

Mechanisms of Resistance

Ferre and Van Rie (2002) reviewed the current knowledge on the biochemical mechanisms and genetics of resistance to *Bt* products and insecticidal crystal proteins. Monitoring the early phases of *Bt* resistance evolution in the field has been viewed as crucial, but extremely difficult, especially when resistance is recessive. Mechanisms associated with development of resistance in insect populations to *Bt* are discussed below.

Reduced Protoxin-Toxin Conversion

Reduced protoxin-toxin conversion is one of the mechanisms of resistance to *Bt*. A resistant colony of *P. interpunctella* displayed reduced capacity to activate Cry1Ac (Oppert et al., 1996), linked to a genetic linkage between absence of major gut proteases and susceptibility of Cry1Ac (Oppert et al., 1997). Altered proteolytic activity is also one of the mechanisms of resistance in this insect to *Bt* toxins (Johnson et al., 1990). However, altered proteolytic activity is not associated with development of resistance to *Bt* in *H. virescens* (Marrone and MacIntosh, 1993), and *P. xylostella* (Liu and Tabashnik, 1997; Tabashnik, Finson, and Johnson, 1992). Slower activation and faster degradation of Cry1Ab toxin in midgut extract in a resistant strain *of H. virescens* is associated with resistance to this toxin (Forcada et al., 1996). Reduced protoxin-toxin conversion is also one of the mechanisms of resistance to Cry1Ca in *P. xylostella* (MacIntosh et al., 1991).

Reduced Binding to Receptor Proteins

Reduced binding is one of the mechanisms of resistance in *P. interpunctella* to Cry1Ab (Van Rie et al., 1990). However, there were no differences between the resistant and susceptible strains in binding levels for Cry1Ca, to which the insect does not display any resistance. Nearly 60-fold reduction in binding has been observed in another colony to Cry1Ab (Herrero, Oppert, and Ferre, 2001). However, there were no differences in binding to Cry1Ac. Resistance to Cry1Ac in *H. virescens* is not linked to binding to Cry1Ac or Cry1Ab, but binding to Cry1Ca was reduced drastically, to which the insect showed moderate levels of cross-resistance (Van Rie et al., 1990; Lee et al., 1995). Competition binding studies indicated that Cry1Aa binds to receptor A, Cry1Ab to receptors A and B, and Cry1Ac to receptors A, B, and C. Altered Cry1Aa binding site may contribute to resistance to all three Cry1A proteins (Lee et al., 1995). Cry1Fa and Cry1Ja also share the binding site A with Cry1A toxins in *H. virescens* (Jurat-Fuentes and Adang, 2001). In another colony of *H. virescens*, small or no compensatory changes were observed in binding to Cry1Ab or Cry1Ac (MacIntosh et al., 1991; Gould et al., 1992). Reduced binding is one of the mechanisms of resistance to Cry1Ac in *H. armigera* (Akhurst et al., 2003).

Reduced binding is one of the mechanisms of resistance in *P. xylostella* to Cry1Ab (Tabashnik et al., 1997b) and Cry1Ac (Tabashnik et al., 1994b, 1997b). In another colony, binding to Cry1Ab was reduced dramatically, but not to Cry1Aa and Cry1Ac (Tabashnik et al., 1997b; Ballester et al., 1999). Competition experiments have indicated the presence of a site recognized only by Cry1Aa, another site by Cry1Aa, Cry1Ab, Cry1Ac, Cry1Fa, and Cry1Ja, and two additional sites by Cry1Ba and Cry1Ca (Ferre et al., 1991; Ballester et al., 1994, 1999; Granero, Ballester, and Ferre, 1996). In the resistant strain, there was no binding to Cry1Ab and Cry1Ac, while Cry1Aa binds equally well to brush border membrane vesicles (BBMV) of the resistant and susceptible strains (Tabashnik et al., 1997b). In general, patterns of cross resistance correspond to binding patterns of different toxins to receptors in the BBMV. Receptor specificity corresponds to levels of amino acid homology in domain II (Tabashnik et al., 1996). Lack of binding to Cry1Ab has also been demonstrated through histopathological studies (Bravo, Jansens, and Peferoen, 1992). Gahan, Gould, and Heckel (2001) suggested that disruption of a cadherin-superfamily gene by retrotransposon-mediated insertion was linked to high levels of resistance to the *Bt* toxin Cry1Ac in *H. virescens*.

In the susceptible strain of *P. gossypiella*, Cry1Aa, Cry1Ab, Cry1Ac, and Cry1Ja bound to a common binding site that was not shared by the other toxins tested (Gonzalez et al., 2003). Reciprocal competition experiments with Cry1Ab, Cry1Ac, and Cry1Ja showed that these toxins do not bind to any additional binding sites. In the resistant strain, binding of 125I-Cry1Ac was not significantly affected. However, 125I-Cry1Ab did not bind to the BBMV. The results indicated that resistance fits the "mode 1" pattern of resistance described previously in *P. xylostella*, *P. interpunctella*, and *H. virescens*. Griffitts et al. (2001) reported the cloning of a *Bt* toxin resistance gene, *Caenorhabditis elegans bre-5*, which encodes a putative beta-1,3-galactosyltransferase. The lack of *bre-5* in the intestine led to development of resistance to *Bt* toxin Cry5B.

Feeding Behavior

Feeding behavior, survival, and development of insects on transgenic crops will have a major bearing on development of resistance to transgenic crops. Gore et al. (2002) observed a change in behavior of *H. zea* larvae infesting *Bt* cotton (Bollgard®, NuCOTN33B). More larvae moved from the terminal parts of *Bt* cotton than in non-*Bt* cotton within one hour of infestation. Greater infestation was recorded on white flowers and small bolls, necessitating scouting of these plant parts in addition to terminals and squares. The first- and second-instar larvae damage squares and flowers, but mostly feed on cotton bolls from third-instar onwards, and there were no differences in feeding behavior on transgenic and nontransgenic cotton (Wu, Guo, and Wang, 2000). However, Zhao et al. (2000b) suggested that flowers of *Bt* cotton were more preferred by the third- and fourth-instars, since they did not express high doses of toxin. The *H. armigera* larvae have the ability to compensate for food intake, digestibility, and utilization of transgenic crops (Gujar, Kalia, and Kumari, 2001). Neonates of *H. armigera* have the ability to detect and avoid transgenic *Bt* cotton Zhong 30 and transgenic *CpTI-Bt* cotton SGK 321, as compared to the nontransgenic cotton Shiyuan 321 (Zhang et al., 2004). The larvae consumed more food on *CpTI-Bt* transgenic cotton than on *Bt* transgenic cotton, possibly to compensate for reduced nutritional quality of the food. Fourth-instars were found in equal numbers on transgenic and nontransgenic cottons, but food consumption on transgenic cotton was lower than on the nontransgenic cotton. In no-choice tests involving fifth-instars, significantly less time was spent in feeding on the two transgenic cottons. The neonates selectively feed on the nontransgenic cotton or the preferred plant parts. Diamondback moth, *P. xylostella*, resistant to *Bt* toxins may be able to use Cry1Ac as a supplementary food protein (Sayyed, Cerda, and Wright, 2003). *Bt* transgenic crops could therefore have unanticipated nutritionally favorable effects, increasing the fitness of resistant populations, and resulting in evolution of resistance to *Bt* transgenic crops.

Cross Resistance

Akhurst and Liao (1996) studied the susceptibility of *H. armigera* to Cry1Ab, Cry1Ac, Cry2Aa, and Cry2Ab. The Cry1Ac selected insects with 188-fold resistance (in the 18th generation) showed 69-fold resistance to Cry1Ac of MVP formulation, and five-fold resistance to Dipel® that contains Cry1Ab, Cry1Ac, and Cry2Aa. The Cry1Ac selected insects with 82-fold resistance (in the 28th generation) showed as high as 157-fold resistance to Cry1Ab. Cry1Ac selected insects with 204-fold resistance (in the 19th generation) showed only 2.3-fold resistance to Cry2Aa, suggesting that Cry1Ac resistance extended to Cry1Ab, but not to Cry2Aa. *Helicoverpa armigera* resistant to Cry1Ac did not possess cross resistance to Cry1C and Cry2Aa, but was resistant to *Bt* corn expressing Cry1A (Zhao et al., 2000b). Chandrashekar et al. (2005) also observed 31.4-fold resistance to Cry1Ac, 25.4-fold resistance to Cry1Ab, and only 8.4-fold resistance to Cry1Aa in strains of *H. armigera* selected for resistance to these *Bt* toxins.

Resistance to the *Bt* formulation, Javelin increased 1.9 to 4.4 times under field conditions, but was significant only at an application rate of 1.12 kg ha^{-1}, irrespective of the presence or absence of a refuge (Perez, Shelton, and Roush, 1997). Selection with Javelin at 0.3 kg ha^{-1} or Xentari did not cause a significant increase in resistance to *Btk* nor did *P. xylostella* selected with Xentari resistance show resistance to *B. thuringiensis* subsp. *aizawai*. Two strains of the diamondback moth, *P. xylostella*, selected using Cry1C protoxin and transgenic broccoli plants expressing *cry1C*, were resistant to Cry1C, but had different cross-resistance patterns (Zhao et al., 2001). In a study involving 12 protoxins (Cry1Aa, Cry1Ab, Cry1Ac, Cry1Bb, Cry1C, Cry1D, Cry1E, Cry1F, Cry1J, Cry2Ab, Cry9Aa, and Cry9C), the resistance ratio (RR) of one strain (BCS-Cry1C-1) to the Cry1C protoxin was 1,090-fold with a high level of cross-resistance to Cry1Aa, Cry1Ab, Cry1Ac, Cry1F, and Cry1J (RR > 390-fold). The cross resistance to Cry1A, Cry1F, and Cry1J in this strain was probably related to Cry1A resistance gene(s) that came from the initial field population as a result of intensive sprayings of *Bt* products containing Cry1A protoxins. The neonates of this strain survived on transgenic broccoli plants expressing either *cry1Ac* or *cry1C* toxins. The other strain (BCS-Cry1C-2) resistant to Cry1C did not show cross resistance to other *Bt* protoxins. The neonates of this strain survived on transgenic broccoli expressing *cry1C*, but not *cry1Ac*.

Genetics of Resistance

Frequency of Resistance Alleles

Frequency of resistance allele between generations in the insect population is based on the selection intensity, fitness cost, percentage of refugia, mortality in refugia due to insecticide sprays, initial frequency of the resistance allele, and the rate of insect movement (Cerda and Wright, 2004). A simulation model for resistance development based on four different spatial patterns of refugia inside the field (border, central, equidistant, and random), four crop rotation patterns, five different sizes of refuge (5% to 50%), and two levels of non-*Bt* insecticide plus an untreated control in the refugia indicted that:

- The greater the size of refugia, the lower the rate of increase in the frequency of resistance alleles.
- Positioning patches of refugia at random in the field resulted in a higher rate of increase in the frequency of resistance alleles compared with nonrandom positions.
- Positioning refugia along the border resulted in a lower rate of increase in the frequency of resistance alleles compared with equidistant and central patterns.
- The frequency of resistance allele increased markedly when the refugia was less than 10%.

- Use of temporal refugia (rotation with a non-*Bt* insect host crop) led to a decline in the frequency of the resistance allele.
- Higher rates of movements increased the rate at which the frequency of resistance increased.
- The frequency of the resistance allele increased when the refugia were sprayed with insecticide.

The frequency of resistant phenotypes can be estimated as the ratio of density of larvae per plant in *Bt* crops to the density of larvae per plant in an adjacent non-*Bt* crop (Andow and Hutchinson, 1998). Functional dominance of resistance alleles and the initial frequency of those alleles have a major impact on evolution of resistance (Storer et al., 2003). The survival of susceptible insects on the transgenic crops and the population dynamics of the insect, driven by winter survival and reproductive rates, were also important. In addition, agricultural practices, including the proportion of the area planted with maize, and the larval threshold for insecticide sprays affected the resistance-allele frequency.

Initial frequency of resistance genes is the key determinant for predicting evolution of resistance in an insect population. However, only a few estimates are available for resistance gene frequencies under field conditions. Andow and Alstad (1998) proposed a procedure to estimate resistance gene frequencies in field populations based on F_2 screening. It does not require a laboratory resistant population, and is more sensitive than discriminating dose assay for detection of recessive traits. This method estimated no homozygotes in *O. nubilalis*, and the frequency of Cry1Ab resistance alleles ranged from <0.013 to 0.0039 in the United States (Andow et al., 2000). Resistance gene frequency in *H. virescens* has been estimated to be 1.5×10^{-3} in the United States (Gould et al., 1995), 4×10^{-3} for low-level resistance in *H. armigera*, and $<10^{-3}$ for high levels of resistance in *P. xylostella* in Australia. However, high levels of resistance alleles have been estimated for *P. xylostella* in Hawaii (0.12) and Arizona (0.16) (Tabashnik et al., 1997a, 2000b). Gould et al. (1997) reported 10^{-3} frequency of resistant alleles in natural populations of *H. virescens*. Akhurst, James, and Bird (2000) also estimated the frequency of resistance alleles at 10^{-3} in *H. armigera*, which was supposed to be three times more common than normally assumed (10^{-6}) for resistant genes in the field population. At this frequency, resistance is likely to be a significant problem in the field in less than 16 generations (4 to 5 years), if *Bt* resistance management strategies are not implemented. Estimated frequency of a recessive allele conferring resistance to *Bt* toxin Cry1Ac in *P. gossypiella* was 0.16 (Tabashnik et al., 2003). Unexpectedly, the estimated resistance allele frequency did not increase from 1997 to 1999 and *Bt* cotton remained extremely effective against pink bollworm. The finding that 21% of the individuals from a susceptible strain were heterozygous for the multiple-toxin resistance gene indicated that the resistance allele frequency was 10 times higher than the most widely cited estimate of the upper limit for the initial frequency of resistance alleles in susceptible populations. These findings suggest that insect pests may evolve resistance to some groups of toxins much faster than previously expected.

Frequency of resistance in *O. nubilalis* to *Bt* maize was $<3.9 \times 10^{-3}$ in an Iowa population (Andow et al., 2000), indicating that the refuge plus high-dose strategy may be effective for managing resistance in *Bt* maize. Partial resistance to Cry1Ab toxin was found to be common. The 95% CI for the frequency of partial resistance were 8.2×10^4 to 9.4×10^4 for the Iowa population. Variable costs of the method were $14.90 per isofemale line, which was a reduction of 25% compared with the initial estimate. Bourguet et al. (2003) screened >1200 isofemale lines of *O. nubilalis*, and no alleles conferring resistance to *Bt* maize producing the Cry1Ab toxin were detected. Frequency of resistance alleles in France was <9.20×10^{-4}. In the northern U.S. Corn Belt, the frequency of the resistance allele to *Bt* maize was <4.23×10^{-4}. The results suggested that resistance is probably rare in France and the U.S. Corn Belt for the high-dose plus refuge strategy to delay development of resistance to *Bt* maize.

Inheritance of Resistance

Resistance to *Bt* products or Cry toxins, in general, behaves as a completely or partially dominant trait (Liu and Tabashnik, 1997; Tabashnik et al., 1997a, 2000b; Sayyed, Ferre, and Wright, 2000; Sayyed et al., 2000; Liu et al., 2001). Resistance to *Bt* products and Cry toxins at the LC_{50} level is partially recessive, but closer to codominance than to being completely recessive (McGaughey, 1985; McGaughey and Beeman, 1988; Gould et al., 1992, 1995; Martinez-Ramirez et al., 1995; Chaufaux et al., 1997; Imai and Mori, 1999; Sayyed, Ferre, and Wright, 2000; Sayyed et al., 2000). High levels of dominance have been observed in *O. nubilalis* for Dipel (Huang et al., 1999), *L. decemlineata* for Cry3Aa (Rahardja and Whalon, 1995), *P. xylostella* for Cry1Ca (Liu and Tabashnik, 1997), and in *H. virescens* for Cry1Ab and Cry2a (Gould et al., 1995). In general, there is a single locus or tightly linked loci associated with resistance to *Bt*, and resistance is inherited as an autosomal trait. Lack of binding in *P. xylostella* is inherited as an autosomal recessive trait (Martinez-Ramirez et al., 1995). However, in a few cases, sex had a significant influence on the survival of F_1 progeny in *P. xylostella* (Martinez-Ramirez et al., 1995; Sayyed et al., 2000) and *S. littoralis* (Chaufaux et al., 1997). Studies involving the progeny of crosses and backcrosses between resistant and susceptible strains showed that resistance was autosomally inherited, incompletely dominant, and controlled by several genetic factors (Sims and Stone, 1991). Genetic control of resistance was unstable. An unchallenged line with an initial resistance of 69-fold declined to 13-fold by generation 5 of nonselection.

Inheritance of resistance to the *Bt* toxins in transgenic crops is typically recessive, and DNA-based screening for resistance alleles in heterozygotes is potentially much more efficient than detection of resistant homozygotes with bioassays (Morin et al., 2003). Using a 37-fold resistant strain, Liang et al. (2000) reported that inheritance of resistance to *Bt* transgenic cotton in *H. armigera* was controlled by a single autosomal incomplete recessive allele. Daly and Olsen (2000) found that resistance in *H. armigera* was not always recessive, and was probably functionally dominant under field conditions at certain times. Resistance in the *Bx* strains of *H. armigera* is completely recessive (Akhurst et al., 2003). It seems that the dominance of resistance depends on the level of selection, as it increased with a decrease in concentration of Cry1Ac for the pink bollworm (Tabashnik et al., 2000b, 2002a, 2002b). Populations of pink bollworm, *P. gossypiella*, harbored three mutant alleles of a cadherin-encoding gene linked with resistance to *Bt* toxin Cry1Ac and survival on transgenic *Bt* cotton. Each of the three resistance alleles has a deletion expected to eliminate at least eight amino acids upstream of the putative toxin-binding region of the cadherin protein. Larvae with two resistance alleles in any combination were resistant, while those with one or no resistance allele were susceptible to Cry1Ac. Together with previous evidence, the results suggested that cadherin gene is a leading target for DNA-based screening of resistance to *Bt* crops in lepidopteran pests.

An autosomal recessive gene conferred extremely high levels of resistance in *P. xylostella* to four *Bt* toxins (Cry1Aa, Cry1Ab, Cry1Ac, and Cry1F) (Tabashnik et al., 1997a). A recessive autosomal gene confers resistance to at least four *Bt* toxins in *P. xylostella* and enables survival without adverse effects on transgenic plants (Heckel et al., 1999). Allelic variants of this gene confer resistance in strains from Hawaii, Pennsylvania, and Philippines. The biphasic nature of the lepidopteran genetic linkage map has been used to detect this gene in diamondback moth, with 207 amplified fragment length polymorphism (AFLP) DNA markers. An AFLP marker has been cloned and sequenced for the chromosome containing the *Bt* resistance gene. Resistance in *O. nubilalis* to *Bt* appears to be inherited as an incompletely dominant autosomal gene, indicating that if the field resistance turns out to be dominant, then the high dose/refuge strategy may be ineffective for management of resistance to this pest.

Strategies for Resistance Management

To preserve the value of *Bt* endotoxins in crop protection, it will be necessary to implement resistance management measures from the very beginning. Individual farmers have limited incentive to adopt resistance management technologies for *Bt* endotoxins, and the greatest incentive lies with the *Bt* industry (Kennedy and Whalon, 1995). However, the implementation of a coordinated, industry-wide, *Bt* resistance management effort is likely to be constrained by competition among different segments of the *Bt* industry interested in different technologies (sprays versus transgenic plants), and among producers of *Bt* products using the same technology. Successful implementation of resistance management for *Bt* endotoxins will require that the *Bt* industry prepackage resistance management technologies, and that these prepackaged resistance management strategies do not add significantly to the costs or complexity of pest control by the end user. In view of criticism of the *Bt* transgenic crops for their ability to enhance development of resistance, prevention of development of resistance to *Bt* is the best policy. The *Bt* resistance management tactics have been incorporated in the regulation of *Bt* transgenic crops in most countries (Haliday, 2001). Transgenic crops must produce high doses of toxin and there should be nontransgenic crops or other alternative host crops as refuge for the production of susceptible insect populations.

To increase the effectiveness of the transgenic plants, it is important to implement the resistance management strategies from the very beginning. A number of conceptual strategies have been developed for resistance management (McGaughey and Whalon, 1992; Tabashnik, 1994; Gould, 1998). A high level of migration or a sensible reduction of the fitness associated with the change in the genome may lead to a long-lasting efficacy of the transgenic crops (Arpaia, Chiriatti, and Giorio, 1998). Most of the strategies for resistance management are based on mixtures of toxins to be deployed for insect control, tissue-specific production, and induced toxin production. Two or more insecticidal proteins or different genes can also be introduced into the same plant to slow down the rate of development of resistance in the target insects. Plants can also be engineered so that the toxin is produced only in the tissues where the insect feeds. In induced toxin production, the plants can be engineered in such a way that the toxin is produced when the insect starts feeding.

The strategies for resistance management would depend on the number and nature of gene action, insect behavior, and insect-genotype-environment interaction. Spring movement of emerging adults onto wild hosts may delay the development of resistance if the movement of adults is far enough from the field in which the pupae over-wintered. Increase in the summer migration and the distance moved will also delay resistance development. It has been suggested that *Bt* sunflower might not lead to the development of *Bt*-resistant sunflower moth, *Homoeosoma electellum* (Hulst.) populations (Brewer, 1991). Ives and Andow (2002) suggested that at least three processes are involved in explaining the effectiveness of the high-dose/refuge strategy, which depends on: (1) the intensity of selection, (2) assortative (nonrandom) mating due to spatial subdivision, and (3) variation in male mating success due to spatial subdivision. Understanding these processes will lead to a greater range of possible resistance management tactics. For example, efforts to encourage adults to leave their natal fields may have the unwanted effect of speeding rather than slowing resistance evolution. Furthermore, when *Bt* maize causes high mortality of susceptible target pests, spraying insecticides in refuges to reduce pest populations may not greatly disrupt resistance management. The following tactics have been accepted for resistance management in *Bt* transgenic crops.

- Stability of transgene expression
- High level of transgene expression
- Gene pyramiding
- Planting refuge crops
- Destruction of carryover population
- Control of alternate hosts
- Use of planting window
- Following crop rotations
- Use of pesticide application based on ETLs
- Use of IPM from the beginning

Stable Transgene Expression

For efficient pest control, it is important to achieve optimum and stable expression of the transgene that exercises a constant pressure on the pest populations. Selective removal of nucleotides has been used to improve transgene expression (F.J. Perlak et al., 1991). Transgene expression is also influenced by the region in which the transgene integrates in different genotypes. Breeding programs should take cognizance of this fact and design screening procedures to reject plants with unstable transgene expression. Based on the information available on molecular mechanisms of transgene silencing, a few precautions need to be observed in genetic transformation of plants to minimize transgene silencing in the transformants. These include:

- The transformation vector should not contain duplicated sequences that might serve as initiation sites for *de novo* methylation.
- Different genes in the same construct should not contain identical promoters or termination signals, and the transgene should be codon optimized to match the composition of the host genome.
- Bacterial vector sequences should not find a place within the segment that will be integrated into the plant genome.
- If the insert contains two transgenes, the promoters should read in different directions. If the insert contains more than two genes, all the promoters should read in the same direction, and should be separated by appropriate stuffer elements to prevent read through.

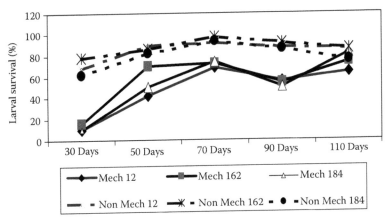

FIGURE 12.3 Survival of neonate larvae of *Helicoverpa armigera* on the leaves of transgenic and nontransgenic cotton hybrids during the crop growing season.

- Expression of the transgene varies across different stages of plant growth, and generally declines towards the end of the crop-growing season (Figure 12.3). The variation in expression of the insecticidal protein in transgenic rape is presumed to be associated with the changes in growing conditions or with integration of the transgene at different loci (Lin et al., 2001).

High Level of Transgene Expression

Expression of toxins at very high levels is one of the strategies to slow the adaptation by insects to a toxin and prevent/delay the evolution of biotypes capable of surviving on the transgenic crops. This strategy can be used effectively if the ecology and genetics of the insect and cropping system fit specific assumptions (Gould, 1994). These assumptions relate to:

- Pattern of inheritance of resistance;
- Ecological costs of resistance;
- Behavioral responses of larvae and adults to the toxins;
- Movement of larvae, adult dispersal, and mating behavior; and
- Distribution of host plants that do and do not produce the toxin(s).

The U.S. Environmental Protection Agency (EPA, 1998) concluded that for a high-dose strategy, the *Bt* transgenic cultivar should produce 25 times the amount of *Bt* toxin needed to kill 99% of the susceptible insects. A large number of plants in the field should be surveyed to make sure that the *Bt* toxin expression in the transgenic cultivar is at the LD_{99} or higher level. High levels of expression of *cry1C* protects transgenic broccoli not only from susceptible or *cry1Ab*-resistant diamondback moth larvae, but also from those selected for moderate levels of resistance to Cry1C (Cao et al., 1999; Liu et al., 2000). High levels of expression of *cryV* in *Phthorimaea operculella* (Zeller)-resistant potato lines resulted in greater mortality of this pest. Transformed tobacco leaves expressing *cry2Aa2* protoxin at 2 to 3% of the total soluble protein (20- to 30-fold higher) are highly effective against the resistant populations of *H. zea*, *H. virescens*, and *S. exigua* (De Cosa et al., 2001).

Gene Pyramiding

Several genes can be inserted in the same plant for effective pest management (L. Chen et al., 1998). To convert transgenics into an effective weapon in pest control, it is important to deploy genes with different modes of action in the same plant (Zhao et al., 1997; L. Chen et al., 1998). Many of the candidate genes that have been used in genetic transformation of crops are highly specific or are only mildly effective against the target insect pests. In addition, crops frequently suffer from a number of primary herbivores. This suggests that single and multiple transgenes will need to be combined in the same variety with other sources, mechanisms, and targets of insect pest resistance in order to generate effective and sustainable seed-based technologies. Several genes such as trypsin inhibitors, secondary plant metabolites, vegetative insecticidal proteins, plant lectins, and enzymes that are selectively toxic to insects can be deployed along with the *Bt* genes to increase the durability of transgenic resistance to insects.

From an evolutionary point of view, the development of multiple toxin systems in transgenic plants will be expected to decrease the ability of insect pests to overcome newly deployed seed-based resistance technologies and thereby prolong the life of such new varieties (Hadi, McMullen, and Finer, 1996; Karim, Raizuddin, and Dean, 1999). Stacking or pyramiding of transgenes is an important strategy and involves introducing more than one insecticidal gene into the same plant. Many advantages can be realized by stacking two different *Bt* genes having a different mode of action. For example, Bollgard II contains two *Bt* genes, *cry1Ac* and *cry2Ab*. The additive effect of both proteins will ensure greater lethality to bollworms, and the product is also expected to have an expanded host range [including the tobacco caterpillar, *Spodoptera litura* (F.)], and delay the development of resistance in bollworms. Commercial introduction of Bollgard II could form a basic component of a resistance management strategy in the future. The second gene can also be a non-*Bt* gene, such as protease or amylase inhibitors. The following strategies can be used for gene pyramiding to increase the effectiveness of transgenic crops for pest management and delay the evolution of insect biotypes capable of surviving on the transgenic crops.

Pyramiding Two or More Bt Genes

Pyramiding two or more *Bt* genes will help in expanding the spectrum of insecticidal activity of *Bt*, thereby providing more protection to crop plants, and also reduce the possibility of development of resistance. The hybrid gene combining *cry1Ac* and *cry2a* in Bollgard II has not only enhanced insecticidal spectrum, but also decreased the possibility of development of resistance in bollworms (J.H. Perlak et al., 2001). Greenplate et al. (2003) studied independent and additive interactive effects of two *Bt* δ-endotoxins expressed in the transgenic cotton variety 15985 by examining the responses of *H. virescens*, *H. zea*, and *S. frugiperda* larvae to field- or greenhouse-grown tissue from near-isolines, which expressed *cry1Ac* only, *cry2Ab* only, or both toxins. In all cases, the *cry2Ab* component was the larger contributor to total toxicity in the two-toxin isoline. Levels of each toxin in tissues of the two-toxin isoline were not statistically different from the levels found in the corresponding tissues of the respective single-toxin isoline. Considering the additive interaction of toxins, a relatively simple insect resistance-monitoring procedure has been proposed for the monitoring of commercial cotton varieties expressing both toxins.

A combination of Cry1Aa and Cry1Ac toxins has a synergistic effect, while a combination of Cry1Aa and Cry1Ab produces an antagonistic effect against the gypsy moth, *Lymantria dispar* L. (Lee et al., 1996). In tobacco plants, a combination of *cry1Ab* and *cry1Ac* genes has been shown to be effective for controlling the lepidopteran insects (Van der Salm et al., 1994). The Cry1Ac-resistant pink bollworm had little or no survival on second-generation transgenic cotton with *cry2Ab* alone or in combination with *cry1Ac* plus *cry2Ab* (Tabashnik et al., 2002a). A mixture of Cry1Ac and Cry1F decreased the EC_{50} to *H. armigera* by 26 times (Chakrabarti et al., 1998). The δ-endotoxins CryIIAa and CryIIAb are ideal candidates for gene pyramiding as they differ in structure and mode of action from Cry1A proteins (Kumar et al., 2004). The expression of *cry1Ab–cry1Ac* genes resulted in increased protection against *S. exigua, Manduca sexta* (L.), and *H. virescens.* In addition, a chitinase gene from *Serratia marcesens* (Bizio) has been shown to act synergistically with *Bt* toxins against *S. littoralis* (Rigev et al., 1996). Broccoli with two *Bt* genes, *cry1Ac* and *cry1C*, is effective for controlling diamondback moth, *P. xylostella*, resistant to Cry1A and Cry1C proteins (Cao et al., 2002).

However, pyramiding different *Bt* events compared with single events did not improve control of fall armyworm whorl damage in maize, but they did prevent more ear damage by corn earworm, *H. zea* (Buntin, Flanders, and Lynch, 2004). The MON 84006 event singly and pyramided with MON 810 had superior control of whorl-stage damage by *S. frugiperda* and ear damage by *H. zea* compared with MON 810. Improved control of whorl and ear infestations by *H. zea* and *S. frugiperda* would increase the flexibility of planting corn and permit double cropping of corn in areas endemic to these pests.

Bollgard II expressing *cry1Ac–cryIIAb* provides better control of bollworms than Bollgard. Survival of *H. zea* larvae was lower on different plant parts of Bollgard II than on the corresponding structures of Bollgard and conventional cotton (Gore, Leonard, and Adamczyk, 2001). Bollgard II significantly reduced larval survival and fruit penetration by bollworm compared to the Bollgard (Jackson, Bradley, and Duyn, 2004). The Cry1Ac-selected bollworm strain displayed increased larval survival, superficial fruit damage, and fruit penetration compared to the feral strain when averaged across genotypes. However, the single *cry2Ab*-producing genotype was similar to both Bollgard and Bollgard II with respect to fruit penetration.

Cry1Ac-resistant pink bollworm had little or no survival on second-generation transgenic cotton with *cry2Ab* alone or with *cry1Ac* plus *cry2Ab* (Tabashnik et al., 2002b). Artificial diet bioassays showed that resistance to Cry1Ac did not confer strong cross resistance to Cry2Aa. Strains with >90% larval survival on a diet with 10 µg of Cry1Ac per milliliter showed no survival on a diet with 3.2 or 10 µg of Cry2Aa per milliliter. However, average survival of the larvae fed on a diet with 1 µg of Cry2Aa per milliliter was greater for Cry1Ac-resistant strains (2% to 10%) than for susceptible strains (0). If plants with *cry1Ac* plus *cry2Ab* are deployed, and if the inheritance of resistance to both toxins is recessive, the efficacy of transgenic cotton might be greatly extended. After 24 generations of selection on broccoli expressing *cry1Ac* and *cry1C* genes alone or in combination, resistance to pyramided two-gene plants was significantly delayed in a *P. xylostella* resistant to *Bt* at frequencies of 0.10 and 0.20 for Cry1Ac and Cry1C, respectively, as compared with resistance to single-gene plants deployed in mosaics, and to Cry1Ac toxin when it was the first used in a sequence (Zhao et al., 2003).

Pyramiding Bt with Protease Inhibitor and Lectin Genes

Activity of *Bt* genes in transgenic plants is also enhanced by protease inhibitors (MacIntosh et al., 1990; Zhao et al., 1997, 1998). *Helicoverpa armigera* developed three-fold resistance on transgenic tobacco containing *cry1Ac* and cowpea trypsin inhibitor (*CpTI*) compared to 13-fold resistance to plants containing only *cry1Ac* (Zhao et al., 1999). This approach has been used in commercialization of *Bt* cotton varieties such as SGK 321 (*cry1Ac* and *CpTI* genes) in China. Zhang, Wang, and Guo (2004) reported >92% mortality of fourth-instar larvae of *H. armigera* in 3 days in *CpTI-Bt* cotton, SGK 321. Larvae reared on *CpTI-Bt* transgenic cotton showed significantly lower approximate digestibility and higher efficiency of conversion of ingested food than those treated with *Bt* transgenic cotton and Shiyuan 321. The chitinase gene from *S. marcesens* has also been shown to act synergistically with *Bt* toxins against *S. littoralis* (Rigev et al., 1996).

Pyramiding Bt Genes with Conventional Host Plant Resistance

Insect-resistant lines derived through conventional host plant resistance and novel genes can be used effectively to achieve high levels of resistance against the target insect pests (Bergvinson, Willcox, and Hoisington, 1997; Sharma et al., 2004). Insect-resistant transgenic plants with different transgenes or with lines derived through conventional breeding with resistance to the target insects can also be deployed as multilines or synthetics (Bergvinson et al., 1997). Sachs et al. (1996) crossed the transformed cotton line MON 81 expressing the *cry1Ab* gene to glandless (terpenoid-free), wild-type (normal terpenoid level), and high-glanded (high-terpenoid) plant isolines derived from the "TAMCOT CAMD-E" and "Stoneville 213" variety backgrounds. Survival of Cry1Ab-susceptible larvae was reduced more by pyramiding the *cry1Ab* with the high-terpenoid trait in a single plant than by either trait alone under no-choice conditions. Pyramiding Cry1Ab with high-terpenoid content should increase plant resistance to *H. virescens* and improve the durability of the Cry1Ab trait in commercial cotton.

Activity of *Bt* genes in transgenic cotton plants is also enhanced by tannic acid (Gibson et al., 1995). Olsen and Daly (2000) observed that plant toxin interactions influence the effectiveness of Cry1Ac protein in transgenic cotton, possibly due to tannins. Differences in LC_{50} varied from 2.4- to 726-fold on different cotton genotypes. Genetic background of the transformed lines exercises considerable effect on the effectiveness of the transgene for controlling the target pests (Li et al., 2001; Kranthi et al., 2005b). Bollworm, *H. armigera* damage in *Bt* transgenic cotton hybrid Mech 184 is much lower than in Mech 12 and Mech 162 (Sharma and Pampapathy, 2006). Resistance levels of *Gossypium arboreum* L. varieties Aravinda and MDL 2450 to the bollworms have been found to be comparable to the transgenic *G. hirsutum* hybrids, suggesting that it would be useful to combine transgenic resistance to *H. armigera* with genotypes derived through conventional plant breeding to improve the effectiveness of transgenic plants for pest management. Protease inhibitors engineered into cotton with high gossypol and/or tannin content may achieve greater protection against *H. armigera* (Wang and Qin, 1996). The resistance of 32B to cotton bollworm showed obvious changes with the developmental stage of the cotton (S. Chen et al., 2002). Resistance during the seedling stage was quite strong. A sharp decrease of resistance occurred from the square stage, which coincided with changes in the toxin content. *Bt* toxin level might be associated with the protein metabolism in *Bt* transgenic cotton. Tannin content was negatively correlated with insect resistance and *Bt* toxin level. It is suggested that the increase in tannin content might contribute to the decrease of *Bt* toxin and the resistance to cotton bollworm.

High levels of expression of *cryV* in potato lines resistant to *P. operculella* resulted in 96% mortality of this pest. These transgenic lines provide a germplasm base to combine insect resistance mechanism from conventional plant resistance and novel genes as a means to achieve durable resistance against this difficult to control pest (Westedt et al., 1998).

A quantitative trait loci (QTL) conditioning maize earworm resistance in soybean PI 229358 and *cry1Ac* transgene from the recurrent parent Jack-*Bt* has been pyramided into BC$_2$F$_3$ plants by marker-assisted selection (Walker et al., 2002). Combining transgene- and QTL-mediated resistance to lepidopteran pests along with *Bt* genes is more effective for insect control than the transgene alone.

Pyramiding Genes for Resistance to Multiple Pests

Development of genetically engineered plants with resistance to more than one pest will be the most ideal strategy to use such plants in pest management. Many of the candidate genes that have been used in genetic transformation of crops are highly specific or are only mildly effective against the target insect pests. In addition, crops frequently suffer from a number of primary herbivores. This suggests that single and multiple transgenes will need to be combined in the same variety with other sources, mechanisms, and targets of insect pest resistance in order to generate highly effective and sustainable seed-based technologies. In this context, it is important to examine whether co-expression of multiple toxins in the same plant will have a synergistic or antagonistic effect. The *Xa21* gene (resistance to bacterial blight), the *Bt* fusion gene (for insect resistance), and the chitinase gene (for tolerance to sheath blight) have been combined in a single rice line by reciprocal crossing of two transgenic homozygous IR 72 lines (Datta et al., 2003). The identified lines showed resistance to bacterial blight, tolerance to sheath blight, and caused 100% mortality of neonate yellow stem borer, *Scirpophaga incertulas* (Walker) larvae. Rice plants with *cry1Ac* and *GNA* genes accumulated high levels of insecticidal gene products (Nguyen et al., 2002; Loc et al., 2002). Transgenic plants expressing *GNA* showed enhanced resistance to brown planthopper, *Nilaparvata lugens* (Stal), while the plants expressing *cry1Ac* were resistant to striped stem borer, *Chilo suppressalis* (Walker). Expression of both transgenes gave protection against both pests, but did not increase protection against either pest significantly over the levels observed in plants containing a single insecticidal transgene.

Regulation of Gene Expression and Gene Deployment

There is a need to develop appropriate strategies for gene deployment for different crops and cropping systems depending on the pest spectrum, their sensitivity to the insecticidal genes, and interaction with the environment. For efficient pest control, it is important that effective levels of insect control proteins are expressed in the plant where the insects feed. Regulation of gene expression by the use of appropriate promoters is important for durability and specificity of resistance. In most cases, resistance genes have been inserted with constitutive promoters such as *CaMV35S*, maize *ubiquitin*, or rice *actin 1*, which direct expression in most plant tissues. Limiting the time and place of gene expression by tissue-specific promoters such as *PHA-L* for seed-specific expression, *RsS1* for phloem-specific expression, or inducible promoters such as potato *pin2* wound-induced promoter might contribute to resistance management, and avoid unfavorable interactions with the beneficial insects. Greater risk of resistance build up would arise from prolonged exposure to sublethal levels of the transgene product. Restricted expression in tissues may also contribute to minimizing the yield penalty associated with transgene expression (Xu et al., 1993; Schular et al., 1998). There are specific situations where specific promoters would have a clear advantage, such as root-feeding insects.

Planting Refuge Crops

A refuge composed of nontransgenic plants of the same crop or other susceptible hosts of the target pest can be used to produce insects that have not been exposed to or selected for resistance to the transgene product. The insects produced on the nontransgenic crop will dilute any resistant genes in the gene pool of the target insect exposed to the transgenic crop (Figure 12.4). This approach can be used to suppress or delay the development of an insect population capable of surviving and multiplying on the transgenic crops. For the refuge strategy to be effective, the insects must emerge from the refuge crop at the same time as those on the transgenic crop so as to dilute the production of insects homozygous for the gene(s) conferring resistance to the transgene. Farmers growing *Bt* transgenic cotton in the United States and Australia are adopting this method. In India, the farmers grow cotton along with sorghum, pigeonpea, sunflower, or tomato in the same season, and chick-pea towards the end of the cotton crop. All these crops are alternate hosts for *H. armigera*. Pigeonpea and chickpea are capable of supporting substantial populations of this pest, and thus serve as a natural refuge. In such situations, there may be no need to undertake plant-ing of refuge crops if transgenic cotton occupies less than 75% of the cropped area.

Refuge Types

The Environment Protection Agency (EPA, 2001a,b) has advocated three types of refuge: (1) 95:5 external structured unsprayed refuge in which 5% of the area is planted as refuge within 0.5 km of the *Bt* cotton field; (2) 80:20 external sprayed refuge in which 20% of the area is sown as border rows of nontransgenic cotton field, which can be treated with insec-ticides, excluding foliar sprays of *Bt*; and (3) 95:5 embedded refuge in which 5% of the area is used as a contiguous block of nontransgenic cotton, which can be treated with any insec-ticide, excluding foliar sprays of *Bt*. As a novelty, a community refuge program based on the first two strategies has also been advocated. The Australian Cotton Cooperative Growers Association (2001) has recommended that the Ingard cotton area should not exceed more than 30% of the total cotton area in Australia.

Patchworks of treated and untreated fields can delay the evolution of pesticide resistance, but the untreated refuge fields are likely to sustain heavy damage (Alstad and Andow, 1995). A strategy that exploits corn borer preferences and movements can eliminate this problem. Shelton et al. (2000) used *Bt*-transgenic broccoli plants and the diamondback

FIGURE 12.4   Planting refuge crops for delaying development of resistance to transgene in insects.

moth, *P. xylostella* as a model system to examine resistance management strategies. The higher number of larvae on refuge plants in field tests indicated that a "separate refuge" is more effective in conserving susceptible larvae than a "mixed refuge," and would thereby reduce the number of homozygous resistant (RR) offspring. Care must be exercised to ensure that refuges, particularly those sprayed with efficacious insecticides, produce adequate numbers of susceptible insects.

A strip configuration is the least costly method of planting a 20% non-*Bt* maize refuge (Hyde et al., 2001). The non-*Bt* plants produce the susceptible insects, which have a probability of mating with insects emerging from the *Bt* crops nearby, and thus dilute the frequency of the resistant individuals. Separate refuges are superior to seed mixtures for delaying resistance. Movement of *H. zea* larvae from nontransgenic to transgenic plants may result in an increase in damage and reduce the yield in mixed stands of *Bt* and non-*Bt* plants (DuRant et al., 1996). In cotton, the numbers of eggs do not differ between mixed and pure stands of transgenic or nontransgenic plants (Halcomb et al., 1996; Lambert, Bradley, and Duyn, 1996). Planting two-row strips may be as good as separate refuges in delaying resistance, but their adoption carries greater risk because of the uncertainty surrounding the movement and survival of neonates (Onstad and Gould, 1998). Transgenic and nontransgenic plants could be grown in separate rows with a wider row spacing (strip planting) to minimize the rate of resistance development (Ramachandran et al., 1998a). As the proportion of nontransgenic plants increases, the number of larvae and amount of injury also increase.

Role of Alternate Hosts as Refugia

The ability of noncrops to support complete development of insect pests is an important factor in determining their impact on resistance management in transgenic crops. For polyphagous pests that feed on several field crops and alternate hosts, there may not be any need to maintain refugia under subsistence farming conditions in the tropics, where several collateral and alternate hosts are available over time. There may not be any need to plant refuge crops for cereal stem borers and *Heliothis/Helicoverpa* in the tropics where they feed on several crop and weed hosts. The probability of European corn borer, *O. nubilalis* to complete development on a plant other than corn is relatively low, and the smaller-stemmed noncorn plants may not make a substantial contribution to the pool of susceptible adults (Losey, Carter, and Silverman, 2002), although some individuals that survive could serve as a source of refuge. Cameron et al. (1997) studied the host range and occupancy of diamondback moth, *P. xylostella*, and potato tuber moth, *P. operculella* to determine what proportion of their populations occur in refugia and would therefore not be exposed to *Bt*-transformed *Brassica* or potato. Although *P. operculella* was found on Solanaceae other than potato in the field, the alternative hosts were rare in areas where vegetables were growing and, therefore, were not a significant source of refugia for this pest. Diamondback moth, *P. xylostella*, was common on Brassicaceae other than vegetable brassicas, particularly on white mustard, *Sinapis alba* L., which was the main refuge. Wild radish, *Raphanus raphanistrum* L., was less preferred, but was the most common alternative host plant. The area covered by wild radish was approximately 6% of the area planted in commercial brassicas and therefore formed a minor refuge for *P. xylostella*. The results suggested that to limit the development of resistance, insect-resistant transgenic potatoes should be less widely used than insect-resistant transgenic brassicas.

A low proportion of potato tuber moths, *P. operculella*, foraged beyond 100 to 250 m to infest tubers or plants (Cameron et al., 2002). Light traps indicated that the number of moths moving out from crops diminished over a 40 m distance. Dispersing moths penetrated 30 m into new crops to infest the foliage. Direct movement of potato tuberworm and diamondback moth between crops has been estimated through mark-recapture experiments. Sweep-net collections of potato tuberworm indicated that 17% of moths moved between crops. Only 1.2% of the diamondback moths were recaptured by sweep-net outside the release area, and very few moths were caught in pheromone traps. The results suggested that sufficient numbers of potato tuberworm would forage between adjacent treated and untreated crops to minimize the development of resistance. Use of refuges to conserve susceptible pest populations has also been recommended for managing resistance that may arise from future use of *Bt*-transgenic potatoes or *Brassica* spp. in New Zealand. The refuges that are intended to dilute potential resistance of potato tuberworm to transgenic crops should be placed close to transgenic potato crops.

The *Bt*-transgenic cotton in China covered over 1.46 million ha in 2001. In Shandong and Hebei provinces, *Bt* cotton occupies 80% and 97% of the cotton area, while other provinces such as Anhui and Jiangsu have less than 30% of the *Bt* cotton area (Dong et al., 2004). The high dose and refuge strategy widely accepted in the United States and Australia for *Bt* crops is not feasible in China in view of a lack of a consistent high dose of Cry1Ac toxin in the cotton varieties, and the complexity of small farming by marginal/poor farmers in a region (Ru, Zhao, and Rui, 2002). Although corn provided refuge to *H. armigera* larvae for *Bt* cotton, nevertheless, resistance management is necessary for regulation of *Bt* transgenic crops.

Resistant larvae of *P. gossypiella* on *Bt* cotton took longer to develop than susceptible larvae on non-*Bt* cotton (Liu et al., 1999). This developmental asynchrony favors nonrandom mating that could reduce the expected benefits of refuge strategy. Consistent with one of the assumptions of the refuge strategy, *P. gossypiella* resistance to *Bt* cotton is recessive. Survival of the hybrid F_1 progeny (2%) was not higher than the survival of the susceptible strain (6%), and both were markedly lower than the survival of the resistant strain (37%). Resistant larvae on *Bt* cotton required an average of 5.7 days longer to develop than susceptible larvae on non-*Bt* cotton. Field data suggested that median longevity of male *P. gossypiella* is less than one week, and 80% of moths mate within three days of emergence. This developmental asynchrony, therefore, favors assortative mating among resistant moths from *Bt* plants. In the field, the extent of developmental asynchrony and assortative mating would be affected by variation in toxin expression, weather, and overlap between generations. Deliberate inclusion of a refuge may also reduce the proportion of marketable produce, and may affect use of this resistance management strategy in both *Bt* sprays and transgenic crops expressing *Bt* toxins (Perez, Shelton, and Roush, 1997).

Removal of Alternate Hosts and Destruction of Carryover Population

Removal of alternate hosts in case the alternate hosts play an important role in pest population buildup will also be effective in delaying the development of resistance to *Bt* toxins through reduction in carryover of the post population from one season to another. Efforts should be made to remove the alternate hosts of the pests from the vicinity of the crop. This will help in reducing the density of the pests, and low to moderate levels of insect abundance can be effectively controlled through transgenic plants. However, this approach may reduce the number of susceptible individuals, which otherwise will act as a source of refuge to dilute the frequency of resistance alleles.

Destruction of the carryover population, for example diapausing larvae or pupae, that have been exposed to transgenic crops in the previous generation, is another important component of a resistance management strategy. Ploughing or flooding the fields immediately after the crop harvest will expose the hibernating larvae or pupae to the biotic and abiotic mortality factors. Destruction of stems or burning of stubbles of crops with insect larvae and pupae is also helpful in reducing the carryover of insects from one season to another (Sharma and Ortiz, 2002). The frequency of resistant individuals in a strain is not associated with induction of diapause or emergence from diapause in early winter in pink bollworm, *P. gossypiella* (Carriere et al., 2001). However, emergence from diapause in the spring has been found to be 71% lower in three highly resistant strains than in two heterogeneous strains from which the resistant strains were derived. Therefore, there is a need to take into consideration the proportion of population entering diapause and the numbers that emerge in the following season. Emergence in the spring of hybrid progeny from crosses between the resistant and heterogeneous strains has been found to be greater than that of the resistant strains, but did not differ from susceptible strains, showing that the over-wintering cost was recessive to some extent.

Use of a Planting Window

Following a planting window, that is, sowing crops such that the most susceptible stage of the crop escapes pest damage or avoids peak periods of insect abundance can also be useful in maximizing the benefits of transgenic crops or prolong the life of transgenic crops. Planting the crops with the first good monsoon rains has been found to be effective in controlling the damage by sorghum shoot fly, *Atherigona soccata* (Rondani), and sorghum midge, *Stenodiplosis sorghicola* (Coquillett) (Sharma, 1993; Sharma and Ortiz, 2002). Similar strategies can be employed to avoid insect damage in other crops to prolong the effectiveness of transgenic crops.

Crop Rotations

Crops with *Bt* gene can be rotated with non-*Bt* crops in the following season or year. This strategy would work better if *Bt* resistance in insect populations is not stable and breaks down when selection pressure is removed. This strategy will also be useful in situations where the resistance has a fitness cost.

Integrated Pest Management

Integration of different control measures, including biocontrol agents and insecticides, as components of pest management is important for resistance management strategy (Australian Cotton Growers Research Association, 2001). Developing knowledge-based strategies such as resistance management aimed at achieving long life of transgenic crops should be commensurate with the integrated crop management (Tuli et al., 2000; Sharma and Ortiz, 2000; Pilcher et al., 2002; Shelton, Zhao, and Roush, 2002).

Simulation Models for Resistance Management

The effects of transgenic plants on the populations of insect pests will be similar to those of conventional host plant resistance to insects (Knipling, 1979). Effects of population dynamics and movement of insects between transgenic and refuge fields on evolution of insecticide resistance have been examined in two different simulation models (Caprio, 2001). The two models were developed to test the hypothesis that increasing the habitat from fine-grained to coarse-grained, and the resultant increase in nonrandom mating, would increase the rate of local adaptation and hence evolution of resistance. The first, a stochastic, spatially explicit model, altered habitat by varying adult dispersal rates between habitat patches. In contrast to the expectation that increasing patch isolation and increasing the coarseness of the habitats would increase the rate of resistance evolution, intermediate levels of dispersal actually delayed resistance by as much as five-fold over the range of dispersal levels observed. A simple deterministic model was also developed to separate the impact of mating and ovipositional behavior. This model showed qualitatively the same results. Under similar assumptions, it predicted longer delays in resistance evolution. In this model, nonrandom mating alone increased the rate at which insects adapted to transgenic crops, but nonrandom mating in combination with nonrandom oviposition could significantly delay resistance evolution. Differences between the two models may be due to population regulation incorporated in a spatially explicit model. The models clearly suggested that resistance management programs using untreated refuges should not over-emphasize random mating at the cost of making the habitat too fine-grained.

Since resistance gene(s) do not appear to be completely recessive, the need for a higher proportion of refuge vis-à-vis transgenic crop has often been advocated (Gould and Tabashnik, 1998). The resistant individuals emerging out of *Bt* cotton should mate randomly with those from the refuge where the susceptible individuals breed. However, synchrony of emergence of adults from *Bt* cotton and refuge will be equally essential. *Helicoverpa armigera*, a highly mobile insect, may mate randomly, although closeness of refuge as a policy also contributes to its success. Soybean and peanut as mixed crops could provide natural refuge for the second to fourth generations of *H. armigera* in North China (Wu, Guo, and Gao, 2002). However, function of refuge appeared to depend on the proportion of *Bt* cotton. However, there is a need to protect the natural refuge of noncotton host plants of *H. armigera* by not commercializing both *Bt* cotton and *Bt* corn in the same area (Zhao et al., 2000b). In India, small farm holdings may not necessarily follow refuge norms. As the transgenic cotton area increases, it may lead to almost monocropping without a significant contribution of mixed crops or nontrangenic *Bt* cotton towards the production of susceptible individuals.

Conclusions

The use of crop protection traits through transgenics will continue to expand in the future, and gene pyramiding might become very common. This approach of controlling insects would offer the advantage of allowing some degree of selection for specificity effects, so that pests but not beneficial organisms are targeted. Resistance to *Bt* is a recessive or partially recessive mutation in a major autosomal gene, and cross resistance extends to Cry proteins sharing the same binding site(s). Cry proteins with specificity to different binding sites should be considered for gene pyramiding as a component of a resistance management strategy. In a vast majority of cases, resistance has been found to be quite unstable because of fitness costs associated with resistance, although there are a few exceptions. A detailed understanding of the mechanisms, insect biology, and plant molecular biology can be used to tailor expression of *cry* genes in transgenic plants for pest management in the future. Planting of refuge crops and use of integrated pest management from the very beginning is crucial to delay the development of resistance to insect-resistant transgenic crops, and increasing their usefulness for sustainable crop production.

References

Akhurst, R. and Liao, C. (1996). Protecting an investment: Managing resistance development to transgenic cotton by *Helicoverpa armigera*. *Proceedings of the Eighth Australian Cotton Conference*, 1996, Broadbeach. Brisbane, Queensland, Australia: Cotton Growers Research Association, 299–305.

Akhurst, R., James, W. and Bird, L. (2000). Resistance to Ingard® cotton by the cotton bollworm, *Helicoverpa armigera*. *Proceedings of the 10th Australian Cotton Conference*, 16–18 August, 2000. Brisbane, Australia: Australian Cotton Growers Research Association, 1–5. http://cotton.pi.csiro.au/Publicat/conf/coconf00/Areawide/25/25.htm.

Akhurst, R.J., James, W., Bird, L.J. and Beard, C. (2003). Resistance to the Cry1Ac δ-endotoxin of *Bacillus thuringiensis* in the cotton bollworm, *Helicoverpa armigera* (Lepidoptera: Noctuidae). *Journal of Economic Entomology* 96: 1290–1299.

Alstad, D.N. and Andow, D.A. (1995). Managing the evolution of insect resistance to transgenic plants. *Science* 268: 1894–1896.

Altre, J.A., Grafius, E.J. and Whalon, M.E. (1996). Feeding behavior of CryIIIA-resistant and susceptible Colorado potato beetle (Coleoptera: Chrysomelidae) larvae on *Bacillus thuringiensis tenebrionis*-transgenic CryIIIA-treated and untreated potato foliage. *Journal of Economic Entomology* 89: 311–317.

Andow, D.A. and Alstad, D.N. (1998). F_2 screen for rare resistance alleles. *Journal of Economic Entomology* 91: 572–578.

Andow, D.A. and Hutchinson, W.D. (1998). *Bt*-corn resistance management. In Mellon, M. and Rissler, J. (Eds.), *Now and Never: Serious New Plans to Save a Natural Pest Control*. Cambridge, Massachusetts, USA: Union of Concerned Scientists, 19–66.

Andow, D.A., Olson, D.M., Hellmich, R.L., Alstad, D.N. and Hutchison, W.D. (2000). Frequency of resistance to *Bacillus thuringiensis* toxin Cry1Ab in an Iowa population of European corn borer (Lepidoptera: Crambidae). *Journal of Economic Entomology* 93: 26–30.

Arpaia, S., Chiriatti, K. and Giorio, G. (1998). Predicting the adaptation of Colorado potato beetle (Coleoptera: Chrysomelidae) to transgenic eggplants expressing CryIII toxin: The role of gene dominance, migration, and fitness costs. *Journal of Economic Entomology* 91: 21–29.

Australian Cotton Growers Research Association. (2001). *Transgenics and Insect Management*. Strategy Committee. Brisbane, Australia. http://cotton.crc.org.au/Publicat/Pest.

Ballester, V., Escriche, B., Mensua, J.L., Reithmacher, G.W. and Ferre, J. (1994). Lack of cross-resistance to other *Bacillus thuringiensis* crystal proteins in a population of *Plutella xylostella* highly resistant to Cry1Ab. *Biocontrol Science and Technology* 4: 437–443.

Ballester, V., Granero, F., Tabashnik, B.E., Malvar, T. and Ferre, J. (1999). Integrative model for binding of *Bacillus thuringiensis* toxins in susceptible and resistant larvae of the diamondback moth (*Plutella xylostella*). *Applied Environmental Microbiology* 65: 1413–1419.

Barton, K., Whiteley, H. and Yang, N.S. (1987). *Bacillus thuringiensis* δ-endotoxin in transgenic *Nicotiana tabacum* provides resistance to lepidopteran insects. *Plant Physiology* 85: 1103–1109.

Bauer, L.S. (1995). Resistance: A threat to the insecticidal crystal proteins of *Bacillus thuringiensis*. *Florida Entomologist* 78: 415–443.

Bauer, L.S., Koller, C.M., Miller, D.L. and Hollingworth, R.M. (1994). Laboratory selection of cottonwood leaf beetle, *Chysomela scripta*, to *Bacillus thuringiensis* var. *tenebrionis* delta endotoxin. *Abstracts of 27th Annual Meeting of Society for Invertebrate Pathology*. Montpellier, France: Society for Invertebrate Pathology, 68.

Bergvinson, D., Willcox, M.N. and Hoisington, D. (1997). Efficacy and deployment of transgenic plants for stem borer management. *Insect Science and Its Application* 17: 157–167.

Bolin, P.C., Hutchison, W.D. and Andow, D.A. (1995). Selection for resistance to *Bacillus thuringiensis* Cry1Ac endotoxin in Minnesota population of European corn borer. *Abstracts of 28th Annual Meeting of Society for Invertebrate Pathology*, Cornell University, Ithaca. New York, USA: Society for Invertebrate Pathology.

Bolin, P.C., Hutchison, W.D. and Andow, D.A. (1999). Long-term selection for resistance to *Bacillus thuringiensis* Cry1Ac endotoxin in a Minnesota population of European corn borer (Lepidoptera: Crambidae). *Journal of Economic Entomology* 93: 1588–1595.

Bolin, P.C., Hutchison, W.D. and Davis, D.W. (1996). Resistant hybrids and *Bacillus thuringiensis* for management of European corn borer (Lepidoptera: Pyralidae) in sweet corn. *Journal of Economic Entomology* 89: 82–91.

Bourguet, D., Chaufaux, J., Seguin, M., Buisson, C., Hinton, J.L., Stodola, T.J., Porter, P., Cronholm, G., Buschman, L.L. and Andow, D. (2003). Frequency of alleles conferring resistance to *Bt* maize in French and US corn belt populations of the European corn borer, *Ostrinia nubilalis*. *Theoretical and Applied Genetics* 106: 1225–1233.

Bravo, A., Jansens, S. and Peferoen, M. (1992). Immunocytochemical localization of *Bacillus thuringiensis* crystal proteins in intoxicated insects. *Journal of Invertebrate Pathology* 60: 237–246.

Brewer, G.J. (1991). Resistance to *Bacillus thuringiensis* subsp. *kurstaki* in the sunflower moth (Lepidoptera: Pyralidae). *Environmental Entomology* 20: 316–322.

Broza, M. (1986). Seasonal changes in population of *Heliothis armigera* (Hb.) (Lepidoptera: Noctuidae) in cotton fields in Israel and its control with *Bacillus thuringiensis* preparations. *Journal of Applied Entomology* 102: 365–370.

Buntin, G.D., Flanders, K.L. and Lynch, R.E. (2004). Assessment of experimental *Bt* events against fall armyworm and corn earworm in field corn. *Journal of Economic Entomology* 97: 259–264.

Cameron, P.J., Gatland, A.M., Walker, G.P., Wigley, P.J. and O'Callaghan, M. (1997). Alternative host plants as refugia for diamondback moth and potato tuber moth. *Proceedings of the Fiftieth New Zealand Plant Protection Conference*, 18–21 August, 1997. Canterbury, New Zealand: Lincoln University, 242–246.

Cameron, P.J., Walker, G.P., Penny, G.M. and Wigley, P.J. (2002). Movement of potato tuberworm (Lepidoptera: Gelechiidae) within and between crops, and some comparisons with diamondback moth (Lepidoptera: Plutellidae). *Environmental Entomology* 31: 65–75.

Cao, J., Tang, J.D., Strizhov, N., Shelton, A.M. and Earle, E.D. (1999). Transgenic broccoli with high levels of *Bacillus thuringiensis* Cry1C protein control diamondback moth larvae resistant to Cry1A or Cry1C. *Molecular Breeding* 5: 131–141.

Cao, J., Zhao, Z., Tang, J.D., Shelton, A.M. and Earle, E.D. (2002). Broccoli plants with pyramided *cry1C* and *cry1Ac Bt* genes control diamondback moths resistant to Cry1A and Cry1C proteins. *Theoretical and Applied Genetics* 105: 258–264.

Caprio, M.A. (2001). Source-sink dynamics between transgenic and non-transgenic habitats and their role in the evolution of resistance. *Journal of Economic Entomology* 94: 698–705.

Carriere, Y., Ellers Kirk, C., Patin, A.L., Sims, M.A., Meyer, S., Liu, Y.B., Dennehy, T.J. and Tabashnik, B.E. (2001). Overwintering cost associated with resistance to transgenic cotton in the pink bollworm (Lepidoptera: Gelechiidae). *Journal of Economic Entomology* 94: 935–941.

Cerda, H. and Wright, D.J. (2004). Modeling the spatial and temporal location of refugia to manage resistance in *Bt* transgenic crops. *Agriculture, Ecosystems, and Environment* 102: 163–174.

Chakrabarti, S.K., Mandaokar, A.D., Ananda Kumar, P. and Sharma, R.P. (1998). Synergistic effect of Cry1Ac and Cry1F δ-endotoxin of *Bacillus thuringiensis* on cotton bollworm, *Helicoverpa armigera*. *Current Science* 75: 663–664.

Chandrashekar, K. and Gujar, G.T. (2004). Development and mechanisms of resistance to *Bacillus thuringiensis* endotoxin Cry1Ac in the American bollworm, *Helicoverpa armigera* (Hubner). *Indian Journal of Experimental Biology* 42: 164–173.

Chandrashekar, K., Kumari, A., Kalia, V. and Gujar, G.T. (2005). Base-line susceptibility of the American bollworm, *Helicoverpa armigera* (Hübner) to *Bacillus thuringiensis* Berl. var. *kurstaki* and its endotoxins in India. *Current Science* 88: 167–175.

Chaufaux, J., Muller-Cohn, J., Buisson, C., Sanchis, V., Lereclus, D. and Pasteur, N. (1997). Inheritance of resistance to the *Bacillus thuringiensis* Cry1C toxin in *Spodoptera littoralis* (Lepidoptera: Noctuidae). *Journal of Economic Entomology* 90: 873–878.

Chaufaux, J., Seguin, M., Swanson, J.J., Bourguet, D. and Siegfried, B.D. (2001). Chronic exposure of the European corn borer (Lepidoptera: Crambidae) to Cry1Ab *Bacillus thuringiensis* toxin. *Journal of Economic Entomology* 94: 1564–1570.

Chen, L., Marmey, P., Taylor, N.J., Brizard, J., Espinoza, C., D'Cruz, P., Huet, H., Zhang, S., de Kocho, A., Beachy, R.N. and Fauquet, C.M. (1998). Expression and inheritance of multiple transgenes in rice plants. *Nature Biotechnology* 16: 1060–1064.

Chen, S., Zhou, D.S., Wu, Z.T., Wang, X.L. and Zhou, B.L. (2002). Changes of insect resistance of *Bt* transgenic cotton 32B at different growth stages and its physiological mechanisms. *Jiangsu Journal of Agricultural Sciences* 18: 84.

Crecchio, C. and Stotzky, G. (1998). Insecticidal activity and biodegradation of the toxin from *Bacillus thuringiensis* subsp. *kurstaki* bound to humic acids from soil. *Soil Biology and Biochemistry* 30: 463–470.

Daly, J. and Olsen, K. (2000). Genetics of *Bt* resistance. *Proceedings of the 10th Australian Cotton Conference*, 16–18 August, 2000. Brisbane, Australia: Australian Cotton Growers Research Association, 185–188. http://cotton.pi.csiro.au/Publicat/conf/coconf00/Areawide/23/23.htm.

Datta, K., Baisakh, N., Thet, K.M., Tu, J. and Datta, S.K. (2003). Pyramiding transgenes for multiple resistance in rice against bacterial blight, yellow stem borer and sheath blight. *Theoretical and Applied Genetics* 106: 1–8.

De Cosa, B., Moar, W., Lee, S.B., Miller, M. and Daniel, H. (2001). Over expression of the *Bt cry2Aa2* operon in chloroplasts leads to formation of insecticidal crystals. *Nature Biotechnology* 19: 71–74.

Deeba, F., Nandi, J.N., Anuradha, K., Ravi, K.C., Mohan, K.S. and Manjunath, T.M. (2003). An insect based assay to quantify *Bacillus thuringiensis* insecticidal protein Cry1Ac expressed in planta. *Entomon* 28: 27–31.

Dong, H., Li, W., Tang, W. and Zhang, D. (2004). Development of hybrid *Bt* cotton in China: A successful story of integration of transgenic technology and conventional techniques. *Current Science* 86: 778–782.

DuRant, J.A., Roof, M.E., May, O.L. and Anderson, J.P. (1996). Influence of refugia on movement and distribution of bollworm/tobacco budworm larvae in bollgard cotton. *Proceedings, Beltwide Cotton Conference*, 9–12 January, 1996, Nashville, Tennessee, Volume 2. Memphis, Tennessee, USA: National Cotton Council, 921–923.

EPA (Environmental Protection Agency). (1998). *FIFRA Scientific Advisory Panel. Subpanel on* Bacillus thuringiensis *(Bt) Plant-Pesticides and Resistance Management*, 9–10 February, 1998 (OPP Docket No. 00231). http://epa.gov/pesticides/biopesticides.

EPA (Environmental Protection Agency). (2001a). *Fact Sheet for Bt Corn*. http://epa.gov/pesticides/biopesticides/ingredients/factsheets/006481%20Fact%20fact.pdf.

EPA (Environmental Protection Agency). (2001b). Bt *Cotton Refuge Requirements for the 2001 Growing Season*. http://epa.gov/pesticides/biopesticides/pips/Bt_cotton_refuge_2001.htm.

Estada, U. and Ferré, J. (1994). Binding of insecticidal crystal proteins of *Bacillus thuringiensis* to the midgut brush border of cabbage looper, *Trichoplusia ni* (Hubner) (Lepidoptera: Noctuidae), and selection for resistance to one of the crystal proteins. *Applied Environmental Microbiology* 60: 3840–3846.

Federici, B.A. and Bauer, L.S. (1998). Cry1A protein of *Bacillus thuringiensis* is toxic to the cottonwood leaf beetle, *Chrysomela scripta*, and suppresses high levels of resistance to Cry3Aa. *Applied Environmental Microbiology* 64: 4368–4371.

Ferre, J. and Van Rie, J. (2002). Biochemistry and genetics of insect resistance to *Bacillus thuringiensis*. *Annual Review of Entomology* 47: 501–533.

Ferre, J., Real, M.D., Van Rie, J., Jansens, S. and Peferoen, M. (1991). Resistance to the *Bacillus thuringiensis* bioinsecticide in a field population of *Plutella xylostella* is due to a change in a midgut membrane receptor. *Proceedings National Academy of Sciences USA* 88: 5119–5123.

Forcada, C., Alcacer, E., Garcera, M.D. and Martinez, R. (1996). Differences in the midgut proteolytic activity of two *Heliothis virescens* strains, one susceptible and one resistant to *Bacillus thuringiensis* toxins. *Archives of Insect Biochemistry and Physiology* 31: 257–272.

398 *Biotechnological Approaches for Pest Management and Ecological Sustainability*

Gahan, L.J., Gould, F. and Heckel, D.G. (2001). Identification of a gene associated with *Bt* resistance in *Heliothis virescens*. *Science* 293: 857–860.

Georghiou, G.P. (1990). Overview of insecticide resistance. In Green, M.B., LeBoron, H.M. and Moberg, W.K. (Eds.), *Managing Resistance to Agrochemicals*. Washington, D.C., USA: American Chemical Society, 18–41.

Gibson, D.M., Gallo, L.G., Krasnof, S.B. and Ketchum, R.E.B. (1995). Increased efficiency of *Bacillus thuringiensis* subsp. *kurstaki* in combination with tannic acid. *Journal of Economic Entomology* 88: 270–277.

Gonzalez, C.J., Escriche, B., Tabashnik, B.E. and Ferre, J. (2003). Binding of *Bacillus thuringiensis* toxins in resistant and susceptible strains of pink bollworm (*Pectinophora gossypiella*). *Insect Biochemistry and Molecular Biology* 33: 929–935.

Gore, G., Leonard, B.R. and Adamczyk, J.J. (2001). Bollworm (Lepidoptera: Noctuidae) survival on Bollgard and Bollgard II cotton flower and bud and flower components. *Annals of Entomological Society of America* 94: 1445–1451.

Gore, J., Leonard, B.R., Church, G.E. and Cook, D.R. (2002). Behavior of bollworm (Lepidoptera: Noctuidae) larvae on genetically engineered cotton. *Journal of Economic Entomology* 95: 763–769.

Gould, F. (1994). Potential and problems with high-dose strategies for pesticidal engineered crops. *Biocontrol Science and Technology* 4: 451–461.

Gould, F. (1998). Sustainability of transgenic insecticidal cultivars: Integrating pest genetics and ecology. *Annual Review of Entomology* 43: 701–726.

Gould, F. and Tabashnik, B.E. (1998). *Bt* cotton resistance management. In Mellon, M. and Rissler, J. (Eds.), *Now and Never: Serious New Plans to Save a Natural Pest Control*. Cambridge, Massachusetts, USA: Union of Concerned Scientists, 65–105.

Gould, F., Anderson, A., Jones, A., Summerford, D., Heckel, D.G., Lopez, J., Micinski, S., Leonard, R. and Laster, M. (1997). Initial frequency of alleles for resistance to *Bacillus thuringiensis* toxins in field populations of *Heliothis virescens*. *Proceedings National Academy of Sciences USA* 94: 3519–3523.

Gould, F., Anderson, A., Reynolds, A., Bumgarner, L. and Moar, W. (1995). Selection and genetic analysis of a *Heliothis virescens* (Lepidoptera: Noctuidae) strain with high levels of resistance to *Bacillus thuringiensis* toxins. *Journal of Economic Entomology* 88: 1545–1559.

Gould, F., Martinez-Ramirez, A., Anderson, A., Ferré, J., Silva, F.J. and Moar, W.J. (1992). Broad-spectrum resistance to *Bacillus thuringiensis* toxins in *Heliothis virescens*. *Proceedings National Academy of Sciences USA* 89: 7986–7990.

Granero, F., Ballester, V. and Ferre, J. (1996). *Bacillus thuringiensis* crystal proteins Cry1Ab and Cry1Fa share a high affinity binding site in *Plutella xylostella* (L.). *Biochemical and Biophysical Research Communications* 224: 779–783.

Greenplate, J.T., Mullins, J.W., Penn, S.R., Dahm, A., Reich, B.J., Osborn, J.A., Rahn, P.R., Ruschke, L. and Shappley, Z.W. (2003). Partial characterization of cotton plants expressing two toxin proteins from *Bacillus thuringiensis*: relative toxin contribution, toxin interaction, and resistance management. *Journal of Applied Entomology* 127: 340–347.

Griffitts, J.S., Whitacre, J.L., Stevens, D.E. and Aroian, R.V. (2001). *Bt* toxin resistance from loss of a putative carbohydrate-modifying enzyme. *Science* 293: 860–864.

Gujar, G.T., Kalia, V. and Kumari, A. (2001). Effect of sublethal concentration of *Bacillus thuringiensis* var. *kurstaki* on food and development needs of the American bollworm, *Helicoverpa armigera* (Hubner). *Indian Journal of Experimental Biology* 39: 1130–1135.

Gujar, G.T., Kumari A., Kalia, V. and Chandrashekar, K. (2000). Spatial and temporal variation in susceptibility of the American bollworm, *Helicoverpa armigera* (Hubner) to *Bacillus thuringiensis* var. *kurstaki* in India. *Current Science* 78: 995–1001.

Gujar, G.T., Mittal, A., Kumari A. and Kalia, V. (2004). Host crop influence on the susceptibility of the American bollworm, *Helicoverpa armigera*, to *Bacillus thuringiensis* sap. *kurstaki* HD-73. *Entomologia Experimentalis et Applicata* 113: 165–172.

Hadi, M.Z., McMullen, M.D. and Finer, J.J. (1996). Transformation of 12 different plasmids into soybean via particle bombardment. *Plant Cell Reporter* 15: 500–505.

Halcomb, J.L., Benedict, J.H., Correa, J.C. and Ring, D.R. (1996). Inter-plant movement and suppression of tobacco budworm in mixtures of transgenic *Bt* and non-transgenic cotton. In *Proceedings, Beltwide Cotton Conference*, 9-12 January, 1996, Nashville, TN, Volume 2. Memphis, Tennessee, USA: National Cotton Council, 924–927.

Haliday, R.E. (2001). Geographical baseline susceptibility establishes the natural genetic variation and provides the foundation for future testing of insect populations exposed to increased use of *Bacillus thuringiensis* based crops. *Journal of Economic Entomology* 94: 397–402.

Han, Q. and Caprio, M.A. (2002). Temporal and spatial patterns of allelic frequencies in cotton bollworm (Lepidoptera: Noctuidae). *Environmental Entomology* 31: 462–468.

Harris, A. (1991). Comparison of costs and returns associated with *Heliothis* resistant *Bt* cotton to non-resistant varieties. In *Proceedings, Beltwide Cotton Conference*, 1991. Memphis, Tennessee, USA: National Cotton Council, 249–297.

Heckel, D.G., Gahan, L.J., Liu, Y.B. and Tabashnik, B.E. (1999). Genetic mapping of resistance to *Bacillus thuringiensis* toxins in diamondback moth using biphasic linkage analysis. *Proceedings National Academy of Sciences USA* 96: 8373–8377.

Herrero, S., Oppert, B. and Ferre, J. (2001). Different mechanisms of resistance to *Bacillus thuringiensis* toxins in the Indianmeal moth. *Applied Environmental Microbiology* 67: 1085–1089.

Hilder, V.A. and Boulter, D. (1999). Genetic engineering of crop plants for insect resistance: A critical review. *Crop Protection* 18: 177–191.

Holloway, J.W. and Dang, H. (2000). Monitoring susceptibility to *Bt* toxins in Australian *Helicoverpa* species. In *Proceedings of 10th Australian Cotton Conference*, 16–18 August, 2000. Brisbane, Queensland, Australia: Cotton Growers Research Association, 189–194.

Huang, F., Buschman, S.S., Higgins, R.A. and McGaughey, W.H. (1999). Inheritance of resistance to *Bacillus thuringiensis* toxin (Dipel ES) in European corn borer. *Science* 284: 965–967.

Hyde, J., Martin, M.A., Preckel, P.V., Dobbins, C.L. and Edwards, C.R. (2001). An economic analysis of non-*Bt* corn refuges. *Crop Protection* 20: 167–171.

Imai, K. and Mori, Y. (1999). Levels, inheritance and stability of resistance to *Bacillus thuringiensis* formulation in a field population of the diamondback moth, *Plutella xylostella* (Lepidoptera: Plutellidae) from Thailand. *Applied Entomology and Zoology* 34: 23–29.

Ives, A.R. and Andow, D.A. (2002). Evolution of resistance to *Bt* crops: Directional selection in structured environments. *Ecology Letters* 5: 792–801.

Jackson, R.E., Bradley, J.R. Jr. and Duyn, J.W. (2004). Performance of feral and Cry1Ac-selected *Helicoverpa zea* (Lepidoptera: Noctuidae) strains on transgenic cottons expressing one or two *Bacillus thuringiensis* ssp. *kurstaki* proteins under greenhouse conditions. *Journal of Entomological Science* 39: 46–55.

James, C. (2007). *Global Status of Commercialized Biotech/GM Crops: 2006*. ISAAA Briefs no. 35. Ithaca, New York, USA: International Service for Acquisition of Agri-Biotech Applications (ISAAA). http://www.isaaa.org/resorces/publications/briefs/35.

Johnson, D.E., Brookhart, G.L., Kramer, F.J., Barnett, B.D. and McGaughey, W.H. (1990). Resistance to *Bacillus thuringiensis* by the Indianmeal moth, *Plodia interpunctella*: Comparison of midgut proteinases from susceptible and resistance larvae. *Journal of Invertebrate Pathology* 55: 235–244.

Jurat-Fuentes, J.L. and Adang, M.J. (2001). Importance of Cry1 δ–endotoxin domain II loops for binding specificity in *Heliothis virescens* (F.). *Applied Environmental Microbiology* 67: 323–329.

Karim, S. and Riazuddin, S. (1999). Natural variation among *Helicoverpa armigera* (Hubner) populations to *Bacillus thuringiensis* Cry1Ac toxin. *Punjab University Journal of Zoology* (Pakistan) 14: 33–39.

Karim, S., Riazuddin, S. and Dean, D.H. (1999). Interaction of *Bacillus thuringiensis* delta-endotoxins with midgut brush border membrane vesicles of *Helicoverpa armigera*. *Journal of Asia Pacific Entomology* 2: 153–162.

Kennedy, G.G. and Whalon, M.E. (1995). Managing pest resistance to *Bacillus thuringiensis* endotoxins: Constraints and incentives to implementation. *Journal of Economic Entomology* 88: 454–460.

Knipling, E.F. (1979). The basic principles of insect population suppression. *Bulletin of Entomological Society of America* 12: 7–15.

Kota, M., Daniell, H., Varma, S., Garczynski, S.F., Gould, F. and Moar, W.J. (1999). Overexpression of the (*Bt*) Cry2Aa2 protein in chloroplasts confers resistance to plants against susceptible and *Bt*-resistant insects. *Proceedings National Academy of Sciences USA* 96: 840–1845.

Kranthi, K.R. and Kranthi, S. (2000). A sensitive bioassay for the detection of Cry1A toxin expression in transgenic cotton. *Biocontrol Science and Technology* 10: 1096–1107.

Kranthi, K.R. and Kranthi, N.R. (2004). Modeling adaptability of cotton bollworm, *Helicoverpa armigera* (Hubner) to *Bt*-cotton in India. *Current Science* 87: 669–675.

Kranthi, K.R., Kranthi, S. and Wanjari, R.R. (2001). Baseline susceptibility of Cry1A toxins to *Helicoverpa armigera* (Hubner) (Lepidoptera: Noctuidae) in India. *International Journal of Pest Management* 45: 141–145.

Kranthi, K.R., Dhawad, C.S., Naidu, K., Mate, K., Patil, E. and Kranthi, S. (2005a). *Bt*-cotton seed as a source of *Bacillus thuringiensis* insecticidal Cry1Ac toxin for bioassays to detect and monitor bollworm resistance to *Bt*-cotton. *Current Science* 88: 796–800.

Kranthi, K.R., Kranthi, S., Ali, S. and Banerjee, S.K. (2000). Resistance to Cry1Ac δ-endotoxin of *Bacillus thuringiensis* in a laboratory selected strain of *Helicoverpa armigera* (Hübner). *Current Science* 78: 1001–1004.

Kranthi, S., Kranthi, K.R., Siddhabathi, P.M. and Dhepe, V.R. (2004). Baseline toxicity of Cry1Ac toxin against spotted bollworm, *Earias vittella* (Fab.) using a diet based bioassay. *Current Science* 87: 1593–1597.

Kranthi, K.R., Naidu, K., Dhawad, C.S., Tatwawadi, A., Mate, K., Patil, E., Bharose, A.A., Behere, G.T., Wadaskar, R.M. and Kranthi, S. (2005b). Temporal and intra-plant variability of Cry1Ac expression in *Bt*-cotton and its influence on the survival of the cotton bollworm, *Helicoverpa armigera* (Hubner) (Noctuidae: Lepidoptera). *Current Science* 89: 291–298.

Kumar, S., Udayasuriyan, V., Sangeetha, P. and Bharathi, M. (2004). Analysis of Cry2A proteins encoded by genes cloned from indigenous isolates of *Bacillus thuringiensis* for toxicity against *Helicoverpa armigera*. *Current Science* 86: 566–570.

Lambert, A.L., Bradley, J.R. Jr. and van Duyn, J.W. (1996). Effects of natural enemy conservation and planting date on the susceptibility of *Bt* cotton to *Helicoverpa zea* in North Carolina. In *Proceedings, Beltwide Cotton Conference*, 9–12 January, 1996, Volume 2. Memphis, Tennessee, USA: National Cotton Council, 931–935.

Lang, B.A., Moellenbeck, D.J., Isenhour, D.J. and Wall, S.J. (1996). Evaluating resistance to CryIA(b) in European corn borer (Lepidoptera: Pyralidae) with artificial diet. *Resistant Pest Management* 8: 29–31.

Lee, M.K., Curtiss, A., Alcantara, E. and Dean, D.H. (1996). Synergistic effect of the *Bacillus thuringiensis* toxins CryIAa and CryIAc on the gypsy moth, *Lymantria dispar*. *Applied Environmental Microbiology* 62: 583–586.

Lee, M.K., Rajamohan, F., Gould, F. and Dean, D.H. (1995). Resistance to *Bacillus thuringiensis* Cry1A δ-endotoxins in a laboratory-selected *Heliothis virescens* strain is related to receptor alteration. *Applied Environmental Microbiology* 61: 3836–3842.

Li, G.L., Liu, Z.H., Hu, Y.M. and Ji, H.Q. (2001). Quantitative analysis of insect-resistant plant with *Bt* transformed gene in maize. *Journal of Henan Agricultural University* 35: 206–208.

Liang, G.M., Tan, W.J. and Guo, Y.Y. (2000). Study on screening and inheritance mode of resistance to *Bt* transgenic cotton in cotton bollworm. *Acta Entomologica Sinica* 43: 57–62.

Liao, C., Heckel, D.G. and Akhurst, R. (2002). Toxicity of *Bacillus thuringiensis* insecticidal proteins for *Helicoverpa armigera* and *Helicoverpa punctigera* (Lepidoptera: Noctuidae), major pests of cotton. *Journal of Invertebrate Pathology* 80: 55–63.

Lin, L.B., Guan, C.Y., Yang, H.W., Li, Y.Z. and Yang, Z.X. (2001). Dynamics of expression of *Bt* insecticidal protein gene in the transgenic plants of rapeseed and its insect-resistance activity. *Journal of Southwest Agricultural University* 23: 100–103.

Liu, Y.B. and Tabashnik, B.E. (1997). Inheritance of resistance to the *Bacillus thuringiensis* toxin Cry1C in the diamondback moth. *Applied Environmental Microbiology* 63: 2218–2223.

Liu Y.B. and Tabashnik, B.E. (1998). Elimination of a recessive allele conferring resistance to *Bacillus thuringiensis* from a heterogeneous strain of diamondback moth (Lepidoptera: Plutellidae). *Journal of Economic Entomology* 91: 1031–1037.

Liu Y.B., Tabashnik, B.E., Dennehy, T.J., Patin, A.L., Bartlett, A.C. and Liu, Y.B. (1999). Development time and resistance to *Bt* crops. *Nature* 400: 519.

Liu, Y.B., Tabashink, B.E., Masson, L., Escriche, B. and Ferre, J. (2000). Binding and toxicity of *Bacillus thuringiensis* protein Cry1C to susceptible and resistant diamondback moth (Lepidoptera: Plutellidae). *Journal of Economic Entomology* 93: 1–6.

Liu, Y.B., Tabashink, B.E., Meyer, S.K., Carriere, Y. and Bartlett, A.C. (2001). Genetics of pink bollworm resistance to *Bacillus thuringiensis* toxin Cry1Ac. *Journal of Economic Entomology* 94: 248–252.

Loc, N.T., Tinjuangjun, P., Gatehouse, A.M.R., Christou, P. and Gatehouse, J.A. (2002). Linear transgene constructs lacking vector backbone sequences generate transgenic rice plants which accumulate higher levels of proteins conferring insect resistance. *Molecular Breeding* 9: 231–244.

Losey, J.E., Carter, M.E. and Silverman, S.A. (2002). The effect of stem diameter on European corn borer behavior and survival: Potential consequences for IRM in *Bt*-corn. *Entomologia Experimentalis et Applicata* 105: 89–96.

Luttrell, R.G., Wan, L. and Knighten, K. (1999). Variation in susceptibility of noctuid (Lepidoptera) larvae attacking cotton and soybean to purified endotoxin proteins and commercial formulations of *Bacillus thuringiensis*. *Journal of Economic Entomology* 92: 21–32.

Lynch, R.E., Hamm, J.J., Myers, R.E., Guyer, D. and Stein, J. (2003). Baseline susceptibility of the fall armyworm (Lepidoptera: Noctuidae) to Cry1Ab toxin: 1998–2000. *Journal of Entomological Science* 38: 377–385.

MacIntosh, S.C., Kishore, G.M., Perlak, F.J., Marrone, P.G., Stone, T.B., Sims, S.R. and Fuchs, R.L. (1990). Potentiation of *Bacillus thuringiensis* insecticidal activity by serine protease inhibitors. *Journal of Agriculture and Food Chemistry.* 38: 1145–1152.

MacIntosh, S.C., Stone, T.B., Jokerst, R.S. and Fuchs, R.L. (1991). Binding of *Bacillus thuringiensis* proteins to a laboratory selected strain of *Heliothis virescens*. *Proceedings National Academy of Sciences USA* 188: 8930–8933.

Marcon, P.C.R.G., Siegfried, B.D., Spencer, T. and Hutchinson, W.D. (2000). Development of diagnostic concentrations for monitoring *Bacillus thuringiensis* resistance in European corn borer (Lepidoptera: Noctuidae). *Journal of Economic Entomology* 93: 925–930.

Marrone, P.G. and MacIntosh, S.C. (1993). Resistance to *Bacillus thuringiensis* and resistance management. In Entwistle, P.F., Cory, J.S., Bailey, M.J. and Higgs, S. (Eds.), Bacillus thuringiensis, *an Environmental Biopesticide: Theory and Practice*. Chichester, UK: Wiley Science Publishers, 221–235.

Martinez-Ramirez, A.C., Escriche, B., Real, M.D., Silva, F.J. and Ferre, J. (1995). Inheritance of resistance to a *Bacillus thuringiensis* toxin in a field population of diamond-back moth (*Plutella xylostella*). *Pesticide Science* 43: 115–120.

Mascarenhas, R.N., Boethel, D.J., Leonard, B.R., Boyd, M.L. and Clemens, C.G. (1998). Resistance monitoring to *Bacillus thuringiensis* insecticides for soybean loopers (Lepidoptera: Noctuidae) collected from soybean and transgenic *Bt*-cotton. *Journal of Economic Entomology* 91: 1044–1050.

McGaughey, W.H. (1985). Insect resistance to the biological insecticide *Bacillus thuringiensis*. *Science* 229: 193–195.

McGaughey, W.H. and Beeman, R.W. (1988). Resistance to *Bacillus thuringiensis* in colonies of Indian meal moth and almond moth (Lepidoptera: Pyralidae). *Journal of Economic Entomology* 81: 28–33.

McGaughey, W.H. and Johnson, D.E. (1987). Toxicity of different serotypes and toxins of *Bacillus thuringiensis* to resistant and susceptible Indianmeal moths (Lepidoptera: Pyralidae). *Journal of Economic Entomology* 80: 1122–1126.

McGaughey, W.H. and Johnson, D.E. (1992). Indianmeal moth (Lepidoptera: Pyralidae) resistance to different strains and mixtures of *Bacillus thuringiensis*. *Journal of Economic Entomology* 85: 1594–1600.

McGaughey, W.H. and Johnson, D.E. (1994). Influence of crystal protein composition of *Bacillus thuringiensis* strains on cross-resistance in Indianmeal moths (Lepidoptera: Pyralidae). *Journal of Economic Entomology* 87: 535–540.

McGaughey, W.H. and Whalon, M.E. (1992). Managing insect resistance to *Bacillus thuringiensis* toxins. *Science* 258:1451–1455.

Meng, F.X., Shen, J.L., Zhou, W.J. and Cen, H.M. (2004). Long-term selection for resistance to transgenic cotton expressing *Bacillus thuringiensis* toxin in *Helicoverpa armigera* (Hubner) (Lepidoptera: Noctuidae). *Pest Management Science* 60: 167–172.

Moar, W.J., Pusztai-Carey, M., Van Faassen, H., Bosch, D., Frutos, R., Rang, C., Luo, K. and Adang, M. (1995). Development of *Bacillus thuringiensis* Cry1C resistance by *Spodoptera exigua* (Hubner) (Lepidoptera: Noctuidae). *Applied Environmental Microbiology* 61: 2086–2092.

Moberg, W.K. (1990). Understanding and combating agrochemical resistance. In Green, M.B., LeBaron, H.M. and Moberg, W.K. (Eds.), *Managing Resistance to Agrochemicals*. ACS Symposium Series. Washington, D.C., USA: American Chemical Society, 3–16.

Mohan, M. and Gujar, G.T. (2000). Susceptibility pattern and development of resistance in the diamondback moth, *Plutella xylostella* L. to *Bacillus thuringiensis* Berl. var *kurstaki* in India. *Pest Management Science* 56: 189–194.

Mohan, M. and Gujar, G.T. (2002). Geographical variation in susceptibility of the diamondback moth, *Plutella xylostella* L. (Lepidoptera: Plutellidae) to *Bacillus thuringiensis* strains and purified toxins and associated resistance development in India. *Bulletin of Entomological Research* 92: 489–498.

Morin, S., Biggs, R.W., Sisterson, M.S., Shriver, L., Ellers Kirk, C., Higginson, D., Holley, D., Gahan, L.J., Heckel, D.G., Carriere, Y., Dennehy, T.J., Brown, J.K. and Tabashnik, B.E. (2003). Three cadherin alleles associated with resistance to *Bacillus thuringiensis* in pink bollworm. *Proceedings National Academy of Sciences USA* 100: 5004–5009.

Muller-Cohn, J., Chaufaux, J., Buisson, C., Gilois, N., Sanchis, V. and Lereclus, D. (1996). *Spodoptera littoralis* (Lepidoptera: Noctuidae) resistance to Cry1C and cross-resistance to other *Bacillus thuringiensis* crystal toxins. *Journal of Economic Entomology* 87: 791–797.

Nguyen, T.L., Tinjuangjun, P., Gatehouse, A.M.R., Christou, P. and Gatehouse, J.A. (2002). Linear transgene constructs lacking vector backbone sequences generate transgenic rice plants which accumulate higher levels of proteins conferring insect resistance. *Molecular Breeding* 9: 231–244.

Olsen, K.M. and Daly, J.C. (2000). Plant-toxin interactions in transgenic *Bt* cotton and their effect on mortality of *Helicoverpa armigera* (Lepidoptera: Noctuidae). *Journal of Economic Entomology* 93: 1294–1299.

Onstad, D.W. and Gould, F. (1998). Modeling the dynamics of adaptation to transgenic maize by European corn borer (Lepidoptera: Pyralidae). *Journal of Economic Entomology* 91: 585–593.

Oppert, B., Kramer, K.J., Beeman, R.W., Johnson, D. and McGaughey, W.H. (1997). Proteinase-mediated insect resistance to *Bacillus thuringiensis* toxins. *Journal of Biological Chemistry* 272: 23473–23476.

Oppert, B., Kramer, K.J., Johnson, D., Upton, S. and McGaughey, W.H. (1996). Luminal proteinases from *Plodia interpunctella* and the hydrolysis of *Bacillus thuringiensis* Cry1Ac protoxin. *Insect Biochemistry and Molecular Biology* 26: 571–583.

Perez, C.J., Shelton, A.M. and Roush, R.T. (1997). Managing diamondback moth (Lepidoptera: Plutellidae) resistance to foliar applications of *Bacillus thuringiensis*: Testing strategies in field cages. *Journal of Economic Entomology* 90: 1462–1470.

Perlak, F.J., Fuchs, T.A., Dean, D.A., McPherson, S.I. and Fischoff, D.A. (1991). Modification of the coding sequences for plant expression of insect-control protein genes. *Proceedings National Academy of Sciences USA* 88: 3324–3328.

Perlak, J.H., Oppenhuizen, M., Gustafson, K., Voth, R., Sivasubramaniam, S., Heering, D., Carey, B., Ihrig, R.A. and Roberts, J.A. (2001). Development and commercial use of Bollgard cotton in the USA: Early promises versus today's reality. *The Plant Journal* 27: 489–501.

Pilcher, C.D., Rice, M.E., Higgins, R.A., Steffey, K.L., Hellmich, R.L., Witkowski, J., Calvin, D., Ostlie, K.R. and Gray, M. (2002). Biotechnology and the European corn borer: Measuring historical farmer perceptions and adoption of transgenic *Bt* corn as a pest management strategy. *Journal of Economic Entomology* 95: 878–892.

Rahardja, U. and Whalon, M.E. (1995). Inheritance of resistance to *Bacillus thuringiensis* subsp. *tenebrionis CryIIIA* δ-endotoxin in Colorado potato beetle (Coleoptera: Chrysomelidae). *Journal of Economic Entomology* 88: 21–26.

Rajmohan, N. (1998). Pesticides resistance: A global scenario. *Pesticide World* 3: 34–40.

Ramachandran, S., Buntin, G.D., All, J.N., Raymer, P.L. and Stewart, C.N. Jr. (1998a). Movement and survival of diamondback moth (Lepidoptera: Plutellidae) larvae in mixtures of nontransgenic and transgenic canola containing a cryIA(c) gene of *Bacillus thuringiensis*. *Environmental Entomology* 27: 649–656.

Ramachandran, S., Buntin, G.D., All, J.N., Tabashnik, B.E. and Raymer, P.L., Adang, M.J., Pulliam, D.A. and Stewart, C.N. Jr. (1998b). Survival, development, and oviposition of resistant diamondback moth (Lepidoptera: Plutellidae) on transgenic canola producing a *Bacillus thuringiensis* toxin. *Journal of Economic Entomology* 91: 1239–1244.

Reed, J.P. and Halliday, W.R (2001). Establishment of Cry9C susceptibility baselines for European corn borer and southwestern corn borer (Lepidoptera: Crambidae). *Journal of Economic Entomology* 94: 397–402.

Rigev, A., Keller, M., Strizhov, M., Sneh, B., Prudovsky, E., Chet, I., Ginzberg, I., Koncz, Z., Schell, J. and Zilberstein, A. (1996). Synergistic activity of a *Bacillus thuringiensis* δ-endotoxin and bacterial endochitinase against *Spodoptera littoralis* larvae. *Applied Environmental Microbiology* 62: 3581–3586.

Ru, L.J., Zhao, J.Z. and Rui, C.H. (2002). A simulation model for adaptation of cotton bollworm to transgenic *Bt* cotton in North China. *Acta Entomologica Sinica* 45: 153–159.

Sachs, E.S., Benedict, J.H., Taylor, J.F., Stelly, D.M., Davis, S.K. and Altman, D.W. (1996). Pyramiding CryIA(b) insecticidal protein and terpenoids in cotton to resist tobacco budworm (Lepidoptera: Noctuidae). *Environmental Entomology* 25: 1257–1266.

Sayyed, A.H., Cerda, H. and Wright, D.J. (2003). Could *Bt* transgenic crops have nutritionally favourable effects on resistant insects? *Ecology Letters* 6: 167–169.

Sayyed, A.H., Ferre, J. and Wright, D.J. (2000). Mode of inheritance and stability of resistance to *Bacillus thuringiensis* var *kurstaki* in diamondback moth (*Plutella xylostella*) population from Malaysia. *Pest Management Science* 56: 743–748.

Sayyed, A.H., Haward, R., Herrero, S., Ferre J. and Wright, D.J. (2000). Genetic and biochemical approach for characterization of resistance to *Bacillus thuringiensis* toxin Cry1Ac in a field population of the diamondback moth, *Plutella xylostella*. *Applied Environmental Microbiology* 66: 1509–1516.

Schular, T.H., Poppy, G.M., Kerry, B.R. and Donholm, L. (1998). Insect resistant transgenic plants. *Trends in Biotechnology* 16: 168–175.

Sharma, H.C. (1993). Host plant resistance to insects in sorghum and its role in integrated pest management. *Crop Protection* 12: 11–34.

Sharma, H.C. and Pampapathy, G. (2006). Influence of transgenic cotton on the relative abundance and damage by target and non-target insect pests under different protection regimes in India. *Crop Protection* 25: 800–813.

Sharma, H.C. and Ortiz, R. (2000). Transgenics, pest management, and the environment. *Current Science* 79: 421–437.

Sharma, H.C. and Ortiz, R. (2002). Host plant resistance to insects: An eco-friendly approach for pest management and environment conservation. *Journal of Environmental Biology* 23: 111–135.

Sharma, H.C., Sharma, K.K, and Crouch, J.H. (2004). Genetic transformation of crops for insect resistance: Potential and limitations. *Critical Reviews in Plant Sciences* 23: 47–72.

Shelton, A.M., Tang, J.D., Roush, R.T., Metz, T.D. and Earle, E.D. (2000). Field tests on managing resistance to *Bt*-engineered plants. *Nature Biotechnology* 18: 339–342.

Shelton, A.M., Zhao, J.H. and Roush, R.T. (2002). Economic, ecological, food, safety and social consequences of the deployment of *Bt* transgenic plants. *Annual Review of Entomology* 47: 845–881.

Shen, J.L., Zhou, W.J., Wu, Y.D., Lin, X.W. and Zhu, X.F. (1998). Early resistance of *Helicoverpa armigera* (Hubner) to *Bacillus thuringiensis* and its relation to the effect of transgenic cotton lines expressing *Bt* toxin on the insect. *Acta Entomologica Sinica* 41: 8–14.

Sims, S.R. and Stone, T.B. (1991). Genetic basis of tobacco budworm resistance to an engineered *Pseudomonas fluorescens* expressing the delta-endotoxin of *Bacillus thuringiensis* var. *kurstaki*. *Journal of Invertebrate Pathology* 57: 206–210.

Stone, T.B. and Sims, S.R. (1993). Geographic susceptibility of *Heliothis virescens* and *Helicoverpa zea* (Lepidoptera: Noctuidae) to *Bacillus thuringiensis*. *Journal of Economic Entomology* 86: 989–994.

Stone, T.B., Sims, S.R. and Marrone, P.G. (1989). Selection of tobacco budworm for resistance to genetically engineered *Pseudomonas fluorescens* containing the δ-endotoxin of *Bacillus thuringiensis* subspecies *kurstaki*. *Journal of Invertebrate Pathology* 53: 228–234.

Storer, N.P., Peck, S.L., Gould, F., van Duyn, J.W. and Kennedy, G.G. (2003). Sensitivity analysis of a spatially-explicit stochastic simulation model of the evolution of resistance in *Helicoverpa zea* (Lepidoptera: Noctuidae) to *Bt* transgenic corn and cotton. *Journal of Economic Entomology* 96: 173–187.

Sumerford, D.V. (2002). Larval development of *Spodoptera exigua* (Lepidoptera: Noctuidae) larvae on artificial diet and cotton leaves containing a *Bacillus thuringiensis* toxin: Heritable variation to tolerate Cry1Ac. *Florida Entomologist* 86: 295–299.

Tabashnik, B.E. (1994). Evolution of resistance to *Bacillus thuringiensis*. *Annual Review of Entomology* 39: 47–79.

Tabashnik, B.E., Finson, N. and Johnson, M.W. (1992). Two protease inhibitors fail to synergize *Bacillus thuringiensis* in diamondback moth (Lepidoptera: Plutellidae). *Journal of Economic Entomology* 84: 49–55.

Tabashnik, B.E., Carriere, Y., Dennehy, T.J., Morin, S., Sisterson, M.S., Roush, R.T., Shelton, A.M. and Zhao, J.Z. (2003). Insect resistance to transgenic *Bt* crops: Lessons from the laboratory and field. *Journal of Economic Entomology* 96: 1031–1038.

Tabashnik, B.E., Cushing, N.L., Finson, N. and Johnson, M.W. (1990). Field development of resistance to *Bacillus thuringiensis* in diamondback moth (Lepidoptera: Plutellidae). *Journal of Economic Entomology* 83: 1671–1676.

Tabashnik, B.E., Dennehy, T.J., Sims, M.A., Larkin, K., Head, G.P., Moar, W.J. and Carriere, Y. (2002a). Control of resistant pink bollworm (*Pectinophora gossypiella*) by transgenic cotton that produces *Bacillus thuringiensis* toxin Cry2Ab. *Applied Environmental Microbiology* 68: 3790–3794.

Tabashnik, B.E., Finson, N., Groeters, F.R., Moar, W.J., Johnson, M.W., Luo, K. and Adang, M.J. (1994a). Reversal of resistance to *Bacillus thuringiensis* in *Plutella xylostella*. *Proceedings National Academy of Sciences USA* 91: 4120–4124.

Tabashnik, B.E., Finson, N., Johnson, M.W. and Heckel, D.G. (1994b). Cross-resistance to *Bacillus thuringiensis* toxin Cry1F in the diamondback moth (Lepidoptera: Plutellidae). *Applied Environmental Microbiology* 60: 4627–4629.

Tabashnik, B.E., Finson, N., Johnson, M.W. and Moar, W.J. (1993). Resistance to toxins from *Bacillus thuringiensis* subsp. *kurstaki* causes minimal cross-resistance to *Bacillus thuringiensis* subsp. *aizawai* in the diamondback moth (Lepidoptera: Plutellidae). *Applied Environmental Microbiology* 59: 1332–1335.

Tabashnik, B.E., Liu Y.B., Finson, N., Masson, L. and Heckel, D.G. (1997a). One gene in diamondback moth confers resistance to four *Bacillus thuringiensis* toxins. *Proceedings National Academy of Sciences USA* 94: 1640–1644.

Tabashnik, B.E., Liu, Y.B., de Maagd, R.A. and Dennehy, T.J. (2000a). Cross-resistance of pink bollworm (*Pectinophora gossypiella*) to *Bacillus thuringiensis* toxins. *Applied Environmental Microbiology* 66: 4582–4584.

Tabashnik, B.E., Liu, Y.B., Dennehy, T.J., Sims, M.A., Sisterson, M.S., Biggs, R.W. and Carrière, Y. (2002b). Inheritance of resistance to *Bt* toxin Cry1Ac in a field-derived strain of pink bollworm (Lepidoptera: Gelechiidae). *Journal of Economic Entomology* 95: 1018–1026.

Tabashnik, B.E., Liu, Y.B., Malvar, T., Heckel, D.G., Masson, L. Ballester, V., Granero, F., Mensua, J.L. and Ferre, J. (1997b). Global variation in the genetic and biochemical basis of diamondback moth resistance to *Bacillus thuringiensis*. *Proceedings National Academy of Sciences USA* 94: 12780–12785.

Tabashnik, B.E., Malvar, T., Liu, Y.B., Finson, N., Borthakur, D., Shin, B.S., Park, S.H., Masson, L., Maagd, R.A. and Bosch, D. (1996). Cross-resistance of diamondback moth indicates altered interactions with domain II of *Bacillus thuringiensis* toxins. *Applied Environmental Microbiology* 62: 2839–2844.

Tabashnik, B.E., Patin, A.L., Dennehy, T.J., Liu, Y.B., Carriere, Y., Sims, M.A. and Antilla, L. (2000b). Frequency of resistance to *Bacillus thuringiensis* in field populations of pink bollworm. *Proceedings National Academy of Sciences USA* 97: 12980–12984.

Tang, J.D., Shelton, A.M., Van Rie, J., de Roeck, S., Moaer, W.J., Roush, R.T. and Peferoen, M. (1996). Toxicity of *Bacillus thuringiensis* spore and crystal protein to resistant diamond black moth (*Plutella xylostella*). *Applied Environmental Microbiology* 62: 564–569.

Tuli, R., Bhatia, C.R., Singh, P.K. and Chaturvedi, R. (2000). Release of insecticidal transgenic crops and gap areas in developing approaches for more durable resistance. *Current Science* 79: 163–169.

Van der Salm, T., Bosch, D., Honee, G., Feng, I., Munsterman, E., Bakker, P., Stiekema, W.J. and Visser, B. (1994). Insect resistance of transgenic plants that express modified Cry1A(b) and Cry1C genes: A resistance management strategy. *Plant Molecular Biology* 26: 51–59.

Van Rie, J., McGaughey, W.H., Johnson, D.E., Barnett, B.D. and Van Mellaert, H. (1990). Mechanism of insect resistance to the microbial insecticide *Bacillus thuringiensis*. *Science* 247: 72–74.

Vaeck, M., Reynaerts, A., Hofte, H., Jansens, S., DeBeuckleer, M., Dean, C., Zabeau, M., Van Montagu, M. and Leemans, J. (1987). Transgenic plants protected from insect attack. *Nature* 327: 33–37.

Venette, R.C., Hutchison, W.D. and Andow, D.A. (2000). An in-field screen for early detection and monitoring of insect resistance to *Bacillus thuringiensis* in transgenic crops. *Journal of Economic Entomology* 93: 1055–1064.

Walker, D., Boerma, H.R., All, J. and Parrott, W.L. (2002). Combining cry1Ac with QTL alleles from PI 229358 to improve soybean resistance to lepidopteran pests. *Molecular Breeding* 9: 43–51.

Wang, C.Z. and Qin, J.D. (1996). Effect of soybean trypsin inhibitor, gossypol and tannic acid on the midgut protease activities and growth of *Helicoverpa armigera* larvae. *Acta Entomologica Sínica* 39: 337–341.

Westedt, A.L., Douches, D.S., Pett, W. and Grafius, E.J. (1998). Evaluation of natural and engineered resistance mechanisms in *Solanum tuberosum* for resistance to *Phthorimaea operculella* (Lepidoptera: Gelechiidae). *Journal of Economic Entomology* 91: 552–556.

Whalon, M.E., Miller, D.L., Hollingworth, R.M., Grafius, E.J. and Miller, J.R. (1993). Selection of a Colorado potato beetle (Coleoptera: Chrysomellidae) strain resistant to *Bacillus thuringiensis*. *Journal of Economic Entomology* 86: 226–223.

Wu, K.M., Guo, Y.Y. and Gao, S.S. (2002). Evaluation of the natural refuge function for *Helicoverpa armigera* (Lepidoptera: Noctuidae) within *Bacillus thuringiensis* transgenic cotton growing areas in North China. *Journal of Economic Entomology* 95: 826–831.

Wu, K.M., Guo, Y.Y. and Nan, L.V. (1999). Geographical variation in susceptibility of *Helicoverpa armigera* (Lepidoptera: Noctuidae) to *Bacillus thuringiensis* insecticidal protein in China. *Journal of Economic Entomology* 92: 273–278.

Wu, K.M., Guo, Y.Y. and Wang, W.G. (2000). Field resistance evaluations of *Bt* transgenic cotton GK series to cotton bollworm. *Acta Phytophylacica Sinica* 27: 317–321.

Xu, D., McElroy, D., Thoraburg, R.W. and Wu, R. (1993). Systemic induction of a potato pin 2 promoter by wounding methyl jasmonate and abscisic acid in transgenic rice plants. *Plant Molecular Biology* 22: 573–588.

Zhang, J.H., Wang, C.Z. and Guo, S.D. (2004). Effects of CpTI-*Bt* transgenic cotton and *Bt* transgenic cotton on survival, growth and nutrition utilization of *Helicoverpa armigera* (Hubner). *Acta Entomologica Sinica* 47: 146–151.

Zhang, J.H., Wang, C.Z., Qin, J.D. and Guo, S.D. (2004). Feeding behaviour of *Helicoverpa armigera* larvae on insect-resistant transgenic cotton and non-transgenic cotton. *Journal of Applied Entomology* 128: 218–225.

Zhao, J.Z., Cao, J., Li, Y.X., Collins, H.L., Roush, R.T., Earle, E.D. and Shelton, A.M. (2003). Transgenic plants expressing two *Bacillus thuringiensis* toxins delay insect resistance evolution. *Nature Biotechnology* 21: 1493–1497.

Zhao, J.Z., Collins, H.L., Tang, J.D., Cao, J. and Earle, E.D., Rousch, R.T., Herrero, H., Escriche, B., Ferre, J. and Shelton, A.M. (2000a). Development and characterization of diamond back moth resistance to transgenic *broccoli* expressing high levels of Cry1C. *Applied Environmental Microbiology* 66: 3784–3789.

Zhao, J.Z., Fan, Y.L., Fan, X.L., Shi, P. and Lu, M.G. (1999). Evaluation of transgenic tobacco express-ing two insecticidal genes to delay resistance development of *Helicoverpa armigera*. *Chinese Science Bulletin* 44: 1871–1873.

Zhao, J.Z., Fan, X.L., Shi, X.P., Zhao, R.M. and Fan, Y.L. (1997). Gene pyramiding: An effective strategy of resistance management for *Helicoverpa armigera* and *Bacillus thuringiensis*. *Resistant Pest Management* 9: 19–21.

Zhao, J.Z., Li, Y.X., Collins, H.L., Cao, J., Earle, E.D. and Shelton, A.M. (2001). Different cross-resistance patterns in the diamondback moth (Lepidoptera: Plutellidae) resistant to *Bacillus thuringiensis* toxin Cry1C. *Journal of Economic Entomology* 94: 1547–1552.

Zhao, J.Z., Lu, M.G., Fan, X.L., Wei, C., Liang, G.M. and Zhu, C.C. (1996). Resistance monitoring of *Helicoverpa armigera* to *Bacillus thuringiensis* in North China. *Resistant Pest Management* 8: 20–21.

Zhao, J.Z., Rui, C., Lu, M.G., Fan, X.L., Ru, L. and Meng, X.Q. (2000b). Monitoring and management of *Helicoverpa armigera* resistance to transgenic *Bt* cotton in Northern China. *Resistant Pest Management* 11: 28–31.

Zhao, J.Z., Shi, X.P., Fan, X.L., Zhang, C.Y., Zhao, R.M. and Fan, Y.L. (1998). Insecticidal activity of transgenic tobacco co-expressing *Bt* and CpTI genes on *Helicoverpa armigera* and its role in delaying the development of pest resistance. *Rice Biotechnology Quarterly* 34: 9–10.

13

Transgenic Resistance to Insects: Gene Flow

Introduction

There is a need for planned introduction of genetically modified organisms into the environment. There are concerns that the deployment of transgenic plants will lead to development of resistance in insects and gene flow between closely related species and the unrelated species in the ecosystem. While some of these concerns may be real, others seem to be highly exaggerated. Therefore, careful thought should be given to the production and release of transgenic crops in different agroecosystems. A number of ecological issues, including gene flow, need to be addressed while considering the production and deployment of transgenic crops for insect control. The greatest risk of a transgenic plant released into the environment is its potential spread to other areas to become a weed. Genes from unrelated sources may change the fitness and population dynamics of the hybrids between native plants and the wild species. However, there are no records of a plant becoming a weed as a result of plant breeding (Cook, 2000). This is mainly because of:

- Low risk of crop plants to the environment;
- Extensive testing of the crop varieties before release; and
- Adequate management practices to mitigate risks inherent in crop plants.

Plant breeding efforts have tended to decrease rather than increase the toxic substances, as a result making the improved varieties more susceptible to insect pests. However, there is a feeling that genes introduced from outside the range of sexual compatibility might present new risks to the environment and human beings. However, many such apprehensions are not supported by data. A study conducted by the National Academy of Sciences (NAS, 1987) has concluded that:

- There is no evidence of hazards associated with DNA techniques.
- The risks, if any, are similar to those with conventional breeding techniques.
- The risks involved are related to the nature of the organism rather than the process.

Role of Pollinators in Gene Flow

There is a need for risk assessment of the likely spread of transgenes via pollen through insect pollinators, largely bees (honeybees and bumblebees) to other crops, or to sexually compatible wild relatives. Insect-mediated pollen and gene flow is a function of the deposition of viable and compatible pollen from donor to recipient plants along insect foraging routes, and of the spatial dynamics of those foraging routes within the foraging areas. Knowledge of bee foraging behavior will be useful to design appropriate confinement strategies for experimental releases, monitoring protocols, and mitigation procedures for risk management. Crucial data are lacking for the development of a credible scientific basis to confirm or deny environmental risks associated with ecological roles of lepidopterans exploiting specific wild relatives of *Bt* transgenic crops and escape of *Bt* transgene constructs to wild relatives (Letourneau, Robinson, and Hagen, 2003).

Insects do not discriminate between conventional and transgenic oilseed rape resistant to glufosinate (Pierre et al., 2003) and chitinase (Picard Nizou et al., 1995), although some differences have been observed in the nectar between the transgenic and nontransgenic lines. Bees (Apidae), hoverflies (Syrphidae), and sawflies (Symphyta) visited glufosinate-resistant oilseed rape, of which bees were most abundant (Saure, Kuhne, and Hommel, 2001). They were involved in pollination of oilseed rape and the transgene was found in *Brassica juncea* (L.) Czern. Oilseed rape pollen was spread through pollen loads of different types (Ramsay et al., 1999). Presence of pollen from transgenic plant in largely nontransgenic *Brassica* pollen loads, and generation of mixed transgenic and nontransgenic progeny following pollination of male-sterile plants with pollen from single bees indicated that bees can readily spread pollen around the foraging area. In rape, a total of 94 bee species and 48 hoverfly species have been recorded. Mark-recapture experiments with bumblebees and solitary bees showed that the same individuals visited the flowers of transgenic rape and later flowers of other cruciferous plants (Saure et al., 2003).

The foraging distances of *Bombus muscorum* (L.) are more restricted to the neighborhood of the nesting habitat than those of *B. terrestris* (L.) and *B. lapidarius* (L.) (Hellwig and Frankl, 2000). High percentages of *B. terrestris* workers were recaptured while foraging on super-abundant resources at distances up to 1750 m from the nest. Isolated patches of highly rewarding forage crops in agricultural landscapes are probably accessed by bumblebee species with large mean foraging distances, such as the short-tongued *B. terrestris*. The long-tongued *B. muscorum*, which is relatively rare, depends on a close connection between nesting and foraging habitat. Long-distance flights of bumblebee pollinators need to be considered for gene flow from transgenic plants on a landscape scale. Consistent and significant reduction in pollen dissemination by insects has been observed in cotton as the distance from the test plot increased (Umbeck et al., 1991). Outcrossing decreased from 5% to <1% at 7 m away from the test plot. The leafcutter bees, *Megachile* spp. used in commercial seed production, show a directional nonrandom bias when pollinating within fields, primarily resulting in the movement of pollen directly towards and away from the bee domicile (Amand, Skinner, and Peaden, 2000). Within-field pollen movement was detected only over distances of 4 m or less. In pigeonpea, the most important insect pollinators are *Megachile* spp. (Saxena, Singh, and Gupta, 1990). The flower type, the abundance of insect pollinators, and weather conditions at flowering influence the degree of cross-pollination. Dispersal of pollen from alfalfa hay and seed production plots occurred up to 1000 m. Random amplified polymorphic DNA (RAPD) markers detected gene movement up to 230 m. The outcrossing frequency for large fields was nearly 10 times greater than that of small-sized plots.

Role of Viruses in Gene Flow

Plants expressing viral sequences have potential risks such as heterologous encapsidation, which can temporarily confer the ability to a virus to be transmitted by a novel vector. However, this is not different from the nontransgenic plants infected by viruses, except when a coat protein gene is introduced into a nonhost plant (Tepfer, 2002). To overcome this problem, genes that code for a coat protein that no longer interacts with the vector may be used. Other options are to delete the motif recognized by the vector or delete the motifs allowing particle assembly. Plant to plant gene flow by sexual outcrossing may lead to creation of virus-resistant weeds, and can be overcome by using terminator technology, and carrying out detailed studies of the potential ecological impact. Plant to virus gene flow by recombination results from covalent joining of DNA or RNA sequences that are not normally adjacent (Hammond, Lecoq, and Raccah, 1999). It can occur by cleavage and ligation (cut and paste) or in the case of RNA recombination, by the RNA polymerase changing matrix during strand synthesis. Plant to virus gene flow by recombination could create a virus with novel properties, worsening of symptoms, and changes in other properties such as host range. Recombination can also occur in nontransgenic plants infected by more than one virus. Under conditions of minimum selection pressure, there is a need to examine the nature of recombinants produced in single-infected transgenic plants, and doubly infected nontransgenic plants. If the nature of recombinants produced is the same, it is reasonable to predict that, selection pressures being equivalent, the fitness of recombinants occurring in transgenic plants will be the same as in doubly infected plants. In nontransgenic plants infected with cauliflower mosaic virus and tomato aspermy virus, recombination has been observed frequently. All crossovers occurred in a region of high sequence identity (Gibs, 1994; Aaziz and Tepfer, 1999; Hammond, Lecoq, and Raccah, 1999). In all plants where recombination occurred, several types of recombinants have been observed.

Gene Flow into the Wild Relatives of Crops: Vertical Gene Flow

To assess the potential ecological impact of commercial release of transgenic crops in a given region, the likelihood and impact of vertical gene flow for that crop in that region should be taken into consideration. To guide this assessment, the concept of gene flow indices or botanical files (Frietema de Vries, van der Meijden, and Brandenburg, 1992, 1994; Frietema de Vries, 1996; Amman, 2001) has been developed. Gene flow indices give an indication of the likelihood of a given species to hybridize with the wild relatives and its impact on the ecosystem. Botanical profiles should be established for each region, consisting of data on plant species, and should provide an index of the likelihood for dispersal of pollen, dispersal of reproductive plant parts such as seeds or fruit, and distribution frequency of wild relatives. Each of these factors can be subdivided into different levels of potential (or unknown) risk.

Botanical profiles indicate the likelihood of gene flow from a particular transgenic crop plant to its wild relatives, but ignore the potential impact of the transgene on crops and recipient wild relatives. Therefore, botanical profiles should be combined with information on the transgene used for transformation, and the transformation event (Amman, 2001).

This will allow the evaluation of issues such as contribution of the transgene to the weed-iness and fitness of the host plant. For each transgene, a "transgene file" with all relevant information should be put together, and encompass all aspects of a "transgene-centered approach to bio-safety" (Metz, Stiekema, and Nap, 1998). Assessing ecological impact of the transgene in a given crop in a region is an important component of risk assessment. Where the transgenic crops are used as food for human beings, then there is also a need to include food profile, in which all nutritional and food safety aspects of a given plant product should be evaluated and classified. There is a need to provide information to the consumers in a way that is equivalent to current labeling of the presence of food addi-tives, and define a unique key for each product (OECD, 2002).

Introgression of transgenes into the crop plants and wild relatives is of potential concern (Gregorius and Steiner, 1993; Serratos, Willcox, and Castillo-Gonzalez, 1997). Transgenes conferring resistance to insect pests, diseases, and herbicides may result in enhanced fit-ness, survival, and spread of weeds (Ellstrand, 2001). This may add to farmers' weed man-agement burden, and/or result in invasion of natural habitats and compromise the biodiversity of these habitats. The potential for a crop plant to hybridize with a weed depends on sexual compatibility and relatedness between the species. The opportunity for natural hybridization between two species depends on many pre- and post-zygotic factors. While plant breeders have repeatedly crossed crops with wild relatives to introgress a wide range of beneficial traits (Harlan, 1976), many of the hybrid combinations developed in this manner would not occur in nature because of the barriers within the plants to prevent normal embryo or endosperm development. The occurrence of interspecific and intergeneric hybrids as a result of manual hybridization only gives a possible indication of sexual compatibility and potential for hybridization in nature. If an interspecific or inter-generic hybrid develops in nature, the success of such hybrids and their progeny depends on a number of factors. Gene introgression from one species to another or from a crop to a weed requires repeated backcrossing to effect the incorporation of alleles from one gene pool to another. The possibility of repeated hybridization leading to allele introgression from cultivated crops to weedy relatives has been recognized for a long time (Anderson, 1949), and is considered to have played an important role in both the domestication of crops and the evolution of weeds (de Wet and Harlan, 1975). These events occur where the distribution of a wild species overlaps with the cultivation of the related crop, and are especially common in the centers of diversity for a specific crop.

The characteristics influencing gene transfer between species are complex and will not, in general, change as a consequence of transgene expression. Changes in flower color as a result of genetic transformation may have either a positive or negative influence on insect pollina-tors. Depending on crop management, male sterility may remove pollen competition and provide a better opportunity for hybridization. For a majority of transgenic traits, transgenic crops are no more likely to transfer either their transgenes or any other gene to other species than through crop cultivars developed by conventional methods in the past. If gene intro-gression from a crop to natural populations does occur, the key issue to consider is whether the impact is any different for cultivars derived from genetic modification as compared to those derived through traditional breeding. When considering the ecological concerns about transfer of transgenes to weedy species, it is the phenotype conferred by a gene that is important, not whether it was derived by genetic engineering or traditional plant breeding (Metz and Nap, 1997). However, extension of the botanical profile approach to include a description of the transgene to assess the ecological impact of a gene is quite important.

If transgenes conferring resistance to insect pests, diseases, and environmental stress are introgressed into weedy relatives of crops, there is a concern that they may enhance particular fitness components of the weed (Ellstrand, Prentice, and Hancock, 1999). However, plant breeders have released many cultivars with new genes for resistance to insect pests, diseases, and environmental stress over many years. Any impacts resulting from the introgression of such traits into weedy species are equally likely for the products of plant breeding and genetic modification. The risks are no different and the use of resistance genes in cultivars from traditional breeding has not enhanced the survival and spread of weeds in the past. When serious weeds have arisen following hybridization of crops and wild species, their aggressive nature has arisen from a coupling of morphological traits conferring weedy attributes and the synchrony of development with the crop rather than a gain in fitness from resistance genes (de Wet and Harlan, 1975).

Another problem highlighted by natural hybridization between crops and their wild relatives is the increased potential of extinction of wild taxa. Some "genetically aggressive" species, referred to as compilospecies (Harlan and de Wet, 1963), may completely assimilate another locally rare species through repeated cycles of hybridization and introgression, causing it to become extinct. The highly successful wheat crop is considered to have assimilated genes from more than one species of *Aegilops* (Harlan and de Wet, 1963). Extinction by hybridization does not depend on relative fitness, but on patterns of mating (Ellstrand, Prentice, and Hancock, 1999). The impact of the release of transgenic crops will be no different than the impact of existing nontransgenic crops. Interspecific hybridization is a common process, but hybrids are rare, and most are sterile, and there is a very low chance of gene introgression into the wild relatives (Fitter, Perrins, and Williamson, 1990). Transgenic plants may also become weeds, except in the context of their normal agricultural environment. Gene escape may occur when a plant species invades a seminatural habitat or the gene is transferred into the wild relative, and persists in uncultivated land. Its spread can be checked by methods similar to any other single trait. There are differences among plant species to disperse from the environment other than the one in which they are released, and their ability to establish feral populations. Such an event has to be compared with that of the original plant.

The process of introgression between a transgenic crop modified for better agronomic characters and a wild relative could potentially lead to increased weediness and adaptation to the environment (Gueritaine et al., 2002). However, formation of a hybrid and hybrid progeny could be associated with functional imbalance and low fitness, which reduces the risk of gene escape and establishment of the wild species in the field. Inter- and intra-specific gene flow between transgenic and nontransgenic plants leads to development of new genetic materials in cultivated or wild populations. Introgression of novel genes into wild relatives of crops may have a substantial impact on crop evolution or wild populations leading either to more desirable cultivars, more aggressive weeds, and/or extinction of rare and endangered species. A better understanding of gene flow is therefore essential for deployment of transgenic plants. Such information would be helpful in assessing the ecological risk of transgene escape from cultivated germplasm to closely related wild relatives. The consequences of gene flow from crop to wild relatives include genetic assimilation, wherein alien genes replace host genes, and demographic swamping, wherein hybrids are less fertile than their wild parents and as a result the wild populations shrink (Haygood, Ives, and Andow, 2003). Factors influencing gene flow include mating system, mode of pollination, mode of seed dispersal, and the habitat characteristics where the crops are grown (Messeguer, 2003). Hybridization is perhaps the most serious genetic threat to endangered species, with extinction often taking place in less than five generations (Wolf, Takebayashi, and Rieseberg, 2000). Pollen can function as a vehicle to disseminate introduced, genetically engineered genes throughout a plant population or into a related species (Amand, Skinner, and Peaden, 2000). The measurement of the risk of inadvertent dispersal of transgenes must include the assessment of accidental dispersion of pollen. Factors to be considered include the rate of pollen spread, the maximal dispersion distance of pollen, and the spatial dynamics of pollen movement within seed production fields, none of which are known for insect-pollinated crop species.

Assessment of realistic risk for gene transfer through pollen is available for many crops (Raybould and Gray, 1993), and agriculturally sound procedures need to be developed for different regions (Boulter, 1995). Evidence of gene flow has been reported in sugarbeets (Desplanque et al., 1999; Bartsch et al., 1999, 2003; Arnaud et al., 2003), cassava (Nassar, 2002), rice (Song et al., 2003; Chen et al., 2004), *Brassica* (Chevre et al., 1997), sunflower (Arias and Rieseberg, 1994; Linder et al., 1998), and sorghum (Paterson et al., 1995; Arriola and Ellstrand, 1996, 1997). Crop to weed gene flow has been implicated in the evolution of enhanced weediness in wild relatives of wheat, rye, rice, soybean, sorghum, millet, beans, and sunflower, and extinction of wild relatives in rice and cotton (Sun and Corke, 1992; Ellstrand, Prentice, and Hancock, 1999). Information on pollen dispersals and gene flow in different crops is discussed below.

Cotton

In cotton, pollen dispersal increases with an increase in the size of the source plot (Llewellyn and Fitt, 1996). A buffer zone of 20 m would limit the dispersal of transgenic pollen from small-scale field tests. In China, the potential for gene introgression from *Bt* cotton lines into wild or cultivated sexually compatible plants is very low, and such events are highly unlikely to increase the weediness potential of any resulting progeny. There is no pollen transfer between *Gossypium hirsuturm* L. and *G. arboreum* L. although these species have been grown together in India for several decades (Figure 13.1).

Cereals

In rice, pollination of recipient plants with pollen of the transgenic plant occurs at a significant frequency (Messeguer et al., 2001). A gene flow slightly lower than 0.1% has been detected in a normal side-by-side plot design. Similar results have been obtained in a

Risk of outcrossing is almost zero. There are numerous genetic barriers to outcrossing.

Not true of all GM crops.

Tetraploid - AD genome
Gossypium hirsuturm

Diploid - A2 genome
Gossypium arboreum

FIGURE 13.1   There are several barriers for gene flow and survival of interspecific hybrids. No intermediates have been observed between cultivated *Gossypium hirsutum* and *Gossypium arboreum* cottons under natural conditions.

circular plot when the plants were placed at 1 m distance from the transgenic plants. Circular-field trial designs could also prove to be useful for studying the gene flow to commercial cultivars of rice and other crops to develop strategies to prevent pollen dispersal from transgenic fields to the neighboring nontransgenic fields. Studies involving allozyme and chloroplast DNA analyses have shown that gene flow does take place between maize and teosinte [*Zea diploperennis* (Iltis et al.), *Z. perennis* (Hitch.) Reeves et Mangelsd., *Z. luxurians* (Durieuet Ash) Bid., and *Z. mays* (L.) subspecies *parviglumis* and subsp. *mexicana* (Hitch.)]. Introgression is in both directions, but at a low level, allowing the species to maintain their distinct genetic constitutions (Doebley, 1990). It has been suggested that an engineered gene in maize could spread to teosinte, and if it were to confer some adaptive advantage, might spread throughout the teosinte population. As teosinte is restricted to certain areas of Mexico, the solution would be not to grow transgenic maize in those regions.

Seefeldt et al. (1999) discovered two imazamox-resistant hybrids from a cross between *Aegilops cylindrica* (Host) and an imazamox-resistant wheat (induced mutant FS-4IR). Six seedlings from BC$_1$ survived an application of 72 g *ai* ha^{-1} of imazamox. Management strategies to reduce the occurrence of herbicide-resistant *A. cylindrica* × wheat hybrids can be incorporated before, during, and after the cropping season when the herbicide-resistant wheat is planted. Arriola and Ellstrand (1996) detected crop weed hybrids in sorghum at distances of 0.5 to 100 m from the crop. Interspecific hybridization can and does occur in this system at a substantial and measurable rate. Transgenes introduced into sorghum can be expected to have the opportunity to escape through interspecific hybridization with Johnson grass, *Sorghum halepense* (L.) Pers. (Figure 13.2). Traits that prove to be beneficial to weeds can be expected to persist and spread. This is an issue that needs to be addressed when developing biosafety guidelines for commercial release of transgenic sorghum. Chromosome counts of plants intermediate in appearance between *Setaria italica* (L.) Beauv. and *S. verticillata* (L.) Beauv., collected from a millet field at Maine et Loire, France, revealed that outcrossing rates within the genus are very low

FIGURE 13.2 Wild relatives of sorghum. (A) *Sorghum plumosum*, (B) *Sorghum halepense*, and (C) *Sorghum bicolor*. There is frequent gene flow between cultivated and wild sorghums, which is a source of tremendous genetic diversity observed in this crop.

(Bottraud et al., 1992). However, examination of plants intermediate between *S. italica* and *S. viridis* (L.) Beauv., collected from two fields in Maine et Loire on which *S. italica* was cultivated, revealed differences in the EcoR1 patterns of chloroplast DNA between cultivated and wild plants, indicating that reciprocal crosses occur in the field. The results suggested that even a largely self-pollinated cultivated species at times may exchange genetic information with wild relatives at rates that may cause problems if transgenic cultivars are released.

Brassicas

Genes from the conventionally bred *Brassica napus* L. have been moving to the wild turnip, *B. rapa* L. (Raybould and Gray, 1993). Transgenic herbicide tolerance is capable of intro- gressing into populations of *B. rapa* and persisting, even in the absence of selection due to herbicide application (Snow and Jorgensen, 1999). However, there were no significant differences between transgenic and nontransgenic plants in survival or the number of seeds per plant, indicating that fitness costs associated with the transgene are likely to be negligible. Pollen fertility and seed production of BC_3 plants were as great as those of *B. rapa* raised in the same growth rooms. Microscopic studies have shown polymorphism within the population of hoary mustard for pollen germination on oilseed rape flowers (Lefol, Fleury, and Darmency, 1996). The transgenic herbicide-resistant and a commercial oilseed rape cultivar did not differ in pollen behavior and ovule fertilization. Pollen tube growth was slow and erratic in interspecific crosses. Fertilization efficiency of oilseed rape, *B. napus* in relation to interspecific gene flow with hoary mustard, *Hirschfeldia incana* (L.) Lagreze-Fossat crosses was 15% and 1.3%, respectively, of that in intraspecific crosses. There were no post-zygotic barriers to the development of hybrid embryos in hoary mustard pods. Up to 26 spontaneous hybrids per male-sterile oilseed rape plant and one per hoary mustard plant were obtained in field experiments. All hybrids were triploid, with 26 chromosomes, and had low fertility. They produced 0.5 seeds per plant after spontaneous backcrossing with hoary mustard. Some of these descendants were produced from unreduced gametes. The results suggested that gene flow is likely to occur, but its actual frequency under crop grow- ing conditions remains to be estimated. Outcrossing studies have also revealed high poten- tial of transgenic pollen transfer from rape to *B. juncea* (Saure et al., 2003). The backcross progenies with oilseed rape, *B. napus* var. *oleifera* cytoplasm (OBC) have a fitness value 100 times lower than that of the backcross with wild radish, *Raphanus raphanistrum* L. cytoplasm (RBC) (Gueritaine et al., 2002). The herbicide-resistant RBC has similar growth to the susceptible RBC, but final male and female fitness values were two times lower. In turn, susceptible RBC exhibited similar fitness to the control wild radishes, and the relative fitness of the different types is the same, whether or not they grow under competitive conditions.

Soybean

The seedlings gathered from seeds from individual plants of a wild accession have been used for isozyme analysis to identify whether they were hybrids or not. In 23 plants of the wild accession, four plants produced hybrids (incidence of hybridization = 17.4%) (Nakayama and Yamaguchi, 2002). There was no specific direction in hybridization. The hybridization rate per maternal plant varied from 0 to 5.89% with a mean of 0.73% for all maternal plants. The results indicated that natural hybrids are easily produced in a certain frequency by pollen flow from the cultivated soybean to the wild soybean when they flow- ered simultaneously, and there was adequate pollinator population.

Sunflower

Wild sunflower occurs in disturbed habitats such as roadsides and agricultural areas, often reaching heights of 2 to 3 m in optimal growing conditions. In contrast to wild populations, commercial varieties lack several fitness-related traits such as branching, extended flowering period, self-incompatibility, and seed dormancy, and often exhibit greater resistance to diseases (Seiler, 1992). Isozyme and molecular studies indicated that gene flow among populations has been extensive (Rieseberg and Seiler, 1990; Arias and Rieseberg, 1995), and populations near commercially grown sunflower harbor crop-specific DNA markers due to crop-wild hybridization. The marginal wild populations (3 m from the cultivar) showed the highest percentage (27%) of gene flow (Arias and Rieseberg, 1994). Cultivated sunflower hybridizes spontaneously with wild/weedy populations. F_1 wild-crop hybrids had lower fitness than wild progeny, especially when grown under favorable conditions, but the F_1 barrier to the introgression of crop genes into wild populations is quite "permeable." Therefore, episodes of crop-to-wild hybridization are likely to lead to long-term persistence of nondeleterious crop genes in wild populations.

Vegetables

Most domesticated species are associated with a closely related wild or weedy species, for example, *Cucurbita pepo* (L.) with *C. texana* (Scheele) Decker, with which they are fully compatible, indicating the possibility of introgressive hybridization. Evidence for this is based on genetic homogeneity [as shown by electrophoretic data from *C. pepo* and its associated forms *C. texana* and *C. fraterna* (Bailey) Andres] and introgressive hybridization (Wilson, 1990).

Beets

Weed beets pose a serious problem for sugarbeet, *Beta vulgaris* var. *saccharifera* Alef. (Desplanque, Hautekeete, and van Dijk, 2002). Spreading of transgenic traits into wild beet can occur if genetically engineered biennial plants survive in the winter, flower in spring, and spread their pollen (Pohl Orf et al., 1999). Survival of sugar beet, transgenic as well as conventional beets, in Germany and along the Dutch border is possible. Survival rates were well correlated with temperature. Differences between sugar beet hybrids and breeding lines have been detected, but not within different breeding lines or hybrids. There were no detectable differences between transgenic and nontransgenic plants. Traditionally, the only efficient method of weed control has been manual removal, but the introduction of transgenic herbicide-tolerant sugarbeets may provide an alternative solution because nontolerant weed beets can be destroyed by herbicide. In northern France, weed beets are present in variable densities in sugarbeet fields of up to 80 weed beet plants m^{-2}. Diploid F_1 crop-wild hybrids and triploid variety bolters (individuals with a low vernalization requirement) have been found in low densities in virtually all sugarbeet fields. Gene flow is possible between all forms, as illustrated by overlapping flowering periods in the field and successful controlled cross-pollination. The F_1 wild hybrids result from pollination in the seed-production region by wild plants possessing the dominant bolting allele B for flowering without experiencing a period of cold. In the case of a transgene for herbicide tolerance incorporated into male-sterile seed-bearer plants, such hybrids will contain both the herbicide tolerance and the bolting allele. Appearance of transgenic weed beets is possible, but can be retarded if the transgene for herbicide tolerance is incorporated into the tetraploid pollinator breeding line.

Development of Resistance to Antibiotics: Horizontal Gene Flow

Horizontal gene transfer (HGT) is the transfer of genetic material from one organism (the donor) to another organism (the recipient), which is not sexually compatible with the donor (Gay, 2001). HGT between bacterial species is particularly common when it involves plasmids and transposons (Courvalin, 1994; Lorenz and Wackernagel, 1994; Landis, Lenart and Spromberg, 2000). With the availability of full genomic sequences of organisms, more and more potential candidates for HGT between species, genera, and even kingdoms are being identified. HGT is considered a significant source of genome variation in bacteria (Ochma, Lawrence, and Groisman, 2000), and may be a common route for evolution of bacterial populations and possibly eukaryotes (De la Cruz and Davies, 2000). Detailed phylogenetic analyses based on the presence of specific DNA sequences does not necessarily support the involvement of HGT (Stanhope et al., 2001). The general concern with respect to transgenic crops is that the novel genes may result in a transfer of genetic material to other species and cause harm. Of particular concern are putative recipient microorganisms in soil or in the digestive tract of humans and livestock (Dröge, Pühler, and Selbitschka, 1998, 1999). The initial debate on HGT from transgenic crops focused on the presence of antibiotic marker genes in the plants. Due to the strong selection pressure of prescription regimes in human and animal therapy, as well as use in farming (as feed additives), spontaneous resistance through mutation, coupled with some HGT between bacteria, has resulted in the spread of antibiotic resistance to such an extent that the medical and veterinary use of antibiotics as therapeutic agents is being seriously compromised (Austin et al., 1999).

Development of techniques to remove antibiotic and herbicide selectable markers has shifted the HGT debate to transgenes used in transgenic plants. The most popular method of gene transfer to plants using *Agrobacterium tumefaciens* (Smith and Townsend) is based on HGT. Whereas the mechanism of HGT from *A. tumefaciens* to plant cells is known in considerable detail, there is no known mechanism for HGT from plants to other organisms. HGT depends on existence of free DNA that should be of sufficient length, and persist long enough for uptake during growth, decay, herbivory, or consumption of transgenic crops. A (bacterial) recipient should be competent for DNA uptake, and a mechanism for uptake should be in place. Bacterial strains may be naturally competent for DNA uptake during some stage of development, such as *Ralstonia solanacearum* (Smith) (Gay, 2001). Transformation mechanisms developed in the laboratory for *Escherichia coli* Escherich such as those operating during PEG-mediated DNA uptake or electroporation (Sambrook, Fritsch, and Maniatis, 1989) may have counterparts in nature. As transgenic-crop DNA will not be released as plasmids, and therefore, conjugal transfer can be totally ruled out. The recipient cell should incorporate, maintain, and use the incoming DNA. This genetic alteration should pose no selection against the recipient organism. Although DNA in decaying plant cells is rapidly degraded, DNA of appropriate length can survive in some soils and aquatic environments (Lorenz and Wackernagel, 1992), or the digestive tract of mice (Schubbert et al., 1997) long enough to be available for uptake. The intestinal tract of cows and other ruminants is likely to be more hostile towards free DNA (Duggan et al., 2000; Gay, 2001). Competence of bacteria in natural surroundings is difficult to assess, but is unlikely to approach the efficiencies reached in optimized laboratory conditions. The maintenance and integration of incoming DNA is mediated by (and may require?) sequence homology with DNA of the recipient bacterium (de Vries and Wackernagel, 1998; de Vries, Meier, and Wackernagel, 2001). The given transfer of a plant gene to a bacterium does not imply functionality in the bacterium. Regulatory sequences (promoters and enhancers) may not work, and introns, if any, may not be recognized in the recipient.

Several studies have failed to demonstrate HGT from transgenic plants to bacteria (Schlüter, Fütterer, and Potrykus, 1995; Nielsen et al., 1998; Bertolla and Simonet, 1999; Gebhard and Smalla, 1999). The kanamycin resistance gene from transgenic maize could be retrieved in an *Acinetobacter* strain (de Vries and Wackernagel, 1998). Without the artificially introduced homology in the recipient strain, no HGT was detected, indicating that transformation frequency is very low. Such systems confirm that HGT can occur, but at exceptionally low frequencies. Another route for HGT could be a plant virus in a process equivalent to transduction in bacteria. It is known that plant viruses can acquire host sequences, but in the case of RNA viruses, it would seem highly unlikely that such DNA will become integrated in the genome of a related plant (Aaziz and Tepfer, 1999; Tepfer, 2002), although RNA recombination itself may be an HGT issue (Malnoe et al., 1999). In the case of DNA viruses, there is some evidence for transfer of genetic material from virus to plants (Bejarano et al., 1996; Harper et al., 1999; Jakowitsch et al., 1999). The transgene of genetically modified crops constitutes only a fraction of the total plant DNA, whereas all plant-derived DNA will be subjected to the same likelihood of decay and HGT. If the average length of a transgene is 3 kb, three transgenes in *Arabidopsis thaliana* (L.) Hegnh. would constitute $7 \times 10^{-6}\%$ of the total genome. The statistical likelihood that "a" piece of DNA undergoes HGT is obviously considerably higher than the likelihood that a given piece of transgene DNA undergoes HGT.

Widespread occurrence of kanamycin resistance in the microbial soil and intestinal flora (Smalla et al., 1993), combined with the low occurrence of HGT, suggests that the likelihood of a bacterium in conditions selecting for kanamycin resistance receiving the gene from another (bacterial) source is much greater than the likelihood that a bacterium receives the gene from a transgenic plant. In combination with the limited therapeutic value of kanamycin as an antibiotic, the concerns about kanamycin resistance in plants compromising human therapy would seem to be not sufficiently supported by scientific evidence. Similar arguments hold for hygromycin resistance as a selectable marker in transgenic crops, which is too toxic as an antibiotic for any therapeutic use (Gay, 2001). The *nptIII* gene confers some resistance against amikacin, an antibiotic in use to combat nosocomial infections. A potato fortuitously containing this gene was withdrawn from commercial application because of perceived problems with HGT (Gay, 2001). The likelihood and impact of HGT with parental plant DNA compared to transgenic plant DNA would seem to indicate that HGT deserves less attention in the regulatory process compared to other concerns. Unless there is strong evidence for impact from HGT of a plant gene, as in the case of antibiotic resistance, HGT from GM plants to other organisms should not be considered a serious risk.

The introduction of bacterial genes into plant genomes does increase the probability of horizontal gene transfer from plant DNA to bacteria due to the presence of homologous sequences. Natural transformation is the most probable mechanism to transfer plant DNA to bacteria. DNA transfer from one organism to another depends on: (1) availability of free DNA, (2) presence of bacteria in a component state (not for lightning), (3) stable integration of the captured DNA, and (4) selection of transformants. Different abiotic and biotic factors, such as: (1) content and type of clay minerals, (2) pH, (3) temperature, (4) humidity, and (5) microbial activity, affect the persistence of free DNA. Persistence of transgenic DNA in soil has been shown in several studies, and recombinant DNA enables specific and sensitive detection of the DNA. Recent research has shown that *Pseudomonas stutzeri* (Leh. and Neum.) Sijderius is naturally transformable in nonsterile soil. Different levels of transformation have been recorded within a population of *P. stutzeri*. During infection of its host, the plant pathogen, *R. solanacearum* naturally develops a state of competence. *Agrobacterium tumefaciens* and *P. fluorescens* (Migula) can undergo natural transformation in soil microcosms. A survey of a collection of rhizobacteria associated with potato indicated that the kind and diversity of bacteria capable of taking up DNA is much larger than previously known. The number of transformants decreases proportionally with the size of the DNA. Experiments following the fate of DNA in tobacco leaves subjected to different decaying conditions (grinding, enzymatic attack, etc.) have shown a rapid decrease of high molecular weight DNA in the plant and of transformation frequencies within a few days.

The antibiotic gene used as a marker to select for gene transfer may lead to resistance in pathogens infecting human beings. However, the general scientific view is that the risk of compromising the therapeutic value of antibiotics is almost negligible. Most genetically engineered plants contain a gene for antibiotic resistance as an easily identifiable marker. Hypothetically, antibiotic resistance genes may move from a crop into bacteria in the environment. Since bacteria readily exchange antibiotic resistance genes, the antibiotic resistance genes also have the potential to move into disease-causing bacteria. Gene transfer from plants to microorganisms is possible in laboratory studies (Gebhard and Samalla, 1998), and possibly has happened during evolution (Doolittle, 1999). The probability of movement of genes from plants to human pathogens (antibiotics) is negligible. Under laboratory conditions, plasmid transfer between *B. thuringiensis* subsp. *tenebrionis* and *B. thuringiensis* subsp. *kurstaki* HD 1 (resistant to streptomycin) strains occurs at a frequency of 10^{-2} (Thomas et al., 1997). However, no plasmid transfer has been observed in soil release experiments, and in insects on leaf discs.

Gene Flow, Selection Pressure, and Enhanced Fitness of Herbivores

Gene Flow and Selection Pressure from Herbivores

Given the absence of information on the identity, level of susceptibility, and ecological roles of insects exploiting specific wild relatives of *Bt* transgenic crops, efforts should be made to assess possible consequences of arthropod mortality on resistant wild relatives (Letourneau, Robinson, and Hagen, 2003). The *Bt* transgenic *B. napus* and *B. rapa* × *B. napus* hybrids are lethal to larvae of diamondback moth, *Plutella xylostella* (L.) (Mason et al., 2003). No measurable plant fitness advantage (reproductive dry weight) was observed for *Bt B. napus* and *Bt* transgenic *B. rapa* × *B. napus* hybrid populations at low insect pressure (one larva per leaf). Establishment of the *Bt* trait in wild *B. rapa* populations may increase its competitive advantage under high insect pressure, although credible scientific basis to confirm the associated risk is lacking.

One of the hazards in gene transfer from the transgenic plants to the wild relatives is the possibility of reduced selection pressure (biological control) on the wild relatives from the pest. If the target pest does not play a significant role in population regulation of the wild hosts, the gene transfer will not constitute any hazard. The build up of resistance in the wild relatives can also act as a component of pest management for the target pest if the wild relative acts as an alternate host, and helps in carryover of the pest population from one season to another. Resistance to insects and diseases can also make the plant more persistent in the wild environment, and confer similar advantages on the wild relatives. The chances for development of resistance in the pest populations also need to be assessed, and strategies devised to use different genes in different crops in the same environment. Transgenic crops containing protease inhibitors may pose similar problems, and their use needs to be carefully planned to avoid the evolution of pest populations capable of withstanding the transgene. Resistance to abiotic stress factors may present additional challenges, as this would enable the plants to grow in environments where they were unable to do well earlier (Fraley, 1992). This confers additional advantage to the transgenic plant, and there are chances for gene transfer through cross-pollination. The exotic species model can be used to assess the risk of introducing transgenic plants with resistance to abiotic stress factors. The risk assessment in such cases requires more information, and should take into account the nature of competitive advantage conferred by the transgene under specific conditions.

Gene Flow and Enhanced Fitness of Herbivores

Widespread cultivation of transgenic crops with resistance to insect pests and diseases will impose intense selection pressure on pest populations to adapt to the transgene. Development of insect pest and disease-resistant cultivars has been one of the primary objectives of plant breeding for many years (Simmonds et al., 1999). The history of plant breeding has clearly established that insect pest and pathogen populations can quickly adapt to crop cultivars with new resistance genes (Bonman, Khush, and Nelson, 1992; McIntosh and Brown, 1997). However, crop improvement is an ongoing process, and plant breeders have not stopped breeding for insect pest and disease resistance simply because the target pest or disease might overcome the resistance. Development of management strategies to minimize the evolution of insect and pathogen populations that might overcome the resistance genes has been ongoing for many years. In order to establish an appropriate resistance management plan, it is important to understand the nature of interactions between the host crop, the insect pest/pathogen population(s), and the mechanisms involved. The key to maintaining an effective management plan is regular monitoring of the response of the insect pest/pathogen populations to the cultivars grown by the farmers.

Although many resistance genes have been identified in crop germplasm, there is no easy way to predict the quality or durability of these resistance genes. The "breakdown" of resistance is usually associated with qualitative resistance conferred by major genes (*R* genes), where resistance versus susceptibility results from a gene-for-gene interaction between the *R* genes in the host and avirulence genes in the pathogen (Flor, 1971). The resistance conferred by many *R* genes has not been durable as a consequence of rapid changes in pathogen populations (Leach et al., 2001). The most widely cited examples of durable resistance against bacterial or fungal pathogens involve multigenic quantitative traits (Johnson, 1984; Parlevliet, 2002). However, there are examples where single *R* genes have conferred highly durable resistance, for example, the *Lr34* gene conferring resistance to leaf rust in wheat (Kolmer, 1996) and the *Xa4* gene conferring bacterial blight resistance in rice (Bonman, Khush, and Nelson, 1992). The experience gained from plant breeding over many years will help define the appropriate management approaches for insect pest- and disease-resistant transgenic crops to prevent or minimize the establishment of insect pests and pathogen populations that may overcome the resistance mechanism underpinning the resistance genes. It is not the transgenic versus nontransgenic status of the crop plants that may result in projected problems, but the way the crops are grown and managed will have a greater influence. The widespread cultivation of genetically modified crops with insect or disease resistance is no more likely to result in the development of difficult-to-control insect pests and diseases than that experienced by traditional breeding in the past.

Gene Flow and Genetic Purity of Crops

One of the concerns about the use of transgenic crops involves the possibility that the transgene will move to the nontransgenic crops, resulting in situations that are either undesirable, for example, presence of Starlink maize (*cry9C*) gene in nontransgenic maize (Dorey, 2000). Inadvertent mixing of transgenic and nontransgenic crops through pollen dispersal and seed is of particular concern for the organic farming, for both economic and emotional reasons (Dale, 1994; Moyes and Dale, 1999). In such cases, liability can become a major issue (Moeller, 2001). Genetic modification does not change the frequency with which admixture of genetic material occurs. However, modern molecular techniques can detect low levels of genetic mixing. Maintenance of seed quality is an important issue in modern agriculture. For commercial cultivars of both nontransgenic and transgenic crops, the genetic purity of seed represents the homogeneity of a single recognized cultivar or its trueness to type (Briggs and Knowles, 1967). Strict management guidelines have been imposed to allow multiplication of sufficient amounts of seed for sowing large areas with commercial crops (Condon, 2001). Without imposing seed production guidelines to maintain the genetic purity of certified seed, a cultivar may quickly deteriorate and become unrecognizable due to factors such as mechanical admixtures, gene flow through natural crossing, mutations, random genetic drift, or selection pressure.

Monitoring of genetic purity and seed certification is based on the phenotype of the plants, and on observing the plant characteristics and ensuring that they match the standards in the cultivar descriptions. The international seed purity standards require the incidence of admixture and genetic instability of cultivars to be maintained above a minimal threshold value that depends on the reproductive characteristics of each crop. Tolerance of a low level of gene transfer by pollen is considered an inherent component of modern day agriculture, especially when growing commercial crops for food production. International seed certification standards require genetic purity levels of 98% to 99% (Leask, 2000). These purity levels represent the compromise between the stringency imposed on seed production, and the market need for affordable seed, especially for crops grown over large areas.

New molecular and biochemical techniques such as polymerase chain reaction (PCR) and enzyme-linked immunosorbent assay (ELISA) allow for more precise testing. Such diagnostic tests allow the testing of a particular gene (or allele) or products of a particular gene (or allele) to be measured. Consequently, cultivar purity can now be estimated on the basis of genotype, rather than on phenotype. An additional advantage is that the environment can substantially influence the latter. Sophisticated diagnostic tests based on genotype reveal higher frequencies of occurrence of adventitious genetic material in commercial seed than previously anticipated. Existing cultivars that are widely used and traded, and thought to be homogeneous, pure, and stable at the phenotypic level, may actually contain considerable variability at the genotypic level. This will present new challenges for commercial seed production, and will require a thorough reassessment of existing quality control paradigms, given the new opportunities to measure genetic purity.

If adventitious genetic content in commercial seed of nontransgenic crops is common, given current seed certification schemes, the appearance of transgenic material in nontransgenic cultivars is unavoidable, except by fully prohibiting the cultivation of transgenic crops. A standard of zero adventitious content in a commercial seed line is unachievable, irrespective of whether the influx is from nontransgenic or transgenic cultivars. Improvements in seed production to minimize the incidence of adventitious seed content are likely to be at least matched by enhanced sensitivity and precision of the modern diagnostic tests. While the use of stringent management approaches will minimize the opportunity for inadvertent admixtures of transgenic with nontransgenic seed, the issue of transgene flow via pollen dispersal presents a more difficult problem. Restricting the movement of pollen between crops is not a new concern, and forms an important basis of seed quality control.

The production of certified seed of specific cultivars requires the maintenance of minimum isolation distances. There are internationally recognized isolation distances that vary depending on the crop, and its reproductive characteristics (Briggs and Knowles, 1967; Simmonds et al., 1999). There may be particular uses of transgenic crops where additional care may be required, for example, in pharmaceuticals, vaccines, biodegradable plastics, or speciality biochemicals. Such products should not be mixed with normal food crops. The environmental release of such transgenic crops will require more stringent levels of containment for keeping these products out of the food chain. To prevent inadvertent admixture of such cultivars with those intended for food use presents special challenges. The economic feasibility and success of molecular farming approaches will depend to a large extent on the ability to meet these challenges. Recent advances in the genetic engineering of chloroplasts can be used to limit gene escape through pollen (Daniell et al., 1998).

Gene Flow and the Centers of Genetic Diversity

There is no evidence that a transgene could alter the genetic structure of a crop. Currently grown transgenic commercial varieties will certainly have a positive effect on agronomic productivity. However, it is important to develop technology aimed at solving specific and pressing socioeconomic problems. Genetic diversity of crops and their wild relatives is and will always be susceptible to displacement by high yielding varieties, whether transgenic or conventional. Therefore, *in situ* conservation methods must be implemented to maintain genetic diversity. It is important to reinforce, characterize, and evaluate the present germplasm collections of landraces, and wild relatives of crops. Information on the impact of transgenic crops on biodiversity is limited or nonexistent. Therefore, it becomes necessary to establish sound monitoring methods, develop basic ecological knowledge of wild relatives of crops, and thoroughly study what has been the effect of hybrids on the landraces and wild relatives of crops, for example, hybrid corn has been grown in Mexico over the past 50 years, and some of the landraces have been replaced in favor of high yielding hybrids. As a result of hybrid production, the identity of landraces has not been lost. The growers continue to select phenotypes that are preferred by the customers or for their own use. Monitoring is underway to determine the presence of transgenes in landraces and test the stability of the transgene in the environment. However, information to date has suggested that: (1) the transgene does not alter the structure of the maize gene, (2) there is no transfer of the transgene to other plants, and (3) there is no effect on human health. There is a better chance of countering the gene flow if the transgene is deployed through hybrids based on male sterility.

Gene Flow and Aquatic Environments

Transgenic crops are intended to reduce the types and quantities of pesticides necessary for production of food, feed, and fiber. Because of decreased reliance on chemical pesticides, extensive cultivation of transgenic crops may result in improved water quality and related ecosystems (Estes et al., 2001). A simple model has been proposed to examine the displacement of insecticides by growing *Bt* cotton and *Bt* maize, and assess the impact on drinking water quality. Based on this module, all the transgenic cropping systems resulted in significantly lower pesticide concentrations in ground and surface waters, thereby reducing whatever impacts these products have on drinking water quality and the aquatic environments.

Management of Gene Flow

There are many ways to prevent transgene introgression from crops to other varieties or to related weeds or wild species (containment strategies), as well as to preclude the impact should containment fail (mitigation strategies) (Gressel and Al Ahmad, 2003). The needs are most acute with rice and sunflowers, which have conspecific weeds, and with oilseed rape, sorghum, and barley, which have closely related weeds. Containment and mitigation are critical for pharmaceutical crops, where gene flow from the crop to edible varieties must be precluded. Some gene flow (leakage) is inevitable with all containment mechanisms, and once leaked, could then move through populations of undesired species, unless their spread is mitigated. Leakage even occurs with chloroplast-encoded genes. A mechanism for mitigation has been proposed where the primary transgene (herbicide resistance, etc.) is tandemly coupled with flanking genes that could be desirable or neutral to the crop, but unfit for the rare weed into which the gene introgresses. Mitigator traits include dwarfing, nonbolting, no secondary dormancy, no seed shattering, and poor seed viability, depending on the crop. Hybrids with the tandem construct are unable to reach maturity when grown interspersed with the wild type. Such mitigation should greatly decrease the risk of transgene movement when coupled with containment technologies, allowing cultivation of transgenic crops having related weeds. As the number of transgenic crops being released for cultivation is increasing, the problem of monitoring such genes also increases geometrically. There is a need for a uniform system, where a small piece of noncoding DNA carrying an assigned variable region is used to mark transgenic crops, allowing monitoring.

Use of Taxonomic Information

Systematic studies can be used for managing gene flow as they provide important information on breeding system and sexual compatibility of related species, likelihood of such crosses, and the factors affecting the successful production and survival of hybrid progeny (Warwick, 1997). They also provide information on the distribution of crop and wild relatives, their ecological requirements, status in natural environments, and if weedy, the areas of infestation and patterns of spread. This information is needed on a case-by-case basis for evaluation of both the likelihood of escape of a transgenic trait and the environmental impact, if such an escape occurs.

Geographic Isolation

Half of the pollen produced by an individual plant falls within 3 m and the probability of fertilization afterwards decreases slowly along a negative exponential of the distance (Lavigne et al., 1998). The percentage of pollen dispersal from *B. napus* in the center of a nontransgenic crop has been estimated to be 4.8% (Scheffler, Parkinson, and Dale, 1993). In rape, pollen trap examination showed a close link between weather and pollen amounts transported from rape fields by wind. The degree of outcrossing depended not only on the pollen concentration in the air, but also on other factors such as the weather and the abundance of flower-visiting insects (Saure et al., 2003). The frequency was estimated to be 1.5% at a distance of 1 m and 0.4% at 3 m. The frequency decreased sharply to 0.02% at 12 m and was only 0.00033% at 47 m. No obvious directional effects were detected that could be ascribed to wind or insect activity. Pollen-dispersal distributions based on dispersal from whole plots instead of individual plants might underestimate the proportion of pollen that dispersed over average or long distances. For insect-pollinated outcrossing crops such as radish, strategies other than distance must be employed to ensure complete isolation (Klinger, Elam and Ellstrand, 1991). Gaussian plume models, which take distance and wind direction into account, have indicated that small conspecific populations of *Lolium perenne* L. might, in some conditions, be swamped by immigrant pollen, even if they are not directly downwind of the source (Giddings, 2000). In sunflower, gene flow decreased with distance. However, gene flow occurred up to distances of 1,000 m from the source population, indicating that physical distance alone may not prevent gene flow between cultivated and wild populations of sunflowers (Arias and Rieseberg, 1994). Pollen dispersal from transgenic cotton is low, but increases with an increase in the size of the source plot (Llewellyn and Fitt, 1996). A 20 m buffer zone has been suggested to limit dispersal of transgenic pollen from small-scale field tests. The border rows fulfilled the purpose of serving as a pollen sink to significantly reduce the amount of pollen dissemination from the test plot of cotton (Umbeck et al., 1991). In China, gene flow between cultivars of *G. hirsutum* is up to 36 m. A buffer zone of at least 72 m has been proposed as a safer distance to avoid gene flow. A minimum isolation distance of 1,557 m has been recommended to prevent gene flow in alfalfa. Complete containment of transgenes within alfalfa seed or hay production fields would be highly unlikely using current production practices.

Use of Border Rows

Barren zones of 4 to 8 m may actually increase seed contamination over what would be expected if the intervening ground were instead planted entirely with a trap crop. When trap crops occupied a limited portion of the isolation zone separating transgenic and nontransgenic varieties, the effectiveness of the trap crop depended on the width of the isolation zone. Gene escape was reduced when the two varieties were separated by 8 m, but increased the gene escape across a 4 m isolation zone (Morris, Kareiva, and Raymer, 1994). For relatively short isolation distances, the most effective strategy for reducing the escape of transgenic pollen is to devote the entire region between transgenic and nontransgenic varieties to a trap crop. There is a significant reduction in pollen dissemination as distance from the test plot increased (Umbeck et al., 1991). Outcrossing decreased from 5% to <1% by 7 m away from the test plot. A low level (<1%) of pollen dispersal was recorded to a distance of 25 m. The border rows reduced the amount of pollen dissemination from the test plot. Economic profit can be maximized by removing field borders after flowering, rather than by leaving a surrounding gap, which would need to occupy up to threefold as much field surface to achieve the same level of containment (Reboud, 2003).

Use of Cytoplasmic Male Sterility

Cytoplasmic male sterility (CMS) can also be used to restrict pollen flow (Feil and Stamp, 2002; Feil, Weingartner, and Stamp, 2003). Such a system can produce grain yields as high as or even higher than those produced by pure male-fertile maize crops, especially when the male-sterile component is pollinated nonisogenically. Growing 80%:20% mixtures of CMS-transgenic hybrids and male-fertile nontransgenic hybrids, whereby the latter component acts as pollen donor for the entire stand, can be used for pollen management. Since the CMS-transgenic plants release no pollen, the transgenes cannot escape from the transgenic maize field (Feil and Stamp, 2002). Blends of male-sterile *Bt* maize and male-fertile nontransgenic maize will help in delaying the development of *Bt* toxin-resistant insect populations. The A_3 CMS system can be used in sorghum to control transgene flow (Pedersen, Marx, and Funnell, 2003). Seed set on A_1F_2 individuals averaged 74%, and on A_3F_2 individuals averaged 0.04%. Upper confidence limits for seed set were 1.32% or less for all A_3 hybrids. PCR analysis detected four individuals with outcrossing (from a population of 1007) in A_3 cytoplasm.

Conclusions

The use of crop protection traits through transgenic plants for pest management will continue to expand in the future and, therefore, there will be a continuing need to understand the degree of gene flow and the likely consequences of such a phenomenon. The consequences of gene transfer to and between bacteria and viruses will be a major concern, particularly in relation to use of antibiotic and herbicide resistance genes as selection markers. Efforts would have to be made to devise appropriate measures to contain gene flow where its likely consequences may be deleterious to the environment. There is considerable evidence that current transgenic crops in conjunction with conventional agricultural practices offer a sufficiently safe and effective technology that may contribute to a better, cost-effective, sustainable, and productive agriculture. Experience has shown that the promise of transgenic crops has met the expectations of large and small farmers, in both industrialized and developing countries, and established an appreciable market share. The risk of not using transgenic crops, particularly in developing countries where the technology may have most to offer, should also be considered more explicitly. Governments, supported by the global scientific and development community, must ensure continued safe and effective testing, and implement harmonized regulatory programs that inspire public confidence. Many of the crop traits being modified through genetic modification are the same as those targeted through conventional plant breeding for many years. The impact of transgenic crops would therefore be similar to the impact of cultivars derived through traditional breeding, which have been an integral part of agriculture for many years. Consequently, the risks of growing most transgenic crops on the environment or ecosystems will be similar to the effects of growing new cultivars from traditional breeding. Whenever unresolved questions arise concerning undesirable effect of transgenic crops, science-based evaluations should be used on a case-by-case basis to the best of our ability. Increased knowledge underpinning transgenic crops provides a greater confidence in the assurances that science can give when evaluating and monitoring the impact of transgenic crops relative to traditional methods of crop improvement. The resulting regulation is not a static activity, but needs continuous revisiting based on increased knowledge and experience.

References

Aaziz, R. and Tepfer, M. (1999). Recombination in RNA viruses and in virus-resistant transgenic plants. *Journal of General Virology* 80: 1339–1346.

Amand, P.C., Skinner, D.Z. and Peaden, R.N. (2000). Risk of alfalfa transgene dissemination and scale-dependent effects. *Theoretical and Applied Genetics* 101: 107–114.

Amman, K. (2001). Safety of genetically engineered plants: An ecological risk assessment of vertical gene flow. In Custers, C. (Ed.), *Safety of Genetically Engineered Crops*. Ghent, Belgium: VIB, 61–87. http://www.vib.be.frame.cfm.

Anderson, E. (1949). *Introgressive Hybridisation*. New York, USA: John Wiley & Sons.

Arias, D.M. and Rieseberg, L.H. (1994). Gene flow between cultivated and wild sunflower. *Theoretical and Applied Genetics* 89: 655–660.

Arias, D.M. and Rieseberg, L.H. (1995). Genetic relationships among domesticated and wild sunflowers (*Helianthus annuus*, Asteraceae). *Economic Botany* 49: 239–248.

Arnaud, J.F., Viard, F., Delescluse, M. and Cuguen, J. (2003). Evidence for gene flow via seed dispersal from crop to wild relatives in *Beta vulgaris* (Chenopodiaceae): Consequences for the release of genetically modified species with weedy lineages. *Proceedings Royal Society of London, Series B, Biological Sciences* 270: 1565–1571.

Arriola, P.E. and Ellstrand, N.C. (1996). Crop-to-weed gene flow in the genus *Sorghum* (Poaceae): Spontaneous interspecific hybridization between Jonhsongrass, *Sorghum halepense* and crop sorghum, *S. bicolor. American Journal of Botany* 83: 1153–1160.

Arriola, P.E. and Ellstrand, N.C. (1997). Fitness of interspecific hybrids in the genus *Sorghum*: Persistence of crop genes in wild populations. *Ecological Applications* 7: 512–518.

Austin, D.J., Kristinsson, K.G. and Anderson, R. (1999). The relationship between the volume of antimicrobial consumption in human communities and the frequency of resistance. *Proceedings National Academy of Sciences USA* 96: 1152–1156.

Bartsch, D., Cuguen, J., Biancardi, E. and Sweet, J. (2003). Environmental implications of gene flow from sugar beet to wild beet: Current status and future research needs. *Environmental Bio-Safety Research* 2: 105–115.

Bartsch, D., Lehnen, M., Clegg, J., Pohl-Orf, M., Schuphan, I. and Ellstrand, N.C. (1999). Impact of gene flow from cultivated beet on genetic diversity of wild sea beet populations. *Molecular Ecology* 8: 1733–1741.

Bejarano, E.R., Khashoggi, A., Witty, M. and Lichtenstein, C. (1996). Integration of multiple repeats of geminiviral DNA into the nuclear genome of tobacco during evolution. *Proceedings National Academy of Sciences USA* 93: 759–764.

Bertolla, F. and Simonet, P. (1999). Horizontal gene transfers in the environment: Natural transformation as a putative process for gene transfers between transgenic plants and micro-organisms. *Research in Microbiology* 150: 375–384.

Bonman, J.M., Khush, G.S. and Nelson, R.J. (1992). Breeding rice for resistance to pests. *Annual Review of Phytopathology* 30: 507–528.

Bottraud, I., Lavigne, C., Reboud, X., Vedel, F., Rherissi, B. and Lefranc, M. (1992). Gene flow between wild and cultivated *Setaria*: Consequences for the release of transgenic crops. In *IXe Colloque international sur la biologie des mauvaises herbes*, 16–18 September, 1992, Dijon, France, 507–512.

Boulter, D. (1995). Plant biotechnology. Facts and public perception. *Phytochemistry* 40: 1–9.

Briggs, F.N. and Knowles, P.F. (1967). *Introduction to Plant Breeding.* New York, USA: Reinhold.

Chen, L.J., Lee, D.S., Song, Z.P., Suh, H.S. and Lu, B.R. (2004). Gene flow from cultivated rice (*Oryza sativa*) to its weedy and wild relatives. *Annals of Botany* 93: 67–73.

Chevre, A.M., Eber, F., Baranger, A. and Renard, M.I. (1997). Gene flow from transgenic crops. *Theoretical and Applied Genetics* 97: 90–98.

Condon, M.S. (2001). Seed genetic purity in the pre and post biotechnology eras. http://pepwagbiotech. org/events/0911/speakersCondon.pdf.

Cook, R.J. (2000). Science-based risk assessment for the approval and use of plants in agricultural and other environments. In Persley, G.J. and Lantin, M.M. (Eds.), *Agricultural Biotechnology and the Poor.* Washington, D.C., USA: Consultative Group on International Agricultural Research, 123–130.

Courvalin, P. (1994). Transfer of antibiotic resistance genes between Gram-positive and Gram-negative bacteria. *Antimicrobial Agents and Chemotherapy* 38: 1447–1451.

Dale, P.J. (1994). The impact of hybrids between genetically modified crop plants and their related species: General considerations. *Molecular Ecology* 3: 31–36.

Daniell, H., Datta, R., Varma, S., Gray, S. and Lee, S.B. (1998). Containment of herbicide resistance through genetic engineering of the chloroplast genome. *Nature Biotechnology* 16: 345–348.

De la Cruz, F. and Davies, J. (2000). Horizontal gene transfer and the origin of species: Lessons from bacteria. *Trends in Microbiology* 8: 128–133.

Desplanque, B., Hautekeete, N. and van Dijk, H. (2002). Transgenic weed beets: Possible, probable, avoidable? *Journal of Applied Ecology* 39: 561–571.

Desplanque, B., Boundry, P., Broomberg, K., Saumitou-Laprade, P., Cuguen, J. and van Dijk, H. (1999). Genetic diversity and gene flow between wild, cultivated and weedy forms of *Beta vulgaris* L. (Chenopodiaceae), assessed by RFLP and microsatellite markers. *Theoretical and Applied Genetics* 98:1194–1201.

de Vries, J., Meier, P. and Wackernagel, W. (2001). The natural transformation of the soil bacteria *Pseudomonas stutzeri* and *Acinetobacter* sp. by transgenic plant DNA strictly depends on homologous sequences in the recipient cells. *FEMS Microbiology Letters* 195: 211–215.

de Vries, J. and Wackernagel, W. (1998). Detection of npt II (kanamycin resistance) gene in genomes of transgenic plants by marker-rescue transformation. *Molecular and General Genetics* 257: 606–613.

de Wet, J.M.J. and Harlan, J.R. (1975). Weeds and domesticates: Evolution in the man-made habitat. *Economic Botany* 29: 99–107.

Doebley, G.F.J. (1990). Molecular evidence for gene flow among *Zea* species. *BioScience* 40: 443–448.

Doolittle, W.F. (1999). Phytogenic classification and the universal tree. *Science* 284: 2124–2127.

Dorey, E. (2000) Taco dispute underscores need for standardized tests. *Nature Biotechnology* 18: 1136–1137.

Dröge, M., Pühler, A. and Selbitschka, W. (1998). Horizontal gene transfer as a bio-safety issue: A natural phenomenon of public concern. *Journal of Biotechnology* 64: 75–90.

Dröge, M., Pühler, A. and Selbitschka, W. (1999). Horizontal gene transfer among bacteria in terrestrial and aquatic habitats as assessed by microcosm and field studies. *Biology and Fertility of Soils* 29: 221–245.

Duggan, P.S., Chambers, P.A., Heritage, J. and Forbes, J.M. (2000). Survival of free DNA encoding antibiotic resistance from transgenic maize and the transformation activity of DNA in bovine saliva rumen fluid and silage effluent. *FEMS Microbiology Letters* 191: 71–77.

Ellstrand, N.C. (2001). When transgenes wander, should we worry? *Plant Physiology* 125: 1543–1545.

Ellstrand, N.C., Prentice, H.C. and Hancock, J.F. (1999). Gene flow and introgression from domesticated plants into their wild relatives. *Annual Review of Ecology and Systematics* 30: 539–563.

Estes, T.L., Allen, R., Jones, R.L., Buckler, D.R., Carr, K.H., Gustafson, D.I., Gustin, C., McKee, J., Hornsby, A.G. and Richards, R.P. (2001). Predicted impact of transgenic crops on water quality and related ecosystems in vulnerable watersheds of the United States. In *Proceedings, Pesticide Behaviour in Soils and Water,* 13 to 15 November, 2001. Brighton, UK: British Crop Protection Council, 357–366.

Feil, B. and Stamp, P. (2002). The pollen-mediated flow of transgenes in maize can already be controlled by cytoplasmic male sterility. *AgBiotechNet* 4: 1–4.

Feil, B., Weingartner, U. and Stamp, P. (2003). Controlling the release of pollen from genetically modified maize and increasing its grain yield by growing mixtures of male-sterile and male-fertile plants. *Euphytica* 130: 163–165.

Fitter, A., Perrins, J. and Williamson, M. (1990). Weed probability challenged. *Bio/Technology* 8: 473–474.

Flor, H.H. (1971). Current status of the gene-for-gene concept. *Annual Review of Phytopathology* 9: 275–296.

Fraley, R. (1992). Field testing genetically engineered plants. In Fraley, R., Frey, N.M. and Schell, J. (Eds.), *Current Combinations in Molecular Biology: Improvement of Agriculturally Important Crops.* New York, USA: Cold Spring Harbor Press, 83–86.

Frietema de Vries, F.T. (1996). *Cultivated Plants and the Wild Flora: Effect Analysis by Dispersal Codes.* Ph.D. thesis, Hortus Botanicus, Leiden, The Netherlands.

Frietema de Vries, F.T., van der Meijden, R. and Brandenburg, W.A. (1992). Botanical files: A case study of the real chances for spontaneous gene flow from cultivated plants to the wild flora of the Netherlands. *Gorteria* (Supplement) 1: 1–100.

Frietema de Vries, F.T., van der Meijden, R. and Brandenburg, W.A. (1994). Botanical files on lettuce (*Lactuca sativa*). On the chance for gene flow between wild and cultivated lettuce (*Lactuca sativa* L. including *L. serriola* L., Compositae) and the generalized implications for risk-assessments on genetically modified plants. *Gorteria* (Supplement) 2: 1–44.

Gay, P. (2001). The biosafety of antibiotic resistance markers in plant transformation and the dissemination of genes through horizontal gene flow. In Custers, R. (Ed.), *Safety of Genetically Engineered Crops.* Zwijnaarde, Belgium: Flanders Interuniversity Institute for Biotechnology, 135–159.

Gebhard, F. and Smalla, K. (1998). Transformation of *Acinetobacter* sp. strain BD413 by transgenic sugarbeet DNA. *Applied Environmental Microbiology* 64: 1550–1554.

Gebhard, F. and Smalla, K. (1999). Monitoring field releases of genetically modified sugar beets for persistence of transgenic plant DNA and horizontal gene transfer. *FEMS Microbiology and Ecology* 28: 261–272.

Gibs, M. (1994). Risks in using transgenic plants? *Science* 264: 1650–1651.

Giddings, G. (2000). Modeling the spread of pollen from *Lolium perenne.* The implications for the release of wind-pollinated transgenics. *Theoretical and Applied Genetics* 100: 971–974.

Gregorius, H.R. and Steiner, W. (1993). Gene transfer in plants as a potential agent of introgression. In Workman, K. and Tommie, J. (Eds.), *Transgenic Organisms.* Basel, Switzerland: Birkhauser Verlag, 83–107.

Gressel, J. and Al Ahmad, H.I. (2003). Containment and mitigation of transgene flow from crops. In *Crop Science and Technology, the BCPC International Congress,* 10–12 November, 2003. Glasgow, Scotland, UK: British Crop Protection Council, 1175–1180.

Gueritaine, G., Sester, M., Eber, F., Chevre, A.M. and Darmency, H. (2002). Fitness of backcross six of hybrids between transgenic oilseed rape (*Brassica napus* var. *oleifera*) and wild radish (*Raphanus raphanistrum*). *Molecular Ecology* 11: 1419–1426.

Hammond, J., Lecoq, H. and Raccah, B. (1999). Epidemiological risks from mixed virus infections and transgenic plants expressing viral genes. *Advances in Virus Research* 54: 189–314.

Harlan, J.R. (1976). Genetic resources in wild relatives of crops. *Crop Science* 16: 329–333.

Harlan, J.R. and de Wet, J.M.J. (1963). The compilospecies concept. *Evolution* 17: 497–501.

Harper, G., Osuji, J.O., Heslop-Harrison, J.S. and Hull, R. (1999). Integration of banana streak badnavirus into the *Musa* genome: Molecular and cytogenetic evidence. *Virology* 255: 207–213.

Haygood, R., Ives, A.R. and Andow, D.A. (2003). Consequences of recurrent gene flow from crops to wild relatives. *Proceedings Royal Society of London. Series B, Biological Sciences* 270: 1879–1886.

Hellwig, W.K. and Frankl, R. (2000). Foraging distances of *Bombus muscorum, Bombus lapidarius* and *Bombus terrestris* (Hymenoptera, Apidae). *Journal of Insect Behavior* 13: 239–246.

Jakowitsch, J., Mette, M.F., van der Winden, J., Matzke, M.A. and Matzke, A.J.M. (1999). Integrated pararetroviral sequences define a unique class of dispersed repetitive DNA in plants. *Proceedings National Academy of Sciences USA* 96: 13241–13246.

Johnson, R. (1984). A critical analysis of durable resistance. *Annual Review of Phytopathology* 22: 309–330.

Klinger, T., Elam, D.R. and Ellstrand, N.C. (1991). Radish as a model system for the study of engineered gene escape rates via crop-weed mating. *Conservation Biology* 5: 531–535.

Kolmer, J.A. (1996). Genetics of resistance to wheat leaf rust. *Annual Review of Phytopathology* 34: 435–455.

Landis, W.G., Lenart, L.A. and Spromberg, J.A. (2000). Dynamics of horizontal gene transfer and the ecological risk assessment of genetically engineered organisms. *Human and Ecological Risk Assessment* 6: 875–899.

Lavigne, C., Klein, E.K., Vallee, P., Pierre, J., Godelle, B. and Renard, M. (1998). A pollen-dispersal experiment with transgenic oilseed rape. Estimation of the average pollen dispersal of an individual plant within a field. *Theoretical and Applied Genetics* 96: 886–896.

Leach, J.E., Vera Cruz, C.M., Bai, J. and Leung, H. (2001). Pathogen fitness penalty as a predictor of durability of disease resistance genes. *Annual Review of Phytopathology* 39: 187–224.

Leask, B. (2000). *Troubles with Thresholds*. Canadian Seed Trade Association. http://cdnseed.ord/press.Troubles%20with%20-Thresholds.pdf.

Lefol, E., Fleury, A. and Darmency, H. (1996). Gene dispersal from transgenic crops. II. Hybridization between oilseed rape and the wild hoary mustard. *Sexual Plant Reproduction* 9: 189–196.

Letourneau, D.K., Robinson, G.S. and Hagen, J.A. (2003). Bt crops: Predicting effects of escaped transgenes on the fitness of wild plants and their herbivores. *Environmental Biosafety Research* 2: 219–246.

Linder, C.R., Taha, I., Seiler, G.J., Snow, A.A. and Rieseberg, L.H. (1998). Long-term introgression of crop genes into wild sunflower populations. *Theoretical and Applied Genetics* 96: 339–347.

Llewellyn, D. and Fitt, G. (1996). Pollen dispersal from two field trials of transgenic cotton in the Namoi Valley, Australia. *Molecular Breeding* 2: 157–166.

Lorenz, M.G. and Wackernagel, W. (1992). DNA binding to various clay minerals and retarded enzymatic degradation of DNA in a sand/clay microcosm. In Gauthier, M.J. (Ed.), *Gene Transfers and Environment*. Berlin, Germany: Springer–Verlag, 103–113.

Lorenz, M.G. and Wackernagel, W. (1994). Bacterial gene-transfer by natural genetic transformation in the environment. *Microbiology Research* 58: 563–602.

Malnoe, P., Jakab, G., Droz, E. and Vaistij, F. (1999). RNA recombination in transgenic virus resistant plants. In Amman, K., Jacot, Y., Kjellsson, G. and Simonsen, V. (Eds.), *Methods for Risk Assessment of Transgenic Plants. Part III. Ecological Risks and Prospects of Transgenic Plants, Where Do We Go From Here? A Dialogue between Biotech Industry and Science*. Basel, Switzerland: Birkhäuser-Verlag, 145–147.

Mason, P., Braun, L., Warwick, S.I., Zhu B. and Stewart, C.N. Jr. (2003). Transgenic Bt-producing *Brassica napus: Plutella xylostella* selection pressure and fitness of weedy relatives. *Environmental Biosafety Research* 2: 263–276.

McIntosh, R.A. and Brown, G.N. (1997). Anticipatory breeding for resistance to rust diseases in wheat. *Annual Review of Phytopathology* 35: 311–326.

Messeguer, J. (2003). Gene flow assessment in transgenic plants. *Plant Cell, Tissue and Organ Culture* 73: 201–212.

Messeguer, J., Fogher, C., Guiderdoni, E., Marfa, V., Catala, M.M., Baldi, G. and Mele, E. (2001). Field assessments of gene flow from transgenic to cultivated rice (*Oryza sativa* L.) using a herbicide resistance gene as tracer marker. *Theoretical and Applied Genetics* 103: 1151–1159.

Metz, P.L.J. and Nap, J.P. (1997). A transgene-centred approach to the biosafety of transgenic plants: Overview of selection and reporter genes. *Acta Botanica Neerlandica* 46: 25–50.

Metz, P.L.J., Stiekema, W.J. and Nap, J.P. (1998). A transgene-centred approach to the biosafety of transgenic phosphoinothricin-tolerant plants. *Molecular Breeding* 4: 335–341.

Moeller, D.R. (2001). *GMO Liability Threats for Farmers. Legal Issues Surrounding the Planting of Genetically Modified Crops*. Minneapolis, Minnesota, USA: Institute for Agriculture and Trade Policy.

Morris, W.F., Kareiva, P.M. and Raymer, P.L. (1994). Do barren zones and pollen traps reduce gene escape from transgenic crops? *Ecological Applications* 4: 157–165.

Moyes, C.L. and Dale, P.J. (1999). *Organic Farming and Gene Transfer from Genetically Modified Crops*. Norwich, UK: John Innes Centre.

Nakayama, Y. and Yamaguchi, H. (2002). Natural hybridization in wild soybean (*Glycine max* subsp. *soja*) by pollen flow from cultivated soybean (*Glycine max* subsp. *max*) in a designed population. *Weed Biology and Management* 2: 25–30.

NAS (National Academy of Sciences). (1987). *Introduction of Recombinant DNA-Engineered Organisms into the Environment: Key Issues*. Washington, D.C., USA: National Academy of Sciences.

Nassar, N.M.A. (2002). Gene flow between cassava, *Manihot esculenta* Crantz and wild relatives. *Gene Conservation* 4: 64–79.

Nielsen, K.M., Bones, A.M., Smalla, K. and van Elsas, J.D. (1998). Horizontal gene transfer from transgenic plants to terrestrial bacteria: Rare event? *FEMS Microbiology Research* 22: 79–103.

Ochman, H., Lawrence, J.G. and Groisman, E.A. (2000) Lateral gene transfer and the nature of bacterial innovation. *Nature* 405: 299–304.

OECD (Organisation for Economic Co-operation and Development). (2002). *OECD Guidance for the Designation of a Unique Identifier for Transgenic Plants. Series on Harmonisation of Regulatory Oversight in Biotechnology*. No. 23. Paris, France: Organisation for Economic Co-operation and Development.

Parlevliet, J.E. (2002). Durability of resistance against fungal, bacterial and viral pathogens: Present situation. *Euphytica* 124: 147–156.

Paterson, A.H., Schertz, K.F., Lin, Y.R., Liu, S.C. and Chang, Y.L. (1995). The weediness of wild plants: Molecular analysis of genes influencing dispersal and persistence of Jonhsongrass, *Sorghum halepense* (L.) Pers. *Proceedings National Academy of Sciences USA* 92: 6127–6131.

Pedersen, J.F., Marx, D.B. and Funnell, D.L. (2003). A_3 cytoplasm to reduce risk of gene flow through sorghum pollen. *Crop Science* 43: 1506–1509.

Picard Nizou, A.L., Pham Delegue, M.H., Kerguelen, V., Douault, P., Marilleau, R., Olsen, L., Grison, R., Toppan, A. and Masson, C. (1995). Foraging behaviour of honey bees (*Apis mellifera* L.) on transgenic oilseed rape (*Brassica napus* L. var. *oleifera*). *Transgenic Research* 4: 270–276.

Pierre, J., Marsault, D., Genecque, E., Renard, M., Champolivier, J. and Pham Delegue, M.H. (2003). Effects of herbicide-tolerant transgenic oilseed rape genotypes on honey bees and other pollinating insects under field conditions. *Entomologia Experimentalis et Applicata* 108: 159–168.

Pohl Orf, M., Brand, U., Driessen, S., Hesse, P.R., Lehnen, M., Morak, C., Mucher, T., Saeglitz, C., von Soosten, C. and Bartsch, D. (1999). Overwintering of genetically modified sugar beet, *Beta vulgaris* L. subsp. *vulgaris*, as a source for dispersal of transgenic pollen. *Euphytica* 108: 181–186.

Ramsay, G., Thompson, C.E., Neilson, S. and Mackay, G.R. (1999). Honeybees as vectors of GM oilseed rape pollen. In *Proceedings, Gene Flow and Agriculture: Relevance for Transgenic Crops*, 12–14 April, 1999, Keele, UK, 209–214.

Raybould, A.F. and Gray, A.J.C. (1993). Genetically modified crops and hybridization with wild relatives: A UK perspective. *Journal of Applied Ecology* 30: 119–219.

Reboud, X. (2003). Effect of a gap on gene flow between otherwise adjacent transgenic *Brassica napus* crops. *Theoretical and Applied Genetics* 106: 1048–1058.

Rieseberg, L. H. and Seiler, G. (1990). Molecular evidence and the origin and development of the domesticated sunflower (*H. annuus* L.). *Economic Botany* 44: 79–91.

Sambrook, J., Fritsch, E.F. and Maniatis, T. (1989). *Molecular Cloning: A Laboratory Manual*, 2nd Edition. New York, USA: Cold Spring Harbor Press.

Saure, C., Kuhne, S. and Hommel, B. (2001). Pollen transfer by insects from oilseed rape to other crucifers: Contribution to the risk assessment of genetically modified plants. Vortrage der Entomologentagung in Dusseldorf vom 26. bis. 31. Marz 2001. *Mitteilungen der Deutschen Gesellschaft fur allgemeine und Angewandte Entomologie* 13: 265–268.

Saure, C., Kuhne, S., Hommel, B. and Bellin, U. (2003). Transgenic herbicide-resistant rape: Flower-visiting insects, pollen dispersal and outcrossing. *Agrarokologie* 44: 1–103.

Saxena, K.B., Singh, L. and Gupta, M.D. (1990). Variation for natural out-crossing in pigeonpea. *Euphytica* 46: 143–148.

Scheffler, J.A., Parkinson, R. and Dale, P.J. (1993). Frequency and distance of pollen dispersal from transgenic oilseed rape (*Brassica napus*). *Transgenic Research* 2: 356–364.

Schlüter, K., Fütterer, J. and Potrykus, I. (1995). Horizontal gene transfer from a transgenic potato line to a bacterial pathogen (*Erwinia chrysanthemi*) occurs, if at all, at an extremely low frequency. *Bio/Technology* 13: 1094–1098.

Schubbert, R., Renz, D., Schmitz, B. and Doerfler, W. (1997). Foreign (M13) DNA ingested by mice reaches peripheral leukocytes, spleen, and liver via the intestinal wall mucosa and can be covalently linked to mouse DNA. *Proceedings National Academy of Sciences USA* 94: 961–966.

Seefeldt, S.S., Young, F.L., Zemetra, R.S. and Jones, S.S. (1999). The production of herbicide-resistant jointed goatgrass (*Aegilops cylindrica*) × wheat (*Triticum aestivum*) hybrids in the field by natural hybridization and management strategies to reduce their occurrence. *Proceedings, Gene Flow and Agriculture: Relevance for Transgenic Crops*, 12–14 April, 1999, Keele, UK.

Serratos, J.A., Willcox, M.C. and Castillo-Gonzalez, F. (Eds.). (1997). *Gene Flow Among Maize Landraces, Improved Maize Varieties, and Teosinte: Implications for Transgenic Maize*. Mexico, DF, Mexico: International Wheat and Maize Research Institute.

Seiler, G. J. (1992). Utilization of wild sunflower species for the improvement of cultivated sunflower. *Field Crops Research* 30: 195–230.

Simmonds, N.W., Smartt, J., Millam, S. and Spoor, W. (1999). *Principles of Crop Improvement*, 2nd Edition. Oxford, UK: Blackwell Publishers.

Smalla, K., van Overbeek, L.S., Pukall, R. and van Elsas, J.D. (1993). Prevalence of *npt*II and Tn5 in kanamycin-resistant bacteria from different environments. *FEMS Microbiology and Ecology* 13: 47–58.

Snow, A.A. and Jorgensen, R.B. (1999). Fitness costs associated with transgenic glufosinate tolerance introgressed from *Brassica napus* ssp. *oleifera* (oilseed rape) into weedy *Brassica rapa*. *Proceedings, Gene Flow and Agriculture: Relevance for Transgenic Crops*, 12–14 April, 1999, Keele, UK, 137–142.

Song, Z.P., Lu, B.R., Zhu, Y.G. and Chen, J.K. (2003). Gene flow from cultivated rice to the wild species *Oryza rufipogon* under experimental field conditions. *New Phytologist* 157: 657–665.

Stanhope, M.J., Lupas, A., Italia, M.J., Koretke, K.K., Volker, C. and Brown, J.R. (2001). Phylogenetic analyses do not support horizontal gene transfers from bacteria to vertebrates. *Nature* 411: 940–944.

Sun, M. and Corke, H. (1992). Population genetics of colonizing success of weedy rye in northern California. *Theoretical and Applied Genetics* 83: 321–329.

Tepfer, M. (2002). Risk assessment of virus resistant transgenic plants. *Annual Review of Phytopathology* 40: 467–491.

Thomas, D.J.I., Morgan, J.A.W., Whipps, J.M. and Saunders, J.R. (1997). Plasmid transfer by the insect pathogen *Bacillus thuringiensis* in the environment. *Proceedings, Microbial Insecticides: Novelty or Necessity?* 16–18 April, 1997, University of Warwick. Coventry. Farnham, UK: British Crop Protection Council, 261–265.

Umbeck, P.F., Barton, K.A., Nordheim, E.V., McCarty, J.C., Parrott, W.L. and Jenkins, J.N. (1991). Degree of pollen dispersal by insects from a field test of genetically engineered cotton. *Journal of Economic Entomology* 84: 1943–1950.

Warwick, S.I. (1997). Use of biosystematic data, including molecular phylogenies, for biosafety evaluation. In *The 4th International Symposium on Biosafety. Results of Field Tests of Genetically Modified Plants and Microorganisms*. Ohwashi, Japan: Japanese International Center for Agricultural Sciences (JIRCAS), 53–63.

Wilson, H.D. (1990). Gene flow in squash species. *BioScience* 40: 449–455.

Wolf, D.E., Takebayashi, N. and Rieseberg, L.H. (2000). Predicting the risk of extinction through hybridization. *Conservation Biology* 15: 1039–1053.

14

Transgenic Resistance to Insects: Nature of Risk and Risk Management

Introduction

Genetically modified plants have been released worldwide, but the regulations governing the use of transgenic plants vary considerably in different countries. Existing regulations may be applied to the production and release of genetically modified plants, but some of these regulations may not be adequate to address the potential environmental effects of the transgenic plants (FAO, 1986; UNIDO, 1991; Tzotzos, 1995). For genetically engineered plants with resistance to insects to gain acceptance, decisions that address the concerns associated with the application of biotechnology in agriculture must be science based. Regulatory agencies should assure credibility, and use a rational basis for decision making (Tiedje et al., 1989; Tzotzos, 1995; Sharma et al., 2002a, 2002b; Clark, 2006). Science and legal processes are inextricably linked for regulations that evaluate biological products for human and animal consumption. Review of any particular product should be based on scientific criteria relevant to the product. Advances in biotechnology should determine the changes that might have occurred at the molecular and biochemical levels to scrutinize the product safety, and the effect of product modification on food safety. The approach to review biosafety of transgenic plants is constantly evolving due to new types of products and the availability of scientific information. New technologies have regularly been utilized to develop new gene pools for crop improvement, including artificial manipulation of chromosome number, development of addition and substitution lines for specific chromosomes, chemical and radiation treatments to induce mutations and chromosome rearrangements, and cell and tissue culture (embryo rescue, *in vitro* fertilization, and protoplast fusion) to allow recovery of interspecific and generic hybrids (Simmonds et al., 1999). The genetic gains from the integration of these technologies in crop improvement have resulted in improved yield, environmental adaptation, resistance to specific diseases and insect pests, and quality attributes preferred by farmers, the food industry, and the consumers. Scientific advances in cell and molecular biology have now culminated in genetic modification of crops resulting in novel germplasm that allows plant breeders to respond more quickly to increasing consumer demands. Despite numerous promises, there is a multitude of concerns about the impact of genetically modified crops on the environment. Key issues in the environmental assessment of genetically modified crops are: non-target effects on natural enemies, development of resistance, invasiveness, vertical or horizontal gene flow, effects on biodiversity, and the biosafety of food and food products (Conner, Glare, and Nap, 2003; Clark, 2006). The biosafety issues arising as a result of deployment of transgenic plants include:

Risk to animal and human health: Toxicity and food quality/safety, allergenicity, and resistance to antibiotics.

Risk for agriculture: Loss of biodiversity, evolution of weeds or super weeds, possible alteration of nutritional value, reduction in diversity of the cultivars grown, and development of resistance or tolerance in the target organisms.

Risk for the environment: Persistence of gene or transgene products, unpredictable gene expression or transgene instability, impact on nontarget organisms, and increased use of chemicals in herbicide-resistant crops.

Risk of horizontal transfer: Interaction among different genetically modified organisms, genetic pollution through pollen or seed dispersal (transgene or promoter dispersion), and transfer of the transgene to microorganisms through DNA uptake or generation of new viruses by recombination—transcapsidation, complementation, etc.

A crucial component of ecological risk assessment is defining the baseline for comparison and decision making. For genetically modified crops, the most appropriate reference point is the impact of plants developed through traditional breeding. An extensive knowledge base underpinning the development of genetically modified crops should provide greater confidence in plant science for deployment of genetically modified crops for sustainable food production. It is suspected that the genetically modified crops may have some undesirable impact on the environment and the people (Tiedje et al., 1989; Tzotzos, 1995). The public concern is becoming increasingly vocal, and at times violent. In Europe, consumer acceptance of genetically modified products seems to be quite distant, while consumers in the United States are awakening to the controversy (McHughen, 2000). The coming years may, therefore, prove to be quite decisive for deployment of genetically modified crops. The United Nations and other international organizations have indicated that the world is facing serious problems with food and nutrition security globally, and that it cannot afford to turn away from genetically modified crops (Chrispeels, 2000; NAS, 2000; Schrope, 2001; Leisinger, Schmitt, and Pandya-Lorch, 2002). In such assessments, genetically modified crops are not presented as the only solution, but as an important component of an array of measures and incentives to solve the problem of food security. Many of the concerns raised about genetically modified crops are a reflection of the changing nature of agriculture (Beringer, 2000), and draws on values and philosophical positions that do not change readily upon presentation of technical information, but indicates the importance of socioeconomic and other issues for technology assessment (Bruce and Bruce, 1998). A prudent and transparent linking of science and politics would be the biggest challenge for evaluation and deployment of genetically modified crops for sustainable food production (Levidow et al., 1996; Levidow and Carr, 2000).

Assessment of Risk to Agriculture

The importance of assessing the invasiveness of genetically modified crops has been the subject of much debate (Fitter, Perrins, and Williamson, 1990). While it is important to be cautious when predicting weediness based on plant attributes, an assessment of such characteristics can be made with much greater confidence when assessing whether well-established crops can invade agricultural or natural ecosystems. Modern crop cultivars no longer possess the ability to become weeds, as this ability has been severely reduced in the absence of gene introgression from wild relatives. Common distinctive attributes of weeds, such as seed dormancy, phenotypic plasticity, indeterminate growth, continuous flowering and seed production, and seed dispersal have been bred out of the crop plants over thousands of generations (Baker, 1965). Such changes appeared early in the domestication of crop plants as a consequence of repeated sowing and harvesting cycles (Harlan, 1992). These characters are not candidates for gene transfer back into crops, whether by genetic modification or traditional breeding, because they would severely reduce the agronomic performance of a crop for modern farming practices. Furthermore, these attributes do not arise from a single or few gene transfers. Therefore, genetically modified crops are no more likely to become weeds outside farming situations than the crop cultivars developed through traditional breeding in the past.

Selection for Weedy Traits

Weeds generally exhibit a preference for habitats disturbed by human activities, such as cultivated fields, field margins, gardens, roadsides, soil dumps, and waste sites (Harlan and de Wet, 1965). One feature that all weeds tend to share is high phenotypic plasticity that allows continuous adaptation to changing environments. It is suspected that genetic modification for adaptation to drought or heat stress may change some of these characteristics of the wild and weedy relatives of crops through gene flow. Closely related species can sometimes be highly successful as weeds as they possess different combinations of weedy characteristics (Williamson, 1993). Whether a genetically modified plant species becomes a serious weed in a new environment may relate more to its ability to grow in the new environment coupled with the absence of effective natural enemies such as insect herbivores and diseases (Williamson, 1994). There has been a close association between the evolution of weeds and the domestication of crop plants (de Wet and Harlan, 1975; Harlan, 1992), and some weeds mimic crop characteristics such as growth form and maturity cycle (Barrett, 1983). Selection away from cultivated attributes and gene introgression between genetically modified crops and wild relatives may lead to evolution of weeds with better adaptation to the environment, and this aspect needs to be carefully assessed for each transgene and crop combination.

Invasiveness

An appropriate measure of invasiveness of a plant is its finite rate of increase (Crawley, 1986), the constant at which a population increases over time, assuming a stable age distribution, and the absence of density-dependent constraints. Risk assessment encompasses all demographic processes regulating population growth of a species, such as rate of growth and maturity, fecundity, seed survival, seed germination, and seedling survival to maturity (Parker and Kareiva, 1996). Transfer of a gene conferring a particular character, whether through genetic modification or traditional breeding, may have a positive influence on one component of the overall demographic processes under some environmental conditions, and a negative influence under other conditions, for example, genetically modified seeds of oilseed rape with modified oil content (high-stearate) can have enhanced longevity in soil (Linder, 1998), but the high-stearate gene confers reduced vigor on seedlings (Linder and Schmitt, 1995). Therefore, changes in a single demographic process may not be a good measure of plant invasion, a methodological concern often overlooked when enhanced invasiveness of genetically modified crops is considered.

When assessing the invasiveness potential of genetically modified crops, the key issue to address is whether their weedy characteristics are likely to be different when the expression of the transgene is taken into account. In this context, the transgene-centered approach to biosafety is important (Metz and Nap, 1997). Experimental studies investigating invasiveness need to be established in appropriate and well-defined environments, and measure parameters that encompass all demographic processes. The demographic parameters of genetically modified oilseed rape with resistance to the herbicide glufosinate and conventional oilseed rape have been estimated over a three-year period involving a range of climatic conditions (Crawley et al., 1993). There was no evidence to indicate that genetically modified oilseed rape may become invasive in undisturbed natural habitats. The genetically modified lines tended to be less invasive and less persistent than the nongenetically modified counterparts. However, a few studies have also indicated that herbicide-resistant oilseed rape, sugarbeet, and maize, and *Bt* and pea lectin transgenic potato were found to be no more invasive, or more persistent than their conventional counterparts (Crawley et al., 2001).

Assessment of Nature of Risk to the Environment

Risk Assessment

"Risk" means many different things to different people. It depends on social, cultural, and economic background and value system (Kaplan and Garrick, 1981). A common description of risk is "probability of harm." When taking the magnitude of the potential harm into account, risk is expressed as:

Risk = Probability × Consequence = Likelihood of event × Negative impact of the event.

Probability of Harm

The negative or undesirable impact of an event is commonly referred to as hazard (Wachbroit, 1991). Managing probability, or consequence, or both influence risk. Risk is determined by the following questions. What can go wrong (possibility of harm)? How likely is that to happen (probability of that harm occurring), and what are the consequences if it happens (consequence of harm)? The concept of risk is therefore taken as the possibility of harm that may occur, rather than the probability of harm. It may be wrong to assume that any effect automatically results in an undesired negative impact. If the probability of a given harm is not zero, for worst-case scenarios, the probability of occurrence could be taken as one and the focus should be on the potential detrimental consequences of the event.

One of the concerns related to transgenic plants is the potential environmental harm if these organisms escape or are released into the environment. Harm can take many different forms from transient to permanent in time and space, and from local to global in scope. To define harm, it is necessary to distinguish between the terms risk and hazard, which are often confused. Risk is the probability that a transgene will spread into natural populations once released, and hazard is the probability of species extinction, displacement, or ecosystem disruption given that the transgene will spread into the population (Muir and Howard, 2001a). Long-term hazards to the ecosystem are difficult to predict because not all nontarget organisms may be identified, and the species can evolve in response to the hazard. An infinite number of direct and indirect biotic interactions can occur in nature. The only way to ensure that there is no harm to the environment is to release only those transgenic organisms whose fitness is such that the transgene will not spread, in which case the hazard, becomes irrelevant because the transgene is lost from the population.

Hazard to the Environment

Hazards posed by genetically modified plants include the tendency of a self-pollinated crop to outcross because of self-sterility or any other factors. Or a plant may have a tendency to become a weed, yield toxic substances in the product, or change in the toxin(s) produced by the plant. Any of these attributes may pose a risk to people consuming the product, working with it, or to the environment. In the case of toxin genes inserted into crop plants, the range of the expressed toxin may be much wider than expected, with adverse consequences for nontarget organisms and the environment. Plants may also display a change in appearance, reaction to other biotic and abiotic stress factors, or the end use characteristics. However, it is not easy to measure weediness, as such characteristics are not easily defined. The ecological consequences in most cases are only qualitative and, therefore, risk assessment for genetically modified plants requires a detailed assessment of the modified plant in comparison to the plant from which it has been derived. The procedures adopted should also take cognizance of the environment where the plant is to be released. Pleiotropic effects of a transgene that have antagonistic effects on net fitness can result in unexpected hazards, such as local extinction of the species containing the transgene (Muir and Howard, 2001b). If a transgene enhances mating success while reducing juvenile viability, the least fit individuals obtain the majority of the matings, while the resulting transgenic offspring have low probability of survival, resulting in a gradual spiraling down of population size until both wild-type and transgenic genotypes become extinct (Hedrick, 2001). Transgenes for pest resistance or stress tolerance can increase offspring viability and can also reduce male fertility (Rahman and Maclean, 1999). Extinction hazards in such cases are similar to the use of sterile males to eradicate insect pests. However, in the sterile insect technique, the sterile males have to be introduced repeatedly to cause extinction. Antagonistic pleiotropic effects of transgenes on viability and fertility represent a new class of "Trojan Genes," which persist in the environment. Attempts to reduce transgenic male fertility that do not result in complete male sterility may increase hazard rather than reduce it. If any of the net fitness components are improved by the transgene, while having no adverse side effects, the transgene will invade a population (Muir and Howard, 2001a). However, advantages in one fitness component can offset disadvantages in another and may still increase invasion risk.

Probability of Horizontal and Vertical Gene Flow

Long-term ecological risk can be determined from the probability that an initially rare transgene might spread into the ecosystem, resulting in vertical gene transfer as a result of gene introgression into feral populations, invasion of new territories as a result of introduction of an exotic species, and horizontal gene transfer mediated by microbial agents, or a combination of these. The relative importance of each factor depends on the species, transgene, and the method used to insert the transgene. Vertical gene transfer depends on the species modified. If feral populations are locally available, then adaptation is not a major barrier to gene spread, as the introgression of the transgene may take place into highly adapted native populations. Although feral populations do not exist locally for every domesticated species, if the transgenic plant has an economic advantage, we must assume that human intervention will transport such organisms to area(s) of the world where native populations exist. Another mechanism is the spread of the transgene into new territories, depending on the functionality of the transgene. The anthropogenic introduction of any exotic organisms into natural communities is a serious ecological concern because exotics may adversely affect the local communities, including eliminating populations of other species. Transgenic individuals retain most of the characteristics of their wild-type counterparts, but also possess some novel advantages. A transgene for enhanced environmental adaptation, such as heat tolerance, would allow the species to invade cool and warm environments, while maintaining populations in current habitats (Tzotzos, 1995). They may reproduce at a faster rate and their populations may increase unchecked and thus adversely affect the other species. As a consequence, transgenic organisms might threaten the survival of wild-type conspecifics as well as other species in the community. The third mechanism of spread involves horizontal gene transfer, which occurs through viruses and transposons (Tefper, 2002). However, horizontal gene transfer may occur at such low rates that it would not normally be an additional concern. If a virus or transposon is used to insert the transgene, then even if the virus is disabled, it may be possible for the transposon element to recombine with other naturally occurring viruses and spread into new hosts.

Regardless of the mechanism of gene spread, the ultimate fate of a transgene will be determined by the same forces that direct evolution, that is, natural selection acting on fitness. Thus, risk assessment can be accomplished by determining the outcome of natural selection for increased fitness. This conclusion assumes that the natural populations are large enough to recover from such introductions, that is, natural selection will have time to readjust the population to its previous state. Fitness is not simply survival, but all aspects of the organism that result in spread of the transgene.

Risk Management

A comprehensive strategy for risk management is necessary once a plant has to be released for small-scale experiments and commercial production (NAS, 1989; CEC, 1990a, 1990b; UNIDO, 1991; OECD, 1992, 1993a, 1993b; Sharma et al., 2002b). The scientists concerned, the biosafety committee, and national and international regulatory authorities should determine whether it is acceptable to release the specific transgenic plants and, if needed, restrictions to be imposed. Field containment should be in place to limit the possible environmental impact of the release experiment. This may include isolation from sexually compatible species, prevention of flowering, use of male-sterile lines, and subsequent monitoring protocols. Data required for risk management includes:

- Organization and the people involved;
- DNA donor, receiving species, and the transgene;
- Target environment and the conditions of release;
- Transgenic plant-environment interaction; and
- Control, monitoring, and waste treatment.

The focus of biosafety regulations needs to be on safety, quality, and efficacy. Management, interpretation, and utilization of information are important components of risk assessment, and determine the effectiveness and reliability of biotechnology (Williamson, Perrins, and Fitter, 1990). Various approaches addressing the risks are concerned with establishing good standards of laboratory practice, efficiency and security of the containment facilities, and effects of modified organisms on the human health and the environment (Levin, 1988; Levin and Strauss, 1993). The risk is assessed in the form of access, as a measure of the probability that a modified organism (or the DNA inserted in it) will be able to enter other organisms and survive, and expression, and anticipated or known level of expression of the inserted DNA. Risk also measures damage in the form of harm likely to be caused by exposure to the modified organism.

General Information

To develop effective strategies for risk management, it is important to have information on the institution and the people involved in development, and the people or organizations to be responsible for field testing and containment, monitoring, and waste treatment of the transgenic event.

DNA Donor and Receiving Species

There is a need to have complete information about the donor and the species receiving the gene. The receiving plant species forms the baseline with which the transgenic plant should be compared. Information on the donor species indicates the type of information needed about the transgene. Information is also needed about the vector used in the transformation, and antibiotic or herbicide resistance genes used as markers. Finally, there should be complete information about the transgenic plant, molecular data on the genes inserted, stability of gene expression, and changes in allergenicity, toxicity, persistence in particular environmental conditions, and ability to invade new habitats. The changes in the transplant should be measured against the unmodified control genotype.

Conditions of Release and the Target Environment

The risk to the target environment requires a qualitative judgment, and should be based on a case-by-case study, depending on experience. Information about the purpose of the release, size, design, and agronomic requirements is important for risk assessment and risk management at the national and international levels. Ecological information about the release site, survey of plant species growing in the target region, and the nature of pollen dissemination are important. The anticipated target and nontarget organisms with which the transgenic plant will interact need to be determined. Information should be also recorded whether the transgenic plant would become a better or worse host. The risk to the environment includes harmful effects on the beneficial nontarget organisms.

Interaction Between the Transgenic Plants and the Environment

It is important to describe the invasiveness of the transgenic plant in the wild habitat, ability to propagate sexually and asexually, possibility of transferring the transgene to the same or related species, or to microorganisms, and the consequences of gene transfer.

Control, Monitoring, and Waste Treatment

The containment of the transgenic plants from the tissue culture, growth room, and green-house are covered by good laboratory practice. Care should be exercised so that pollen and seed from the laboratory-produced plants do not escape outside the facilities. The plants should be labeled properly, and there should be no mixing between the transgenic plants. A high level of quality control is needed over the DNA sequences, gene constructs, trans-genic plants, and the experimental results. Growing plants in the greenhouse involves the same level of controls as in the laboratory. The greenhouse should be properly designed to keep out insects and pollen. The facilities should be run under the control of a biosafety committee, and the level of containment should depend on the type of transgenic plants. The greenhouse should have a controlled and filtered airflow system, control of water out-lets, and sterilization. Autoclaving of plant and soil material coming out of the greenhouse is very important.

Post Release Monitoring of the Transgene

Once the transgenic plants are released into the environment, there is a movement of pollen, seed, and the plants outside the immediate environment of release (Sharma et al., 2002a, 2002b). It is important to monitor the transgenes in the environment after the release for its efficacy in controlling the target pest, emergence of secondary pest problems, development of resistance, and the efficiency with which it is possible to destroy the plant material, in case it becomes necessary. Efficient methods of detecting the transgenic plants and the transgene in nontarget species is necessary. It can be done by visual marker (e.g., β-glucorinidase) or a selectable marker (e.g., herbicide/antibiotic resistance, or molecular analysis (e.g., PCR, southern hybridization, and ELISA). If necessary, methods of destroying the plant material at the end of the experiment should be described.

Commercial releases follow once the results of experimental releases have been found to be satisfactory. The aim of risk analysis is to find out the changes in experimental protocol and the method by which transgenic plants may be confined in order to minimize risk to the environment and human health. The risk assessment should be carried out by the multidisciplinary biosafety committee with expertise in molecular biology, environmental science, entomology, pathology, and any other field as appropriate. At this stage, the governmental authorities, environmental groups, social activists, NGOs, and progressive farmers may be involved to make the process transparent, and assure the public that care is being taken to minimize risks to the environment and human health. The institute biosafety committee should have the responsibility for monitoring the site after release for risk assessment to enhance accountability. All the unexpected events should be documented and reported. The national biosafety committee should look into the proposal, and recommend any additional information that may be necessary. The oversight responsibility of the national biosafety committee should also involve interaction with the field experimentation staff, and the release team or institution. Finally, the release proposal may be examined by regional or international agencies as appropriate to ensure a proper and transparent release process.

It is difficult to conclude that release of a transgenic plant would or would not pose a risk to the environment or human health. And it is not possible to put a value on the degree of risk. The essential point is to determine whether and how the transgene might alter the risk compared to the nontransgenic counterpart. There will certainly be questions that cannot be answered fully. Not much can be said about the nature and extent of gene flow. In situations where enough knowledge is not available, it is important to use the knowledge available through conventional plant breeding to aid in risk assessment. Plant breeding has been carried out for several thousand years, and many of the genes that are being inserted through the recombinant technology fall into the same classes as those manipulated by conventional plant breeding.

Once the transgenic plants are released for commercial cultivation, measures such as prevention of flower production and destroying all plant parts are not possible (Sharma et al., 2002a, 2002b). Therefore, the risk assessment should take into account pollen transfer between the nontransgenic crop and wild relatives. There may also be possibilities for taking the transgenic plants into areas where the sexually compatible wild relatives of the crop are present in large numbers. Transgene instability may be another cause of concern when the transgenic crops are grown on a large scale. Strategies for introducing herbicide resistance into several crops or different toxins against the prevalent pests and diseases have to be properly devised. The options for containment after large-scale cultivation of a transgenic plant are limited. Therefore, risk assessment must take these factors into account, and consider all information available from small-scale experiments. To overcome such problems, it may be useful to use tissue-specific expression (target site for insect feeding or infection by the pathogen) or use of male-sterile lines to limit the dispersal of pollen, as is the case with hybrids produced by conventional plant breeding. The cultivation of genetically modified crops may be stopped if there is a risk to human health and the environment.

Many agricultural practices are in place for risk management. The risk of gene transfer by outcrossing from a herbicide-resistant crop to a weed, for example, from canola to weedy mustard, can be managed by spraying a herbicide with a different mode of action. Crop rotations can also be used to control such weeds (Sharma et al., 2002b). The risk of introducing a fertile hybrid between the transgenic plants and weedy relatives can be managed by seeds grown under strict certification procedures to identify crop weed hybrids in the seed production plots. Gene transfer within the same species can be avoided by keeping a safe distance between the adjacent plots. Such information to avoid outcrossing is available for most of the cultivated crops. In areas where there is a greater chance of gene transfer, for example, in the center of origin of a crop plant, serious thought should be given before introducing a transgenic crop with certain genes. Varieties or crops that are likely to be carried to the next crop season or contaminate the same crop next season can be replaced by crop varieties with less or no carryover of seed to the next season.

Conclusions

Genetically modified plants have been released in several countries, but the regulations governing the use of transgenic plants vary considerably in different countries. The decisions that address the concerns associated with the application of biotechnology to agriculture must be science based. The regulatory agencies should assure credibility and use a rational basis for decision making. The approach to risk assessment and risk management is constantly evolving due to new types of products and the availability of scientific information. Long-term ecological risk can be determined from the probability that an initially rare transgene might spread into the ecosystem, resulting in vertical gene transfer as a result of gene introgression into feral populations, invasion of new territories as a result of introduction of an exotic species, and horizontal gene transfer mediated by microbial agents, or a combination of these.

References

Baker, H.G. (1965). Characteristics and modes of origin of weeds. In Baker, H.G. and Stebbins, G.L. (Eds.), *The Genetics of Colonising Species*. New York, USA: Academic Press, 147–172.

Barrett, S.C.H. (1983). Crop mimicry in weeds. *Economic Botany* 37: 255–282.

Beringer, J. (2000). Releasing genetically modified organisms: Will any harm outweigh any advantage? *Journal of Applied ecology* 37: 207–214.

Bruce, D. and Bruce. A. (Eds.). (1998). *Engineering Genesis. The Ethics of Genetic Engineering in Non-Human Species*. London, UK: Earthscan Publications.

Chrispeels, M.J. (2000). Biotechnology and the poor. *Plant Physiology* 124: 3–6.

Clark, E.A. (2006). Environmental risks of genetic engineering. *Euphytica* 148: 47–60.

CEC (Commission of the European Communities). (1990a). Council Directive of 23rd April 1990 on the Contained use of Genetically Modified Microorganisms. Reference no 90/219/EEC. *Official Journal L117*, Volume 33, May 1990. Brussels, Belgium: Commission of the European Communities.

CEC (Commission of the European Communities). (1990b). Council Directive of 23rd April 1990 on the Deliberate Release into the Environment of Genetically Modified Organisms. Reference no 90/220/EEC. *Official Journal L117*, Volume 33, May 1990. Brussels, Belgium: Commission of the European Communities.

Conner, A.J., Glare, T.R. and Nap, J.P. (2003). The release of genetically modified crops into the environment. II. Overview of ecological risk assessment. *The Plant Journal* 33: 19–46.

Crawley, M.J. (1986). The population biology of invaders. *Philosophical Transactions of Royal Society, London (Series B)* 314: 711–731.

Crawley, M.J., Brown, S.L., Hails, R.S., Kohn, D.D. and Rees, M. (2001). Transgenic crops in natural habitats. *Nature* 409: 682–683.

Crawley, M.J., Hails, R.S., Rees, M., Kohn, D. and Buxton, J. (1993). Ecology of transgenic oilseed rape in natural habitats. *Nature* 363: 620–623.

de Wet, J.M.J. and Harlan, J.R. (1975). Weeds and domesticates: Evolution in the man-made habitat. *Economic Botany* 29: 99–107.

FAO (Food and Agriculture Organization). (1986). *International Code of Conduct on the Distribution and Use of Pesticides*. Rome, Italy: Food and Agriculture Organization.

Fitter, A., Perrins, J. and Williamson, M. (1990). Weed probability challenged. *Bio/Technology* 8: 473–474.

Harlan, J.R. (1992). *Crops and Man*, 2nd Edition. Madison, Wisconsin, USA: Agronomy and Crop Science Society of America.

Harlan, J.R. and de Wet, J.M.J. (1965). Some thoughts about weeds. *Economic Botany* 19: 16–24.

Hedrick, P.W. (2001). Invasion of transgenes from salmon or other genetically modified organisms into natural populations. *Canadian Journal of Fisheries and Aquatic Sciences* 58: 841–844.

Kaplan, S. and Garrick, B.J. (1981). On the quantitative definition of risk. *Risk Analysis* 1: 11–27.

Leisinger, K.M., Schmitt, K. and Pandya-Lorch, R. (2002). *Six Billion and Counting. Population and Food Security in the 21st Century*. Washington, D.C., USA: John Hopkins University Press and International Food Policy Research Institute.

Levidow, L. and Carr, S. (2000). Unsound science? Transatlantic regulatory disputes over GM crops. *International Journal of Biotechnology* 2: 257–273.

Levidow, L., Carr, S., von Schomberg, R. and Wield, D. (1996). Regulating agricultural biotechnology in Europe: Harmonization difficulties, opportunities, dilemmas. *Science and Public Policy* 23: 135–157.

Levin, S.A. (1988). Safety standards for the release of genetically engineered organisms. *Trends in Biotechnology* 6: 547–549.

Levin, M. and Strauss, H.S. (1993). Overview of risk assessment and regulation of environmental biotechnology. In Levin, M. and Strauss, H.S. (Eds.), *Risk Assessment in Genetic Engineering. Environmental Release of Organisms*. New York, USA: McGraw-Hill, 1–17.

Linder, C.R. (1998). Potential persistence of transgenes: seed performance of transgenic canola × wild canola hybrids. *Ecological Applications* 8: 1180–1195.

Linder, C.R. and Schmitt, J. (1995). Potential persistence of escaped transgenes: Performance of transgenic oil-modified Brassica seeds and seedlings. *Ecological Applications* 5: 1050–1068.

McHughen, A. (2000). *A Consumer's Guide to GM food. From Green Genes to Red Herrings*. London, UK: Oxford University Press.

Metz, P.L.J. and Nap, J.P. (1997). A transgene-centred approach to the biosafety of transgenic plants: Overview of selection and reporter genes. *Acta Botanica Neerlandica* 46: 25–50.

Muir, W.M. and Howard, R.D. (2001a). Methods to assess ecological risks of transgenic fish releases. In Letourneau, D.K. and Burrows, B.E. (Eds.), *Genetically Engineered Organisms: Assessing Environmental and Human Health Effects*. New York, USA: CRC Press, 355–383.

Muir, W.M. and Howard, R.D. (2001b). Fitness components and ecological risk of transgenic release: A model using Japanese medaka (*Oryzias latipes*). *American Naturalist* 158: 1–16.

NAS (National Academy of Sciences). (1989). *Field Testing of Genetically Modified Organisms*. Washington, D.C., USA: National Academy of Sciences.

NAS (National Academy of Sciences). (2000). *Transgenic Plants and World Agriculture*. Washington, D.C., USA: National Academy Press.

OECD (Organisation for Economic Co-operation and Development). (1992). *Report of OECD Workshop on the Monitoring of Organisms Introduced into the Environment*. Paris, France: Organisation for Economic Co-operation and Development.

OECD (Organisation for Economic Co-operation and Development). (1993a). *Safety Considerations for Biotechnology: Scale-up of Crop Plants*. Paris, France: Organisation for Economic Co-operation and Development.

OECD (Organisation for Economic Co-operation and Development). (1993b). *Safety Considerations of Foods Derived by Modern Biotechnology: Concepts and Principles*. Paris, France: Organisation for Economic Co-operation and Development.

Parker, I.M. and Kareiva, P. (1996). Assessing risks of invasion for genetically engineered plants: Acceptable evidence and reasonable doubt. *Biological Conservation* 78: 193–203.

Rahman, M.A. and Maclean, N. (1999). Growth performance of transgenic tilapia containing an exogenous piscine growth hormone gene. *Aquaculture* 173: 333–346.

Schrope, M. (2001). UN backs transgenic crops for poorer nations. *Nature* 412: 109.

Sharma, H.C., Seetharama, N., Sharma, K.K and Ortiz, R. (2002a). Transgenic plants: Environmental concerns. In Singh, R.P. and Jaiwal, P.K. (Eds.), *Plant Genetic Engineering, Vol. 1. Applications and Limitations*. Houston, Texas, USA: Sci-Tech Publishing, 387–428.

Sharma, K.K., Sharma, H.C., Seetharama, N. and Ortiz, R. (2002b). Development and deployment of transgenic plants: *Biosafety considerations. In Vitro Cellular and Developmental Biology—Plant* 38: 106–115.

Simmonds, N.W., Smartt, J., Millam, S. and Spoor, W. (1999). *Principles of Crop Improvement*, 2nd Edition. Oxford, UK: Blackwell Publishers.

Tepfer, M. (2002). Risk assessment of virus resistant transgenic plants. *Annual Review of Phytopathology* 40: 467–491.

Tiedje, J.M., Colwell, R.K., Grossman, W.L., Hodson, R.E., Lenski, R.E., Mack, R.N. and Regal, P.J. (1989). The planned introduction of genetically modified organisms: Ecological considerations and recommendations. *Ecology* 70: 298–315.

Tzotzos, T. (Ed.). (1995). *Genetically Modified Microorganisms: A Guide to Biosafety*. Wallingford, Oxon, UK: Commonwealth Agricultural Bureau, International.

UNIDO (United Nations Industrial Development Organization). (1991). *Voluntary Code of Conduct for the Release of Organisms into the Environment*. Vienna, Austria: United Nations Industrial Development Organization.

Wachbroit, R. (1991). Describing risk. In Levin, M.A. and Strauss, H.S. (Eds.), *Risk Assessment in Genetic Engineering. Environmental Release of Organisms*. New York, USA: McGraw-Hill, 368–377.

Williamson, M. (1993). Invaders, weeds and the risks from GMOs. *Experientia* 49: 219–224.

Williamson, M. (1994). Community response to transgenic plant release: Predictions from British experience of invasive plants and feral crop plants. *Molecular Ecology* 3: 75–79.

Williamson, M., Perrins, J. and Fitter, A. (1990). Releasing genetically engineered plants. Present proposal and possible hazards. *Trends in Ecology and Evolution* 5: 417–419.

15

Biosafety of Food from Genetically Modified Crops

Introduction

The human population is expected to exceed 8 billion by 2025, and most of this increase will occur in the developing countries. World food production capacity is quite substantial, and yet millions of people are too poor to meet their basic need for food. Agriculture is the primary interface between people and the environment and, therefore, agricultural transformation will be essential to meet the global challenges of reducing poverty and environmental pollution, and enhancing food security. Agricultural growth is central to economic growth in developing countries, and very few low-income countries have achieved rapid growth in nonagricultural sectors without a corresponding increase in agricultural production. Socioeconomic transformation will have to occur at the level of smallholder farmers so that their complex farming systems can be made more productive and efficient in resource use. The "Green Revolution" led to a rapid increase in food production between the 1950s and the 1990s, but the total food production and per capita availability of food have become almost stagnant for the past decade. Therefore, there is a need to harness all the technologies, including biotechnology, for a sustainable growth in agriculture for food security (Serageldin, 1999).

Because of the potential benefits of growing genetically modified crops, their cultivation has increased from 1.97 million hectares in 1996 to over 100 million hectares in 2006 (James, 2007). Large-scale planting of insect-resistant transgenic crops has resulted in a drastic reduction in pesticide use and pesticide residues in food and food products (Conway, 2000; NAS, 2000; Pray et al., 2001). However, the promise of genetically modified crops for increasing crop production has been dimmed by concerns related to their possible impact on nontarget organisms and food biosafety (Williamson, Perrins, and Fitter, 1990; Miller and Flamm, 1993; Sharma, Sharma, and Crouch, 2004). In developed countries, social and environmental groups have raised a hue and cry about the real or conjectural issues related to biosafety of transgenic crops, while in the developing countries, the caution has given rise to fear because of lack of adequate information. In response to these concerns, biosafety working groups have been formed by the Food and Agricultural Organization (FAO), the

United Nations Environment Program (UNEP), the United Nations Industrial Organiza-tion (UNIDO), and the World Health Organization (WHO), and guidelines for handling and release of genetically modified organisms have been published by several agencies (UNIDO, 1991; Tzotzos, 1995).

Potential effects of the transformation process should continue to be taken into consid-eration in the safety assessment, and phenotypic characteristics should be compared between foods derived from genetically modified plants and their comparators. These will include, but may not be limited to, composition, nutritional value, allergenicity, and toxicity. It will be important to define the growing conditions of the comparative plants, the scope of the comparisons, and the acceptable margins of measured differences. It may not be nec-essary or feasible to subject all genetically modified foods to the full range of evaluations, but the conditions that have to be satisfied should be defined. Safety assessments should make use of new profiling techniques such as microarray technology for detailed studies of mRNA expression, quantitative two-dimensional gel electrophoresis and mass spec-trometry for protein analysis, and metabolomic analyses to look at changes in all meta-bolites and metabolic intermediates. Application of such techniques to characterize the differences between the genetically modified crop and the appropriate comparator should help provide a rigorous scientific basis for biosafety of food from transgenic plants. Sometimes, the genetically modified food products include components that are deliberately introduced by genetic modification. In this case, the genetically modified food product might be regarded as "substantially equivalent" to its conventional counterpart, except for a small number of clearly defined differences. Assessment is then limited to examining the implications of the difference(s), perhaps by testing the novel components of the genetically modified plant in isolation.

Biosafety Assessment of Food from Genetically Modified Crops

Historically, crop plants improved by cross-pollination have not been subjected to risk assessment, safety analysis, or risk management. Genetically modified plants have a better predictability of gene expression than the conventional breeding methods, and transgenes are not conceptually different than the use of native genes or plants modified by conven-tional technologies. The ability to synthesize genes in the laboratory and insert them into plants has raised some concerns by the general public. The potential of recombinant tech-nologies to allow a greater modification than is possible with the conventional technologies may have a greater bearing on the environment (Tiedje et al., 1989) and, therefore, the focus of biosafety regulations needs to be on safety, quality, and efficacy (Table 15.1) (Levin, 1988). The need and extent of safety evaluation may be based on the comparison of the new food and the analogous food, if any, and the interaction of the transgene with the environ-ment (Figure 15.1). The lack of any adverse effects resulting from the production and con-sumption of transgenic crops grown on more than 100 million hectares supports these conclusions. Before food derived from a genetically modified crop can be approved for human consumption or an animal feed, it must satisfy three criteria (Moseley, 2002).

- It must be safe.
- It must satisfy the consumer.
- It must be nutritionally adequate.

TABLE 15.1

Assessment of Biosafety of Food Derived from Transgenic Crops

Effects	General/Specific Tests
Human health assessment	Mammalian toxicity
	Digestibility
	Allergenicity
	Homology with known food allergens
Food safety assessment	Composition
	Nutritional quality
	Substantial equivalence
	Animal feed value
	Unanticipated effects
Nonfood safety	Occupational health effects and environmental contamination (via pollen, water, etc.)

Source: Cockburn, A. (2002). *Journal of Biotechnology* 98: 79–106.

A rigorous safety-testing paradigm has been developed for transgenic crops, which utilizes a systematic, stepwise, and holistic approach. The resultant science-based process focuses on a classical evaluation of the toxic potential of the novel trait introduced and the wholesomeness of the transformed crop. In addition, detailed consideration should be given to the history and safe use of the parent crop as well as that of the transgene product. Despite application of rigorous procedures, there is no evidence of harm resulting from the consumption of food derived from transgenic crops. Direct and indirect health effects of genetically modified foods should be used as a guideline for safety assessment, including the tendency to provoke allergic reactions, specific components thought to have nutritional or toxic properties, stability of the inserted gene, and nutritional effects that could result from gene insertion (Haslberger, 2003). Several studies have focused on understanding the intended alterations in composition of food crops that may occur as a result of genetic modification (Kuiper, Kok, and Engel, 2003). The available information on biological effects,

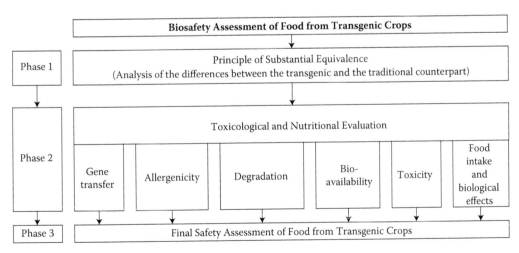

FIGURE 15.1 Biosafety assessment of food from genetically modified crops.

nutritional quality, and chemical composition of food derived from genetically modified crops is discussed in the following pages.

Comparative Effects of Traditional Breeding and Genetic Engineering on Food Quality

There is a need for the new technologies to be tested rigorously for potential allergenic, toxic, and antimetabolic effects in a transparent manner (Gillard, Flynn, and Rowell, 1999). In the context of food safety, it is useful to consider the genetic consequences of plant breeding approaches that provide the opportunity to introduce or modify a plant trait. Conventional plant breeding may also cause rearrangements of the genome and result in production of previously unknown toxins, antinutrients, or allergens. Examples, though uncommon, include insect-resistant celery, which accumulates psoralen in response to light and thereby causes skin burns (Ames and Gold, 1999), and the Magnum Bonum potato, which accumulates toxic levels of solanine in cool weather (Van Gelder, Vinke, and Scheffer, 1988). This raises the question of whether the same safety assessment criteria should be applied to conventionally modified foods as to genetically modified foods. In tomato, it has been possible to evaluate the precision of the genetic change in relation to the size and position of introduced DNA, whereby a trait that modifies carbohydrate composition was introgressed from a wild tomato or engineered by the introduction of an antisense transgene into the cultivated tomato. The results indicated that both classical and molecular approaches introduce a degree of uncertainty in the final genetic makeup, but the source of this uncertainty is quite different in each case (Bennett, Chetelat, and Klann, 1995), and, therefore, it is reasonable to assume that analysis of the safety of genetically engineered food products be evaluated relative to that of traditional approaches that achieve the same goal.

Substantial Equivalence to the Nontransgenic Food

The overall safety evaluation is conducted under the concept known as "substantial equivalence," which is enshrined in all international crop biotechnology guidelines. This provides the framework for a comparative approach to identify the similarities and differences between the transgene product and its comparator, which has a known history of safe use. By building a detailed profile on each step in the transformation process, from parent to new crop, and by thoroughly evaluating the significance of change from a safety perspective of any differences that may be detected, a comprehensive matrix of information is constructed to enable decision making as to whether the food or feed derived from a transgenic crop is as safe as its traditional counterpart. Using this approach, more than 100 transgenic crops have been approved worldwide based on the conclusion that foods and feeds derived from genetically modified crops are as safe and nutritious as those derived from traditional crops (Cockburn, 2002).

Substantial equivalence is based on the principle that if genetically modified food can be shown to be equivalent in composition to an existing food, then it can be considered as safe as its conventional equivalent. Composition and nutritional values of many crops will depend, among other things, on growth conditions, climate, and time of harvesting. Toxicological testing of whole foods has limitations due to bulkiness (difficulties in ingesting sufficient quantities of the whole food in the diet) as compared to food additives or medicines. However, application of such tests to conventional crops with a "history of safe use" may define them as unsafe. In recognition of these difficulties, the principle of substantial equivalence requires that genetically modified plants be compared with its conventional counterpart (FAO/WHO, 1996).

Substantial equivalence is a starting point in the safety evaluation, rather than an end-point of biosafety assessment. Comparative analysis of the transgenic crop as compared to the conventional nontransgenic counterpart, including biological, nutritional, chemical, DNA/RNA microarray profiling, proteomics, and metabolomics, should be carried out scientifically for valid health risk assessment (Kuiper et al., 2001; Pusztai, 2002). The potential of chemical profiling methods is obvious, but further exploration of specificity, sensitivity, and validation is needed. Moreover, the successful application of profiling techniques to the safety evaluation of transgenic foods will require databases that contain information on variations in profiles associated with differences in developmental stages and environmental conditions. An issue that may gain importance in the near future is that of post-marketing surveillance of the foods derived from genetically modified crops. Application of the principle of substantial equivalence has been found to be fairly adequate and no alternative safety assessment strategies that are free of any risk have been proposed so far. It has been suggested that the approach of substantial equivalence is subjective and inconsistent (Millstone, Brunner, and Mayer, 1999). One particular concern has been that the application of substantial equivalence may not reveal any unexpected effects of genetic modification. For instance, the introduction of a gene or multiple genes into a plant species may result in production of low levels of previously unknown toxins, antinutrients, or allergens.

Nutritional Quality

Nutritional assessments are made as part of the safety assessment of a genetically modified food. The assessment reviews the composition of the novel food, its preparation, and the role it is expected to have in the diet. The novel food is compared to traditional counterparts and the significance of any differences is assessed. This may include the use of animal models to establish some aspects of nutritional quality. Nutrient composition data take into account the effects of storage, processing, and cooking. Attention is paid to the particular physiological characteristics and metabolic requirements of vulnerable groups such as infants, children, pregnant and lactating women, the elderly, and those with chronic disease.

Over the last few years, many studies have determined the nutrient value of transgenic feeds compared to their conventional counterparts and some have followed the fate of DNA and the novel protein. The results available to date are reassuring and have revealed no significant differences in the safety and nutritional value of food derived from the first generation of genetically modified plants in comparison with nontransgenic varieties (Flachowsky, Chesson, and Aulrich, 2005). In addition, no residues of recombinant DNA or novel proteins have been found in any organ or tissue samples obtained from animals fed with genetically modified plants, suggesting that compositionally equivalent feeding with transgenic plants generally added little to nutritional and safety assessment. However, the strategies devised for the nutritional and safety assessment of the first-generation products will be much more difficult to apply to second-generation genetically modified plants, in which significant changes in constituents may have been deliberately introduced, for example, increased fatty acid and amino acid content, or reduced concentration of undesirable constituents.

Soybean meal obtained from herbicide-tolerant lines and insect-resistant maize are nutritionally equivalent to their conventional lines (Zdunczyk, 2001). Oil derived from genetically modified crops such as maize or soybean is equivalent to the conventional nontransgenic crops as it does not include detectable protein or DNA derived from the genetically modified plant. Grain and forage of event MON 863 are comparable in their nutritional content to the control maize hybrid and the conventional nontransgenic maize (Berberich et al., 1996; George et al., 2004). These comparisons, together with the history of the safe use of corn as a common component of animal feed and human food, support the conclusion that corn event MON 863 is compositionally equivalent to, and as safe and nutritious as, conventional corn hybrids grown commercially. Very little *Bt* toxins remain in plant parts to be consumed by human beings or dairy cattle, for example, the raw seed of line 81 (with the *cryIAb* gene) showed 14.00 µg per gram of active protein, while the line 531 (with the *cryIAc* gene) contained 2.22 µg per gram of active protein as detected by ELISA (Berberich et al., 1996). Chemical analysis has not shown major changes in nutritional composition of transgenic *Bt* tomatoes compared to that of the control lines. Moreover, levels of the glycoalkaloid α-tomatine were similar to the parent variety. Preliminary analysis of the results of a 91-day feeding trial of transgenic *Bt* tomatoes in rats did not reveal any signs of adverse effects (Noteborn et al., 1995). Russet Burbank potato plants, which resist damage by Colorado potato beetle, *Leptinotarsa decemlineata* (Say), produce two additional proteins, the *Bacillus thuringiensis* subsp. *tenebrionis* protein and the neomycin phosphotransferase II protein (Lavrik et al., 1995). The presence of these proteins and the process used to produce these plants did not alter the quality, composition, or safety of the tubers from insect-resistant plants as compared to commercial Russet Burbank potatoes.

Toxicological Effects and Metabolism of the Transgene Products

Safety evaluations include a detailed consideration of genetic modification with respect to both the DNA sequences that are introduced and the site of integration in the genome of the recipient species. Phenotypic data and detailed chemical composition covering a wide range of nutritionally important parameters are usually considered. Assessment is also made of allergic potential arising from foreign DNA (transgenes) in genetically modified foods. If there are differences between the genetically modified product and its conventional counterpart, then further investigations are carried out. These may include toxicological assessments of the introduced protein and animal feeding studies. In animal studies, any changes in tissue structure or metabolic function of various organs (liver, kidney, lungs, brain, and cardiovascular organs) should be assessed. In human studies, measures of general health, development, and psychological well-being should be included in biosafety evaluations.

Biosafety of Transgene Products

Bt Cry *Proteins*

Based on published information, it has been concluded that *Bt* is safer to humans and other nontarget organisms (McClintock, Schaffer, and Sjoblad, 1995; Siegel, 2001). There are no specific receptors for *Bt* protein in the gastrointestinal tract of mammals, including man (Kuiper and Noteborn, 1994). The toxin proteins produced by *Bt* are toxic to only a few species of insects based on brush border membrane receptors and the structure of the proteins. The species that lack the protein receptors tolerate *Bt* exposure without any injury.

TABLE 15.2

No Observable Effect Level (NOEL) for Mortality of Rats Following Exposure to Purified
Bt Proteins

Event/Transgene	NOEL (μg kg^{-1})	Digestibility in Simulated Gastric Juices	Homology to Known Food Allergens
Mon810 (Cry1Ab)	>4000	Rapidly degraded	None
Bt 11	>4000	Rapidly degraded	None
Cry1F	>5080	Rapidly degraded	None
Bt 176 (Cry1Ab)	>3280	Rapidly degraded	None
Cry9C	>3760	Stable	None
Bt spray	>5050	Stable	None

Source: http://www.epa.gov/appbppdl/biopesticides/ai/plant-pesticides.html.

The plant-expressed Cry1Ab protein has not shown any toxic effects on rats (Table 15.2). The levels of *Bt* protein produced in the currently registered genetically modified plants are quite low (<0.05 μg g^{-1} of plant tissue). A person would have to consume 27.5 kg of corn to reach an equivalent dose of 5000 mg protein kg^{-1} of rat body weight. There is no production of secondary metabolites due to pleiotropic effects (Pedersen, Eriksen, and Knudsen, 2001), and there is no change in the products of chymotrypsin inhibitors (Novak and Haslberger, 2000).

The *Bt* proteins are rapidly degraded by the stomach juices of vertebrates. Most of the *Bt* toxins are specific to insects as they are activated in the alkaline medium of the insect gut. Processing removes >97% of the active proteins in transgenic cottonseed (Sims and Berberich, 1996). CryIAb protein as a component of post-harvest transgenic maize plants dissipates readily on the surface of, or cultivated into, soil (Sims and Holden, 1996), and has not been detected in silage prepared from transgenic plants (Fearing et al., 1997). The *nptII* gene used as a selectable marker is also readily degraded like other dietary proteins, and does not compromise the efficacy of aminoglycoside antibiotics, and presents no risk to humans (Fuchs et al., 1993). The *Bt* proteins Cry34Ab1/Cry35Ab1 conferring resistance to western corn rootworm, *Diabrotica virgifera virgifera* LeConte, in maize are rapidly degraded in simulated gastric fluid, comparable to other registered plant-incorporated protectants (Herman et al., 2003).

Histopathological effects of *Bt* proteins have been observed in the gut mucosa, but no systemic effects have been observed in mice and rabbits following oral administration. Transgenic *Bt* tomatoes pose no additional risk to human and animal health (Noteborn et al., 1996). *In vivo* and *in vitro* experiments involving gastrointestinal tissues from rodents, rhesus monkey, and humans have indicated the absence of specific binding sites for Cry1Ab protein (Noteborn et al., 1995). Short-term toxicity testing revealed no adverse effects in laboratory animals, and no evidence was found for immunotoxicity of the protein. There were no apparent differences in percentages of testicular cell populations (haploid, diploid, and tetraploid) between the mice fed the *Bt*-corn diet and those fed on conventional corn diet (Brake, Thaler, and Evenson, 2004). Because of the high rate of cell proliferation and extensive differentiation that makes testicular germ cells highly susceptible to some toxic agents, it was concluded that the *Bt* corn diet had no measurable or observable effect on fetal, postnatal, pubertal, or adult testicular development, and therefore, *Bt* corn is not harmful to human reproductive development.

The fate of Cry1Ab protein has also been examined in the gastrointestinal contents and visceral organs of calves fed insect-resistant genetically modified maize *Bt11*. Trace amounts of Cry1Ab protein were detected in the gastrointestinal contents, but not in the liver, spleen, kidney, muscle, or mesenteric lymph nodes (Chowdhury et al., 2003). No lesions were observed pathologically. Cry1Ab protein in the feces was degraded quickly at atmospheric temperature, suggesting that only trace amounts of Cry1Ab protein survived passage through the gastrointestinal tract, but were not transferred to liver, spleen, kidney, lymph nodes, or muscles. Fragments of transgenic and endogenous plant DNA as well as *Bt* Cry1Ab protein were not detected in chicken breast muscle samples from animals fed MON 810 (YieldGuard) (Jennings et al., 2003). A 396-bp fragment of the chicken ovalbumin (*ov*) gene used as a positive control was amplified from all the samples showing that the DNA preparations were amenable to PCR amplification. By using a competitive immunoassay with a limit of detection of approximately 60 ng of Cry1Ab protein per gram of chicken muscle, neither the Cry1Ab protein nor immunoreactive peptide fragments were detectable in the breast muscle homogenates from chickens fed YieldGuard.

Protease Inhibitors

Rats fed on purified cowpea trypsin (EC 3.4.21.4) inhibitor in a semisynthetic diet based on lactoalbumin (10 g inhibitor kg^{-1}) for 10 days showed a moderate reduction in weight gain in comparison with controls, despite identical food intake (Pusztai et al., 1992). Although most of the cowpea trypsin inhibitor (CpTI) was rapidly broken down in the digestive tract, its inclusion in the diet led to a slight, though significant, increase in the nitrogen content of feces, but not of urine. Accordingly, the net protein utilization in rats fed on inhibitor-containing diets was also slightly lower, while their energy expenditure was elevated. The slight antinutritional effects of CpTI were probably due mainly to the stimulation of growth and metabolism of the pancreas. Thus, the nutritional penalty for increased insect resistance after the transfer of the cowpea trypsin inhibitor gene into food plants is quite low in the short term.

Lectins

The level of *Galanthus nivalis* L. lectin expression that provides insecticidal protection for plants did not reduce the growth of young rats, and showed negligible effect on weight and length of the small intestine, even though there was a slight hypertrophy of this tissue (Pusztai et al., 1996). However, the activity of brush border enzymes was affected. Sucrase-isomaltase activity was nearly halved, and those of alkaline phosphatase and aminopeptidase increased significantly. Incorporation of N-acetylglucosamine-specific agglutinins from wheat germ (WGA), thorn apple, *Datura stramonium* L., or nettle, *Urtica dioica* L. rhizomes in the diet at the level of 7 g kg^{-1} reduced the apparent digestibility and utilization of dietary proteins and the growth of rats, with WGA being the most damaging (Pusztai et al., 1993). As a result of their binding and endocytosis of the epithelial cells of the small intestine, all three lectins interfered with metabolism and function to varying degrees. WGA also induced hypertrophic growth of the pancreas and caused thymus atrophy. Transfer of WGA genes into crop plants has been advocated to confer resistance to insect pests. However, the presence of this lectin in the diet may harm higher animals at concentrations required to be effective against most insect pests.

Changes in Chemical Profile/Secondary Metabolites

Comparative analysis of genetically modified crops and the conventionally bred crop plants has been suggested to monitor the changes in secondary metabolites, assuming that these products have a long history of safe use, and a hierarchical approach has been proposed based on comparing the chemical fingerprints of the transgenic crop plant to those of isogenic parental or closely related lines bred at identical and multiple sites, extended range of commercial varieties, and downstream processing effects (Noteborn et al., 2000). It is also important to know the likelihood that some of the statistical differences in a transgenic crop plant may be false positives due to chance alone or genetic, environmental, and physiological effects. The levels of antinutrients such as gossypol, cyclopropenoid fatty acids, and aflatoxins in the seed from the *Bt* cotton have been found to be similar to or lower than the levels present in the parent and other commercial varieties. The levels of MDA (malonaldehyde), proline, chlorophyll, and endogenous ABA (abscisic acid) in the main leaves of the transgenic and control lines have been found to be similar (Li, 2003). However, there was a fluctuation in levels over the years because of climatic conditions. The level of endogenous IAA (indole-3-acetic acid) in transgenic plants was low during the seedling stages, but increased with plant age. The activity of SOD (superoxide dismutase) was consistent over the years, being lowest during the seedling stage. Transgenic Kangchong 931 tobacco had the same 22 flavor constituents in its leaves as the nontransgenic NC 89 tobacco (Ding et al., 2001). The differences in aroma contents and the main chemical components (Cl, total sugar, nitrogen, and nicotine content) in the leaves were not significant between the two varieties. Transgenic Kangchong 931 had lower total residues of BHC in the leaves as compared to the conventional variety, NC 89. Potato tubers of insect- and virus-protected varieties are substantially equivalent to tubers of conventional potato varieties in terms of total solids, vitamin C, dextrose, sucrose, soluble protein, glycoalkaloids, fat, ash, calories, total protein, crude fiber, vitamin B6, niacin, copper, magnesium, potassium, and amino acids (Rogan et al., 2000). The information available so far has indicated that there are no substantial differences in chemical fingerprints, secondary metabolites, and nutritional quality of food from transgenic crops as compared to that of the nontransgenic parental lines.

Biosafety of Transgenic Feed/Forage for Animals

Nearly 75% to 80% of genetically modified crops are used as animal feed. The equivalence in digestible energy and crude protein between isogenic and transformed plants expressing a wide range of modifications (insect resistance, herbicide tolerance, or barnase/barstar system of sterility/fertility restoration genes) has been clearly demonstrated in different species. In none of these experiments was animal performance affected by feeding transformed plants compared to animals fed control or isogenic plants, whether measured in terms of growth rate, feed efficiency, and carcass merit in beef cattle; egg mass in laying hens; milk production, composition, and quality in dairy cows; or digestibility in rabbits. Laboratory animals have been used in toxicological studies of the products of introduced genes, but they are rarely fed the entire transformed plants, or their by-products (Aumaitre et al., 2002). Despite not being required to or recommended by existing legislation, many new products have been tested intensively with farm animals to measure effects on performance and animal health, digestibility of key nutrients, wholesomeness, and feeding value. Detection of chloroplasts-specific gene fragments by polymerase chain reaction has shown the presence of plant DNA fragments (199 base pairs) in lymphocytes and duodenal juice of the dairy cow, and in muscle, liver, kidney and spleen of broilers. However, tDNA expressing *Bt* genes usually found as a single copy has not been detected in milk, tissues, or eggs in livestock fed *Bt* maize.

Transgenic soybean, rapeseed, cottonseed, maize, and whole plant maize are used as cattle feed/forage (Kosieradzka, 2002). Several studies have confirmed that their nutritive value is similar to the isogenic varieties, and no adverse effects on growth and production parameters have been observed. Transgenic protein and DNA are degraded in the gut, and no transgenic protein or DNA has been found in animal-origin products. Cultivation of transgenic crops for cattle feed may also reduce the use of pesticides and decrease the contamination of crops with mycotoxins. Higher amounts of α-amylase and protease inhibitors, lectins, or alkaloids may increase plant resistance to insects, but may also decrease the nutritional value of grain. Subchronic *in vivo* experiments as well as comparison of nutritional equivalence of transgenic and conventional crops should be carried out to rule out any adverse effects of transgenic crops as feed/forage.

There were no differences in the survival and body weight of broilers reared on mashed or pelleted diet prepared with *Bt* transgenic maize and similar diets prepared using control maize (Brake and Vlachos, 1998). Comparative feeding studies with broilers and layers in which conventional maize (50% to 78%) or soybean (27%) in feeds was replaced with transgenic varieties did not show any significant differences in production parameters (Chesson and Flachowsky, 2003). Comparative growth studies with broiler chicks, particularly sensitive to any change in nutrient supply or the presence of toxic elements in their feed, can be used to screen for any unintended adverse consequence of the recombinant event not detected by compositional analysis. This does, however, depend on whether the transgenic plant can be matched to the parental line or another suitable control, and its suitability for inclusion in broiler diets. Plant DNA derived from feed has been detected in the muscle, liver, spleen, and kidneys of broilers and layers, but not in eggs. However, no fragments of transgenic DNA or its expressed protein have been found to date in poultry meat or eggs or in any other animal tissues examined.

DNA Transfer from Transgenic Food to Microorganisms/Humans

Potential Effects on Human Health Resulting from the Use of Viral DNA in Human Food

Two types of plant viral DNA sequences are commonly used in construction of genes inserted into genetically modified plants. The first includes "promoters," usually short sequences of DNA that are required for the expression of all genes. In genetically modified plants, the inserted gene is often combined with a promoter derived from the cauliflower mosaic virus (*CaMV35S*). The second type of sequence comprises genes that encode the outer protective coat proteins of viruses, which when expressed in the host plant, interferes with the infecting viruses and confers resistance. However, to date, no genetically modified crops using this second type of gene are grown commercially. It has been suggested that the introduction of viral DNA sequences into genetically modified plants could produce new viruses through recombination ("gene exchange"), either with the remnants of viral DNA sequences that are commonly found in the genomes of all species or with naturally infecting plant and animal viruses. There are, however, natural barriers to this process (Aaziz and Tepfer, 1999; Worobey and Holmes, 1999). Most importantly, viruses generally infect a limited range of species, although there are genetic similarities between some viruses that infect plants and animals, suggesting that they may have jumped between these kingdoms in their evolutionary past, but such events must be rare.

Plant and animal viruses are quite dissimilar and plant viruses cannot infect animal cells. There is only one reported case of recombination between a plant and an animal virus (Gibbs and Weiller, 1999). Humans have eaten virally infected plants for millennia, and there is no evidence that this has created new viruses by recombination, or has caused serious disease. In the extremely unlikely case of recombination producing a novel virus, this would probably be defective, because most recombination events disrupt functional genes. The suboptimal viruses would be removed from the population by natural selection as is the case for most recombinant viruses produced naturally. There is a potential risk that the use of complete viral genes to create transgenic plants resistant to viral infections may create new plant viruses. These novel recombinants could result in plant diseases, but this process has not been documented to date. Concern has been expressed over the use of the *CaMV35S* promoter (Ho, Ryan, and Cummins, 1999, 2000) because this functions in a wide variety of species, including some vertebrates, and has been shown to undergo recombination in laboratory studies (Kohli et al., 1999; Morel and Tepfer, 2000). However, the promoter sequences used in genetically modified plants are a normal constituent of common plant viruses that frequently infect food plants and there is no evidence that these sequences have been involved in the creation of new viruses. Studies have shown that 10% of cabbages and 50% of cauliflowers are infected with *CaMV35S* virus (Hull, Covey, and Dale, 2000), and it has never been shown to cause disease in humans or to recombine with human viruses. It is also highly unlikely that the *CaMV35S* promoter could reactivate the remnants of viruses that are integrated into the genomes of most species. Although there is a great variation among species, most integrated viruses are inert because they contain multiple mutations and cannot be reactivated by the simple acquisition of *CaMV35S* or any other promoter.

Transposable Elements

It has been suggested that genetic modification may activate transposable elements already present in the human genome. Like viruses, transposable elements (short DNA sequences that have the ability to move around the genomes of eukaryotes and bacteria) have been associated with host organisms since early evolution. Because of their mobility, transposable elements have the ability to insert themselves into and thereby damage host genes, and thus potentially lead to pathological effects such as tumors (Hiom, Melek, and Gellert, 1998). These elements comprise up to 40% of the total DNA of higher animals and plants. There is strong evidence that transposable elements have repeatedly been transferred among different species during evolution (Kidwell, 1993; Silva and Kidwell, 2000). Consequently, it seems improbable that the accidental mobilization of transposable elements during the construction and use of genetically modified plants would have any broad impact on the biology of humans, animals, or plants compared with what has happened under natural conditions. Therefore, the risks to human health associated with the use of specific viral DNA sequences in genetically modified plants are negligible.

Fate of Genetically Modified Plant DNA in the Digestive System

One concern associated with genetically modified foods is the possibility that genes introduced into the plant might be incorporated into the consumer's genetic makeup. There is no evidence for transfer of intact genes to humans either from bacteria in the gut or from foodstuffs, despite daily consumption of DNA in the diet. Most ingested DNA is rapidly broken down in the intestinal tract, although it can persist for some time in saliva (Schubbert et al., 1994). Nevertheless, low levels of uptake of gene-sized DNA into cells of the gastrointestinal tract have been detected (Schubbert et al., 1996; Doerfler, 2000; Duggan et al., 2000; Flachowsky, 2000; Einspanier et al., 2001). The uptake may be due to specialized cells of the lining of the gastrointestinal tract, which actively sample gut contents as part of the process of protecting the body from infection (Nicoletti, 2000). There have been no reports of transgenes detected in the cells of cows fed genetically modified maize, although the presence of plant chloroplast genes, which are present at about 1000 times higher concentration than any transgene, could be detected (Flachowsky, 2000; Einspanier et al., 2001). There is no obvious reason that a cell with altered biological properties due to foreign DNA uptake could transmit this effect to other cells or affect the germline of the host organism.

Any untoward consequences of DNA consumption would probably be due to ingestion and transmission of intact autonomous genetic elements rather than to transfer of fragments of DNA. Such elements might include the complete genomes of viruses or transposable elements, or large pieces of DNA from normal intestinal microbial flora. There is strong evidence that gene transfer events of this sort have occurred during evolution (Kidwell, 1993; Capy, Anxolabehere, and Langin, 1994). Uptake of fragments of transgenic DNA from genetically modified food should therefore be seen in the context of an ongoing biological process involving intact autonomous genetic elements, which has had no detectable negative consequences. An alternative scenario might involve the entry of a novel DNA sequence into gastrointestinal microbial flora, where it would replicate and persist in its new host and deliver a product into its surroundings. This has occurred throughout mammalian evolution and has apparently had little biological consequence (Stanhope et al., 2001).

All the consequences of use of genetically modified crops as food need to be viewed in the context of a normal diet, which for humans and animals comprises large amounts of DNA. For example, a 600 kg cow is estimated to ingest about 600 mg of DNA a day (Beever and Kemp, 2000), and digestion of DNA in the gastrointestinal tract may make a significant contribution to nutrition. The DNA is derived not only from the cells of food sources, but also from any contaminating microbes and viruses. Therefore, it is unlikely that the ingestion of well-characterized transgenes in normal food and their possible transfer to mammalian cells would have any significant deleterious biological effects.

Transgenic DNA is no different than the DNA consumed as a part of daily diet, and that transgenic DNA will have the same effects on humans as the DNA from nontransgenic food. Ingested DNA is subjected to degradation, but the degradative process is not necessarily complete (Heritage, 2004; Netherwood et al., 2004). People fed on transgenic soybean meal containing 5-enol pyruvyl shikimate-3-phosphate synthase (*epsps*) transgene, which provides resistance to herbicide glyphosate, degrades completely when passing through the human gut (Netherwood et al., 2004). However, transgenic DNA sequences have been detected in digesta of ileostomists (the microbes found in the terminal ileum and diversion of digests to a colostomy bag, although such microbes may not be present in the gut of normal people). Gene flow from transgenic plants to the gut microbes does occur. However, such a gene transfer is unlikely to influence gastrointestinal function or endanger human health.

Allergenic Effects

Expression of a new gene in a crop could also introduce new allergens, normally not present in the nontransformed plants (Lehrer, 2000). Allergic reactions to foods are hard to predict, but can be life threatening. Virtually every gene transfer in crops results in some protein production, and proteins trigger allergic reactions. Biotechnology can introduce new proteins into food crops, not only from the known sources (such as peanuts and shellfish), but also from plants, bacteria, and viruses whose allergenicity is unknown. This might lead people to avoid foods that are actually safe. If the indigenous proteins or the new proteins are from the known sources of allergens, then assessing the allergens within the genetically modified plants is easier. If the source of the allergenic protein is known, and is related to the introduced gene from sources that have not been used as a human food, then one has to rely on criteria with which to assess their potential allergenicity. Eight commonly allergenic and 160 less allergenic foods have been identified and scientists can avoid the transfer of genes with known allergenic effect (Lehrer, 2000).

Food allergies occur in 1 to 2% of adults and 6 to 8% of children (Metcalfe et al., 1996; Sampson, 1997), although severe allergic reactions (anaphylaxis) to foods are relatively rare. The introduction of a new gene into a plant, or a change in the expression of an existing gene, may cause that plant to become allergenic. Therefore, known allergens should not be introduced into food crops. At present, there is no evidence that genetically modified foods that are commercially available cause any clinical manifestations of allergenicity, and assertions to the contrary have not been supported by systematic analysis. The allergenic risks posed by genetically modified plants are in principle no greater than those posed by conventionally derived crops or by plants introduced from other parts of the world. Nevertheless, it is important to consider potential allergenic risks posed by genetically modified plants and to place them in the context of risks posed by introduced plants and plants produced by conventional or organic means.

Allergic sensitization to a genetically modified plant, as with a conventionally derived plant, could also occur via the lungs (perhaps through inhaling pollen or dust created during milling) or through skin contact (for example, during handling), as well as via the gastrointestinal tract following ingestion of foods. Occupational allergies to conventional plants can take the form of either immediate hypersensitivity or delayed hypersensitivity reactions. The latter occur as a consequence of handling plant materials and generally express themselves as contact dermatitis. Of immediate hypersensitivity allergies to plants, those that involve inhalation of particulates are quite common. Those at risk from pollen include the general population and, in particular, farm workers. Individuals involved in the harvesting of crops and in food processing that generate dusts are at risk of sensitization through both inhalation and skin contact. Therefore, in order to adequately assess any risks, it is important to evaluate the allergenic potential of genetically modified plants through inhalation and skin contact, as well as via ingestion.

Guidelines for assessing the allergenic risks of genetically modified foods have been developed by FAO/WHO (2001). This approach includes determining whether the source of the introduced gene is from an allergenic plant, whether genetically modified foods react with antibodies in the sera of patients with known allergies, and whether the product encoded by the new gene has similar chemical and biological properties to known allergens. It also involves animal models that can be used to screen genetically modified foods for allergenic reaction. Ongoing research to develop adequate animal models for allergenicity testing will further increase the assurance of this procedure.

Several protein families that contribute to the defense mechanisms of food plants are allergens or putative allergens, and some of these proteins have been used in molecular approaches to increase insect resistance in crop plants (Franck Oberaspach and Keller, 2002). These include α-amylase and trypsin inhibitors, lectins, and pathogenesis-related proteins. Several "self defense" substances produced by plants may be toxic to mammals, including humans. Food safety can also be severely influenced by invading pathogens and their metabolic products, for example, mycotoxins. This may result in a trade-off between "nature's pesticides" produced by transgenic plants and varieties from traditional breeding programs, synthetic pesticides, mycotoxins, or other poisonous products of insect pests. Allergenicity assessment must consider several factors, including the source of the transferred protein, expression levels, the physical and chemical properties of the protein, and similarity to known allergens. Although no single factor can be considered definitive, consideration of all these factors together may provide some indication of potential allergenicity (Gendel, 1998).

All *Bt* proteins have been subjected to allergenicity tests. With the exception of Cry9C, all are degraded quickly in the stomach, and are unstable when heated. The Cry9C is stable under simulated stomach conditions, but it is not glycosylated and has not caused any adverse effects or shown characteristics of immune system responses in mammalian toxicity studies (Carpenter et al., 2002). It has been suggested that ingredients in vegetative cells and parasporal bodies of *Bt* may trigger immune reactions rather than crystalline proteins expressed in plants (Bernstein et al., 1999).

Bollgard cotton lacks the characteristics found in potential allergenic proteins (Perlak et al., 2001). Genetically modified maize with an amaranth globulin protein termed amarantin showed 4% to 35% more protein and 0% to 44% higher amounts of specific essential amino acids as compared to nontransgenic maize (Sinagawa Garcia et al., 2004). Individual sequence analysis with known amino acid sequences reported as allergens showed that none of these *IgE* elicitors were identified in amarantin. Amarantin was digested in 15 minutes by simulated gastric fluid as observed by Western blot. Expressed amarantin did not induce levels of specific *IgE* antibodies in BALB/c mice, as analyzed by ELISA, and therefore, transgenic maize with amarantin is not an important allergenicity inducer, just as nontransgenic maize.

Public Attitude to Food from Transgenic Plants

Public perception is likely to have a great impact on innovation, introduction, and diffusion of products of biotechnology. And negative public perception is likely to keep the products of biotechnology away from the marketplace. Public perception is influenced by a broad range of issues, including environmental safety, ethics, legal repercussions, economic gain, and socioeconomic impact. Public opinion can be influenced by nonscientific considerations based on impressions created by the media and pressure groups. Concerns about recombinant technology were raised in the early 1970s, and even led to protests over setting up of biotechnology laboratories in developed countries (Leopold, 1993). As a result, regulations and guidelines for handling DNA-based technologies were formulated. With increasing stakes for economic gains and potential risks, public debate over the need to develop regulatory mechanisms became all the more important. This also led to intense public debate among scientists, particularly between molecular geneticists and ecologists. As a result, the politicians, the industry, and environmental groups began to take a more active stance. Much of this debate is likely to play itself out in the regulatory arena, with the shift of focus from the laboratory to commercial applications. This role needs to be taken up by international and national research organizations for agriculture, health, food, and the environment.

Introduction of Food from Genetically Modified Crops to the General Public

There is a need for introducing the products of biotechnology through the peer-reviewed press rather than through the general media, balanced presentation of the benefits of biotechnology to the general public, and an understanding of the trade-offs. The role of transgenics in reducing the load of pesticides in the environment needs to be considered seriously. The first agricultural products of biotechnology have already reached the world markets, and have received a frosty response in some parts of the world. Despite some sensational headlines in the press, people in North America have not reacted to foods containing ingredients developed through biotechnology. It is clear from consumer surveys that perceptions about biotech foods are strongly influenced by the type of information, confidence in government, and cultural preferences. Most of the people are cautiously optimistic about the benefits of biotechnology. They will accept the products if they see a benefit to themselves or to the society.

Consumer Response to Food from Genetically Modified Crops

Consumer response to food from genetically modified plants varies across regions. In general, the response is similar to any other food, and the major considerations are taste, nutrition, price, safety, and convenience. How seeds and food ingredients are produced is relevant only for a small group of concerned "organic" consumers. In general, consumers see considerable value in human genetic testing, development of new medicines, and the use of biotechnology to develop insect-resistant crop plants. Because of ethical considerations, consumers are less likely to accept the use of biotechnology with animals and food products compared to crop plants. The most acceptable applications are those that offer a clear consumer benefit, as well as those perceived to be ethical and safe. However, public attitudes about agricultural biotechnology vary considerably across countries.

European consumers generally are not against the pharmaceutical products of biotechnology, but are much less willing to accept food and food ingredients, especially when derived from genetically modified plants (Moses, 1999). Objections are mainly based on fears for the health and safety of the consumer, worries about the possibility of deleterious effects on the environment, and a range of moral and ethical concerns often deriving from a distaste, however expressed, for the concept of interfering with nature. Transgenic maize with *Bt* genes has been banned in the European Union due to the presence of the antibiotic (ampicillin) resistance gene. The end result is a serious setback to the public acceptability of transgenic crops. Consumer understanding of the science underlying biotechnology is patchy. In no country does more than a small proportion of the population claim a good grasp. As a consequence of these attitudes, the introduction of genetically modified foods into Europe has occurred slowly.

In the developing countries, there is a need for an environment that is institutionally, socially, culturally, politically, and educationally favorable. Developing countries that are entering biotechnology also have a large proportion of population capable of making rational decisions. If large sections of the population are under the poverty line and illiterate, the nongovernmental organizations become the advocates of public opinion, which at times may be guided by several extraneous factors. Under such circumstances, international organizations such as FAO, WHO, UNIDO, and international agriculture research centers (IARCs) can be expected to play an important role in enhancing the public perception of the usefulness and the risks associated with the introduction of genetically modified organisms. In the Philippines, almost 80% of the people were willing to buy genetically modified food with improved characteristics (Mendoza and Ables, 2003). Sixty percent believed they would benefit from biotechnology in the near future, and 82% would strongly support policy on labeling a product only if there is an allergen or substantial change in the composition of foods. Cloning of animals for food was acceptable to 37% of the respondents, while cloning of humans was unacceptable to 79% of the people.

Role of the Scientific Community, NGOs, and the Media

There is a continued need to educate the general public about the benefits of biotechnology and the perceived risks, if any. Environmentalists often focus on possible ecological impact from the use of biotechnology. In countries where consumers are more negative about biotechnology, media coverage and public opposition have been more pronounced. As a result, a discussion on the benefits of biotechnology has been ignored, while the perceived risks have been emphasized.

Most of the scientific community and the public agree that the risks of genetic engineering are largely exaggerated, but there is a need for strict regulatory mechanisms. The farmers and the public believe that biotechnology will lead to increased food production and improved nutrition (Pilcher and Rice, 1998). Level of education, religion, socioeconomic factors, pressure by the environmental groups, and governmental policy are likely to shape public opinion about biotechnology. Scientific literacy, scientific proof of lack of evidence for presumed risks, informal dissemination of information through the public media, clear standards, food labeling, reducing the extent of exaggerated expectations arising from science fiction, making the public part of the decision-making process, and reliability of information are important to provide a clear picture of the benefits and risks of biotechnology. Farmers' willingness to use or consume a range of crops modified genetically for on-farm production or input benefits (such as insect pest and herbicide resistance) is quite high (McDougall et al., 2001). Acceptability of products specifying cross-species or cross-kingdom gene transfer was quite low. Concerns have also been raised about corporate ownership and marketing of the technology, and the potential lack of demand for food derived from genetically modified plants.

In light of public concerns with the safety of genetically modified crops, the issue of food labeling, and governmental regulation of transgenic plants, quality control, and monitoring of the process of transgenic seed product development has been advocated for ensuring customer satisfaction (Mumm and Walters, 2001). The primary goal of quality control monitoring is to ensure the authenticity (transgenic event identity and purity) of seed materials used in product testing, in the development of regulatory data packages for governmental review, and to produce seed for commercial release. Monitoring is performed to confirm the presence of the presumed transgenic event(s) and the absence of all others. Sophisticated quality control strategies formulated to monitor the product development process and to maintain quality standards in the manufacturing industry can serve as a foundation in devising efficient strategies tailored to meet the needs of the seed industry. However, it is impossible to follow food labeling in village markets in developed and developing countries, where food and food products from both transgenic and non-transgenic sources is bought by local traders and sold in small quantities to consumers in unpacked and unprocessed form.

Conclusions

Concerns have been raised about the biosafety of food produced through genetically modified organisms. Genetically modified organisms have a better predictability of gene expression than the conventional breeding methods, and transgenes are not conceptually different than the use of native genes or organisms modified by conventional technologies. The focus of biosafety regulations needs to be on safety, quality, and efficacy. The need and extent of safety evaluation may be based on the comparison of the new food and the analogous food, if any, and the interaction of the transgene with the environment. The overall safety evaluation is conducted under the concept known as "substantial equivalence," which is enshrined in all international crop biotechnology guidelines. This provides a framework for a comparative approach to identify the similarities and differences between the transgene product and its comparator, which has a known history of safe use. A rigorous safety-testing paradigm has been developed for transgenic crops, which utilizes a systematic, stepwise, and holistic approach. The resultant science-based process focuses on a classical evaluation of the toxic potential of the introduced novel trait and the wholesomeness of the transformed crop. The lack of any adverse effects resulting from the production and consumption of transgenic crops grown on more than 100 million hectares supports the conclusions that food derived from genetically modified plants is safe for human consumption. Risks to human health associated with the use of specific viral DNA sequences in genetically modified plants are negligible. Given the long history of DNA consumption from a wide variety of sources by animals and humans, it is likely that such consumption poses no significant risk to human health, and that additional ingestion of DNA of genetically modified crops will have no adverse effect. It may not be necessary or feasible to subject all genetically modified foods to the full range of evaluations, but those conditions that have to be satisfied should be defined. Research should be undertaken to develop modern profiling techniques and to define the "normal" composition of conventional plants. The procedures to assess allergy should be expanded to encompass inhalant as well as allergy to food. Labeling of food derived from transgenic or other sources will remain a challenge, as much of the food produced in developing countries is consumed in unprocessed and unpackaged form locally.

References

Aaziz, R. and Tepfer, M. (1999). Recombination in RNA viruses in virus-resistant transgenic plants. *Journal of General Virology* 80: 1339–1346.

Ames, B.N. and Gold, L.S. (1999). Pollution, pesticides and cancer misconceptions. In Morris, J. and Bate, R. (Eds.), *Fearing Food: Risk, Health and Environment.* Oxford, UK: Butterworth Heinemann, 18–38.

Aumaitre, A., Aulrich, K., Chesson, A., Flachowsky, G. and Piva, G. (2002). New feeds from genetically modified plants: Substantial equivalence, nutritional equivalence, digestibility, and safety for animals and the food chain. *Livestock Production Science* 74: 223–238.

Beever, D.E. and Kemp, C.F. (2000). Safety issues associated with DNA in animal feed derived from genetically modified crop: A review of scientific and regulatory procedures. *Nutrition Abstracts and Reviews* 70: 175–182.

Bennett, A.B., Chetelat, R. and Klann, E. (1995). Exotic germplasm or engineered genes: Comparison of genetic strategies to improve fruit quality. In *Genetically Modified Foods: Safety Issues.* Washington, D.C., USA: American Chemical Society, 88–99.

Berberich, S.A., Ream, J.E., Jackson, T.L., Wood, R., Stipanovic, R., Harvey, P., Patzer, S. and Fuchs, R.L. (1996). The composition of insect-protected cottonseed is equivalent to that of conventional cottonseed. *Journal of Agricultural and Food Chemistry* 44: 365–371.

Bernstein, I.L., Bernstein, J.A., Miller, M., Tierzieva, S., Bernstein, D.I., Lummus, Z., Selgrade, M.K., Deerfler, D.L. and Seligy, V.L. (1999). Immune responses in farm workers after exposure to *Bacillus thuringiensis* pesticides. *Environmental and Health Perspectives* 107: 575–582.

Brake, D.G., Thaler, R. and Evenson, D.P. (2004). Evaluation of Bt (*Bacillus thuringiensis*) corn on mouse testicular development by dual parameter flow cytometry. *Journal of Agricultural and Food Chemistry* 52: 2097–2102.

Brake, J. and Vlachos, D. (1998). Evaluation of transgenic event 176 "Bt" corn in broiler chickens. *Poultry Science* 77: 648–653.

Capy, P., Anxolabehere, D. and Langin, T. (1994). The strange phylogenies of transposable elements: Are transfers the only explanation? *Trends in Genetics* 10: 7–12.

Carpenter, J., Felsot, A., Goode, T., Hamming, M., Onstad, D. and Sankula, S. (2002). *Comparative Environmental Impacts of Biotechnology Derived and Traditional Soybean, Corn, and Cotton Crops.* Ames, Iowa, USA: Council for Agricultural Science and Technology.

Chesson, A. and Flachowsky, G. (2003). Transgenic plants in poultry nutrition. *World's Poultry Science Journal* 59: 201–207.

Chowdhury, E.H., Shimada, N., Murata, H., Mikami, O., Sultana, P., Miyazaki, S., Yoshioka, M., Yamanaka, N., Hirai, N. and Nakajima, Y. (2003). Detection of Cry1Ab protein in gastrointestinal contents but not visceral organs of genetically modified Bt11-fed calves. *Veterinary and Human Toxicology* 45: 72–75.

Cockburn, A. (2002). Assuring the safety of genetically modified (GM) foods: The importance of an holistic, integrative approach. *Journal of Biotechnology* 98: 79–106.

Conway, G. (2000). *Crop Biotechnology. Benefits, Risks, and Ownership.* The Rockfeller Foundation. http://www.rockfound.org/news/03062000.cropbiotech.html.

Ding, Y.L., Xu, Z.C., Yang, T.Z., Yan, X.F., Liu, X.Z., Yie, H.C., Zhang, H. and Chen, G.L. (2001). Analysis on quality indexes of insect-resistant transgenic tobacco (*Nicotiana tabacum* L.). *Journal of Huazhong Agricultural University* 20: 81–84.

Doerfler, W. (2000). Consequences of foreign DNA integration and persistence. In *Foreign DNA in Mammalian Systems.* New York, USA: Wiley-VCH Verlag, 129–146.

Duggan, P.S., Chambers, P.A., Heritage, J. and Forbes, J.M. (2000). Survival of free DNA encoding antibiotic resistance from transgenic maize and the transformation activity of DNA in bovine saliva, ovine rumen fluid and silage effluent. *FEMS Microbiology Letters* 191: 71–77.

Einspanier, R., Klotz, A., Kraft, A., Aulrich, K., Poser, R., Schwagele, F., Jahreis, G. and Flaschowsky, G. (2001). The fate of forage plant DNA in farm animals: A collaborative case study investigating cattle and chicken fed recombinant plant material. *European Food Research Technology* 212: 2–12.

FAO (Food and Agriculture Organization)/WHO (World Health Organization). (1996). *Biotechnology and Food Safety.* Report of a Joint Food and Agriculture Organization/World Health Organization Consultation. Rome, Italy: Food and Agriculture Organization/World Health Organization.

FAO (Food and Agriculture Organization)/WHO (World Health Organization). (2001). *Evaluation of Allergenicity of Genetically Modified Foods.* Report of a Joint Food and Agriculture Organization/World Health Organization Consultation. Rome, Italy: Food and Agriculture Organization/World Health Organization.

Fearing, P.L., Brown, D., Vlachos, D., Meghji, M. and Privalle, L. (1997). Quantitative analysis of CryIA(b) expression in Bt maize plants, tissues, and silage and stability of expression over successive generations. *Molecular Breeding* 3: 169–176.

Flachowsky, G. (2000). GMO in animal nutrition: Results of experiments at our institute. In *Proceedings, 6th International Feed Production Conference*, 27–28 November, 2000, Piacenza, Italy, 291–307.

Flachowsky, G., Chesson, A. and Aulrich, K. (2005). Animal nutrition with feeds from genetically modified plants. *Archives of Animal Nutrition* 59: 1–40.

Franck Oberaspach, S.L. and Keller, B. (1997). Consequences of classical and biotechnological resistance breeding for food toxicology and allergenicity. *Plant Breeding* 116: 1–17.

Fuchs, R., Ream, J., Hammad, B., Naylor, M., Leimgruber, R. and Berberich, S. (1993). Safety assessment of the neomycin phosphotransferase Ii (*Nptii*) protein. *Bio/Technology* 11: 1543–1547.

Gendel, S.M. (1998). Assessing the potential allergenicity of new food proteins. *Food Biotechnology* 12: 175–185.

George, C., Ridley, W.P., Obert, J.C., Nemeth, M.A., Breeze, M.L. and Astwood, J.D. (2004). Composition of grain and forage from corn rootworm-protected corn event MON 863 is equivalent to that of conventional corn (*Zea mays* L.). *Journal of Agricultural and Food Chemistry* 52: 4149–4158.

Gibbs, M.J. and Weiller, G.F. (1999). Evidence that a plant virus switched hosts to infect a vertebrate and then recombined with a vertebrate-infecting virus. *Proceedings National Academy of Sciences USA* 96: 8022–8027.

Gillard, M.S., Flynn, L. and Rowell, A. (1999). Food scandal exposed. *The Guardian*, December 2, p. 1.

Haslberger, A.G. (2003). Codex guidelines for GM foods include the analysis of unintended effects. *Nature Biotechnology* 21: 739–741.

Heritage, J. (2004). The fate of transgenes in the human gut. *Nature Biotechnology* 22: 170–172.

Herman, R.A., Schafer, B.W., Korjagin, V.A. and Ernest, A.D. (2003). Rapid digestion of Cry34Ab1 and Cry35Ab1 in simulated gastric fluid. *Journal of Agricultural and Food Chemistry* 51: 6823–6827.

Hiom, K., Melek, M. and Gellert, M. (1998). DNA transposition by RAG1 and RAG2 proteins: A possible source of oncogenic translocations. *Cell* 94: 463–470.

Ho, M.W., Ryan, A. and Cummins, J. (1999). Cauliflower mosaic virus promoter: A recipe for disaster. *Microbial Ecology in Health and Disease* 10: 33–59.

Ho, M.W., Ryan, A. and Cummins, J. (2000). Hazards of transgenic plants containing the cauliflower mosaic virus promoter. *Microbial Ecology in Health and Disease* 12: 6–11.

Hull, R., Covey, S. and Dale, P. (2000). Genetically modified plants and the 35S promoter: Assessing the risks and enhancing the debate. *Microbial Ecology in Health and Disease* 12: 1–5.

James, C. (2007). *Global Status of Commercialized Biotech/GM Crops: 2006*. ISAAA Briefs no. 35. Ithaca, New York, USA: International Service for Acquisition on Agri-Biotech Applications (ISAAA). http://www.isaaa.org/resources/publications/briefs/35.

Jennings, J.C., Albee, L.D., Kolwyck, D.C., Suber, J.B., Taylor, M.L., Hartnell, G.F., Lirette, R.P. and Glenn, K.C. (2003). Attempts to detect transgenic and endogenous plant DNA and transgenic protein in muscle from broilers fed YieldGard corn borer corn. *Poultry Science* 82: 371–380.

Kidwell, M. (1993). Lateral transfer in natural populations of eukaryotes. *Annual Review of Genetics* 27: 235–256.

Kohli, A., Griffiths, S., Palacios, N., Twyman, R.M., Vain, P., Laurie, D.A. and Christou, P. (1999). Molecular characterization of transforming plasmid rearrangements in transgenic rice reveals a recombination hotspot in the CaMV 35S promoter and confirms the predominance of microhomology mediated recombination. *The Plant Journal* 17: 591–601.

Kosieradzka, I. (2002). Genetically modified crops in cattle feeding. *Biuletyn Informacyjny Instytut Zootechniki* 40: 237–248.

Kuiper, H.A., Kok, E.J. and Engel, K.H. (2003). Exploitation of molecular profiling techniques for GM food safety assessment. *Current Opinion in Biotechnology* 14: 238–243.

Kuiper, H.A. and Noteborn, H.P.J.M. (1994). Food safety assessment of transgenic insect-resistant Bt tomatoes. Food safety evaluation. In *Proceedings of an OECD-Sponsored Workshop*, 12–15 September, 1994, Oxford, UK. Paris, France: Organisation for Economic Co-operation and Development, 50–57.

Kuiper, H.A., Kleter, G.A., Noteborn, H.P.J.M. and Kok, E.J. (2001). Assessment of the food safety issues related to genetically modified foods. *Plant Journal* 27: 503–528.

Lavrik, P.B., Bartnicki, D.E., Feldman, J., Hammond, B.G., Keck, P.J., Love, S.L., Naylor, M.W., Rogan, G.J., Sims, S.R. and Fuchs, R.L. (1995). Safety assessment of potatoes resistant to Colorado potato beetle. In *Genetically Modified Foods: Safety Issues*. Washington, D.C., USA: American Chemical Society, 148–158.

Lehrer, S.B. (2000). Potential health risks of genetically modified organisms: How can allergens be assessed and minimized. In Persley, G.J. and Lantin, M.M. (Eds.), *Agricultural Biotechnology and the Poor*. Washington, D.C., USA: Consultative Group on International Agricultural Research, 149–155.

Leopold, M. (1993). The commercialization of biotechnology. *Annals of the New York Academy of Sciences* 700: 214–231.

Levin, S.A. (1988). Safety standards for the release of genetically engineered organisms. *Trends in Biotechnology* 6: 547–549.

Li, F.G. (2003). Study on the physiological and biochemical characters of insect-resistant transgenic cotton harbouring double-gene. *Cotton Science* 15: 131–137.

McClintock, J.T., Schaffer, C.R. and Sjoblad, R.D. (1995). A comparative review of the mammalian toxicity of *Bacillus thuringiensis* based pesticides. *Pesticide Science* 45: 95–105.

McDougall, D.J., Longnecker, N.E., Marsh, S.P. and Smith, F.P. (2001). Attitudes of pulse farmers in Western Australia towards genetically modified organisms in agriculture. *Australasian Biotechnology* 11: 36–39.

Mendoza, T.L.T. and Ables, H.A. (2003). Awareness of and attitudes towards modern biotechnology of selected UPLB (Philippines University, Los Banos, Laguna, Philippines) personnel. *Philippine Journal of Crop Science* 25: 35–44.

Metcalfe, D.D., Astwood, J.D., Townsend, R., Sampson, H.A., Taylor, S.L. and Fuchs, R.L. (1996). Assessment of the allergenic potential of foods from genetically engineered crop plants. *Critical Reviews in Food Science and Nutrition* 36: S165–186.

Miller, H.I. and Flamm, E.L. (1993). Biotechnology and food regulation. *Current Opinion in Biotechnology* 4: 265–268.

Millstone, E.P., Brunner, E.J. and Mayer, S. (1999). Beyond "substantial equivalence." *Nature* 401: 25–26.

Morel, J.B and Tepfer, M. (2000). Pour une évaluation scientifique des risques: le cas du promoteur 35S. *Biofutur* 201: 32–35.

Moseley, B.E.B. (2002). Safety assessment and public concern for genetically modified food products: the European view. *Toxicologic Pathology* 30: 129–131.

Moses, V. (1999). Biotechnology products and European consumers. *Biotechnology Advances* 17: 647–678.

Mumm, R.H. and Walters, D.S. (2001). Quality control in the development of transgenic crop seed products. *Crop Science* 41: 1381–1389.

NAS (National Academy of Sciences). (2000). *Transgenic Plants and World Agriculture*. Washington, D.C., USA: National Academy Press.

Netherwood, T., Martin-Orue, S.M., O'Donnell, A.G., Gockling, S., Graham, J., Mathers, J.C. and Gilbert, H.J. (2004). Assessing the survival of transgenic plant DNA in the human gastrointestinal tract. *Nature Biotechnology* 22: 204–209.

Nicoletti, C. (2000). Unsolved mysteries of intestinal M cells. *Gut* 47: 735–739.

Noteborn, H.P.J.M., Bienenmann Ploum, M.E., Alink, G.M., Zolla, L., Reynaerts, A., Pensa, M., Kuiper, H.A. and Fenwick, G.R. (1996). Safety assessment of the *Bacillus thuringiensis* insecticidal crystal protein CRYIA(b) expressed in transgenic tomatoes. In Hedley, C., Richards, R.L. and Khokhar, S. (Eds.), *Agri Food Quality: An Interdisciplinary Approach*. Special Publication No. 179. Cambridge, UK: Royal Society of Chemistry, 23–26.

Noteborn, H.P.J.M., Bienenmann Ploum, M.E., van den Berg, J.H.J., Alink, G.M., Zolla, L., Reynaerts, A., Pensa, M. and Kuiper, H.A. (1995). Safety assessment of the *Bacillus thuringiensis* insecticidal crystal protein CRYIA(b) expressed in transgenic tomatoes. In *Genetically Modified Foods: Safety Issues*. Washington D.C., USA: American Chemical Society, 134–147.

Noteborn, H.P.J.M., Lommen, A., van der Jagt, R.C. and Wiseman, J.M. (2000). Chemical fingerprinting for the evaluation of unintended secondary metabolic changes in transgenic food crops. *Journal of Biotechnology* 77: 103–114.

Novak, W.K. and Haslberger, A.G. (2000). Substantial equivalence of anti-nutrients and inherent plant toxin in genetically modified novel foods. *Foods and Chemical Toxicology* 38: 473–483.

Pedersen, J., Eriksen, F.D. and Knudsen, I. (2001). Toxicity and food safety of genetically engineered crops. In *Safety of Genetically Engineered Crops*. VIB. Zwisnearde, Belgium: Flanders Institute of Biotechnology, 27–59.

Perlak, F.J., Oppenhuizen, M., Gustafson, K., Voth, R., Sivasubramaniam, S., Heering, D., Carey, B., Ihrig, R.A. and Roberts, J.K. (2001). Development and commercial use of Bollgard(R) cotton in the USA: Early promises versus today's reality. *Plant Journal* 27: 489–501.

Pilcher, C.D. and Rice, M.E. (1998). Management of European corn borer (Lepidoptera: Crambidae) and corn rootworms (Coleoptera: Chrysomelidae) with transgenic corn: A survey of farmer perceptions. *American Entomologist* 44: 36–44.

Pray, C., Ma, D., Huang, J. and Qiao, F. (2001). Impact of *Bt* cotton in China. *World Development* 29: 813–825.

Pusztai, A. (2002). Can science give us the tools for recognizing possible health risks of GM food? *Nutrition and Health* 16: 73–84.

Pusztai, A., Ewen, S.W.B., Grant, G., Brown, D.S., Stewart, J.C., Peumans, W.J., Damme, E.J.M. and Bardocz, S. (1993). Antinutritive effects of wheat-germ agglutinin and other N-acetylglucosamine-specific lectins. *British Journal of Nutrition* 70: 313–321.

Pusztai, A., Grant, G., Brown, D.J., Stewart, J.C., Bardocz, S., Ewen, S.W.B., Gatehouse, A.M.R. and Hilder, V. (1992). Nutritional evaluation of the trypsin (EC 3.4.21.4) inhibitor from cowpea (*Vigna unguiculata* Walp.). *British Journal of Nutrition* 68: 783–791.

Pusztai, A., Koninkx, J., Hendriks, H., Kok, W., Hulscher, S., Damme, E.J.M. van, Peumans, W.J., Grant, G. and Bardocz, S. (1996). Effect of the insecticidal *Galanthus nivalis* agglutinin on metabolism and the activities of brush border enzymes in the rat small intestine. *Journal of Nutrition and Biochemical Science* 7: 677–682.

Rogan, G.J., Bookout, J.T., Duncan, D.R., Fuchs, R.L., Lavrik, P.B., Love, S.L., Mueth, M., Olson, T., Owens, E.D., Raymond, P.J. and Zalewski, J. (2000). Compositional analysis of tubers from insect and virus resistant potato plants. *Journal of Agricultural and Food Chemistry* 48: 5936–5945.

Sampson, H.A. (1997). Immediate reactions to foods in infants and children. In Metcalfe, D.D., Sampson, H.A. and Simon, R.A. (Eds.), *Food Allergy: Adverse Reactions to Food and Food Additives*. Cambridge, UK: Blackwell Science, 169–182.

Schubbert, R., Lettmann, C. and Doerfler, W. (1994). Ingested foreign phage M13 DNA survives transiently in the gastrointestinal tract and enters the bloodstream of mice. *Molecular and General Genetics* 242: 495–504.

Schubbert, R., Renz, D., Schmitz, B. and Doerfler, W. (1996). Foreign M13 DNA ingested by mice reaches peripheral lymphocytes, spleen and liver via the intestinal wall mucosa and can be covalently linked to mouse DNA. *Proceedings National Academy of Sciences USA* 94: 961–966.

Serageldin, I. (1999). Biotechnology and food security in the 21st Century. *Science* 387–389.

Sharma, H.C., Sharma, K.K, and Crouch, J.H. (2004). Genetic transformation of crops for insect resistance: Potential and limitations. *CRC Critical Reviews in Plant Sciences* 23: 47–72.

Siegel, J.P. (2001). A review of the mammalian toxicity of *Bacillus thuringiensis* var. *israelensis* for mammals. *Journal of Invertebrate Pathology* 77: 13–21.

Silva, J. and Kidwell, M. (2000). Horizontal transfer and selection in the evolution of P elements. *Molecular Biology Evolution* 17: 1542–1557.

Sims, S.R. and Berberich, S.A. (1996). *Bacillus thuringiensis* Cry IA protein levels in raw and processed seed on transgenic cotton: determination using insect bioassay and ELISA. *Journal of Economic Entomology* 89: 247–251.

Sims, S.R. and Holden, L.R. (1996). Insect bioassay for determining soil degradation of *Bacillus thuringiensis* subsp. *kurstaki* CryIA(b) protein in corn tissue. *Environmental Entomology* 25: 659–664.

Sinagawa Garcia, S.R., Rascon Cruz, Q., Valdez Ortiz, A., Medina Godoy, S., Gutierrez, E. and Paredes Lopez, O. (2004). Safety assessment by *in vitro* digestibility and allergenicity of genetically modified maize with an amaranth 11S globulin. *Journal of Agricultural and Food Chemistry* 52: 2709–2714.

Stanhope, M.J., Lupas, A., Italia, M.J., Koretke, K.K., Volker, C. and Brown, J.R. (2001). Phylogenetic analyses do not support horizontal gene transfers from bacteria to vertebrates. *Nature* 411: 940–944.

Tiedje, J.M., Colwell, R.K., Grossman, W.L., Hodson, R.E., Lenski, R.E., Mack, R.N. and Regal, P.J. (1989). The planned introduction of genetically modified organisms: Ecological considerations and recommendations. *Ecology* 70: 298–315.

Tzotzos, T. (Ed.). (1995). *Genetically Modified Microorganisms: A Guide to Biosafety.* Wallingford, Oxon, UK: Commonwealth Agricultural Bureau, International.

UNIDO (United Nations Industrial Development Organization). (1991). *Voluntary Code of Conduct for the Release of Organisms into the Environment.* July 1991. Vienna, Austria: United Nations Industrial Development Organization.

Van Gelder, W.M.J., Vinke, J.H. and Scheffer, J.J.C. (1988). Steroidal glycoalkaloids in tubers and leaves of solanum species used in potato breeding. *Euphytica* 38: 147–158.

Williamson, M., Perrins, J. and Fitter, A. (1990). Releasing genetically engineered plants. Present proposal and possible hazards. *Trends in Ecology and Evolution* 5: 417–419.

Worobey, M. and Holmes, E.C. (1999). Evolutionary aspects of recombination in RNA viruses. *Journal of General Virology* 80: 2535–2544.

Zdunczyk, Z. (2001). *In vivo* experiments on the safety evaluation of GM components of feeds and foods. *Journal of Animal and Feed Sciences* 10: 195–210.

16

Detection and Monitoring of Food and Food Products Derived from Genetically Modified Crops

Introduction

The global area under genetically modified crops has increased from 1.7 million hectares in 1996 to over 100 million hectares in 2006, and nearly one-third of the area under transgenic crops was in developing countries (James, 2007). The major transgenic crops include soybean (60%), corn (23%), cotton (12%), canola (5%), and potato (~1%). The major traits include herbicide tolerance (71%), insect resistance (28%), and quality traits (1%). More than 8 million farmers have benefited from this technology, and 90% of the beneficiaries are the resource-poor farmers in developing countries. Although considerable progress has been achieved in the development of analytical methods for detection of genetically modified food based on polymerase chain reaction (PCR), several other analytical technologies, including mass spectrometry, chromatography, near infrared spectroscopy, micro fabricated devices, and DNA chip technology (microarrays), can also be used for monitoring and detection of genetically modified food. So far, only PCR-based methods have been accepted for regulatory purposes. Monitoring and detection of genetically modified crops consists of three distinct steps, detection, identification, and quantification.

Detection: To determine whether a product has genetically modified plant material or not, a general screening method can be used, which could be positive or negative. The screening methods are usually based on the PCR, immunoassays, or bioassays. Analytical methods for detection must be sensitive and reliable enough to obtain accurate and precise results.

Identification: To find out which genetically modified crops or transgenes are present in food, and whether they are authorized or not in a particular country. DNA-based methods or immunoassays can be used for this purpose.

Quantification: If a crop or its product has been shown to contain genetically modified material, then it becomes necessary to assess the amount of each genetically modified variety or the transgene product present. Normally, quantification is performed using real-time PCR.

Sampling for Detection of Genetically Modified Food

One of the major considerations in analytical testing of almost any transgenic crop or its product is the sampling procedure. The sample analyzed must be representative of the material from which the sample is derived; otherwise, the testing regime becomes flawed. Sample preparation for both DNA-based and protein-based methods is critical for detection and/or quantification. It is important to know the limitations of each procedure, as well as the purpose of detection. Both the sample size and sampling procedures dramatically impact the conclusions that may be drawn from any of the testing methods. The first step is to collect a sample for analysis, which should be representative of the materials to be analyzed. The sample should then be homogenized for isolation and purification of RNA, DNA, or protein. The quality, purity, and amount of DNA, RNA, or protein are important for quantification and detection. Variance associated with sampling is the most likely factor to influence estimates of the proportion of genetically modified material present in food and food products. The sampling strategies should take into account the nature of the analyte/foodstuff, and distribution of the analyte in the bulk food. The main parameters to be taken into account should include lot size and uniformity, accepted testing methods, and the preparation of the sample prior to analysis (Paoletti et al., 2003).

Detection of Genetically Modified Foods

Foods derived from genetically modified plants are now appearing in the market and many more are likely to emerge in the future. The safety, regulation, and labeling of these foods are still contentious issues in most countries and recent surveys highlight consumer concerns about the safety and labeling of genetically modified foods. In most countries, it is necessary to obtain approval for the use of genetically manipulated organisms in food production (MacCormick et al., 1998). In order to implement the regulations governing the production and marketing of food derived from genetically modified organisms, there is a need to develop and standardize technology to detect such foods. In addition, a requirement to label approved genetically modified food would necessitate a constant monitoring system. One solution is to "tag" approved genetically modified food from transgenic plants with some form of biological or genetic marker, permitting the surveillance of foods for the presence of approved products of genetic engineering. While genetically modified transgenic plants that are not approved would not be detected by such surveillance, they might be detected by a screen for DNA sequences common to all or most modified transgenic plants.

Identification of vector sequences, plant transcription terminators, and marker genes by PCR and hybridization techniques can be used to detect the presence of genetically modified plant material in food and food products (Table 16.1). Detection methods can be based on detection of DNA, RNA, or the transgene product associated with the genetic modification. The majority of the methods developed for detection of genetically modified products are based on DNA, while a few are based on RNA. The DNA can be purified in billions of copies while the multiplication of RNA or the protein is more complicated and a slow process. The RNA and the protein molecules are also less stable. The relationship between quantity of genetically modified foods and DNA is linear if genetically modified DNA is nuclear, but there is no such correlation between the quantity of genetically modified food and the RNA or protein (Holst-Jensen, 2006). The most commonly used methods are gel electrophoresis and hybridization techniques that allow the size and amount of DNA to be estimated. This can be coupled with digestion of DNA with restriction enzymes that are known to cut a PCR fragment into segments of specified sizes. Determination of melting point profiles, which is characteristic of a specific DNA sequence, can also be used for detection of food derived from genetically modified crops. Another alternative is to use short synthetic molecules called probes (similar to, but smaller than primers) and allow

TABLE 16.1

Methods for Detection of Genetically Modified Food and Food Products

Method	Target Sequence	References
Plant-derived DNA	Chloroplast tRNA^{Leu} gene (*trnL*) intron	Taberlet et al. (1991)
Specific plant species	Maize single copy invertase (*Ivr*) gene	Ehlers et al. (1997)
	Soybean single copy lectin gene	Meyer et al. (1996)
	Tomato single copy polygalacturonase gene	Busch et al. (1999)
Gene-specific methods	Cauliflower mosaic virus promoter (*P-35S*)	Pietsch et al. (1997)
	Nopaline synthase terminator (*T-Nos*)	Pietsch et al. (1997)
	Bar (phosphoinothricin acetyltransferase) gene	Ehlers et al. (1997)
	Synthetic *cry1Ab* gene	Ehlers et al. (1997), Vaitilingom et al. (1999)
Construct-specific methods	*Bt*11 maize: junction alcohol dehydrogenase 1S intron *IVS6* (enhancer)—*cry1Ab* gene	Matsuoka et al. (2001)
	*Bt*176 maize: junction *CDPK* (calcium dependent protein-kinase) promoter—synthetic *cry1Ab* gene	Hupfer et al. (1998)
	GA21 maize: OTP (enhancer)—*epsps* gene (RoundupReady tolerance)	Matsuoka et al. (2001)
	Mon810 maize: junction *P-35S*—heat shock protein (*hsp*) 70 intron I (enhancer)	Zimmermann et al. (1998)
	Mon810 maize: junction *hsp 70* intron—*cry1Ab* gene	Matsuoka et al. (2001)
	RoundupReady®: junction *P-35S*—*Petunia hybrida* CTP (chloroplast transit peptide)	Wurz and Willmund (1997)
	T25 maize: junction *pat* (phosphoinothricin acetyltransferase) gene—*T-35S*	Matsuoka et al. (2001)
	Zeneca tomato: junction *T-Nos*-truncated tomato polygalacturonase gene	Busch et al. (1999)
Event-specific methods	*Bt*11 maize: junction host plant genome—integrated recombinant DNA	Zimmermann, Luthy, and Pauli (2000)
	Roundup Ready® soybean: junction host plant genome—integrated recombinant DNA	Berdal and Holst-Jensen (2006), Taverniers et al. (2001), Terry and Harris (2001)

these to bind to DNA or RNA. If designed properly, these can be used to discriminate between the right molecule (sequence) and any other DNA/RNA (Holst-Jensen, 2006). Labeling of molecules with fluorescence, radioactivity, antibodies, or dyes can also be used to facilitate the detection of genetically modified foods.

DNA-Based Methods

The DNA-based methods involve detection of the specific genes or DNA genetically engineered into the crop. Although, there are several DNA-based methodologies, the most commercial testing is conducted using PCR technology. The PCR technique is based on multiplying a specific target DNA, allowing the million- or billion-fold amplification by two synthetic oligonucleotide primers. The process consists of extraction and purification of DNA, amplification of the inserted DNA by PCR, and confirmation of the amplified PCR product. In principle, PCR can detect a single target molecule in a complex DNA mixture. The first step involves separation of the two strands of the original DNA molecule. The first primer matches the start of the coding strand of the DNA, while the second primer matches the end and the noncoding strand of the DNA to be multiplied. The second step involves binding of the two primers to their oligonucleotide primers. The third step involves making two perfect copies of the original double-stranded DNA molecule by adding the right nucleotides to the end of each primer, using the strands as templates. Once the cycle is completed, it can be repeated, and in each cycle, the number of copies is doubled, resulting in an exponential amplification. After 20 cycles, the copy number is 1 million times higher than at the beginning of the first cycle. The amplified fragment can be detected by gel electrophoresis or hybridization techniques.

Qualitative PCR Analysis

For general screening purposes, PCR-based methods have been widely used for the detection of genetically modified crops (Holst-Jensen et al., 2003; Holst-Jensen, 2006). The focus should be on target sequences that are characteristic for the group to be screened. Genetic control elements such as the cauliflower mosaic virus 35S promoter and the *Agrobacterium tumefaciens* (Townsend) *nos* terminator (*nos3'*) are present in many genetically modified crops. The PCR detects the presence of the genetically modified crop, which then needs to be identified. Primer selection has to be based on target sequences that are characteristic of individual transgenic crops. The junction sequences between two adjoining DNA segments can be the target for specific detection of the genetic construct such as cross-border regions between integration site and transformed genetic element of a specific transgenic variety or specific sequence alterations. The junction sequences in the integration site (plant-construct junction fragment) can be used to detect a specific transformation event. When the same gene construct is used to produce different transgenic crops, this will be the only strategy to distinguish between crops containing the same gene construct. The scheme to test for the presence of transgenic plants includes initial screening of samples for species-specific DNA, known as housekeeping genes, for example, lectin in soybean or invertase gene (*ivrl*) in maize to determine whether DNA from that species can be detected. If DNA is detectable, samples are then screened using the general genetic elements for the detection of transgenic crops. Positive indication from initial screening can be followed with screens for specific genes or constructs used in the most common transgenic crops, followed by identification tests depending on the DNA sample (e.g., *cry* genes, *EPSPS* gene, *Pat* gene, etc.), or more ideally, for the plant-construct junction fragments.

Hupfer et al. (2000) developed a method for quantification of transgenic insect-resistant *Bt* maize by single- and dual-competitive PCR. The analysis of mixtures of DNA solutions, as well as of maize flours containing defined amounts of *Bt* maize, demonstrated the usefulness of single-competitive PCR based on co-amplification of the *CDPK* promoter/*cryIAb* gene region of *Bt* maize and an internal standard. Upon heat treatment of DNA solutions and maize flour, the recovery of the *Bt* proportion compared to the starting material determined by single-competitive PCR decreased significantly. This systematic error could be compensated for by using a dual-competitive approach based on PCR quantification of the transgenic target sequence of *Bt* maize and of the maize specific invertase gene.

Multiplex PCR-Based Detection Methods

Several target DNA sequences can be screened with multiplex PCR-based methods, and detected in a single reaction. In the multiplex method, fewer reactions are needed to test a sample for potential presence of transgenic plant-derived DNA. Development of multiplex assays requires careful testing and validation. After the PCR, the resulting pool of amplified fragments needs to be analyzed further to distinguish between the amplicons. Multiplex assay for detection of five transgenic maize varieties (*Bt*11, *Bt*176, Mon810, T25, and GA21) has been well standardized (Matsuoka et al., 2001).

Quantitative PCR

In principle, PCR-based quantification can be performed either after completion of the PCR (end-point analysis), or during the PCR (real-time analysis) (Table 16.2). Conventional PCR measures the products of the PCR reaction at the end point in the reaction profile. End-point analyses are based on comparison of the final amount of amplified DNA of two DNA targets, the one to be quantified and a competitor (an artificially constructed DNA that is added in a small and known quantity prior to the PCR amplification, which is co-amplified with the target to be quantified). The competitor has the same binding sites for the same primer pair, but is of a different size. This is called competitive quantitative PCR, and the two DNA targets are amplified with equal efficiency. A series of dilutions

TABLE 16.2

Methods for Quantification of Derivatives in Genetically Modified Food and Food Products

Method	Genetically Modified Material	References
Double competitive PCR methods	Roundup Ready soybean (raw materials)	Van den Eede et al. (2000)
	*Bt*176 maize (raw materials)	Van den Eede et al. (2000)
	P-35S (screening)	Hardegger, Brodmann, and Herrmann (1999)
	T-Nos (screening)	Hardegger, Brodmann, and Herrmann (1999)
	Epsps gene (RoundupReady)	Studer et al. (1998)
	Synthetic *cry1Ab* gene	Studer et al. (1998)
	*Bt*11 maize (event specific)	Zimmermann, Luthy, and Pauli (2000)
Real-time PCR methods	Roundup Ready soybean (event specific)	Berdal and Holst-Jensen (2001), Terry and Harris (2001), Taverniers et al. (2001)
	Synthetic *cry1Ab* gene	Vaitilingom et al. (1999)

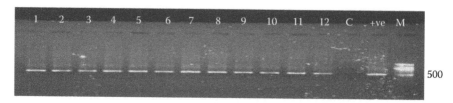

FIGURE 16.1   RT-PCR analysis of genetically engineered pigeonpea plants for *cry1Ac* gene. Lanes 1 to 12 are pigeonpea transgenic events with *cry1Ac* gene, lane C is a negative control, lane positive is the plasmid DNA having the *cry1Ac* gene, and lane M is the 1 kb ladder as marker.

of the DNA to be analyzed is prepared, and a constant amount of the competitor added. After completion of the PCR, the resulting amplification products are visualized through gel electrophoresis. When both DNA targets yield the same amount of product, it is assumed that the starting amount was also the same. By setting up two competitive PCRs, one for the transgenic crop (*Bt* maize) and one for the species of interest (nontransgenic maize), and including competitors in both, the quantity of transgenic crop relative to the species can be estimated by extrapolation from the degree of dilution and concentration of the competitors. The competitive PCR methods are semiquantitative.

Real-time PCR is based on continuous monitoring of PCR product. This is done via fluorometric measurement of an internal probe during the reaction. In real-time analyses, the amount of product synthesized during PCR is estimated directly by measuring fluorescence in the PCR reaction. Several types of hybridization probes are available that emit fluorescent light corresponding to the amount of synthesized DNA. The quantitative estimation is based on extrapolation by comparing the transgenic crop sequence relative to the reference, for example, gene sequence *cry1Ac* (Figure 16.1) and *ivr1* gene from maize. With the use of fluorescence, it becomes possible to measure exactly the number of cycles that are needed to produce a certain amount of PCR product. Since one cycle corresponds to doubling the amount of product, a simple formula can be used to estimate the ratio. While real-time PCR requires more sophisticated and expensive equipment than competitive PCR, it is faster, automated, and more specific. Presently, real-time PCR can be considered as the most powerful tool for the detection and quantification of food derived from transgenic crops.

Detection systems for monitoring the presence of genetically modified food could also be based on the presence of the 35S promotor, which has been used in many insect protected crops. However, the detection limit based on 35S has been found to vary by a factor of 20 in different laboratories, and is not suited to distinguish between genetically modified food mixers and co-mingling of produce during harvest, transport, and storage. It allows the detection of 35S promotor in the range of 0.01% to 0.1%. The fixing of a limit of 1% as the basis for labeling food as genetically modified has necessitated the use of quantitative PCR (Hardegger, Brodmann, and Herrmann, 1999), and real-time PCR (Heid et al., 1996). Hubner, Studer, and Luthy (1999) suggested that QC-PCR could be used for survey of threshold values of 1% for the labeling of genetically modified food.

Microarray Technology

In this technique, many selected probes are bound in an array format to a solid surface, with each spot containing numerous copies of the probe. The array is then hybridized with the isolated DNA sample of interest labeled with a fluorescent marker. During the hybridization phase, the labeled fragments associate with the spotted probes on the basis of complementary sequences. The larger the number of complementary sequences, the stronger the bond will be. After the hybridization phase, the free labeled sequences and weakly bound probes are washed off, and the array can be scanned for fluorescence intensity of each spot. Analysis of the resulting patterns and intensity gives an idea of the presence of genetically modified food. It can also be used for protein-based methods. Promising new developments in this area are the development of an electroarray system, where the fluorescently labeled negative DNA fragments are guided to individual spots that are positively charged in order to increase the rate of hybridization events (Nanogen, 2003). Systems that increase the surface where hybridization may occur by using three-dimensional spot structures such as gel-based chips have also been developed (Miraglia et al., 2004).

RNA-Based Methods

The RNA-based methods rely on specific binding between RNA molecule and the synthetic RNA molecule called a primer. The primer should be complementary to the nucleotide sequence at the start of the RNA molecule. The binding between the RNA molecule and the primer is followed by conversion of RNA to a DNA molecule through a process called reverse transcription. Finally, the DNA can be multiplied by PCR or translated into over 100 copies of the original RNA molecule, and the procedure repeated by using each copy as a template using the NASBA (nucleic acid sequence-based amplification) technique. The specific primers needed for the process can be developed without prior knowledge of the composition of the RNA molecule to be detected (Holst-Jensen, 2006).

Protein-Based Methods

Immunoassay can also be used for detection and quantification of foreign proteins introduced through genetic transformation of plants. Immunoassay is based on the specific binding between an antigen and an antibody. Thus, the availability of antibodies with the desired affinity and specificity is the most important factor for setting up an immunoassay. Immunoassays can be used qualitatively or quantitatively over a wide range of concentrations. Western blot, enzyme-linked immunosorbent assay (ELISA), and lateral flow sticks are typical protein-based methods. The antibodies can be polyclonal (raised in animals) or monoclonal (produced by cell cultures). Commercially available polyclonal antiserum is often produced in rabbits, goats, or sheep. Monoclonal antibodies offer some advantages over polyclonal antibodies because they express uniform affinity and specificity against a single epitope or antigenic determinant and can be produced in vast quantities. Both polyclonal and monoclonal antibodies may require further purification to enhance sensitivity. The specificity of the antibodies must be checked carefully to elucidate any cross-reactivity with similar substances, which might cause false positive results.

Enzyme-Linked Immunosorbent Assay

In ELISA, the antigen–antibody reaction takes place on a solid phase (microtiter plates). Antigen and antibody react and produce a stable complex, which can be visualized by the addition of a second antibody linked to an enzyme. Addition of a substrate for that enzyme results in color formation, which can be measured photometrically or recognized by the naked eye. ELISA test kits provide semiquantitative results in a short time with detection limits less than 0.1%. ELISA has been designed to detect a novel transgene protein or trait.

An antibody-sandwich ELISA has been developed to detect and quantify *Bt* insecticidal protein expressed in transgenic insect-resistant cotton, using a purified *Bt* toxin (67 kDa) as the standard protein and antigen (Chen et al., 1999). The minimal detectable concentration of purified *Bt* toxin was 1 to 15 g L^{-1}. The mean recovery rate was 94.94%. Quality control experiments indicate that the ELISA method is stable and reliable, and is characterized by its strong specificity and high sensitivity. A sandwich-type ELISA method has also been developed for Cry1Ab in maize (Walschus, Witt, and Wittmann, 2002). Monoclonal antibodies were developed by immunization of mice. The secondary antibody was labeled with horseradish peroxidase and 3,3',5,5'-tetramethylbenzidine (TMB) used as a chromogene. The detection limit was 0.4 ng mL^{-1} of Cry1Ab. There was no Cry1Ac, Cry1C, Cry2A, and Cry3A, or little Cry9C cross-reactivity with other Cry proteins. This assay has been applied to protein extracts from maize powder containing different amounts of *Bt* maize from three different lines (*Bt* 176, *Bt* 11, and MON 810). Using 100 mg maize powder and Tris-borate buffer for rapid protein extraction, the minimum detection limit was 0.10% for *Bt* 11 and 0.23% for MON 810. In contrast, *Bt*176 maize could not be detected due to the fact that the Cry1Ab gene is not expressed in the seed tissue of *Bt* 176 maize.

Lateral Flow Sticks

The lateral flow test (dipstick format) uses a membrane-based detection system. The membrane contains two capture zones, one captures the bound transgene protein, while the other captures color reagent. Paper strips or plastic paddles can be used as support for capturing the antibody immobilized onto a test strip in the specific zone. The lateral flow test strip is dipped in the sample in extraction solution, and the sample migrates up the strip by capillary action. As the sample flows through the detection antibody strip and the capture antibody strip, the protein of interest will accumulate and give a high intensity band. These tests generally provide qualitative or semiquantitative results using antibodies and color reagents incorporated into a lateral flow strip. By following appropriate sampling procedures, it is possible to obtain a 99% confidence level of less than 0.15% transgenic material.

Phenotypic Characterization

Phenotypic characterization allows detection of the presence or absence of a specific trait. Such methods can be used to test for presence or absence of herbicide-resistant transgenic varieties. They consist of conducting germination tests on solid germination media in the presence of a specific herbicide, where the nontransgenic and transgenic seeds show distinct characteristics. The detection level depends on the efficiency of seed germination. Seeds testing positive should be exposed to subsequent tests for confirmation. The herbicide bioassay tests are accurate, inexpensive, and useful as preliminary tests. Negative and positive trait seeds should be included as controls with every test. In the future, bioassays for insect-resistant or other transgenic crops may also be developed.

Mass Spectrometry

Spectrometrical methods (matrix-assisted laser desorption/ionization) can also be used for analysis of oligonucleotides. In this system, the analyte is embedded in a UV absorbing matrix in vacuum on a carrier between electrodes. When the UV light is applied, the UV energy is absorbed by the matrix, and also carried over to the sample such that it is ionized. The ionized ions move towards oppositely charged electrode, and enter the flight tube towards the detector. During this process, the molecules are separated based on mass and charge ratio. This technique has been used successfully in genotyping of single nucleotide polymorphism (SNP) of genes, and can also be used for detection of genetically modified food (Miraglia et al., 2004).

Surface Plasmon Resonance

Surface plasmon resonance (SPR) uses thin-layered metal films (plasmon) on a sensor chip, usually coated with gold, to which the molecules (protein, DNA, or oligonucleotides) are bound. The surface is rinsed with a fluid that contains a binding partner to the surface attached molecules. The interaction between the molecules is studied by polarized light, which is reflected from the plasmon surface. If the molecules from the fluid and the chip bind to each other, the reflected light intensity is reduced, and the size of change in the surface plasmon resonance signal is directly proportional to the mass. This system has been used for detection of herbicide-resistant soybean by immobilizing biotinylated PCR products of the target oligonucleotides on the chip and hybridizing them with respective probes (Feriotto et al., 2002). This approach of detecting genetically modified foods needs to be evaluated further. Electrochemiluminescence PCR (ECL-PCR) combined with the hybridization (probes labeled with biotin and TBR) technique has also been used for detecting genetically modified food (Liu et al., 2004). This method can rapidly detect genetically modified food with high sensitivity, over a wide range.

At the moment, only PCR offers a general screening for transgenic crops, and detection of particular "events." Phenotypic characterization and immunoassays detect particular traits that may be present in several transgenic crops (e.g., the Cry1a protein and genes conferring insect resistance are present in a number of maize cultivars).

Monitoring of Genetically Modified Food and Food Labeling

The current policies on authorization and labeling of products derived from genetically modified plants in different countries are not clear, and often hampers international trade. Therefore, there is a need to harmonize these regulations globally. Harmonization of national requirements for labeling of genetically modified products will lead to increase in transparency in international trade in food and food products. Harmonized data requirements are essential for internationally accepted analytical methods and reference materials and implementing current and future regulations governing genetically modified food. At the moment, the detection methods are DNA based in Europe, while the United States follows a protein-based system, and the two approaches may give different results depending on the matrix under analysis.

Despite lack of regulations to provide information on food and food products derived from genetically modified plants, many products carry positive or negative labels with regard to genetic modification. A maize product not labeled as genetically modified has been found with ELISA to contain the transgene product (Urbanek Karlowska et al., 2003). In another study involving 58 soybean and maize products in the food chain in South Africa, 44 tested positive for the presence of genetically modified plants. Of the 20 products labeled as genetically modified, only 14 tested positive, suggesting the need for effective regulations for production and labeling of food derived from genetically modified plants

(Viljoen, Dajee, and Botha, 2006). Detection of genetically modified food varies across different foods and food processing (Cazolla and Petrucelli, 2006). Based on PCR screening for *CaMV35S* and transgene specific primers (CRY and EPSPS), the quantification within the European Union limits was possible only in case of ice-cream, flours, soybean, and starch. Samples with high lipid content or subjected to intense thermal treatments, such as snacks, mayonnaise, and creamy soup, etc., could not be amplified mainly due to the presence of PCR inhibitors.

There are practical difficulties in labeling the food and food products derived from genetically modified plants when they are produced and consumed at the village level without processing and packaging, particularly in the developing countries. Because of the difficulties involved in monitoring and detection of genetically modified food and unregulated mixing, marketing, and consumption of food in the developing countries, it may not be practical to follow a harmonized system of leveling genetically modified food the world over. Therefore, it may be more pertinent to ensure the safety and quality of the genetically modified food during product development and de-regulation stages.

Conclusions

The need for identification and detection of transgenic crops and the food products derived from them has increased with the rapid expansion in the cultivation of transgenic crops over the past decade. Labeling and traceability of transgenic material is important to address the concerns of the consumer. Establishment of reliable and economical methods for detection, identification, and quantification of genetically modified food continues to be a challenge at the international level. It is important to know the limitations of each procedure as well as the purpose of detection. Both the sample size and sampling procedures dramatically impact the conclusions that may be drawn from any of the testing methods. Currently, available methods for detecting transgenic crops and their products are almost exclusively based on PCR, because of sensitivity, specificity, and the need for only a small amount of DNA. Real-time PCR has been regarded as a powerful tool for detection and quantification of transgenic material despite its high cost. There is a need to refine other methods for detection of genetically modified food that are economic, reliable, and sensitive to meet the future needs for monitoring food and food products for the presence of transgene or transgene products. In view of the difficulties involved in detection and monitoring of genetically modified food, it will be useful to ensure the safety and quality of food derived from genetically modified plants during the development and testing phases.

References

Berdal, K.G. and Holst-Jensen, A. (2001). RoundupReady® soybean event specific real-time quantitative PCR assay and estimation of the practical detection and quantification limits in GMO analyses. *European Food Research and Technology* 213: 432–438.

Busch, U., Muhlbauer, B., Schulze, M. and Zagon, J. (1999). Screening und spezifische Nachweismethode fur transgene Tomaten (Zeneca) mit der Polymeraskettenreaktion. *Deutsche Lebensmittelrundschau, Heft* 2: 52–56.

Cazzolla, M.L. and Petrucelli, S. (2006). Semiquantitative analysis of modified maize and soybean in food. *Electronic Journal of Biotechnology*. http://www.ejbiotechnology.info.content/vol9/issue3/index.htm.

Chen, S., Wu, J.Y., Cheng, D.R., Zhang, R.X. and Huang, J.Q. (1999). Study on the enzyme-linked immunosorbent assay of *Bacillus thuringiensis* insecticidal protein expressed in transgenic cotton. *Acta Gossypii Sinica* 11: 259–267.

Ehlers, B., Strauch, E., Goltz, M., Kubsch, D., Wagner, H., Maidhof, H., Bendiek, J., Appel, B. and Buhk, H.J. (1997). Nachweis genetechnischer Veranderungen in Mais mittels PCR. *Bundesgesundhblogy* 4: 118–121.

Feriotto, G., Borgatti, M., Bianchi, N. and Gambari, R. (2002). Biosensor technology and surface plasmon resonance for real-time detection of genetically modified round up ready soybean sequences. *Journal of Agricultural and Food Chemistry* 50: 955–962.

Hardegger, M., Brodmann, P. and Herrmann, A. (1999). Quantitative detection of the 35S promoter and the NOS terminator using quantitative competitive PCR. *European Food Research and Technology* 209: 83–87.

Heid, C.A., Stevens, J., Livak, K.J. and Williams, P.M. (1996). Real time quantitative PCR. *Genome Research* 6: 986–994.

Holst-Jensen, A. (2006). *GMO Detection Methods and Validation: DNA-Based Methods*. http://www.entransfood.nl/workinggroups/wg4TQA/GMO%20detection%20methods%20.htm.

Holst-Jensen, A., Ronning, S.B., Lovseth, A. and Berdal K.G. (2003). PCR technology for screening and quantification of genetically modified organisms (GMOs). *Analytical and Bioanalytical Chemistry* 375: 985–993.

Hubner, P., Studer, E. and Luthy, J. (1999). Quantitation of genetically modified organisms in food. *Nature Biotechnology* 17: 1137–1138.

Hupfer, C., Hotzel, H., Sachse, K. and Engel, K.H. (1998). Detection of the genetic modification in heat treated products of *Bt* maize by polymerase chain reaction. *Zeitschrift für Lebensmitteluntersuchung und Forschung* 206(A): 203–207.

Hupfer, C., Hotzel, H., Sachse, K., Moreano, F. and Engel, K.H. (2000). PCR-based quantification of genetically modified Bt maize: Single-competitive versus dual-competitive approach. *European Food Research and Technology* 212: 95–99.

James, C. (2007). *Global Status of Commercialized Biotech/GM Crops: 2006*. ISAAA Briefs no. 35. Ithaca, New York, USA: International Service for Acquisition on Agri-Biotech Applications (ISAAA). http://www.isaaa.org/resources/publications/briefs/35.

Liu, J., Xing, D., Shen, X. and Zhu, D. (2004). Detection of genetically modified organisms by electroluminescence PCR method. *Biosensors and Bioelectronics* 20: 436–441.

MacCormick, C.A., Griffin, H.G., Underwood, H.M. and Gasson, M.J. (1998). Common DNA sequences with potential for detection of genetically manipulated organisms in food. *Journal of Applied Microbiology* 84: 969–980.

Matsuoka, T., Kuribara, H., Akiyama, M.H., Goda, Y., Kusakabe, Y., Isshiki, K., Toyoda, M. and Hino, A. (2001). A multiplex PCR method of detecting recombinant DNAs from five lines of genetically modified maize. *Journal of Food Hygiene Society of Japan* 42: 24–32.

Meyer, R., Chardonnes, F., Hubner, P. and Luthy, J. (1996). Polymerase chain reaction (PCR) in the quality and safety assurance of food: Detection of soya in processed meat products. *Zeitschrift für Lebensmitteluntersuchung und Forschung* 203: 339–344.

Miraglia, M., Berdal, K.G., Brera, C., Corbisier, P., Holst-Jensen, C., Kok, E.J., Marvin, H.J.P., Schimmel, J., van Rie, J.P.P.F. and Zagon, J. (2004). Detection and traceability of genetically modified organisms in the food production chain. *Food Chemistry and Toxicology* 42: 1157–1180.

Nanogen. (2003). *Core Technology*. San Diego, California, USA: Nanogen. http://www.nanogen.com/technology/core_technology.htm.

Paoletti, C., Donatelli, M., Kay, S. and Van den Eede, G. (2003). Simulating kernel lot sampling: The effect of heterogeneity on the detection of GMO contamination. *Seed Science and Technology* 31: 629–638.

Pietsch, K., Waiblinger, H.U., Brodmann, P. and Wurz, A. (1997). Screeningverfahren zur Identifizierung gentechnisch veranderter pflanzlicher Lebensmittel. *Deutsche Lebensmittel-Rundschau* 93: 35–38.

Studer, E., Rhyner, C., Luthy, J. and Hubner, P. (1998). Quantitative competitive PCR for the detection of genetically modified soybean and maize. *Zeitschrift für Lebensmitteluntersuchung und Forschung* 207A: 207–213.

Taberlet, P., Gielly, L., Pautou, G. and Bouvet, J. (1991). Universal primers for amplification of three non-coding regions of chloroplast DNA. *Plant Molecular Biology* 17: 1105–1109.

Taverniers, I., Wiendels, P., Van Bockstaele, E. and De Loose, M. (2001). Use of cloned DNA fragments for event specific quantification of genetically modified organisms in pure and mixed food products. *European Food Research and Technology* 213: 417–424.

Terry, C. and Harris, N. (2001). Event specific detection of RoundupReady soya using two different real time PCR detection systems. *European Food Research and Technology* 213: 425–431.

Urbanek Karlowska, B., Sawilska Rautenstrauch, D., Jedra, M. and Badowski, P. (2003). Detection of genetic modification in maize and maize products by ELISA test. *Roczniki Panstwowego Zakladu Higieny* 54: 345–353.

Vaitilingom, M., Pijnenburg, H., Gendre, F. and Brignon, P. (1999). Real-time quantitative PCR detection of genetically modified maximizer maize and RoundupRedy soybean in some representative foods. *Journal of Agriculture and Food Chemistry* 47: 5261–5266.

Van den Eede, G., Lipp, M., Eyquem, F. and Enklam, E. (2000). *Validation of a Double Competitive Polymerase Chain Reaction Method for the Quantification of GMOs in Raw Materials*. IHCP Eur 19676. Ispra, Italy: European Commission, Joint Research Centre.

Viljoen, C.D., Dajee, B.K. and Botha, G.M. (2006). Detection of GMO in food products in South Africa: Implications of GMO labeling. *African Journal of Biotechnology* 20: 73–82.

Walschus, U., Witt, S. and Wittmann, C. (2002). Development of monoclonal antibodies against Cry1Ab protein from *Bacillus thuringiensis* and their application in an ELISA for detection of transgenic Bt-maize. *Food and Agricultural Immunology* 14: 231–240.

Wurz, A. and Willmund, R. (1997). Identification of transgenic glyphosate-resistance soybeans. In Schreiber, G.A. and Vogel, K.W. (Eds.), *Foods Produced by Genetic Engineering*. 2nd Status Report. Berlin, Germany: Bundesinstitut fur gesundheitlichen Verbraucherschutz und Vetrinarmedizin, 115–117.

Zimmermann, A., Liniger, M., Luthy, J. and Pauli, U. (1998). A sensitive detection method for genetically modified MaisGard™ corn using a nested PCR-system. *Lebensmittel Wissenschaft und Technologie* 31: 664–667.

Zimmermann, A., Luthy, J. and Pauli, U. (2000). Event specific transgene detection in Bt11 corn by quantitative PCR at the integration site. *Lebensmittel Wissenschaft und Technologie* 33: 210–216.

17

Molecular Markers for Diagnosis of Insect Pests and Their Natural Enemies

Introduction

Many insect and mite species possess an astounding potential to cause damage in field crops and storage. For developing appropriate management strategies, it is important to have a correct identification of the pest species. Correct taxonomic identification is also important for import and export of plant material and food grains to implement appropriate quarantine procedures. Quarantine measures are critical to prevent or delay the introduction of exotic pests into newer areas. Identification of insect pests of quarantine importance has primarily relied on morphological characters of adult life stages. However, intercepted specimens often are not in the adult stage and may be damaged, which seriously handicaps correct identification. The molecular tools now enable precise and rapid identification of insect pests irrespective of the developmental stage and condition of the samples. The modern tools of biotechnology can be used for detection and identification of insect pests, insect biotypes, and understand genetic diversity, population structure, tritrophic interactions, and insect plant relationships (Caterino, Cho, and Sperling, 2000; Heckel, 2003). Molecular markers can also be used to gain a basic understanding of insect metabolism, development, interaction with the environment, and for developing sound strategies for pest management. Molecular techniques have also been found to be useful for studying insect phylogeny and predator-prey relationships (Hillis, Martiz, and Noble, 1996; Crampton, Beard, and Louis, 1997; Osborne, Loxadale, and Woiwod, 2002; Hoy, 2003). Molecular tools also enable genetic characterization of specific attributes of an organism, and if combined with a high-throughput technology such as microarrays, they are ideally suited for rapid molecular screening. Molecular diagnostics will play a vital role in detection and identification of insect pests. Common molecular techniques that can be used for detection of insect pests include:

- Hydrocarbons for species recognition;
- Secondary metabolites;

- Isozyme and protein profiles;
- Immunodiagnostic methods; and
- DNA-based methods.

Molecular Tools for Diagnosis of Insect Pests

Polymerase Chain Reaction

Current molecular systematics depends on polymerase chain reaction (PCR) amplification of a few "universal" genes to provide phylogenetic data. However, the need for sequencing is increasing quite fast (Murphy et al., 2001; Wheeler et al., 2001; Philippe et al., 2004; Teeling et al., 2005), and expanding PCR approaches to a wider selection of genes becomes difficult because of the need to develop new degenerate primers for the amplification of single-copy loci.

Random Amplified Polymorphic DNA

Random amplified polymorphic DNA (RAPDs)-based DNA fingerprinting has been used for insect diagnosis that does not require prior knowledge of DNA sequence. RAPDs provide a rapid means of identifying genetic markers to distinguish closely related species (Alvarez and Hoy, 2002). These markers are inexpensive, rapid, and easy to use when studies involve many insects. However, results at times are difficult to reproduce (Ellsworth, Rittenhouse, and Honeycut, 1993).

Expressed Sequence Tags

The expressed sequence tags (ESTs) provide a more readily available resource of genomic data (Rudd, 2003). Most publicly available ESTs have been generated for gene discovery or to complement genome sequencing efforts. However, species for EST analyses have rarely been selected based on taxonomic criteria, which limits their use for phylogenetic analyses and comparative genomics. A concerted effort to enlarge EST databases to encompass disparate taxa may alleviate these problems (Bapteste et al., 2002; Theodorides et al., 2002). The EST databases for the taxa specifically selected to obtain comprehensive coverage of Coleoptera, a group that includes nearly one-third of all known species of animals, have also been generated. A critical problem for comparative studies is that ESTs from different taxa may not contain overlapping sets of genes. The challenge of matching orthologous genes between taxa is amplified by the low expression of many transcripts. Sequencing of tens of thousands of ESTs in *Drosophila melanogaster* Meigen (Rubin et al., 2000) and *Bombyx mori* L. (Mita et al., 2003) has fallen short of reaching a full complement of predicted genes. The EST databases are growing rapidly, with approximately 27.6 million entries in the GenBank as of June 2005 (http://www.ncbi.nlm.nih.gov/dbEST/). However, taxonomic coverage of the Class Insecta has been limited to eight of the 25 insect orders. The EST representation in insects has been severely biased towards Diptera, comprising 15 of 47 holometabolous insects.

Restriction Fragment Length Polymorphisms

The restriction fragment length polymorphisms (RFLP) probes allow very fine mapping of loci, and have several advantages over morphological markers. These include the ability to behave in a codominant manner and detect heterozygotes, whereas the morphological markers do not detect heterozygotes. The RFLP markers detect greater allelic variation in natural populations than morphological markers and are unaffected by environmental effects. The disadvantages of RFLP linkage analysis include the additional time required to complete an analysis (7 to 10 days) and the use of radioactive isotopes.

Inter-Simple Sequence Repeats

The inter-simple sequence repeats (ISSRs) have shown good promise for studying the population biology of plants. The ISSRs can be advantageous when time and material costs preclude development of more robust markers such as locus-specific SSRs. ISSRs can reveal polymorphisms and elaborate detection protocols. These are a valuable addition to PCR-based markers in studies where genomic fingerprinting is appropriate for large-scale screening of genetic variation in animal populations (Abbot, Withgott, and Moran, 2001). ISSRs have been used to study variation in aphids, *Acyrthosiphon pisum* Harris and *Pemphigus obesinymphae* Moran, and yellow fever mosquito, *Aedes aegypti* (L.) (Abbot, 2001).

Application of Molecular Markers for Insect Diagnosis

Diagnosis of Insect Pests and Their Natural Enemies

The addition of molecular techniques to the taxonomist's arsenal has been one of the major advances for insect identification. It has shown a great potential to expedite the identification of cryptic species, unidentifiable life stages, and critical taxa such as disease vectors. This has led to development of diagnostic kits, enabling nonspecialists to discriminate rapidly between closely related species of insect pests, which are difficult to separate using conventional taxonomic approaches. It has also made it possible to discriminate between different isolates of microbial pathogens such as entomopathogenic fungi, which vary in virulence against the target insect pests, allowing monitoring of the spread and survival of released strains in the crop environment. Similarly, the spread of exotic or introduced strains of crop pests and diseases can also be monitored using these techniques.

Correct taxonomic identification is essential for effective pest management, particularly for identification of natural enemies. RAPDs, SSRs, and allozymes have been widely used for such studies. RAPDs have been used for distinguishing populations of coccinnellid beetles from different geographical areas (Roehrdanz and Flanders, 1993), while RFLP and mtDNA markers have been used as diagnostic markers for distinguishing insect populations or migration (Hall, 1998; Nielsen et al., 2000) (Table 17.1). RAPDs have also been used to differentiate between colonies of *Ageniaspis citricola* Logvin., which showed differences in life cycle and behavior (Hoy et al., 2000). Allozymes have been used for differentiating *Aphidius ervi* Haliday attacking pea aphid, *A. pisum*, and nettle aphids, *Microlophium carnosum* (Buckton) (Atanassova et al., 1998). Enzyme electrophoresis has been used to monitor parasitism of the aphid, *Sitobion avenae* (F.) (Walton, Loxdale, and Allen-Williams, 1990), while Black et al. (1992) used RAPDs to identify two endoparasitic wasps, *Diaeretiella rapae*

TABLE 17.1

PCR Primers Used for Detecting Parasitoids

Species	Family	Region	References
Ageniaspis citricola	Encyrtidae	Actin genes	Hoy et al. (2000)
Aphelinus abdominalis (Dalman)	Aphelinidae	Microsatellite	Vanlerberghe-Masutti and Chavigny (1997)
Aphelinus hordei		ITS-2	Zhu et al. (2000)
		ITS-2, 16s rDNA	Prinsloo et al. (2002)
Aphidius ervi		CO I-II, CO B	Daza-Bustamante et al. (2002)
		Microsatellite	Hufbauer, Bogdanowicz, and Perez (2001)
Cotesia congregata (Say)	Braconidae	Microsatellite	Jensen et al. (2002)
Diaeretiella rapae		Microsatellite	Loxdale and MacDonald (2004)
Lydella thompsoni	Tachinidae	CO I	Agusti et al. (2005)
Lysiphlebus testaceipes	Aphidiidae	ITS-2	Persad, Jayaprakash, and Hoy (2004)
Nasonia vitripennis (Walker)	Pteromalidae	ITS-1	Ratcliffe et al. (2002)
Peristenus digoneutis Loan	Pteromalidae	ITS-1, ITS-2	Erlandson et al. (2003)
Pseudoperichaeta nigrolineata	Tachinidae	CO I	Agusti et al. (2005)
Stomodys calcitrans (L.)	Pteromalidae	ITS-1	Ratcliffe et al. (2002)
Trichomalopsis sarcophagae (Gahan)	Pteromalidae	ITS-1	Ratcliffe et al. (2002)
Trichogramma australicum (Girault)	Trichogrammatidae	ITS-2	Amornsak, Gordh, and Graham (1998)
		ITS-1, ITS-2	Sappal et al. (1995)
Trichogramma deion Pinto and Oatman		ITS-2	Stouthamer et al. (2000)
Trichogramma dendrolimi Matsumura		ITS-2	Li and Shen (2002)
Trichogramma minutum Riley		ITS-1, ITS-2	Sappal et al. (1995)
Trichogramma ostriniae Pang et Chen		ITS-1	Chang et al. (2001)
Trichogramma pretiosum Riley		ITS-2	Stouthamer et al. (2000)
Trichogramma turkestanica Meyer		ITS-2	Silva et al. (1999)

McIntosh and *Lysiphlebius testaceips* (Cresson), within the bodies of their aphid hosts. Tilmon et al. (2000) used PCR-RFLP of mitochondrial *COI* gene to detect parasitism of mirid bugs, *Lygus lineolaris* (Palisot de Beauvois) by *Peristenus* spp. Species-specific *ITS2* and *16S* primers have been used to determine the establishment of the parasitoid, *Aphelinus hordei* Kurd. for the control of Russian wheat aphid, *Diuraphis noxia* (Mord.) (Prinsloo et al., 2002), and of the eggs of *L. lineolaris* by *Anaphes iole* Girault (Zhu and Williams, 2002).

Detection of Insect Biotypes

Molecular tools have been used for solving routine taxonomic and ecological problems regarding biotype or cryptic status of insect pests. Sequences of *ITS2* region have been used to separate the Australian and Taiwanese *Ageniaspis fuscicollis* (Dalman) populations (Alvarez and Hoy, 2002). Molecular markers have been found to be useful for characterizing strains to determine genetic origin and identity of the parasitoids for use in pest management.

They have also been found to be useful for detection of alien strains or species, which may be unadapted to hosts or genotypes present, and lead to failure of control operations (Hufbauer, Bogdanowicz, and Harrison, 2004). Allozymes, RFLP, RAPD, microsatellite, and mtDNA-based markers have been used for differentiating biotypes and sympatric species (Laroche et al., 1996; Hoy et al., 2000; Hufbauer, Bogdanowicz, and Harrison, 2004). The tephritid fruit flies, *Bactrocera philippinensis* (Drew and Hancock) and *B. occipitalis* (Bezzi) are of significant importance as quarantine pests (Yu et al., 2005). Real-time qualitative PCR using SYBR green assay with melting curve analysis has been developed to identify these two symptomatic species, and can be used as a rapid detection technique in quarantine inspection.

Phylogenetics and Population Structure

The advent of molecular techniques to allow discrimination of species and identification of genetic polymorphism within and between populations has added a new dimension for understanding taxonomic relationships, evolution, and epidemiology of insect pests and their natural enemies. Molecular techniques are being used in ecological research to elucidate the structure of populations and to estimate gene flow between populations occupying discrete habitats or utilizing different host species. Such population studies are valuable for assessing the effects of habitat fragmentation, and may be useful to develop ecological modification strategies for maintaining biodiversity on farmland and promoting sustainable pest management. Assessing gene flow among populations in agricultural ecosystems will also aid risk assessment of the release of genetically modified organisms into the environment. Ribosomal DNA and mtDNA have been used widely in phylogenetic studies. Several studies have used molecular phylogenetic approaches to ascertain the phylogenetic position of Aphidinae and the relationship of the taxa within the family (Belshaw and Quicke, 1997; Downton and Austin, 1998; Smith et al., 1999). Simultaneous application of morphological and molecular data has also been used to resolve the relationship among the apocritan wasps (Dowton and Austin, 2001).

Studies on basal relationships in Coleoptera have been based on the mitochondrial *cox1* (Howland and Hewitt, 1995) and the nuclear small subunit rRNA genes (Caterino et al., 2002), but the use of a single locus in these cases has been found to be insufficient to resolve the main phylogenetic questions. Using EST-based approaches that do not rely on degenerate PCR would be of great advantage, and the utility of this approach has been tested by producing phylogenetic trees for the basal groups of Coleoptera from 66 genes coding for ribosomal proteins (RP). The use of nuclear genes as a source of phylogenetic data requires an appreciation of the complex nature of genome evolution, involving gene loss, duplications, expansion of gene families, and functional diversification. Assignment of gene orthology is difficult even between fairly closely related groups such as the dipteran, *Anopheles gambiae* (Giles) and *D. melanogaster*, where genes diversified independently in each lineage (Zdobnov et al., 2002). Increased taxon sampling can improve the confidence of orthology assignments by identifying the origin of gene copies, facilitating inferences on gene duplications, clarifying the relationship between gene content, and the diversity of lineages (Parkinson et al., 2004).

The ISS-PCR markers have been used to study geographic variation in population structure of the egg parasitoid, *Gonatocerus ashmeadi* Girault of the grassy winged sharp-shooter, *Homalodisca coagulata* (Say) (de Leon and Jones, 2005). The molecular data indicated restricted gene flow using six populations from the United States and Argentina. Ji et al. (2003) characterized five polymorphic loci in cotton bollworm, *Helicoverpa armigera* (Hubner), from two genomic DNA libraries in China. The expected heterozygocity of a novel set of five poly-morphic di- or trinucleotide microsatellite loci suitable for population genetic studies were developed from an enriched genomic library, and the cross amplifiability of these and other published loci was tested in a closely related species, *Helicoverpa assulta* Guen. (Ji, Wu, and Zhang, 2005). The expected heterozygocity at these loci ranged from 0.62 to 0.91, and the allele number varied from 4 to 12, and these can be used for population studies of *H. armigera* in the future.

Phylogenetic studies have been carried out using allozymes (Richardson, Baverstock, and Adams, 1996) or DNA-based markers, for example, regions of the nuclear ribosomal RNA cluster (rDNA) such as *18S* (Sanchis et al., 2000), and *28S* rDNA (Mardulyn and Whitfield, 1999), mitochondrial DNA (mDNA) including *12S* and *16S* (Cameron and Williams, 2003), Cytochrome oxidase subunit1 (*CO1*) (Machado et al., 2001), Cytochrome oxidase subunit II (*COII*) (Despres et al., 2002), and Cytochrome B (Kerdelhue, Le Clainche, and Rasplus, 1999). Nuclear coding genes such as elongation factor 1α (Belshaw and Quicke, 1997) and DNA internal transcribed spacer regions (*ITS*) (Thomson et al., 2003) have also been found to be useful. Microsatellites are useful for detecting relationships between closely related taxa, for example, taxonomy of host races of *A. pisum* (Simon et al., 2003). A number of studies have used rDNA and mtDNA markers for parasitic Hymenoptera (*18S*, *28S*, *16S*, *CO1*, Cytochrome B, and adenine dehydrogenase subunit1 (*ND1*) (Quicke and Belshaw, 1999; Smith et al., 1999; Belshaw et al., 2000; Schmidt, Naumann, and De Barro, 2001). Some phylogenetic studies have also been carried out on predatory beetles using *28S* and nuclear wingless gene (Ober, 2002) and *ND5* (Su, Imura, and Osawa, 2005), and on vespid wasps using *16S* and *28S* (Schmitz and Moritz, 1998; Carpenter, 2003).

Application of Molecular Markers for Studying Population Genetics

Using RAPD-PCR markers, population structure and dynamics of the parasitoid, *D. rapae* within the agricultural systems has been studied by Vaughn and Antolin (1998) on cabbage aphid, *Brevicoryne brassicae* (L.) and Russian wheat aphid, *D. noxia*. *Diaeretiella rapae* popula-tions <1 km apart have been genetically differentiated on a spatial scale corresponding to host use patterns. Using microsatellite markers, heterogeneity in the populations was detected even on the same plant (Loxdale and MacDonald, 2004), and populations at two sites (40 km apart) were found to be quite distinct (Vaughn and Antolin, 1998). Allozyme markers have been used to study population structure in relation to habitat heterogeneity or attitude of beetle predators (Liebherr, 1986, 1988). Genetic structure of the carabids, *Carabus nemoralis* Muller and *C. punctatoauratus* Germar has been found to be quite distinct at distances of 13.6 km through the use of microsatellite markers (Brouat et al., 2003). Populations of the specialist predator, *C. punctatoauratus* were more structured spatially than those of the generalist predator, *C. nemoralis*.

Several molecular markers have been used for tracking insects in space and time (Caterino, Cho, and Sperling, 2000). Microsatellites, mtDNA sequences, RAPDs, AFLPs, and introns have been used widely. Microsatellites have been used in studies on parasitoid wasps (Hufbauer, Bogdanowicz, and Harrison, 2004), carabid beetles (Brouat et al., 2003), and hon-eybees, *Apis mellifera* L. (De la Rua et al., 2003). The use of mitochondrial DNA in population genetic studies has been quite effective due to intraspecific polymorphism. This has allowed analysis of spatial distribution of geographical lineages in the case of braconid parasitoids (Hufbauer, Bogdanowicz, and Harrison, 2004), honeybees (De la Rua et al., 2001), bumblebees (Widmer et al., 1998), and yucca moth, *Prodoxus quinquepunctella* (Chambers) (Althoff and Thomson, 2001). RAPDs have been used for studies on asexual insects such as aphids (Lushai et al., 1997), and the parasitoids, *A. ervi* (Daza-Bustamante et al., 2002) and *D. rapae* (Vaughn and Antolin, 1998).

Application of Molecular Markers for Studying Social Behavior of Insects

Molecular techniques have been used to study behavioral aspects of insect populations (Hughes, 1997). Application of molecular techniques has suggested that behavioral dynamics of insect populations is much more complex than had been known previously. Use of molecular techniques has allowed a better understanding of kinship, colony sex status, and cooperative breeding. Molecular analysis of insect behavior will lead to an understanding of how changes in species environment have been associated with evolutionary changes in social behavior. Molecular markers have also been used to study the behavior of insects in relation to their hosts/environment. Mitochondrial DNA has been used to detect an alien inseminated female in the nest of the polygynous ant, *Leptothorax acervorum* F. (Stille and Stille, 1992). Synthetic probes have been used in studies on beneficial Hymenoptera for studying colony structure and mating behavior, for example, paper wasp, *Polistes annularis* (L.) (Queller et al., 1997). Microsatellite markers have been used to genotype wingless colony queen and sperms in the spermatheca (Hammond, Bourke, and Bruford, 2001) to understand colony structure and mating preference. The results showed that 95% of the females mated only once, while the female offspring attributable to 31 queens were full sisters, and that *L. acervorum* adults mate at random. Using microsatellite markers, Keller and Fournier (2002) suggested that kin recognition mediated by genetic cues in the Argentine ant, *Linepithema humile* (Mayr) may be intrinsically error prone within colonies of social insects.

Molecular Basis of Insect–Plant Interactions

Functional genomics can also be used to study the interaction of insects with their host plants. *Nicotiana attenuata* Torrey ex Watson genes that respond to feeding by *Manduca sexta* L. have been identified through differential display PCR, and then spotted onto microarray (Halitschke et al., 2003; Hui et al., 2003). These have been used to identify genes that responded to wounding as opposed to insect secretions, as well as to different components in insect secretions. Over half of the plant genes that were induced by *M. sexta* secretions could also be induced by fatty acid–amino acid conjugates in the regurgitant, suggesting that plants recognize the herbivores through the composition of their oral secretions (Halitschke et al., 2003). Microarrays have also been used to understand the response of cowpea weevil, *Callosobruchus maculatus* (F.), to defenses produced by cowpea (Moon et al., 2004). Of the 1,920 cDNAs on the array, 151 were induced or suppressed by soyacystatin N (protease inhibitor). Proteases, carbohydrate digestive enzymes, and microbial defense and detoxification genes were also induced. An advantage of this approach is that sequencing is limited to cDNAs representing responsive genes while the disadvantage is that it is limited to responsive genes, since these are most likely to be retained during prescreening, and several physiologically important genes may be missed.

Application of Molecular Markers to Understand Functional Genomics of Insects

Functional genomics can play an important role in understanding the chemical ecology of insects (Tittiger, 2004). The ESTs have been used to identify pheromone biosynthetic genes in *B. mori*. A database of ESTs from pheromone glands has been used to catalogue transcription of *B. mori* (Mita et al., 2003). The Δ-9 desaturase has been confirmed to be specific to pheromone glands (Yoshiga et al., 2000). The EST databases have also been used to isolate pheromone fatty acyl-denaturases from *D. melanogaster* (Jallon and Wicker-Thomas, 2003). In *Ips pini* (Say), HMGGA reductase gene (HMG-R) expression and enzyme activity levels are simulated by JH III (Tillman et al., 2004). HMG-R plays an important role in regulation of the mevalonate pathway (Goldstein and Brown, 1990). Though not directly related to pheromone synthesis, these genes could be the targets to develop future control strategies to disrupt pheromone synthesis (through RNAi or specific inhibitors).

Feeding coordinately stimulates the mevalonate pathway in male *I. pini* (Keeling, Blomquist, and Tittiger, 2004), but is not expected in female *I. pini*, as they do not produce ipsdienol (Seybold et al., 1995). However, early steps showing isomerization of isoprenyl phosphate to dimethylallyl phosphate has been observed in the females, while the later steps were not. Basal transcript levels for all mevalonate pathway genes, and GPPS in particular, were significantly higher in males compared to females. Some genes are strongly induced by feeding, while others are already highly expressed, probably in response to developmental or environmental cues (Keeling, Blomquist, and Tittiger, 2004). Most mevalonate pathway genes are coordinately upregulated by JH III in the males.

The molecular basis of perception of pheromones and other odors includes binding proteins, receptors, degrading enzymes, and the signal transduction pathway (Jacquin-Joly and Merlin, 2004). The cDNAs synthesized from male antennal RNA have revealed the presence of four new odor-binding proteins in tobacco hornworm, *M. sexta* (Robertson et al., 1999). Microarrays have been used to study pheromone reception in honeybee, *A. mellifera* (Grozinger et al., 2003; Whitfield, Cziko, and Robinson, 2003). Genes with the strongest change in expression levels were not as consistently regulated to the less strongly responding genes (Whitfield, Cziko, and Robinson, 2003). Queen mandibular pheromone (QMP) is a multicomponent pheromone that regulates worker development and behavior in *A. mellifera* (Plettner et al., 1996). Microarray analysis has shown that 1,200 genes were upregulated, and nearly 1,300 were downregulated in QMP exposed workers (Grozinger et al., 2003). Transcription factors were more highly upregulated than the other functional groups such as phosphatases, kinases, receptors, etc., suggesting that pheromones can trigger developmental programs by activating one or more transcription factors, which in turn regulate the downstream genes.

Detection of Natural Enemy–Insect Host Interactions

Identification of gut contents of predatory insects provides useful information on predator-prey interactions (Symondson, 2002). Direct observations in the field are quite difficult and cumbersome as both prey and predator can be quite small, and at times hide away from visual detection. Analysis of gut contents is useful for chewing types of insects, but not suitable for piercing-sucking types of insects. Immunological assays based on prey-specific protein antibodies have been used widely (Hagler and Naranjo, 1997; Symondson et al., 1999), but their efficacy is influenced by several factors. To overcome these problems, several molecular marker-based techniques can be used to study the interrelationships between the natural enemies and their insect hosts (Symondson, 2002; Greenstone, 2006).

Specific probes and markers have been used to detect the prey in gut contents of the predators (Symondson, 2002). PCR-based probes have been found to be more effective than protein electrophoresis (Murray, Solomon, and Fitzgerald, 1989; Solomon, Fitzgerald, and Murray, 1996). SCAR markers derived from RAPD bands have been used to detect *H. armigera* (Agusti, De Vicente, and Gabarra, 1999a) and whitefly, *Trialeurodes vaporariorum* (West.) (Agusti, De Vicente, and Gabarra, 1999b) in predator grubs. Multiple copy esterase genes from *Culex quinquefasciatus* Say have been detected in carabid, *Pterostichus cuprea* (L.) (Zaidi et al., 1999), while Cuthbertson, Fleming, and Murchie (2003) used species-specific mitochondrial primers to detect *Rhopalosiphum insertum* (Walker) in the gut of the predatory mite, *Anystis baccarum* (L.).

A number of studies have used PCRs for analyzing predator gut contents (Agusti, De Vicente, and Gabarra, 1999a, 1999b; Chen et al., 2000; Hoogendoorn and Heimpel, 2001). Primer sets that amplify fragments of different lengths, which have a characteristic detection time, have also been used to estimate the time since feeding from the number of bands that were detectable (Hoogendoorn and Heimpel, 2001). Shorter fragments can be detected for a longer time span after ingestion of the prey than the longer fragments in *Coleomegilla maculata* DeGeer and *Harmonia axyridis* (Pallas). Detection time has been found to be independent of meal size, sex, or predator weight. Primers amplifying fragments of mitochondrial *COI* have been used for identification and distribution of the corn borer, *Ostrinia nubilalis* (Hubner), larval parasitoids, *Lydella thompsoni* (Herting) and *Pseudoperichaeta nigrolineata* (Walker). Molecular markers indicated three times higher levels of parasitism than with the conventional methods (Agusti et al., 2005). There were no differences in parasitism by the two tachinids across geographical regions.

Conclusions

Molecular markers can be used for insect identification, understanding population structure, and ecological processes at the micro level. They are also useful for understanding the characteristics of natural enemies, host preference, biotypes, movement, invasion, sex ratio, population genetics, and interaction with other insects and insecticides. Such information can be used as a critical input for the success of pest management programs. Information generated through molecular markers can be used to build a detailed picture of insect population structure, behavior, and the interactions that will lead to a more effective pest management strategy, and have a bearing on the environment and human health. The use of high-resolution molecular markers is likely to increase their application for diagnosis of insect pests, pest invasions, traceability of introduction of parasitoids and predators, direction of insect migration, host selection behavior, and molecular ecology.

References

Abbot, P. (2001). Individual and population variations in invertebrates revealed by inter-simple sequence repeats (ISSRS). *Journal of Insect Science.* http://insectscience.org//8.

Abbot, P., Withgott, J.H. and Moran, M.A. (2001). Genetic conflict and conditional altruism in social aphid colonies. *Proceedings National Academy of Sciences USA* 98: 12068–12071.

Agusti, N., De Vicente, M.C. and Gabarra, R. (1999a). Development of sequence amplified characterized region (SCAR) markers of *Helicoverpa armigera*: A new polymerase chain reaction-based technique for predator gut analysis. *Molecular Ecology* 8: 1467–1474.

Agusti, N., De Vicente, M.C. and Gabarra, R. (1999b). Developing SCAR markers to study predation on *Trialeurodes vaporariorum. Insect Molecular Biology* 9: 263–268.

Agusti, N., Bourguet, D., Spataro, T., Delos, M., Eychenne, N., Folcher, L. and Arditi, R. (2005). Detection, identification and geographical distribution of European corn borer larval parasitoids using molecular markers. *Molecular Ecology* 14: 3267–3274.

Althoff, D.M. and Thompson, J.N. (2001). Geographic structure in the searching behaviour of a specialist parasitoid: Combining molecular and behavioral approaches. *Journal of Evolutionary Biology* 14: 406–417.

Alvarez, J.M. and Hoy, M.A. (2002). Evaluation of the ribosomal ITS2 DNA sequences in separating the closely related populations of the parasitoid *Ageniaspis* (Hymenoptera: Encyrtidae). *Annals of the Entomological Society of America* 95: 250–256.

Amornsak, W., Gordh, G. and Graham, G. (1998). Detection of parasitized eggs with polymerase chain reaction and DNA sequence of *Trichogramma australicum* Girault (Hymenoptera: Trichogrammatidae). *Australian Journal of Entomology* 37: 174–179.

Atanassova, P., Brookes, C.P., Loxdale, H.D. and Powell, W. (1998). Electrophoretic study of five aphid parasitoid species of the genus *Aphidius* (Hymenoptera: Braconidae), including evidence for reproductively isolated sympatric populations and a cryptic species. *Bulletin of Entomological Research* 88: 3–13.

Bapteste, E., Brinkmann, H., Lee, J.A., Moore, D.V., Sensen, C.W., Gordon, P., Durufle, L., Gaasterland, T., Lopez, P., Muller, M. and Philippe, H. (2002). The analysis of 100 genes supports the grouping of three highly divergent amoebae: *Dictyostelium, Entamoeba*, and *Mastigamoeba. Proceedings National Academy of Sciences USA* 99: 1414–1419.

Belshaw, R. and Quicke, D.L.J. (1997). A molecular phylogeny of the Aphidinae (Hymenoptera: Braconidae). *Molecular Phylogenetics and Evolution* 7: 281–293.

Belshaw, R., Downton, M., Quicke, D.L.J. and Austin, A.D. (2000). Estimating ancestral geographical distributions: A Gondwanan origin for aphid parasitoids? *Proceedings Royal Society of London (Series B), Biological Sciences* 267: 491–496.

Black, W.C., Duteau, N.M., Puterka, G.J., Nechols, J.R. and Pettorini, J.M. (1992). Use of the random amplified polymorphic DNA-polymerase chain reaction (RAPD-PCR) to detect DNA polymorphism in aphids (Homoptera, Aphididae). *Bulletin of Entomological Research* 82: 151–159.

Brouat, C., Sennedot, F., Audiot, P., Leblosis, R. and Rasplus, J.Y. (2003). Fine-scale genetic structure of two carabid species with contrasted levels of habitat specialization. *Molecular Ecology* 12: 1731–1745.

Cameron, S.A. and Williams, P.H. (2003). Phylogeny of bumble bees in the New World subgenus *Fervidobombus* (Hymenoptera: Apidae): Congruence of molecular and morphological data. *Molecular Phylogenetics and Evolution* 28: 552–563.

Carpenter, J.M. (2003). On molecular phylogeny of Vespidae (Hymenoptera) and the evolution of sociality in wasps. *American Museum Novitates* 3389: 1–20.

Caterino, M.S., Shull, V.L., Hammond, P.M. and Vogler, A.P. (2002). Basal relationships of Coleoptera inferred from 18S rDNA sequences. *Zoologica Scripta* 31: 41–49.

Caterino, M.S., Cho, S. and Sperling, F.A. (2000). The current state of insect molecular systematics: A thriving Tower of Babel. *Annual Review of Entomology* 45: 1–54.

Chang, S.C., Hu, N.T., Hsin, C.V. and Sun, C.N. (2001). Characterization of differences between two *Trichogramma* wasps by molecular markers. *Biological Control* 21: 75–78.

Chen, Y., Giles, K.L., Payton, M.E. and Greenstone, M.H. (2000). Identifying key cereal aphid predators by molecular gut analysis. *Molecular Ecology* 9: 1887–1898.

Crampton, J.M., Beard, C.B. and Louis, C. (Ed.). (1997). *The Molecular Biology of Insect Disease Vectors.* London, UK: Chapman and Hall.

Cuthbertson, A.G.S., Fleming, C.C. and Murchie, A.K. (2003). Detection of *Rhopalosiphum insertum* (Apple grass aphid) predation by the predatory mite, *Anystis baccarum* using molecular gut analysis. *Agricultural and Forest Entomology* 5: 219–225.

Daza-Bustamante, P., Fuentes-Contreras, E., Rodriguez, L.C., Figueroa, C.C. and Niemeye, H.M. (2002). Behavioral differences between *Aphidius ervi* populations from two tritrophic systems are due to phenotypic plasticity. *Entomologia Experimentalis et Applicata* 104: 321–328.

De La Rua, P., Galian, J., Serrano, J. and Moritz, R.F.A. (2001). Molecular characterization and population structure of the honeybees from the B'alearic Islands (Spain). *Apidologie* 32: 417–427.

De La Rua, P., Galian, J., Serrano, J. and Moritz, R.F.A. (2003). Genetic structure of Balearic honeybee populations based on microsatellite polymorphism. *Genetics, Selection, Evolution* 35: 339–350.

de Leon, J.H. and Jones, W.A. (2005). Genetic differentiation among geographic populations of *Gonatocerus ashmeadi*, the predominant egg parasitoid of glassy sharp-shooter, *Homalodisca coagualata. Journal of Insect Science* 5: 1–9.

Despres, L., Pettex, E., Plaisance, V. and Pompanon, F. (2002). Speciation in the globeflower fly *Chiastocheta* spp. (Diptera: Anthomyiidae) in relation to host plant species, biogeography, and morphology. *Molecular Phylogenetics and Evolution* 22: 258–268.

Downton, M. and Austin, A.D. (1998). Phylogenetic relationships among the microgastroid wasps (Hymenoptera: Braconidae): Combined analysis of 16S and 28S rDNA genes and morphological data. *Molecular Phylogenetics and Evolution* 10: 354–366.

Dowton, M. and Austin, A.D. (2001). Simultaneous analysis of 16S, 28S, COI and morphology in the Hymenoptera: Apocrita: Evolutionary transitions among parasitic wasps. *Biological Journal of the Linnean Society* 74: 87–111.

Ellsworth, D.L., Rittenhouse, K.D. and Honeycut, R.L. (1993). Artificial variation in randomly amplified polymorphic DNA banding patterns. *Biotechniques* 14: 214–217.

Erlandson, M., Braun, L., Baldwin, D., Soroka, J., Ashfaq, M. and Hegedus, D. (2003). Molecular markers for *Peristenus* spp. (Hymenoptera: Braconidae), parasitoids associated with *Lygus* spp. (Hemiptera: Miridae). *Canadian Entomologist* 135: 71–85.

Goldstein, J.I. and Brown, M.S. (1990). Regulation of the mevalonate pathway. *Nature* 343: 425–430.

Greenstone, M.H. (2006). Molecular methods for assessing insect parasitism. *Bulletin of Entomological Research* 96: 1–13.

Grozinger, C.M., Sharabash, N.M., Whitfield, C.W. and Robinson, G.E. (2003). Pheromone-mediated gene expression in the honey bee brain. *Proceedings National Academy of Sciences USA* 100: 14519–14525.

Hagler, J.R. and Naranjo, S.E. (1997). Measuring the sensitivity of an indirect predator gut content by ELISA: Detectability of prey remains in relation to predator species, temperature, time, and meal size. *Biological Control* 9: 112–119.

Halitschke, R., Gase, K., Hui, D., Schmidt, D.D. and Baldwin, I.T. (2003). Molecular interactions between the specialist herbivore *Manduca sexta* (Lepidoptera, Sphingidae) and its natural host *Nicotiana attenuata*. VI. Microarray analysis reveals that most herbivore-specific transcriptional changes are mediated by fatty acid-amino acid conjugates. *Plant Physiology* 131: 1894–1902.

Hall, H.G. (1998). PCR amplification of a locus with RFLP alleles specific to African honey bees. *Biochemical Genetics* 36: 351–361.

Hammond, R.L., Bourke, A.F.G. and Bruford, M.W. (2001). Mating frequency and mating system of the polygynous ant, *Leptothorax acervorum. Molecular Ecology* 10: 2719–2728.

Heckel, D.G. (2003). Genomics in pure and applied entomology. *Annual Review of Entomology* 48: 235–260.

Hillis, D.M., Mortiz, C. and Mable, B.K. (Eds.). (1996). *Molecular Systematics.* Sunderland, Massachusetts, USA: Sinauer Associates Inc.

Hoogendoorn, M. and Heimpel, G.E. (2001). PCR-based gut content analysis of insect predators using ribosomal ITS-1 fragments from prey to estimate predation frequency. *Molecular Ecology* 10: 2059–2067.

Howland, D.E. and Hewitt, G.M. (1995). Phylogeny of the Coleoptera based on mitochondrial cytochrome oxidase I sequence data. *Insect Molecular Biology* 4: 203–215.

Hoy, M.A. (2003). *Insect Molecular Genetics*, 2nd edition. London, UK: Academic Press.

Hoy, M.A., Jeyaprakash, A., Morakote, R., Lo, P.K.C. and Nguyen, R. (2000). Genomic analyses of two populations of *Ageniaspis citricola* (Hymenoptera: Encyrtidae) suggest that a cryptic species may exist. *Biological Control* 17: 1–10.

Hufbauer, R.A., Bogdanowicz, S.M. and Harrison, R.G. (2004). The population genetics of a biological control introduction: Mitochondrial DNA and microsatellite variation in native and introduced populations of *Aphidius ervi*, a parasitoid wasp. *Molecular Ecology* 13: 337–348.

Hufbauer, R., Bogdanowicz, S.M. and Perez, Z. (2001). Isolation and characterization of microsatellites in *Aphidius ervi* (Hymenoptera: Braconidae) and their applicability to related species. *Molecular Ecology Notes* 1: 197–199.

Hughes, C. (1997). Interacting molecular techniques with field methods in studies of social behavior: A revolution results. *Evolution* 79: 383–399.

Hui, D., Iqbal, J., Lehmann, K., Gase, K., Saluz, H.P. and Baldwin, I.T. (2003). Molecular interactions between the specialist herbivore *Manduca sexta* (Lepidoptera, Sphingidae) and its natural host *Nicotiana attenuata*. V. Microarray analysis and further characterization of large-scale changes in herbivore-induced mRNAs. *Plant Physiology* 131: 1877–1893.

Jacquin-Joly, E. and Merlin, C. (2004). Insect olfactory receptors: Contributions of molecular biology to chemical ecology. *Journal of Chemical Ecology* 33: 575–578.

Jallon, J.M. and Wicker-Thomas, C. (2003). Genetic studies on pheromone production in *Drosophila*. In Blomquist, G.J. and Vogt, R.G. (Eds.), *Insect Pheromone Biochemistry and Molecular Biology*. Amsterdam, The Netherlands: Elsevier Press, 253–282.

Jensen, M.K., Kester, K.M., Kankare, M. and Brown, B.L. (2002). Characterization of microsatellite loci in the parasitoid *Cotesia congregata* (Say) (Hymenoptera: Braconidae). *Molecular Ecology Notes* 2: 346–348.

Ji, Y.J., Wu, Y.U.C. and Zhang, D.X. (2005). Novel polymorphic microsatellite markers developed in the cotton bollworm, *Helicoverpa armigera* (Lepidoptera: Noctuidae). *Insect Science* 12: 331–334.

Ji, Y.J., Zhang, D.X., Hewitt, G.M., Karg, L. and Li, D.M. (2003). Polymorphic microsatellite loci for the cotton bollworm, *Helicoverpa armigera* (Lepidoptera: Noctuidae) and some remarks on their isolation. *Molecular Ecology Notes* 3: 102–104.

Keeling, C.I., Blomquist, G.J. and Tittiger, C. (2004). Coordinated gene expression for pheromone biosynthesis in the pine engraver beetle, *Ips pini* (Coleoptera: Scolytidae). *Naturwissenschaften* 91: 324–328.

Keller, L. and Fournier, D. (2002). Lack of inbreeding avoidance in the Argentine ant, *Linepithema humile. Behavioral Ecology* 13: 28–31.

Kerdelhue, C., Le Clainche, I. and Rasplus, J.Y. (1999). Molecular phyllogeny of *Leratosolen* species pollinating *Ficus* of the subgenus *Sycomorus* senu stricto. Biogeographical history and origins of the species specificity breakdown cases. *Phyllogenetics and Evolution* 11: 401–414.

Laroche, A., Declerck-Floate, R.A., Lesage, L., Floate, K.D. and Demeke, T. (1996). Are *Altica carduorum* and *Altica cirsicola* (Coleoptera: Chrysomelidae) different species? Implications for the release of *A. cirsicola* for the biocontrol of Canada thistle in Canada. *Biological Control* 6: 306–314.

Li, X.Z. and Shen, Z.R. (2002). PCR-based technique for identification and detection of *Trichogramma* spp. (Hymenoptera: Trichogrammattidae) with specific primers. *Entomologia Sinica* 9: 9–16.

Liebherr, J.K. (1986). Comparison of genetic variation in two carabid beetles (Coleoptera) of differing vagility. *Annals of the Entomological Society of America* 79: 424–433.

Liebherr, J.K. (1988). Gene flow in ground beetles (Coleoptera, Carabidae) of differing habitat preference and flight-wing development. *Evolution* 42: 129–137.

Loxdale, H.D. and MacDonald, C. (2004). Tracking parasitoids at the farm and field scale using microsatellite markers. In Wemer, D. (Ed.), *Biological Resources and Migration. Proceedings of the International Conference and OECD Workshop*, 1–8 October, 2003. Philippips University, Marburg, Germany. Springer-Verlag, 107–126.

Lushai, G., Loxdale, H.D., Brookes, C.P., Von Mende, N., Harrington, R. and Hardie, J. (1997). Genotypic variation among different phenotypes within aphid clones. *Proceedings of the Royal Society, London, Biological Sciences (Series B)* 264: 725–730.

Machado, C.A., Jousselin, E., Kjellberg, F., Compton, S.G. and Herre, E.A. (2001). Polygenetic relationships, historical biogeography and character evolution of fig pollinating wasps. *Proceedings of the Royal Society of London Biological Sciences (Series B)* 268: 685–694.

Mardulyn, P. and Whitfield, J.B. (1999). Phylogenetic signal in the COI, 16S, and 28S genes for inferring relationships among genera of *Microgastrinae* (Hymenoptera; Braconidae): Evidence of a high diversification rate in this group of parasitoids. *Molecular Phylogenetics and Evolution* 12: 262–294.

Mita, K., Morimyo, M., Okano, K., Koike, Y., Nohata, J., Kawasaki, H., Kadono-Okuda, K., Yamamoto, K., Suzuki, M.G., Shimada, T., Goldsmith, M.R. and Maeda, S. (2003). The construction of an EST database for *Bombyx mori* and its application. *Proceedings National Academy of Sciences USA* 100: 14121–14126.

Moon, J., Salzman, R.A., Ahn, J.E., Koiwa, H. and Zhu-Salzman, K. (2004). Transcriptional regulation in cowpea bruchid guts during adaptation to a plant defense protease inhibitor. *Insect Molecular Biology* 13: 283–291.

Murphy, W.J., Eizirik, E., O'Brien, S.J., Madsen, O., Scally, M., Douady, C.J., Teeling, E., Ryder, O.A., Stanhope, M.J., deJong W.W. and Springer, M.S. (2001). Resolution of the early placental mammal radiation using Bayesian phylogenetics. *Science* 294: 2348–2351.

Murray, R.A., Solomon, M.G. and Fitzgerald, J.D. (1989). The use of electrophoresis for determining patterns of predation in arthropods. In Loxdale, H.D. and den Hollander, J. (Eds.), *Electrophoretic Studies on Agricultural Pests*. Oxford, UK: Clarendon Press, 467–483.

Nielsen, D.I., Ebert, P.R., Page, R.E., Hunt, G.J. and Guzmannovoa, E. (2000). Improved polymerase chain reaction-based mitochondrial genotype assay for identification of the Africanized honeybee (Hymenoptera: Apidae). *Annals of the Entomological Society of America* 93: 1–6.

Ober, K.A. (2002). Phylogenetic relationships of the carabid subfamily Harpalinae (Coleoptera) based on molecular sequence data. *Molecular Phylogenetics and Evolution* 24: 228–248.

Osborne, J.L., Loxadale, H.D. and Woiwod, I.P. (2002). Monitoring insect dispersal: Methods and approaches. In Bullock, J.M., Kenward, R.E. and Halls, R.S. (Eds.), *Dispersal Ecology. British Ecological Society Symposium*, 3–5 April, 2001, Reading University. Oxford, UK: Blackwell Publishing, 24–49.

Parkinson, J., Mitreva, M., Whitton, C., Thomson, M., Daub, J., Martin, J., Schmid, R., Hall, N., Barrell, B., Waterston, R.H., McCarter, J.P. and Blaxter, M.L. (2004). A transcriptomic analysis of the phylum Nematoda. *Nature Genetics* 36: 1259–1267.

Persad, A.P., Jayaprakash, A. and Hoy, M.A. (2004). PCR assay discriminates between immature *Lipolexis orygmae* and *Lysiphlebus testaceips* (Hymenoptera: Aphidiidae) within their aphid hosts. *Florida Entomologist* 87: 18–24.

Philippe, H., Snell, E.A., Bapteste, E., Lopez, P., Holland, P.W. and Casane, D. (2004). Phylogenomics of eukaryotes: Impact of missing data on large alignments. *Molecular Biology and Evolution* 21: 1740–1752.

Plettner, E., Slessor, K.N., Winston, M.L. and Oliver, J.E. (1996). Caste-selective pheromone biosynthesis in honey bees. *Science* 271: 1851–1853.

Prinsloo, G., Chen, Y., Giles, K.L. and Greenstone, M.H. (2002). Release and recovery in South Africa of the exotic aphid parasitoid *Aphelinus hordei* verified by the polymerase chain reaction. *Biocontrol* 47: 127–136.

Queller, D.C., Peters, J.M., Solis, C.R. and Strassmann, J.E. (1997). Control of reproduction in social insect colonies: Individual and collective relatedness preferences in the paper wasp, *Polistes annularis*. *Behavioral Ecology and Sociobiology* 40: 3–16.

Quicke, D.L.J. and Belshaw, R. (1999). Incongruence between morphological data sets: An example from the evolution of endoparasitism among parasitic wasps (Hymenoptera: Braconidae). *Systematic Biology* 48: 436–454.

Ratcliffe, S.T., Robertson, H.M., Jones, C.J., Bollero, G.A. and Weinzerl, R.A. (2002). Assessment of parasitism of housefly and stable fly (Diptera: Muscidae) pupae by pteromalid (Hymenoptera: Ptermalidae) parasitoids by using a polymerase chain reaction. *Journal of Medical Entomology* 39: 52–60.

Richardson, B.J., Baverstock, P.R. and Adams, M. (1996). *Alloyzyme Electrophoresis. A Handbook for Animal Systematics and Population Studies*. London, UK: Academic Press.

Robertson, H.M., Martos, R., Sears, C.R., Todes, E.Z., Walden, K.K. and Nardi, J.B. (1999). Diversity of odorant binding proteins revealed by an expressed sequence tag project on male *Manduca sexta* moth antennae. *Insect Molecular Biology* 8: 501–518.

Roehrdanz, R.L. and Flanders, R.V. (1993). Detection of DNA polymorphisms in predatory Coccinellids using polymerase chain reaction and arbitrary primers (RAPD-PCR). *Entomophaga* 38: 479–491.

Rubin, G.M., Hong, L., Brokstein, P., Evans-Holm, M., Frise, E., Stapleton, M. and Harvey, D.A. (2000). A *Drosophila* complementary DNA resource. *Science* 287: 2222–2224.

Rudd, S. (2003). Expressed sequence tags: Alternative or complement to whole genome sequences? *Trends in Plant Science* 8: 321–329.

Sanchis, A., Latorre, A., Gonzalez-Candelas, F. and Michelena, J.M. (2000). An 18S rDNA-based molecular phylogeny of Aphidiinae (Hymenoptera: Braconidae). *Molecular Phylogenetics and Evolution* 14: 180–194.

Sappal, N.P., Jeng, R.S., Hubbel, M. and Liu, F. (1995). Restriction fragment length polymorphisms in polymerase chain reaction amplified ribosomal DNAs of three *Trichogramma* species (Hymenoptera: Trichogrammatidae). *Genome* 38: 419–425.

Schmidt, S., Naumann, I.D. and De Barro, P.J. (2001). *Encarsia* species (Hymenoptera: Aphelinidae) of Australia and the Pacific Islands attacking *Bemisia tabaci* and *Trialeurodes vaporariorum* (Hemiptera: Aleyrodidae): A pictorial key and descriptions of four new species. *Bulletin of Entomological Research* 91: 369–387.

Schmitz, J. and Moritz, R.F.A. (1998). Molecular phylogeny of Vespidae (Hymenoptera) and the evolution of sociality in wasps. *Molecular Phylogenetics and Evolution* 9: 183–191.

Seybold, S.J., Ohtsuka, T., Wood, D.L. and Kubo, I. (1995). The enantiomeric composition of ipsdienol: A chemotaxonomic character for North American populations of *Ips* spp. in the *pini* subgeneric group (Coleoptera: Scolytidae). *Journal of Chemical Ecology* 21: 995–1016.

Silva, I.M.M.S., Honda, J., van Kan, F., Hu, J., Neto, L., Pinturequ, B. and Stouthamer, R. (1999). Molecular differentiation of five *Trichogramma* species occurring in Portugal. *Biological Control* 16: 177–184.

Simon, J.C., Carre, S., Boutin, M., Prunier-Leterme, N., Sabatermunoz, B., Latorre, A. and Bournoville, R. (2003). Host-based divergence in populations of the pea aphid: Insights from nuclear markers and the prevalence of facultative symbionts. *Proceedings of the Royal Society of London (Series B), Biological Sciences* 270: 1703–1712.

Smith, P.T., Kambhampati, S., Volkl, W. and Mackauer, M. (1999). A phylogeny of aphid parasitoids (Hymenoptera: Braconidae: Aphidiinae) inferred from mitochondrial NADH 1 dehydrogenase gene sequence. *Molecular Phylogenetics and Evolution* 11: 236–245.

Solomon, M.G., Fitzgerald, J.D. and Murray, R.A. (1996). Electrophoretic approaches to predator-prey interactions. In Symondson, W.O.C. and Liddell, J.D. (Eds.), *The Ecology of Agricultural Pests: Biochemical Approaches, Systematics Association*. Special Volume No. 53. London, UK: Chapman and Hall, 457–468.

Stille, M. and Stille, B. (1992). Intranest and internest variation in mitochondrial DNA in the polygynous ant *Leptothorax acervorum* (Hymenoptera, Formicidae). *Insectes Sociaux* 39: 335–340.

Stouthamer, R., Hu, J., van Kan, F.J.P.M., Platner, G.R. and Pinto, J.D. (2000). The utility of internally transcribed spacer 2DNA sequences of the nuclear ribosomal gene for distinguishing sibling species of *Trichogramma*. *Biocontrol* 43: 421–440.

Su, Z.H., Imura, Y. and Osawa, S. (2005). Evolutionary history of *Calosomina* ground beetles (Coleoptera, Carabidae, Carabinae) of the world as deduced from sequence comparisons of the mitochondrial ND5 gene. *Gene* 360: 140–150.

Symondson, W.O.C. (2002). Molecular identification of prey in predator diets. *Molecular Ecology* 11: 627–641.

Symondson, W.O.C., Erickson, M.L., Liddell, J.E. and Jayawardena, K.G.I. (1999). Amplified detection using a monoclonal antibody of an aphid specific epitope exposed during digestion in the gut of a predator. *Insect Biochemistry and Molecular Biology* 29: 873–882.

Teeling, E.C., Springer, M.S., Madsen, O., Bates, P., O'Brien, S.J. and Murphy, W.J. (2005). A molecular phylogeny for bats illuminates biogeography and the fossil record. *Science* 307: 580–584.

Theodorides, K., De Riva, A., Gomez-Zurita, J., Foster, P.G. and Vogler, A.P. (2002). Comparison of EST libraries from seven beetle species: Towards a framework for phylogenomics of the Coleoptera. *Insect Molecular Biology* 11: 467–475.

Thomson, L.J., Rundle, B.J., Carew, M.E. and Hoffmann, A.A. (2003). Identification and characterization of *Trichogramma* species from southeastern Australia using the internal transcribed spacer 2 (ITS-2) region of the ribosomal gene complex. *Entomologia Experimentalis et Applicata* 106: 235–240.

Tillman, J.A., Lu, F., Goddard, L., Donaldson, Z., Dwinell, S.C., Tittiger, C., Hall, G.M., Storer, A.J., Blomquist, G.J. and Seybold, S.J. (2004). Juvenile hormone regulates *de novo* isoprenoid aggregation pheromone biosynthesis in pine bark beetles, *Ips* spp. (Coleoptera: Scolytidae), through transcriptional control of HMG-CoA reductase. *Journal of Chemical Ecology*. 30: 2459–2494.

Tilmon, K.J., Danforth, B.N., Day, W.H. and Hoffmann, M.P. (2000). Determining parasitoid species composition in a host population: A molecular approach. *Annals of the Entomological Society of America* 93: 640–657.

Tittiger, C. (2004). Functional genomics and insect chemical ecology. *Journal of Chemical Ecology* 30: 2335–2357.

Vanlerberghe-Masutti, F. and Chavigny, P. (1997). Characterization of a microsatellite locus in the parasitoid wasp, *Aphelinus abdominalis* (Hymenoptera: Aphelinidae). *Bulletin of Entomological Research* 87: 313–318.

Vaughn, T.T. and Antolin, M.F. (1998). Population genetics of an opportunistic parasitoid in an agricultural landscape. *Heredity* 80: 152–162.

Walton, M.P., Loxdale, H.D. and Allen-Williams, L. (1990). Electrophoretic keys for the identification of parasitoids (Hymenoptera, Braconidae, Aphelinidae) attacking *Sitobion avenae* (F.) (Hemiptera, Aphididae). *Biological Journal of the Linnean Society* 40: 333–346.

Wheeler, W.C., Whiting, M., Wheeler, Q.D. and Carpenter, J.M. (2001). The phylogeny of the extant hexapod orders. *Cladistics* 17: 113–169.

Whitfield, C.W., Cziko, A.M. and Robinson, G.E. (2003). Gene expression profiles in the brain predict behaviour in individual honeybees. *Science* 302: 296–299.

Widmer, A., Schmid-Hempel, P., Estoup, A. and Scholl, A. (1998). Population genetic structure and colonization history of *Bombus terrestris s.l.* (Hymenoptera: Apidae) from the Canary Islands and Madeira. *Heredity* 81: 563–572.

Yoshiga, T., Okano, K., Mita, K., Shimada, T. and Matsumoto, S. (2000). cDNA cloning of acyl-coA desaturase homologs in the silkworm, *Bombyx mori*. *Gene* 246: 345.

Yu, D.J., Chen, Z.L., Zhang, R.J. and Yim, W.J. (2005). Real time qualitative PCR for the inspection and identification of *Bactrocera philippinensis* and *Bactrocera occipitalis* (Diptera: Tephritidae) using SYBR green assay. *The Raffles Bulletin of Zoology* 53: 73–78.

Zaidi, R.H., Jaal, Z., Hawkes, N.J., Hemingway, J. and Symondson, W.O.C. (1999). Can multiple copy sequences of prey DNA be detected amongst the gut contents of invertebrate predators? *Molecular Ecology* 8: 2081–2087.

Zdobnov, E.M., von Mering, C., Letunic, I., et al. (2002). Comparative genome and proteome analysis of *Anopheles gambiae* and *Drosophila melanogaster*. *Science* 298: 149–159.

Zhu, Y.C. and Williams, L. (2002). Detecting the egg parasitoid *Anaphes iole* (Hymenoptera: Mymaridae) in tarnished plant bug (Heteroptera: Miridae) eggs by using a molecular approach. *Annals of the Entomological Society of America* 95: 359–365.

Zhu, Y.C., Burd, J.D., Elliot, N.C. and Greenstone, M.H. (2000). Specific ribosomal DNA marker for early polymerase chain reaction detection of *Aphelinus hordei* (Hymenoptera: Aphelinidae) and *Aphidius colemani* (Hymenoptera: Aphidiidae) from *Diuraphis noxia* (Homoptera: Aphididae). *Annals of the Entomological Society of America* 93: 486–491.

18

Molecular Techniques for Developing New Insecticide Molecules and Monitoring Insect Resistance to Insecticides

Introduction

Crop protection is still dominated by conventional chemical control, and this approach will continue to be important in crop protection in the future. Therefore, there will be a continued need for new insecticides either to be used alone or in combination with insect-resistant transgenic crops for integrated pest management. Overuse of insecticides has led to development of resistance to insecticides in many insect species. As a result, there are concerns that under outbreak situations, it may be difficult to control certain pests or vectors of human and plant diseases that have developed high levels of resistance to the commonly used insecticides. Therefore, there is a continuing need to monitor insect resistance to insecticides and develop and identify insecticide molecules with novel modes of action. It is in this context that molecular techniques can be employed to detect different receptor sites, and for screening of the molecules for receptor specificity and mode of action.

Development of New Insecticide Molecules

Traditionally, the discovery of new agrochemicals has used *in vivo* screens to identify new compounds, and has been very successful. Functional genomics offers the opportunity to acquire in-depth knowledge of the genetic makeup and gene function of insect pests that may lead to the discovery of new processes that could be the targets for novel chemistry. Advances in pest control are now being aided through rapid synthesis of novel compounds using combinational chemistry, genomics, proteomics, and molecular modeling.

Combining genomics with high-throughput biochemical screening and combinatorial chemical approaches to generate extensive arrays of compounds for screening will lead to a range of new chemicals for pest control. The adoption of such approaches by the pharmaceutical industry has permitted screening rates of up to 10,000 molecules a day.

The new tools involving combinational synthesis, high throughput, and *in vitro* screening has become an integral part of discovering new chemicals for agriculture (Hess, Anderson, and Reagan, 2001). Depending on the synthesis design, the products of combinational synthesis, referred to as a library, can be biased toward an intended target. Unbiased libraries are prepared to maximize chemical diversity around a central core. Compounds in biased libraries are rationally designed to contain structural motifs for pharmacophores that are presumed to be beneficial for activity on the intended target. The compounds can be screened on microtiter plates having 96 to 864 wells. For *in vitro* assays, high-density formats are preferred that allow for testing higher concentrations. This leads to rapid identification of active molecules. A wide range of *in vitro* or *in vivo* assays can be conducted for a range of pesticides using microtiter plates.

Another application of molecular biology is the development of novel strains of entomopathogenic bacteria, fungi, nuclear polyhedrosis viruses, and nematodes. Recombinant *Bacillus thuringiensis* (Berliner) strains with enhanced toxicity and broad insecticidal spectrum have been developed for pest management (Kaur, 2004). To increase the persistence of insecticidal crystal proteins (ICPs), alternative modes of delivery through *Pseudomona* sp. and endophytes have also been developed. The ICPs have been modified by site-directed mutagenesis to improve their insecticidal efficacy, while the yield of ICPs has been increased through the use of strong expression promoters and other regulatory elements. Gene disabling of the sporulation-specific proteases has resulted in increased yield of ICPs. Development of a ligand-mediated system can be used for structure and function analysis of pesticides based on ecdysone (a hormone that regulates molting in insects) (Tran et al., 2001). Such a system could provide a tool for structure function analysis of ecdysone receptor (ECR) in relation to ecdysteroids and other known analogs, and can be used as an effective means for screening new chemicals, and to validate and improve potential insecticidal molecules.

Molecular Markers for Monitoring Insect Resistance to Insecticides

Genomic technologies are now allowing investigation of some previously intractable resistance mechanisms. These cover resistance to both synthetic insecticides and biopesticides. The molecular techniques permit fundamental insights into the nature of mutations and genetic processes (gene amplification, altered gene transcription, and amino acid substitution) underpinning resistance and other adaptive traits. This in turn will lead to high-resolution diagnostics for resistance alleles, in both homozygous and heterozygous forms, especially for insect pests with multiple resistance mechanisms, or for resistance mechanisms not amenable to biochemical assays. Evolution of insecticide-resistant insects provides evolutionary biologists an ideal model system for studying how new adaptations can be rapidly acquired. There is, therefore, a great interest in the use of tools of molecular biology to elucidate the mechanisms of resistance to insecticides.

Recent studies have shown how genomic techniques can access mechanisms that had previously proven intractable to molecular analysis (Gahan, Gould, and Heckel, 2001;

Daborn et al., 2002; Ranson et al., 2002). Resistance mechanisms involving enhanced detoxification of insecticides and rendering the target insensitive to insecticides have been detected earlier. One of the detoxification mechanisms involving sequestration has been reported in the case of resistance to organophosphates and carbamates in aphids (Field, Devonshire, and Forde, 1988) and culicine mosquitoes (Raymond et al., 1998). The other detoxification mechanism, involving active degradation of the insecticides, has been observed in two species of flies as a result of structural mutations in specific carboxyl-esterases that converted them to kinetically inefficient but physiologically efficient organo-phosphate hydrolases (Newcomb et al., 1997; Campbell et al., 1998; Claudianos, Russel, and Oakeshott, 1999). The third mechanism, in which the target molecule mutated in such a way that it became insensitive to insecticides, has now been found in several species involving different types of chemicals (Mutero et al., 1994; Williamson et al., 1996; Vaughan, Rocheleau, and Ffrench-Constant, 1997; Ffrench-Constant et al., 2000; Martin et al., 2000). The mutant target molecules included acetylcholinesterase for organophos-phates, γ-aminobutyric acid (GABA) receptors for cyclodienes, and voltage-gated sodium channels for the synthetic pyrethroids and dichlorodiphenyltrichloroethane (DDT).

The knock down resistance (*kdr*) gene in the housefly, *Musca domestica* L., confers resis-tance to rapid paralysis (knockdown) and lethal effects of DDT and pyrethroids. Flies with the *kdr* trait exhibit reduced neuronal sensitivity to these compounds, which are known to act at voltage-sensitive sodium channels of nerve membranes. The *kdr* trait is tightly linked (within about 1 map unit) to the voltage-sensitive sodium channel gene segment exhibit-ing the DNA sequence polymorphism. Resistance to pyrethroids has been well described at the molecular level for several insect species. Consensus polymerase chain reaction (PCR) primers that amplify a segment of the *para*-like sodium channel gene, which is the molecular target of pyrethroids, have also been developed (Knipple et al., 1994). This segment is known to carry pyrethroid resistance conferring point mutations (*kdr* and *superkdr*) in Diptera and other insects, and although homologous, differed from published genomic sequences of *Blatella germanica* (L.) and *M. domestica* for intron position and size. The *kdr* mutation was found at a high frequency and could be associated to a low level of pyrethroid resistance. In contrast, the *superkdr* mutation that confers high levels of insecti-cide resistance in Diptera was not found in *Frankliniella occidentalis* (Perg.). However, another point mutation close to the *superkdr* position was significantly linked to pyrethroid resistance in *F. occidentalis*. Combined occurrence of *kdr* and "thrips-*superkdr*" allows for a correct diagnosis of pyrethroid resistance in 94% of the individuals tested.

Comparative Genomics and Divergent Evolution of Detoxification Genes

One of the major advances in the study of insecticide resistance enabled by genomics has been the cataloguing of relevant gene families. The lack of such information has been a major constraint in studying resistance based on sequestration and degradation. Three gene families that have been involved in insecticide detoxification [cytochrome P450s, carboxylesterases, and glutathione-S-transferases (GSTs)] have been fully sequenced in *Drosophila melanogaster* Meigen and *Anopheles gambiae* (Giles) (Ranson et al., 2002). Most of the P450s and GSTs are thought to have detoxification or related digestive and/or meta-bolic functions (Stevens et al., 2000; Tijet, Helvig, and Feyereisen, 2001), although many of the esterases are also expected to have specialist nondetoxification functions (Claudianos et al., 2002). There is substantial scope for secondment of various members of these fami-lies to resistance-related functions. The P450s seem to be the main resource for evolution of resistance through enhanced detoxification (Feyereisen, 1999). Although *D. melanogaster* and *A. gambiae* are in the same order, only a small number of the members of the three gene families have been identified as orthologs between the two species (Ranson et al., 2002). Most clades of genes within each of the three families were represented in both species, but the origins of most genes within each clade were explained by independent duplication events within each species. Therefore, finding similar mutations in orthologous genes may be an exception rather than the rule.

Microarrays and Regulatory Mutations in Cytochrome P450s

Mutation in a P450 gene leading to insecticide resistance has been elucidated at the molecular level (Daborn et al., 2002). Expression profiling with microarrays has shown that high levels of DDT resistance in strains of *D. melanogaster* is due to 100-fold upregulation of a specific P450 enzyme (*Cyp6g1*), owing to insertion of a transposable element into its promoter. Expression profiling identified a specific causal change in a specific member of a large gene family without any prior knowledge or assumption regarding the identity of that gene within the family. Several enzymes that metabolize pesticides but do not belong to the three major detoxification gene families (P450s, carboxylesterases, and GSTs) have been found in soil bacteria, and several of these have homologs of unknown function in insects (Claudianos et al., 2002), suggesting that additional resistance mechanisms may be discovered as the power of genomic technologies is applied to understand the resistance mechanisms. Proteomics along with positional cloning using quantitative trait loci (QTLs) will be quite useful for this purpose. There may be some mechanisms that will be difficult to resolve even with genomic technologies, for example, upregulation mediated by changes to *trans*-acting factors, a mechanism that appears to underlie some cases of resistance involving P450s, carboxylesterases, and GSTs (Feyereisen, 1999; Hemingway, 2000).

Quantitative Trait Loci, Positional Cloning, and Multiple Resistance to *Bt* Toxins

The *Bt* toxins differ from most other insecticides in that they are proteins and are not neurotoxins. In fact, their mode of action is quite complex and not properly understood, involving binding to sites on at least four different protein and carbohydrate targets in the insect midgut (Tabashnik et al., 1997; Marroquin et al., 2000; Ferre and van Rie, 2002). Resistance to *Bt* toxins has become a critical concern with the expression of *Bt* toxins in transgenic crops (Ferre and van Rie, 2002). Resistance has already been reported in natural populations of the diamondback moth, *Plutella xylostella* (L.), and it has been quite easy to select for resistance in laboratory populations of several insect species. Several *Bt* toxin resistance genes have been reported from *P. xylostella*, most of which probably encode the proteins that act as toxin-binding sites. Similarly, most of the laboratory-selected strains of insects with resistance to *Bt* toxins involve multiple genes (Ferre and van Rie, 2002). Some individual genes underlying *Bt* toxin resistance have now been mapped onto high-density linkage maps using QTL mapping (Heckel et al., 1999; Marroquin et al., 2000). Two laboratory-selected *Bt* toxin resistance genes have been identified using genomic technologies. The first involved positional cloning in *Heliothis virescens* (F.) (Gahan, Gould, and Heckel, 2001). The coding region of this gene was apparently disrupted in resistant individuals by insertion of a transposable element. The mutant target site appears to be insensitive to *Bt*, although the native function of the target molecule has probably been lost, leading to a significant fitness penalty in the absence of insecticide—a phenomenon that is also characteristic of the resistant *P. xylostella* found in the field (Tabashnik et al., 1997).

Selective Sweeps and Genomic Consequences

One feature of the evolution of insecticide resistance in the field is the rapid spread of resistance alleles after the initial outbreak. Since it happens so quickly, there is little time for recombination to separate the favored resistance gene from the particular combination (haplo-type) of closely linked genes where it appeared first. It is likely that the sweep of mutant esterase/organophosphate hydrolase genes in organophosphate-resistant blowflies (Newcomb et al., 1997; Campbell et al., 1998; Smyth et al., 2000) replaced most of the variation throughout the cluster of ten esterase genes in which it lies with a couple of whole-cluster haplotypes. This cluster makes up a large proportion of the detoxifying esterase genes in the blowfly's genome (Claudianos et al., 2002). Probably, similar sweeps have occurred in clusters containing resistant mutants of P450s and GSTs (Russell et al., 1990). A major reduction may have occurred in the genetic variation in this species' chemical defense system in a short span of time. There may also be additional detoxification systems beyond the three major gene families so far studied. In addition to direct fitness costs in the absence of insecticide that may occur for some resistance mutants, there may also be a substantial "opportunity cost" in terms of lost variation with which the species can respond (Batterham et al., 1996).

Conclusions

There is a continued need for identifying new insecticides either to be used alone or in combination with insect-resistant transgenic crops for integrated pest management. It is in this context that molecular techniques can be employed to detect different receptor sites, and screening of the molecules for receptor specificity, and mode of action. Functional genomics offers the opportunity to acquire in-depth knowledge of the genetic makeup and gene function of insect pests that may lead to the discovery of new processes that could be the targets for novel chemistry. Advances in pest control are now being aided through rapid synthesis of novel compounds using combinational chemistry, genomics, proteomics, and molecular modeling. Combining genomics with high-throughput biochemical screening and combinatorial chemical approaches to generate extensive arrays of compounds for screening will lead to a range of new chemicals for pest control. Another application of molecular biology is the development of novel strains of entomopathogenic bacteria, fungi, nuclear polyhedrosis viruses, and nematodes. Genomic technologies will allow investigation of some previously intractable resistance mechanisms to both synthetic insecticides and biopesticides. The molecular techniques will be useful to understand the nature of mutations and genetic processes such as gene amplification, altered gene transcription, and amino acid substitution. This in turn will lead to high-resolution diagnostics for resistance alleles, in both homozygous and heterozygous forms, especially for insect pests with multiple resistance mechanisms, or for resistance mechanisms not amenable to biochemical assays. The evolution of insecticide-resistant insects also provides evolutionary biologists an ideal model system for studying how new adaptations can be rapidly acquired. There is, therefore, a great interest in the use of the tools of molecular biology to develop new insecticide molecules and elucidate the mechanisms of resistance to insecticides.

References

Batterham, P., Davies, A.G., Game A.Y. and Mckenzie J.A. (1996). Asymmetry: Where evolutionary and developmental genetics meet. *Bioassays* 18: 841–845.

Campbell, P.M., Newcomb, R.D., Russell, R.J. and Oakeshott, J.G. (1998). Two different amino acid substitutions in the ali-esterase, E3, confer alternative types of organophosphorus insecticide resistance in the sheep blowfly, *Lucilia cuprina*. *Insect Biochemistry and Molecular Biology* 28: 139–150.

Claudianos, C., Russell, R.J. and Oakeshott, J.G. (1999). The same amino acid substitution in ortholo-gous esterases confers organophosphate resistance on the housefly and a blowfly. *Insect Biochemistry and Molecular Biology* 29: 675–686.

Claudianos, C., Crone, E., Coppin, C., Russell, R. and Oakeshott, J. (2002). A genomic perspective on mutant aliesterases and metabolic resistance to organophosphates. In Marshall Clark, J. and Yamaguchi, I. (Eds.), *Agrochemical Resistance: Extent, Mechanism and Detection*. Washington, D.C., USA: American Chemical Society, 90–101.

Daborn, P.J., Yen, J.L., Bogwitz, M.R., Le Goff, G., Feil, E., Jeffers, S., Tijet, N., Perry, T., Heckel, D., Batterham, P., Feyereisen, R., Wilson, T.G. and Ffrench-Constant, R.H. (2002). A single P450 allele associated with insecticide resistance in *Drosophila*. *Science* 297: 2253–2256.

Ferre, J. and Van Rie, J. (2002). Biochemistry and genetics of insect resistance to *Bacillus thuringiensis*. *Annual Review of Entomology* 47: 501–533.

Feyereisen, R. (1999). Insect P450 enzymes. *Annual Review of Entomology* 44: 507–533.

Ffrench-Constant, R.H., Anthony, N., Aronstein, K., Rocheleau, T. and Stilwell, G. (2000). Cyclodiene insecticide resistance: From molecular to population genetics. *Annual Review of Entomology* 45: 449–466.

Field, L.M., Devonshire, A.L. and Forde, B.G. (1988). Molecular evidence that insecticide resistance in peach-potato aphids (*Myzus persicae* Sulz.) results from amplification of an esterase gene. *Biochemistry Journal* 251: 309–312.

Gahan, L.J., Gould, F. and Heckel, D.G. (2001). Identification of a gene associated with *Bt* resistance in *Heliothis virescens*. *Science* 293: 857–860.

Heckel, D.G., Gahan, L.J., Liu, Y.B. and Tabashnik, B.E. (1999). Genetic mapping of resistance to *Bacillus thuringiensis* toxins in diamondback moth using biphasic linkage analysis. *Proceedings National Academy of Sciences USA* 96: 8373–8377.

Hemingway, J. (2000). The molecular basis of two contrasting metabolic mechanisms of insecticide resistance. *Insect Biochemistry and Molecular Biology* 30: 1009–1015.

Hess, F.D., Anderson, R.J. and Reagan, J.D. (2001). High throughput synthesis and screening: The partner of genomics for discovering of new chemicals for agriculture. *Weed Science* 49: 249–256.

Kaur, S. (2004). Molecular approaches towards development of novel *Bacillus thuringiensis* biopesti-cides. World Journal of Microbiology and Biotechnology 16: 781–793.

Knipple, D.C., Doyle, K.E., Marsella-Henrick, P.A. and Soderland, D.M. (1994). Tight genetic linkage between the Kdr insecticide resistance trait and a voltage sensitive sodium channel gene in the house fly. *Proceedings National Academy of Sciences USA* 91: 2483–2487.

Marroquin, L.D., Elyassnia, D., Griffitts, J.S., Feitelson, J.S. and Aroian, R.V. (2000). *Bacillus thuringi-ensis* (*Bt*) toxin susceptibility and isolation of resistance mutants in the nematode *Caenorhabditis elegans*. *Genetics* 155: 1693–1699.

Martin, R.L., Pittendrigh, B., Liu, J., Reenan, R., Ffrench-Constant, R. and Hanck, D.A. (2000). Point mutations in domain III of a *Drosophila* neuronal Na channel confer resistance to allethrin. *Insect Biochemistry and Molecular Biology* 30: 1051–1059.

Mutero, A., Pralavorio, M., Bride, J.M. and Fournier, D. (1994). Resistance-associated point mutations in insecticide-insensitive acetyl-cholinesterase. *Proceedings National Academy of Sciences USA* 91: 5922–5926.

Newcomb, R.D., Campbell, P.M., Ollis, D.L., Cheah, E., Russell, R.J. and Oakeshott, J.G. (1997). A single amino acid substitution converts a carboxylesterase to an organophosphorus hydrolase and confers insecticide resistance on a blowfly. *Proceedings National Academy of Sciences USA* 94: 7464–7468.

Ranson, H., Claudianos, C., Ortelli, F., Abgrall, C., Hemingway, J., Sharakhova, M.V., Unger, M.F., Collins, F.H. and Feyereisen, R. (2002). Evolution of supergene families associated with insecticide resistance. *Science* 298: 179–181.

Raymond, M., Chevillon, C., Guillemaud, T., Lenormand, T. and Pasteur, N. (1998). An overview of the evolution of overproduced esterases in the mosquito *Culex pipiens*. *Philosophical Transactions of Royal Society, London, Biological Sciences (B)* 353: 1707–1711.

Russell, R.J., Dumancic, M.M., Foster, G.G., Weller, G.L., Healy, M.J. and Oakeshott, J.G. (1990). Insecticide resistance as a model system for studying molecular evolution. In Barker, J., Starmer, W. and MacIntyre, R. (Eds.), *Ecological and Evolutionary Genetics of Drosophila*. New York, USA: Plenum Press, 293–314.

Smyth, K.A., Boyce, T.M., Russell, R.J. and Oakeshott, J.G. (2000). MCE activities and malathion resistances in field populations of the Australian sheep blowfly (*Lucilia cuprina*). *Heredity* 84: 63–72.

Stevens, J.L., Snyder, M.J., Koener, J.F. and Feyereisen, R. (2000). Inducible P450s of the CYP9 family from larval *Manduca sexta* midgut. *Insect Biochemistry and Molecular Biology* 30: 559–568.

Tabashnik, B.E., Liu, Y.B., Malvar, T., Heckel, D.G., Masson, L., Ballester, V., Granero, F., Mensua, J.L. and Ferre, J. (1997). Global variation in the genetic and biochemical basis of diamondback moth resistance to *Bacillus thuringiensis*. *Proceedings National Academy of Sciences USA* 94: 12780–12785.

Tijet, N., Helvig, C. and Feyereisen, R. (2001). The cytochrome P450 gene superfamily in *Drosophila melanogaster*: Annotation, intron-exon organization and phylogeny. *Gene* 262: 189–198.

Tran, H.T., Askari, H.B., Shaaban, S., Prince, L., Palli, S.R., Dhadialla, T.S., Carlson, G.R. and Butt, T.R. (2001). Reconstruction of ligand dependant transactivation of *Choristoneura fumiferana* ecdysone receptor in yeast. *Molecular Endocrinology* 15: 1140–1153.

Vaughan, A., Rocheleau, T. and Ffrench-Constant, R. (1997). Site-directed mutagenesis of an acetylcholinesterase gene from the yellow fever mosquito *Aedes aegypti* confers insecticide insensitivity. *Experimental Parasitology* 87: 237–244.

Williamson, M.S., Martinez Torres, D., Hick, C.A. and Devonshire, A.L. (1996). Identification of mutations in the housefly para-type sodium channel gene associated with knockdown resistance (kdr) to pyrethroid insecticides. *Molecular and General Genetics* 252: 51–60.

19

Biotechnology, Pest Management, and the Environment: The Future

Introduction

The world population is expected to exceed 8.0 billion by 2025, and over 800 million people are food insecure. The availability of cropland is decreasing over time, and therefore there is a need to increase crop productivity on the available arable land to meet the increasing demand for food, fodder, and fuel in the future. Productivity gains are essential not only for increasing food availability, but also for economic growth. One of the areas where a substantial increase in food production can be realized is through the reduction in crop losses due to biotic stresses, currently valued at US$243.4 billion (Oerke et al., 1994). Among these, insect pests cause an estimated loss of US$90.4 billion. Massive application of pesticides to minimize losses due to insect pests has resulted in adverse effects on beneficial organisms, pesticide residues in food, and environmental pollution. A large number of insects have also developed high levels of resistance to commonly used insecticides (Rajmohan, 1998). Development of resistance to insecticides has necessitated the application of higher dosages of the same pesticide or more pesticide applications. It is in this context that the modern tools of biotechnology can be used for pest management and sustainable crop production (Table 19.1).

Genetic Engineering of Crops for Resistance to Insect Pests

Recombinant DNA technology offers the possibility of developing novel biological insecticides that retain the advantages of classical biological control agents, but have fewer or none of their drawbacks. In addition to widening the pool of useful genes, genetic

TABLE 19.1

Potential Applications of Biotechnology to Improve Resistance to Insect Pests, Yield, and Quality of Major Field Crops

Crops	Areas of Improvement	WH	MAS	Trans
Rice	Drought and salinity tolerance		X	
	Resistance to stem borers, plant hoppers, gall midge, and leaf sheath blight	X	X	X
	Nutritional and table quality of grains		X	X
	Resistance to lodging		X	
Wheat	Yield, quality, and adaptation	X	X	X
	Resistance to rusts, Karnal bunt, and insects	X	X	X
Maize	Yield and quality		X	X
	Resistance to lodging and stem borers	X	X	X
Sorghum	Yield, quality, and adaptation to drought	X	X	X
	Resistance to shoot fly, stem borer, midge, and head bugs	X	X	X
Pearl millet	Yield and adaptation to drought		X	
	Resistance to downy mildew, stem borers, and head miner		X	X
Pigeonpea	Yield and adaptation to drought		X	X
	Resistance to *Helicoverpa* and *Fusarium* wilt	X	X	X
Chickpea	Adaptation to drought and chilling tolerance		X	X
	Resistance to wilt, *Ascochyta* blight, and *Helicoverpa*	X	X	X
Mustard	Yield and adaptation to drought		X	
	Oil content and quality		X	
	Resistance to insects	X	X	X
Groundnut	Yield, oil content, and adaptation to drought		X	X
	Resistance to foliar diseases, aflatoxins, and leaf miner	X	X	X
Cotton	Yield, fiber quality, and oil content		X	X
	Resistance to jassids and bollworms	X	X	X
	Flushing pattern		X	X
Sugarcane	Resistance to stem borers	X	X	X
	Yield and induction of early maturity		X	X
Tobacco	Yield and quality	X	X	X
	Resistance to aphids, tobacco caterpillar, and viruses	X	X	X

WH = wide hybridization; MAS = marker-assisted selection; Trans = transgenics.

engineering has also allowed the introduction of several desirable genes into a single plant, and reduce the time to introgress the novel genes into elite backgrounds. Genes conferring resistance to insects have been inserted into several crop plants, such as maize, cotton, potato, tobacco, rice, chickpea, pigeonpea, sorghum, broccoli, lettuce, walnut, apple, alfalfa, and soybean (Sharma, Sharma, and Crouch, 2004). There has been a rapid increase in the area planted to transgenic crops, from 1.7 million hectares in 1996 to over 100 million hectares in 2006 (James, 2007). Development and deployment of insect-resistant transgenic plants for pest control will lead to:

- Reduction in insecticide usage;
- Reduced exposure of farm labor to insecticides;

- Reduction in harmful effects of insecticides on nontarget organisms; and
- Reduced amounts of insecticide residues in food and food products.

The benefits to growers of transgenic crops have been higher yield, lower input costs in terms of pesticide use, and easier crop management (James, 2007). These factors are likely to have substantial impact on the livelihoods of farmers in both developed and developing countries. Transgenic corn and cotton with resistance to insects have reduced pest damage and pesticide use, and have increased crop yield (Qaim and Zilberman, 2003; James, 2007). However, there is considerable debate on this issue, and several claims to the contrary have also been published in the media. In many developing countries, small-scale farmers suffer pest-related yield losses because of technical and economic constraints. Pest-resistant genetically modified crops can contribute to increased yields and agricultural growth in such situations. Adopting transgenic crops also offers the additional advantage of controlling insect pests that have become resistant to commonly used insecticides. However, deployment of insect-resistant transgenic plants should be based on an overall philosophy of integrated pest management, and consider not only gene construct, but alternate mortality factors, reduction of selection pressure, and development of resistance to design effective pest management strategies. Future efforts in developing insect-resistant transgenic plants should focus on developing:

- Transgenic plants with a wide spectrum of activity against the insect pests feeding on a crop, but harmless to natural enemies and other nontarget organisms;
- Technologies that limit transgene expression to plant parts or growth stage of plants that are vulnerable to insect damage;
- Transgenes that target the receptor sites in insects that have developed resistance to conventional insecticides;
- Delivery systems for toxins from the transgenic plants to the insect hemolymph, removing a key constraint in exploiting the transgenic approach to crop protection;
- Pyramiding of novel genes with different modes of action (Table 19.2) and with conventional host plant resistance, and multiple resistances to insect pests and diseases; and
- Insect-resistant plants in cereals, legumes, and oil seed crops that are a source of sustenance for poorer sections of the society, particularly in the developing countries.

TABLE 19.2

Candidate Genes for Use in Genetic Transformation in Crops for Resistance to Insect Pests

Toxins	Candidate Genes
Cry toxins	Cry1Ab, Cry1Ac, Cry IIa, Cry9c, Cry IIB, Vip I, and Vip II
Plant metabolites	Flavonoids, alkaloids, and terpenoids
Enzyme inhibitors	SBTI, CpTi, and GPTi
Enzymes	Chitinase and lipoxygenase
Plant lectins	GNA, ACAL, WAA, and avidin
Toxins from predators	Scorpion, spiders, and ants
Insect hormones	Neuropeptides and peptidic hormones

Genetic Improvement of Natural Enemies

Some of the major problems in using natural enemies in pest control are the difficulties involved in mass rearing and their inability to withstand adverse conditions in the field. Traditional approaches have been used successfully to select natural enemies for resistance to pesticides and enhanced tolerance to high temperatures. Genetic improvement can be useful when the natural enemy is known to be a potentially effective biological control agent, except for one limiting factor; the limiting trait is primarily influenced by a major gene; and the gene can be obtained by selection, mutagenesis, or cloning. The manipulated strain should be fit, effective, and able to be maintained in some form of reproductive isolation (Headley and Hoy, 1987). Some of the desirable characteristics for transgenic insects include resistance to pathogens, adaptation to different environmental conditions, high fecundity, and improved host-seeking ability (Atkinson and O'Brochta, 1999).

Biotechnological interventions can also be used to broaden the host range of natural enemies and enable their production on artificial diet or nonhost insect species that are easy to multiply under laboratory conditions. There is a tremendous scope for developing natural enemies with genes for resistance to pesticides (Hoy, 1992). Genetic transformation can also be used to improve commercial production of materials such as silk, honey, and lac (Mori and Tsukada, 2000), or production of pharmaceuticals and biomolecules (Yang et al., 2002). However, release of genetically modified insects might have a potential risk to the environment (Spielman, Beier, and Kiszewski, 2002). This is of particular concern when the same vector transmits several disease-causing pathogens, as it might be difficult to develop transgenic individuals incapable of transmitting different pathogens. Future research on genetic manipulation of insects should focus on:

- Production of insects for improved silk, honey, lac, and biomaterials;
- Genetically modifying insect vectors for disease control;
- Production of robust natural enemies with resistance to pesticides or adaptation to extremes of climatic conditions;
- Increased host range and efficient rearing systems;
- Mechanisms to drive the gene of interest (such as sex-linked sterility or lethal genes) through natural populations; and
- Procedures for testing, deployment, and risk management of genetically modified insects.

Genetic Improvement of Entomopathogenic Microorganisms

Entomopathogenic bacteria, viruses, fungi, nematodes, and protozoa have a good potential as a component of integrated pest management. However, they still account for <3% of the total pesticide market, and formulations based on the bacterium, *Bacillus thuringiensis* (Berliner) account for 80 to 90% of the commercial microbial pesticides. The major constraint to the use of biopesticides for pest management is the need for simultaneous management of three biological systems: the pathogen, the prey, and the crop. Despite several advantages of biological insecticides, many factors have hindered their commercial success and practical effectiveness. Some of the problems associated with the use of microbial pesticides for pest management are:

- Quality and effectiveness;
- Unstable formulations and poor delivery systems;
- Sensitivity to light, relative humidity, and heat;
- Short shelf-life, especially under hot and humid conditions; and
- Limited host range and specificity to a particular stage of the insect.

Genetic engineering can be used to improve the efficacy of entomopathogenic microorganisms. Efforts to improve *Bt* have largely been focused on increasing its host range and stability. Work on *baculoviruses* is largely focused on incorporation of genes that produce the proteins, which kill the insects at a faster rate (Bonning and Hammock, 1992), and removal of the polyhedrin gene, which produces the protective viral-coat protein, and its persistence in the field (Corey, 1991). Neurotoxins produced by spiders and scorpions have also been expressed in transgenic organisms (Barton and Miller, 1991). Incorporation of benomyl resistance into *Metarhizium anisopliae* (Metsch.) and other entomopathogenic fungi could make entomopathogenic fungi more useful in integrated pest management (Goettel et al., 1990). The role of neurotoxins from insects and spiders needs to be studied in greater detail before they are deployed in other organisms because of their possible toxicity to mammals. Future attention on improvement of microorganisms through genetic engineering should focus on:

- Improved shelf-life;
- Stability under UV light and extremes of environmental conditions;
- Strains of entomopathogenic fungi with resistance to commonly used fungicides;
- Virulent and effective strains of entomopathogenic bacteria, viruses, fungi, and nematodes; and
- Interactions with wild relatives and nontarget organisms.

Molecular Markers, Biosystematics, and Diagnostics

Diagnosis and characterization of insect pests and their natural enemies through molecular approaches will play a major role in pest management. Characterization of natural enemies is particularly challenging in the tropics, where there is enormous diversity (Waage, 1991). Biotechnological tools based on molecular markers offer a great promise for improving biosystematics, and identification of insect pests and natural enemies across a range on taxonomic groups (Hawksworth, 1994). Molecular markers can be used for rapid diagnosis even by nontaxonomists in far-flung regions. Future research efforts in this area should focus attention on the development of:

- Molecular markers effective against a wide range of taxa;
- Kits that are based on both the marker and the gene product;

- Genetic linkage maps of economically important insects; and
- Quantitative trait loci (QTLs) associated with insect behavior and insect-plant interactions.

Development of New Insecticide Molecules and Monitoring Insect Resistance to Insecticides

Crop protection is still dominated by chemical control, and this approach will continue to be important in crop protection. Traditionally, the discovery of new agrochemicals has used *in vivo* screens to identify new compounds. Functional genomics offers the opportunity to acquire in-depth knowledge of the genetic makeup and gene function of insect pests that may lead to the discovery of new processes that could be the targets for novel chemistry (Hess, Anderson, and Reagan, 2001). Combining genomics with high-throughput bio-chemical screening can be used for a range of new chemicals for pest control. Genomic technologies are now allowing investigation of some previously intractable mechanisms involved in insect resistance to insecticides. New molecular techniques permit fundamental insights into the nature of mutations and genetic processes such as gene amplification, altered gene transcription, and amino acid substitution to underpin insecticide resistance mechanisms. This, in turn, will lead to high-resolution diagnostics for resistance alleles in homozygous and heterozygous forms, especially for insect pests with multiple resistance mechanisms, or for resistance mechanisms not amenable to biochemical assays. Future efforts in this area should focus on:

- Functional genomics that may lead to development of molecules with different modes of action;
- Genetic linkage maps of insects, and QTLs associated with different resistance mechanisms; and
- Diagnostic kits for detecting resistance to a group of insecticides with the same mechanism of resistance or with different resistance mechanisms.

Marker-Assisted Selection and the Genomics Revolution

The last decade has seen the whole genome sequencing of a number of model organisms (Chalfie, 1998). Systematic whole genome sequencing will provide critical information on gene and genome organization and function, which will revolutionize our understanding of crop production and provide an ability to manipulate traits contributing to high productivity (Pereira, 2000). Advances in plant genomics will fuel the mapping of QTLs associated with resistance to insect pests. Use of QTLs will also facilitate rapid and efficient transfer of genes conferring resistance to insects from wild relatives of crops. Some of the secondary plant metabolites, such as flavonoids, terpenoids, and alkaloids, have been implicated in host plant resistance to insects, and some of these metabolites accumulate in plants in response to biotic and abiotic stresses. Biotechnology also offers the promise to increase the production of secondary metabolites in plants that are used in medicine, aromatic industry, or confer resistance to insect pests and diseases. Future research on marker-assisted selection to breed for insect resistance should focus on:

- Developing genetic linkage maps of crops;
- Identification of robust QTLs for use in marker-assisted selection for resistance to insect pests;
- Improving accuracy and precision of phenotyping for insect resistance;
- Development of cultivars with resistance to insects combining transgenic and marker-assisted selection; and
- Precise mapping of the QTLs associated with resistance to insects to develop new paradigms in crop breeding.

Transgenic Crops and the Environment

There are a number of ecological and economic issues that need to be addressed when considering the development and deployment of transgenic crops for pest management (NRC, 2000; Sharma and Ortiz, 2000). Deployment of insect-resistant transgenic plants has raised some concerns about the real or conjectural effects of transgenic plants on nontarget organisms, including human beings (Miller and Flamm, 1993), and evolution of resistant strains of insects (Williamson, 1992; Tabashnik, 1994; Shelton et al., 2000). As a result, caution has given rise to doubt because of lack of adequate information. However, genetically modified organisms have a better predictability of gene expression than the conventional breeding methods, and transgenes are not conceptually different than the use of native genes or organisms modified by conventional technologies (Table 19.3).

One of the risks of growing transgenic plants for pest management is the potential spread of the transgene beyond the target area (Chevre et al., 1997). Genes from unrelated sources may change the fitness and population dynamics of hybrids between native plants and the wild species. Plant breeding efforts, in general, have tended to decrease rather than increase the toxic substances, as a result, making the improved varieties more susceptible to insect pests. However, there is a feeling that genes introduced from outside the range of sexual compatibility might present new risks to the environment and humans, and will lead to development of resistance to herbicides in weeds, and to antibiotics. While some of

TABLE 19.3

Comparison of Conventional Breeding and Biotechnological Approaches for Crop Improvement

Conventional Breeding	Genetic Engineering
Uses limited gene pool	Unlimited gene pool
Simple skills sufficient	Requires highly technical skills
Gene regulation native	Gene regulation modified
Imprecise transfer of genes	Precise gene transfer
Integration homologous	Gene integration random
Function of genes unknown	Known gene function

these concerns may be real, the others seem to be highly exaggerated. There are no records of a plant becoming a weed as a result of plant breeding (Cook, 2000). This may be because of low risk of crop plants to the environment, extensive testing of the crop varieties before release, and adequate management practices to mitigate risks inherent in crop plants. Future research on environmental effects of transgenic crops should focus attention on:

- Effects of transgenic plants on the activity and abundance of nontarget herbivore arthropods, natural enemies, and fauna and flora in the rhizosphere and aquatic systems;
- Development of transgenes with specificity effects, so that pests but not the beneficial organisms are targeted;
- Considering Cry proteins with specificity to different binding sites for gene pyramiding as a component of the resistance management strategy;
- Using a detailed understanding of resistance mechanisms, insect biology, and plant molecular biology to tailor gene expression in transgenic plants for efficient pest management;
- Planting of refuge crops and use of integrated pest management from the very beginning to delay development of resistance;
- Studying the extent and implications of gene transfer to and between bacteria and viruses;
- Devising appropriate measures to contain gene flow where its likely consequences may be deleterious to the environment; and
- Considering the risk of not using transgenic crops, particularly in developing countries, where the technology may have most to offer.

In addition, governments, supported by the global scientific and development community, must ensure safe and effective testing, and implement harmonized regulatory programs that inspire public confidence.

Biosafety of Food from Transgenic Crops

Genetically modified plants have been released in different countries, but the regulations governing the use of transgenic plants vary considerably. Existing regulations have been applied to the production and release of genetically modified plants, but have not been adequate to address the potential environmental effects from the transgenic plants. The biosafety issues related to the deployment of transgenic plants include risks for animal and human health, such as toxicity and food quality and safety, allergies, and resistance to antibiotics. The need for and extent of safety evaluation may be based on the comparison of the new food and the analogous food, if any, and the interaction of the transgene with the environment (Cook, 2000). Foods derived from genetically modified plants are now appearing in the market and many more are likely to emerge in the future. The safety, regulation, and labeling of these foods are still contentious issues in most countries (MacCormick et al., 1998). In order to implement the regulations governing the production and marketing of food derived from genetically modified organisms, there is a need to develop and standardize technology to detect such foods (Holst-Jensen, 2006). In addition, a requirement to label approved genetically modified food would necessitate a constant monitoring system, and the cost of such monitoring will be formidable for many developing countries. This will also take precious resources away from their main agenda of using science for development. Future research efforts in this area should focus on:

- Biosafety regulations based on safety, quality, and efficacy;
- Expanding procedures to assess allergy to encompass inhalants as well as allergy to food;
- Establishment of reliable methods for detection and quantification of genetically modified food; and
- Address the issue of food labeling in the unorganized sector, and ensure food safety at the testing stage.

Conclusions

The products of biotechnology should be commercially viable, environmentally benign, easy to use in diverse agroecosystems, and have a wide spectrum of activity against the target insect pests, but be harmless to nontarget organisms. There is a need to pursue a management strategy that takes into account the insect biology, insect plant interactions, and their influence on the natural enemies. Emphasis should be placed on combining exotic genes with conventional host plant resistance, and also with traits conferring resistance to other insect pests and diseases of importance in the target region. It is important to follow the biosafety regulations and make this technology available to farmers who cannot afford the high cost of seeds and chemical pesticides. Use of biotechnological tools for diagnosis of insect pests and their natural enemies, and for gaining an understanding of their population genetics, behavior, and interactions with the host plant will provide a sound foundation for pest management. There is a need to use genetic engineering to produce robust natural enemies, and more stable and virulent strains of entomopathogenic bacteria, fungi, viruses, protozoa, and nematodes for use in integrated pest management. Molecular makers can be used for development of newer pesticide molecules with different modes of action, and monitoring insect resistance to insecticides. Augmentation of conventional breeding with the use of marker-assisted selection and transgenic plants promises to facilitate substantial increases in food production. Rapid and cost-effective development and adoption of biotechnology-derived products will depend on developing a full understanding of the interaction of genes within their genomic environment, and with the environment in which their conferred phenotype must interact for sustainable crop protection.

References

Atkinson, P.W. and O'Brochta, D.A. (1999). Genetic transformation of non-drosophilid insects by transposable elements. *Annals of the Entomological Society of America* 92: 930–936.
Barton, K.A. and Miller, M.J. (1991). Insecticidal toxins in plants. *EPA* 0431829 A1 910612.

Bonning, B.C. and Hammock, B.D. (1992). Development and potential of genetically engineered viral insecticides. *Biotechnology and Genetic Engineering Reviews* 10: 455–489.

Chalfie, M. (1998). Genome sequencing. The worm revealed. *Nature* 396: 620–621.

Chevre, A.M., Eberm F., Baranger, A. and Renard, M. (1997). Gene flow from transgenic crops. *Nature* 389: 924.

Cook, R.J. (2000). Science-based risk assessment for the approval and use of plants in agricultural and other environments. In Persley, G.J., and Lantin, M.M. (Eds.), *Agricultural Biotechnology and the Poor*. Washington, D.C., USA: Consultative Group on International Agricultural Research, 123–130.

Corey, J.S. (1991). Releases of genetically modified viruses. *Medical Virology* 1: 79–88.

Goettel, M.S., St. Leger, R.J., Bhairi, S., Jung, M.K., Oakley, B.R., Roberts, D.W. and Staples, R.C. (1990). Pathogenicity and growth of *Metarhizium anisopliae* stably transformed to benomyl resistance. *Current Genetics* 17: 129–132.

Hawksworth, D.L. (Ed.). (1994). *The Identification and Characterization of Pest Organisms*. Wallingford, Oxon, UK: CAB international.

Headley, J.C. and Hoy, M.A. (1987). Benefit/cost analysis of an integrated mite management program for almonds. *Journal of Economic Entomology* 80: 555–559.

Hess, F.D., Anderson, R.J. and Reagan, J.D. (2001). High throughput synthesis and screening: The partner of genomics for discovering of new chemicals for agriculture. *Weed Science* 49: 249–256.

Holst-Jensen, A. (2006). *GMO Detection Methods and Validation: DNA-based Methods*. http://www.entransfood.nl/workinggroups/wg4TQA/GMO%20detection%20methods%20.htm.

Hoy, M.A. (1992). Biological control of arthropods: Genetic engineering and environmental risks. *Biological Control* 2: 166–170.

James, C. (2007). *Global Status of Commercialized Biotech/GM Crops: 2006*. ISAAA Briefs no. 35. Ithaca, New York, USA: International Service for Acquisition on Agri-Biotech Applications (ISAAA). http://www.isaaa.org/resources/publications/briefs/35.

MacCormick, C.A., Griffin, H.G., Underwood, H.M. and Gasson, M.J. (1998). Common DNA sequences with potential for detection of genetically manipulated organisms in food. *Journal of Applied Microbiology* 84: 969–980.

Miller, H.I. and Flamm, E.L. (1993). Biotechnology and food regulation. *Current Opinion in Biotechnology* 4: 265–268.

Mori, H. and Tsukada, M. (2000). New silk protein: Modification of silk protein by gene engineering for production of biomaterials. *Reviews in Molecular Biotechnology* 74: 95–103.

NRC (National Research Council). (2000). *Genetically Modified Pest-Protected Plants: Science and Regulation*. Washington, D.C., USA: National Research Council, 33–35.

Oerke, E.C., Dehne, H.W., Schonbeck, F. and Weber, A. (1994). *Crop Production and Crop Protection: Estimated Losses in Major Food and Cash Crops*. Amsterdam, The Netherlands: Elsevier.

Pereira, A. (2000). Plant genomics is revolutionising agricultural research. *Biotechnology Development Monitor* 40: 2–7.

Qaim, M. and Zilberman, D. (2003). Yield effects of genetically modified crops in developing countries. *Science* 299: 900–902.

Rajmohan, N. (1998). Pesticide resistance: A global scenario. *Pesticides World* 3: 34–40.

Sharma, H.C. and Ortiz, R. (2000). Transgenics, pest management, and the environment. *Current Science* 79: 421–437.

Sharma, H.C., Sharma, K.K. and Crouch, J.H. (2004). Genetic transformation of crops for insect resistance: Potential and limitations. *CRC Critical Reviews in Plant Sciences* 23: 47–72.

Shelton, A.M., Tang, J.D., Roush, R.T., Metz, T.D. and Earle, E.D. (2000). Field tests on managing resistance to *Bt*-engineered plants. *Nature Biotechnology* 18: 339–342.

Spielman, A., Beier, J.C. and Kiszewski, A.E. (2002). Ecological and community considerations in engineering arthropods to suppress vector-borne disease. In Letourneau, D.K. and Burrows, B.E. (Eds.), *Genetically Engineered Organisms: Assessing Environmental and Human Health Effects*. Boca Raton, Florida, USA: CRC Press, 315–329.

Tabashnik, B.E. (1994). Evolution of resistance to *Bacillus thuringiensis. Annual Review of Entomology* 39: 47–79.

Waage, J.K. (1991). Biodiversity as a resource for biological control. In Hawksworth, D.L. (Ed.), *The Biodiversity of Microorganisms and Invertebrates: Its Role in Sustainable Agriculture.* Wallingford, Oxon, UK: CAB international, 149–161.

Williamson, M. (1992). Environmental risks from the release of genetically modified organisms (GMOs): The need for molecular ecology. *Molecular Ecology* 1: 3–8.

Yang, G., Chen, Z., Cui, D., Li, B. and Wu, X. (2002). Production of recombinant human calcitonin from silkworm (*B. mori*) larvae infected by baculovirus. *Current Pharmaceutical Design* 9: 323–329.

Species Index

Subject Index

A

Acylsugar, 178
Additive gene effects, 138
Agrobacterium, 31, 210, 213, 416, 418, 468
Allelochemicals, 64, 164, 326
    allomones, 69, 99
    antifeedants, 65, 66, 68, 126, 160, 163, 164,
        225, 232, 234
    attractants, 162, 163, 256
    kairomones, 44, 47, 69, 99
    phagostimulants, 160, 162, 165
    repellents, 126, 160, 162
Allozymes, 179, 413, 479, 481, 482
Alternate hosts, 5, 305, 326, 384, 390, 391,
        392, 418
Aminopeptidase, 228, 266, 450
Antibodies, 24, 32, 235, 236, 239, 266, 278, 455,
        456, 468, 471, 472, 485
Apomixes, 34
Associate resistance, 85

B

Baculoviruses, 6, 7, 33, 234, 257–263, 278,
        298, 299, 303, 307, 505
    AcAaIT, 260
    Ac-NPV, 258, 259
    AcMNPV, 260, 261
    Af-MNPV, 258, 259
    Ag-NPV, 258
    H-NPV, 258
    HaNPV, 258
    HaSNPV, 259, 261

    Mb-NPV, 258
    Sl-NPV, 258
Bazooka applicator, 49, 50, 51
Biochemical markers, 65, 162, 163
Biological control, 4–7, 14, 33, 94, 98–104,
        111, 263, 272, 275–277, 294, 295,
        299, 301–306, 325, 326, 343, 344,
        348, 418, 501, 504
Biopesticides, 3, 18, 83, 255–278, 324, 369,
        449, 494, 497, 504
Biosafety–transgenic food, 443–456
    allergenicity, 276, 432, 437, 444, 445,
        455, 456
    chemical profile, 450
    nutritional quality, 445–451
    substantial equivalence, 445, 446,
        447, 459
    toxicological effects, 448
Biosafety–transgenic food, 448–451
    Bt Cry proteins, 448
    lectins, 450
    protease inhibitors, 450
Biosafety-transgenic feed/forage, 451
Biosafety-microbial pesticides, 276–278
Biotypes, 92, 104, 110–112, 130, 133, 134,
        136, 138, 172, 174, 294, 305, 328,
        330, 385, 386, 477, 480, 481, 485
Breeding methods, 133, 340, 444, 459, 507
    backcross breeding, 26, 132, 133, 181
    cytoplasmic male-sterility, 133, 134,
        135, 138, 141, 424
    mass selection, 131
    pedigree breeding, 26, 131, 132, 134
    recurrent selection, 131, 132, 141, 176

Toxin genes for insect resistance (*continued*)
   protease inhibitors, 225–230
   secondary plant metabolites, 224–225
   vegetative insecticidal proteins, 17, 224,
      340, 386
   viruses, 234
Transgene instability, 307, 432, 439
Transducing particles, 257
Translocation lines, 154
Transposable elements, 263, 297, 307, 453, 454
   hobo, 297, 307
   mariner, 297, 306, 307
   piggyback, 297
Transposons, 212, 257, 263, 271, 300, 416, 436
Tridecanone, 59, 65, 111, 162, 163, 164, 178

V

Vaccines, 24, 32, 294, 308, 421
Vertical gene flow, 409, 436
Visual stimuli, 157

W

Waste treatment, 437, 438
Weediness, 410, 411, 412, 433, 435

Y

Yellow stem borer, 53, 89, 136, 172, 215, 220, 269, 322, 389